T0189478

Lecture Notes on Data Engineering and Communications Technologies

Volume 38

Series Editor

Fatos Xhafa, Technical University of Catalonia, Barcelona, Spain

The aim of the book series is to present cutting edge engineering approaches to data technologies and communications. It will publish latest advances on the engineering task of building and deploying distributed, scalable and reliable data infrastructures and communication systems.

The series will have a prominent applied focus on data technologies and communications with aim to promote the bridging from fundamental research on data science and networking to data engineering and communications that lead to industry products, business knowledge and standardisation.

** Indexing: The books of this series are submitted to ISI Proceedings, MetaPress, Springerlink and DBLP **

More information about this series at http://www.springer.com/series/15362

D. Jude Hemanth · Subarna Shakya · Zubair Baig
Editors

Intelligent Data Communication Technologies and Internet of Things

ICICI 2019

 Springer

Editors
D. Jude Hemanth
Department of ECE
Karunya University
Coimbatore, India

Subarna Shakya
Department of Electronics and Computer
Engineering
Tribhuvan University
Kirtipur, Nepal

Zubair Baig
School of Science
Edith Cowan University, Joondalup Campus
Joondalup, WA, Australia

ISSN 2367-4512   ISSN 2367-4520   (electronic)
Lecture Notes on Data Engineering and Communications Technologies
ISBN 978-3-030-34079-7   ISBN 978-3-030-34080-3   (eBook)
https://doi.org/10.1007/978-3-030-34080-3

This Springer imprint is published by the registered company Springer Nature Switzerland AG
The registered company address is: Gewerbestrasse 11, 6330 Cham, Switzerland

We are honored to dedicate the proceedings of ICICI 2019 to all the participants and editors of ICICI 2019.

Foreword

It is with deep satisfaction that I write this foreword to the proceedings of the ICICI 2019 held in JCT College of Engineering and Technology, Coimbatore, Tamil Nadu, on September 12–13, 2019.

This conference was bringing together researchers, academics, and professionals from all over the world, experts in Data Communication Technologies and Internet of Things.

This conference particularly encouraged the interaction of research students and developing academics with the more established academic community in an informal setting to present and to discuss new and current work. The papers contributed the most recent scientific knowledge known in the fields of data communication and computer networking, communication technologies and its applications like IoT, Big Data, and cloud computing. Their contributions helped to make the conference as outstanding as it has been. The local organizing committee members and their helpers put much effort into ensuring the success of the day-to-day operation of the meeting.

We hope that this program will further stimulate research in Intelligent Data Communication Technologies and Internet of Things, and provide practitioners with better techniques, algorithms, and tools for deployment. We feel honored and privileged to serve the best recent developments in the field of Intelligent Data Communication Technologies and Internet of Things to you through this exciting program.

We thank all authors and participants for their contributions.

K. Geetha
Conference Chair

Preface

This conference proceedings volume contains the written versions of most of the contributions presented during the conference of ICICI 2019. The conference provided a setting for discussing recent developments in a wide variety of topics including data communication, computer networking, communicational technologies, wireless and ad hoc network, cryptography, Big Data, cloud computing, IoT, and healthcare informatics. The conference has been a good opportunity for participants coming from various destinations to present and discuss topics in their respective research areas.

ICICI 2019 conference tends to collect the latest research results and applications on Intelligent Data Communication Technologies and Internet of Things. It includes a selection of 86 papers from 284 papers submitted to the conference from universities and industries all over the world. All of accepted papers were subjected to strict peer-reviewing by 2–4 expert referees. The papers have been selected for this volume because of quality and the relevance to the conference.

ICICI 2019 would like to express our sincere appreciation to all authors for their contributions to this book. We would like to extend our thanks to all the referees for their constructive comments on all papers, especially; we would like to thank to organizing committee for their hard work. Finally, we would like to thank the Springer publications for producing this volume.

K. Geetha
Conference Chair

Acknowledgements

ICICI 2019 would like to acknowledge the excellent work of our conference organizing committee, keynote speakers for their presentation on September 12–13, 2019. The organizers also wish to acknowledge publicly the valuable services provided by the reviewers.

On behalf of the editors, organizers, authors, and readers of this conference, we wish to thank the keynote speakers and the reviewers for their time, hard work, and dedication to this conference. The organizers wish to acknowledge Dr. D. Jude Hemanth and Dr. K. Geetha for the discussion, suggestion, and cooperation to organize the keynote speakers of this conference. The organizers also wish to acknowledge for speakers and participants who attend this conference. Many thanks given for all persons who help and support this conference. ICICI 2019 would like to acknowledge the contribution made to the organization by its many volunteers. Members contribute their time, energy, and knowledge at a local, regional, and international level.

We also thank all the Chair Persons and conference committee members for their support.

Contents

Summarization of Research Publications Using Automatic Extraction

Nikhil Alampalli Ramu[(⊠)], Mohana Sai Bandarupalli[(⊠)],
Manoj Sri Surya Nekkanti[(⊠)], and Gowtham Ramesh

Department of Computer Science and Engineering,
Amrita School of Engineering, Amrita Vishwa Vidyapeetham, Coimbatore, India
{CB.EN.U4CSE14103, CB.EN.U4CSE14110,
CB.EN.U4CSE14130}@cb.students.amrita.edu,
r_gowtham@cb.amrita.edu

Abstract. The past two decades have witnessed the significant proliferation of technologies, which has laid a strong foundation for the scientific research in different fields. However, the changes across every field have also created new challenges related to the management of large chunks of data present in the process of converting, storing, searching and providing the user with relevant data. Extracting and transforming the data from one form to another remains as an important task in the current era. It becomes challenging when we focus on the particular extraction instances. Finding the proper research paper from the huge number of papers that includes navigation through the data is not an easy task. It includes huge amount of time search to provide the user with the most appropriate scientific paper of search. This paper concentrates on the extraction of problem statement from one research paper and will be further used to find the related papers. The use of phrases makes the search to considerably reduce the number of search across the Internet and at the same time, it yields a high performance.

Keywords: Summarization · Keywords · Phrases · Automatic extraction · Keyphrases extraction · Tagging · Information retrieval · Pattern identification

1 Introduction

With expansion of growth in the knowledge in all directions, the development in the quantity of records and research paper has additionally extended, this is because of the numerous chains of reasons like exponential development in innovation, which brought about the enormous volume of papers distributed over the world. This makes the creators or the clients hard to look the related the papers which are distributed in Internet.

To sort this trouble, various outline procedures came into the image yet those methodologies clarify about condensing the record or making a lot of numerous reports into a solitary archive yet these methodologies doesn't assist us with minimizing the pursuit time and doesn't deliver the well-suited paper. In this manner, the yield doesn't meet the desires for the user. This paper clarifies an option, where a summarization

D. J. Hemanth et al. (Eds.): ICICI 2019, LNDECT 38, pp. 1–10, 2020.
https://doi.org/10.1007/978-3-030-34080-3_1

procedure is utilized which makes the report to gather the proper phrases dependent on the phrases proclamation and these phrases are utilized in the research paper and expelling the superfluous information from the archive. This methodology just focuses on the significant expressions over the paper and recovers the data which is suitable to user search across the internet.

The model proposed in this paper explains how to extract the Problem Statement of the paper, which is utilized as contribution for the further technique pursued. Extracting the problem statement can't be done through summarization technique so there is a need to develop the approach to bring out the problem statement. Once the problem statement is extracted it is used to extract the related papers.

This methodology forms the paper in different ways and channels the information in the paper to extract the Problem Statement from the paper. In this methodology where the user needs to encourage the program with the contribution of a research paper. The program keeps running on the paper which is given as the input with the numerous methods for a way to deal with the problem statement extracted and giving a yield of appropriate research papers to the user.

2   Literature Survey

Mihalcea [9] proposed a method to identify the similar sentences from the natural text using Graph-based Ranking Algorithms. This work considers all sentences or phrases into vertex where the similarity between these vertexes is represented with the help of edges. The weight of the edges states that weight of similarity match between the sentences. The author also has compared his work with HITS (Hyperlinked Induced Topic Search), Positional Power Function, PageRank, Undirected Graphs, Weighted Graphs methods.

Li et al. [10] developed a method to find similarity between the short texts. In the short text, the reoccurrence of the words are not as frequent as the larger texts, so the traditional techniques may not be directly applied. In order to overcome this issue, this method uses the syntactic and semantic information to find the similarity between the short texts.

Gong et al. [11] proposed approaches for text summarization using Information Retrieval (IR) and Latent Semantic Analyses (LSA) methods. The first summarization method works based on the weights calculated using the rank sentence relevance method. The other method calculates the score for the sentences based on the sentence semantics that adds value to the sentences. Finally, the authors have concluded that both the methods made their mark by producing acceptable performance scores.

Gupta et al. [8] explored the characteristics of various text summarization technique that uses Term Frequency-Inverse Document Frequency (TF-IDF), clustering, graph-theoretic approach, Machine Learning approach, Latent Semantic Analysis, Concept-Obtained approach, neural networks, fuzzy logic, regression technique for estimating feature weights, query-based extractive methods in the text summarization. This study provides insights into various summarization techniques.

Allahyari et al. [3] performed a survey on the summarization techniques specifically on the non-traditional documents such as web documents, scientific articles, and e-mails. They evaluated the effectiveness of the summarization approaches that use graph-based methods, and machine learning approaches. They also detailed the metrics that are used to assess the performance of the summarization techniques.

Kang et al. [7] proposed the method CFinder for extracting the key concepts from the texts. The CFinder primarily POS tags the words in the sentence and further the tagged words are used in finding the required patterns. The texts with the JJ*NN+ pattern are considered as the core technical terms. The core technical terms are further processed to find the key concepts in the text.

Batouche et al. [5] describe an RDF based method for querying the content of the sentence. This method splits the sentences in the text into RDF triples using the NLP tool. The RDF documents are further queried using SPARQL queries.

3 System Design

See Fig. 1.

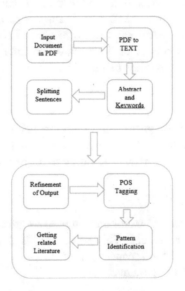

Fig. 1. System design

3.1 PDF to TEXT Conversion

The study papers are downloaded in pdf format from online journal systems, which is the study document's most frequently accessible format. For further processing this requires the transformation of study papers from the pdf format into text format. The JAVA library pdfbox [12] was used for the transformation (Fig. 2).

Fig. 2. PDF to Text

3.2 Abstract and Keywords Extraction

The module identifies and separates the abstract and keyword sections of the converted research documents. Though this module gets the input in the form of text file, the alignment of the sections as in the source files are not maintained. This is due to various reasons like the way that the authors have publishes their documents as well as the various formats followed in the article publishers [4]. So, this requires a pre-processing to make sure that only the abstract and keyword sections are isolated from the paper. We have created section pointers for an abstract section and keyword section individually based on the formatting pattern learned from the various publications. The first index points to a text region from the start of an abstract section to start of the keyword section. The next points to a text region from the start of Keywords section to start of the Introduction section of the paper. The region covered by each of the section is extracted given as input to the other module [6].

3.3 Sentences Extraction

Each of the sentences is distinguished and disencumbered from the portioned unique area. This module uses Stanford-NLP-parser to accurately distinguish the sentences. These sentences are used for further procedures to acquire significant keywords.

3.4 Refinement of Output

From the "Sentences extraction" module the sentences are further processed to obtain the minimum number of sentences. This processing is done through some patterns which are observed from multiple papers [1].

The two approaches using patterns:

- Authors use some phrases, using this as the primary way of approach to extract the PS.
- And the alternative of approach are used by using the keywords count pattern in the keyword section.

PS identification using Phrases: This approach is based on author usage of phrases. Authors use these phrases more frequently which signifies this statement has more weight to justify it has most probable to be the problem statement. The phrases that the author uses more frequently are shown in Fig. 3, these phrases are purely based on observation.

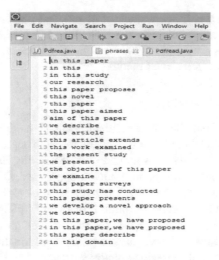

Fig. 3. Key phrases

PS identification using Probability count: This approach uses Keywords has the tool to extract the problem statement. Using these keywords, we use them to count the number of occurrences in all the sentences of the abstract. Here we can choose a threshold count, such that the sentences with count more than threshold count are selected as the problem statement. With some alteration to this approach instead of a threshold value, we can make a statement with maximum count.

This methodology comes vigorously if there is no outcome from the First Approach. In some cases, we acquire the outcome from both the ways to deal with confirm the announcements are available in both the methodologies.

3.5 POS Tagging

To further process the extracted sentence we have to POS tag each word in the sentence, using these tags we make some filtration. POS Tagging means assigning corresponding Parts of Speech to each word. So here, we assign parts of speech to each word in the problem statement. Using a set of defined parts of speech notations, we filter the words with JJ and NN tags. JJ means Adjective and NN means Noun.

3.6 Pattern Identification

After POS Tagging is done, we will be having a set of notations for the pre-defined parts of speech which are assigned to the extracted sentence. The main challenge is to find the technical terms in the extracted sentence. A pattern is used to find the technical terms. This extraction of technical terms is done by using "CFinder: An intelligent key concept finder from the text for Ontology Development". The CFinder says that any POS Tagged sentence which is having a pattern of JJ*NN+ is said to be the technical terms. JJ*NN+ means zero or more repetitions of an adjective (JJ) and one or more repetitions of the noun (NN). Once the tagging of each word is done, the tags are traversed from the first tag. It keeps track of blocks that possess tag from JJ to NN+ . This collection of blocks with this pattern is done till it reaches the last tag. The outcome of this process gives a set of blocks which signifies the technical terms. These blocks are used as input in the next module.

3.7   Getting Related Literature

The blocks extracted from the "pattern identification" module is processed in this module. Here, we use JSON file to get the related research papers. The separated blocks in the previous module are given as input to the JSON file. Each block is separately processed and the output is taken. In JSON file we used DBLP, API.SPRINGER, DOAJ websites to extract related research papers. Each block is given as an input and the technical terms in it are searched through DBLP, API.SPRINGER, DOAJ websites recursively and the unique set of papers are retrieved as an output. Eventually, we extract a set of related research papers that possess information about the technical terms that are extracted in the previous module.

4   Experiment Design

4.1   Description of Data

We collected over 2500 research papers from various publications over the period of 10 years from 2008 to 2018. Specifically, the corpus consists of 1840 collected from different sciences and 660 which comprises other integrated streams. The 1840 papers considered for the evaluation are obtained from the known sciences like Artificial Intelligence, Machine Learning, Environmental Sciences. Most of these papers are popular and most of the authors publish their papers in those areas. The papers for the evaluation are downloaded from the various sources are listed in Table 1.

Table 1. Research paper data sources.

Sciences	No of papers	Link
Machine learning	212	https://www.ieee.org/
	222	https://www.icret.org/
	236	http://iaescore.com/journals/index.php/IJECE
Artificial intelligence	146	https://www.ieee.org/
	220	https://www.icret.org/
	210	http://iaescore.com/journals/index.php/IJECE
Environmental	331	https://www.ieee.org/
	118	https://www.icret.org/
	145	http://iaescore.com/journals/index.php/IJECE

4.2   Metrics Used in Evaluation

In our experiments, we used true positive rate, false negative rate and accuracy for evaluating the performance of our proposed system.

The True Positive Rate (TPR) measures the rate of correctly detected Research papers in relation with the given research paper.

$$TPR = \frac{M_{R \to L}}{M_R} = \frac{M_{R \to L}}{M_{R \to R} + M_{R \to L}}$$

The False Negative Rate (FNR) measures the rate of research paper incorrectly classified as not related research paper.

$$FNR = 1 - TPR$$

The False Positive Rate (FPR) measures the rate of research papers classified as not a related research papers.

$$FPR = \frac{M_{L \to L}}{M_L} = \frac{M_{L \to L}}{M_{L \to L} + M_{L \to R}}$$

The True Negative Rate (TNR) measures the rate of research papers classified as not a related research papers.

$$TNR = 1 - FPR$$

Accuracy indicates how often the detection of related research paper is correct.

$$ACC = \frac{M_{L \to L} + M_{R \to R}}{M_{L \to L} + M_{L \to R} + M_{R \to R} + M_{R \to L}}$$

The notations $M_{L \to L}, M_{R \to R}$ represent correctly classified samples and the notations $M_{R \to L}, M_{L \to R}$ represent wrongly classified Research papers by our proposed system.

4.3   Experiment Results

See Table 2, Figs. 4, 5, 6.

Table 2. Confusion matrix of the experimental results.

Confusion matrix	Related research paper	Non related research paper
Classified as related research paper	TP = 1916	FP = 89
Classified as non-related research paper	FN = 57	TN = 438

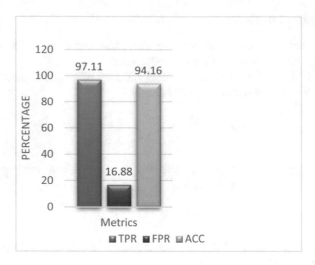

Fig. 4. Assessment of experiment

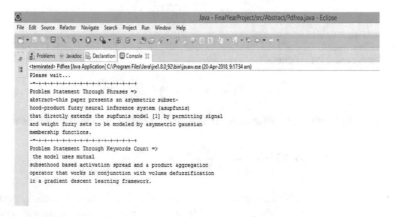

Fig. 5. Output using both the approaches

5 Discussion

In the context of extracting research papers, there is a necessity to address some limitations as it purely based on human evaluation, and where we can face some issues in the scenarios like use of the phrases which are restricted to samples we collected from various papers, if the unique paper comes for evaluation then it may not be expected to produce most efficient outcome. This approach fails to produce expected outcome if the problem statement content is written in multiple lines.

Fig. 6. Related papers output

6   Conclusion

This project helps the user to know more about the related papers for his/her publications that are available in the online and also helps in extracting the related publication from online data sources which includes multiple publications and saves a lot of time to the user in retrieving data. The approach followed here is passing the paper (assuming in pdf format) through different phases like "pdf to text conversion" and followed by an important phase "Abstract and keywords Extraction" and "Sentence Extraction", the output from the above pharses are refined with the help of patterns, then it is moved to final phases called "POS tagging" and "Pattern Identification" to get related literatures. Sometimes approach may fail due to some different format followed in the paper such as alignment of the topics and unique way in representing the paper.

References

1. Gambhir, M., Gupta, V.: Recent automatic text summarization techniques: a survey. Artifi. Intell. Rev. **47**(1), 1–66 (2017)
2. Yao, J.-g., Wan, X., Xiao, J.: Recent advances in document summarization. Knowl. Inf. Syst. **53**(2), 297–336 (2017)
3. Allahyari, M., Pouriyeh, S., Assefi, M., Safaei, S., Trippe, E.D., Gutierrez, J.B., Kochut, K.: Text summarization techniques: a brief survey. arXiv preprint arXiv:1707.02268 (2017)
4. Bharti, S.K., Babu, K.S.: Automatic keyword extraction for text summarization: a survey. arXiv preprint arXiv:1704.03242 (2017)
5. Batouche, B., Gardent, C., Monceaux, A.: Parsing text into RDF graphs. In: Proceedings of the XXXI Congress of the Spanish Society for the Processing of Natural Language (2014)
6. Alami, N., Meknassi, M., Rais, N.: Automatic texts summarization: Current state of the art. J. Asian Sci. Res. **5**(1), 1–15 (2015)

7. Kang, Y.-B., Haghighi, P.D., Burstein, F.: CFinder: an intelligent key concept finder from text for ontology development. Expert Syst. Appl. **41**(9), 4494–4504 (2014)
8. Gupta, V., Lehal, G.S.: A survey of text summarization extractive techniques. J. Emerg. Technol. Web Intell. **2**(3), 258–268 (2010)
9. Mihalcea, R.: Graph-based ranking algorithms for sentence extraction, applied to text summarization. In: Proceedings of the ACL 2004 on Interactive Poster and Demonstration Sessions, p. 20. Association for Computational Linguistics (2004)
10. Li, Y., McLean, D., Bandar, Z.A., O'shea, J.D., Crockett, K.: Sentence similarity based on semantic nets and corpus statistics. IEEE Trans. Knowl. Data Eng. **18**(8), 1138–1150 (2006)
11. Gong, Y., Liu, X.: Generic text summarization using relevance measure and latent semantic analysis. In: Proceedings of the 24th Annual International ACM SIGIR Conference on Research and Development in İnformation Retrieval, pp. 19–25. ACM (2001)
12. Apache PDFBox - A Java PDF Library: https://pdfbox.apache.org/

Cluster Head Enhance Selection Using Type-II Fuzzy Logic for Multi-hop Wireless Sensor Network

K. M. Ramya$^{(\boxtimes)}$ and S. N. Hanumanthappa

Department of Electronics and Communication, UBDT College of Engineering, Davangere, India
ramyakm.95@gmail.com, hanumanthappasn456@gmail.com

Abstract. Wireless sensor network is a remote network of spatially distributed small, lightweight sensors to observe physical and environment status by the measurement of temperature, pressure, vibration and to co-operatively pass their information via network to a base station (BS). In designing wireless sensor network routing protocol, enhancing energy efficiency and lifetime of remote sensor systems are critical issues as a large portion of the remote sensor systems work in unattended condition where accessing and observing are not easy. Low energy adaptive clustering hierarchy (LEACH) is a randomized probabilistic model which is not advisable in practice because it consider energy only to elect cluster head (CH) and it follows the single-hop communication which burdens the CH and may not scale well for bigger applications. Wireless sensor network has routing chain which is of requested grouping of the considerable number of nodes in the system framing a chain structure to convey a collected information to BS.

Clustering techniques arranges the framework activity in related way to go to the system versatility, limit energy utilization and accomplish delayed system lifetime. Orchestrate the framework activity in related way to go to the system versatility, limit energy utilization and accomplish delayed system lifetime. Most of the algorithms overburden the CH during cluster formation. An idea of fuzzy logic is come up as decision maker in applied wireless sensor network (WSN). A large portion of the algorithms use type-1 fuzzy logic (T1FL) model, but uncertain level decisions are handled by type-2 fuzzy logic (T2FL) model superior to T1FL model. The performance is analysed using NS2 simulator.

Keywords: WSN · LEACH · Type-2 fuzzy logic · Aggregated cluster head (ACH)

1 Introduction

The huge uses of wireless sensor system bring numerous difficulties in dangerous places where conventional framework based network is basically infeasible because they are facing problems like energy resources limitation, computing capacity restriction, open condition and remote availability leads the sensor network disappointment more often than not. The sensor nodes communicate through wireless

© Springer Nature Switzerland AG 2020
D. J. Hemanth et al. (Eds.): ICICI 2019, LNDECT 38, pp. 11–24, 2020.
https://doi.org/10.1007/978-3-030-34080-3_2

channels to share the information. Two way of communication like single or multi-hop takes place between CH and BS depending on distance. So as to adjust the utilization of energy among nodes, clustering algorithm for networks was introduced. It is a development over the LEACH protocol. LEACH uses a probabilistic approach to select the CH. Those methods use only local information to make decisions to choose CH. In practical sense, considering only energy factor is not enough to select the CH. This is because other conditions such as distances to sink concentration entire dissipation during transmission here. Backup CH is chosen rather than CH to send data to the BS if there should arise an occurrence of any energy drop out happens to ready CH. Fuzzy logic takes continuous choices without requiring total data about the environment. It is basic and adaptable to take continuous choices under dubious condition. T2FL model can deal with the vulnerability condition more exactness than T1FL model on the grounds that the enrolment inputs of T2FL are fuzzy sets themselves. The irregular vulnerabilities are identified with probabilistic hypothesis and etymological haphazardness is identified with fuzzy sets.

The proposed protocol is in light of three fuzzy parameters they are concentration, distance to sink and remaining battery power concentration. To address the problem of LEACH protocol, fuzzy logic is used which is suitable for real time uncertainties. This algorithm improves the scalability, network lifetime prolongation and increase efficiency of energy. Usage of T2FL in remote sensor system helps message to convey for far away base station in larger application.

The paper is sorted out as pursues: Sect. 2 point to related work. Section 3 discusses overview of WSN and its issues. Existing model in Sect. 4, proposed work over WSN is explained in Sect. 5 and in the final section we present the summary.

2   Related Work

In this segment, we first make a short rundown about the broadly utilized clustering protocols, and afterward examine related work on Fuzzy Logic. In addition, we delineate the advantage of our adaptive T2FL mechanisms in WSNs.

Fu, Jiang and others proposed notes on wireless sensor as the basic important factor to consider is energy. To overcome node energy limitation the following routing protocol is designed that is LEACH protocol which follows typical hierarchy process. CH selection in LEACH is random and it provides over burden to selected cluster head because selection does not depend on parameters like distance and energy. So, distance may be far or near and similarly energy may less or more not considered [6]. Hence it creates overhead communication in the sensor network. Lindsey briefs Power efficient gathering in sensor information systems (PEGASIS) protocol that the base station is at far distance from the sensor node and it is fixed. Depend on the distance of packet transmission cost of energy varies [3]. LEACH is a clustering protocol. In chain PEGASIS each node communicates with few neighbour nodes which are close to it and takes next step to communicate with BS and nodes have capability to aggregate the message and this fused data is transmitted to BS. Dasgupta, Dutta et al. proposes A WSN which contains number of clusters which has hundreds or thousands of sensor nodes communicates through cluster head. Every cluster has cluster head to create a link between sensor node and base station. Usually, effective decision of CH is a significant issue in the presentation of the devoured WSN [15]. A new approach is proposed, that is traditional fuzzy algorithm follows the criteria of minimum distance and maximum energy. This statistical strategy improves the CH choice contrast with LEACH and provides long network life time. Nayak and Anurag defines it's unrealistic and once in a while difficult to supplant the battery and to keep up longer framework life time. The challenging issue is battery power lifetime restriction and consumption of energy [19]. LEACH is the for the most part renowned various leveled steering convention. In this methodology a standby cluster head is chosen beside the CH considering fuzzy logic. Heinzelman et al. states application specific protocol which explains about specific data aggregation for great execution as far as system lifetime and latency. This architecture improves the self- adaptability and nodes in the cluster acts as self-organized. It depends on the function of application as important parameter for simplicity of arrangement and spare vitality limitation of the hubs. The highlights of convention design lead to perform locally to lessen the measure of transmitted information and network system arrangement [2]. Ran, Zhang and Gong [11]: In WSN, the important elements to consider are energy and lifetime of WSN. To make proficiency of these, clustering approach is demanded. The CH which is at edge regions needs more energy [11]. Mamdani and Assilian initiates the learning machines which are programmed with all cases of possibilities. Synthesis of new parameters considering real-time actions was developed with linguistic variables [14]. Human check the machines in a while for its proper working, these actions can be automated using logic of fuzzy was recommended.

3   Overview of Wireless Sensor Network and Its Issues

Wireless sensor node is an electronic device which is used to recognize or quantify a physical amount and changes over it into electronic signal. In the other way, these gadgets decipher parts of physical reality into a representation understandable and process able by the computer. The basic requirements of the sensor are microcontroller, a power supply unit, transceiver, memory unit and sensor unit.

WSN consist sensor nodes has ability speak with one another by methods for remote. The sensor nodes are commonly dispersed in the field and have capacity to gather the information and course the data to BS. The sink may speak with assignment director or end-client through web or satellite. Sensor nodes goes about as the two information originators and information switches also, every hub in any system has a predefined objective. WSN ought to have an incorporated and synchronized structure for imparting and information sharing. The sensor hubs are set in an associated system as per a specific topology, for example, straight, star and work. In WSNs, information accumulation and information move are cultivated in 4 stages: gathering the information, handling the information, bundling the information and moving the data [20]. A WSN is made out of a few quantities of sensors and a door to give association with the Internet (Figs. 1 and 2).

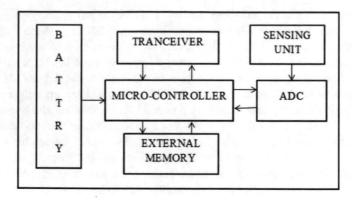

Fig. 1. Architecture of sensor node

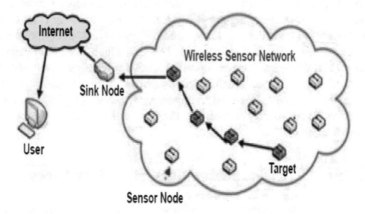

Fig. 2. WSN Architecture

Issues: WSN design issue is different for different wireless networks. It depends on the application, and for what purpose WSN required. Initially, WSN was mostly motivated by the military application and afterwards it has great scope in application domains like environmental monitoring, species monitoring, healthcare canter, smart homes and etc.

Fault Tolerance: It is the capacity to keep up sensor network utilities with no disturbance because of sensor node disappointment. The sensor node may become failure when there is lack of power, environmental interference and by any physical damage. Because sensor node failure, the whole system event of the sensor network influenced which cause fault tolerance issues.

Scalability: The hundreds, thousands or many more sensor nodes are organized in the detecting area. As network size is increasing, routine events which are carrying like constructing the path to sink should not be affected and it should not introduce the overhead communication.

Self-organizing Capability: After the organization of sensor nodes in the hazardous places or remote places where the human monitoring the system is not so easy, there sensor nodes must communicate with one another without human interference to monitor the system is known as self-organizing capability.

Load Balancing: So as to maintain and increase the network lifetime, load balancing plays an important role. Routing protocols should design in a way to adjust the load in the communication network.

Latency: The very important design issue in the real time applications. Information got from the WSN is time delicate. The computing time to produce data should be less to achieve real time applications.

Clustering: Grouping of sensor nodes is known as clustering. It is powerful technique to achieve objectives of scalability, improves energy efficiency and prolonged network lifetime in huge scale conditions (Fig. 3).

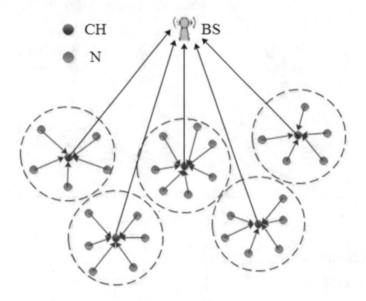

Fig. 3. Clustering in WSN

4   Existing Model

LEACH is a self-adaptive, self-organized and low energy protocol. It has application explicit information preparing, for example, collection or compression. In LEACH protocol, CH is selected on rotation bases and each is comprised of two stage for the reduction of pointless vitality costs [8]. In the set-up stage cluster arrangement is done and genuine information is transmitted in consistent state stage. Each sensor node picks an arbitrary number somewhere in the range of 0 and 1 to be the cluster head. LEACH group head is chosen based on of limit esteem. The selected irregular number by the sensor node is compared with T(n) threshold value which is nothing but limit esteem. In the event that the number chose is not as much as T(n), then node turns into CH for the present rotation. T(n) is controlled by condition:

$$T(n) = p/1 - p * (r mod 1/p) \; if \; n \in G$$
$$0, \; otherwise$$

Where p is the CHs percentage in network, r indicates round number which already ended, G represents collection of nodes which have not at been chosen as CH (Fig. 4).

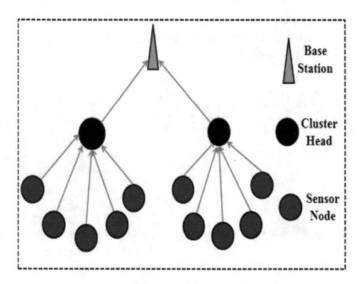

Fig. 4. Overview of LEACH

The following two stages of LEACH are:

1. **Set-up stage:**

   - Based on the limit esteem T(n), each node will pick whether to turn into CH or not.
   - After the CH choice, every one of residual node will pick its very own CH and depending on energy nodes join the cluster.
   - Every node chooses closest CH.

2. **Steady-state stage:**

   - CH adds the information packets and transfer the collected to BS through single-hop.
   - It randomly selects the CH for uniform energy utilization.
   - It balance the load in sensor organize.

Energy is consumed by "L" bits to move a separation d from transmitter or collector. This energy consumption is analysed by the radio model. The wireless system is isolated and forms layers, the count of sensor nodes forms the different size of clusters at each layer of network and the CH is selected at each clusters [7]. CH collects data and it is transmitted to BS. During this transmission a certain amount of energy is dissipated.

$$E_{TX}(L, d) = E_{TX-elec}(L) + E_{TX-amp}(L,d) \tag{1}$$

Eelec represent the energy disseminated per bit.

- εfs is the characteristic of the amplifier in the transmitter for free space.
- εmp is the characteristic of the transmitter amplifier for multipath.

Model of free space is utilized, at the point when the separation between the transmitter and recipient is not exactly the limit esteem d0. The d0 threshold value is written as:

$$d0 = \varepsilon\!f\!s / \sqrt{\varepsilon m p} \tag{2}$$

If distance d is more prominent than or equivalent to threshold value d0, at that point it pursues multipath. The power loss done by model of free space or multipath model can be controlled by the power control which inverts this loss by appropriate power amplifier adjustments. The measure of energy devoured by the recipient to get L bits of information is figured as

$$E_{RX}(L) = E_{elec} * L \tag{3}$$

5 Proposed Work

Fuzzy logic is valuable for settling on on-going choices with inadequate data about nature yet by and large regular control components need exact and complete data about the environment. Logic of the fuzzy is used for analysis of experimental model, human intervention and it can deal with continuous application more precisely than the stochastic model. Fuzzy logic acts as leader dependent on various ecological parameters by mixing them as indicated by predefined rules and these bunching calculations utilize fuzzy logic to deal with vulnerabilities in the remote sensor systems. T2FL handle the uncertain level decisions better than T2FL model. The parameters used in the T2FL for the enhanced selection of CH are remaining battery power, distance and concentration [19]. These parameters described as:

- Remaining battery power: residual power of every sensor node.
- Distance: separation between sensor nodes to sink.
- Node concentration: nodes number in the particular locality.

There are 27 fuzzy rules considered, that guidelines are predefined at the BS. The BS chooses the group head as indicated by these fuzzy guidelines[19]. The main benefit of T2FL is it evaluates energy efficiency and relieves network traffic by exploring base station. The proposed approach simulation is better than LEACH protocol and carried out using NS2 simulator (Fig. 5).

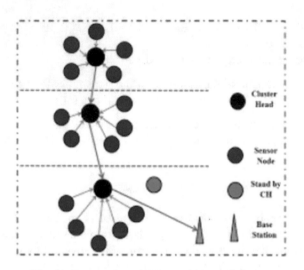

Fig. 5. Proposed model for multi hop clustering

/* for each rounds */

(1) 81 sensor nodes appropriated randomly over 550×550 locate zone and 9 is the cluster number expected.
(2) 81 sensor hubs are separated into various levels.
(3) Level ought to be masterminded by separation.
(4) Select cluster head at each dimension dependent on T2FL.
(5) Apply fuzzy rules, it is rule of if-then to choose group head.

  /* for "k" CHs */

(6) Transmit the information from one cluster head to another group head till it achieves the base station yet information must originate from the upper layer.
(7) Sensor another node with more energy is chosen as a aggregated CH which is close to BS to keep up the availability, if any failure happens by ready CH.

  /* one round ends */

(8) Base station gathers the collected information from CH in the chain.

  /* rounds ends */

The clustered T2FL algorithm as following step:

- Initialization phase
- splitting into number of clusters
- Cluster head selection
- Aggregated cluster head selection
- Self-sorted out information gathering and transmission
- Data exchange phase

Initialization Phase. In this phase, parameters of the systems are presented. The considered parameters are area required, nodes number, radio model, initial energy, packet size, routing protocol etc. After the assumption of these parameters, the same amount of energy is given to all wireless sensor nodes. Since, assumed network is homogeneous. Wireless network stars its initialization process by communication with each other sensor nodes.

Splitting into Clusters. To accomplish scalability, wireless sensor nodes in the wireless sensor system are assembled to frame groups. Important of all sensor nodes are splitting into number of clusters is that, the data packets send within the cluster through which the energy consumption is reduced.

Cluster Head Selection. T2FL algorithm is used to choose the CH within cluster. The input linguistic variables for type2 fuzzy logic taken in current algorithm are battery power, distance and concentration. Data packets to BS or to another sensor node transmitted through CH only.

Aggregated Cluster Head Selection. It acts as Stand by CH to CH in the cluster, through which data packets are send to BS when cluster head fails to transmit.

Selforganized Data Collection and Transmission. As per the self-organized capability of the sensor node, each sensor node collects or gathers the data and it will deliver information packet. These data packet reaches the base station through cluster head. It follows multi-hop communication using Viterbi algorithm. The sensor nodes has self-organized energy level, concentration and distance.

Information Exchange Phase. After the recognition of reduction of energy level in current CH, it is exchanged in next round. All nodes are again monitored and automatically update the change. Every one of the packets is sent, then goes to next round.

6   Simulation Setup

Network simulator is a discrete instance test system, which gives significant help to recreate bundle of conventions like transmission control protocol, record move protocol, user datagram protocol, hypertext move convention and dynamic source routing. It mimics both wired and remote system. It is a bundle of devices that reproduce organize practices, for example, making system topology examines the occasions and comprehend the system. Network simulator of version 2 is called as NS2which is utilized to break down the presentation of proposed protocol (Table 1, Figs. 6 and 7).

Table 1. Simulation parameters

Simulation parameters	Value
Simulator	Ns-2 (2.35)
Topology	550*550
No. of clusters	9
No. of nodes	81
Nodes distribution	Random
Channel	Wireless
Packet size	50 bytes
Simulation time	50.5 s
Initial energy	100 Joules
Routing protocols	DSR
Mac	IEEE 802.11
Model	Two ray ground

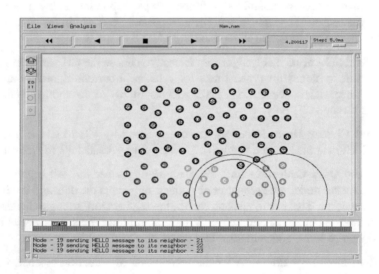

Fig. 6. Random deployment of sensor nodes with cluster forming in WSN

Performance Analysis

The performance is analysed between Type2 fuzzy logic clustering algorithm and LEACH with parameters:

- Throughput
- PDR
- Packet drop
- Throughput
- Average energy

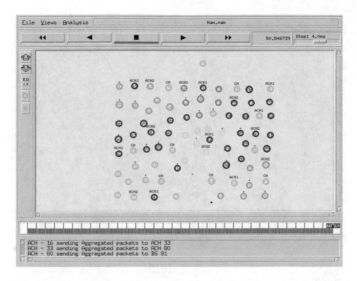

Fig. 7. Data packets are transmitted to BS through aggregated CH.

Throughput: Average packet rate is increased using fuzzy logic in a network system (Fig. 8).

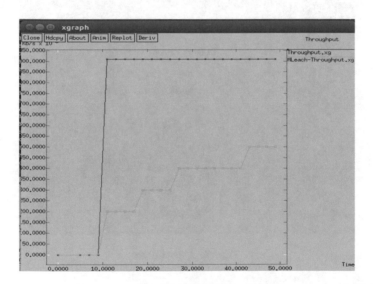

Fig. 8. Throughput

Packet Delivery Ratio: It increases the complete number of packets effectively at BS to the number of packets sent by the source (Fig. 9).

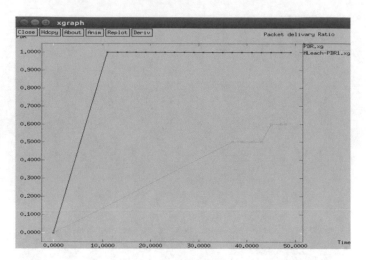

Fig. 9. PDR

Packet Drop: Packet drop is nothing but packet loss, it occurs when one or more packets of data are moving to BS in WSN (Fig. 10).

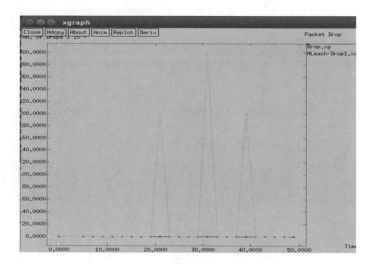

Fig. 10. Packet drop

Energy Consumption: The average energy consumption by network is minimized using fuzzy logic (Fig. 11).

Fig. 11. Average energy

7 Conclusion

The T2FL clustering algorithm is compared with LEACH protocol. From the outcomes obtained, it is analysed that T2FL clustering algorithm provides better performance with regard to average energy consumption, PDR, packet drop and throughput. The T2FL inputs are linguistic variables which are fuzzy itself. Hence, it can handle real time uncertainties better than other protocols. Uncertain parameters are considered to select the CH in the fuzzy logic. Implementation of fuzzy logic in WSN is an enhanced and improved method to elect CH.

References

1. Heinzelman, W.R., Chandrakasan, A., Balakrishnan, H.: Energy-efficient communication protocol for wireless microsensor networks. In: Proceedings of 33rd Hawaii International Conference on System Science (HICSS), Washington, DC, USA, pp. 1–10, January 2000
2. Heinzelman, W.B., Chandrakasan, A.P., Balakrishnan, H.: An application-specific protocol architecture for wireless microsensor net-works. IEEE Trans. Wirel. Commun. **1**(4), 660–670 (2002)
3. Lindsey, S., Raghabendra, C.S.: PEGASIS: power efficient gathering in sensor information systems. In: Proceedings of IEEE Aerospace Conference, pp. 3-1125–3-1130, March 2002
4. Gupta, I., Riordan, D., Sampalli, S.: Cluster-head election using fuzzy logic for wireless sensor networks. In: Proceeding of Communication Networks and Services Research Conference, pp. 255–260, May 2005

5. Kim, J.-M., Park, S.H., Han, Y.J., Chung, T.: CHEF: cluster head election mechanism using fuzzy logic in wireless sensor networks. In Proceedings of ICACT, pp. 654–659, Februray 2008
6. Fu, C., Jiang, Z., Wei, W., Wei, A.: An energy balanced algorithm of LEACH protocol in WSN. IJCSI Int. J. Comput. Sci 10, 354 (2012)
7. Taheri, H., Neamatollahi, P., Younis, O.M., Naghibzadeh, S., Yaghmaee, M.H.: An energy-aware distributed clustering protocol in wireless sensor networks using fuzzy logic. Ad Hoc Netw. 10(7), 1469–1481 (2012)
8. Sharma, T., Kumar, B.: F-MCHEL: fuzzy based master cluster head election leach protocol in wireless sensor network. Int. J. Comput. Sci. Telecommun. 3(10), 8–13 (2012)
9. Siew, Z.W., Liau, C.F., Kiring, A., Arifianto, M.S., Teo, K.T.K.: Fuzzy logic based cluster head election for wireless sensor network. In: Proceedings of 3rd CUTSE International Conference, Miri, Malaysia, pp. 301–306 November 2011
10. Nehra, V., Pal, R., Sharma, A.K.: Fuzzy based leader selection for topology controlled PEGASIS protocol for lifetime enhancement in wireless sensor network. Int. J. Comput. Technol. 4(3), 755–764 (2013)
11. Ran, G., Zhang, H., Gong, S.: Improving on LEACH protocol of wireless sensor networks using fuzzy logic. J. Inf. Comput. Sci. 7(3), 767–775 (2010)
12. Ando, H., Barolli, L., Durresi, A., Xhafa, F., Koyama, A.: An intelligent fuzzy-based cluster head selection system for WSNs and its performance evaluation for D3N parameter. In: Proceedings of International Conference on Broadband, Wireless Computing, Communication and Applications, pp. 648–653, November 2010
13. Arabi, Z.: HERF: a hybrid energy efficient routing using a fuzzy method in wireless sensor networks. In: Proceedings International Conference on Intelligent and Advanced Systems (ICIAS), June 2010
14. Mamdani, E.H., Assilian, S.: Implementation of mamdani fuzzy method in employee promotion system. In: IOP Conference 2017
15. Dasgupta, S., Dutta, P.: An improved leach approach for head selection stratergy in a fuzzy-c means induced clustering of a WIN, 16 December 2010
16. Nayak, P., Anurag, D., Bhargavi, V.V.N.A.: Fuzzy based method super cluster head election for wireless sensor network with mobile base station (FM-SCHM). In: Proceedings of 2nd International Conference on Advanced Computation, Hyderabad, India, pp. 422–427 (2013)
17. Wang, Y.-C., Wu, F.J., Tseng, Y.C.: Mobility management algorithms and applications for mobile sensor networks. Wirel. Commun. Mobile Comput. 12(1), 7–21 (2012)
18. Handy, M.J., Haase, M., Timmermann, D.: Low energy adaptive clustering hierarchy with deterministic cluster-head selection. In: Proceedings of International Workshop Mobile Wireless Communication Networks, pp. 368–372, September 2002
19. Nayak, P., Anurag, D.: A fuzzy logic-based clustering algorithm for WSNto extend the network lifetime. IEEE Sensor J. 16(1), 137–144 (2016)
20. Kocakulak, M., Butun, I.: An overview of WSNs towards internet of things. In: IEEE 7th Annual Computing Workshop and Conference (2017)

Information Diffusion: Survey to Models and Approaches, a Way to Capture Online Social Networks

Mohd Abas Shiekh[1](✉), Kalpana Sharma[1],
and Aaquib Hussain Ganai[2]

[1] Bhagwant University, Sikar Road, Ajmer, Rajasthan, India
Er.abasshiekh2014@gmail.com, kalpanasharma56@gmail.com
[2] University of Kashmir, Hazaratbal, Srinagar, J&K, India
hussainaaquib332@gmail.com

Abstract. Since the emergence of the Online Social Networks, people have been increasingly involved in the online mode of communications. They have been more or less influenced by these online communications and the idea of having or making an influence has been a key specialization for Online Social Network users. To control or spread something that is good or bad, we always have to point out the key players on Online Social Networks, so that we can take out remedial steps in order to deal with good or bad diffusions around the Globe in the modern era of Online Social Networks. These key players are called as influential users. To model the extraction of these influential users and to model information influence in the Online Social Networks the researchers have modeled as information diffusion model. In this paper, we are going to explore some of the models and the approaches that have an advantage to tackle the problems relating to the information diffusion in Online Social Networks.

Keywords: Social networks · Online Social Networks · Influential user · Information diffusion · Online communications

1 Introduction

The term "social networks" was used by Barnes [11] and now an Online Social Network is always any web place that involves the interactions of people [9]. The idea of taking Online Social Networks under the scanner of researchers was made by Girvan and Newman in 2002 [10, 12], since then the field have emerged as a new domain of research and this field has come across a long journey of problem which are carried in the Online Social Networking domain. A social network is represented by a graph, called a social graph G (V, E) [10], V is a set of nodes representing the individuals or users in the given Online Social Network and E is a set of edges representing the communications in the given Online Social Network. Social influence is the act of making other user's behavior dependent on our communications in an Online Social Network [13]. The information flows in an Online Social Network [11, 15], with users changing their activities, behaviors and perceptions, these characteristics are changed not only by the content of the information but also by the frequency of the mentions

© Springer Nature Switzerland AG 2020
D. J. Hemanth et al. (Eds.): ICICI 2019, LNDECT 38, pp. 25–32, 2020.
https://doi.org/10.1007/978-3-030-34080-3_3

(messages) among the users in an Online Social Network. It is clear from the real Online Social Networks that every user in an Online Social Network has a different degree of influencing on others. As the Online Social Networks are implicitly complex so solving the problems of finding the influential users, influential information diffusions and content popularity are very hard. These problems are to be captured by many state of the art models and approaches, so as to have their impressions in the minds of the researchers will act as the milestones for the further journey of carrying out the field to the further expectations. These problems in the hand can be modeled by the information diffusion models [6].

2 Models

Online Social Networks are diverse in nature and can give the users a wide domain of perceptions to take these fatty networks in their own means according to the problem set for the given Online Social Networks in their hands. Every model has a means to process the Online Social Networks, to the tackle some related queries. Information diffusion in Online Social Networks can be best understood by Fig. 1 below:

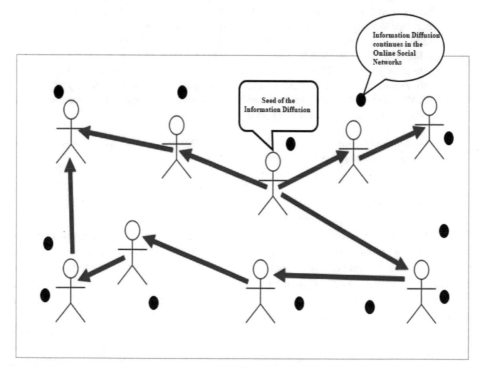

Fig. 1. Information Diffusion in Online Social Networks

Every model takes advantage of some implicit parameter or characteristic to devise model that acts as a healthy vehicle in providing conceptual understanding of the domain of Information Diffusion in Online Social Networks. Ever model helps in providing some sort of insight into the field of the information diffusion in Online Social Networks, so we should always take each and every model as a fruit bearing model to capture the queries relating to information Diffusion in Online Social Networks. If one is truly thirsty about handling the different aspects on Information Diffusion in Online Social Networks, he should first get an insight into every model so that if he has to deal with the Information Diffusion in Online Social Networks whose domain is vast, his dealing will naturally match to the truth about the field of Information Diffusion in Online Social Networks. The most giant models of information diffusion [14] are:

Epidemic model, Cascading model and the *Threshold model* [5] and are shown in the Fig. 2 given below.

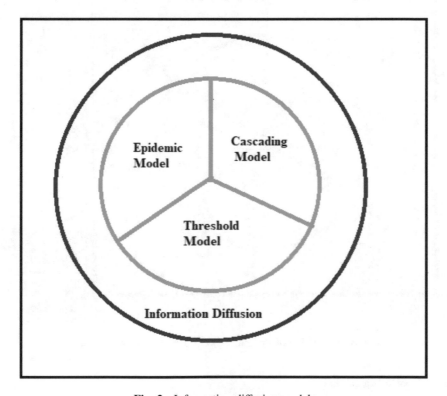

Fig. 2. Information diffusion models

2.1 Epidemic Model

These models are inherited from the biological sciences and these models are very popular in the field of medical science to capture how diseases were getting replicating at a very rapid pace. These models are now used for the Online Social Networks to tackle the problem of how some contents are getting popular and why others are not getting popular. These models are best suited for finding content influence in Online Social Networks. Epidemics of some contents in Online Social Networks need to be captured, in order to patternise which contents peacefully created bad percepts and which one created the good percept among the users of Online Social Networks. These epidemics can help to devise such tools that can help in dumping the bad contents or blowing the contents that can spread the good. In this model, we capture the pace at which the Information gets Diffused in the given Online Social Network. To work on this model of information diffusion we have to carry out the snapshots of Online Social Networks in the frame of temporal standards that are defined for the given model (Fig. 3).

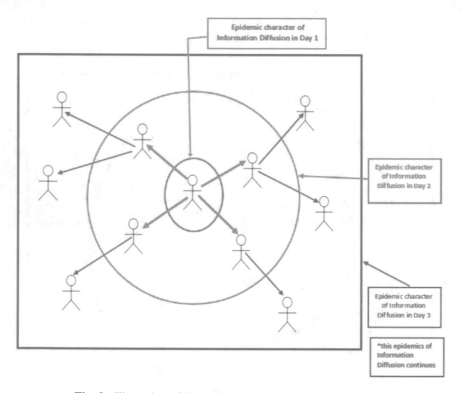

Fig. 3. Illustration of Epidemic model for Information diffusion

2.2 Cascading Model

This model was pioneered by Goldenberg et al. [7, 8]. These models are just using probability of activating a near neighbor at time t+1, if it has been activated by its previous member at t; which has been activated by its predecessor at time t−1. It just works on covering the activation chains by working on the model of induction of activation on temporal bases. In this model missing links has to be taken in the restrictive manner. This model works as of taking the advantage of mathematical induction; dividing the online communications as of being carried out in the pipeline stages and making sure that if earlier stage is active then only this pipeline cascade can proceed otherwise it can't. This pipeline of information cascades or activation cascades are not as usual but they are of giant stage pipelines so to the unknown stages from known stages we should use the probability of occurrence of current snapshots in the future stages as well. In this cascading model of information diffusion we must capture the present communications paths in the natural order so as to make the prediction for unknown paths very precise. This cascading model of Information Diffusion is depicted in the given Fig. 4 as:

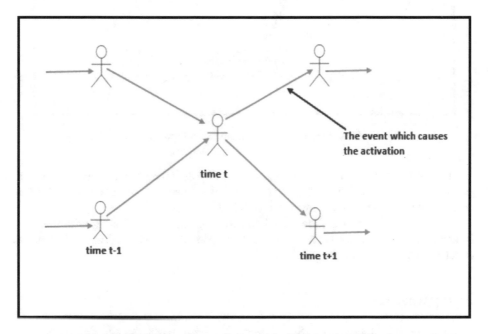

Fig. 4. Illustration of cascading model of Information Diffussion

2.3 Threshold Model

The model made its steps the work of Granovetter [9]. Here a threshold that is a number which is used to state whether a user get activated or not that is it gets influenced by its neighbors: by taking these influence values in an integral manner. If the threshold of the node under processing gets more influence value that its threshold. It is definite that it is influenced or activated. This Threshold Model is depicted in the Fig. 5 given:

These model form the basis to capture the Information Diffusion in Online Social Networks, although there are many variations of these models but the basic framework remains same. Depending on the context of problem to be solved respective model should be taken in an advantageous manner.

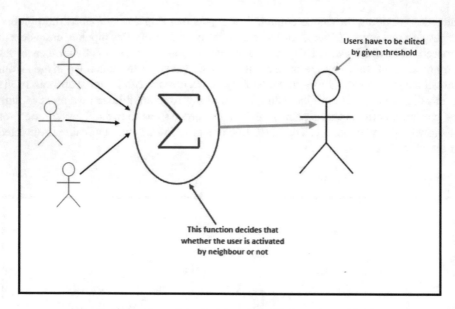

Fig. 5. Illustration of threshold model of Information Diffussion

This model gives users a good categorization by assigning them thresholds that are just weights and to activate a user which is wholly dependent on its threshold and its neighboring communications must overweigh it.

The real challenge of Online Social Networks when we are dealing with them are isolated nodes, so it is always best practice to get rid of these isolated nodes in order to lower the computational complexities.

3   Approaches

Many researchers have taken the advantage of these basic models to use them in proposing the approaches to solve the queries that are getting evolved to them from the perceptions of the Online Social Networks.

Some of the important approaches are:

3.1   Profile Rank Approach

Profile Rank Approach [1] finds the influential users and their content relevance, taking the advantage of *Page Rank Algorithm* with *Random Walks*. This approach applies Page Rank algorithm with Random Walks on the user's content Graph. This approach has all its bases on user content with temporal framework.

There are *many drawbacks* to this approach; it does not take the real temporal feature into consideration, lacks the real dynamics of the Online Social Network that is implicit in Online Social Networks.

3.2   Logistic Mathematical Approach

For Logistic Mathematical Approach [2] logistics are made possible by using differential equations. This approach uses two distance metrics:

I. Friendship Hops.
II. Shared Interests.

It uses two processes to model the diffusion:

I. Social Process.
II. Growth Process.

The *main drawback* of this approach is that most parameters are known only when the Online Social Network graph gets synthesized.

3.3   Information Cascading Approach

Information Cascading Approach [3] learns how the information gets propagated from user to user. In this the Information Diffusion is based on the message streams in the underlying Online Social Network and in the respective social graph of the online network under investigation. This approach learns the influential users by uncovering the influence paths in the given network.

The *measure drawback* of this approach are the *missing links* and *messages,* which are halting the cascade paths starting from a sender, therefore needs to be handled in a restrictive case.

3.4   General Diffusion Threshold Approach

General Diffusion Threshold Approach [4] takes the advantages of the two well-known models of information diffusion; linear threshold model and information cascade model. This approach works by getting trained from activities and after training classifies the further interactions in Online Social Networks.

The measure drawbacks of this approaches are:

Giving threshold for the activation testing to a user is random and is very hard to offer concretely, and another disadvantage is to enumerate human emotions is very hard.

These are the some of the approaches which provide enough means to tackle the real queries that are implicitly asserted by Online Social Networks.

4   Conclusion

Information diffusion and its handlings in an Online Social Networks is a diverse domain of discourse in the contemporary world. We tried to precisely summarize the various models and approaches used to find out the influential nodes in the Online Social Networks and found that what exists that is what has been done already is best and provides the better insights in the era of information diffusion in Online Social

Networks, so we must use the state of the art models and approaches to urge for further findings and ours have the aim to refine the field further by taking existing toil in hand as a base.

References

1. Silva, A., Guimaraes, S., Meira Jr., W., Zaki, M.: Profile rank: finding relevant content and influential users based on information diffusion. In: ACM. 978-1-4503-2330-7-7/06/13
2. Wang, F., Wang, H., Xu, K.: Dynamic mathematical modelling of information diffusion in online social networks
3. Taxidou, I., Fischer, P.M.: Online analysis of information diffusion in Twitter. In: ACM. 978-1-4503-2745-9/14/04
4. Cole, W.D.: An information diffusion approach for detecting emotional contagion in online social networks. A thesis for MS degree
5. Zhang, H., Mishra, S., Thai, M.T.: Recent advances in information diffusion and influence maximization of complex social networks. Opportunistic Mobile Soc. Netw. **37**, 37 (2014)
6. Mottin, D., Lazaridou, K.: Graph mining: social network analysis and information diffusion. Hasso Plattner Institute
7. Goldenberg, J., Libai, B., Muller, E.: Talk of the network: a complex systems look at the underlying process of word-of-mouth. Mark. Lett. **3**, 211–223 (2001)
8. Goldenberg, J., Libai, B., Muller, E.: Using complex systems analysis to advance marketing theory development. Acad. Mark. Sci. Rev. **9**, 1–18 (2001)
9. Granovetter, M.S.: Threshold models of collective behaviour. Am. J. Sociol. **83**(6), 1420–1443 (1978)
10. Singh, A.K., Gambhir, S.: A comprehensive review of overlapping community detection algorithms for social networks. Int. J. Eng. Res. Appl. (IJERA) ISSN: 2248-9622
11. Li, M., Wang, X., Gao, K., Zhang, S.: A survey on information diffusion in online social networks: models and methods. Information **8**, 118 (2017)
12. Choudhury, D., Paul, A.: Community detection in social networks: an overview. IJRET: Int. J. Res. Eng. Technol. eISSN: 2319-1163
13. Tang, J.: Social influence and information diffusion. Department of Computer Science and Technology Tsinghua University
14. Weng, L.: Information diffusion in online social networks, April 2014
15. Wani, M., Ahmad, M.: Survey of information diffusion over interaction networks of Twitter. Int. J. Comput. Appl. **3**, 310–313 (2014)

A Miniaturized Circular Split Ring Embedded Monopole for WiMAX Applications

S. Imaculate Rosaline[(⊠)], M. B. Akkamahadevi, N. Pooja,
and Rajani S. Kallur

Ramaiah Institute of Technology, Bangalore, Karnataka, India
rosaline@msrit.edu

Abstract. Two concentric circular split rings embedded onto a rectangular monopole radiator is designed for 3.5 GHz (WiMAX) applications. The antenna is designed on a $20 \times 20 \times 1.6$ mm^3 FR4 dielectric with a slotted ground plane structure. The monopole is designed as a $\lambda/2$ radiator. The addition of splits in the two concentric circular rings supports in bringing down the frequency to a lower band of 3.5 GHz, which is nearly half of the monopole's resonance. This yields a miniaturization of about 50%. The slotted ground plane aids in achieving wider impedance matching at the desired mode of operation. The antenna radiates uniformly well in all directions with a peak gain of 1.5 dBi, making it suitable for wireless communication devices.

Keywords: Circular split ring · Slotted ground · Monopole radiator · WiMAX · Miniaturization

1 Introduction

Antenna is the vital component of any communication device. Most of the portable devices require miniaturized and compact antennas. Microstrip technology serves as the best option in such cases. Many antennas employing several techniques like fractal geometries, defected ground plane structures, parasitic elements, stubs and slots, metamaterial inspired structures are proposed in the past for WiMAX applications. Several fractals based methodologies for WiMAX applications is discussed in [1]. A monopole with fractal Koch [2] is utilized for UWB application. A cambered shaped defected ground plane [3] provides a wideband resonance. Three slots on a half solid ground excites additional resonance as proposed in [4], A patch in the shape of fan along with an additional radiator is discussed in [5] for WLAN/WiMAX applications. A patch antenna along with parasitic components [6] operates on various wireless bands. A modified folded dipole with dual elements and a monopole [7] is employed for WLAN/WiMAX applications. T shaped and L shaped stubs [8] are used in achieving triple band resonance. Metamaterial inspired antenna [9] yields dual band resonance. A compact MNG material loaded antenna for WLAN/WiMAX applications is proposed in [10]. A monopole antenna with an asymmetrical meander lines SRR is

© Springer Nature Switzerland AG 2020
D. J. Hemanth et al. (Eds.): ICICI 2019, LNDECT 38, pp. 33–39, 2020.
https://doi.org/10.1007/978-3-030-34080-3_4

proposed [11] to enhance the antenna bandwidth and radiation properties. Different ways to achieve size reduction in antennas is discussed in [12]. An annular slot ring antenna loaded with Split Ring Resonators which are also electromagnetically coupled is proposed in [13] to yield miniaturization of about 30%.

This paper presents a compact split ring embedded monopole radiator with slotted ground plane for 3.5 GHz application. The antenna achieves a broad impedance bandwidth of about 500 MHz, which is 142% of centre frequency. The split in the rings is capable of achieving miniaturization of about 50%. The radiation pattern is nearly omnidirectional at the operating frequency making it suitable for wireless communication devices.

2 Structure of the Proposed Circular Split Ring Antenna

The design process evolved in obtaining the circular split ring antenna is shown in Fig. 1. Configuration 1 shows a simple circular ring embedded rectangular monopole with a partial ground plane. The rectangular monopole is 17 mm in length which corresponds to $\lambda_g/2$ at 4 GHz. Embedding circular ring onto the monopole shifts the resonance to 7 GHz. Here, the radius of the ring plays a vital role in choosing the operating frequency. The width of the ring is optimized to yield the desired band of resonance. Further, a concentric inner ring is added to realize configuration 2. The distance between the rings is carefully chosen so that the microstrip feed and the antenna are properly matched. Subsequently, four splits are introduced along both the circular rings at an angle of 45° to the horizontal axis of the antenna to develop configuration 3. The splits in the rings contribute to the size reduction of the antenna. The position and width of the splits acts as controlling parameter of the resonating frequency. The proposed antenna is then obtained by introducing a slotted ground plane. With a partial ground structure, an impedance bandwidth of merely 200 MHz was achieved. But with the slotted ground plane, the significance of the antenna resonance enhances and hence yields a wide impedance bandwidth. Detailed geometrical layout of the proposed antenna is displayed in Fig. 2. The antenna is printed on a $20 \times 20 \times 1.6$ mm^3 FR 4 substrate. The dimensions of the antenna are chosen carefully after several parametric studies. The finalized dimensions are as follows: L1 = 20 mm; L2 = 0.5 mm; L3 = 0.5 mm; L4 = 3 mm; L5 = 5 mm; L6 = 2 mm; W1 = 20 mm; W2 = 2 mm; W3 = 0.2 mm; R = 7 mm; θ = 45°.

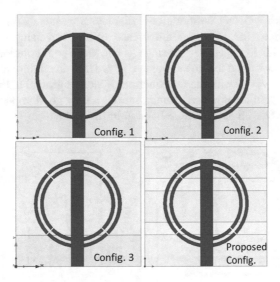

Fig. 1. Evolution of the proposed circular split ring antenna

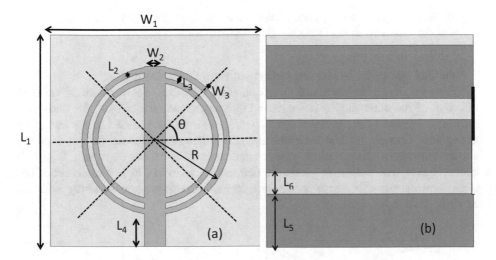

Fig. 2. Layout of the proposed antenna design (a) Top plane (b) Bottom plane

3 Results and Discussion

The simulations are carried out using Ansys Electronics 19.1. Figure 3 illustrates the comparison of S_{11} characteristics of different antenna configurations proposed in Fig. 1. It is inferred that the configuration 1 shows a resonance around 7 GHz. The radius of the circular ring is chosen based on the circular patch antenna formula given by $a = \dfrac{F}{\sqrt{\left\{1 + \frac{2h}{\pi \varepsilon_{rF}}\left[ln\frac{\pi F}{2h} + 1.7726\right]\right\}}}$ where $F = \dfrac{8.791 \times 10^9}{f_r \sqrt{\varepsilon_r}}$. So, for the radius of 7 mm, we infer a resonance around 7 GHz. Configuration 2 displays a frequency dip around 8.5 GHz. This is attributed to the path length increase of the traversing current. With the introduction of splits in the concentric rings as shown in configuration 3, the operating frequency is further shifted to a lower value of 3.8 GHz, in additional to a higher band around 9 GHz. However, with the introduction of slotted ground at the rear side of the substrate, as shown in the proposed configuration, a very good impedance matching is realized at 3.5 GHz with an extensive bandwidth of 500 MHz.

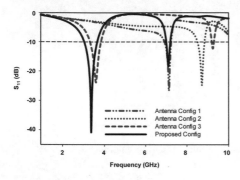

Fig. 3. Simulated S11 characteristics of the different configurations shown in Fig. 1

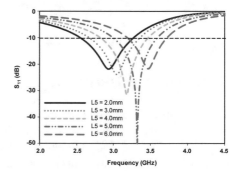

Fig. 4. Simulated S11 characteristics for some values of R

Various parametric studies are carried out to optimize the results.

Figure 4 shows the simulated reflection (dB) characteristics for various value of radius R of the outer ring. It is concluded that as the radius increases, the frequency moves to a lower value, because, the frequency is in reverse related to the antenna's dimension. Thus, R = 7 mm is chosen as the optimum value for our design. Figure 5 shows the similar characteristics for various values of monopole width W_2. It is observed that, with the rise in width, the frequency moves to a higher value. Since this dimension corresponds to the feed width, it loses its impedance matching with the antenna as its value increases.

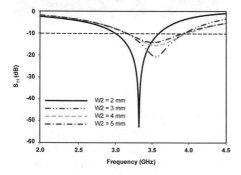

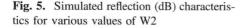

Fig. 5. Simulated reflection (dB) characteristics for various values of W2

Fig. 6. Simulated reflection (dB) characteristics for various values of L5

Figures 6 and 7 shows the simulated characteristics of the return loss in dB, for various values of L5 and L6 in the ground plane. L5 denotes the partial ground plane's length and L6 denotes the slot length in the ground plane. As L5 increases, resonant frequency shifts from lower range to higher range is observed, whereas, increasing the length L6, exactly depicts the increase in the impedance bandwidth. Here, L6 = 0 mm means there is no slot between the three strips of ground planes. As, this is increased, the lower frequency range gets shifted still lower, while maintain the upper frequency range. Thus, L6 = 2 mm is chosen as optimum for our design. Beyond 2 mm is not possible, as the ground plane will exceed out of the substrate.

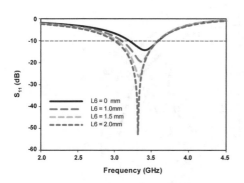

Fig. 7. Simulated S11 characteristics for various values of L6

Fig. 8. Simulated S11 characteristics for various values of W3

W3, width of the splits is an important parameter in achieving miniaturization of our antenna. A parametric study on this parameter is shown in Fig. 8. It is inferred that as this value is increased the resonance is shifting to a higher frequency value, which means the percentage of miniaturization is increased which is not desirable. Hence, an optimum value of W3 = 0.2 mm is selected for our design.

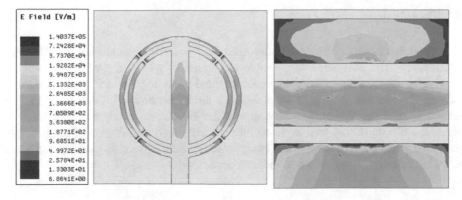

Fig. 9. Simulated Electric field distribution of the proposed antenna

Figure 9 shows the simulated electric field distribution on the antenna's top and bottom plane at the operating frequency 3.5 GHz. It is seen that the filed distribution is more at the splits and at the edges of the ground plane justifying our parametric analysis which contributes to the miniaturization and impedance matching concepts. Figure 10 shows the computer-generated azimuthal plane and elevation plane pattern of the proposed antenna at 3.5 GHz. It is clear that the pattern is omnidirectional and bidirectional in the H plane and E planes. It also exhibits a peak gain of 1.5 dBi at the operating frequency.

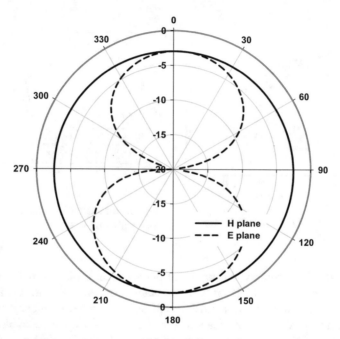

Fig. 10. Simulated E plane and H plane pattern of the proposed antenna

4 Conclusion

The paper proposes a simple circular split ring embedded monopole for WiMAX applications. The antenna is very simple and compact with a size of $20 \times 20 \times 1.6$ mm^3. The use of splits in the rings which is tilted at an angle of $45°$ to the horizontal axis of the antenna is utilized for achieving miniaturization of about 50% and the slotted ground plane aids in achieving a wider bandwidth of about 500 MHz. The antenna emits uniformly well in the azimuthal plane with a peak gain of 1.5 dBi. The proposed antenna is very simple and compact to reproduce in a bulk quantity, making it more suitable for all wireless communication devices.

References

1. Azaro, R., et al.: Fractal-based methodologies for WiMAX antenna synthesis. In: WiMAX, pp. 35–54. CRC Press (2018)
2. Zarrabi, F.B., et al.: Triple-notch UWB monopole antenna with fractal Koch and T-shaped stub. AEU-Int. J. Electron. Commun. **70**(1), 64–69 (2016)
3. Gautam, A.K., Bisht, A., Kanaujia, B.K.: A wideband antenna with defected ground plane for WLAN/WiMAX applications. AEU-Int. J. Electron. Commun. **70**(3), 354–358 (2016)
4. Kunwar, A., Gautam, A.K., Kanaujia, B.K.: Inverted L-slot triple-band antenna with defected ground structure for WLAN and WiMAX applications. Int. J. Microw. Wirel. Technol. **9**(1), 191–196 (2017)
5. Verma, M.K., Kanaujia, B.K., Saini, J.P.: Design of fan-shaped stacked triple-band antenna for WLAN/WiMAX applications. Electromagnetics **38**(7), 469–477 (2018)
6. Singh, S., Bharti, G.: A multi-band patch antenna with parasitic strips for wireless applications. In: 2018 IEEE International Students' Conference on Electrical, Electronics and Computer Science (SCEECS). IEEE (2018)
7. Park, J., et al.: Design and fabrication of triple-band folded dipole antenna for GPS/DCS/WLAN/WiMAX applications. Microw. Opt. Technol. Lett. **61**(5), 1328–1332 (2019)
8. Kumar, A., Sharma, M.M.: Compact triple-band stubs-loaded rectangular monopole antenna for WiMAX/WLAN applications. In: Optical and Wireless Technologies, pp. 429–435. Springer, Singapore (2018)
9. Hasan, M.M., Faruque, M.R.I., Islam, M.T.: Dual band metamaterial antenna for LTE/bluetooth/WiMAX system. Sci. Rep. **8**(1), 1240 (2018)
10. Patel, U., Upadhyaya, T.K.: Design and analysis of compact µ-negative material loaded wideband electrically compact antenna for WLAN/WiMAX applications. Prog. Electromagnet. Res. **79**, 11–22 (2019)
11. Alibakhshi-Kenari, M., Naser-Moghadasi, M., Sadeghzadeh, R.A.: Bandwidth and radiation specifications enhancement of monopole antennas loaded with split ring resonators. IET Microwav. Antennas Propag. **9**(14), 1487–1496 (2015)
12. Fallahpour, M., Zoughi, R.: Antenna miniaturization techniques: a review of topology-and material-based methods. IEEE Antennas Propag. Mag. **60**(1), 38–50 (2017)
13. Mulla, S.S., Deshpande, S.S.: Miniaturization of multiband annular slot ring antenna using reactive loading. J. Electromagn. Waves Appl. **32**(14), 1779–1790 (2018)

Extended Fibonacci Series for Selection of Carrier Samples in Data Hiding and Extraction

Virendra P. Nikam[✉] and Shital S. Dhande

Department of Computer Science and Engineering, SCET, Amravati, India
virendranikamphd@gmail.com,
sheetaldhandedandge@gmail.com

Abstract. Now a days data sensitivity and information security remains as a key challenge for every IT industry. Most of the industries have their focus on data security. The existing data security techniques like cryptography, steganography are used only to protect the data from unauthorized access. Recently, it is found that cryptography is not enough to protect the data when it is transmitted over wireless network. Proposed concept focuses on the Data Hiding mechanism, which enables the proper use of sample selections of carrier channel through the use of extended Fibonacci series. Proposed concept have achieved better results and generates less noise in carrier object after the concealing of secret information in it.

Keywords: Data security · Carrier object · Wireless media · Steganography · Cryptography · Extended Fibonacci series · psnr · Mean square error · Data Hiding Capacity

1 Introduction

Most of the Information Technology industries will store their data on server system. Data security is the primary task for any IT industry. Almost all the companies paid millions of dollars on security of data stored on server system. So many techniques are available to protect data on server. Some of the common techniques are cryptography, steganography, watermarking etc. Cryptography is a art of conversion of readable data into unreadable format. This is a very strong technique for data protection on storage system. So many cryptography techniques are available like public and private cryptography. Lots of algorithms are implemented as a concept of cryptography. Cryptography is a good art protecting data from unauthorized users. many of the IT industries are using cryptography techniques for data protection. Major drawback of

---

Virendra P. Nikam is a PhD Scholar doing his research work at Sipna College of Engineering and Technology. His major area of interest is developing cloud security through the use of Cryptography, Stegnography, Visible Watermarking. Prof. Dr. Shital S. Dhande is a Professor at Sipna College of Engineering and Technology in the Department of Computer Science. She is having more than 20 years of research experience especially in cloud computing and database management.

D. J. Hemanth et al. (Eds.): ICICI 2019, LNDECT 38, pp. 40–50, 2020.
https://doi.org/10.1007/978-3-030-34080-3_5

cryptography techniques is that it occupies either equal or more space than the original plain text. The drawbacks of cryptography technique are removed by concealing secret information in one of the object called as carrier object. This carrier object hides secret information without modifying original visual appearance of carrier object. This technique is called as steganography. Steganography is simply a art of hiding data behind another object. This removes the existence of secret data and creates an illusion for Intruder who always looks for sensitive data over transmission network. As steganography not shows any clue to Intruder, intruder may get difficulty in hacking process. However, not all the steganography technique hide secret information without modifying carrier object. If an expert scan carrier object, it may find change in carrier object. Change in carrier object can be located with the factors like psnr, mean square error, absolute difference, mean difference, structural content etc. This change in carrier object after hiding of sensitive data can be possible to dominate by the proper selection of samples of carrier object. From literature analysis, it is observed that possibility of suppressing noise in carrier object is almost 100% possible. This can be achieved by two ways

1. Random selection of pixels
2. Sequential selection of pixels

Random selection of carrier pixels generate very less noise, but it is too complicated.

Random sample selection of carrier achieved better results than sequential sample selections approach. Propose technique of sample selection through the use of extended Fibonacci series comes under random sample selection approach. Today, many programming languages directly provides functions for calculating random values within a range Like in C Sharp, rand (1, n) function is used to find random value from 1 to n numbers. One drawback associated with random function is that, the probability of occurrence of collision is more i.e. generating same number in a list. Random number is good to achieve suppression of noise occurs in carrier object after concealing of data but it is too complex and have an implications of generating large size key file. While extracting data on receiver side, receiver should know position of sample components from where secret information is to be extracted. Length of key file generated by random sample selection approach, is directly proportional to amount of secret information. Sequential sample selection approach not generate any key. Due to the simplicity of sequential approach, possibility of intruder attack is more. This is one of the simplest sample selection method in steganography techniques.

Sequential approach is a simple and generates more noise. It creates bands of noisy samples on carrier. Sequential approach is not popular sample selections of carrier in steganography techniques.

To use the proposed concept for selection of samples from carrier media, may generate excellent results than any other existing steganographic techniques. carrier media generally referred as either images, audio, video etc. If carrier media is an image, steganographic technique is called as image steganography. Similarly if carrier object is audio or video, steganographic technique called as audio or video steganography respectively. For image, carrier samples are consist of RGB channels. For audio, samples are consists of left and right stereo channels. One good thing associated with every sample is that, all sample are consist of 8 bits. With 8 bits sample, which have minimum value 0 and maximum value 255. In order to process steganography technique, maximum difference generated by hiding secret bits in carrier sample is vary from 0 to 255. Here with extended Fibonacci series, one cannot guaranteed the minimization of error difference between original samples and result samples because it depends on steganographic techniques where secret bits get hidden. But with proposed concept, noise get equally distributed throughout the carrier object and hence not easily visible to end user. For any steganography technique, quantization error is a big problem. So many approaches are given by authors to minimize quantization error [1] like flipping of sample of carrier object. Quantization error can be expressed as difference between result simple and original sample $Q_e = P_{Original} \, P_{Result}$. Weight of quantization error depends on at what position secret bit get hide in samples of carrier object. If bit hidden at first LSB position, it will generate quantization error of 1. As one should go from LSB to Most significant bit (MSB), quantization error increases. Weight of Quantization error can be Express $2BitPosition$ [3].

2   Background History

Steganography comes from Greek word steganos (means covered or secret) writing for drawing. Steganography can be defined as hiding of information by embedding message with another format without harming the original message. First stenographic technique was developed in ancient Greece around 440 BC. Greek ruler employed earlier version of steganography which in world shaving head of slave, tattoo team the message on slave scalp.

Steganography and Cryptography are guided by the same principle. Both hide message in specific media. One distinct difference among them is that, cryptography is dependent on hiding the meaning of message whereas steganography is dependent on hiding the presence of message to another. The most of the common techniques of steganography implemented using image and sound as a carrier file. Famous steganography technique is called as least significant bit (LSB) substitution for writing. As its name implies, LSB involves writing in lowest arithmetic value position

10010101   00001101   11001001   10010110   00001111   11001011   10011111 00010000

Although the above given example would be a perfectly viable way to use **LSB** substitution method, it is too basic to be practical. The main goal of steganography is to hide the presence of hidden message from human senses. Now a days modern steganography detection applications has alter those goals to include securing the hidden message from both human senses and digital applications. Due to this reason, meaning of the modern steganography techniques prefer to use randomize sample selection method for securing secret data behind carrier object.

2.1   Overview of Data Hiding and Extraction

Figure 1, architecture diagram gives overall view of transmission of sensitive data through the concept of steganography. Sender hide the data behind the carrier object by using steganography technique. Almost all the steganography technique requires key for sample selection from carrier object. Role of key is for selection of samples from carrier media. In random approach size of key varies from 1 to length of secret data. Steganography technique takes care of originality of carrier object after concealing of secret information. If result stego object is not similar with original carrier object, technique is not called as steganography. All algorithms of steganography take cares for noise suppression generated due to concealing of secret data. At receiver end, exact revers process is performed where receiver can extract data from resultant stego object with the help of key.

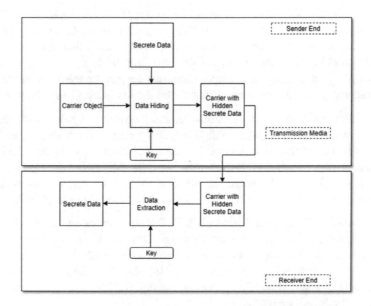

Fig. 1. Architecture view of Stegnography data hiding and Extraction

Key use on sender side is exactly equal to key at receiver side. This kind of steganography is called as symmetric steganography. One drawback associated with this approach is that, there is need to share key between sender and receiver. However it may create possibility to hack key while transmitting it over wireless network. Propose concept take care for not generating such types of key during data hiding process. It also suppressed noise created due to data Hiding process.

Proposed approach reduces the length of key used at the time of Data Hiding and extraction. With the help of extended Fibonacci series, it possible to reduce the length of key file with only size of two values that is current and previous index values. Fibonacci series have a drawback of limited samples selection as number of samples selected by series is always less than total number of item present in series. With propose extended Fibonacci series, it is possible to select almost all samples present in carrier object. As it minimizes key size, it may be a future choice for many developers in steganography techniques.

3   Literature Review

Purpose of literature review is to aware the readers about the existing techniques available with the given concept. Meaning of the concepts as written in introduction section, uses simple sample selection process.

Fu and Au [23]. In 2002, proposed a novel method for Data Hiding in halftone images. They proposes a method that enables to hide huge amount of data when the original multitone images are unavailable. The resulting stego images that are generated have high quality and virtually indistinguishable from original image. This approach also concentrate on hiding data in multiple channels orphan image with random pixel selection approach. Digest mentioned that secret data is hidden in halftone images.

Dey, Abraham and Sanyal [20]. In 2007 present a novel approach to hide the data in carrier images and then it is added to high frequency signal. The high frequency signal are replaced by one middle channel and then data is embedded. The detection of data is done by employing correlations between the sum and difference of stereo signal. This proposed approach directly replace middle one channel by using sequential sample selection approach due to which the exact data is not extracted through this algorithm.

Naseem, Hussain, Khan and Ajmal [17]. In 2011 presented an optimized bit plane slicing algorithm to Hide data in carrier image. This method suggest to hide data in carrier image pixels intensity instead of hiding it sequentially in pixel by pixel and plane by plane. They uses random approach to Hide data in plane instead of hiding it in pixels adjacent to each other. This approach making it difficult to guess and intercept the hidden data during transmission.

Mahjabin, Hossain and Haque [22]. In 2012 proposes Data Hiding method on pixel value differences and LSB substitution. This method achieved increase in embedding capacity and lower image degradation. Data Hiding capacity of this approach is high but creates noise in carrier object.

Fallahpour, Shirmohammadi, Semsarzadeh, and Zhao [21]. In 2014 use non zero DCT coefficient for hiding data. The proposed Method can hold more data, increases the Data Hiding capacity. After Data Hiding and re-encoding, which affect the data extraction process and therefore this method reduces robustness. They uses randomize pixel selection approach where only those pixels are selected which are not closer to zero and not closer to 255.

Shiu, Chen and Hong [25]. In 2015 proposed reversing order of encryption and vacating room technique prior to image encryption. In this approach, space is created for Secret data and then secret data is combined with surrounding pixel so as to minimize quantization error. This approach even though minimize quantization error, sequential pixel combination of surrounding generate noise that affect the quality of carrier image.

Qian, Dai, Jiang and Zhang [3, 6, 10, 11]. In 2016 Propose an improved method of reversible Data Hiding in encrypted images. Original image get encrypted using cipher stream algorithm and then upload it to remote server. On receiver side data get extracted from image using swapping/shifting based algorithm. With this technique large number of secret data get hidden in carrier image. Process of samples selection on encrypted image is done using sequential approach. Obtained result sample is flipped so as to minimize quantization error. **Huang, Huang and Shi** [12] in 2016 suggest Simple but effective Framework for RDH in encrypted domain. In this proposed work, plane image is first divided into multiple blocks and then with an encryption key, stream of block are combined with plain text message to produce an encrypted blocks. After stream encryption, encrypted block are randomly permuted with a permutation key. Since correlations between neighbouring pixels in each encrypted block can be preserved in an encrypted domain, data recovery is possible with exactly same as that of hidden data. This project approach uses sequential sample selection strategy for combining divided blocks with plain text message. After encrypting entire block, all blocks are assembled so as to generate final encrypted stego image.

4 Proposed Concept

Propose extended Fibonacci series is an extension of existing Fibonacci series where prior objective is to select almost all samples available in carrier object. As mentioned in introduction, Fibonacci series requires base number F_n and F_{n-1}. Let's illustrate this concept with an example of series as given below 1, 2, 3, 4, 5, 6, 7, …… n Let's initial index selected are 1 and 0. So first sample selected is 1 and second sample selected is 2. Resultant sample selected with simple Fibonacci series are 1, 2, 3, 5, 8, 13, 21……m. Where m ™ n.

Sample selected with extended Fibonacci series as below 1, 2, 3, 5, 8, ….. m. if $m > n$ New two indices are chosen so as to select remaining sample in a list. So new indices will be selected as 3 and 2. Item present at position 3 is 3 which is already present in a chosen list. It means selected choice of index is somewhere wrong. So again new indices are selected like 4 and 3. Item present at index 4 is 4, which is not present in sample list. It means selected index choice is correct. So 4 is added in list and then 7 etc. With this one, result obtained from extended Fibonacci series is as below

  1, 2, 3, 5, 8, ….. m
  4, 7, 11, 18, 29, …… m

Process of new index selections will be continued until required number of samples are not selected. Below **algorithm** implemented for extended Fibonacci series can be written as

1. Start
2. Input carrier object.
3. Extract Samples of Carrier Object L_i .
4. Count length of Extracted Samples n .
5. Select Start Index (S_i) and End Index (S_{i-1})
6. For i=S_i to n
Add Sample from L_i (S_i)
Add Sample from $L_i(S_{i-1})$ E_i
=S_i
$S_i = S_i + S_{i-1}$
If S_i ¿ n
Set new S_i if L_i (new S_i) is not in list
End
End
7. Stop

5   Result and Discussion

Result analysis and discussion is used to compare the obtained results of propose work with other existing techniques results. Results gives an idea about the performance of proposed system.

Table 1. Sample selected with EFS

1	2	3	4	5	6	7	8	9	10	Original list
1	2	3	5	8						Round 1
4	7									Round 2
5	9									Round 3
6										Round 4
10										Round 5

As shown in Table 1, number of rounds required for extended Fibonacci series is more than existing Fibonacci series. To select with EFS, almost _n_ iterations are required. Where n represents to total number of items present in list. Even though number of iterations are required to be higher than existing Fibonacci series, it does not effect on the key size. Only two points need to be remember on extraction site. This implies that, with EFS, performance of Fibonacci series get extend its applicability.

Comparative analysis shown in Table 2 implies that, propose extended Fibonacci series (EFS) achieves good result over other existing sample selection techniques. However popularity even mentioned in Table 2 is present but it cannot be measure with any specific method. Propose sample selection technique is new and hence cannot be perfectly measured with popularity parameter.

Table 2. Comparative analysis

Parameters	Sequential	Random	Fibonacci	EFS
Sample selection	Sequential	Random	Random	Random
All samples selected	Yes	No	No	Yes
Noise level	High	Low	Low	Low
Implications	No	Yes	No	No
Complexity	No	Yes	No	No
Popularity	No	No	No	Yes
Key size	No key	High	Less	Less

Aim of propose work is not to concentrate on data Hiding technique but to provide a sample selection approach so as to make available Data Hiding technique more efficient (Fig. 2).

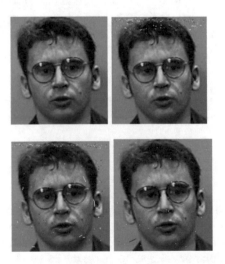

Fig. 2. 1, 2, 3, 4

As shown in above results, stego images are obtained from above discuss steganography techniques with different sample selection approach. First image represent to original image, second image obtain from sequential sample selection, third image obtained from random sample selection approach and fourth image from extended Fibonacci sample selection approach. From above Images, one can conclude that extended Fibonacci series generate less noise and maintain visual perceptual quality of original carrier object. Sequential random approach generates noise that concentrate on specific region of carrier object. Whereas random selection approach which distributes noise throughout the carrier, disturb originality of carrier object from original object.

Performance of above mentioned sample selection techniques can also be measured with the parameters like peak signal to noise ratio (psnr), mean square error, absolute error, maximum difference, structural content, cross correlation etc.

Psnr, mean square error, cross correlation, average difference and other parameters signifies the difference between two images. If two images are similar with each other, psnr becomes closer to infinity. psnr and mean square error are inversely proportional to each other. Lowering the value of psnr means it increases the value of mean square error.

Table 3. Image parameters comparison

Parameters	Sequential	Random	Fibonacci	EFS
PSNR	Low	Low	Low	High
MSE	High	High	High	Low
SC	0 to 1	0 to 1	0 to 1	Closer to 1
CC	0 to 1	0 to 1	0 to 1	Closer to 1
Min diff	0 to 255	0 to 255	0 to 255	Closer to 0
Max diff	0 to 255	0 to 255	0 to 255	Closer to 0
Avg diff	0 to 255	0 to 255	0 to 255	Closer to 0

From Table 3, parameter comparison suggest acceptability of extended Fibonacci series in future. As work is going on and expected to get better results than result mention in Table 3.

6   Future Scope

Every application will have some drawbacks, which should be removed by someone so as to maintain the growth in research domain. Even though extended Fibonacci series suppresses the noise, sample selection approach needs even more simplification. Every time new indices require to be selected to cover the required number of samples from carrier sample list. One may also need to focus on minimization of quantization error obtained from random selection approach. If sample selected by extended Fibonacci series are optimized samples i.e. selected sample will have the exact value with secrete decimal value, then it is possible to minimize the quantization error. Also with extended Fibonacci series, there is a need to expand the Data Hiding capacity.

7 Conclusion

Extended Fibonacci series provides good results over other steganographic sample selection approach. From result analysis, one May conclude that, it will be a good choice for developer to implement a sample selection process. One should get compared with more parameters so as to measure its efficiency, popularity, simplicity etc. One advantage to propose EFS is that, it can be possible to implement on any kinds of carrier object like image and audio. Size and type of carrier does not matter the performance of EFS.

References

1. Ni, Z., Shi, Y.-Q., Ansari, N., Su, W.: Reversible data hiding. IEEE Trans. Circuits Syst. Video Technol. **16**(3), 354–362 (2006)
2. Tai, W.-L., Yeh, C.-M., Chang, C.-C.: Reversible data hiding based on histogram modification of pixel differences. IEEE Trans. Circuits Syst. Video Technol. **19**(6), 906–910 (2009)
3. Zhang, X.: Reversible data hiding in encrypted images. IEEE Signal Process. Lett. **18**(4), 255–258 (2011)
4. Hong, W., Chen, T., Wu, H.: An improved reversible data hiding in encrypted. IEEE Signal Process. Lett. **19**, 199–202 (2012)
5. Liao, X., Shu, C.: Reversible data hiding in encrypted images based on absolute mean difference of multiple neighbouring pixels. J. Vis. Commun. Image Representation **28**, 21–27 (2015)
6. Qian, Z., Dai, S., Jiang, F., Zhang, X.: Improved joint reversible data hiding in encrypted images. J. Vis. Commun. Image Representation **40**, 732–738 (2016)
7. Zhou, J., Sun, W., Dong, L., et al.: Secure reversible image data hiding over encrypted domain via key modulation. IEEE Trans. Circuits Syst. Video Technol. **26**(3), 441–452 (2016)
8. Zhang, X.: Separable reversible data hiding in encrypted image. IEEE Trans. Inf. Forensics Secur. **7**(2), 826 832 (2012)
9. Wu, X., Sun, W.: High-capacity reversible data hiding in encrypted images by prediction error. Signal Process. **104**, 387–400 (2014)
10. Qian, Z., Zhang, X.: Reversible data hiding in encrypted image with distributed source encoding. IEEE Trans. Circuits Syst. Video Technol. **26**(4), 636–646 (2016)
11. Qian, Z., Zhang, X., Feng, G.: Reversible data hiding in encrypted images based on progressive recovery. IEEE Signal Process. Lett. **23**(11), 1672–1676 (2016)
12. Huang, F., Huang, J., Shi, Y.Q.: New framework for reversible data hiding in encrypted domain. IEEE Trans. Inf. Forensics Secur. **11**(12), 2777–2789 (2016)
13. Ma, K., Zhang, W., Zhao, X., et al.: Reversible data hiding in encrypted images by reserving room before encryption. IEEE Trans. Inf. Forensics Secur. **8**(3), 553–562 (2013)
14. Zhang, W., Ma, K., Yu, N.: Reversibility improved data hiding in encrypted images. Signal Process. **94**, 118–127 (2014)
15. Cao, X., Du, L., Wei, X.: High capacity reversible data hiding in encrypted images by patch-level sparse representation. IEEE Trans. Cybern. **46**(5), 1132–1143 (2016)
16. Kayarkar, H., Sanyal, S.: A survey on various data hiding techniques and their comparative analysis, pp. 1–9 (2012)

17. Naseem, M., Hussain, I.M., Khan, M.K., Ajmal, A.: An optimum modified bit plane splicing LSB algorithm for secret data hiding. Int. J. Comput. Appl. **29**(120), 36–38 (2011)
18. Nosrati, M., Karimi, R., Hariri, M.: An introduction to steganography methods. World Appl. Program. **1**(3), 191–195 (2011). ISSN 2222-2510
19. Thampi, S.M.: Information hiding techniques: a tutorial review. ISTE-STTP on Network Security and Cryptography, LBSCE, pp. 1–19 (2004)
20. Dey, S., Abraham, A., Sanyal, S.: An LSB data hiding technique using natural numbers. In: IEEE Third International Conference on Intelligent Information Hiding and Multimedia Signal Processing, IIHMSP 2007, 26–28 November 2007, Kaohsiung City, Taiwan, pp. 473–476. IEEE Computer Society press, USA (2007). ISBN 0-7695-2994-1
21. Fallahpour, M., Shirmohammadi, S., Semsarzadeh, M., Zhao, J.: Tampering detection in compressed digital video using watermarking. IEEE Trans. Instrum. Meas. **63**(5), 1057–1072 (2014)
22. Mahjabin, T., Hossain, S.M., Haque, M.S.: A block based data hiding method in images using pixel value differencing and LSB substitution method. IEEE (2012)
23. Fu, M.S., Au, O.C.: Data hiding watermarking for halftone images. IEEE Trans. Image Process. **11**(4), 477–484 (2002)
24. Kuo, W.-C., Jiang, D.-J., Huang, Y.-C.: A reversible data hiding scheme based on block division. In: Congress on Image and Signal Processing, vol. 1, , pp. 365–369, 27–30 May 2008
25. Shiu, C.-W., Chen, Y.-C., Hong, W.: Encrypted image-based reversible data hiding with public key cryptography from difference expansion. Signal Process Image Commun. **39**, 226–233 (2015)

FAFinder: Friend Suggestion System for Social Networking

Navoneel Chakrabarty$^{(\boxtimes)}$, Siddhartha Chowdhury, Sangita D. Kanni, and Swarnakeshar Mukherjee

Jalpaiguri Government Engineering College, Jalpaiguri, West Bengal, India
{nc2012,sc2024,sdk2081,sm2084}@cse.jgec.ac.in

Abstract. The emergence of social networking has led people to stay connected with friends, family, customers, colleagues or clients. Social networking can have social purposes, business purposes or both through sites such as Facebook, Instagram, LinkedIn, Twitter and many more. Recently, a large active social involvement have been seen from all echelons of society which keeps the friend circle increasing than never before. But, the friend suggestions based on one's friend list or profile may not be appropriate in some situations. Considering this problem, in this paper, a Friend Suggestion System, FAFinder (Friend Affinity Finder) based on 5 major dimensions (attributes): Agreeableness, Conscientiousness, Extraversion, Emotional range and Openness is proposed. This will help in understanding more about the commonalities that one shares with the other on the basis of their behaviour, choices, likes and dislikes etc. The suggested list of friends are extracted from the People Database (containing details of the 5 dimensions of different people) by deploying the concept of Hellinger-Bhattacharyya Distance (H-B Distance) as a measure of dissimilarity between two people.

Keywords: Social networking · Friend · Agreeableness · Conscientiousness · Extraversion · Emotional range · Openness · H B Distance

1 Introduction

Distanced from friends in olden days, people stood no chance in finding and getting connected to their friends again. Time passed and development of new ways of communication gave birth to internet. This constant pursuit of reliable connectivity gradually led to the development of social media. In recent times, social media plays a very crucial role in the field of communication. One of the important entities is finding or making new friends on social media. But, it is tougher than it seems because there are great chances of mismatch between the traits of different people in terms of behavioural attributes. With the growth of social media, social networking sites are facing the challenge of developing a proper and stable approach for recommending friends to the users. In order to

© Springer Nature Switzerland AG 2020
D. J. Hemanth et al. (Eds.): ICICI 2019, LNDECT 38, pp. 51–58, 2020.
https://doi.org/10.1007/978-3-030-34080-3_6

overcome this problem and provide a more justified way of suggesting friends, a simple methodological web application has been developed to determine a good suggestive set of friends based on the attributes used: Agreeableness, Conscientiousness, Extraversion, Emotional range and Openness. In this paper, different users and their attribute ratings prime to define his/her behavioural qualities are considered.

2   Literature Review

Some of the earlier works relevant to Friend Suggestion or Recommendation include:

Sharma et al. designed a hybrid friend recommender model using k-means Clustering and Similarity Index Calculation [1].

Kaviya et al. presented a friend recommendation system based on social networks (user log) considering network correlations [2].

Shahane et al. developed a friend recommender system that suggested friends to users on the basis of their life-styles using Collaborative Filtering [3].

Phepale et al. [4] performed text mining and Bhandari et al. [5] developed an Android Application for recommending friends on the basis of similarities in life-styles.

Likewise, Veeramani et al. proposed a life-style analyzing friend recommender, which is query-based, using k-Means Algorithm for analyzing the life-style and extracting the similarities in between [6].

Wang et al. implemented a life-style based friend recommender system as developed by Shahane et al. [3], Phepale et al. [4] and Veeramani et al. [6] via sensor-rich smartphones [7].

Bian et al. developed a Friend Recommendation System (named as Match-Maker) by employing Collaborative Filtering. In MatchMaker, a real-time user is mapped to TV Character(s) based on the similarities among them using Collaborative Filtering. Now, on considering one of the mapped TV Character(s), if the friend(s) of that TV Character is/are similar to any of the other users on social network, then they are hence recommended [8].

Agarwal et al. modelled a trust-intensified friend recommendation system by implementing Genetic Algorithms for learning user preferences [9].

Silva et al. proposed a friend recommender based on graphs by deploying Genetic Algorithm [10].

Kwon et al. proposed a friend recommender that is based on similarities among users in social and physical context [11].

The article is organized as Introduction and Literature Review, followed by Proposed Methodology and Implementation Details. Finally, the article is concluded in Sect. 5.

3 Proposed Methodology

3.1 The Dataset

The dataset for this work is created manually with 49 entries and saved as (**/display/Custom_user.csv**). The dataset contains 7 columns namely: Username, Name and the self-ratings (rated by people themselves) of the 5 dimensions: Agreeableness, Conscientiousness, Extraversion, Emotional range and Openness on a scale of 5.

The literal explanation of the 5 dimensions follows:

1. Agreeableness: A person who is highly agreeable will exhibit pro-social forms of behaviour. They are more sociable, aim to please other people and are willing to help those in need. As a result, agreeable people tend to work well as part of a team. During arguments or times of conflict, they will seek to resolve, rather than prolong confrontation.
2. Conscientiousness: A person scoring high in conscientiousness usually has a high level of self-discipline. These individuals prefer to follow a plan, rather than acting spontaneously. Their methodical planning and perseverance usually make them highly successful in their chosen occupation.
3. Extraversion: It is a person's behavior that makes him/her joyful around people rather than being alone.
4. Emotional range: It is the natural reaction to which a person's feelings are sensitive to his/her surrounding.
5. Openness: It is the person's experience and joy in trying new things.

3.2 System Outline

The System Architecture is actually based on the Client-Server Model. The proposed model generates the details (name, username, self-ratings of the 5 dimensions and similarity scores) of the five most suggested friends at the end of the process as final output, banking on the self-ratings of the five dimensions input by the corresponding user using the Web Application.

Fig. 1. Snapshot of Registration Page(**/go/register.html**) [LEFT] and Login Page(**/go/login.html**) [RIGHT]

Fig. 2. Entering the self-rated values of the 5 attributes

Fig. 3. Snapshot of the corresponding display page (**/display/run.py**)

1. At the very beginning, each new user needs to register for a new account with details: First Name, Last Name, username and password whereas old users need to login with their necessary input details: username and password. A snapshot for the Registration Page and Login Page is given in Fig. 1.

   Now, there are two possibilities for each case where registration (**/go/register.html**) and login (**/go/login.html**) are the 2 cases:

   (a) Successful

       i. After registration is successfully done, the database of users using the web application (**/go/database.csv**) is updated with the current user details and the page is redirected to the (**/go/register.py**) page where user needs to rate himself/herself on the five dimensions(attributes): Agreeableness, Conscientiousness, Extraversion, Emotional Range and Openness on a scale of 5.

       ii. After login is successfully done, the page is redirected to the (**/go/login.py**) page where user needs to do the same as after successful registration because of the fact that the behavioural aspects of people may change.

   (b) Failed

       i. Registration: If the username already exists in (**/go/database.csv**) i.e., if a user is already registered with that username, it is a case of failed registration and the system is redirected to (**/go/register.py**) having action buttons connected to the Registration Page (**/go/register.html**), Login Page (**/go/login.html**) and Home Page (**/index.html**).

       ii. Login:

           A. If a non-existant username (new username) is entered, then the user is a new user and hence, system is redirected to (**/go/login.py**) having action buttons connected to the Registration Page (**/go/register.html**), Login Page (**/go/login.html**) and Home Page (**/index.html**).

           B. If the username already exists (registered) and a wrong password corresponding to that username is entered, the system is similarly redirected to (**/go/login.py**) having action buttons connected to the Registration Page (**/go/register.html**), Login Page (**/go/login.html**) and Home Page (**/index.html**).

   So, the pages (**/go/register.py**) and (**/go/login.py**) manage the cases of Successful/Failed Registration and Login respectively in the Back-End.

2. After successfully entering the self-rated values of the five attributes, system will direct to the display page (**/display/run.py**) that will dynamically generate the list of five suggested friends from the People Dataset/Database (**/display/Custom_user.csv**) on their ratings. A snapshot showing the self-ratings of the 5 attributes being entered and the corresponding snapshot of the display page (**/display/run.py**) are given in Figs. 2 and 3 respectively.

The Architectural Diagram of the System Model is given in Fig. 4.

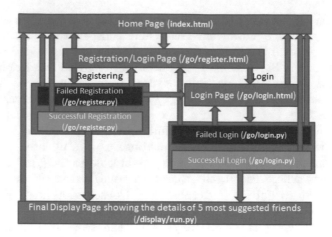

Fig. 4. Architectural diagram of the system model

3.3 Working Mechanism of Friend Suggestion:

1. The self-rated attribute ratings of a user is visualized as a point in a 5-Dimensional Space where the 5 attributes are the dimensions. Say, the point is named as O.
2. Similarly, each person's set of self-ratings in the dataset (**/display/ Custom_user.csv**) is treated as a point in a 5-Dimensional Space where the 5 attributes are the dimensions. Say, the points are named as P_1, P_2, P_3, ... ,P_{49}.
3. Hellinger-Bhattacharya Distance is calculated between O and each of P_1, P_2, P_3, ... ,P_{49} by the formula given in Eq. (1).

$$HB(X,Y) = \frac{1}{\sqrt{2}} * \sqrt{\sum_{i=1}^{5}(\sqrt{x_i} - \sqrt{y_i})^2} \tag{1}$$

where X and Y are points in 5-dimensional space and $X = (x_1, x_2, x_3, ... ,x_5)$ & $Y = (y_1, y_2, y_3, ... ,y_5)$ with $HB(X,Y)$ as Hellinger-Bhattacharya Distance between points X and Y
4. 5 points among P_1, P_2, ... ,P_{49} from which O is the nearest (minimum H-B Distance) are selected in ascending order of H-B Distance to be the most suggested friends with self-ratings characterized by each of the 5 points.
5. The maximum H-B Distance between any 2 points in a 5 dimensional space, given the maximum value representing a dimension is 5 (rating is done on a scale of 5), is approximately 3.54 as when $X = (0,0,0,0,0)$ & $Y = (5,5,5,5,5)$. Hence, the similarity/affinity score (in %) is calculated by the formula given in Eq. (2).

$$s = (100 - \frac{H * 100}{3.54})\% \tag{2}$$

where s is the similarity or affinity score between 2 people and H is the corresponding H-B Distance between them calculated in Step 3.

4   Implementation Details

The Front-End Development of the Web Application is done using Hyper Text Markup Language (HTML) and Cascading Style Sheets (CSS). The form designing and handling are done using CSS integrated with Javascript (JS) and jQuery. The frameworks used in UI/UX Development are shown in Fig. 5.

Fig. 5. Frameworks used for UI/UX Development

The Back-End Development involving the registered user database (**/go/database.csv**) maintenance, registration & login process execution and the working mechanism of Friend Suggestion, is done using Python Programming Language with Common Gateway Interface (CGI) technology incorporated in WAMP Server as the local server for testing purposes (shown in Fig. 6).

Fig. 6. Python 3.7 Programming Language and WAMP Server 2.1 are used in the Web Application Development

The web application can be downloaded online at the referenced link [12].

5   Conclusion

In this paper, a web application has been penned in which H-B Distance based System Model is deployed for suggesting friends to the user based on his/her 5 characteristics: Agreeableness, Conscientiousness, Extraversion, Emotional range and Openness. This also has a high probability of ensuring that after getting connected as friends, their friendship will last longer due to the similarity between their characteristic traits related to the aforementioned 5 major dimensions.

References

1. Sharma, S.K.: Hybrid friend recommendation approach based on clustering and similarity index. Int. J. Res. Appl. Sci. Eng. Technol. **6**, 1528–1534 (2018). https://doi.org/10.22214/ijraset.2018.5248
2. Kaviya, R., et al.: Friend suggestion in social network based on user log. In: Materials Science and Engineering Conference Series, vol. 263, no. 4 (2017)
3. Shahane, A., Galgali, R.: Friend recommendation system for social networks (2016)
4. Phepale, P., Longani, C.V.: A friend suggestion system for social networks. Int. J. Eng. Dev. Res. (IJEDR) **4**(3), 221–226 (2016). http://www.ijedr.org/papers/IJEDR1603037.pdf
5. Bhandari, C., Ingle, M.D.: FriendFinder: a lifestyle based friend recommender app for smart phone users. Int. J. Comput. Appl. **145**(6) (2016)
6. Veeramani, S., Jeba, L.: A query based friend recommendation system with de-trop message detection. Global J. Pure Appl. Math. **12**(2), 1293–1298 (2016)
7. Wang, Z., et al.: Friendbook: a semantic-based friend recommendation system for social networks. IEEE Trans. Mobile Comput. **14**(3), 538–551 (2014)
8. Bian, L., et al.: MatchMaker: a friend recommendation system through TV character matching. In: 2012 IEEE Consumer Communications and Networking Conference (CCNC). IEEE (2012)
9. Agarwal, V., Bharadwaj, K.K.: Trust-enhanced recommendation of friends in web based social networks using genetic algorithms to learn user preferences. In: International Conference on Computational Science, Engineering and Information Technology. Springer, Berlin (2011)
10. Silva, N.B., et al.: A graph-based friend recommendation system using genetic algorithm. In: IEEE Congress on Evolutionary Computation. IEEE (2010)
11. Kwon, J., Kim, S.: Friend recommendation method using physical and social context. Int. J. Comput. Sci. Netw. Secur. **10**(11), 116–120 (2010)
12. https://github.com/ns3-jalpaiguri/Friend_Affinity_Finder.git

Forensic Analysis on IoT Devices

A. R. Jayakrishnan[1]($\boxtimes$) and V. Vasanthi[2]

[1] RCAS, Bharathiar University, Coimbatore, Tamilnadu, India
anilakkad.jk@gmail.com
[2] Department of IT, SKASC, Coimbatore, Tamilnadu, India
vasanthiv@skacas.ac.in

Abstract. When the number of connected devices increases, the chances of getting hacked also increases. IoT is the interconnection of smart devices to usher automation in most of the areas. It has been identified that the study of attack patterns on IoT networks are limited, hence it becomes difficult to recognize a methodology yet. This paper demonstrates the experiments carried out to analyse the IoT attacks on Wi-Fi camera and SCA attack to get AES key. The aim of this paper is to show intrusion to the network as a hacker through the unprotected Things (IoT) connected to the network for performing illegal activities. These experiments would be an eye-opener to the investigators and also leads to study the attack patterns in the IoT devices and then to derive the methodology for forensics analysis.

Keywords: IoT attack · Network forensics · SCA · Digital security · Cyber forensics · Wi-Fi camera intrusion

1 Introduction

Connected devices incessantly emit a stream of data as the part of their communication in the network. In a way, people create this massive amount of data, fuelled by their profound hunger for inevitable connectivity. As the technology grows day by day, humans thought process are also getting changed. Their life style gets changed, this also leads for getting involved in smarter systems, which have the capability to connect to the internet. Network of smart systems, mobile phones, home appliances, vehicles, medical equipments and other electronic devices that can connect, interact and interchange data is known as IOT, the Internet of things [1]. In general, the current incalculable gadgets shape a pervasive communication network ceaselessly transmitting data. The swift growth of the Internet has become responsible for the advent of the IoT, with protruding examples as smart homes, smart cities, healthcare systems, and other cyber physical systems.

2 Background

2.1 Definition

The Internet of Things is a collection of connected devices made of light processors and network cards that are capable of being managed through web, app or other

D. J. Hemanth et al. (Eds.): ICICI 2019, LNDECT 38, pp. 59–68, 2020.
https://doi.org/10.1007/978-3-030-34080-3_7

interfaces [5]. However, the boundless evolution gives abundant chance for the attackers to intrude the network since the surface for attack grows. This is a nightmare to the forensics analysts as to finding the source of intrusion, type of attacks, method of attacks always remains as big questions in front of them. Bigger the network, the more complex it acts and much tougher it becomes to safeguard [8].

2.2   Challenges

Any device connected to the internet grudgingly exposes to the attackers waiting for users to make mistakes, by ignoring security warnings, leaving vulnerable services open or by making a visit to malicious site unprotected. Massive effort is spent on protecting large networks in corporates quarantining infected hosts, alerting admins, preventing future similar attacks by implementing protective measures [6]. Nowadays, cheap cost devices are made and distributed in the market which has many capabilities and features to make it a smart device. Device specifications of such "Thing" are not standardised across the industry. Controlling such device is again not standardised too. Connectivity is always a challenge. With the notion of IoT, a device is becoming a "THING". Not all devices are things even though they are connected to the internet. A device become a thing if it is at least self-described, reconfigurable, autonomous, interacting with some decision systems, safe and secure, uniquely identified, connected, reliable and resilient [2].

Protecting a network from attacks and retrospective attack analysis in addition to the continuous process of establishing secure network are some of the major tasks of security analysts. Network activity trail would be an essential data for investigators to narrow down the root cause of issues. Key questions that come in their mind are How did the intruders get in? What did they do once inside network? Where did they originate from? What attack patterns they follow? How to prevent such attacks? Administrators can just answer such inquiries by analysing logs of past action recorded over a period of time. IoT devices are at huge risk, malicious apps target Internet of Things. Currently, over 8 billion of smart devices are connected globally. It was when the mirai botnet emerged in 2016 that the whole world learned how dangerous such devices may become in the hands of cyber criminals [2]. However, the history of malware attacking IoT devices began much earlier. Symantec has observed a 159% increase in IoT attacks from July 2017 to July 2018. It has been predicted that, by 2020, the security of Internet-of-Things will be 20% of the annual security budget and will continue to grow in due course.

Devices are rarely patched, leaving security holes open. Some devices have default username, password combinations unchanged. They are often plugged in forgotten about, so infections go unnoticed. IoT devices are proving to be a soft target for worms.

2.3   IoT Device

Typically, an IoT device contains a general purpose microcontroller MCU as the heart of it. Tons of features like Wifi, Bluetooth, Zigbee, memory etc are provided for extremely low cost MCUs. Many interfaces and sensors are added along with the feature packed devkit too. Majority of the general purpose MCUs are quite vulnerable to attacks [12]. The truth is that vendors know this as they also offer secure MCUs for a pricier one. Secure MCUs, protocols, coding standards cost money and time.

Bad security = Limited money + Design trade-offs + Time pressure [11].

Attackers search for secrets and sensitive information about a device. What if that information affect all/majority of devices? It becomes worse if the secret information exposes a remote attack path. The attack scales very fast in such scenarios (Fig. 1, Tables 1 and 2).

Fig. 1. Hackers intrusion

Table 1. Most attacked devices

Router	DVR	Camera	Alarm system	Soho router	Set topbox	Satellite dish	P. logic controller
54.6%	21.7%	7.1%	6.1%	3.9%	3.1%	2.1%	1.5%

Table 2. Most attacked ports

Port 23	2323	80	445	22	443
82%	8%	6%	2%	1%	1%

2.4   IoT Attack Metrics

If there is an inadequately configured IoT gadget in your home system, which may contain vulnerabilities too, it can create critical issues. The most critically recognized situation is that the Thing could end up as a part of the botnet [10]. For its owner, this situation could be perhaps inoffensive; however other circumstances are increasingly risky. For instance, your house system appliances could be utilized to accomplish illicit actions and crimes, or a cyber-criminal who accessed an IoT gadget could keep an eye on and later bully its owner. There are many such cases reported in the newspapers these days. Eventually, the infected IoT thing can be simply damaged, however this is in no way the worst things that could occur [9]. Usually, device manufacturers are a bit slow to come up with firmware patches and updates for smart devices. In the worst scenario, the firmware does not get patched at all, and many devices do not have the ability to install the patch updates of firmwares [3]. The protocol Telnet which allows connecting remote computers has been around since 1969, however it is not that broadly used as it was before. There are lots of embedded system applications in Internet of Things devices such as IP phones, routers, industrial control systems, smart televisions, DVRs and others that leverage its distant access capabilities. In fact, a Shodan search conducted during this research returned more than 17 million devices connected globally with an accessible port and telnet server. Telnet has been always an easy method and target for hackers to sniff for user names and passwords, since Telnet does not encrypt communications. Attacker can easily determine information shared between connected devices.

3   Experiments

The below two experiments illustrate how weak the IoT gadgets are and it shows how easily they are vulnerable to attackers.

3.1   Experiment 1 - IoT Camera

- Made in RPC
- Costs around Rs2000
- Lots of features: -
  - Wi-Fi Connection
  - 2-way audio
  - HD Image/Video
  - Motors for rotating camera
  - Night imaging IR lights
  - Records data to the SD card
  - Interface apps for Android/iOS etc.

Camera was easily opened and the interfaces were identified as seen in below image (Figs. 2 and 3).

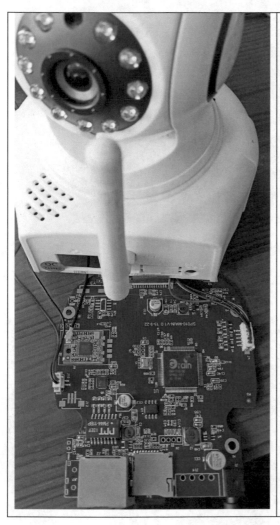

It has the following configurations

- Grain Media's IP Camera SoC GM8135S embedded with ARM CPU Core which is the heart of the system.
- MediaTek MT7601 SoM (Wi-Fi communications)
- DoSilicon FM25Q64A (Storage for Linux)
- Atmel AT24C02 I2C flash (Storing camera model, MAC address)
- Transistor array for powering the camera motor to turn 360 degrees.

Fig. 2. Dismantled camera and its parts

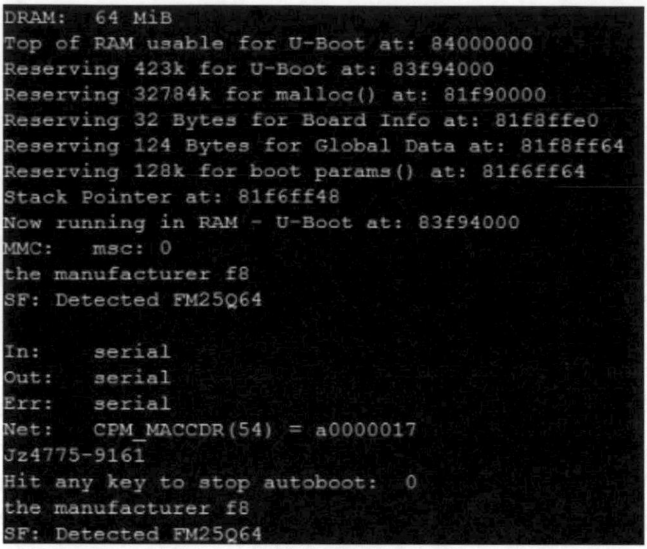

Fig. 3. Connected the camera through UART Rx/Tx and powered it to boot up the system.

The camera prints all sorts of debug information – Ports of camera, config files, users of camera and **passwords**. Wi-Fi configuration, SSID + **passwords** (Fig. 4).

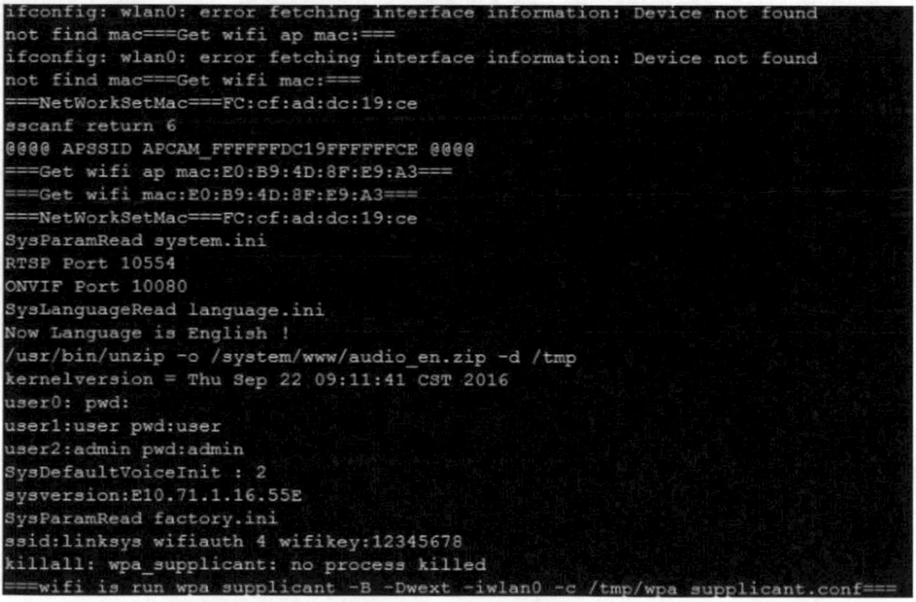

Fig. 4. Debug information

Camera has a root password for Linux, however almost all cameras have same root passwords. We can even boot camera without root password by using the following script.

- Stop u-boot procedure, and print bootargs:
- printenv
- Then append to the bootargs init = / bin / sh:
- $setenv bootargs 'console=ttyS1115200,n8 mem=39@M0x0 ispmem=5M@0x27 00000 rmem=20M@0x2c00000 init=/linuxrc rootfstype=squashfs rw root=/dev/ mtdblock2 rw mtdparts=jz_sfc:256k(boot),2176k(kernel),3584k(rootfs),2176k(sys- tem) init=/bin/sh'
- $ boot

Shockingly, reverse engineering on different Wi-Fi cameras show that they have same type of configurations.
Crucial files:

- /etc/password

  root:$1y$bdHbPDn@ii9aEIFNio@lBbM9QxW9mr0:0:0::/root:/bin/sh

- / etc /shadow does not exist. hash above is an MD5 hash

Any password cracking program can be used to crack the salt$hash string or even google the string gives results.
ybdHbPDn$ii9aEIFNiolBbM9QxW9mr0 = md5("ybdHbPDn" + "hslwificam")
Same weak password was found in every device investigated. If we find any remote vulnerability, we have the root access password on all these cameras.
One main observation here is that always use strong, per-device passwords/keys.
Reverse engineering on such cameras reveals more and more threats. Telnetd is running default backdoor on all cameras. Interestingly, there is a RSTP (Rapid Spanning Tree Protocol) port in 10554 published by camera and if we try to access it, we could see the camera streaming. The port 81 ONVIF interface can even move the camera. The panicky situation is that lots of cameras are connected to internet and are available to be hacked easily [4]. Searching in Shodan.io returns alarming results (Figs. 5 and 6).

Fig. 5. Shodan search results for Telnet

Fig. 6. Shodan search results for WiFiCam

The consequences are

- Camera security fully bypassed & backdoor for free
- Hardware attack: Serial port reveals root password
- Telnet - users and configs exposed, remote access to video stream.
- These cameras are used typically as baby monitors - privacy violation
- Linux system can be used for illicit activities, e.g. bitcoin miners
- IoT botnet Mirai almost brought down DNS on various parts of the world. In the IoT realm, consequences can scale fast due to the networking of things.

3.2   Experiment 2: AES on IoT Device

AES - Advanced Encryption Standard is the most generally used symmetric key cryptographic systems in IoT edge devices [7]. The goal in this experiment is to get the AES key from the IoT device. The device has no logic flaws, it performs AES encryption of message and replies the encrypted message. Note, there will be side channel attacks on all sort of crypto.

Side-Channel Attacks (SCA) recipe:

- Communicate to a device performing crypto (e.g. AES)
- Measure the device's power consumption while doing crypto
- SCA program "computes math" with collected data
- You get the crypto key (Fig. 7).

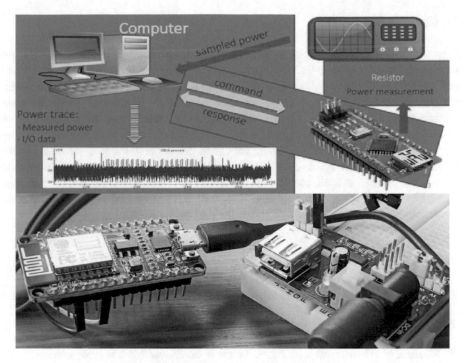

Fig. 7. Device connected for power measurement.

The consequences:

- Hardware attack: SCA reveals AES master key from devices
- All devices share the same master key
- This allows to impersonate manufacturer/authorised users
- Malicious updates will be indistinguishable from legit updates, updating a malicious firmware in IoT device as attack tool.
- This can also allow to decrypt the IoT device communications.

It's evident that even with good practices, there can be security issues. However, good practices make attack scalability much harder.

4 Future Work

With the diversified things connecting the internet, it's a big challenge for analysts to monitor and spot the intrusions since most of the so called smart devices doesn't follow the security standards. There are no such common standards set across the globe, however a methodology for network forensics on IoT can be derived by studying the various attack patterns on IoT devices. With the help of honeypots, the attackers can be attracted to the IoT network and the pattern of attacks could be studied. This will help to formulate a methodology for network forensics on IoT.

5   Conclusion

Network forensics is an art of monitoring, capturing, analysing and reporting the network audit trails so as to find the root of security breaches. It takes tremendous effort and patience to study the patterns of attacks, how intruders breach through networks weaknesses which a skilful hacker can use, as well as advantages which forensic investigators can turn into their favour. This paper demonstrated the possibilities of attacking unsecured IoT devices and how it can be done. In the current market, where lots of Wi-Fi smart devices are coming for cheap price, people are not bothered about the security of such devices and the repercussions when compromised. Experiments were carried out to exhibit hacking an IoT camera bought from the market and side channel attack (SCA) to get AES key. This will pave way further for investigators/ researchers to study the patterns of attack for deriving a methodology.

References

1. Davidoff, S., Ham, J.: Network Forensics – Tracking Hackers Through Cyberspace. Prentice Hall (2012). ISBN-13 978-0-13-256471-7
2. Carpi, R.B.: Cheapsgate attacking IoT in less dollars. In: International Conference on Cyber Security, Data Privacy and Hacking, Cocon XI, October 2018
3. Khadshe, S.: New tools and trends in cyber forensics. In: The Proceedings of the International Conference by CSI Cyber FIF-C (2019) – Cyber Frauds, Investigations & Forensics Conference, January 2019
4. Djemaiel, Y.: Tracing, detecting and tolerating attacks in communication networks. In: Proceedings of Sup'Com Research Thesis CNAS, January 2018
5. Pfleeger, C.P., Pfleeger, S.L.: Security in Computing, 4th edn. (2013). ISBN 978-81-317-2725-6
6. Fadia, A.: Network Security, a Hackers' Perspective, 2nd edn. (2009). ISBN 1403-93088-0
7. Stallings, W.: Cryptography and Network Security – Principles and Practice, 6th edn. (2014). ISBN 978-93-325-1877-3
8. IEEE 5G Summit Trivandrum: IoT & Cyber security (2017)
9. Anchit, B., Wazid, M., Pilli, E.S.: Forensics of random-UDP flooding attacks. J. Netw. **10**, 287–293 (2015)
10. Cato Networks: Global industry report, top networking and security challenges in the enterprise, November 2016
11. Liu, C., Singhal, A., Wijesekera, D.: A logic based network forensics model for evidence analysis. In: IFIP International Conference on Digital Forensics (2015)
12. Pollitt, M.: A history of digital forensics. In: Chow, K.-P., Shenoi, S. (eds.) Advances in Digital Forensics VI, vol. 337, pp. 3–15. Springer, Berlin (2010)

Design and Analysis of Wideband Decagon Microstrip Antenna

Subba Rao Chalasani[✉], Srinivasa Rao Mannava,
and Madhavi Devi Lanka

Department of Electronics and Communication Engineering,
Prasad V Potluri Siddhartha Institute of Technology, Vijayawada 520007, India
csr949@gmail.com, madhavidevi.kodali@gmail.com,
msrao@pvpsiddhartha.ac.in

Abstract. In this paper a decagon microstrip antenna for Ultra-Wide Band operation is proposed. A microstrip feed is used to feed the radiating patch. The ground plane of the antenna is defected with length below the patch. The proposed antenna is compact in size of $28 \times 26 \times 1.6$ mm. This antenna is designed using FR4 substrate material that is readily available and cost effective. It's reflection coefficient $S_{11} < -10$ dB 3–40 GHz. This proposed antenna operates from 3 GHz to 40 GHz covering UWB, Ku, K, Ka bands.

Keywords: Decagon · Return loss · Wideband · Microstrip · Compact

1 Introduction

Modern wireless communication systems are operating with multimedia data. These systems are operating in more than one frequency band. Using separate antenna for each frequency band increases the size of the system and simultaneously its cost. It is always preferable to integrate and fabricate the antenna used with the circuit inside the system instead of mounting separately and this greatly reduces its size and makes it portable. For microwave range microstrip antennas are preferable to be used for various frequency bands such as L, S, C, and Ultra-Wide Band (UWB) above. Microstrip antennas have advantages of very thin structure, low power consumption, easy to integrate with circuit and low cost. Microstrip antennas are limited by their narrow bandwidth of operation. Many varieties microstrip antennas have been developed in recent times for different applications.

There are many patch antennas designed and reported for UWB applications. Today's wireless communication involves transmission of multimedia data consisting of Tera Bytes. To meet this demand the U.S Federal Communication Commission (FCC) has permitted the use of 3.1 to 10.6 GHz (7.5 GHz bandwidth) for wireless communication for various needs [1]. Microstrip antennas are compact in size and dissipate low power. These antennas are bandwidth limited. Various techniques have been proposed for the enhancement of bandwidth of these antennas. The radiating patch is fed by different feed types. The feed for the patch plays a very important role for the enhancement of the bandwidth [2]. This has created the need for broad band,

© Springer Nature Switzerland AG 2020
D. J. Hemanth et al. (Eds.): ICICI 2019, LNDECT 38, pp. 69–76, 2020.
https://doi.org/10.1007/978-3-030-34080-3_8

low power and compact antennas. Li and Ye has reported the design and analysis of a rectangular slot antenna with two rejected frequency bands for WLAN/WiMax applications has been reported [3]. Mandal and Das designed and analyzed a hexagonal patch with a rectangular cut for the Ultra-Wide Band frequency range [4]. Azim and Islam has reported the designed an annular slot ring antenna for short range communications with [5]. A hexagonal patch antenna is designed and its analysis is reported for circular polarization. The antenna is narrow band for operation at 2.45 GHz with circular polarization [6]. A microstrip antenna with patch of twelve sides is designed for circular polarization at 2.45 GHz. A hexagonal slot is made in the middle of the patch to get the circular polarization. This is a narrow band antenna [7]. A Hexadecagon for Ku band operation is designed and analyzed. This antenna operates at 13.67 GHz and 15.28 GHz only [8]. A wide band octagonal patch antenna with tapered microstrip line feed is designed and analyzed. This antenna operates over the frequency 3.1 to 16 GHz [9]. A hexagonal printed antenna is proposed and analyzed for UWB operation. Hexagonal slots are inserted to give notched band characteristics [10]. A regular pentagon antenna with slots and irregular pentagon antenna has been designed and the gain is improved. The proposed antenna is fabricated and its parameters are measured in the two frequency bands 2.6–7.5 & 8.7–20 GHz [11]. A microstrip antenna for multi frequency operation in L and S band is designed fabricated and the simulated and measured results are compared and analyzed [12]. A band notched microstrip fed patch antenna for WiMAX operation is designed fabricated and tested for Ultra- Wide Band Operation [13].

All the above papers reported wide band antennas of around 8 GHz frequencies. But the proposed antenna in this paper is much compact in size and provides much wider bandwidth from 3 GHz to 40 GHz. Microstrip antennas of various shapes such as circular, rectangular, hexagon, pentagon and sizes have been analyzed and are widely reported in the literature. But the decagon shape patch is not analyzed for its wide band performance. In this paper radiating patch of decagon shape of the antenna is considered for design and analysis. The antenna is designed, modeled and simulated in CST antenna software to evaluate its performance numerically.

2   Design of the Printed Decagon Antenna

The proposed decagon antenna consists of ten sides and its radius is the main parameter to be determined for its design.

$$R \approx \frac{c}{8f_l\sqrt{\frac{(1+\varepsilon_r)}{2}}} \tag{1}$$

Where R is the radius of the patch and f_l is the lowest operating frequency. The radius of the patch is calculated using the Eq. (1) and is optimized around 3.1 GHz. The radius of the patch is small and is only 7.1 mm obtained using Eq. (1). The dielectric constant is ε_r and 'c' is the velocity of the light. All the design parameters are listed in the Table 1. The antenna is compact for its overall size of 28.5 × 26 mm only.

The substrate used is FR4 which is readily available and is of low cost. The ground plane is split around the strip feed to enhance its bandwidth. There is no ground directly below the feed and is defected. The radiating patch of this antenna consists of 10 sides. A small gap is maintained between the patch and the ground plane to reduce the reflection coefficient at high frequencies. The strip feed is used to excite the antenna as it is simple and is easy to fabricate and its dimensions are optimized to match with the radiating patch for 50 Ohms.

Table 1. Dimensions of the decagon microstrip antenna

Parameter	Material	Symbol	Dimensions(mm)
Patch radius	Conductor	R	7.1
Substrate	FR4	$L_s \times W_s \times h_s$	$28.5 \times 26 \times 1.6$
Permittivity of substrate		ε	4.4
Ground	Conductor	$L_g \times W_g$	10.75×6.3
Ground and patch gap			0.75
Feed length	Conductor	Lf	7.5
Feed width	Conductor	W_f	2.0

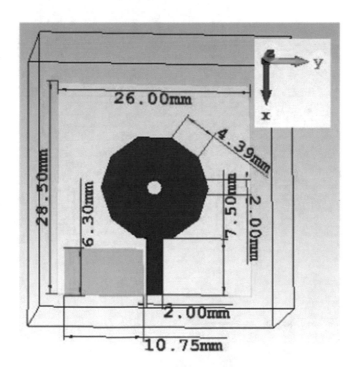

Fig. 1. Decagon antenna modeled in CST

Using the lower frequency of 3 GHz the radius of the patch is determined. While the substrate is FR4 with permittivity 4.4 and loss tangent 0.02 its length and width are selected to be minimum possible as shown in the Table 1 above. The ground plane dimensions are optimized to get best performance. The proposed decagon microstrip antenna modeled in the CST software and its optimized dimensions are shown in Fig. 1 above. All the dimensions of the proposed design are listed in the Table 1 above along with the materials selected. The antenna is modeled and simulated in Computer Simulation Technology (CST) software. This is a Method of Moments (MOM) numerical technique for simulation of radiating systems.

3 Radiation Characteristics

The surface current distribution on the patch and the feed is shown in Fig. 2. The current is uniform on the patch and also on the feed at 20.5 GHz and also at other frequencies. In particular the current is dense at the edges than at the centre of the patch. Along the periphery the current is maximum it is minimum at the centre of the patch. A circular hole is made at the centre of the patch and exhibits no predominant effect on the radiation properties (Fig. 3).

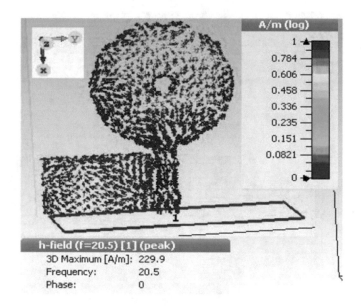

Fig. 2. Current distribution of surface antenna

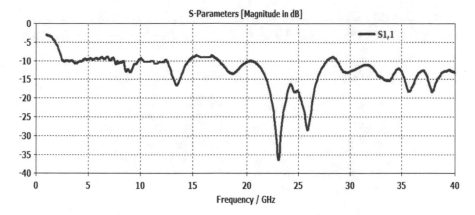

Fig. 3. Reflection coefficient vs. frequency

Reflection coefficient of the antenna exhibits the bandwidth performance of the element. Less the reflection coefficient the less is the return loss of the antenna and a perfect match of the patch with the feed attached to excite the patch. For this antenna the each design parameter is optimized to get wide band performance for wide frequency range. With full ground plane the low frequency response up to 20 GHz is poor and the reflection coefficient is higher. To improve the low frequency response the ground plane is defected and reduced. As the ground plane is reduced the low frequency response improved and reflection coefficient is reduced to –10 dB. The gap between the ground plane and the feed is varied from 0.1 mm to 0.5 mm and is fixed at 0.25 mm. Also the gap between the radiating patch and the ground plane is varied from 1 to 1.5 mm and fixed at 1.25 mm. These two gaps have improved considerably the high frequency response above 20 GHz and reduced the reflection coefficient. A circular hole at the center of the patch is introduced as the current is less at the center of the patch. The reflection coefficient 3 to 40 GHz is less than -10 dB and the return loss is less than 10 dB. The decagonal shaped patch is designed for wide band operation providing coverage for UWB and K bands.

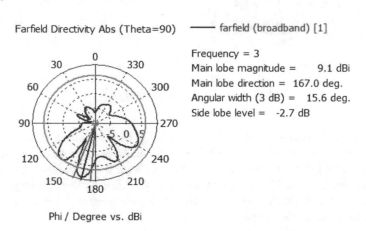

Fig. 4. Radiation pattern at 3 GHz

At 3 GHz the radiation pattern is shown in Fig. 4. Its maximum is directed away from the feed. The side lobes and back lobes are less but the main beam has ripples.

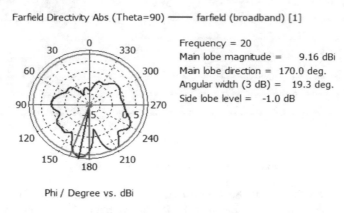

Farfield Directivity Abs (Theta=90) ——— farfield (broadband) [1]

Frequency = 20
Main lobe magnitude =   9.16 dBi
Main lobe direction =   170.0 deg.
Angular width (3 dB) =   19.3 deg.
Side lobe level =   -1.0 dB

Phi / Degree vs. dBi

Fig. 5. Radiation pattern at 20 GHz

The radiation pattern at 20 GHz is depicted in Fig. 5. At this frequency there are no side lobes and ripples are also less. The main beam is also smooth.

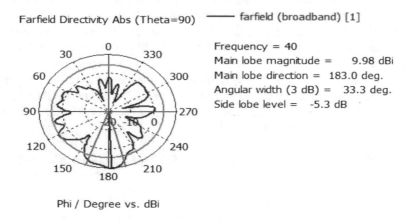

Farfield Directivity Abs (Theta=90) ——— farfield (broadband) [1]

Frequency = 40
Main lobe magnitude =   9.98 dBi
Main lobe direction =   183.0 deg.
Angular width (3 dB) =   33.3 deg.
Side lobe level =   -5.3 dB

Phi / Degree vs. dBi

Fig. 6. Radiation pattern at 40 GHz

At 40 GHz the radiation pattern is shown in Fig. 6. At this frequency there are side lobes and ripples for the pattern. The width of the main beam is wide. The main beam maximum of the pattern is directed away from the feed. Beam width of the main beam is changing from 15° to 33° at high frequency at 40 GHz. At low frequencies the side lobes are less and the main beam is sharp but at high frequencies the side lobes and beam width increase considerably.

The proposed decagon antenna is compared with other similar elements as shown in Table 2. Its size is small as compared to other antennas and also bandwidth is wide. Many of the designs reported in the literature are for UWB. But the proposed antenna provides satisfactory operation above UWB.

Table 2. Comparison of the proposed antenna with those presented

Paper	Dimensions (mm)	Bandwidth
Reference [5]	50 × 50	0.33 GHz
Reference [6]	50 × 50 × 1.6	2.45 GHz
Reference [7]	40 × 48	13.67 &15.28 GHz
Reference [8]	60 × 70 × 1.6	3–16 GHz
Reference [9]	30 × 30 × 0.8	3.1–14 GHz
Reference [10]	33 × 30 × 1.6	8.7–20 GHz
Proposed antenna	28.5 × 26 × 1.6	3–40 GHz

4   Conclusion

The decagon antenna with ten sides is designed and fed by the microstrip feed. The antenna is compact in size of only 28.5 × 26 × 1.6 mm as compared to other similar designs. Ground plane of the antenna is reduced below the patch. The gap between the patch and the ground plane has enhanced the band width at high frequency range. The circular hole at the center of the patch doesn't improve the bandwidth. At the center of the patch the current is very less. Reducing the ground plane to one side of the feed also improved the band width at high frequency side. Dimensions of the antenna are optimized to operate above 3 GHz and up to 40 GHz. It provides satisfactory coverage for UWB, Ku, K and Ka bands. Main lobe of the radiation pattern is stable for the bands but there are significant side lobes at high frequencies. Beam width of the main lobe is narrow around 16–19 . The proposed decagon antenna is compact and provides wide band operation as compared to other similar designs.

References

1. First Report and Order: Revision of Part 15 of the commission's Rule Regarding Ultra-Wideband Transmission System FCC 02 -48, Federal Communications Commission (2002)
2. Balanis, C.A.: Antenna Theory Analysis and Design, 3rd edn, p. 811. Wiley, Hoboken (2009)
3. Li, C.M., Ye, L.H.: Improved dual band-notched UWB slot antenna with controllable notched band widths. Prog. Electromagnet. Res. **115**, 477–493 (2011)
4. Mandal, T., Das, S.: Microstrip feed spanner shape monopole antenna for ultra wide band applications. J. Microwaves Optoelectron. Electromagnet. Appl. **12**(1), 15–22 (2012)
5. Azim, R., Islam, M.T.: Compact planar UWB Antenna with Band notch characteristics for WLAN and DSRC. Prog. Electromagnet. Res. **133**, 391–406 (2013)

6. Kushwaha, N., Kumar, R.: Design of slotted ground hexagonal microstrip patch antenna and gain improvement with FSS screen. Prog. Electromagnet. Res. B **51**, 177–199 (2013)
7. Prakash, K.C., Mathew, S., et al.: Circularly polarized dodecagonal patch antenna with polygonal slot for RFID applications. Prog. Electromagnet. Res. C **61**, 9–15 (2016)
8. Kumar Naik, K., Amala Vijaya Sri, P.: Design of hexadecagon circular patch antenna with DGS at Ku band for satellite communications. Progress In Electromagnetics Research M **63**, 163–173 (2018)
9. Parit, S.S., Kabegoudar, V.G.: Ultra wideband antenna with reduced radar cross section. Int. J. Electromagn. Appl. **6**(2), 23–30 (2016)
10. Elfergani, I., Rodriguez, J., et al.: Slotted printed monopole UWB antennas with tunable rejection bands for WLAN/WiMAX and X-Band coexistence. J. RadioEng. **27**(3), 694–702 (2018)
11. Thaher, R.H., Alsaidy, S.N.: New compact pentagonal microstrip patch antenna for wireless communications applications. Am. J. Electromagn. Appl. **3**(6), 53–64 (2015)
12. Murugun, S., Rohini, B., Muthumari, P., Padma Priya, M.: Multi-frequency T-slot loaded elliptical patch antenna for wireless applications. Assoc. Comput. Electromagn. Soc. (ACES) J. **33**(2), 247–250 (2018)
13. Ammal, M.N., Ramachandran, B., Hanumantha Rao, P.: A Microstrip-fed UWB printed monopole antenna for WiMAX band rejection. Int. J. Microw. Opt. Technol. **11**(5) (2016)

Lifetime Improvement for Hierarchical Routing with Distance and Energy Based Threshold

Remika Ngangbam[1(✉)], Ashraf Hossain[2], and Alok Shukla[3]

[1] Department of Electronics and Communication Engineering,
National Institute of Technology, Aizawl 796012, Mizoram, India
ngangbamremika7@gmail.com
[2] Department of Electronics and Communication Engineering,
National Institute of Technology, Silchar 788010, India
ashrafiit@gmail.com
[3] Department of Physics, National Institute of Technology,
Aizawl 796012, Mizoram, India
aloks.nitmz@gmail.com

Abstract. Hundreds or thousands of sensor nodes are used in the region of interest for observing and examining the regional environmental condition and are irreplaceable once it has been deployed. Much larger energy is spent by cluster head nodes for transmitting data directly to the destination node and lesser energy is required for computational activities and energy spent by group member nodes in a clustering protocol is also lesser comparing to that of head nodes. Efficient energy consumption for network lifetime extension can be accomplished with proper cluster head selection. This paper proposed an improved threshold condition with energy and distance parameters for electing cluster head and the simulated results indicate better overall lifetime performance than the conventional LEACH and some other LEACH related protocols for different network sizes and different density of nodes.

Keywords: Distance · Network lifetime · Cluster head · Stability · Nodes

1 Introduction

Due to the modern day advancement in integrated circuits technology (IC technology), radio networking systems and high computation ability sensors, Wireless Sensor Network (WSN) has engraved a remarkable position in the networking scenario. Basically, it contains peculiar number of sensor nodes that have computational abilities, data transmission and reception among the nodes which have been set out in the area of interest. The power source of their operation depends on their own battery or from some other renewable sources like solar systems. Embedding some other power source in the sensor nodes becomes a trade-off factor between system value and its complexity. Hence, optimising the utilisation of limited power source of these tiny nodes becomes the main consideration among the various problems faced by WSN [5, 11]. Amidst the

D. J. Hemanth et al. (Eds.): ICICI 2019, LNDECT 38, pp. 77–83, 2020.
https://doi.org/10.1007/978-3-030-34080-3_9

energy efficient routing protocols, clustering based protocol is the most prominent protocol in the current scenario [8, 10].

İn clustering based protocol transmission of data is done by leader nodes and integration process of data is also performed by cluster heads (leader). Our proposed work is based on hierarchical based protocol basically known as Low energy adaptive clustering hierarchy [2, 10, 11] focussing mainly in the threshold condition for electing leader node to extend lifetime of the network.

The other content of the paper are divided in the order as described here. Section 2 contains few existing WSN clustering protocols and other related routing protocols. Section 3 includes LEACH network model and the problem statement as a routing protocol. Section 4 marks out the proposed work in details. Simulated results are represented in Sect. 5. Conclusion is included in Sect. 6 and also future outlook. References are listed as the last Section.

2   Related Protocols

The present protocols and few of the associated routing protocols which are cluster based are briefly outlined in this section. LEACH [1, 2] is the utmost hierarchical based clustering routing protocol and leader node election is completely random in nature. This randomness in CH selection is overcome by LEACH-C (centralized LEACH) [9] protocol in which base station (BS) controls the clustering process in the network into clusters to ensure uniform CH distribution.

The authors proposed a modified LEACH [12] where the optimum cluster head (leader node) selection and present energy of the node are considered in the threshold condition to have extended network lifetime. In [3] Xu et al. developed a new threshold condition where they have included parameters like remaining energy of the nodes after each clustering process and new probability for leader node election. In Hybrid Energy Efficient distributed (HEED) [7] the authors proposed a routing protocol that takes into account nodes' communication power strength and separation between the nodes for clustering process. İt also considers number of neighboring nodes as the main deciding factors for selecting cluster head.

İn [4] the authors presented a routing protocol that employs the number of nodes neighbor to a node, nodes to BS distance alongwith present energy of sensor nodes as the determining factors for election of cluster head and also it is based on centralised data transmission. In [6] the author proposed new clustering protocol for WSNs where the energy of nodes are not homogeneous and consists of two types of clustering protocols namely single-hop based type and multihop based type.

3   Conventional LEACH Protocol

LEACH comprises of set up stage and steady state stage for each clustering process. Among the network nodes CH is chosen depending on the random number generated between [0, 1] by any nodes in the set up stage. The node will become CH for that particular clustering process (round) if the number generated randomly is less than the value of threshold condition (1).

$$T(n) = \begin{cases} \frac{p}{1-p*\left(r\,mod\frac{1}{p}\right)}, & n \in G \\ 0, & otherwise \end{cases} \tag{1}$$

where p represents likelihood of nodes to become CH, r is the present round, G are nodes that behave as member of the clusters in recent rounds. Data transmission takes place in the steady phase after accumulating the data transmitted by the member nodes.

4 Proposed Approach

The proposed LEACH (I-LEACH) uses the same energy model as the typical LEACH and has similar clustering operation. This proposed work is basically meant for application specific purpose for extending network lifetime. This paper focuses on improving the threshold condition with the inclusion of factor R_c for smooth cluster head selection till the end of the network lifetime and has additional energy and distance factors to the conventional LEACH threshold condition. The inclusion of these factors provide preferences to the nodes nearer to BS and higher energy to become cluster head more number of times without much computation than the far away one since the expenditure of energy is more with distance and chances of quicker death of far away nodes can also be reduced to some extent. As the improved threshold includes additional determinant rather than just random number generated by nodes the possibility of electing nodes acted as cluster members recently have better chances to become CH in next clustering process. The modified threshold condition is given as:

$$T(n) = \begin{cases} \frac{p_r}{1-p_r*\left(R_i\,mod\frac{1}{p_r}\right)}\left(\frac{1}{\log d_n} * E_{CR} * R_C * p_r\right), & n \in C \\ 0, & otherwise \end{cases} \tag{2}$$

where p_r is the likelihood of nodes to become CH, R_i is the current round, C denotes nodes that behave as member of the clusters in recent rounds, E_{CR} signifies fraction of sensor nodes' present energy to original starting energy of each sensor node, R_C is number of successive rounds for nodes of not being selected as cluster head, d_n represents separation between nodes and BS and logarithm function is performed in base 10.

With this improved threshold condition overall energy expenditure of the network can also be minimised to some extent as only appropriate nodes are elected as CH. It can also reduce the problem of rapid energy depletion of the cluster head. The proposed approach considers some assumptions for setting up the network:

- All nodes are homogeneous in the initial stage.
- Each node is stationary and location aware and can communicate with BS.
- Every node can perform the same function.
- Communication between nodes is symmetric in nature.

5 Simulation Results

The network lifetime performance analysis for the modified threshold is executed using Matlab software and the comparison is done for different network sizes and density of nodes with conventional LEACH and other relevant protocols based on the threshold condition for the same base station location. For simulation purpose the given specifications are used (Table. 1):

Table 1. Specifications for simulation

Specifications	Values
Initial energy (sensor nodes)	0.5 (Joules)
Number of nodes	100,50
P_r	0.1
$Eelec$	50×10^{-9} J/bit
E_{DA}	5×10^{-9} J/bit
ε_{fs}	10×10^{-12} J/bit/m^2
ε_{mp}	0.0013×10^{-12} J/bit/m^4
Packets size (data)	4000 bits
Packets size (control)	200 bits
BS location	(50,50) centre of the network
Network area	(100 $\times$ 100 & 200 $\times$ 200) sq. m.

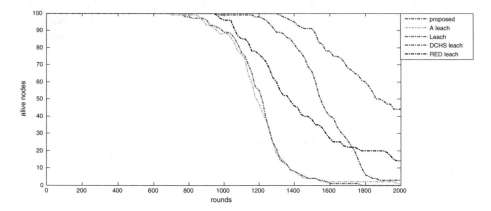

Fig. 1. Alive nodes vs rounds for 100 $\times$ 100 network size (100 nodes)

In Fig. 1 it is shown that life expectancy of improved LEACH (I-LEACH) is extended than the regular LEACH and some other LEACH derivative protocols. The proposed LEACH FND (first node dead) takes place at 1300 rounds as compared to FND at rounds 780, 840, 960, 962 of LEACH [1], A LEACH [13], DCHS [14], RED LEACH [15] respectively showing better stability period in the proposed I-LEACH protocol. This better stability is basically due to the eligible nodes selection as cluster head. Figure 2 shows that I-LEACH (proposed) protocol exhibit elongated network lifetime than conventional LEACH and other related protocols when the network size is increased showing that it is also applicable for larger size network lifetime extension.

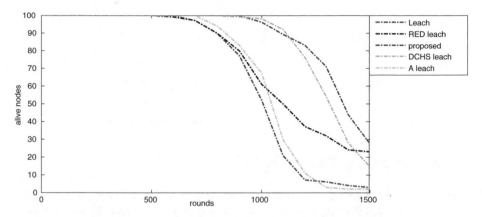

Fig. 2. Alive nodes vs rounds for 200 × 200 network size (nodes density with 100)

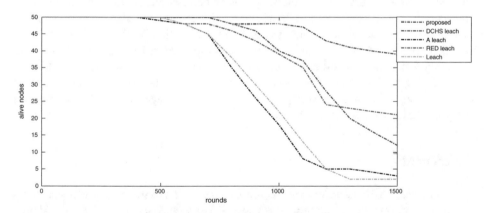

Fig. 3. Alive nodes vs rounds for 200 × 200 network size (nodes density with 50)

From Figs. 3 and 4 it is seen that I-LEACH protocol exhibits longer lifetime of network even when the nodes density varies in the network. The new threshold value still holds better lifetime existence since it is independent of number of nodes deployed. This proposed work chooses appropriate nodes to be CH depending on its distance and its energy to avoid rapid death of nodes by keeping away of choosing farther and low energy nodes to be CH.

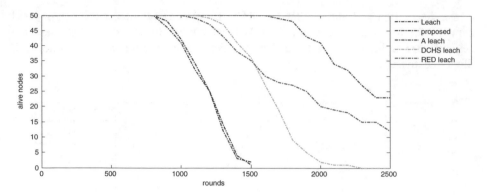

Fig. 4. Alive nodes vs rounds for 100×100 network size (50 nodes)

6   Conclusion and Future Outlook

This proposed work provides an improved threshold condition for cluster head election for each clustering process to further extend network lifetime and the comparison is performed for different network sizes and density of nodes. Even though conventional LEACH and other LEACH related protocols impart longer lifetime, both distance and energy factors consideration for electing cluster head provides better extension. The consideration of sensor nodes to base station distance along with energy of nodes avoids quicker death of nodes and non uniform distribution of alive nodes after each clustering process. As for the future outlook this proposed work can be further explored by combining with some other advanced coverage techniques as the additional factor for analysing performance of network in regards to connectivity, throughput, overall service quality of sensor network and effects on the average energy consumption in every clustering process of the nodes.

References

1. Heinzelman, W.B., Chandrakasan, A., Balakrishnan, H.: Energy-efficient communication protocol for wireless microsensor networks. In: Proceedings of the 33rd Annual Hawaii International Conference on System Sciences, HICSS 2000 (2000)
2. Heinzelman, W.B., Chandrakasan, A.P., Balakrishnan, H.: An application-specific protocol architecture for wireless microsensor networks. IEEE Trans. Wireless Commun. **1**(4), 660–670 (2002)
3. Xu, J., Jin, N., Lou, X., Peng, T., Zhou, Q., Chen, Y.: Improvement of LEACH protocol for WSN. In: International Conference on Fuzzy Systems and Knowledge Discovery FSKD (2012)
4. Haseeb, K., Abu, B.K., Abdullah, A.H., Darwish, T.: Adaptive energy aware cluster-based routing protocol for wireless sensor networks. Wireless Netw. **23**, 1953–1966 (2017)
5. Singh S.P., Sharma S.C.: A survey on cluster based routing protocol in wireless sensor networks. In: International conference on advanced computing technologies and applications (2015)

6. Kumar, D.: Performance analysis of energy efficient clustering protocols for maximising lifetime of wireless sensor networks. IET Wireless Sens. Syst. **4**(1), 9–16 (2014)
7. Younis, O., Fahmy, S.: Distributed clustering in ad-hoc sensor networks: a hybrid, energy-efficient approach. In: IEEE Transactions on Mobile Computing, pp. 909–914. IEEE Press, Bangladesh (2008)
8. Akkaya, K., Younis, M.: A survey on routing protocols for wireless sensor networks. Ad Hoc Netw. **3**(3), 325–349 (2003)
9. Pantazis, N.A., Nikolidakis, S.A., Vergados, D.D.: Energy-efficient routing protocols in wireless sensor networks: a survey. IEEE Commun. Surv. Tutorials **15**(2), 551–591 (2013)
10. Ajay, K.T., Suraj, G.P., Nilkanth, B.C.: A survey on data routing and aggregation techniques for wireless sensor networks. In: International Conference On Pervasive Computing (ICPC) (2015)
11. Karl, H., Willig, A.: Protocols and Architecture for Wireless sensor networks. Wiley, Hoboken (2005). ISBN 0470095105
12. Thein, M.C.M., Thein, T.: An energy efficient cluster-head selection for wireless sensor networks. In: International Conference on Intelligent Systems, Modeling and Simulation, pp. 287–291. IEEE Press, UK (2010)
13. Ali Md., S., Dey, T., Biswas, R.: ALEACH: advanced leach routing protocol for wireless microsensor networks. In: 5th IEEE International Conference on Electrical and Computer Engineering, pp. 909–914 (2008)
14. Handy, M., Haase, M., Timmermann, D.: Low energy adaptive clustering hierarchy with deterministic cluster-head selection. In: 4th IEEE International Workshop on Mobile and Wireless Communications Network, pp. 368–372 (2002)
15. Chit, T.A., Zar, K.T.: Lifetime improvement of wireless sensor network using residual energy and distance parameters on LEACH protocol. In: 18th International Symposium on Communications and Information Technologies, IEEE Press, Yangon (2018)

Improved WiFi Based Real-Time Indoor Localization Strategy

T. Rajasundari[1]([✉]), A. Balaji Ganesh[1], A. Hari Prakash[1], V. Ramji[1],
and A. Lakshmi Sangeetha[2]

[1] Electronic System Design Laboratory, Velammal Engineering College,
Chennai, India
`trajigopi@gmail.com, abganesh@velammal.edu.in,`
`harish091198@gmail.com, ramjivijayv@gmail.com`
[2] Department of Electronics and Instrumentation Engineering,
Velammal Engineering College, Chennai, India
`lakshmisangeetha@velammal.edu.in`

Abstract. Accurate mapping and localization of an environment have been improved owing to the advancements in mobile internet technology. However, indoor localization still requires more intelligent algorithms to keep continuous track of a mobile user because of presence of obstacles and satellite's incapability. The paper presents an indoor wifi based algorithm that combines fingerprint and least square algorithms to track location of a mobile user. The fingerprint data is computed in terms of received signal strength (RSS) acquired from different access points at predefined locations, dynamically. Similarly, the mobile user coordinate is estimated by involving digital filtering process followed by least square technique. The variance in RSS is observed between fingerprint and least square algorithm and applied to Kalman filter for the estimation of weightage value. Thus, the combined mechanism helps to result the value with more accuracy leaving behind low accurate value. The efficiency of the proposed method is evaluated by involving both MATLAB simulation environment covers up to 30 m × 35 m and also hardware resources in real time environment covers up to 13 m × 21 m. The results have showed the accuracy of less than a meter.

Keywords: Localization · Wifi · Real-time · Finger print · Least square method

1 Introduction

In recent times, applications that provide location based services are huge in demand. In the Internet of Things, track and identify a moving object are becoming increasingly essential feature. In various applications, including health monitoring, environmental monitoring and home office automation, location information finds an important role. Since access to wireless information is currently widespread, there is an intense demand to position the wireless systems, accurately. Indoor positioning is used to position objects or devices in an environment where Global Positioning System (GPS) is not governed [1–3].

© Springer Nature Switzerland AG 2020
D. J. Hemanth et al. (Eds.): ICICI 2019, LNDECT 38, pp. 84–95, 2020.
https://doi.org/10.1007/978-3-030-34080-3_10

For indoor navigation, the system needs effective placement of the transmitter device, dynamic mapping structures based on the user location and energy efficient solutions. The Indoor Positioning System (IPS) consists of a network of devices that are used to find objects or persons carrying handheld devices that use Bluetooth, ZigBee or Wifi as positioning signs. A number of indoor traceability technologies, infrared sensing, ultrasound scanning, time of flight, angle of arrival and magnetic sensor, have been adapted in the current situation. Various localization techniques and indoors tracking systems for indoor localization have been proposed [7–9]. Received signal strength (RSS) is the favoured parameter for evaluating indoor location.

The rest of this paper is organized as follows: the related works on wireless technologies are described in Sect. 2. The localization system is presented in Sect. 3. Section 4 discusses about experimental methodology and results in Sect. 5. Section 6 describes the conclusion and future work.

2 Related Works

Trilateration technique is considered as one of the most normally used Wi-Fi based localization techniques. Trilateration uses the distances of an object from three known point, that are usually fixed factors to determine the position of an object. In order to discover a point in space, triangulation uses the geometric property of the triangle. The lateration and angulation approaches are the two categories. The propagation time system measurements (e.g. TOA, RTOF, and TDOA) are called as angulation technique. And the lateration techniques are RSS and received signal phase based methods.

Mok et al. [5] proposed a fingerprinting approach, when an individual enters the environment, without prior understanding they set a fingerprint database. Existing databases also can be updated with the aid of gathering data from neighbour user. The replacement of the database and the reduction of training expenses had been introduced with both historical and new information [6]. The precept of fingerprinting positioning algorithm based on wifi is described in two forms, specifically offline and on-line. In online phase, the primary job is to calculate the target node position on the basis of current fingerprint position data. Some of the commonly used methods are the k nearest neighbourhood (KNN), and another is probabilistic method. In offline, the monitoring area is segmented into numerous virtual grids and every grid vertices are defined as sampling points [4, 15, 19].

In both trilateration and fingerprint algorithm, the significant problem to be addressed is the path loss due to the presence of obstacles in dynamic environments. To overcome path loss model, we combine least square algorithm and fingerprint algorithm. Kalman filter provides excellent signal processing result. It is presumed that applying Kalman filter for indoor localization might give more accurate results as it is known for its immunity towards noise signals. The variance of least square and fingerprint results used in the kalman filter reflects path loss and obstacle nature, respectively. Due to the simple and robust nature of changes in wireless propagation properties, including path loss there was a lot of interest in Fingerprint and least square algorithm [11–14].

3 Localization System

Localization of a user based on merely RSS value deviates from accurate location. Least square is one of the algorithms used to locate user with good prediction. It is implemented by involving the Eqs. (1–4). The equations, 2 and 3 are formed by considering 3 Access Points.

Distance d is calculated by,

$$d_i = 10^{(1mRSS_i - RSS_i) \div (10 \times N)} \tag{1}$$

$$A = \begin{bmatrix} (x_2 - x_1) & (y_2 - y_1) \\ (x_3 - x_1) & (y_3 - y_1) \end{bmatrix} \tag{2}$$

$$B = \begin{bmatrix} x_2^2 - x_1^2 + y_2^2 - y_1^2 - d_2^2 + d_1^2 \\ x_3^2 - x_1^2 + y_3^2 - y_1^2 + d_2^2 - d_3^2 \end{bmatrix} \tag{3}$$

$$X = \frac{1}{2} \times (A^T A)^{-1} (A^T B) \tag{4}$$

Which results the user coordinates Eq. (5) represented by,

$$X = \begin{bmatrix} x \\ y \end{bmatrix} \tag{5}$$

3.1 Filtration

The measured RSS value varies a lot due to many reasons like obstacle and multipath effect. In order to reduce these variations a digital filter is applied before localization algorithm. This is achieved by the method of regression using successive sub-sets of adjacent data points.

For localization, varying RSS values are smoothed and mean value of the filtered result is considered for the conversion of RSS to distance d, which is further processed by localization algorithm. Since filtering process also may contain variations, variance is observed and the same is quantified by,

$$Variance = (1 \div n) \sum_{i=1}^{n} (X_i - \mu)^2 \tag{6}$$

$$Mean = \mu = (1 \div n) \sum_{i=1}^{n} X_i \tag{7}$$

Where n is the size of the set and X is set of filtered RSS values. This variance is then converted into their respective distances with a unit of meter by

$$d_{var} = 10^{(1mRSS - 1mVariance - RSS + Variance) \div (10 \times N)} \tag{8}$$

The one meter of variance denotes the respective AP measured at 1 m distance. Distance d_{var} is given as an input to least square equations that estimates X and Y position. The user position calculated by using d_i and d_{var} variance and it is represented as sensor variance in Kalman filter.

The weightage enables the Kalman filter to evaluate the guess and sensor values obtained from the algorithms, to predict the best accurate value. The initial weight is calculated from guess variance (fingerprint) and sensor variance (Least square). The variance values are calculated in meters for *abscissa* and ordinates, respectively. The guess variance of the fingerprint database is calculated by the variance observed in the database at periodic instances.

To locate user position, this algorithm goes through some iteration based on the number of samples.

$$X_{weight} = X_{guessvariance} \div \left(X_{guessvariance} + X_{sensorvariance} \right) \tag{9}$$

By using the above weights, X and Y can be estimated by

$$X_{estimate} = X_{guess} + X_{weight} \times \left(X_{measurement} - X_{guess} \right) \tag{10}$$

$$Y_{estimate} = Y_{guess} + Y_{weight} \times \left(Y_{measurement} - Y_{guess} \right) \tag{11}$$

The variance in the above estimation is calculated by

$$X_{est\_var} = \left(X_{guess\_var} \times X_{sensor\_var} \right) \div \left(X_{guess\_var} + X_{sensor\_var} \right) \tag{12}$$

$$Y_{est\_var} = \left(Y_{guess\_var} \times Y_{sensor\_var} \right) \div \left(Y_{guess\_var} + Y_{sensor\_var} \right) \tag{13}$$

Now the guess value and guess variance for the next iteration is predicted. These values are updated in the next iteration as guess and guess variance, $X_{measurement}$ and $Y_{measurement}$ updated at the end of each iteration by using least square method and the iteration is continued on the basis of number of samples to be taken in a particular position.

4 Experimental Methodology

The experiment is performed in an indoor laboratory environment that covers an area of 13×21 m^2 as shown in Fig. 1. The scenario involves both indoor localization and indoor navigation. The experimental setup consists of hardware nodes with wifi communication protocol to constitute a wireless sensor network. A wireless sensor network is established by using wireless nodes to monitor environmental parameters. The nodes are grouped as data nodes and access point nodes meant for data acquisition and for navigation, respectively. The nodes are grouped into clusters for effective usage of the resources. In this scenario the number of available access points is same as the number of clusters formed. The access point nodes act as a cluster heads and acquire the data from the sub-nodes within the cluster. The data nodes which act as a sub-node transmit the environment monitoring sensors data to the cluster head and further, it transmits the data to the centralized node, which tops in the hierarchy of the wireless sensor network.

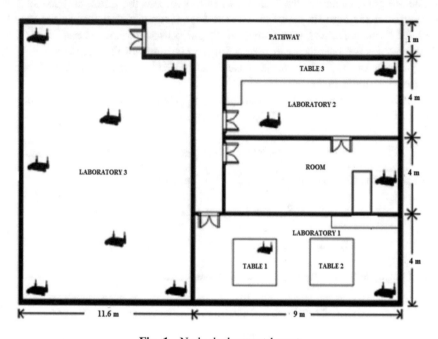

Fig. 1. Node deployment layout

The presence of access points makes it possible to keep track of user location, which helps to provide accurate navigation facility to the end-user. The cluster formation makes the data handling much better than individual data transmission. The involvement of a greater number of data nodes involves high traffic in data transmission and handling. To avoid the chaos in the data processing, the data is received for predefined time limit from each cluster heads in the Round-Robin method by using the wifi switching and parallel processing as shown in Fig. 2, and are stored in the cloud server. The centralized node gets connected to each of the cluster networks for the predefined time and switches to next network. The server data are accessed by the API and are displayed appropriately based on the user requirement. The user can access the live data of the nodes in the deployed environment, remotely. The user inside the indoor environment can easily navigate to any deployed node location by using the developed android application, which has a mapping of the indoor environment.

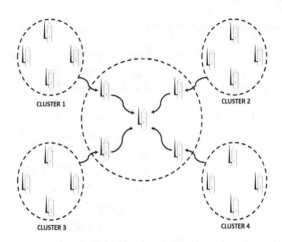

Fig. 2. Data Node cluster formation

The received signal strength (RSS) obtained from the wifi access point is considered as an input for the indoor localization algorithm to estimate the user location.

Input parameters: Node Position point (x_1,y_1), (x_2, y_2) & (x_3, y_3) , 1 meter RSS data, 1 meter Variance of RSS and Fingerprint database.

Event: **Localization**

1. Measure RSS values and create a sample list with the RSS values.

2. Filter is applied to each sample.

3. With the filtered results, the normalized value is given to fingerprint algorithm.

4. Fingerprint algorithm gives (X, Y) and assigned as X_{guess} and Y_{guess} respectively and its respective variance as $X_{guessvariance}$ and $Y_{guessvariance}$.

5. The filtered RSS values along with its variance are given to least square algorithm.

    5.1. For i in range of Access Points
         The mean value of the filtered RSS values is considered and d is calculated from each access points
         $$d_i = 10^{(1mRSS_i - RSS_i) \div (10 \times N)}$$
    5.2. The results are substituted in the least square equation
         $$X = (xy)$$
         By solving these equations (x, y) is obtained.

6. The (x, y) co-ordinates obtained from least square algorithm results $X_{measurment}$ and $Y_{Measurment}$ and its variance results $X_{sensorvariance}$ and $Y_{sensorvariance}$.

7. These results are given as input to the Kalman Filter.

7.1. Calculate weight based on the variance for X and Y by using

$$X_{weight} = X_{guessvariance} \div \left(X_{guessvariance} + X_{sensorvariance}\right)$$
$$Y_{weight} = Y_{guessvariance} \div \left(Y_{guessvariance} + Y_{sensorvariance}\right)$$

7.2. By using weight an estimate of X and Y is found by

$$X_{estimate} = X_{guess} + X_{weight} \times \left(X_{measurement} - X_{guess}\right)$$
$$Y_{estimate} = Y_{guess} + Y_{weight} \times \left(Y_{measurement} - Y_{guess}\right)$$

7.3. The variance for the estimate value is also calculated by,

$$X_{est\_var} = \frac{\left(X_{guess\_var} \times X_{sensor\_var}\right)}{\left(X_{guess\_var} + X_{sensor\_var}\right)}$$

$$Y_{est\_var} = \frac{\left(Y_{guess\_var} \times Y_{sensor\_var}\right)}{\left(Y_{guess\_var} + Y_{sensor\_var}\right)}$$

8.   The resultant output estimates the user coordinates.

9.   Jump to step 1 and start the next iterations. As the number of iterations increases the accuracy increases.

Event: **Calculating the X and Y Variance by Least square algorithm**

1.   Variance for the filtered RSS is found by
2. $Variance = (1 \div n) \sum_{i=1}^{n}(X_i - \mu)^2$

$$Mean = \mu = (1 \div n) \sum_{i=1}^{n} X_i$$

Were X_i is the filtered RSS datum value.

3.   For i in range of access points
$$d_i = 10^{(1mRSS_i - 1mVariance_i - RSS_i + Variance_i) \div (10 \times N)}$$

4.   These results are substituted in the least square equation
$$X = (xy)$$
By solving these equations varied (x, y) is obtained.

5.   Difference between (x, y) obtained from previous event and varied (x, y) gives us variance in x and y, updated as sensor variance.

5 Results and Discussions

The current research examines the localization method by the way of simulating it in the MATLAB platform and also in real time by using wireless nodes. The simulation deployment strategy is displayed in Fig. 3. The overall region covers as a whole lot as 35 m × 3 0 m and 50 sink nodes are distributed with the tx power of −31 dBm. It can be observed that the active nodes near the user are chosen for estimating the user position.

The neighbouring nodes selection is based on the strength of the received RSS values by the user. The neighbouring node RSS values are trusted to be more reliable than the RSS from other nodes due to the environment that both the user and neighbouring nodes have less distance and transmission losses. The Fig. 3 proves to be a valid support for the above theory.

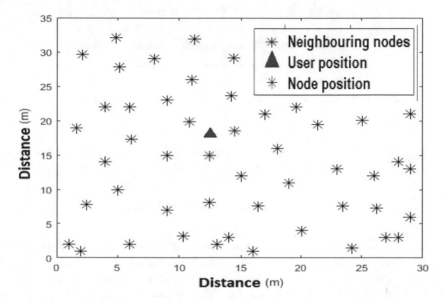

Fig. 3. MATLAB simulation-graph plot of node position and user position

A real-time experiment is performed with the proposed localization algorithm in an indoor environment. The variations in the parameters such as temperature and wind, affect the propagation of wifi signal, causing the RSS value to oscillate around the absolute value with large magnitude variations. The presence of obstacles also adds up in causing distortion in RSS values. The obstacles result in inducing of multi-path effect and wave bouncing, which limits the wifi signal to reach the destination. In order to normalize the obtained RSS values, large numbers of data are collected and filtered.

Figures 4 describe corresponding access points 1, respectively which based on the RSS readings considered for 400 s at a rate of one data per second. The data is given to the filtering mechanism which approximates the values with a moving window by a polynomial of order zero (a constant). The mechanism filters out the small scale variation over both time and space caused due to Rayleigh fading (multipath fading) and it is introduced by the obstacles. The resultant graph shows that the raw value of RSS received by the mobile node varies a lot and filtering has almost reduced those variations.

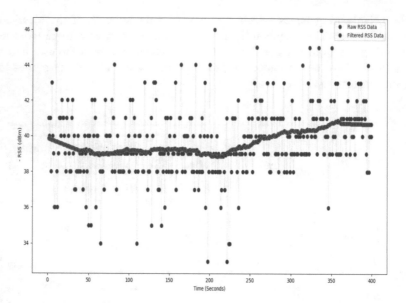

Fig. 4. Normalization value of Node 1

As the localization algorithm uses the log-normal shadowing model unrefined RSS data results in exponential variation. This variation is suppressed by the filter, thereby increasing the precision. Despite using filter, there is a variance even after filtering which can be observed in the resultant data. This variance is quantified by the Eq. (6). The proposed algorithm is applied over the filtered data to estimate a better result.

In order to discover the position of the mobile device, the study has examined the usability of the localization algorithm with the WiFi channels at 2.4 GHz range. The strategy consists of following stages, distance calculation using the RSS value followed by localization. Distance (d) is estimated using filtered RSS value, the resultant (d) is given as an input to position estimation algorithm.

Position estimation is based on least squares and it is implemented to further reduce the estimation bias which has been compared to fingerprint database. In order to use fingerprint data in a dynamic environment, a database is to be maintained regularly. The resultant (x, y) from both algorithm is further processed through Kalman filter which results accurate (x, y) by applying weightage Eqs. (10) and (11). Table 1 describes the real time data (RSS) from the user is processed, estimated and also corrected by the combining strategy of the proposed algorithm.

Table 1. Results of real time estimated position

S. no	Actual location (x, y) (m)	Estimated location using proposed algorithm (x, y) (m)
1	(0, 4.5)	(0, 3.8)
2	(3.15, 4.75)	(3.02, 4.5)
3	(6.3, 14.75)	(6, 13.7)
4	(9.45, 10)	(8.56, 11.03)
5	(12.6, 20.5)	(12.4, 20)
6	(5.4, 1.8)	(4.8, 1.4)
7	(7.8, 4.2)	(8, 4.6)
8	(6.6, 3)	(6, 3)
9	(3, 0.6)	(3.1, 1)
10	(5.4, 4.2)	(5.3, 4.5)

Comparison between the exactness of the least square and fingerprint techniques, our set of rules brings out accuracy & precision less than a meter. Table 2 tabulates the accuracy results of existing works along with the proposed research.

Table 2. Result comparison table

Author	Technique used	Accuracy
İlçi [16]	Least square Algorithm	5 m
Mok [5]	Fingerprint Algorithm	3–5 m
Xiao [15]	Fingerprint Algorithm	3.25 m
Chai [18]	Kalman Filter and least square estimation	2.6 m
Liu [9]	Applying Matrices for RSS value	2 m
Santhosh [17]	Fingerprint Algorithm	1.5 m–2.5 m
Proposed work	**Least square and fingerprint method**	**0.5 m–0.6 m**

To monitor the sensor node and indoor navigation, we develop an android application. The application interface displays the deployment of sensor map along with sensory data. The user can monitor the data at anytime irrespective of the user's presence in the test environment. It has featured with a login page to provide access only to the authorized individuals for security reasons. The privileged users can enter their login credentials to view the data and indoor map which are shown in Figs. 5 and 6.

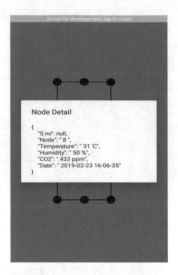

Fig. 5. Application login screen   **Fig. 6.** Node data display screen

6   Conclusion and Future Work

In this proposed research work, indoor localization is developed with wireless sensor nodes, Wifi access points, and mobile phone. In this paper an algorithm is proposed to combine least square and fingerprint by using sensor fusion technique and developed model tested in simulation and in a real time test bed. The simulation output indicates 0.3 m accurate results and real time results precise error measurements less than one meter. The algorithm proposed is 96.7% precise compared to the existing tracking techniques. The combination of these methods enables the creation of a complete localization implementation and framework for a practical wireless sensor network. In addition this research identified a number of indoor localization problems and offered overall alternatives which can assist to address these problems. The future work involves by implementing the present study in large dynamic environments such as warehouse or in shopping malls giving rise to potential applications like indoor localization in the areas of wireless sensor networks.

Acknowledgments. The authors gratefully acknowledge the financial support from Department of Science and Technology by sanctioning a project (File No: DST/SSTP/TN/29/2017-18) to Velammal Engineering College, Chennai, under SSTP scheme.

References

1. Liu, Y., Yang, Z.: Location, Localization, and Localizability: Location Awareness Technology for Wireless Networks. Springer, New York (2010)
2. Xiao, J., Zhou, Z., Yi, Y.: A survey on wireless indoor localization from the device perspective. ACM Comput. Surv. **49**(2), Article no. 25 (2016)

3. Au, A.W.S., Feng, C., Valaee, S., Reyes, S., Sorour, S., Markowitz, S.N., Gold, D., Gordon, K., Eizenman, M.: Indoor tracking and navigation using received signal strength and compressive sensing on a mobile device. IEEE Trans. Mob. Comput. **12**(10), 2050–2062 (2013)
4. Torres-Sospedra, J., Moreira, A.: Analysis of sources of large localization errors in deterministic fingerprinting. Sensors **17**, 2736 (2017)
5. Mok, E., Retscher, G.: Location determination using WiFi fingerprinting versus WiFi trilateration. 145–159 (2007)
6. Emery, M., Denko, M.K.: IEEE 802.11 WLAN based real-time location tracking in indoor and outdoor environments, 0840-7789 (2007)
7. Sakpere, W., Adeyeye Oshin, M., Mlitwa, N.B.W.: A survey on a state-of-the-art survey of indoor positioning and navigation systems and technologies. S. Afr. Comput. J. SACJ **29**(3), 145–197 (2017)
8. Liu, K., Motta, G., Ma, T.: XYZ indoor navigation through augmented reality: a research in progress. In: IEEE International Conference on Services Computing (SCC), San Francisco, CA, pp. 299–306 (2016)
9. Liu, K., Motta, G., Ma, T., Guo, T.: Multi-floor indoor navigation with geo-magnetic field positioning and ant colony optimization algorithm. In: 2016 IEEE Symposium on Service-Oriented System Engineering (SOSE), Oxford, pp. 314–323 (2016)
10. Mendoza-Silva, G.M., Torres-Sospedra, J., Huerta, J., Montoliu, R., Benítez, F., Belmonte, O.: Situation goodness method for weighted centroid-based Wi-Fi APs localization. Springer (2017). https://doi.org/10.1007/978-3-319-47289-8_2
11. Wen-jian, W., Jin, L., He-lin, L., Bing, K.: An improved weighted trilateration localization algorithm. J. Zhengzhou Univ. Light. Ind. (Nat. Sci.) **3**(6), 84–85 (2012)
12. Adler, S., Schmitt, S., Kyas, M.: Path loss and multipath effects in a real world indoor localization scenario. In: 2014 11th Workshop on Positioning, Navigation and Communication (WPNC), pp. 1–7, March 2014
13. Mahiddin, N.: Indoor position detection using WiFi and trilateration technique. In: The International Conference on Informatics and Applications (2012)
14. Yim, J., Jeong, S., Gwon, K., Joo, J.: Improvement of Kalman filters for WLAN based Indoor Tracking. Expert Syst. Appl. **37**, 426–433 (2010). https://doi.org/10.1016/j.eswa.2009.05.047
15. Xiao, T.-T., Liao, X.-Y., Hu, K., Yu, M.: Study of fingerprint location algorithm based on WiFi technology for indoor localization. In. International Conference on Wireless Communications, Networking and Mobile Computing (WiCOM 2014), 26–28 September 2014
16. İlçi, V., Gülal, E., Çizmeci, H., Coşar, M.: RSS-based indoor positioning with weighted iterative nonlinear least square algorithm. In: The Twelfth International Conference on Wireless and Mobile Communication, ICWMC 2016 (2016)
17. Subedi, S., Pyun, J.-Y.: Practical fingerprinting localization for indoor positioning system by using beacons. J. Sens. **2017**, 16 (2017). Article ID 9742170
18. Chai, S., An, R., Du, Z.: An indoor positioning algorithm using Bluetooth low energy RSSIS. In: International Conference on Advanced Material Science and Environmental Engineering (2016)
19. Honkavirta, V., Perala, T., Ali-Loytty, S., Piche, R.: A comparative survey of WLAN location fingerprinting methods. In: Proceedings of the 6th Workshop on Positioning, Navigation and Communication 2009, WPNC 2009, pp. 243–251, March 2009

Smart Monitoring for Waste Management Using Internet of Things

Jagadevi N. Kalshetty$^{(\boxtimes)}$, Akash Anil, Harsh Jain, Sawan Beli,
and Abhishek Kumar Singh

Nitte Meenakshi Institute of Technology, Bengaluru 560064, India
jagadevi.n.kalshetty@nmit.ac.in

Abstract. As the second most populated nation on the planet, India faces a noteworthy issue in garbage management. In the majority of the urban area, the flooded garbage bins are making an unhygienic environment. This additionally leads to promotion of various sorts of anonymous diseases. Starting at now there are customary waste administration frameworks like intermittent and routine clearing by the different urban bodies like the municipality. In any case, despite the fact that these normal systems of support are done we frequently go over flooding refuse containers from which the trash spills on to the lanes. This happens in light of the fact that starting at now there is no framework set up that can screen the trash canisters and demonstrate the same to the company. We introduce a waste gathering arrangement in view of giving knowledge to dustbins, by utilizing an IoT model implanted with sensors, which can read, gather, isolate the waste and transmit dustbin volume information and area over the Internet. This framework screens the dustbins and educates about the level of trash gathered in the waste canisters by means of a site page. For this the framework utilizes infrared sensors and MQ-6 sensor set inside the receptacles to quantify the status of the dustbin. At the point when the dustbin is being filled, the tallness of the gathered misuse of the receptacle will be shown and transmitted.

Keywords: Arduino-MEGA · Rain-drop sensor · IR sensor · Smart bins · Wi-Fi module · Gas sensor · GPS module · Web page · Android app

1 Introduction

Web and its applications have transformed into a fundamental bit of the present human lifestyle. It has transformed into a fundamental gadget in every perspective. In light of the gigantic solicitation and need, experts went past interfacing just PCs into the web. These request about provoked the presentation of an amazing gizmo, Internet of Things (IoT). Correspondence over the web has created from customer - customer collaboration to contraption interchanges these days. The IoT thoughts were proposed quite a while back yet in the meantime it's in the basic period of business association. Things (Embedded devices) that are related with Internet and sometimes these devices can be controlled from the web is typically called as Internet of Things.

© Springer Nature Switzerland AG 2020
D. J. Hemanth et al. (Eds.): ICICI 2019, LNDECT 38, pp. 96–103, 2020.
https://doi.org/10.1007/978-3-030-34080-3_11

In our system, the Smart clean containers are related with the web to get the continuous information of the insightful dustbins. In the present years, there was a quick improvement in masses which prompts increasingly waste exchange. Along these lines, an authentic waste organization structure is essential to refrain from spreading some deadly diseases. Managing the splendid repositories by checking the status through page and android application in like way taking the decision.

2 Related Background

India is a developing country and has a diverse climate and topography. The population of India was 1.32 billion in 2017 to that of1.15 billion in 2010. This large Populations a major contributor of increase in Municipal Solid Waste (MSW) in India. Advances in common infrastructures are obligatory for India to wind up a world driving economy. Scaling up of smart infrastructures directs the issues of the masses and ensures that the earth needs to address viable monetary development as a basic issue. Waste management plays an important part in conveying manageable improvement. Fast population development in India has led to growing extinction of characteristic assets. Major extrication from waste can be materials, vitality or supplements. A new industry can be set up to extract these; also this offers numerous employment opportunities to individuals.

The development of assets from squanders must be accomplished by advancing in fields like Smart Waste Management as this depends upon a planned arrangement of activities to create reusable/recyclable materials. Materials, vitality and supplement extraction must be a major point of future Smart Waste Management framework improvement in India. Assets can be extricated from squanders utilizing existing advances and India has a great success in reusing these assets. The proposed strategy for our paper is an exceptional instance of presenting an android application utilizing which would status be able to the different canister conditions like half, full and carbon dioxide levels can be checked, related information will be put away in the page.

3 Proposed Method

Numerous installed modules in the framework have unique purpose as indicated by their capacities and utilities. The framework is primarily made out of a solitary microcontroller (Arduino Mega), IR sensor, rain drop sensor, MQ-6 sensor, LCD Display, GSM module, Wi-Fi module, Android application and site page. Aside from the database, the whole unit is put inside clean receptacle. The microcontroller (Arduino Mega) is situated at the focal point of the framework. In view of the condition and information sources the microcontroller makes a move. The IR sensor will identify the question at the underlying stage and helps in finding the levels of canister. Rain drop sensor helps in recognizes the kind of waste. The driver circuit will move the loss to specific containers by turning the transport line to the separate receptacles using the information from the rain drop sensor. LCD will help in showing the canister status and significantly more. MQ-6 sensor will help in distinguishing proof of wet waste deterioration (or) foul smell. GPS module gets the present area of the container. Wi-Fi module transfers the information to the page and android application.

In the square outline in Fig. 1 LCD is used to exhibit the working of the whole unit.

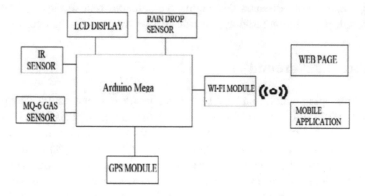

Fig. 1. System architecture

The Hardware Requirements for the project are microcontroller (Arduino Mega), IR Sensor, MQ-6 Sensor, Rain drop Sensor, Wi-Fi module, GSM module, Android cell phone, LCD display.

a. Hardware

Code is executed in embedded c programming and the outcome will be shown in the driven screen.

When the sensor esteem crosses the limit esteem the notice is sent to phone also, information is put away in MySQL.

b. A mobile user should check the availability of the following:

The application should be installed and should be connected to the same network as the PC. Android phone should be able to display the app.

User interface such as login screen, registration screen should be displayed.

c. Database:

The data received should be stored in MySQL database.

The database should be available to receive data and send data

These are the requirements that are not functional in nature; these are the restrictions within which the system must work.

a. Scalability, Capacity and Availability: The system will achieve 100% accessibility at all times. The system should be scalable for surplus hardware requirements.

b. Maintainability: The system should be enhanced for ease of maintenance or supportability, as far as possible. The software requirements of the systems are:

  Arduino IDE
  Embedded C
  Android Version 4.0

Arduino board used is Mega-2560 which is based upon the microcontroller AT Mega 2560. It consists of fifty-four digital I/O pins out of those fifteen pins have the capability to be used as PWM Outputs, 4 UARTs (equipment serial ports), 16 simple info, a 16 MHz gem-oscillator, a USB associate in, a reset button, an ICSP header and a power-jack. It has every functionality needed to help the microcontroller. An IR sensor is used to sense the surroundings by discharging a IR radiation and by also distinguishing them. IR sensors are similarly armed for estimating the heat being produced by a protest and recognizing movement. An Infrared (IR) sensor is used to identify hindrances. An IR sensor consists of a producer, identifier and related hardware. All the things which has temperature even slightly above 0 K has a warm vitality and acts as a IR source.

Wellsprings of infrared-radiation integrate blackbody radiators, silicon carbide and tungsten lights. Infrared- sensors commonly use infrared lasers and LEDs with particular infrared wavelengths as sources. A rain sensor is used to detect rainfall. It can even be described as a switch that activates when water droplets pass or hits the sensor surface. This module consists a rain-board and a control panel which is separate from the board to avoid any mishap and also a potentiometer. The Control board has a LED which glows when the rain-board is dry and also the DO yield is high. A gas sensor (or gas locator) is a gadget that recognizes the nearness of gasses in a region. A sensor is a mechanical gadget that identifies faculties a flag, physical condition and synthetic mixes. They are produced as settled/stationary units and operate by using increased quantity of gases through advancements and are capable of being noticed or noticeable markers, for instance, alerts, flashes or a mixture of signs.

The Moisture-sensor is utilized to quantify the liquid quantity (moisture) in the mud. The moment when the soil encounters a limitation of moisture content, the device yield is at abnormal state; otherwise the yield is at a minor state. This sensor prompts the client to water their floras and furthermore screens the dampness substance of earth. It has been broadly utilized as a part of farming, isolation of sort of waste, arrive water system and natural cultivating. ESP8266 Wi-Fi module is Wi-Fi serial handset module, in light of ESP8266. Little size and minimal effort makes it reasonable for sensor hubs. It deals with 3.3 V and devours current up to 250 mA. Current utilization is very huge so it's generally not controlled on battery. On the off chance that you are utilizing 5 V Arduino, at that point read ESP8266 Wi-Fi and 5 V Arduino association. At anywhere and anytime, there are at least 4 GPS satellites are orbiting. Each one of them transmits data about its position and current time at a constant interval of time.

These signals going at the speed of light are received by GPS receiver; the GPS receiver gets a signal from each of these four satellites. The cluster of satellites is then tasked with transmitting at the same time when the signals are being sent. Then the next process is to subtract the time at which the signal was transmitted from the other values i.e. from the time at which it was received. This way, GPS is able say the distance from each satellite.

4 Simulation and Results

The proposed venture utilizes android user board, site page and android application utilizing which a portable android application is produced. The unit which has been inserted is equipped with different sensors. The implanted unit that makes utilization of IR sensor, MQ-6 sensor, Rain Drop sensor, Wi-Fi module and GPS module is given in the figure underneath (Figs. 2, 3, 4, 5 and 6).

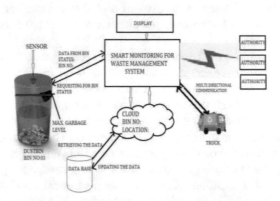

Fig. 2. A general architecture of the model

Fig. 3. A view of the home screen of the application used to monitor the smart bins.

Fig. 4. Log in screen of the application used to monitor the smart bins

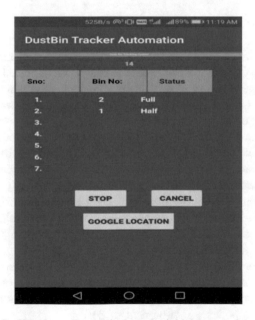

Fig. 5. The above figure displays the initial status of the bins

Fig. 6. The figure displays the current status of the bins

5 Conclusion

Waste Segregation using smart bin has been successfully implemented for the segregation of waste into dry and wet waste at root source. Unregulated waste management is surely one of biggest reason of air, land& water pollution which ultimately causes disastrous effects on health. The Smart Bin provides a simple way to overcome this problem by making entire system automated. The components used in Smart Bin are economic, environmentally friendly and gives accurate results for separating different types of wastes which are commonly produced in day-to-day activities.

Presently there are no device/product available for segregation of waste at root source other than manual segregation. Segregating waste manually is not very accurate and many of us don't like to do that. The device in question has got the task of separating all types of waste and not only spike the economic value of waste but also give a clean and hygienic environment at lesser cost. This could probably the biggest advantage of smart dustbin.

Open dumping of solid waste emits foul smell due to presence of dead or decaying matter. Often animals are found feeding on such dumping place which in turn affects their health. Such dumping sites are generally present in city outskirts or near some village, which in turn pollutes environment of nearby.

In this study, waste segregation and monitoring procedure were reviewed. By using GPS module, Gas sensor and Raindrop sensor, Smart Waste monitoring system has provided the waste segregation operator with a platform to work in real-time to improve the service by optimizing operation time and feasibility of operation.

6   Future Enhancement

At present the proposed model can segregate only two types of waste i.e. dry and wet waste. In future it can be enhanced to segregate more types of municipal solid wastes such as plastic, glass, metals using various high accuracy sensors and devices. It can also be enhanced to segregate combination of solid waste dumped at a time.

Size, weight and cost of the proposed model can also be reduced using more efficient design and 3-D printed parts. Some additional feature can be added such generating biogas source itself and generated biogas can be used for power generation. With the current arrangement the proposed model is confined to a locality. In the future, smaller IoT devices for waste segregation and management can be developed and integrated the garbage chutes of high-rise apartments, widening the implementation to the citizen.

References

1. Catani, V., Venture, D.: An approach for monitoring and smart planning and urban solid waste management using smart-M3 platform. In: Proceeding of the 15th FRUCT Conference, 24 April 2014 (2014)
2. Flora, A.: Towards a clean environment: a proposal on sustainable and integrated solid waste management system for University Kebangsaan Malaysia. Report from Alam Flora (2009)
3. Kumar, A., Malvadkar, A.R.: Automatic arduino-controlled garbage collector. In: 2018 IRAJ ARSSS (2018)
4. Patel, T., Jariwala, K.: Fog computing in Iot. In: 2018 IRAJ Advanced Research Society for Science and Sociology (ARSSS 2018) (2018)
5. Thakur, D., Nikam, N., Harnik, C.: Multi function auto-cleaner. Int. J. Mech. Prod. Eng. **6**, 44–47 (2018)
6. Devadiga, N., Potluri, S.: Design and fabrication of floor cleaning machine. Int. J. Innov. Eng. Res. Technol. (2017)
7. Garach, P.V., Thakkar, R.: Design and implementation of smart waste management system using FOG computing. Int. Res. J. Eng. Technol. (IRJET) **5**, 3888–3891 (2018)
8. Rajaprabha, M.N., Jayalakshmi, P., Vijay Anand, R., Asha, N.: IoT based smart garbage collector for smart cities. Int. J. Civ. Eng. Technol. (IJCIET) **9**, 435–439 (2018)
9. Jadhao, P.A., Sakhare, S.D., Bhaldane, K.G., Narkhede, A.P., Girnale, V.S.: Smart garbage monitoring and collection system using internet of things. Int. J. Adv. Eng. Res. Dev. **5**, 1–7 (2018)
10. Duan, Q., et al.: A survey on service-oriented network virtualization towards convergence of networking and IoT. IEEE Trans. Netw. Serv. Manag. **9**(4), 373–392 (2012)
11. Enevo, Enevo-Optimising Waste Water Collection (2016)
12. Kumar, R.P., Smys, S.: Analysis of dynamic topology wireless sensor networks for the Internet of Things (IOT). Int. J. Innov. Eng. Technol. **8**, 35–41 (2017)

Enhanced Scrambled Prime Key Encryption Using Chaos Theory and Steganography

Shanmukha Shreyas Vedantam, Kushalnath Devaruppala,
and Ravi Shankar Nanduri[✉]

Geethanjali College of Engineering and Technology, Hyderabad,
Telangana, India
shreyas1998vs@gmail.com, kushaldev97@gmail.com,
ravish00@yahoo.com

Abstract. In this paper we have analyzed the strength and weakness of scrambled prime key encryption proposed by Haidar et al. [10] for data encryption using key elements which are prime numbers. However, this scheme exhibits its vulnerabilities in the face of chosen plaintext attack. Hence we, in this paper, extended the algorithm by introducing chaos theory and steganography, which offers better security and robustness. We have taken the partially encrypted text from [10] and embedded the same in a carrier image. In this process, we have chosen a novel chaos generator to select the pixels into which the text is to be embedded.

Keywords: Double rod pendulum · Pendulum velocity · Pendulum acceleration · Encryption · Decryption · Steganography · Chaos theory

1 Introduction

The advent of internet and migration of transactions on to the net necessitated invention of strong measures for ensuring the security of information. Cryptography is widely used for protecting the confidentiality of information [1, 2]. Literature of cryptography is replete with a large number of algorithms to ensure information security. Conventional symmetric key encryption algorithms like DES, Blowfish etc. [3, 4] were widely used in the last decade of the previous century. Modern block ciphers like AES [5] replaced the conventional block ciphers in the 21s't century. In addition, asymmetric key algorithms like RSA [6, 7], ECC [8, 9] were widely used for key exchange.

In this paper we examine a cryptographic algorithm proposed by Haidar et al. [10], which henceforth will be referred to as base algorithm. We found a few vulnerabilities in the base algorithm which might compromise the security. In this paper we propose modifications which enhance the effectiveness of the base algorithm.

Chaos is a concept from non-linear dynamics where generator equation is highly sensitive to initial conditions [11]. Many researchers have focused on application of chaos theory in cryptographic functions [12–21]. Sensitivity to initial conditions, ergodicity and mixing are three important properties of chaos systems.

Steganography is the method of embedding data in various ways to avoid disclosure of hidden information. Steganography is derived from Greek, meaning "covered

© Springer Nature Switzerland AG 2020
D. J. Hemanth et al. (Eds.): ICICI 2019, LNDECT 38, pp. 104–113, 2020.
https://doi.org/10.1007/978-3-030-34080-3_12

writing". It comprises various communication methodologies that can hide the existence of a message [22–25]. Steganography includes methods such as invisible inks, covert channels, Masking and filtering, character arrangement, Redundant Pattern Encoding, digital signatures, etc.,

Industrial applications of steganography such as digital watermarks, confidential communication and secret data storing – are currently in place to track the copyright protection and ownership of electronic media. Convenience and availability of Steganography allowed its implementation in trafficking illicit material through digital images, audio and other files. In order to avoid its usage in illegal activities, researchers are developing methods of message detection in files. Due to its potential, development in the area of steganography will continue, so will be its implementation in covert communication.

Application of Steganography with the help of chaos theory promises high security against attackers. But selection of appropriate chaos concept is important due to compatibility issues. Double rod pendulum is one such chaos concept ideal for usage with image steganography. Double rod pendulum is a system with one pendulum attached to the end of another pendulum. Path traversed by second bob of the double rod pendulum is chaotic where introduction of multiple parameters play a key role in elevating the complexity of the resultant path. With increase in parameter count, an exponential rise in complexity is evident. When the rods are considered massless, double rod pendulum provides a total of 6 parameters. Masses of the two bobs, lengths of the two bobs, initial velocity of the pendulum and acceleration act as the initial constraints. With combined change in the initial conditions one can observe enormous variations in the path traversed by the second pendulum.

Double rod pendulum can be considered as an optimal chaos concept to incorporate in Steganography application. We will be implementing image steganography and since the pendulum swings within the bounds of an image, it satisfies the principles of image steganography. During the first stage of cryptography, the plaintext is encrypted and its cipher binary output is serially embedded in the pixels traversed by the second pendulum. The receiver decodes the image by starting the pendulum motion with identical criterions. The application then extracts the pixels spanned by the second pendulum to grasp the ciphertext. Sequentially perceived ciphertext is decrypted to give rise to the original plaintext.

We present the short coming of the base algorithm and the improvements we propose, in Sect. 2. We illustrate the functioning of the improved algorithm in Sect. 3 and put forth an analysis on the strength of our proposal in Sect. 4. Finally we deal with conclusions and future enhancements in Sect. 5.

2 Design of the Cipher

In the base algorithm, as referred to in Sect. 1, the authors have proposed a scheme where the ASCII values of the plaintext characters are first converted into their binary equivalent. Let us denote the binary string so formed as α. A key comprising prime numbers, in some random order, is generated (the number of key elements will be equal to $|\alpha|$). Each binary bit of plaintext is multiplied with the corresponding prime number of the key. Here if the binary bit is a 1, the bit is replaced with the corresponding key element, 0 otherwise. The authors have then multiplied all the non-zero numbers produced in the previous step to generate a large integer. This integer is converted again into binary equivalent, which then undergoes a series of key dependent permutation.

If we examine the base algorithm closely we realize that the number of key elements (prime numbers) to be generated for encrypting one character will be 8. If each of these prime numbers is in the range of 0–255, 8 bits are required to represent each of these numbers. Thus, for encryption we need a 64-bit key for each character. Assuming around 16 characters to be encrypted as a block, it takes 1024 bit long key, which is way too high for a symmetric key cipher. NIST has recommended a maximum of 256 bit key for a block cipher which encrypts 128 bit plaintext at a time [26].

Further base algorithm also shows its vulnerability against chosen plaintext attack. Let us assume cryptanalyst chooses a plaintext comprising a block of characters in which the ASCII value of the first character is 1 and the ASCII value of the remaining characters is 0. In this case, there will be only 1 bit with a value of '1' and remaining bits are 0s'. When the base algorithm is applied on this chosen plaintext, the resulting ciphertext will contain as many 1s' as in the corresponding key element's binary equivalent (eighth element in this case). Thus we will know how many 1s' are there in the eighth key element's binary equivalent. If we modify the previous chosen plaintext replacing the first character with a character whose ASCII value is two, we will get the number of 1s' in the binary equivalent of the seventh key element. This process can be adopted to find each of the key elements. This is a glaring vulnerability of the base algorithm.

To overcome these vulnerabilities we propose an extension where we employ a chaos based random number generator. However the scrambling of the modified plaintext, for seven iterations, is omitted in our proposal. The values so generated will be transformed into pixel positions of an image. We then hide the ciphertext generated from the base algorithm in the RGB values of the pixels of the image. The size of the key also, as mentioned in the previous section is very high. Hence we limit the size of the key to 32 prime numbers which intern translates into a 256-bit key. Here we use a chaos scheme based on double rod pendulum movement for which 6 parameters, lengths of two rods of pendulum (taken as 40, 50 pixels respectively), the weights at the end of two rods of pendulum (15, 10 units), initial velocity (2 m/s) of the pendulum and acceleration (9.8 m/s^2) given for the pendulum movement. The values mentioned above for the pendulum are taken for the purpose of illustration of the cipher in this paper. These values can be chosen according to the convenience and the size of the carrier image. The six parameters of the pendulum are private and form a part of the key. Figure 1 depicts the structure of the pendulum along with the parameters.

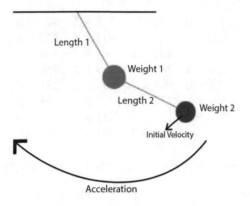

Fig. 1.

It is to be noted that the pendulum movement varies with change in parameters and hence it would not be possible for two double rod pendulums to exhibit same behaviour when their respective parameter differ in numerical values.

We take the pixels that fall on the pendulum's oscillation path. We hide the encrypted data into these pixels, as discussed earlier. The properties of the pendulum will govern the pixels to be selected. If any pixel occurs twice on the path of pendulum's oscillation, its first occurrence will only be taken into consideration. Here, we ignore the first six oscillations of the pendulum for selection of pixels to hide the data.

3 Illustration of the Cipher

In this section we present the functioning of the cipher with illustration. We have a key K, comprising 32 prime numbers, in some random order taken as

$$K = \{479, 11, 757, 827, 607, 2, 977, 223, 503, 887, 661, 839, 137, 743, 433, 769, \\ 463, 907, 199, 281, 509, 151, 239, 191, 373, 13, 271, 379, 491, 733, 523, 23\} \tag{1}$$

Let us take a plaintext, P of 4 characters as

$$P = gold \tag{2}$$

The binary representation of ASCII values of this plaintext will be

$$\alpha = \{01100111\ 01101111\ 01101100\ 01100100\} \tag{3}$$

If we multiply each bit with the corresponding key element we get

$$
\begin{array}{llll}
497 \times 0 & 11 \times 1 & 757 \times 1 & 827 \times 0 \\
607 \times 0 & 2 \times 1 & 977 \times 1 & 223 \times 1 \\
503 \times 0 & 887 \times 1 & 661 \times 1 & 839 \times 0 \\
137 \times 1 & 743 \times 1 & 433 \times 1 & 769 \times 1 \\
463 \times 0 & 907 \times 1 & 199 \times 1 & 281 \times 0 \\
509 \times 1 & 151 \times 1 & 239 \times 0 & 191 \times 0 \\
373 \times 0 & 13 \times 1 & 271 \times 1 & 379 \times 0 \\
491 \times 0 & 733 \times 1 & 523 \times 0 & 23 \times 0
\end{array}
\tag{4}
$$

Multiplying all the non-zero elements, obtained from (4), gives rise to

$$
\begin{aligned}
& 11 \times 757 \times 2 \times 977 \times 223 \times 887 \times 661 \times 137 \times 743 \times 433 \times 769 \times 907 \times 199 \\
& \times 509 \times 151 \times 13 \times 271 \times 733
\end{aligned}
\tag{5}
$$

Which is equal to

$$
2583083197401994229001129201324251797688378
\tag{6}
$$

Converting (6) into its binary equivalent will result in

```
00110010 00110101 00111000 00110011 00110000 00111000 00110011 00110001 00111001 00110111
00110100 00110000 00110001 00111001 00111001 00110100 00110010 00110010 00111001 00110000
00110000 00110001 00110001 00110010 00111001 00110010 00110000 00110001 00110011 00110010
00110100 00110010 00110101 00110001 00110111 00111001 00110111 00110110 00111000 00111000
00110011 00110111 00111000
```

$$
\tag{7}
$$

Here we took individual digits of the number obtained in (6), converted their ASCII equivalent into binary numbers. We then pad End of Text (EoT) character at the end of this string to recognize the end of the input. We now take an image shown in Fig. 2.

Fig. 2.

Fig. 3.

We take 8 bit blocks from the string obtained in (7) and hide them in the RGB values of the pixels by placing them in the three least significant bits of red, three LSBs of green and two LSBs of blue intensities. The image embedded with the message is Fig. 3.

Same plaintext, when embedded into other images, the resultant images appear as shown in Figs. 4, 5 and 6.

Fig. 4.   Fig. 5.

Fig. 6.

It can be noticed that there is no perceptible change in the images after embedding the plaintext.

4 Analysis

Security of any cipher is analyzed through

- Brute force attack
- Known plaintext attack
- Chosen plaintext attack
- Chosen ciphertext attack

Since the size of the key is now going to be 320 bits, exhaustive key search requires $2^{(320)}$ computations. Even if we assume birthday paradox is true, a minimum of $\sqrt{2^{(320)}} = 2^{(160)}$ computations are needed for brute force attack which computationally is infeasible. In case of other attacks like chosen plaintext attack, the ciphertext is hidden in a carrier (image). The pixels in which the ciphertext is hidden are governed by the chaotic maps generated by the double rod pendulum whose parameters are confidential. Hence the cryptanalyst will not be able to identify the pixels in which the information is hidden.

Figure 7 gives the pixels which are on the path of pendulum's oscillation, into which the encrypted data is hidden.

Fig. 7.

5 Results and Conclusion

Messages can be encrypted in many formats such as image, video or voice. Each steganography application has its significance and security strength. Steganography alone does not ensure security. However, if a message is encrypted and then embedded in an image, video or voice then it is well protected. If the embedded message was intercepted, the interceptor will have no knowledge about the message since it is hidden. Usage of steganography along with encryption techniques helps in improving security. Where cryptography and strong encryption are prohibited from using, steganography can circumvent such policies to pass messages covertly.

In this paper, we have introduced a robust cryptosystem coupled with steganography to ensure confidentiality of information. It offers stiff resistance to all kinds of cryptanalytic attacks. Table 1 shows the variation in the pixels when there is a minute change in the parameters of the double rod pendulum, which is a prerequisite for chaos.

Table 1.

Sno	Original	Initial Position Changed by one degree 201, 201	Inner rod length changed by one pixel 41	Outer rod length changed by one pixel 51	Inner weight changed by one unit 16	Outer weight changed by one unit 11
1	(177, 169)	(57, 76)	(87, 201)	(91, 56)	(211, 120)	(108, 47)
2	(96, 73)	(82, 53)	(126, 115)	(41, 160)	(94, 80)	(104, 143)
3	(36, 106)	(102, 61)	(92, 123)	(132, 107)	(115, 89)	(121, 134)
4	(115, 141)	(120, 153)	(140, 172)	(193, 172)	(124, 111)	(132, 132)
5	(121, 186)	(119, 136)	(119, 132)	(158, 126)	(68, 57)	(106, 111)
6	(96, 174)	(133, 94)	(104, 109)	(151, 160)	(206, 125)	(212, 113)
7	(181, 185)	(39, 122)	(45, 91)	(51, 148)	(129, 163)	(154, 96)
8	(103, 128)	(93, 153)	(183, 141)	(185, 84)	(125, 135)	(178, 184)
9	(156, 107)	(121, 77)	(163, 199)	(134, 91)	(129, 164)	(42, 152)
10	(164, 68)	(190, 65)	(115, 125)	(183, 57)	(188, 188)	(146, 209)
11	(131, 146)	(201, 140)	(39, 131)	(154, 51)	(188, 117)	(67, 72)
12	(172, 156)	(110, 112)	(52, 140)	(202, 101)	(93, 81)	(61, 116)
13	(139, 53)	(85, 93)	(115, 110)	(148, 97)	(115, 119)	(149, 121)
14	(100, 211)	(86, 45)	(106, 103)	(92, 192)	(83, 195)	(65, 65)
15	(109, 136)	(139, 105)	(112, 123)	(148, 188)	(67, 119)	(160, 79)

Time required for encrypting and embedding the plaintext in the image is 6.6 ms Time required for that of decryption is 5.83 ms.

In this paper we have presented a novel concept of encrypting a plaintext and embedding the resultant bit stream into an image using chaos theory. The unpredictability and high sensitivity of chaos to initial conditions make it hard to find the pixels in which information is hidden. This scheme can be extended by increasing the plaintext size beyond 256 characters by changing pendulum parameters for every 256 characters.

Acknowledgements. Authors are thankful to the management and the principal of Geethanjali College of Engineering and Technology for all the support, encouragement, both professionally and financially, extended to us in carrying out this work.

References

1. Stallings, W.: Cryptography and Network Security: Principles and Practices, 4th edn. Prentice Hall, Saddle River (2005). ISBN-13 978-0-13-187316-2
2. Schneier, B.: Applied Cryptography: Protocols, Algorithms, and Source Code in C, 2nd edn. Wiley, New York City (1995). ISBN-13 978-0471128458
3. Daley, W.D., Kammer, R.G.: Data encryption standard (DES). U.S. Department of Commerce/National Institute of Standards and Technology, Federal Information Processing Standards Publication No. 46, National Bureau of Standards, 15 January 1977 (1999)
4. Schneier, B.: The blowfish encryption algorithm. Dr. Dobb's J. Softw. Tools **19**(4), 98–99 (1994)
5. Daemen, J., Rijmen, V.: The Design of Rijndael: AES - The Advanced Encryption Standard. Springer, Heidelberg (2002)
6. Daemen, J., Rijmen, V.: FIPS PUB 197, Advanced Encryption Standard (AES), National Institute of Standards and Technology, U.S. Department of Commerce, November 2001. http://csrc.nist.gov/publications/fips/fips197/fips-197.pdf
7. Rivest, R.L., Shamir, A., Adleman, L.: A method for obtaining digital signatures and public key cryptosystems. Commun. ACM **21**, 120–126 (1978)
8. Koblitz, N.: Elliptic curve crypto systems. Math. Comput. **48**(177), 203–209 (1987)
9. Miller, V.S.: Use of elliptic curves in cryptography. In: Williams, H C. (ed.) CRYPTO. LNCS, vol. 218, pp. 417–426. Springer (1986)
10. Haidar, I., Haidar, A.M., Haraty, R.A.: Scrambled prime key encryption. In: Proceedings of the 10th International Conference on Management of Emergent Digital EcoSystems (MEDES 2018, Tokyo, Japan, 25–28 September 2018, 6 p. ACM, New York (2018). https://doi.org/10.1145/3281375.3281376
11. Pecora, L.M., Carroll, T.L.: Synchronization in chaotic systems. Phys. Rev. Lett. **64**, 821 (1990)
12. Raghuwanshi, P., Nair, J.S., Jain, S.: A secure transmission of 2D image using randomized chaotic mapping. In: Symposium on Colossal Data Analysis and Networking (CDAN). IEEE (2016)
13. Mao, Y., Chen, G., Lian, S.: A novel fast image encryption scheme based on 3D chaotic baker maps. Int. J. Bifurc. Chaos **14**(10), 3613–3624 (2004)
14. Li, S., Chen, G., Zheng, X.: Chaos-based encryption for digital images and videos. In: Multimedia Security Handbook, Chap. 4. CRC Press LLC, February 2004
15. Kumar, T., Chauhan, S.: Image cryptography with matrix array symmetric key using chaos based approach. Int. J. Comput. Netw. Inf. Secur. **10**, 60–66 (2018). https://doi.org/10.5815/ijcnis.2018.03.07
16. Bandyopadhyay, D., Dasgupta, K., Mandal, J., Dutta, P.: A novel secure image steganography method based on chaos theory in spatial domain. Int. J. Secur. Priv. Trust. Manag. **3**, 11–22 (2014). https://doi.org/10.5121/ijsptm.2014.3102
17. Luo, Y., et al.: A novel chaotic image encryption algorithm based on improved baker map and logistic map. Multimed. Tools Appl. **78**, 22023–22043 (2019)
18. Tütüncü, K., Demirci, B.: Adaptive LSB steganography based on chaos theory and random distortion. Adv. Electr. Comput. Eng. **18**, 15–22 (2018). https://doi.org/10.4316/aece.2018.03003
19. Li, C., Luo, G., Li, C.: An image encryption scheme based on the three-dimensional chaotic logistic map. Int. J. Netw. Secur. **21**, 22–29 (2019)
20. Krishnaveni, N., Periyasamy, S.: Image steganography using LSB embedding with chaos. Int. J. Pure Appl. Math. **118**, 505–508 (2018)

21. Danforth, C.M.: Chaos in an atmosphere hanging on a wall. Mathematics of Planet Earth, April 2013
22. Rasras, R., Alqadi, Z., Rasmi, M., Sara, A.: A methodology based on steganography and cryptography to protect highly secure messages. Eng. Technol. Appl. Sci. Res. **9**, 3681–3684 (2019)
23. Gupta, A., Ahuja, S.: An improved image steganography technique using block division & least significant bit approach. In: 2018 International Conference on Advances in Computing, Communication Control and Networking (ICACCCN), Greater Noida (UP), India, pp. 335–339 (2018)
24. Sridhar, S., Smys, S.: A hybrid multilevel authentication scheme for private cloud environment. In: 2016 10th International Conference on Intelligent Systems and Control (ISCO), 7 January 2016, pp. 1–5. IEEE (2016)
25. Praveena, A., Smys, S.: Anonymization in social networks: a survey on the issues of data privacy in social network sites. J. Int. J. Eng. Comput. Sci. **5**(3), 15912–15918 (2016)
26. Dworkin, M.J.: SP 800-38A 2001 Edition. Recommendation for Block Cipher Modes of Operation: Methods and Techniques. Technical report, NIST, Gaithersburg, MD, United States (2001)

A Review Paper on Comparison of Different Algorithm Used in Text Summarization

Setu Basak$^{(\boxtimes)}$, MD. Delowar Hossain Gazi,
and S. M. Mazharul Hoque Chowdhury

Department of Computer Science and Engineering, Daffodil International
University, Dhaka, Bangladesh
{setu.cse, mazharul2213}@diu.edu.bd,
delowargazi@yahoo.com

Abstract. At present, Data remains as the most important part of human life. The future of data generation is manipulated through different data analysis techniques. But every day it is becoming much more difficult. Due to current growth of technology, People are generating huge amount of uncontrollable data. Because of that text summarization became important to reduce the volume of the data and extracts the required useful information. This review based paper discusses about different text summarization techniques and algorithms along with their accuracy and efficiency. So that the researchers can easily understand the concept of text summarization and find their expected information in a fast pace.

Keywords: Algorithm · Classification · Sentence · Text summarization · Tokenization

1 Introduction

Content synopsis alludes to the method of shortening long bits of content. The expectation is to make a cognizant and familiar synopsis having just the primary concerns laid out in the archive. The International Data Corporation (IDC) ventures that the aggregate sum of computerized information coursing every year around the globe would grow from 4.4 zettabytes in 2013 to hit 180 zettabytes in 2025. In this way, utilizing programmed content summarizers fit for extricating valuable data that forgets inessential and inconsequential information is getting to be crucial. Actualizing rundown can improve the comprehensibility of records, decrease the time spent in exploring for data, and take into consideration more data to be fitted in a specific territory. Programmed content synopsis is the errand of delivering a brief also, familiar summary while safe-guarding key data content and generally speaking importance. Programmed content rundown is extremely testing, because when we as people abridge a bit of text, we ordinarily perused it altogether to build up our comprehension, and then write a synopsis featuring its central matters. Since PCs need human learning and language ability, it makes programmed content rundown a very troublesome and non-trifling assignment.

D. J. Hemanth et al. (Eds.): ICICI 2019, LNDECT 38, pp. 114–119, 2020.
https://doi.org/10.1007/978-3-030-34080-3_13

2 Literature Review

Bijalwan et al. Proposed a KNN bascd model to categorize documents of same types [1]. They classified the documents in reuter-21578. They first preprocessed the data and after that they have implemented the KNN to train their system using five categories of training data. Finally, they implemented the classification framework to find out the accuracy compared to term graph and Naïve Bayes. Paulus et al. proposed Recurrent Neural Network-based models for abstractive rundown that have accomplished great execution on short information and yield successions [2]. They used neural intra attention model on input sequence. After that they used intra decoder attention followed by the token generation and pointer generation. After that they shared the decoder weights to avoid the repentance at testing time. They have reported the full-length F-1 value of the ROUGE-1, ROUGE-2 and ROUGE-L measurements with the (PTO) Porter stemmer option. After that they have performed qualitative and quantitative analysis to generate the result. Mani and Bloedorn have described the utilization of ML on a preparation corpus of reports and their modified works to find notability capacities which portray what blend of highlights is ideal for a given synopsis task [3]. They treated the summary as the representation of user's need. They spoke to the essential "Boolean-marking" type of this technique, all source sentences over a specific importance edge are treated as "synopsis" sentences. They spoke to the source sentences as far as their component depictions, with "synopsis" sentences being named as positive example. Their intension was to find out rules which can easily be understood by human-being. Their point was to create at both conventional outlines just as client centered synopses, consequently broadening the nonexclusive synopsis direction of the Practical work. After that by regarding the unique as a question, we coordinate the whole theoretical to each sentence in the source, rather than coordinating individual sentences in theory to at least one sentences in the source. Xu et al. have worked with different model and methods to identify the effective algorithms that can automatically classify and summarize music content [4]. They first calculated the Mel Frequency Cepstral Coefficients. After that they calculated the music flux. Then they calculated the distance of every pair of music frames and then embedded the calculated distances. Next they created a normalized matrix. After that they have implemented the algorithm hierarchical clustering to summarize the lyrics of music and lastly, they implemented classification algorithms to classify the summarized music into categories. Shen et al. presented a Conditional Random Fields based method to find the measurements of the supervised and unsupervised approaches while avoiding their disadvantages in summarization [5]. They contrasted their proposed strategy and both administered and unsupervised techniques. Among the managed strategies, they picked Hidden Markov Model, Support Vector Machine, Logistic Regression and Naive Bayes Padma Priya and Duraiswamy have worked with a graphical model for binary random variables effected by Restricted Boltzmann Machine (RBM) method for better efficiency by removing continuing sentences [6]. Their assessment procedure was completed in three diverse report sets. The review, accuracy and f-measure for all the dataset were determined by fluctuating diverse limit esteems. The distinctive edge esteems were utilized to confirm the reactions of the proposed content outline calculation under various condition. They, from their analysis have proposed that text summarization algorithm is sensitive to data. Litvak and Last have presented chart based syntactic portrayal of content and web archives, which upgrades the conventional vector-space model and think about between two novel methodologies, regulated and unsupervised, for distinguishing the catchphrases to be utilized in extractive rundown of content records [7].

3   Text Summarization Concept

In general, there are two different approaches for automatic summarization: extraction and abstraction. Extractive outline techniques work by recognizing significant areas of the content and producing them verbatim; in this way, they depend just on extraction of sentences from the first content. Conversely, abstractive rundown techniques go for creating significant material in another manner. In different words, they translate and look at the content utilizing progressed regular language methods so as to produce another shorter content that passes on the most basic data from the first content. Despite the fact that rundowns made by people are generally not extractive, the greater part of the rundown look into today has centered on extractive synopsis. In the long run, the summarizer framework chooses the top k generally significant sentences to deliver a rundown. A few methodologies utilize eager calculations to choose the significant sentences and a few methodologies may change over the choice of sentences into an improvement issue where a gathering of sentences is picked, thinking about the requirement that it ought to boost generally speaking significance and coherency fur-thermore, limit the excess. Numerous strategies, including regulated and unsupervised calculations, have been created for extractive archive rundown. Generally directed strategies are considered to be the rundown work as two different classes of grouping issue and characterize every sentence independently without utilizing any kind of relationship among sentences. The unsupervised techniques utilize heuristic guidelines to choose the most educational sentences into a synopsis legitimately, which are hard to sum up. Outline development is, by and large, a perplexing undertaking which preferably would include profound normal language handling limits. So as to disen-tangle the issue, momentum inquire about is centered on extractive-synopsis age. The extractive synopsis is essentially makes a smaller set of the sentences of the first content. These synopses do not ensure a decent account lucidness, yet they can advantageously speak to a rough substance of the content for significance judgment. An outline can be utilized in a characteristic way - as a pointer to certain pieces of the first archive, or in a useful way - to cover all applicable data of the content. In the two cases the most significant bit of leeway of utilizing an outline is its decreased perusing time. Synopsis frameworks need to create a brief and familiar rundown passing on the key data in the information. In this segment we force our discussion to extractive summary structures for short, passage length diagrams and explain how these systems perform rundown. The decision about what substance is critical is driven fundamentally by the commitment to the summarizer.

4   Text Summarization

The First we have to convert the paragraph into sentences. Then we have to process the text. After that we need to tokenize each word of the sentences. Next we have to calculate the weighted the frequency of occurring for each word. Finally, we need to substitute the words with their weighted frequencies. In this way we will generate the desired document as a meaningful summary (Fig. 1).

Fig. 1. Phases of text summarization

Different approaches for text summarization are -

Table 1. Graph-based single document summarization

Methods	Accuracy	False positive
J48	**.847**	.022
Naïve Bayes	.839	.011
SMO	.838	**.002**
	Higher is best	Lower is best

According to Table 1, while summarizing the text from a large number of documents the tree based J48 algorithm produces the best result with an accuracy of about 84.7%.

Table 2. Deep learning methods

Method	Recall
Existing system	.72
Restricted Boltzmann machine	.62

Based on the information on Table 2, text summarization from websites, emails and from different documents if the deep learning method is used then the method RBM returns with a recall value of 62% which is pretty good but the existing system performs better here.

Table 3. For music summarization and classification

Method	Accuracy
SVM	**.83**
HMM	.0964
NN	.080

Here, Table 3 showing that, for music classification after the lyrics of music have been summarized the support vector machine can easily classify the music to its actual class with an accuracy of about 83%.

Table 4. Generic and user focused

Method	Accuracy in generic	Accuracy in user focused
SCDF	.64	.88
AQ	.56	.81
C4.5	.69	**.89**

In Summarization of different journals, publications, other websites user-focused model performs better and in this model the C4.5 algorithm can summarize the text better with an accuracy of 89% (Table 4).

Table 5. Temporal summarization

Method	Performance
2APSal	Best
BJUT	Better
uogTR	Good

For temporal summarization of websites like wiki, and emails the 2APSal method performs the best and after that the BJUT and uogTR performs accordingly (Table 5).

Table 6. Deep reinforced model for abstractive summarization (human readability score)

Method	HRS	Relevance
ML	6.76	7.14
RL	4.18	6.32
ML+RL	**7.04**	**7.45**

If summarization of documents uses the ML and RL method together then the readability and relevance of the summarized document is higher with a value of 7.04 and 7.45 which is pretty high than ML or RL anole (Table 6).

5 Conclusion

Text summarization techniques and algorithms are changing and being modified rapidly based on its use and necessity. But there is a lot to do if we want to handle the data generated every day. Current technology is developing in a great speed and this is creating a big wave of data. If new techniques and higher accuracy algorithms are not prepared because it may reduce the growth. On the other hand, as 5G technology is coming up soon, AI will also be available soon and for AI development data is the most important thing. If data is not in a good shape and don't have a clear understanding developed will not be able to do the work. Therefore, a clear understanding of text summarization has been more important now-a-days.

References

1. Bijalwan, V., Kumari, P., Pascual, J., Semwal, V.B.: Machine learning approach for text and document mining. Int. J. Database Theory Appl. (2016). https://doi.org/10.14257/ijdta.2014. 7.1.06
2. Paulus, R., Xiong, C., Socher, R.: A deep reinforced model for abstractive summarization. In: International Conference on Learning Representations, arXiv:1705.04304v3 (2017)
3. Mani, I., Bloedorn, E.: Machine learning of generic and user-focused summarization. In: Proceedings of the Fifteenth National/Tenth Conference on Artificial Intelligence/Innovative Applications of Artificial Intelligence, pp. 820–826 (1998). ISBN 0-262-51098-7
4. Xu, C., Maddage, N.C., Shao, X.: Automatic music classification and summarization. IEEE Trans. Speech Audio Process. **13**(3), 441–450 (2005)
5. Shen, D., Sun, J.T., Li, H., Yang, Q., Chen Z.: Document summarization using conditional random fields. In: Proceedings of the 20th International Joint Conference on Artificial Intelligence, Hyderabad, India, pp. 2862–2867 (2007)
6. Padma Priya, G., Duraiswamy, K.: An approach for text summarization using deep learning algorithm. J. Comput. Sci. (2014). https://doi.org/10.3844/jcssp.2014.1.9
7. Litvak, M., Last, M.: Graph-based keyword extraction for single-document summarization. In: Proceedings of the Workshop on Multi-Source Multilingual Information Extraction and Summarization, Manchester, pp. 17–24 (2008)

Bank Vault Security System Based on Infrared Radiation and GSM Technology

Mithun Dutta[1(✉)], Md. Ashiqul Islam[2], Mehedi Hasan Mamun[2],
Kangkhita Kaem Psyche[3], and Mohammad Al Mamun[1]

[1] Rangamati Science and Technology University, Rangamati, Bangladesh
mithundutta92@gmail.com
[2] Daffodil International University, Savar, Dhaka, Bangladesh
ashiqul15-951@diu.edu.bd
[3] Jahangirnagar University, Savar, Dhaka, Bangladesh

Abstract. Banking security systems are emerging as a matter of great importance nowadays. We are presenting here a smart security system, which has been designed to protect the bank vault from being theft or unauthorized access. This highly secured system is based on Infrared ray, GSM technology, different types of sensors (Motion sensor, Laser sensor, Sound sensor and Gas sensor), IP camera, microcontroller, keyboard, and LCD. GSM technology will be used here to send any type of warning message to the concerned phone number. According to our concept a biometric function, i.e., hand geometry, an iris scanner, a fingerprint scanner or heartbeat rate with a password/PIN can be used here which will be responsible for the fine security of the principle gate of the vault. It may also include Infrared radiation between the walls of bank vault room to provide increased security of the vault. The walls of the vault room is specially designed in two sections maintaining collateral gap where the Infrared radiation will pass from the Infrared radiation sources to prevent any accident like wall break robbery. The proposed security system of the bank vault has been implemented as well as tested several times and found more effective than the other existing security systems.

Keywords: GSM · Sensors · Infrared radiation · IP camera · PIN · Microcontroller

1 Introduction

A secure place provided by a bank is a place where different types of record, documents and valuable assets can be stored is known as a bank voult, primarily designed to give a certain protection to their contents from being theft, illigal access, any types of natural disasters, robbery or any other accidents. In the current system, all branches are monitored at a stretch from the control room under the surveillance of CCTV cameras. Here, an integrated security system for the current Bank vault has been proposed. The system includes IP camera, sensors, Infrared ray & GSM module which can be operated independently.

© Springer Nature Switzerland AG 2020
D. J. Hemanth et al. (Eds.): ICICI 2019, LNDECT 38, pp. 120–127, 2020.
https://doi.org/10.1007/978-3-030-34080-3_14

There is no need of any human being to maintain the mentioned system. In the existing system, a security person, who is employed by the bank, is responsible to control the doors by simply the help of a handle locks operated by a key or sometimes via a PIN code. Our proposed system will stay at the front door of the locker room area, surrounding wall as well as at the gate of the locker room. When an authorized person will insert the password/PIN with verified fingerprint correctly the front door will be opened automatically. Surrounding walls of the room are specially designed into two parts and those will maintain a collateral gap. This proposed system will efficiently work to detect the unauthorized access and protect the bank vault from being theft.

2   Proposed System Design

2.1   Basic Architecture

Proposed vault security system; there have been used multiple sensors, fingerprint scanner, alarm system, microcontroller, LCD, IP camera, GSM Module and also Infrared radiation.

Following Fig. 1 shows the basic architecture of the proposed system which con-structs various hardware modules. If any sensors is activated in the vault room it sent signals to the microcontroller because all sensors are linked up to the microcontroller [2, 5]. Vault room walls are specially designed into two parts maintaining continuous distance where the Infrared radiation will pass from Infrared radiation sources to secure the whole vault walls and areas.

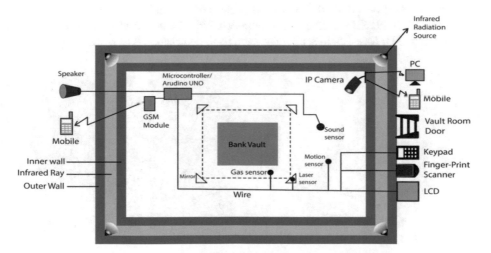

Fig. 1. The architecture of vault security system based on GSM and infrared radiation

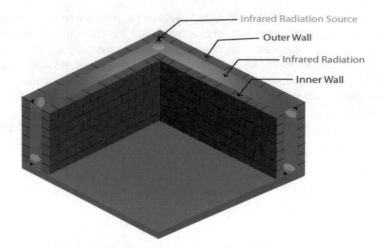

Fig. 2. Design of the protective vault wall using infrared radiation

Above Fig. 2 shows a different type of security for the vault room where harmonized civil engineering and infrared radiation technology. The Infrared radiation system is used to prevent any kind of accident like wall break robbery, disaster and so on.

2.2   Inner Architeture Design

Firstly, we would envisage GSM technology (module) and we used the SIM900 to support a Quad-band850/900/1800/1900 MHz because this module composes the holder which holds SIM Card also composes an RS232 a consecutive port and a GSM antenna [10, 11]. It assists to send a message to the corresponding authorities (Fig. 3).

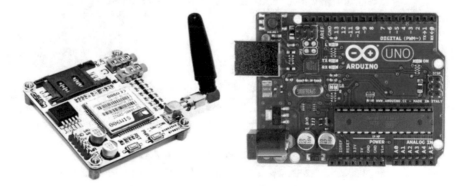

Fig. 3. SIM900 GSM module and Microcontroller Arduino

The microcontroller board namely Arduino UNO based on ATMEGA328P-AN which is high endurance, memory segments is a non-volatile, ISP (Involuntary Separation Planner), flash memory size is 32 KB; 1.8 to 5.5 V ranges and conducts operations [12]. This module is connected to all the sensors as well as the GSM module. It transmits an admonishment to the authorities by the help of the GSM module after getting any wrong allusion. By using the IP camera, we can allow someone to observe the bank vault locally or remotely. Sound sensors can detect sound pressure waves which are not within an audible range [6, 13]. Namely, the sensor PIR is used as motion sensor. All the things are deputizing heat energy in the formation of radiation above at absolute zero temperature. PIR radiates a wavelength namely infrared wavelength which is invisible to the human eye. By detecting entirely closed energy from the objects of this radiation has been working. PIR doesn't detect heat instead it detects IR from the objects [10]. After getting unusual motion then it sent an admonishment to authorities and activated alarm through the help of GSM module.

Generally, LDR (Light Dependent Resistor) is used as a laser sensor which is imperceptible; as a result if someone (any object) can endeavor to come closer to the vault the laser light takes a break through expeditiously activated the alarm system. We used MQ5 or MQ7 in here as gas sensor which is suitable for detecting flammable substance, LPG, CH4 etc. This sensor makes a warning alarm and sends a message to authority when any object pricks gas.

2.3 Formula

$$Vpi = \frac{Vpi}{\sum_i^j Vpi}, for\ j = 1, 2, \ldots, 10.$$

3 System Flow Chart

3.1 Algorithm for the System

The above Flow-chart is described by the following Algorithm

Step 1: Start the procedure.
Step 2: Initialize Keyboard, Scanner, LCD Display, Sensors, Camera, Micro-controller, Radiation and GSM System as inputs etc.
Step 3: Update the functional status of all inputs.
Step 4: Check all the conditions are; Sensors triggered and detect any rarity?
    if 'Yes', then go to Step 5; else go to Step 7;
Step 5: Take snapshots using the camera, alarm will be ringing and send SMS to the authorized phone number.
Step 6: End the procedure.
Step 7: Insert a password/PIN by a user.
Step 8: Check the conditions are; Password/PIN is wrong for three times or Unauthorized fingerprints detected?
    if 'Yes', then repeat Step 5; else go to Step 9;
Step 9: Check the condition; the user wants to change password/PIN?
    if 'Yes', then go to Step 10; else go to Step 11;

Step 10: User insert old password/PIN for confirmation and set a new password and go to Step 6;

Step 11: Unlock the main door of the locker room using the microcontroller & display sensor.

Step 12: After completing the operation, lock the door & enable the security Then Repeat Step 6;

Step 13: End the procedure.

3.2   Flow Chart

The whole security system is automated so no need to human help to operate after once the system is ON.

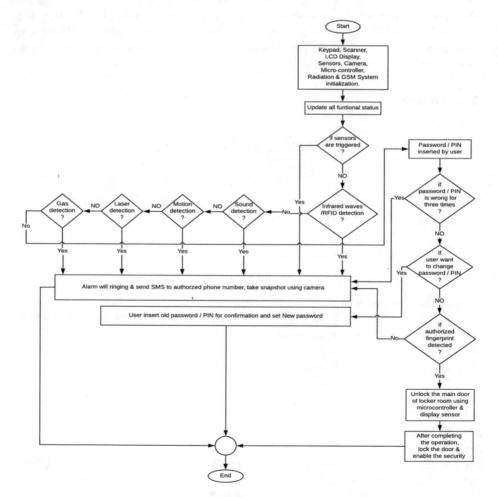

Fig. 4. Flow chart of proposed system

From above Fig. 4, when motion sensor, laser sensor, sound sensor and gas sensor detects respectively any kind of undesirable noise, any object is moving, any unauthorized object/person tries to get close to the vault and leakage of gas inside the vault just then alarm is ON and dispatches a message to the dedicated phone number based on the GSM technology. Through these IP camera vault area can be superintend remotely from anywhere in the world [6].

4   System Components

Principle Components:

- Microcontroller/Arduino UNO
- GSM Technology
- Internet Protocol (IP) Camera
- Alarm system
- Biometric input device (Fingerprint scanner)
- Keypad
- LCD
- Infrared radiation source
- Sensors (Laser, Motion, Gas, Sound)

5   Results and Discussion

Sample Output:

Fig. 5. Sample output of proposed system

From the above Fig. 5 we can see a sample output. In this proposed system many sensors are using to protect the bank vault. If the system gets any wrong allusion from the sensors then it creates an alarm and also sends a warning message to the officials via GSM Module.

A correlative study between the present system and the proposed system is given below (Table 1).

Table 1. A comparative study of existing and proposed system

S. no.	Existing systems	Proposed systems
1	Present system requires manual presence to manage the security	Proposed System is completely automated no need any manual appearance
2	Cameras are used only for monitoring	Camera is not only used for monitoring but also taking action by taking a snapshot
3	Low level of security	Will increase the number of security levels
4	It only focuses on vault room security	It will focus both on the vault room and entrance of the vault area
5	Traditional CCTV camera used	It will use IP Camera wherein the vault area can be controlled from anywhere
6	No sensors	The proposed system will be used different types of sensors
7	No GSM technology	It will be used GSM technology
8	No use of Infrared Radiation	Infrared Radiation will be used

6 Advantages

(a) This system is fully automated.
(b) It contains low cost and reliable.
(c) A high alert banking security system uses GSM technology.
(d) Strong security with fingerprint and password option to more secure the system.
(e) Multilevel security system based on Infrared ray, GSM technology, sensors, IP camera, microcontroller and biometric authentication.
(f) Walls of the bank vault room are specially constructed into two sections with the combination of infrared radiation.
(g) Can be monitored & controlled from distant main office by wireless module.
(h) Also can be used as Home security [3] system or in Museum.
(i) No need to keep eye on surveillance continuously [7, 8].
(j) Auto locking & turning on the Alarm after shutting the vault door.

7 Conclusion

The paper has successfully presented a functional, low complexity Infrared radiation and GSM based low cost strong security system which indicates that it is time to bring a revolution in the bank vault security system by doing the procedure more systematic and little smooth to the bank officials [1, 4]. Only authorized person get entry to vault room through valid password and biometric data input and if these password & fingerprint are correct the microcontroller provides necessary control signal to open the bank vault. If the system finds any unauthorized motion in the vault room through IP camera and sensors, it will close the main vault room door automatically and alarm will be on, send a SMS to dedicated phone number though GSM technology. This security system also can be exercised in high restricted areas likes question paper store room, laboratories, military areas for better security.

References

1. Sridharan, S.: Authenticated secure bio-metric based access to the bank safety lockers. In: IEEE International Conference on Information Communication & Embedded System "ICICES 2014", S.A. Engineering College, Chennai, Tamil Nadu, India, February 2014. ISBN 978-1-4799-3834-6/14/$31.00©2014 IEEE
2. Teja, P.S.R., Kushal, V., Srikar, A.S., Srinivasan, K.: Photosensitive security system for theft detection and control using GSM technology. In: Fifth International Conference on Security, Privacy and Applied Cryptography Engineering, MNIT, Rajasthan, India, pp. 122–125, October 2015
3. Javare, A., Dabhade, J., Ghayal, T., Shelar, A.: Access control and intrusion detection in door lock system using bluetooth technology. In: IEEE International Conference on Energy, Communication, Data Analytics and Soft Computing ICECDS 2017, pp. 2246–2251 (2017)
4. Tams, B., Mihăilescu, P., Munk, A.: Security considerations in minutiae-based fuzzy vaults. IEEE Trans. Inf. Forensics Secur. **10**(5), 985–998 (2015). https://doi.org/10.1109/tifs.2015.2392559
5. Ray, R.K., Uddin, M.A., Islam, S.F.: GSM based bank vault security system. Int. J. Comput. Sci. Inf. Secur. (IJCSIS) **14**(2), 35–38 (2016)
6. Verma, A.: A multi layer bank security system. In: 2013 International Conference on Green Computing, Communication and Conservation of Energy (ICGCE), pp. 914–917, December 2013
7. Ramani, R., Selvaraju, S., Valarmathy, S., Niranjan, P.: Bank locker security system based on RFID and GSM technology. Int. J. Comput. Appl. **57**(18), 15–20 (2012). ISSN 0975-8887
8. Islam, S., Saiduzzaman, Md.: Design of a bank vault security system with password, thermal & physical interrupt alarm. Int. J. Sci. Eng. Res. **4**(8), 70–73 (2013)
9. Bucko, J.: Security of smart banking applications in Slovakia. J. Theor. Appl. Electron. Commer. Res. **12**(1), 42–52 (2017). ISSN 0718-1876
10. Khera, N., Verma, A.: Development of an intelligent system for bank security. In: IEEE 2014 5th International Conference-Confluence the Next Generation Information Technology Summit, pp. 319–322 (2014)
11. Chopkar, T.A., Lahade, S.: Real time detection of moving object based on FPGA. IOSR J. Electron. Commun. Eng. (IOSR-JECE) **11**(1), 37–41 (2016). e-ISSN 2278-2834, p-ISSN 2278-8735
12. Anatoliy, P.N., Kristina, V.A., Elena, A.K., Vagiz, D.G., Aleksandr, V.S.: Technologies of safety in the bank sphere from cyber attacks. In: 2018 IEEE Conference of Russian Young Researchers in Electrical and Electronic Engineering (EIConRus), 29 January–1 February 2018 (2018)
13. Chernyi, S., Zhilenkov, A., Sokolov, S., Nyrkov, A.: Algorithmic approach of destabilizing factors of improving the technical systems efficiency. Vibroeng. Procedia **13**, 261–265 (2017)

Image Processing Based Intelligent Traffic Controlling and Monitoring System Using Internet of Things

M. Sankar[1(✉)], R. Parvathi[1], Sanket C. Mungale[2(✉)],
and Deepak Khot[2(✉)]

[1] Faculty of Engineering, AMGOI Wathar, Kolhapur, Maharashtra, India
ms@amgoi.edu.in
[2] Department of E&TC, AMGOI Wathar, Kolhapur, Maharashtra, India
smungale044@gmail.com

Abstract. In recent years, smart control and management system plays an essential role in our day-to-day life. In this populated world, the smart management of traffic on highways, railways, and subway town is still remaining as a tedious task. This arises the need for developing a smart control and monitoring models for traffic management. With this requirement, the proposed systems uses the image processing model to observe the traffic in the roadways and communicates the information through an efficient IoT network. Furthermore, the proposed model has delivered a stable output, when compared with other existing traffic control systems.

Keywords: Traffic management system · Solar energy system · Internet of things · Crowd management · Real-time image processing

1 Introduction

Traffic management is emerging as an essential requirement to maintain a hassle-free roadways. Some countries uses the direct traffic control and observance method. Whereas, some countries uses IoT based systems for developing a smart and automated traffic management. By using the traditional direct traffic management, man power will be additionally required when the number of vehicle on the roadways increase at an unprecedented rate. The Image processing and IoT based smart traffic management systems enables a user friendly and automated control over the congested roadways. Some traffic signals are sourced with solar power, which provides free power to the traffic signal lights.

2 Proposed Work of Tarffic Management

In our project main objective is to reduce the congestion on roads and establish a smart traffic control models. Recently, the accommodation of passengers traveling is increasing at an unprecedented rate because of the boarding time shortage. We tend to

© Springer Nature Switzerland AG 2020
D. J. Hemanth et al. (Eds.): ICICI 2019, LNDECT 38, pp. 128–138, 2020.
https://doi.org/10.1007/978-3-030-34080-3_15

compare and develop a higher alternative to the present system. In our project we used image processing based intelligent traffic control and monitoring model. We have calculated the density of traffic and introduced a provision to automatically adjust the signal delay according to the density of traffic. We have made this method to reduce traffic congestion on roadways and develop a hassle-free journey on the busy roadways. We have implemented IoT based systems with a better accuracy when compared with the best control solutions for traffic management signal.

The proposed system will work according to the below described workflow process:

First camera will capture the image through the Matlab software and sends it to the computer, where it calculates the density of traffic count in order to display it on the Matlab software screen. Further, the information will be send to an Arduino controller from the main client along with the traffic count. From main client this information will be provided to the remote client via ZigBee module. Also some information will sent from this ZigBee module through serial communication and Wi-Fi network in a secured manner. The main remote client will be more suitable for monitoring the traffic in real time. With this system the data can also be transferred to Wi-Fi or serially connected android smart phone. Both the user and traffic police can analyse the traffic data easily by using particular login details in the proposed system. Figure 1 shows that the proposed workflow block diagram of traffic control system.

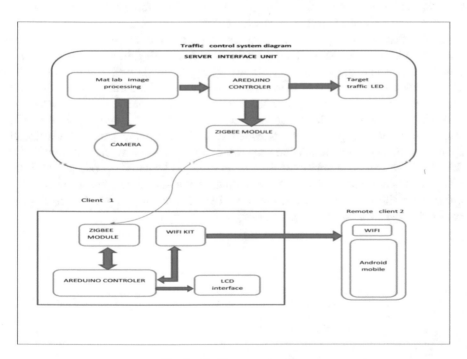

Fig. 1. Traffic control system

3 IoT System Overview in Smart Traffic Management

The internet of things is the network of object devices, vehicles, buildings and alternative things that are embedded with physics, software, device and network property. IoT is restructuring the wireless communication systems with its smart and efficient connectivity. The main aim of most of the developed and developing countries is to develop a smart city that make the lives of human beings to be easier and secured. Smart city Mission, is an innovative program initiated by the Govt. of Asian nation with the mission to smartly automate 100 cities across the country. This method clearly delivers a comprehensive analysis of the services that should be additionally along with a good smart city design.

IoT consists of Physical Object + Controller, Sensor, Actuators + Internet. In this system we have develop a model for traffic management & dominant exploitation IOT & star based system that will be used during this project. Once we have connected the camera through the MATLAB code that will additionally capture the camera pictures. Furthermore, it will send the traffic pictures through the MATLAB software. This is also used to calculate the density of traffic delay and also send the information from controller this data is additionally connected from stoplight. That is additionally connected to traffic signal. Its machine connected the traffic signal. For camera which may be calculated the traffic density and provides space wise calculation on system. And thus that is provides the traffic delay and density. Therefore traffic signal delay are going to be adjusted at instant level. Some data is additionally send from Arduino controller that is serially connect from the zig bee module. Which may send the knowledge to user. As before long as main system provides the traffic information to central system and the count of traffic are going to be showed from liquid crystal display. It's show the count of traffic vehicles on screen. Then this data is additionally send threw the wireless fidelity network from serial communication. This stoplight we will send the information threw wireless fidelity network that name is traffic. As before long as this data are going to be provided sensible automaton phone is additionally connect from serial communication apps. This square measure connected from Badger State WiFi network and appointed the user id, this can be show the output and instant real time system output. Show on the output of the screen and causation real time information threw pictures. Therefore we will assist the higher resolution. Figure 2 shows that the flow chart for traffic management system.

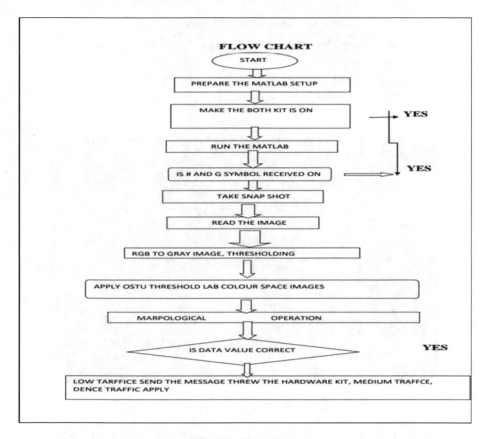

Fig. 2. Flow chart

4 Results and Discussion

In this system we have developed the system from OSTU perform. Density of traffic space is definitely calculated from one space to a different space. Otsu's technique is roughly a one-dimensional, distinct analog of Fisher's Discriminant Analysis. Otsu's technique is directly associated with the Jenks optimization technique. The extension of the initial technique to multi-level 3 holding is spoken because the multi Otsu technique. In ostu technique we can calculate from one dimensional area to multi-dimensional area which will also calculate the density of area. And RGB to GRAY conversion this method to calculate the count of area using ostu technique. So in our project we have used this method from traffic analysis data in easy way.

When we've taking mat research lab threw software package result we have three condition to calculate the density of traffic and it is shown in Table 1.

Table 1. Traffic light signal system output:

Traffic density level	Traffic signal delay	Signal display colour
Low dense traffic	5 s	Green signal
Medium dense traffic	15 s	Green signal
High dense traffic	25 s	Red signal

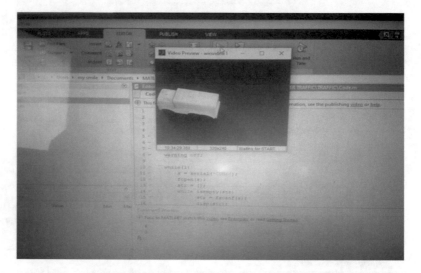

Fig. 3. Low traffic density result count on camera

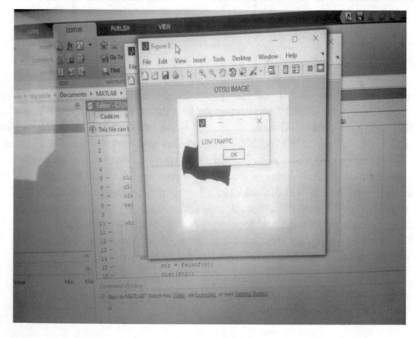

Fig. 4. Low traffic density count display on Matlab

Fig. 5. Low traffic density display result on LCD display.

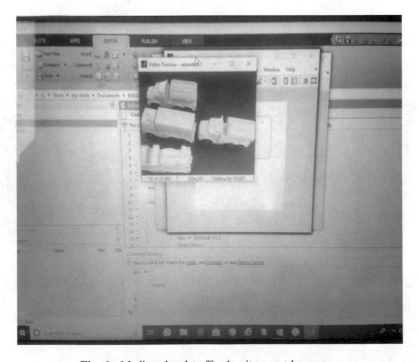

Fig. 6. Medium level traffic density count by camera

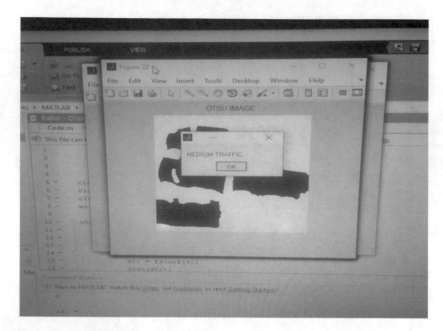

Fig. 7. Medium traffic density display result on Matlab

Fig. 8. Medium traffic density display on LCD display

Fig. 9. High traffic density display on camera

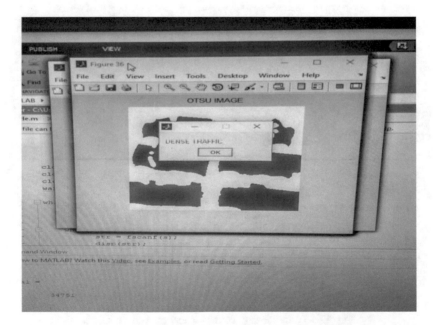

Fig. 10. High dense traffic count display on Matlab

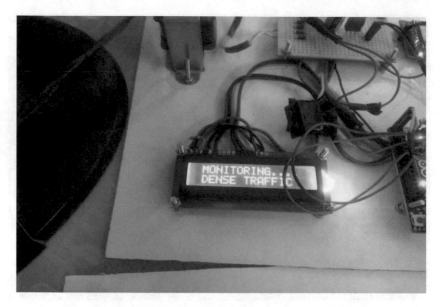

Fig. 11. High dense traffic count display on LCD

Fig. 12. Traffic density count display on WiFi network

We have taken result from MATLAB code. Three conditions that play a significant role in traffic management is:

[1] Low traffic density
[2] Medium traffic density
[3] High traffic density

So this condition has additionally calculated the traffic density from the road space (Figs. 3, 4, 5, 6, 7, 8, 9, 10, 11 and 12).

5 Conclusion

In our project we are able to develop a good control and management system. It has been proven that combination of image processing and IoT has been proven to be the best solution to reduce the traffic congestion in roads and highways. In addition the proposed work increases the systems stability.

Acknowledgement. We thank our guide Prof Dr M. SANKAR who supported to develop this project work and also our collage AMGOI Kolhapur.

References

1. Iszaidy, I., Ngadiran, R., Ahmad, R.B., Jais, M.I., Shuhaizar, D.: Implementation of raspberry Pi for vehicle tracking and travel time information system: a survey. In: 2016 International Conference on Robotics, Automation and Sciences (ICORAS), 09 March 2017 (2017). https://doi.org/10.1109/icoras.2016.7872605
2. Dangi, M.V., Parab, A.: Image processing based intelligent traffic controller. Undergrad. Acad. Res. J. (UARJ) **1**(1), 13–17 (2012)
3. Tubaishat, J.M., Shang, Y., Shi, H.: Adaptive traffic light control with wireless sensor networks. Department of Computer Science University of Missouri – Columbia, MO. IEEE (2007)
4. Gradinescu, V., Gorgorin, C., Diaconescu, R., Cristea, V.: Adaptive traffic lights using car-to-car communication. "Politehnica" University Bucharest, Computer Science Department, Bucharest
5. Nagmode, V.S., Rajbhoj, S.M.: An IoT platform for vehicle traffic monitoring system and controlling system based on priority. Department of Electronics and Telecommunication, Bharati Vidyapeeth's College of Engineering for Women, Pune, India
6. Dorle, S.S., Patel, P.L., Raisoni, G.H.: Design approach for dynamic traffic control system based on radio propagation model in VANET
7. Youseph, K.M., Al-Karaki, J.N., Shatnawi, A.M.: Intelligent traffic light flow control system using wireless sensors networks, Department of Computer Engineering, Jordan University of Science and Technology, Irbid, Jordan
8. de Charette, R., Nashashibi, F.: Traffic light recognition using Image processing compared to learning processes
9. Panjwani, M., Tyagi, N., Shalini, D., Venkata Lakshmi Narayana, K.: Smart traffic control using image processing

10. Kumar, S.: Ubiquitous smart home system using android application. Int. J. Comput. Netw. Commun. (IJCNC) **6**(1) (2014)
11. Fathy, M., Siyal, M.Y.: An image detection technique based on morphological edge detection and background differencing for real time traffic analysis

Internet of Things Based Data Acquisition System of Acoustic Emission Sensor for Structural Health Monitoring Applications

Arpita Mukherjee$^{(\boxtimes)}$, Abhishek Maurya, Pratap Karmakar, and Partha Bhattacharjee

CSIR-Central Mechanical Engineering Research Institute, Durgapur 713209, West Bengal, India
a_mukherjee@cmeri.res.in

Abstract. In this paper a IoT based data acquisition (DAQ) system has been developed for acquisition of the data from acoustic emission sensor for structural health monitoring (SHM) of the steel bridge using acoustic emission technique. The DAQ device is developed by using Raspberry Pi 3 Model B+ interfaced with a high-speed 1 MSPS and 16 bit ADC ADAQ7980 to capture the high frequency acoustic emission signal generated by crack related activities on the surface of the steel structure for real-time monitoring, identification and localisation of any crack happening on the structure through cloud computing.

Keywords: Internet of Things (IoT) · Structural health monitoring · Acoustic emission sensor · Data acquisition system

1 Introduction

With the recent boom in the IOT field, it is now possible to communicate all sort of information in a wireless manner. The IoT is becoming a well-known concept across all the organization present across the globe and also in common man's everyday life. Thus, there is an increasing interest in integrating IoT with structural health monitoring (SHM) system [1–3], which has created the requirement of IoT based real-time data acquisition system, capable of sending automated alert to the monitoring station. The authors in [4] has designed a real-time smart wireless structural health monitoring system (SHM) using Transmission Control Protocol/Internet Protocol (TCP/IP) protocol and Bluetooth technology for communicating the data measured by the sensors. The flexibility of the SHM system can be enhanced by using IPs with WSN. An SHM scheme for environmental effect removal in an IoT environment has been studied by Zhang et al. [5]. Authors in [6], proposed a mathematical model which can integrate with an IoT platform using a piezoelectric sensor for detection of damages in any physical structures. An IoT based SHM system using LabVIEW platform was designed and implemented by Panthati and Kashyap [7]. In [8], a low-cost, flexible platform has been designed and implemented for health monitoring of bridge using IoT by connecting the accelerometer sensors to the Internet.

© Springer Nature Switzerland AG 2020
D. J. Hemanth et al. (Eds.): ICICI 2019, LNDECT 38, pp. 139–146, 2020.
https://doi.org/10.1007/978-3-030-34080-3_16

In this paper, a IoT based data acquisition system (DAQ) is developed for acoustic emission (AE) sensor for the application of structural health monitoring using AE technique [9]. The DAQ device is built by using Raspberry Pi 3 Model B+ interfaced with a high-speed 1 MSPS and 16 bit ADC ADAQ7980 [10]. It is built to capture the high frequency acoustic emission generated by crack related activities on the surface of the structure under monitoring. The sensor at the place of its mounting is connected to a battery-powered DAQ which is made by using a raspberry pi, a pocket-size cost-effective computer, which sends the data it captures from the sensor upon the event of any fracture or crack-related activities on the structure under monitoring. The data is now sent through the internet to a computer having internet connection for complex mathematical analysis through Matlab to know the nature of Acoustic emission being captured and thus deduce the location of crack happening on the structure.

2   Data Acquisition System Components

In order to accommodate large number of acquisition devices required for a big structure, an indigenous data acquisition system is built which can achieve the desired goal of data acquisition at low cost and at optimum performance. The different components used for acquisition of data from AE sensor is discussed below.

2.1   Raspberry Pi

Here a popular, cheap and small single board computer (SBC), Raspberry Pi [11] has been used. It runs on Linux operating system. The control of different electronic components for physical computing and exploration of the Internet of Things (IoT) can be performed by a set of GPIO (general purpose input/output) pins in Raspberry Pi (Fig. 1).

Fig. 1. Raspberry Pi

2.2   Analog to Digital Converter (ADC)

Another component used is ADC ADAQ7980 which is a high accuracy, low power, 16-bit SAR ADC [10] (Fig. 2).

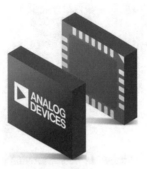

Fig. 2. ADAQ7980 ADC IC form

Evaluation Board. ADC ADAQ7980 comes with an evaluation board having space on the board for additional passive components like resistors and capacitors to be mounted according to needs such as the setting gain of the ADC's in-built amplifier etc. [12] (Fig. 3).

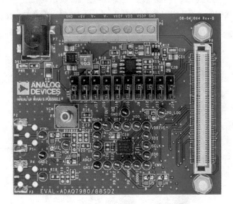

Fig. 3. Evaluation board

Pin Configuration. ADC ADAQ7980 has 24 pins which are used as input and output of the ADC. The pin configuration is shown in Fig. 4.

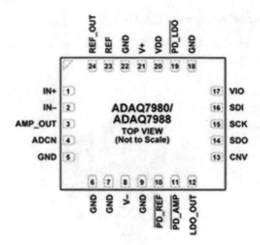

Fig. 4. Pin configuration

Power Supply. The recommended dual-supply, power-on sequence is as follows: The ADC Driver is powered by 7.5 V through V+ pin and 2.5 V through V− pin which is supplied by evaluation board. A logic high needs to applied at VIO and VDD pins which is also provided by the evaluation board then a logic high is applied to PD_LDO, PD_AMP and PD_REF to turn on voltage regulator, ADC Driver and Reference buffer respectively.

Configuring the ADC Driver. The ADC driver is configured for non-inverting unity gain configuration by setting jumper J1 at A, R21 and R23 to zero ohm. The links and components for configuration of the ADC driver is shown in Fig. 5.

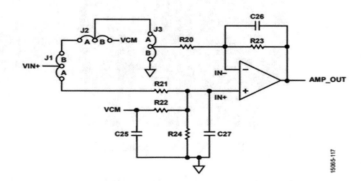

Fig. 5. Relevant links and components for configuring the ADC driver

3 Integration of IoT-Data Acquisition System

In the SHM, it is required to have sensors and data acquisition devices to be mounted at different points on the structure to be monitored, which needs a network of data lines to be drawn towards the computer for processing. But with an increase in the size of the structure and consequently increase in the number of sensors and DAQ devices, all structure may seem to be flooded with data cables and then in order to accommodate such a large number of data lines, the number of processing computers will also increase. This problem can be solved drastically by transmitting data wirelessly through internat from the point of the sensor mounting itself which will lead to termination of all data lines and also reduction in the number of processing computer to one. The process of integrating IoT with data acquisition system is discussed below in details.

3.1 Operation

The SPI (serial peripheral interface) 3-wire $\overline{cs}$ (chip select) mode is used for communication between Raspberry pi and ADC. The circuit diagram and timing diagram of 3-wire $\overline{cs}$ mode is shown in Figs. 6 and 7 respectively. A conversion is initiated on a rising edge of CNV, $\overline{cs}$ mode is selected, and SDO is forced to high impedance, with SDI connected to VIO. After a conversion is initiated, it continues until completion irrespective of the state of CNV. However, before the minimum conversion time elapses, return CNV high and then to avoid the generation of a busy signal indicator, hold it high for the maximum conversion time. When the conversion completes, the ADAQ7980 enters the phase of acquisition and power down. The MSB is output onto SDO, when CNV goes low. Then, the remaining data bits clock out by subsequent SCK falling edges. When CNV goes high or after the 16th SCK falling edge, whichever is earlier, SDO returns to high impedance.

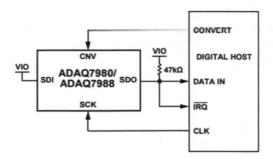

Fig. 6. Circuit diagram for 3-wire $\overline{cs}$ mode

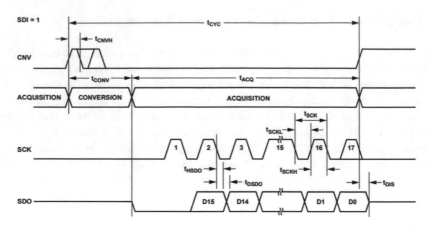

Fig. 7. Timing diagram for 3-wire $\overline{cs}$ mode

3.2   Sending of Data Through Raspberry Pi

Raspberry Pi is used to receive the data through the sensor. It continuously checks for the crack-related activities and upon detecting such activity, it makes an acquisition, saves the data in a file, connects to the internet and sends data to the cloud. First a file hosting service provider, here Dropbox is selected which can provide cloud storage. Dropbox uses OAuth 2, specification which is an open standard for token-based authentication and authorization on the internet. OAuth provides a way for applications to gain credentials to other application without directly using user names and passwords in every request. Dropbox python SDK is installed in the Linux OS of Raspberry Pi to make access Dropbox easier. A Dropbox app is created where parameters such as 'Development Users' which specify the users who can access the Dropbox, 'Permission type' to select the desired folder, 'App folder name' are configured and upon successful configuration, the Access token is granted. Then from the python using the Access Token and libraries installed a request is made to connect to the desired app folder in the Dropbox and upon the establishment of secure connection file containing data is uploaded (Fig. 8).

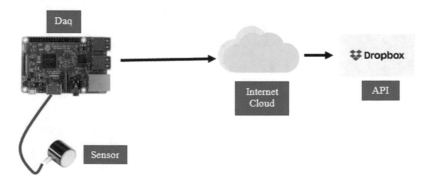

Fig. 8. Wireless transmission of sensor data

3.3   Receiving of Data by Matlab

A program is created in Matlab which detects ADC data in the cloud continuously, if there is new data, it downloads and then performs analysis over it. And this process of detection and analysis is repeated upon its every completion. Again the authorization is done as same by using the Dropbox access token, the parameters such as 'Media Type', 'Character Encoding' and 'Request Method' for reading the web file from the Dropbox is configured then a request is made for downloading the web file from Dropbox (Fig. 9).

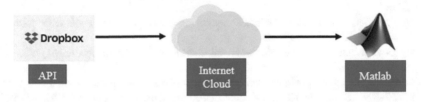

Fig. 9. Access of sensor data

4   Conclusion

A low cost data acquisition (DAQ) system for remote steel bridge health monitoring using IoT to connect the acoustic emission sensors to the internet has been designed and implemented. To capture the high frequency acoustic emission signal data generated by crack related activities on the surface of the steel structure, a high-speed 1 MSPS and 16-bit ADC ADAQ7980 is used. The sensors are connected to the battery-powered DAQ, developed using a pocket-size, cost-effective computer known as raspberry pi which transmits the sensor data upon the event of any fracture or crack-related activities on the structure under monitoring. The sensor data is stored and analysed for real-time monitoring, identification and localisation of any crack happening on the structure through cloud computing.

References

1. Tokognon, C.A., Gao, B., Tian, G.Y., Yan, Y.: Structural health monitoring framework based on internet of things: a survey. IEEE Internet Things J. **4**(3), 619–635 (2017)
2. Myers, A., Mahmud, M.A., Abdelgawad, A., Yelamarthi, K.: Toward integrating structural health monitoring with Internet of Things (IoT). In: 2016 IEEE International Conference on Electro Information Technology (EIT), Grand Forks, ND, pp. 0438–0441 (2016)
3. Heo, G., Son, B., Kim, C., Jeon, S., Jeon, J.: Development of a wireless unified-maintenance system for the structural health monitoring of civil structures. Sensors **18**, 2–16 (2018)
4. Heo, G., Jeon, J.: A smart monitoring system based on ubiquitous computing technique for infra-structural system: centering on identification of dynamic characteristics of self-anchored suspension bridge. KSCE J. Civil Eng. **13**(5), 333–337 (2009)

5. Zhang, H., Guo, J., Xie, X., Bie, R., Sun, Y.: Environmental effect removal based structural health monitoring in the internet of things. In: Proceedings of the IEEE 7th International Conference on Innovative Mobile and Internet Services in Ubiquitous Computing (IMIS), pp. 512–517 (2013)
6. Myers, A., Mahmud, M.A., Abdelgawad, A., Yelamarthi, K.: Toward integrating structural health monitoring with Internet of Things (IoT). In: Proceedings of the IEEE International Conference on Electro Information Technology (EIT), pp. 438–441 (2016)
7. Panthati, S., Kashyap, A.A.: Design and implementation of structural health monitoring based on IoT using lab VIEW. Int. J. Mag. Eng. Technol. Manag. Res. 3(2), 77–82 (2016)
8. Pandey, S., Haider, M., Uddin, N.: Design and implementation of a low-cost wireless platform for remote bridge health monitoring. Int. J. Emerg. Technol. Adv. Eng. 6(6), 57–62 (2016)
9. Nair, A., Cai, C.S.: Acoustic emission monitoring of bridges: review and case studies. Eng. Struct. 32(6), 1704–1714 (2010)
10. https://www.analog.com/media/en/technical-documentation/data-sheets/ADAQ7980-7988.pdf
11. https://www.raspberrypi.org/products/raspberry-pi-3-model-b-plus/
12. https://www.analog.com/media/en/technical-documentation/user-guides/EVAL-ADAQ7980SDZ-UG-1060.pdf

A Data Security Scheme for the Secure Transmission of Images

Parvathy Suresh[(⊠)] and T. K. Ratheesh

Department of IT, Government Engineering College Idukki, Idukki, Kerala, India
parvathysuresh549@gmail.com,
ratheeshtk@gecidukki.ac.in

Abstract. The transfer of image is happening over the unsecured network needs good security mechanisms to keep the image away from the unauthorized access. This paper is suggesting a new approach for providing security of images using Elliptic Curve Cryptography (ECC) and DNA encoding. The RGB image is encoded using DNA encoding followed by DNA addition to increase the randomness. After that a hybrid encryption is performed based on Elliptic Curve Cryptography and Hill Cipher. This will increase the security of the image and thus prevent attacks from hackers.

Keywords: Elliptic Curve Cryptography · DNA encoding · Hill Cipher · Self-invertible key matrix

1 Introduction

The amount of data being transferred over network, which has expanded exponentially these days in the areas such as military, medical, industries and education. It is required to safeguard such data from the access of unauthorized users. Hence, researchers are keep on looking for novel security mechanisms to safeguard such confidential information. To protect the data, different cryptographic algorithms have been suggested with the intent of making data indecipherable for intruders.

Data that are transferring through internet are of various forms such as text, image, and video. Here we propose a mechanism to protect image data so as to resist from attackers. The RGB image will split into three channels which are Red, Green and Blue. Then it will be encode into DNA sequence. DNA addition is then performed to enhance the diffusion property. After that it will undergo a hybrid encryption scheme which consist of Elliptical Curve Cryptography (ECC) followed by Hill Cipher (HC). Then make the pixels random to reduce the correlation.

The remaining part of the paper is structured as follows. Section 2 provides an outline of the recent works and studies that associated to the area. The proposed scheme in explained in Sect. 3. Section 4 shows the results of simulation and Sect. 5 describes the security analysis of simulated result. In Sect. 6, the paper is concluded.

D. J. Hemanth et al. (Eds.): ICICI 2019, LNDECT 38, pp. 147–154, 2020.
https://doi.org/10.1007/978-3-030-34080-3_17

2   Related Works

Image encryption methods are extensively studied in recent years. Many authors have proposed methods for securing images.

Bansal et al. [2] put forward an image enciphering scheme for color images using diffusion and vigenere scheme. Key space was quite large and high sensitive towards initial conditions so that makes the system secure. Chai et al. [3] presented an image enciphering scheme for gray scale images which uses compressive sensing. Plain image is altered using wavelet coefficients and after that it is shuffled in zigzag path. Image is ciphered into compressive cipher image using compressive sensing. A dynamic key selection mechanisms is used in the study proposed by, Yang et al. [4]. The 2D image is stored into 1D array. The hyper chaotic system will be used for generate random number sequence. It is possible to encrypt the image in bidirectional way. But it is only possible in grayscale image. Praveenkumar et al. [5] put forward an image encryption technique based on shuffling and scrambling. There is a large key space which is enough to withstand brute force attack and strong key sensitivity is there. It is possible only on grayscale image. Yao et al. [6] Presented an image encryption scheme which uses Singular Value Decomposition. Original colour image is encrypted into indexed image. The problem is that it cannot resist plain text attack effectively.

A technique is presented by Wu et al. [7] to secure the image data using Elliptic Curve ElGamal scheme. The process is to encrypt the colour image without expanding the pixel scale and the size of image. But long encryption time is needed for the mentioned system. Zhang et al. [8] has developed an image encryption scheme for gray scale images which is based on mixed image element and permutation. It segments the input images into several pure image elements. Then mess up all the pure image elements with the arrangement produced by the piece wise linear chaotic map system to get a scrambled images.

However aforesaid methods proposed by different authors have some downside. Those methods are endure by different attacks and applicable only for grayscale images. So it results in thought of new techniques and strong encryption methods for color images.

3   Proposed Method

3.1   Encryption

In the proposed scheme, an RGB image of size (m × n × 3) is used for encryption, where m and n are the row and column size, respectively. The image is split into three channels (Red, Green, Blue) and encodes each channel into sequence of DNA after selecting one of the eight complimentary rules. To increase the diffusion property of pixels, DNA addition is then performed. For DNA addition, DNA encoded red component and green component will be added by following the rule that selected at the time of DNA encoding to get corresponding image for red channel. Similarly green component will be added to blue component to get resultant image for green channel. For blue channel, the resultant will be sum of the resultant image of green and blue channel. These three intermediate images are the input of ECCHC (Elliptical Curve Cryptography and Hill Cipher) for the encryption.

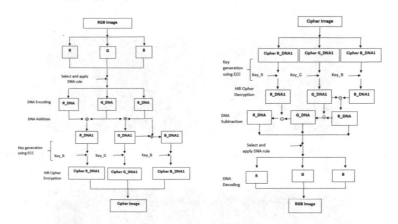

Fig. 1. The process of encryption and decryption.

In ECCHC [1], ECC is used to generate a key for Hill Cipher which is a self-invertible matrix. For key generation, both the sender and receiver select the same elliptic curve function for each channel and exchange the parameters such as the coefficients of the EC function, a prime number which is enough large so that it is hard to break and a generator point G which is a point in that curve. Both users chooses his secret key. It is from the interval [1, p − 1]; n_A for sender A and n_B for receiver B, and generates their public keys as follows:

$$P_A = n_A \cdot G \text{ and } P_B = n_B \cdot G \tag{1}$$

Both sender and receiver will calculate the product of their secret key and the public key of the other user to obtain the first key $K_1 = (x, y)$. i.e.

$$K_1 = x \cdot G = (k_{11}, k_{12}) \text{ and } K_2 = y \cdot G = (k_{21}, k_{22}) \tag{2}$$

Both users will generate a matrix called self-invertible matrix [9] which is the secret key matrix. $K = K^{-1}$. Four pixels from each channel represented as a 4×1 matrix is multiplied with 4×4 self-invertible matrix and take modulo 256 to get ciphered pixel for each channel. Repeat it for all quadruples of each channels. These three image is then combined to form the ciphered image. The enciphering and deciphering processes are shown in Fig. 1 algorithm for the same is given in Fig. 4.

3.2 Decryption

The ciphered image is first separated into three channels Cipher R_DNA1, Cipher G_DNA1 and Cipher B_DNA1. After extracting the four ciphered pixels from each channel the decryption process with ECCHC is performed which is multiplying the components with self-invertible matrix and take modulo 256. This is followed by DNA subtraction process in each channel. DNA sequences in each channel are then decoded to binary using the same complimentary rule used during encryption.

Fig. 2. The original image, enciphered image, and deciphered image of Lena image.

Fig. 3. The red, blue, and green components of Lena image after DNA addition.

Input: RGB image
Output: Ciphered image

1. Split RGB image into Red, Green and Blue channels.
2. Select any on complimentary rule of DNA.
3. Encode each pixel into DNA sequence.
4. Perform DNA addition.
5. Repeat step 4 for three channels.
6. Generation of key using ECC for each channel.
7. Hill Cipher encryption for each four pixels.
8. Merge the channels.

Input: Ciphered image
Output: RGB image

1. Split the ciphered image to three channels.
2. Extract the four ciphered pixels from each channel.
3. Generate the self-invertible key matrix for each channel.
4. Multiply four pixels as 4X1 matrix with self-invertible matrix and taking modulo 256.
5. Replace each four pixels with ciphered pixels in each channel.
6. Apply DNA subtraction in each channel after retrieving rule.
7. Decoding DNA sequences to binary.
8. Combine three channels to form the original image.

Fig. 4. The algorithm for encryption and decryption.

4   Simulation Results

The system have been implemented on Python 3.6, 64- bit software on Core i3 computer with CPU 2.00 GHz and RAM 4 GB. Figure 2 shows the original, encrypted and decrypted image (lena) of size 256×256 with DNA rule 7. Intermediate result after DNA addition of red, green and blue is shown in Fig. 3.

5 Security Analysis

5.1 Statistical Analysis

Statistical analysis on the proposed method, exposes its confusion and diffusion properties to prevent the statistical attacks. From Figs. 2 and 3 itself it is evident that there are considerable difference in the image after the DNA addition and the whole encryption, when each result is compared with corresponding component images. Figure 5 shows the histograms of Lena and Baboon before and after encryption. It is clear that there is no similarities among the histogram of original image and its enciphered image.

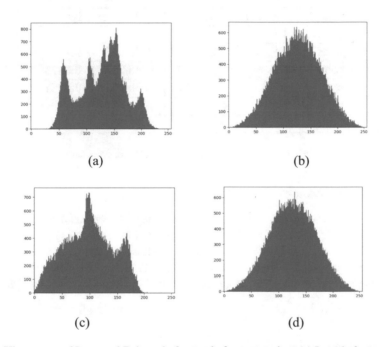

(a)   (b)

(c)   (d)

Fig. 5. Histograms of Lena and Baboon before and after encryption. (a) Lena before encryption (b) Lena after encryption (c) Baboon before encryption (d) Baboon after encryption

5.2 Peak Signal to Noise Ratio (PSNR)

PSNR is a measure for inspecting the efficiency of an image enciphering algorithm. If the PSNR value is high which shows the distortions in the deciphered image is less. That implies deciphered image will be similar to the exact image. This shows the high performance of the enciphering technique. The Figs. 6 and 7 shows the MSE and PSNR values of simulated results. If MSE increases the PSNR values will decrease and the values show that the encrypted image is more random than the original one. The results indicates that the original and encrypted images are not identical which shows the efficiency of the encryption technique.

Image	Red Component		Green Component		Blue Component	
	MSE	PSNR	MSE	PSNR	MSE	PSNR
Peppers	7406.566	9.434635	3348.429	12.88239	4089.103	12.01452
Baboon	6903.747	9.739955	4657.354	11.44941	1593.937	16.10609
Lena	8168.824	9.009208	8286.579	8.947051	866.9193	18.75102
Airplane	26414	3.912462	29697.04	3.403672	1245.005	17.17909
Barbara	8757.461	8.707022	8465.72	8.854165	502.5839	21.11872

Fig. 6. MSE and PSNR after DNA addition

Image	Existing System		Proposed System	
	MSE	PSNR	MSE	PSNR
Peppers	7769.014	9.22714	7846.022	9.184308
Baboon	9472.292	8.366253	9546.558	8.332335
Lena	10594.3	7.88008	10606.41	7.875119
Airplane	10183.16	8.051979	10026.07	8.119497
Barbara	8643.529	8.763893	8516.105	8.828393

Fig. 7. MSE and PSNR after ECCHC encryption

5.3   Differential Attacks

Any minor change if any applied to the plain image which results in predominant changes in the encrypted image, then the system could be assumed to resist differential attacks. Using UACI and NPCR it is possible to measure these changes.

When any one pixel of the plain image is altered, the rate in which the number of pixels in the encrypted image get modified is known as Number of Pixels Change Rate (NPCR).

Image	Red Component		Green Component		Blue Component	
	NPCR (%)	UACI (%)	NPCR (%)	UACI (%)	NPCR (%)	UACI (%)
Peppers	95.42847	27.75938	95.5246	17.43938	94.89288	19.63897
Baboon	99.84436	27.68008	99.80011	21.85303	98.76862	12.18023
Lena	99.823	28.17117	100	32.32916	98.2254	8.698701
Airplane	99.94659	59.46799	100	64.19996	94.36493	7.061217
Barbara	99.97253	31.54167	99.99084	8.854165	96.80634	5.981529

Fig. 8. NPCR and UACI after DNA addition

Image	Existing System		Proposed System	
	NPCR (%)	UACI (%)	NPCR (%)	UACI (%)
Peppers	99.14703	28.59961	99.63989	28.82821
Baboon	99.4751	31.20443	99.646	31.37924
Lena	98.95935	32.92723	99.63989	32.96388
Airplane	98.69537	32.30215	99.6109	32.05761
Barbara	99.31183	30.00272	99.57733	29.79703

Fig. 9. NPCR and UACI after ECCHC encryption

The Unified Average Change Intensity (UACI) computes the average intensity of differences among two enciphered images. Its value have dependency on the size and format of that image.

The Fig. 8 and 9 shows NPCR and UACI analysis. The simulation results shows that the proposed method works fairly well in defending differential attacks.

6 Conclusion

There are several encryption algorithms and methods to ensure the confidentiality of images. But it is noticeable that there is not much systems to ensure security of color images. The paper proposed a method which effectively makes the color images more secured. The system uses DNA encoding and its operations such as DNA addition and subtraction. A second level encryption & decryption is performed using Hill cipher and Elliptic Curve Cryptography. The proposed method give efficient result and thereby enhancing the security of the system compared to existing system. From the analysis of proposed system it have been obtained a better results in histograms, PSNR, NPCR and UACI values, when it is compared with existing system.

References

1. Dawahdeh, Z.E., Yaakob, S.N., bin Othman, R.R.: A new image encryption technique combining elliptic curve cryptosystem with Hill Cipher. J. King Saud Univ. Comput. Inf. Sci. **30**, 349–355 (2017)
2. Bansal, R., Gupta, S., Sharma, G.: An innovative image encryption scheme based on chaoticmap and Vigenere scheme. Multimed. Tools Appl. **76**, 16529–16562 (2017)
3. Chai, X., Gan, Z., Chen, Y., Zhang, Y.: A visually secure image encryption scheme based on compressive sensing (2017). https://doi.org/10.1016/j.sigpro.2016.11.016
4. Chai, X., Yang, K., Gan, Z.: A new chaos-based image encryption algorithm with dynamic key selection mechanisms. Multimed. Tools Appl. **76**, 9907–9927 (2016)
5. Praveenkumar, P., Amirtharajan, R., Thenmozhi, K., Rayappan, J.B.B.: Fusion of confusion and diffusion: a novel image encryption approach (2017). https://doi.org/10.1007/s11235-016-0212-0

6. Yao, L., Yuan, C., Qiang, J., Feng, S., Nie, S.: Asymmetric color image encryption based on singular value decomposition. Opt. Lasers Eng. **89**, 80–87 (2017)
7. Wu, J., Liao, X., Yang, B.: Color image encryption based on chaotic systems and elliptic curve ElGamal scheme. Signal Process. **141**, 109–124 (2017)
8. Zhang, X., Wang, X.: Multiple-image encryption algorithm based on mixed image element and permutation. Opt. Lasers Eng. **92**, 6–16 (2017)
9. Acharya, B., Rath, G.S., Patra, S.K., Panigrahy, S.K.: Novel methods of generating self-invertible matrix for Hill Cipher algorithm. Int. J. Secur. **1**(1), 14–21 (2007)
10. Xu, L., Gou, X., Li, Z., Li, J.: A novel chaotic image encryption algorithm using block scrambling and dynamic index based diffusion. Opt. Lasers Eng. **91**, 41–52 (2017)
11. Gehani, A., LaBean, T.H., Reif, J.H.: DNA-based cryptography. DI MACS Ser. Discret. Math. Theor. Comput. Sci. **54**(1), 233–249 (2000)
12. Wasiewicz, P., Mulawka, J.J., Rudnicki, W.R., Lesyng, B.: Adding numbers with DNA. In: IEEE International Conference on Systems Man and Cybernetics, vol. 1 (2000)

Development of Chronic Kidney Disease Prediction System (CKDPS) Using Machine Learning Technique

Sumana De$^{(\boxtimes)}$ and Baisakhi Chakraborty

Department of Computer Science and Engineering,
National Institute of Technology, Durgapur, India
sumanade@gmail.com, baisakhichak@yahoo.co.in

Abstract. Chronic Kidney Disease (CKD) should be diagnosed earlier before kidneys fail to work. To help doctors or medical experts in prediction of CKD among patients easily, this paper has developed an expert system named Chronic Kidney Disease Prediction System (CKDPS) that can predict CKD among patients. The dataset used to develop CKDPS is taken from the Kaggle machine learning database. Before the implementation of CKDPS, different machine learning algorithms such as, k-Nearest Neighbors (KNN), Logistic Regression (LR), Decision Tree (DT), Random Forest (RF), Naïve Bayes (NB), Support Vector Machine (SVM), Multi-Layer Perceptron (MLP) algorithm are applied on the dataset and their performances are compared to the matter of accuracy, precision and recall results. Finally, Random Forest algorithm is chosen to implement CKDPS as it gives 100% accuracy, precision and recall results. This paper also compares the accuracy results of different machine learning algorithms from different previous related works where same or different CKD dataset has been used.

Keywords: Chronic Kidney Disease (CKD) · Chronic Kidney Disease Prediction System (CKDPS) · Machine learning algorithms · Random Forest Algorithm · User input · System feedback

1 Introduction

In today's world chronic kidney disease (CKD) becomes one of the serious public health problems. CKD is the damaging condition of kidneys that can be worse over time. If the kidneys are damaged very badly then they fail to work. This is called kidney failure, or end-stage renal disease (ESRD). As per report from [1], in India 6000 renal transplants are done annually. It is also reported that, CKD among people is increasing rapidly all over the world. Diabetes, heart disease, heredity and high blood pressure are the most common causes of kidney disease. However, it has been observed that if CKD is diagnosed in early stage then kidney failure can be prevented with proper medical treatment. Medical experts or doctors diagnose CKD by patient's symptoms, physical exam, urine test and blood tests. So, if an expert system is able to predict the kidney disease in patients, it will be very helpful for the medical experts or doctors to give proper treatment to the patients in time. In the recent world, many researches are done in

© Springer Nature Switzerland AG 2020
D. J. Hemanth et al. (Eds.): ICICI 2019, LNDECT 38, pp. 155–164, 2020.
https://doi.org/10.1007/978-3-030-34080-3_18

the medical field. Nowadays, in our society machine learning algorithms are used widely for effective prediction of various diseases as referred to in [2]. In this paper, to help medical experts or doctors in prediction of CKD, Chronic Kidney Disease Prediction System (CKDPS) using Random Forest algorithm has been developed. Random Forest algorithm is one of the machine learning techniques. An application of artificial intelligence (AI) is machine learning that provides the able systems to learn automatically and improve from experience without being explicitly programmed. In medical field machine learning analyzes massive quantities of medical data and delivers accurate results in order to identify the disease. In this paper, the dataset used to develop CKDPS is taken from the Kaggle machine learning database. The detail about dataset is discussed in Sect. 3. The machine learning algorithms such as, k-Nearest Neighbors (KNN), Logistic Regression (LR), Decision Tree (DT), Random Forest (RF), Naïve Bayes (NB), Support Vector Machine (SVM), Multi-Layer Perceptron (MLP) algorithm are applied on the dataset. Experimental result shows that only Random Forest Classifier algorithm gives better classification performance with 100% accuracy, precision and recall results. So, here, only Random Forest Classifier has been used to predict CKD in CKDPS. As a screening tool the graphical user interface(GUI) of CKDPS has been developed with the help of python so that doctors or medical experts diagnose CKD among patients very easily. The rest of the paper is arranged as follows: Sect. 2 focuses on previous related works, Sect. 3 on methodology and system implementation Sect. 4 discusses about the proposed model of CKDPS, Sect. 5 shows accuracy comparison between previous related papers and the experimental results of this paper, Sect. 6 depicts the simulation result, and at last Sect. 7 ends up with conclusion.

2 Previous Related Works

Nowadays, machine learning algorithms are widely used in the field of medicine. Numerous works have been done where machine learning techniques are used to predict disease. The paper [2] shows the usage of machine learning in disease prediction over big data analysis. In the paper [3], machine learning (ML) techniques are used to investigate how CKD can be diagnosed. In another research work [4], classification of CKD is done using Logistic Regression, Wide & Deep Learning and Feed forward Neural Network. Various kernel-based Extreme Learning Machines are evaluated to predict CKD in [5]. In [6], Naive Bayes, Decision Tree, K-Nearest Neighbour and Support Vector Machine are applied to predict CKD. In the paper [7], Back-Propagation Neural Network, Radial Basis Function and Random Forest are used to predict CKD. For predicting the CKD in [8] support vector machine (SVM), K-nearest neighbors (KNN), decision tree classifiers and logistic regression (LR) are used. Multiclass Decision forest algorithm performed best in [9] to predict CKD. After using Adaboost, Bagging and Random Subspaces ensemble learning algorithms for the diagnosis of CKD, the paper [10] suggests that ensemble learning classifiers provide better classification performance. Decision tree and Support Vector Machine algorithm are used in [11]. XGBoost based model is developed in [12] for CKD prediction with better accuracy. In [13], J48 and random forest works better than Naive Bayes (NB), minimal sequential optimization (SMO), bagging, AdaBoost algorithm.

3   Methodology and System Implementation

This paper presents three consecutive steps to implement CKDPS and diagnose CKD in a patient through the system. Steps are shown in Fig. 1.

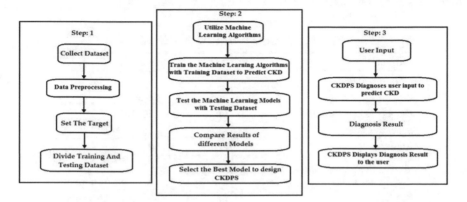

Fig. 1.   Steps to implement CKDPS and diagnose CKD in a patient through the system.

Processing Detail in Step 1

(a) Collect Dataset: The dataset is collected from Kaggle machine learning database, obtained by the survey of CKD in India that contains laboratory results of both positive and negative cases of CKD. It contains cases of 400 patients with 25 attributes (eg, red blood cell count, white blood cell count, etc.), detail in Table 1. Input attributes are used to take input from user in CKDPS and output attribute is for diagnosis result.

(b) Data Preprocessing: CKD dataset contains some attributes with Nominal data, some attributes with Numerical data and some attribute with Null values. Data pre-processing is a data mining technique that is used to transform incomplete raw data in a useful and efficient format. Three steps in Data Preprocessing are shown below:

(i) Data Transformation: All of the nominal data are converted into numerical data. For example, Red Blood Cells & Pus Cell values: Normal = 1; Abnormal = 0. Pus Cell Clumps & Bacteria Values: Present = 1, Not present = 0.

(ii) Missing data handle: The null values of an attribute are replaced with the calculated mean value of the attribute. This strategy is applied on each attributes that contain null value.

(iii) Rearrange the dataset: Each patient's records are repositioned haphazardly.

(c) Set the Target: To classify that patient has CKD or Not CKD according to the values of input attributes the output attribute "Classification" is set as the target class.

(d) Divide Training and Testing Dataset: Except the target class column, the entire dataset is divided into two sets in 7:3 ratios. 70% of which is used for training the machine learning algorithms and the 30% is used to test their accuracy, precision and recall. Table 2 shows the dataset division details.

Processing Detail in Step 2

(a) Utilize Machine Learning Algorithms: The basic idea of machine learning is to build Machine Learning Models that can receive input data and analyze statistically to predict an output more accurately. When a machine learning algorithm is trained from training dataset then Machine Learning Model is generated. This paper uses following different supervised machine learning algorithms to predict CKD.

Table 1. Dataset-attributes with their values.

Attribute	Abbr	Values
Input attributes		
Age	age	2–90
Bloodpressure (mm/Hg)	bp	50–100
Specific Gravity	sg	1.005–1.025
Albumin	al	0–4
Sugar Degree	su	0–4
Red Blood Cells	rbc	Normal, Abnormal
Pus Cell	pc	Normal, Abnormal
Pus Cell Clumps	pcc	Present, Not present
Bacteria	ba	Present, Not present
Blood Glucose Random (mgs/dl)	bgr	22–490
Blood Urea (mgs/dl)	bu	1.5–391
SerumCreatinine (mgs/dl)	sc	0.4–7.6
Packed Cell Volume	pcv	0–54
Potassium (mEq/L)	pot	0–4.7
Sodium (mEq/L)	sod	0–163
Hemoglobin (gms)	hemo	0–17.8
White Blood Count (cells/cumm)	wbcc	0–26400
Red blood cell count (millions/cmm)	rbcc	0–8
Hypertension	htn	Yes, No
Diabetes Mellitus	dm	Yes, No
Coronary Artery Disease	cad	Yes, No
Appetite	appet	Good, Poor
Pedal edema	pe	Yes, No
Anemia	ane	Yes, No
Output attribute		
Classification	Classification	ckd, notckd

Table 2. Dataset-division

Dataset	Total patient's records	Class	
		CKD	Not CKD
Training set	280	143	137
Testing set	120	106	14

(i) K-Nearest Neighbors (KNN): After the training phase, whenever a testing sample is given to K-Nearest Neighbors classifier then firstly it calculates distance between inputted testing sample and training samples then searches for the K-nearest neighbors; next it checks the majority class to which its neighbors belong and lastly assigns that majority output class for the testing sample, referred to in [14].

(ii) Logistic regression (LR): It is a regression model that is used to predict a binary outcome (1/ 0, Yes/ No, True/ False). Using logistic function, this model estimates the probabilities; and measures the relationship between the output variable and the predictor variables as referred to in [4].

(iii) Decision Tree (DT): There are two entities in decision tree; these are decision nodes and leaves. The decision nodes are where the data are split and the leaves are the decisions or the final outcomes. In this tree, the topmost node is called root node; a test condition on an attribute is denoted by the internal node; each result of the test condition is denoted by each branch, and each terminal node (or leaf node) is assigned with a class label as referred to in [14].

(iv) Random Forest (RF): It is an ensemble of decision tree classifiers and each tree is built from bootstrap sample of the data as referred to in [7]. Random Forest combines decision trees to get more accurate and stable prediction. Here, each decision tree predicts a certain class, finally based on the majority votes for class prediction from each of the decision trees; Random Forest Classifier makes its final prediction.

(v) Naïve Bayes (NB): It is a simple probabilistic classifier based on applying bayes theorem with strong independence assumptions between the features. As stated in [6], Bayes theorem calculates the posterior probability, $P(c|x)$, from $P(c)$, $P(x)$ and $P(x|c)$. Naive Bayes classifier assumes that the effect of the value of a predictor (x) on a given class (c) is independent of the values of other predictors.

(vi) Support Vector Machine (SVM): It is a discriminative classifier that is formally defined by a separating hyperplane. It performs classification by finding the hyperplane that maximizes the margin between the classes as referred to in [6]. In 2D space this hyperplane is a separating line that separates the plane in two parts wherein lie two classes separately.

(vii) Multi-layer Perceptron (MLP): MLP is a deep, artificial neural network that is the combination of more than one perceptron. MLP has the input layers to receive the input signal; as computational engine it has an arbitrary number of hidden layers and to make a decision or prediction about the input, it has an output layer.

(b) Train the Machine Learning Algorithms with Training Dataset to Predict CKD: Each of the Machine Learning Algorithms is trained with Training Dataset and Machine Learning Models are generated. The learning algorithms find the patterns in the training dataset that map the input data attributes to the target class and generates a machine learning model that captures these patterns.

(c) Test the Machine Learning Models with Testing Dataset: Testing Dataset is applied to each Machine Learning Models to check their performances to the matter of accuracy, precision and recall.

d) Compare results from different Models: Accuracy, precision and recall results, obtained from different Machine Learning Models are compared.

(e) Select the best Model to develop CKDPS: After comparing the results from the different models, the best model with best accuracy, precision and recall result is selected to develop CKDPS.

Processing Detail in Step 3
Finally, Random Forest model is selected to develop CKDPS, as it gives best accuracy, precision and recall result. CKDPS is a diagnosis application that is developed using python.
(a) User Input: Through GUI of the system doctors or medical experts submit a patient's health condition detail as an input.
(b) CKDPS Diagnoses user input to predict CKD: According to that new input values CKDPS using Random Forest model predicts whether the patient has CKD or not.
(c) Diagnosis Result: After the prediction, CKDPS gets a diagnosis result.
(d) CKDPS Displays Diagnosis Result to the user: Finally, user gets feedback result from the System.

4 Proposed CKDPS Model

This section provides detail about the model of Chronic Kidney Disease Prediction System (CKDPS). It is an inquiry-response model. Figure 2 shows the architectural components of CKDPS model. Three main modules of this model are, Administrator Module, Computational Module, and Data Processing Module. CKD Dataset contains 400 samples of the disease diagnosis. User and Administrator are two components of the model, who have interaction with the system through system's User Interface. The work of individual components is given below:

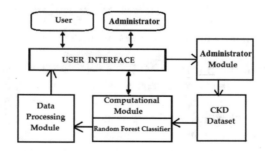

Fig. 2. Architectural components of CKDPS model

1. User Interface: User Interface is responsible to make a connection between user and the system. Through the User Interface, CKDPS displays a query form to the user and user fills up the form with values and submits it to the system. **2. User:** User may be doctors or the medical experts. They provide patient's age, urine and blood test related data to the system through the User Interface. **3. Computational Module:** Using Random Forest model Computational Module classifies whether the newly posted inputted data is in CKD class or Not CKD class. And it also calculates the system accuracy to perform the diagnosis.

4. Data Processing Module: After the complete classification performed by the Computational Module, Data Processing Module checks the diagnosis result. If it finds CKD class then it shows a message to the user that "Patient is suffering from CKD" otherwise it shows "Patient is not suffering from CKD". This module also displays the system accuracy to the user. **5. Administrator Module:** Administrator Module assists the Administrators for administering CKDPS. Only Administrators have the permission to add, delete, update and modify the CKD Dataset records. **6. Administrator:** Administrator should be doctors or medical experts, who should have proper knowledge about CKD. They can update the Dataset with valid data or delete unnecessary data from Dataset. **7. CKD Dataset:** In this model CKD Dataset is stored after following the step-1 of Sect. 3.

5 Accuracy Comparison and Experimental Results

Table 3 shows a comparison of best accuracy results of different machine learning algorithms from different previous related works that used either same or different CKD-dataset for CKD prediction. It shows that Random Forest Classifier algorithm works best. As discussed in Sect. 3, before the implementation of CKDPS, different machine learning algorithms are applied on the dataset and their performances are compared to the matter of accuracy, precision and recall results. Table 4 shows different accuracy, precision, recall results obtained from different machine learning algorithms. This experiment also shows Random Forest Classifier algorithm gives better performance.

Table 3. Comparison of different accuracy results of machine learning algorithms obtained from different previous related works

Previous work	Machine learning algorithm with highest performance	Accuracy
From [3]	Random Forest classifier algorithm	100%
From [6]	K-Nearest Neighbour algorithm	98%
From [7]	Radial basis function algorithm	85.3%
From [8]	SVM classifier	98.3%
Form [9]	Multiclass Decision Forest	99.1%
From [11]	Support Vector Machine	96.75%
From [12]	With three feature selection technique Extreme Gradient Boosting (XGBoost) model	97.6%
From [13]	Random Forest (RF)	100%

Table 4. Accuracy, Precision, Recall results from different machine learning algorithms.

Machine learning algorithms	Accuracy	Precision	Recall
KNN	80%	74%	69%
LR	99.2%	98%	98%
DT	95.8%	98%	97%
RF	**100%**	**100%**	**100%**
NB	98.3%	97%	97%
SVM	62%	36%	60%
MLP	62%	58%	58%

6 Simulation Results

As the simulation tool, in this paper Jupyter Notebook is used in the Python Environment. To create the GUI, Tkinter method in Python has been used. GUI of CKDPS is presented in the Figs. 3 and 4.

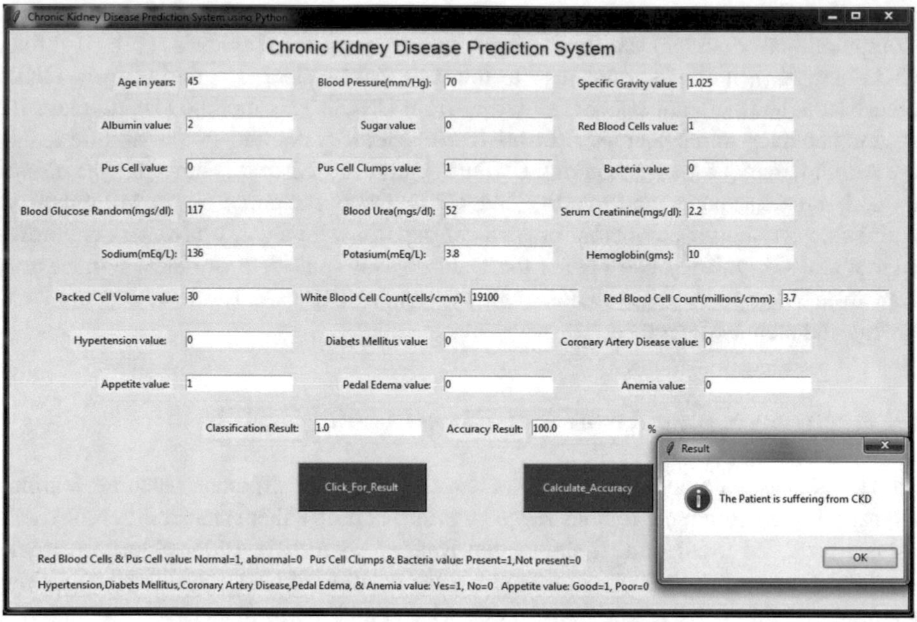

Fig. 3. According to user input patient has CKD

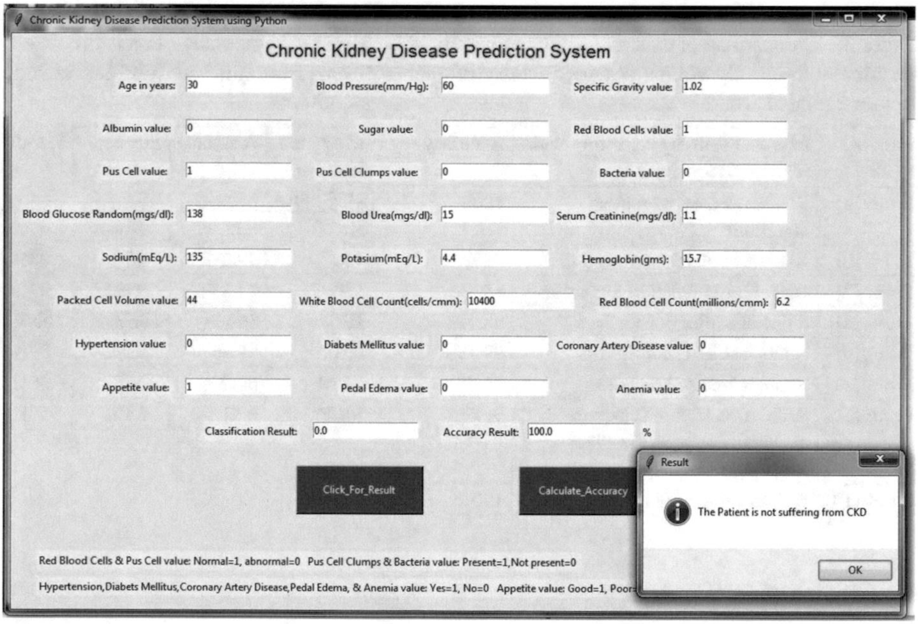

Fig. 4. According to user input patient has no CKD

7 Conclusion

Chronic Kidney Disease (CKD) should be diagnosed earlier before kidneys fail to work. To help doctors or medical experts in prediction of CKD among patients easily, this paper has developed a Chronic Kidney Disease Prediction System (CKDPS) using Random Forest Algorithm. Random Forest Algorithm is a machine learning algorithm that combines decision trees to get more accurate and stable prediction. To design the CKDPS, CKD dataset is taken from Kaggle machine learning database. The dataset contains cases of 400 patients with 25 attributes (e.g., red blood cell count, white blood cell count, etc.). The method of system implementation follows three steps; the primary step to implement CKDPS is Collection of Dataset, Preprocessing of Dataset, Setting of Target class and division of Dataset into Training and Testing Datasets. In the second step machine learning algorithms such as, k-Nearest Neighbors (KNN), Logistic Regression (LR), Decision Tree (DT), Random Forest (RF), Naïve Bayes (NB), Support Vector Machine (SVM), Multi-Layer Perceptron (MLP) algorithm are applied on the CKD Dataset. Each of the different machine learning algorithms provides different accuracy, precision and recall results. From which Random Forest algorithm is selected to develop CKDPS as it provides 100% accuracy, precision and recall results. At the last step, user posts query to the system and system feedbacks the user with classification result. There are seven architectural components in CKDPS model such as User Interface, Administrator Module, Computational Module, Data Processing Module, User, Administrator and CKD Dataset. In this paper, accuracy results from the previous related works are compared. Comparison shows Random Forest algorithm works better to predict CKD.

References

1. Agarwal, S.K.: Chronic kidney disease in India - magnitude and issues involved (2009)
2. Vinitha, S., Sweetlin, S., Vinusha, H., Sajini, S.: Disease prediction using machine learning over big data. Comput. Sci. Eng. Int. J. (CSEIJ) **8**, 1–8 (2018)
3. Subas, A., Alickovic, E., Kevric, J.: Diagnosis of chronic kidney disease by using random forest, pp. 589–594 (2017)
4. Imran, A.A., Amin, M.N., Johora, F.T.: Classification of chronic kidney disease using logistic regression, feedforward neural network and wide and deep learning. In: 2018 International Conference on Innovation in Engineering and Technology (ICIET), pp. 1–6 (2018)
5. Wibawa, H.A., Malik, I., Bahtiar, N.: Evaluation of kernel-based extreme learning machine performance for prediction of chronic kidney disease. In: 2018 2nd International Conference on Informatics and Computational Sciences (ICICoS), pp. 1–4 (2018)
6. Radha, N., Ramya, S.: Performance analysis of machine learning algorithms for predicting chronic kidney disease. Int. J. Comput. Sci. Eng. Open Access **3**, 72–76 (2015)
7. Ramya, S., Radha, N.: Diagnosis of chronic kidney disease using machine learning algorithms. Int. J. Innov. Res. Comput. Commun. Eng. **4**, 812–820 (2016)
8. Charleonnan, A., Fufaung, T., Niyomwong, T., Chokchueypattanakit, W., Suwannawach, S., Ninchawee, N.: Predictive analytics for chronic kidney disease using machine learning

techniques. In: 2016 Management and Innovation Technology International Conference (MITicon), pp. 80–83 (2016)

9. Gunarathne, W.H.S.D., Perera, K.D.M., Kahandawaarachchi, K.A.D.C.P.: Performance evaluation on machine learning classification techniques for disease classification and forecasting through data analytics for chronic kidney disease (CKD). In: 2017 IEEE 17th International Conference on Bioinformatics and Bioengineering (BIBE), pp. 291–296 (2017)

10. Basar, M.D., Akan, A.: Detection of chronic kidney disease by using ensemble classifiers. In: 2017 10th International Conference on Electrical and Electronics Engineering (ELECO), pp. 544–547 (2017)

11. Tekale, S., Shingavi, P., Wandhekar, S., Chatorikar, A.: Prediction of chronic kidney disease using machine learning algorithm. Int. J. Adv. Res. Comput. Commun. Eng. **7**, 92–96 (2018)

12. Ogunleye, A., Wang, Q.: Enhanced XGBoost-based automatic diagnosis system for chronic kidney disease. In: 2018 IEEE 14th International Conference on Control and Automation (ICCA), pp. 805–810 (2018)

13. Sisodia, D.S., Verma, A.: Prediction performance of individual and ensemble learners for chronic kidney disease. In: 2017 International Conference on Inventive Computing and Informatics (ICICI), pp. 1027–1031 (2017)

14. Jadhav, S.D., Channe, H.P.: Comparative study of K-NN, Naive Bayes and decision tree classification techniques. Int. J. Sci. Res. (IJSR) **5**, 1842–1845 (2016)

Developing Classifier Model for Hand Gesture Recognition for Application of Human Computer Interaction (HCI)

Suvarna Nandyal[1] and B. Sangeeta[2(✉)]

[1] CSE Department, PDA College of Engineering, Kalaburgi, Karnataka, India
suvarna_nandyal@yahoo.com
[2] PDA College of Engineering, Kalaburgi, Karnataka, India
sangeeta.bingi3@gmail.com

Abstract. The hand gesture technique that is regarded as the natural and easy method for the human-machine interaction, has paved way for the development of the multitudes of applications. The hand gestures basically employed in most of the application are either sensor based or the vision based. In case of verbal communication the gesture depiction involves the application of the natural and the bare hand gestures. So the paper proposes a bare hand gesture recognition with the light in variance conditions, involving the image cropping algorithm in the preprocessing, considering only the region of interest. The mapping of the image oriented histogram is primarily done utilizing the Euclidean distance method and further supervised neural network are trained using the images mapped, to have a better recognition of images with the same gestures under different light intensities.

Keywords: Hand image segmentation · Light invariant hand gesture recognition · Fingertips detection · Bent fingers' angles calculation · Both hands' angles calculation

1 Introduction

As a piece of HCI, joining hand motions into specialized strategies is a significant research zone. Hand motion acknowledgment (HGR) is the normal method for human device connection. Today numerous scientists in the scholarly world n industry are concentrating various procedures that create such associations simpler, characteristic and helpful without the prerequisite for any extra gadgets. The reason of executing such a framework is motion acknowledgment. Motion acknowledgment has turned into an intriguing issue for a considerable length of time. These days two strategies are utilized essentially to perform motion acknowledgment. One depends on expert, wearable electromagnetic gadgets, similar to exceptional gloves. The other one uses PC vision. The previous one is basically utilized in the film business. It makes well yet is expensive n in some condition. The last one includes picture preparing. In any case, the presentation of signal acknowledgment straightforwardly dependent on the highlights separated by picture preparing is generally restricted. Despite the fact that the presentation has improved as the presence of cutting edge devices, as Microsoft Kinect

© Springer Nature Switzerland AG 2020
D. J. Hemanth et al. (Eds.): ICICI 2019, LNDECT 38, pp. 165–178, 2020.
https://doi.org/10.1007/978-3-030-34080-3_19

sensors, the generally more expensive amount of such gadgets is as yet a hindrance to the enormous scale use of signal based HCI frameworks. Also, such propelled devices make considerably more problematically than optical cameras in some specific condition. For example, the lessening of ultraviolet beam in water can generally restrict the utilization of those like Microsoft Kinect devices in water with a decent light situation.

The idea of motion acknowledgment is a characterization issue. There are loads of ways to deal with handle 2D motion acknowledgment, with the direction histogram, the concealed Markov ideal, molecule separating, bolster path machine (SVM), and so on. The majority of those methodologies want preprocessing the info motion picture to concentrate highlights. The exhibition of those methodologies depends a great deal on the component learning method.

2   Hand Image Segmentation

It was previously examined in that various forthcoming just as existing applications depend available motion acknowledgment (HGR) procedures including an uncovered hand. Such methods permit a characteristic correspondence with engines. HGR based frameworks face numerous issues in skin division because of luminance and power in pictures which is a noteworthy reason for commotion in the pre-prepared outcomes. The HGR based frameworks for the most part make suspicions about the hand bearing. This causes confinements in the normal articulation of people. Numerous frameworks depend on the supposition that the clients need to demonstrate the turn in honest place, with the finger pointing upward. Handling period is factor that should be viewed as when structuring picture created preparing calculations.

In this section the emphasis is on bearing division of a characteristic hand with ongoing execution. It is viewed as that there is no confinement on the signal bearing. Additionally, with regards to characteristic registering, there is no prerequisite for scarves, devices or shading bands to section the hand. The main supposition that will be that the client shows the hand to the framework with the end goal that framework while the course of the hand isn't confined. The client is allowed to move the turn toward any path normally as the arrows move. This section shows a novel picture trimming pre-handling calculation, which affixes the motion acknowledgment method. This is finished by performing trimming and henceforth lessening the quantity of handling pixels. A point by point discourse available geometry limitations (HGP) for example fingertips and COP discoveries. Figure 1 presents a square graph of the whole procedure.

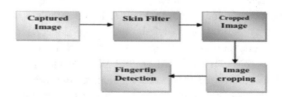

Fig. 1. Algorithm flow for the pre-processing Method. This figure says the flow of preprocessing method

Fig. 2. System prototype. This figure shows how is hand gesture image is captured.

2.1 Hand Segmentation

Video is a succession of picture edges busy at a fixed amount. In this test format all pictures are caught constantly with a basic web camera or laptop camera in 2D and are handled one by one as appeared in Fig. 2. The procedure of hand division is talked about in three stages. Right off the bat, a HSV shading space put together skin channel is connected with respect to the pictures for hand division. Besides, a force created histogram is produced course recognition. Thirdly, the picture is trimmed with the goal that the resulting picture is made out of just the motion pixels.

2.1.1 Skin Filter

The skin channel utilized on the info picture can be founded on HSV or YC shading space to diminish the light impact somewhat. In the HSV shading is sifted utilizing the chrome city (tint and immersion) ethics whereas in the Y shading space the C esteems are utilized for skin separating. A HSV shading channel is connected to the present picture outline for hand division. This shading space isolates three parts: tint (H), immersion (S) and splendor (I, V or L).Basically HSV shading universes are mis-shapenness of the RGB shading solid shape and they can be the RGB space by means of a nonlinear change. The purpose for the determination of this shading space in skin identification is that it enables clients to instinctively indicate the limit of the skin shading session as far as the tint and immersion. As I, V or L provide the brilliance data, they are frequently released to lessen enlightenment reliance of skin shading. The consequence of channel is appeared in Fig. 3. The skin channels are recycled to make a

(a)   **(b)**

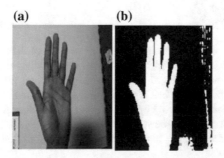

Fig. 3. Skin filtering results (a) Initial hand image (b) Binary silhouette. In this figures it shows how the hand image skin filtering and its binary image is done.

matched picture with the establishment in dull shading and the hand locale in white. Double pictures are bi-level pictures where every pixel is put away as a solitary piece (0 or 1). Smoothening of the ensuing picture is necessary, as the yield picture may cover some barbed ends. This paired picture is smoothened utilizing be around channel.

2.1.2 Hand Direction Detection

In this framework, the client can provide guidance free contribution by displaying a hand signal to the camera. For acquiring better and explicit outcomes, it is important to discover the heading of the hand. For burden as such a 4-route output of the pre-handled picture is executed as appeared in Fig. 4 and histograms are created dependent on the skin shading force toward every path. In every one of the four sweeps the most extreme estimation of is chosen after the histograms. It is absent that the greatest estimation of skin pixels in the picture speaks to the wrist edge and the furthest edge of the output speaks to the finger edge. The capacities utilized for producing the power histograms are introduced in (2.1) and (2.2).

$$H_\chi = \sum_{y=1}^{n} imb(x, y) \tag{2.1}$$

$$H_\gamma = \sum_{y=1}^{m} imb(x, y) \tag{2.2}$$

Where *imb* signifies the binary silhouette and *mn* be the rows and Columns individually of the matrix *imb*.

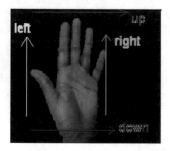 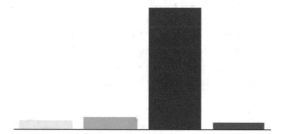

Fig. 4. Image scanning and corresponding bars. In the above figure shown the yellow bar appeared relates to the main skin pixel in the parallel outline filtered after the left course. The green bar relates to the sweep achieved after the correct heading, the red bar compares to the output achieved after the descending bearing n pink bar compares to the output achieved from the rising course. Plainly the red bar has a higher size contrasted with different bars for this specific signal picture. Henceforth, it tends to be surmised that the wrist edge is the descending way of the casing and thusly the course of digits is the rising way. Henceforth, the course after the wrist to the fingers is distinguished.

3   Light Invariant Hand Gesture Recognition

As of late there takes been a developing enthusiasm for the ground of light invariant acknowledgment. For cutting edge uses a framework can be setup in the research center by perfect situations. Be that as it may, in pragmatic circumstances the hand movement affirmation systems can have utilizes in various settings. The light power may not be same everywhere, in this way a solid structure that workings in a wide scope of bright situations is necessary.

In image taking care of employments, the light power accept a critical activity later it basically impacts the division of the ROI after one of a kind image diagram. If the light power variations, by then the breaking point for skin channel moreover should be altered. This rouses the headway of procedures that are significant to powers. As this proposition is committed to the regular communication with engines utilizing hand signals, in this part the examination of the owner on exposed needle motion acknowledgment is made. The light force at various occasions of the day, explicitly, at regular intervals, is measured.

The hand gestures are deciphered by the framework through the needle signal acknowledgment method, regardless of whether it is a pre-defined legitimate gesture or not. On the off chance that the signal is incorporated into the rundown of the gestures, at that point the framework reacts to the relating activity as pre-defined in the framework. The bright power in pre-defined signal and light force in the present framework can be extraordinary, yet the framework should remember it as equal gesture.

3.1   Pattern Recognition

The objective of example acknowledgment is to order the matters of enthusiasm into one of various classifications or lessons (Therrien 1989). The objects of intrigue are by and large termed examples. They can be written cultures or types, natural cells, hardware waveforms or sign, conditions of a framework or whatever other things want to group. Any example acknowledgment framework comprises of two segments, in particular component change and classifier (Therrien 1989) (see Fig. 5).

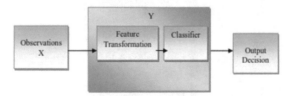

Fig. 5. Approach of feature extraction. Figure shows the flow of feature extracted and classified in pattern recognition.

3.2   Orientation Histogram

Direction Histogram (OH) strategy for highlight removal was created by McConnell (McConnell 1986). The significant bit of leeway of this strategy is that it is straight-forward n hearty to lightvagaries (Freeman and Roth 1995). On the off chance that the pixel-powers method is taken, certain issues emerge because of shifting brightening (Messery 1998). On the off chance that pixel by pixel contrast for a similar signal is taken after two distinct images, whereas the brightening situations are unique, the separation among them can be huge. In such situations the image itself goes about as a component course.

The principle inspiration for utilizing the direction histogram is the necessity for helping and place invariance. Extra significant part of motion acknowledgment is that independent of the direction of the submit various pictures, for the same gesture framework must create a similar yield. This should be possible by shaping a neigh-borhood histogram for nearby directions (Liang and Ouhyoung 1998). Consequently, this methodology need be hearty for enlightenment vagaries it should likewise deal invariance.

3.3   Light Invariant System

The acknowledgment framework deals with the standard of the PC dream in 2D universe. The fundamental strategy takes been appeared in Fig. 6. The framework has an line with a little camera which catches clients' signals. Contribution to the frame-work is picture casing of moving hand before a camera caught as a live video. The pre-processing of picture casing was ended as talked about in Sect. 2. The outcome picture can be ROI for example just motion picture. When the ROI is accessible, following stage is to discover include courses from the information picture to remember it with the assistance of ANN.

Fig. 6. Gesture recognition methodology.

As this framework was for research reason, just 6 unique motions are taken in the informational index the same amount of specialists were additionally have tried their strategies with 6 motions before. These six distinct signals utilized in this examination, are appeared in Fig. 7. The pictures of every motion, utilized for ANN preparing, were distinctive in the skin shading and light force.

When the signal can get perceived the relating move makes place which could be related with it. In created framework the sound depiction of the coordinated motion was appended as relating activity. On acknowledgment of the motion, the sound document comparing to the perceived motion would be played.

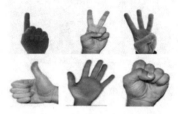

Fig. 7. Hand Gestures to be used in the System. Figures shows the different hand gestures that are used to train the system

3.4   Light Invariant Gesture Recognition

Motions ought to be perceived same paying little mind to where they happen inside the camera's ground of view. This interpretation invariance can be accomplished by the extreme advance of overlooking location by and large, essentially organizing a histogram of how regularly every direction heading happened in the picture. Unmistakably, this tosses out some data, and couple of particular pictures will be confounded by their direction histograms. With the end goal of highlight extraction representation, take one case of pointer signal raised and their relating OH as appeared Figs. 8 and 9. Here it tends to closeness among these direction histograms even the skin shading were especially unique. This skin shading distinction could be a direct outcome of various individuals or a similar individual in various light force. These similitude's would be all the more obviously watched on the off chance that we plan the OH signal. It is a universal suspicion that places of fingertips in the area in respect to the palm and it is quite often adequate to separate a limited amount of various signal (Ahmad and Tresp 1993); (Davis and Shah 1994); (Kuno and Sakamoto+ 1994). We should reflect the motion as every one of the five members open and it's OH as appeared in Fig. 8. After this unmistakably OH plotted for two distinct signals would be especially unique while for same motion it would indicate equal OHs just abundance of course can differ with respect to light force.

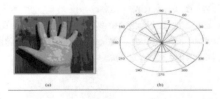

Fig. 8.   (a) Gesture III and (b) OH of Gesture III.

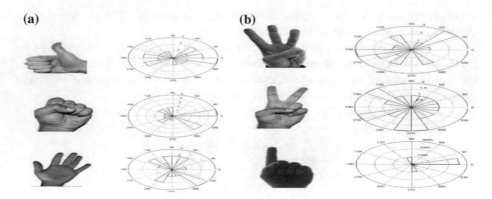

Fig. 9. Gesture and (b) OH of Gesture.

Table 1. Gestures and their target vectors

Gesture	Target Vectors
☝	000001
✌	000010
🖐	000100
👍	001000
✊	010000
🤟	100000

3.5 Experimental Results

The objective directions for six example signals are characterized in Table 1. The signal acknowledgment of trial picture was made with coordinating the pictures put away in DB, first with Euclidean separation technique and after that utilizing neural system to improve outcomes. Euclidean strategy was giving palatable outcomes yet the bogus location amounts for few signals were great.

4 Fingertips Detection

Fingertip identification frames a significant segment of HGR when picture created copies are utilized for building or recognizing hand locations. In this part, the center is on bearing invariant fine focus of palm identification of characteristic hand with genuine period execution. The HGP is distinguished in 2D utilizing a basic webcam and in 3D utilizing KINECT. Low dimension picture preparing techniques are utilized to recognize HGP in 2D, whereas KINECT encourages by giving the profundity data of closer view objects. The motion shares are divided utilizing the profundity vector and the focuses of awards are distinguished utilizing separation change on opposite picture.

4.1 HGP Detections

HGP incorporates fingertips and the focal point of palm. In Sect. 2 the hand is portioned as a piece of pre-processing. Programmed focal point of palm (COP) identification in a continuous info framework is a difficult errand; however it opens another arrangement of utilizations where hand motion acknowledgment can be utilized. This segment proceeds by the consequences of division from part 2 to identify HGP.

4.1.1 Fingertips Detection

Now after the portioned outcomes, one littler picture is accessible which covers for the most part skin pixels (hand signal figure). Later the hand course is recognized, the heading of fingertips is additionally identified. To decide all fingertips in the edited hand picture, a sweep of the trimmed double picture since finishes is started. Therefore, the quantities of pixels are determined for every line or section dependent on the hand heading, regardless of whether it is in the up-down or left-right location. At that point, the power of every pixel is allocated values somewhere in the range of 1 and 255 in expanding request end in equivalent conveyance. The procedure is introduced in Fig. 10 (Figs. 11 and 12).

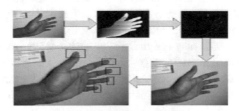

Fig. 10. Fingertip detection process.

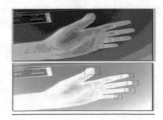

Fig. 11. Results of fingertip detection in the original image frame.

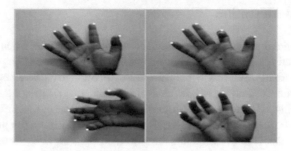

Fig. 12. Enhanced results of fingertips and centre of palm detection.

5   Bent Fingers' Angles Calculation

Hands assume a significant job in the execution of numerous significant errands in the everyday existences of people also for various other uncommon resolves. The state of the is with the end goal that it can without much of a stretch play out various generally repetitive errands. It can twist its fingers to various edges to get or n to relate power by means of fingers or the palm zone. In various situations, a human hand can play out the assignments significantly extra proficiently than a device through. This is because of the capacity of a human hand to work ended various degrees of opportunity and its capacity to curve fingers at various edges.

In any case, in certain situations it cannot be reasonable to utilize a human hand, whereas the utilization of a device might be ideal. For example, in circumstances similar bomb identification/dissemination, finishing of supposed ammo and landmine evacuation, in the event that people are driven into the field, at that point losses may happen. Consequently, there is a requirement for a mechanical hand which can play out indistinguishable activities from a human hand. The automated hand ought to have the option to twist & it ought to be effectively manageable. The automated hand ought to, which it can twist like a human in collaboration way. Henceforth, as a rule, the mechanical hand ought to have the option to play out every one of the tasks of a human turn progressively.

The strategies displayed in this section can be utilized to controller a remotely found mechanical hand which can play out indistinguishable tasks from a human hand. The client demonstrates his common finger (minus trying any powered-electronic gear) to the camera and the palm would confront the camera. The conduct of human finger is recognized by the camera and the automated arm is prepared to work in like manner (Fig. 13).

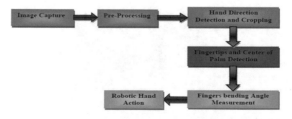

Fig. 13. Block diagram flow of the system.

6 Both Hands' Angles Calculation

In characteristic correspondence individuals utilize two hands to convey what needs be whereas chat. The twisted fingers' plot for one hand can be recognized utilizing the technique examined in Sect. 3. There is a want to recognize the plots for two hands to create the motion created framework progressively powerful. In this part a novel strategy for point guess of two hands' bowed fingers is introduced and its use to a mechanical needle controller is examined. The framework utilizes a straightforward camera and a computer.

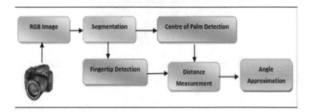

Fig. 14. Algorithmic flow for angle calculation for both hands. Figure shows the algorithm flow for calculating both hand gesture.

Used for ascertaining the plot for two hands, one choice is that the calculation (introduced in Sect. 3 for a solitary hand) can be connected double to the picture outline. This prompts extra time utilization, which isn't attractive progressively uses. It is fundamental that the computational period ought to be low for constant execution. Consequently, another methodology is produced for ascertaining the point of the fingers of two hands. This methodology is displayed by the square chart appeared in Fig. 14. The COP and the are recognized after the sectioned picture and two hands fingers' edge were determined in parallel. Concerning two hands, the identification procedure is completed in parallel. Consequently, the time devoured is littler than that expended when the calculation exhibited is actualized twice, independently for two hands. An itemized exchange on parallel picture preparing is introduced in (Bräunl and Feyrer+ 2001).

6.1 Pre-processing

As there are two turns in the caught gesture picture and it is a want presently to recognize them two. For this, an adjustment in division strategy is required. The HSV shading universe created skin channel is utilized to shape the parallel outline of the information picture. The hands are divided utilizing a similar strategy, where the return for money invested is resolved. After the arrangement of the double picture, two paired connected articles (BLOBs) are gathered and the BLOB examination dependent on 8 availability norms is made. Because of that two hands would be recognized after one another. It is ensure that while computing the separation among COP and fingertips, framework does not commit any error by thinking about then COP of other hand. The fundamental motivation behind the BLOB examination is to extricate the two greatest BLOBs to take out the deficiency identification of skin and the skin shading create out of sight and to separate the two BLOBs from one another. Figure 15 offerings the outcomes for hand division. The more splendid BLOB compares to the correct needle of the fundamental edge and the other BLOB relates to one side hand.

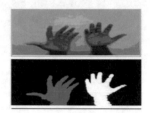

Fig. 15. Result of both hands' segmentation. Figure shows the results of both hand gestures segmentation image.

7   Results

The planned method is evaluated on the following events.

1. Signal to Noise Ratio (SNR): Signal-to-noise ratio is clear as the ratio of the control of a sign (telling info) to the power of related noise (unwelcome signal)

$$SNR = \frac{P_{signal}}{P_{noise}}$$

Where P is normal power. Together sign and commotion power must be estimated at the equivalent or comparable focuses in a framework, and inside a similar framework data transfer capacity.

2. Peak Signal to Noise Ratio (PSNR): PSNR is most effectively characterized through the mean formed mistake (MSE). Given a clamor free m × n monochrome picture I and its loud guess K, MSE is characterized as:

$$MSE = \frac{1}{mn} \sum_{i=0}^{m-1} \sum_{j=0}^{n-1} [I(i,j) - K(i,j)]^2$$

The PSNR (in dB) is defined as:

$$PSNR = 10 \cdot \log_{10}\left(\frac{MAX_I^2}{MSE}\right)$$

Now, MAX_I is the max possible pixel worth of the picture? Once the pixels are signified consuming 8 bits per example, this is 255.

3. Mean Square Absolute error (MSAE): mean total error (MAE) is a measure of change among two continuous variables. The Mean Complete Error is given by:

$$MAE = \frac{\sum_{i=1}^{n} |y_i - x_i|}{n}$$

4. Accuracy: $ACC = \frac{TP+TN}{P+N}$
5. Sensitivity: $TPR = \frac{TP}{TP+FN}$
6. Specificity: $TNR = \frac{TN}{TN+FP}$

Where, TP → True Positive; TN → True Negative; FP → False Positive; FN → False Negative.

8 Conclusions and Future Work

The pre-handling is talked about in hand image segmentation where ROI is getting extricated after the picture edge and picture can be trimmed. The decreased scope picture would create the more procedure quicker than previously. Return for capital invested division is likewise appeared particular gadget MS KINECT, where profundity data is utilized for signal discovery. The hand signal acknowledgment is clarified in light invariant hand gesture recognition which shows light invariant motion acknowledgment. Scarcely any motions were at that point chosen for the framework and their OH was contrasted picture ROI with order the signal. The outcomes are empowering as in two altogether dissimilar light situations, motions were recognized effectively. Signal grouping was finished utilizing Euclidean separation n utilizing ANN.

The ANN execution gave well outcomes and false positives were less. In fingertip detection clarifies the procedure of HGP recognitions utilizing webcam and KINECT. The fingertip recognition is heading invariant in the two situations; client is allowed to demonstrate the turn toward any path. The COP using webcam was detected using sum of area method while using KINECT it was detected after applying inverse transformation on ROI depth image. If user is showing both hand simultaneously, in that case also these methods are effective and could be used to get both hand HGPs. After detecting HGPs, the angles for bent fingers are needed to calculate. In bent finger angles calculation portray it first utilizing a geometrical technique and later utilizing ANN usage. The performance analysis is also given for the both approaches. These angles could be passed to a robotic hand to mimic the human hand gesture. This part considers just one hand activity either right or left, where discussed imagine a scenario where two hands are appeared by user. A new concurrent fingertips method was applied to reduce computational time. The approach described in fingertips detection was modified to minimize the processing time of algorithm. If the same fingertip detection algorithm is applied, it will take 294 ms to compute; whereas the new approach takes 198 ms only . Accuracy of system is around 90–92%.

Future Work: Further examinations in the region of HGR would incorporate shut finger location and gesture invariant strategies. In spite of the fact that KINECT can identify shut finger positions with few obliges however utilizing basic gadgets it is still to accomplish. This proposition is centered around spatial area examination, somebody can likewise explore these matters in recurrence space or model built methodologies. The exhibition on constant implanted framework is as yet a major issue as the cameras will have great goals, more pixels are have to method in space. The handling force recall is expanding with camera goals yet techniques to limit these potentials ought to be examined.

References

Islam, M.M., Siddiqua, S., Afnan, J.: Real time Hand Gesture Recognition using different algorithms based on American Sign Language. In: 2017 IEEE International Conference on Imaging, Vision & Pattern Recognition (icIVPR), pp. 1–6. IEEE (2017)

Haria, A., Subramanian, A.: Hand gesture recognition for human computer interaction. In: 7th International Conference on Advances in Computing &Communication, ICACC 2017, p. 22 (2017)

Grif, S.H., Farcas, C.C.: Mouse cursor control system based on hand gesture. Procedia Technol. **22**, 657–661 (2016)

Thakur, S., Mehra, R., Prakash, B.: Vision based computer mouse control using hand gestures. In: 2015 International Conference on Soft Computing Techniques and Implementations (ICSCTI), pp. 85–89. IEEE (2015)

Freeman, W.T., Roth, M.: Orientation histograms for hand gesture recognition. In: International Workshop on Automatic Face and Gesture Recognition, vol. 12, pp. 296–301 (1995)

Starner, T., Pentland, A.: Real-time American sign language recognition from video using hidden Markov models. In: Motion-Based Recognition, pp. 227–243. Springer, Netherlands (1997)

Bretzner, L., Laptev, I., Lindeberg, T.: Hand gesture recognition using multi-scale colour features, hierarchical models and particle filtering. In: Proceedings of Fifth IEEE International Conference on Automatic Face and Gesture Recognition, pp. 423–428. IEEE (2002)

Dardas, N.H., Georganas, N.D.: Real-time hand gesture detection and recognition using bag-of-features and support vector machine techniques. IEEE Trans. Instrum. Measur. **60**(11), 3592–3607 (2011)

Fritsch, J., Lang, S., Kleinehagenbrock, A., Fink, G.A., Sagerer, G.: Improving adaptive skin color segmentation by incorporating results from face detection. In: 2002 Proceedings of 11th IEEE International Workshop on Robot and Human Interactive Communication, pp. 337–343. IEEE (2002)

Krizhevsky, A., Sutskever, I., Hinton, G.E.: Imagenet classification with deep convolutional neural networks. In: Advances in Neural Information Processing Systems, pp. 1097–1105 (2012)

Banerjee, A., Ghosh, A., Bharadwaj, K., Saikia, H.: Mouse control using a web camera based on colour detection. arXiv preprint arXiv:1403.4722 (2014)

Buehler, P., Everingham, M., Huttenlocher, D.P., Zisserman, A.: Long term arm and hand tracking for continuous sign language TV broadcasts. In: Proceedings of the 19th British Machine Vision Conference, pp. 1105–1114. BMVA Press (2008)

Comprehensive Survey of Algorithms for Sentiment Analysis

V. Seetha Lakshmi$^{(\boxtimes)}$ and B. Subbulakshmi

Department of CSE, Thiagarajar College of Engineering, Madurai, India
seethavenkatesan1997@gmail.com, bscse@tce.edu

Abstract. The growth of the web results with a wider increase in online communications. These online communications include reviews, comments and feedbacks that are posted online by the internet users. It is important to discover and analyse their opinion for a better decision making. These opinionated data are analysed using sentiment analysis. Because of its importance in numerous areas, sentiment analysis was adopted as a subject of increasing research interest in the recent years. It is a technique adopted to extract the useful information and identify the user views either as positive or negative. This paper develops a survey on various approaches used in sentiment analysis and a comparative study has also been made along with the elucidation of recent research trends in the sentiment analysis.

Keywords: Sentiment analysis · Opinions · Reviews · Dictionary · Lexicon · Corpus

1 Introduction

The information from multiple sources is uploaded to web every second. So there occurs an overload of information. Now the way in which the information is seen will make a difference that is, whether it is seen as obstacles or as an opportunity to analyse all the data. If the latter is selected, then it is known as social content analysis, which will aggregate the rough human generated data to get valuable need based findings. The major applications include online brand monitoring, social Customer Relationship Management (CRM), online Commerce, Voice of Customers, Voice of Market, Recommendation Systems, etc. To achieve this, a methodology to automatically associate an opinion to every piece of content is needed and the solution is sentiment analysis, which is an area that analyse people's emotions, reviews, sentiments and opinions towards a lot of categories such as services, products, individuals or a whole organization. It includes polarity detection (i.e., positive or negative) of the given data.

1.1 Sentiment Analysis Levels

Three levels in sentiment analysis are:

© Springer Nature Switzerland AG 2020
D. J. Hemanth et al. (Eds.): ICICI 2019, LNDECT 38, pp. 179–186, 2020.
https://doi.org/10.1007/978-3-030-34080-3_20

1.1.1   Document Level

It is a simple form of classification in which the whole document of the given text is considered as a basic unit of information. In this analysis, the sentiment of a complete document or paragraph is obtained by assuming that the document is having an opinion about a single object. So it is not suitable if document contains opinions about different objects.

1.1.2   Sentence Level

It is a fine- grained analysis of the document in which polarity is calculated for each sentence and every sentence may have different opinions. It has the following tasks:

1.1.2.1. Subjectivity Classification

There are two types of sentences and they are subjective and objective. The objective sentence contains the facts and has no opinion about the entity while subjective sentence has opinions. For example, "I have searched for a digital camera for more than 3 months, and the picture is absolutely amazing with xyz camera". The former sentence is factual and there are no statements that convey the sentiment towards the camera. Since it does not play any role in identifying the polarity of the review it should be removed out.

1.1.2.2. Sentiment Classification

Sentence polarity is classified based on the opinions present in it.

1.1.3   Aspect Level

It is to categorize the sentiment of the aspects (features) in an entity. For example, "The battery works for a long time, but the camera quality is bad". Here there are two aspects (battery, camera) for a single entity phone. Battery aspect has positive sentiment whereas camera aspect has negative sentiment.

1.2   Steps in Sentiment Analysis

Figure 1 illustrates various steps involved in sentiment analysis. Each step is explained below:

Fig. 1. Steps in sentiment analysis

1.2.1   Data Collection

Sentiment analysis requires specialized, large datasets to learn and classify efficiently. Data sources include review sites, blogs and microblogs etc. Datasets can be downloaded from UCI repositories, Kaggle website, etc.

1.2.2   Pre-processing

It includes normalization, tokenization for converting sentences into words, removing unnecessary punctuation & tags, removing stop words which are frequent words such as "is", "the" etc., which does not represent any specific semantic associated with it. Also, stemming reduces the words into a root by removing inflection by dropping unnecessary characters like suffix and lemmatization removes inflection without affecting the meaning of it by determining morphological analysis of each word and utilizing dictionaries.

1.2.3   Feature Extraction

It is the process that encodes categorical and discrete characteristics in a way that is ready to be used by algorithms that map real valued vectors from textual data.

1.2.3.1. Bag of Words
It is one of the simple techniques to represent the text numerically. It makes the list of distinct words in the corpus called vocabulary. Then it is easy to represent every document or sentence as a vector with every word distinguished as 1 and 0 (Presence/Absence) from the vocabulary.

1.2.3.2. Term Frequency-Inverse Document Frequency (TF-IDF)
It computes the count of every words appeared in a text document. Where,

Term Frequency (TF) = (count of a particular term present in a input)/(count of terms present in the input) and

Inverse Document Frequency (IDF) = log (M/n), where, M is the documents count and n is the count of documents a term (t) has occurred in. So, for the rare word IDF is high and for the frequent word IDF is probably low.

The TF-IDF value of a term is calculated as TF * IDF

1.2.1.1. Word Embedding
It is used to represent related words closer in an equivalent system, depending on a relationship from the corpus. Glove and Word2Vec are few examples of word embedding. It is advantageous over Bag Of Words (BOW) since it maintains the context of the words in a document.

1.2.4   Sentiment Classification

This is the most important part in sentiment analysis so it is surveyed deeply in this paper. This phase will take the output of the feature extraction phase and classify the polarity of the sentence which is the final expected outcome.

This paper is arranged in the following way: Sect. 2 discusses the various sentiment analysis approaches, Sect. 3 deals with article summary to compare the discussed approaches and eventually the conclusion and directions for future are given in Sect. 4.

2   Sentiment Analysis Approaches

There are four basic approaches for doing sentiment analysis as depicted in Fig. 2 and they are:

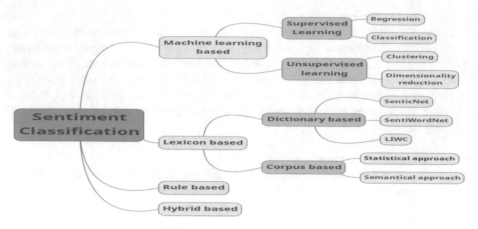

Fig. 2. Approaches of sentiment analysis

2.1   Machine Learning Based Approach

In this technique, two sets say a training and testing set are used. The dataset whose pattern and outputs are known are called as training datasets. In contrast, the datasets whose outputs or behavioural pattern are unknown are termed as test datasets. Different classifiers are trained with training data to build a classification model for sentiment classification and then a test data is given to this model to validate and to get the desired results. The machine learning approach is further categorized as supervised and unsupervised approaches.

2.1.1   Supervised Approach

It allocates labelled data so that a particular function or pattern can be retrieved from that data. It allocates an input object, a vector, at the same time anticipates the most desired output value. Thus, in supervised learning, the input data is known and labelled appropriately. It uses complex computation to obtain accurate and reliable results. Dridi et al. (2018) proposed a fine grained sentiment analysis that learns from feature sets and uses support vector regression for sentiment classification. It is suited only for subjective datasets. Deshmukh and Tripathy (2018) adopts an approach that uses a maximum entropy algorithm and bipartite clustering to extract and classify the sentiments from given domain and predicts the sentiments of the target domain. It uses parts of speech information and word relations as the feature sets. This approach has reduced annotation cost and the n-grams alone are considered.

2.1.2   Unsupervised Approach

In unsupervised learning, inferences are drawn from unlabelled input data. The goal is to determine the unknown patterns. It is frequently used for analysis of exploratory data, the significant characteristics where the input and the output values are not known. It has less computational complexity. Results obtained might be moderate, accurate or reliable. Yang et al. (2018) proposed a sentiment analysis method that employs self-organizing maps to derive the association between the message and keyword clusters. It works better on bipolar messages.

2.2 Lexicon Based Approach

Opinions or sentences are used in this approach. It is the method of defining a negative or positive sentiments to a text that finds the text's opinions with respect to its main subjectivity. To define any favourable details positive word is used, while to define any unfavourable details negative view words are used. It is further divided into two approaches:

2.2.1 Dictionary Based Approach

In this approach a dictionary that contains synonyms and antonyms of a word is used. It uses some seed sentiment words to aid, depending on the synonyms and antonyms arrangement of a dictionary. In this approach, firstly a set of sentiment words that are oriented towards positive or negative is collected manually and it is known as seed. Then, this set is improved by adding new words found by searching in any lexical resource such as WordNet. The process repeatedly adds the words until no new words are found. It requires manual inspection to clean the list at last. It is very time intensive and it has the shortcoming to seek out context and domain specific opinion words. For sample, Teso et al. (2018) suggested a method which provides communication and preference differences of women and men in a linguistic domain. There is also a chance to predict the gender with the help of a frequency key term set when it is not mentioned explicitly. Linguistic Inquiry and Word Count Dictionary was employed here to measure the dimensions of various languages.

2.2.2 Corpus Based Approach

It helps in solving the complications in finding the opinions oriented to context specification. It depends on grammatical patterns that appear in the seed list of opinions to search other opinion words relatively in a corpus. It starts with a fundamental series of opinions and then moving towards different opinion words in a gathering to help searching the opinions on the same text origin. It is again divided into two types, statistical and semantic approaches. The former is used in finding co-occurrence patterns. And the latter assigns relevant sentiment scores to semantically similar words. These words are collected by getting the sentiment words list and iteratively increasing the pioneer seed set with their corresponding meanings and opposites and then determining the polarities for a new input word by the relative count of negative and positive synonyms of that input word. Han et al. (2018) proposed a sentiment lexicon generation method which is a domain specific, overcoming the demerits of dictionary based approach. SentiWordNet 3.0 is the sentiment lexicon used here. This method shows improvement in recognition performance. Precision rate variation occurs since the least amount of reviews leads to poorly generated sentiment lexicon.

2.3 Rule Based Approach

This method uses a collection of rules. On the left hand side, disjunctive normal form representation of the feature set is present and on the right hand side class label is represented. The conditions are usually based on the presence of a term. For rule generation, all the rules are built depending on support and confidence in the training phase. Ali et al. (2017) suggested a fuzzy ontology-based sentiment classification. To monitor the activities of transport and to aid in making a polarity map with city features for travellers, it adopts a Semantic Web Rule Language (SWRL) - a decision making strategy that uses rules and thus requires annotation cost.

2.4 Hybrid Approach

It is the combination of one or more approaches discussed above. This is more advantageous since it will cover the benefits of one or more techniques. So, most of the recent works in sentiment analysis are based on hybrid approaches.

For instance, Wu et al. (2018) suggested a hybrid approach for fine grained sentiment analysis that has adopted set of rules at the chunk level to extract the opinion targets and aspects with high recall. They also exploited high level linguistic features by the use of deep networks. But large memory is required for network training and precision still needs some improvement. Nagarajan et al. (2018) performed a ternary classification by applying particle swarm optimization and genetic algorithm for optimization and decision trees for polarity classification accuracy. Alarifi et al. (2018) proposed a LSTM neural network with a optimization technique for five level polarity classification. Also a greedy algorithm is applied to compute the important features and to reduce the error rates. However, text noise should be minimized for achieving improvement in terms of accuracy. Zainuddin et al. (2018) planned a approach which uses association rule mining based on an apriori algorithm for identifying the frequent aspects, principal component analysis for feature extraction and support vector machine for doing polarity classification.

3   Article Summary

The Table 1 provides a summary of articles discussing the techniques and datasets used in recent literature of sentiment analysis along with their merits and demerits.

Evaluating the classification model used for sentiment analysis is an essential part. Accuracy, F1 Score, Recall and Precision are the most commonly used metrics in sentiment analysis.

Table 1. Comparison of various techniques for sentiment analysis

Ref. No.	Technique used	Merits	Demerits	Dataset
[1]	Hybrid approach: Combination of rules and unsupervised learning.	• There is no dependency on labelled data and the manual annotation rates are reduced. • Implicit sentiments can also be extracted. • Provides new domain portability.	• When the threshold selected is smaller the optimal performance cannot be obtained. • It takes more time and space to parse the sentence. • Still improvement is required in terms of precision.	Restaurant and laptop reviews dataset in SemEval 2014 and Restaurant reviews in SemEval 2015 and 2016 are used.
[2]	Hybrid approach: It is the combination of supervised learning classifiers and optimization techniques.	• It achieves better performance and accuracy than other existing classifiers.	• It is domain dependent.	Tweets collected using twitter streaming API.
[3]	Hybrid approach: It combines both optimization algorithms and supervised learning.	• Ideally important sets can be discovered. • Obtains best classification of words.	• Sometimes it gets stuck with local minimum. • Improvement in accuracy is required.	Dataset collected from Amazon Web 2.0 sites.
[4]	Hybrid approach: It is based on association rule mining, Support vector machine and principal component analysis.	• This is a domain independent approach.	• It deals with the tweets that are subjective.	Hate Crime Twitter Sentiment (HCTS) dataset, Stanford Twitter Sentiment (STS) dataset, Sanders Twitter Corpus (STC) dataset.
[5]	ML-Supervised learning: Support vector regression is applied.	• Achieves good accuracy through replacement and augmentation.	• Better performance is achieved only with spans.	SemEval 2014 task 5 dataset.
[6]	ML-Supervised learning: It uses Maximum entropy algorithm and bipartite graph clustering.	• Achieves Domain adaptability. • Annotation cost is reduced.	• Considers only the n-grams.	Amazon product review dataset.
[7]	Lexicon – Corpus based approach: SentiWordNet 3.0 lexicon is applied.	• Negative reviews are recognized with improved performance.	• Variation in precision rate.	IMDB movie review dataset and the Amazon Product Review Dataset.
[8]	Lexicon – Dictionary based approach: Linguistic Inquiry and Word Count Dictionary is used.	• Applied for gender prediction in lexicon domain.	• Domain dependent.	Dataset is collected from Ciao UK, a well-known eWOM website.
[9]	ML- Unsupervised learning: Self-organizing maps are used.	• There is no dependency on labelled data.	• Accuracy should be improved further.	Corpus of social messages collected from Twitter between January 2012 and March 2012.
[10]	Rule based approach: Semantic web rule language based decision making is adopted.	• Used for preparing city feature polarity.	• It achieves better performance than the existing ontologies.	Using REST and Streaming APIs of e-commerce sites to retrieve reviews and tweets from social network sites.

4 Conclusion and Future Work

As sentiment analysis is used in many applications it is considered as a broad area of research. Contribution on sentiment analysis increases year by year. So this paper covers a brief introduction about sentiment analysis, different levels of sentiment analysis and the basic steps involved in analysing the sentiments of a data. The benefit of this survey is substantial since it provides a refined categorization of current articles depending on the approaches used. This could help the beginners of sentiment analysis to have an overview of the overall area and also help the researchers to choose the suitable technique and benchmark the dataset for a selected application. In future, aspect level sentiment analysis will be considered due to its importance in various applications and then the challenges encountered in previous researches like negation handling, word sense disambiguation, precision and accuracy improvement should be addressed by using deep learning algorithms.

References

Wu, C., Wu, F., Wu, S., Yuan, Z., Huang, Y.: A hybrid unsupervised method for aspect term and opinion target extraction. Knowl.-Based Syst. **148**, 66–73 (2018)

Nagarajan, S.M., Gandhi, U.D.: Classifying streaming of Twitter data based on sentiment analysis using hybridization. Neural Comput. Appl. **31**, 1425–1433 (2018)

Alarifi, A., Tolba, A., Al-Makhadmeh, Z., Said, W.: A big data approach to sentiment analysis using greedy feature selection with cat swarm optimization-based long short-term memory neural networks. J. Supercomputing 1–16 (2018)

Zainuddin, N., Selamat, A., Ibrahim, R.: Hybrid sentiment classification on Twitter aspect-based sentiment analysis. Appl. Intell. **48**, 1218–1232 (2018)

Dridi, A., Atzeni, M., Recupero, D.R.: FineNews: fine-grained semantic sentiment analysis on financial microblogs and news. Int. J. Mach. Learn. Cybern. 1–9 (2018)

Deshmukh, J.S., Tripathy, A.K.: Entropy based classifier for cross-domain opinion mining. Appl. Comput. Inform. **14**(1), 55–64 (2018)

Han, H., Zhang, J., Yang, J., Shen, Y., Zhang, Y.: Generate domain-specific sentiment lexicon for review sentiment analysis. Multimedia Tools Appl. **77**(16), 21265–21280 (2018)

Teso, E., Olmedilla, M., Martínez-Torres, M.R., Toral, S.L.: Application of text mining techniques to the analysis of discourse in eWOM communications from a gender perspective. Technol. Forecasting Soc. Change **129**, 131–142 (2018)

Yang, H.C., Lee, C.H., Wu, C.Y.: Sentiment discovery of social messages using self-organizing maps. Cogn. Comput. **10**(6), 1152–1166 (2018)

Ali, F., Kwak, D., Khan, P., Islam, S.R., Kim, K.H., Kwak, K.S.: Fuzzy ontology-based sentiment analysis of transportation and city feature reviews for safe traveling. Transp. Res. Part C: Emerg. Technol. **77**, 33–48 (2017)

Schouten, K., Van Der Weijde, O., Frasincar, F., Dekker, R.: Supervised and unsupervised aspect category detection for sentiment analysis with co-occurrence data. IEEE Trans. Cybern. **48**(4), 1263–1275 (2017)

Zhao, W., Guan, Z., Chen, L., He, X., Cai, D., Wang, B., Wang, Q.: Weakly-supervised deep embedding for product review sentiment analysis. IEEE Trans. Knowl. Data Eng. **30**(1), 185–197 (2017)

Firefly Based Word Spotting Technique for Searching Keywords from Cursive Document Images

A. Sakila$^{(\boxtimes)}$ and S. Vijayarani

Department of Computer Science, Bharathiar University, Coimbatore, India
sakivani27@gmail.com, vijimohan_2000@yahoo.com

Abstract. In the fast pace development of digitized technologies, document images have become more fashionable for an information management system present in libraries, organization and educational institutions. Searching information from the document image is very difficult to perform as it compared with digital text. Optical Character Recognition (OCR) is employed to detect the characters and converts the images into their text format. OCR system is not properly converts the various fonts, styles, size, symbols, dark background and poor quality of the document images, however it's not an efficient method. For this reason, there is a necessity for a searching strategy to find the user specified keywords from document images. Word spotting is an alternative method, whereas keyword is identified without changing the document images. The primary objective of this research work is to search the keywords from printed cursive English document images using word spotting techniques. In this research work, the Firefly based word spotting technique is proposed to search the keyword based on query given by the user. To estimate the efficiency the Firefly technique is compared with existing Enhanced Dynamic Time Warping (EDTW) technique. From the experimental analysis, the proposed Firefly based word spotting technique has produced high accuracy rate and less execution time compared with existing EDTW technique.

Keywords: Document image · Information retrieval · Optical Character Recognition · Word spotting

1 Introduction

From Past decades the libraries and organizations are maintaining a massive collection of hardcopies such as text books, dissertation, newspapers, magazines, novels, bank statements, invoices, research articles, questionnaire, resume, brochure, letters, advertisement, static data and so on. The difficulties of this system are, hard copies of documents occupies more space and it takes long time for searching particular document. Therefore, at present, most of the libraries and organizations are becoming digitized [1]. Digitization involves the hardcopies of the documents are converted into digital images by using mobile phones, scanners and cameras. These digital images are compactly saved in computer with different image formats. Document images are utilized in several real world applications and it is an emerging research area in the field

D. J. Hemanth et al. (Eds.): ICICI 2019, LNDECT 38, pp. 187–196, 2020.
https://doi.org/10.1007/978-3-030-34080-3_21

of Document Image Analysis (DIA). DIA is used to identify the text and graphical components from the document images. Extracting information from document images becomes a tedious task to perform as compared with digital texts [2]. Optical Character Recognition (OCR) technique is used to identify the characters from image and then transforms into editable text document format [3]. Nowadays numbers of the commercial, free OCR tools and android OCR applications are available in online and offline for converting images into editable text format. OCR is not achieving accurate conversion results, because of different font styles, size, symbols and degraded document images. Therefore, the development of an alternate technique is mandatory for extracting information from document images. An inventive technique for searching keywords from the document images and extracts the similar words without transforming these images using word spotting or word matching techniques [4]. This technique receives the keyword/query from user and the keyword is translated into template image and this process is known as rendering. Distance based matching algorithms are used for searching keyword from document images. Some of the existing word spotting techniques are Euclidian Distance, Dynamic Time Warping (DTW) Cross Correlation (CC) and Normalized Cross Correlation (NCC).

1.1   Contribution

The proposed research work is used to search the keyword from cursive document images. The main contribution of this research work is as follows,

- This research work proposes an innovative approach to word spotting, by extracting information from cursive document images based on the user's query.
- The proposed approach, used 35 × 35 track and sector matrix for character feature extraction. The firefly technique is used for finding the keyword in between the words from cursive document images. Figure 1 depicted the normal font types verses cursive font types.

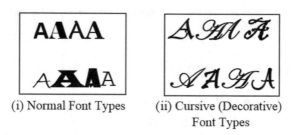

(i) Normal Font Types   (ii) Cursive (Decorative)
                                       Font Types

Fig. 1. Normal font types verses cursive font types

The remaining portion of this paper is organized as follows. Section 2 describes the related works for word spotting techniques, Sect. 3 provides the methodology used for the proposed word spotting technique. Section 4 gives the experimental results and discussions and the conclusion is given in Sect. 5.

2 Related Works

Sakila et al. [5], proposed a word spotting technique for searching keywords from printed English document images. The authors found the template values for each character in the document images by using 15×15 track and sector matrix. Based on the template values, characters are identified using existing templates. If both the values are matched, characters were identified and it was stored in the indexed files. The track and sector matrix is not properly recognize the cursive English document images. Distance based word spotting technique named as Enhanced Dynamic Time Warping (EDTW) was proposed for word spotting. The EDTW supports limited font types such as Arial, book antique, calibre, high tower text and times new roman. The performance of the EDTW is compared with existing Normalized Cross Correlation (NCC) and Dynamic Time Warping (DTW) word spotting techniques. Based on their results EDTW supports different font size, styles and also keywords which are found in between the words from document images, and it is known as partial/inexact keyword spotting. For example 'test' keyword is also find the relevant words like 'test', 'tests', 'testing', 'tested' and so on from the document images.

Harkamaljit et al. [6], analysed a Firefly and neural network algorithm for Content Based Image Retrieval (CBIR) system. The features are extracted from images used radon transform method. Extracted features are reduced by firefly algorithm and trained by the neural network. These trained images are saved in the image database. The proposed CBIR system work by relevant image is retrieved based on the user query image from the image database. Their results are acceptable based on precision and recall measures for image retrieval and this technique is applied on the different image datasets.

Jawahar et al. [7], proposed a novel technique for retrieving similar documents from a large collection of printed document images. Three different languages such as English, Hindi and Amharic document images are used for analysis. In this technique without performing any conversion process for searching the particular word from document images. The similar images are stored in indexed documents based on a query.

3 Methodology

In the proposed method, cursive character features are extracted by using 35×35 track and sector matrix. The ASCII values of the characters are stored in indexed file. Subsequently, word spotting technique is applied for searching keywords from cursive document images. The proposed system has four different steps. They are Preprocessing, Segmentation, Feature Extraction and Word Spotting.

3.1 Textual Query Processing

Whenever, a user gives a textual query for finding the keywords from cursive document images, the query word is initially rendered. Text rendering translates the user's textual query into equivalent query image, it is also known as template image.

3.2   Pre-processing

Pre-processing plays significant role in word spotting system, which improves the recognition rate for information retrieval system. In the digitization process, paper documents are transformed into digital images. In this process the document images may suffer from some level of degradations [5]. Historical document images have a variety of damages like aged effects, sun spots, marks, fungus and stains. For this reason, there is a necessity for pre-processing the input document images [8]. In this research work, binarization technique is used for image enhancement in pre-processing stage. Binarization is used to transforms the document images into binary format, it means foreground (text information) as black color and the background as white color. In this research work has analyzed the performance of existing binarization techniques, namely Otsu, Niblack, Savola, Nick, and Hybrid. The performance has been validated by using PSNR, NRM and execution time. From this analysis it is found that Hybrid binarization technique gives better results than traditional binarization techniques.

3.3   Segmentation

Segmentation is an essential step for information retrieval from document images. It is used to segment the paragraphs, lines, words and characters from document images [9]. Line, Word and Character segmentation techniques are used in this research work for searching keyword from document images. Every text line in the document images are not properly aligned in a horizontal manner. Line segmentation is used in this work to properly align the text lines from document images for correct the skewed text lines. Each line in the original document images may suffer from some level of skew [10]. Hence the lines are straightly aligned using line segmentation [11]. Subsequently split into lines from document images, words are segmented from lines using word segmentation. Words are separated by using character segmentation.

3.4   Feature Extraction

Feature extraction is used to extract the cursive character features from document images. In this research work 35×35 track and sector matrix is used for feature extraction. This technique is used to find the ASCII template values for binarized segmented characters. Then the character is divided into eight tracks (lines) and five sectors (circle), then to find the intersection values for each track and sectors. Current ASCII template values of each character from document images is identified by using existing character template values. If both the values are matched, the characters were identified and it was stored in the indexed files. In the centre of the matrix, there will be 1's in each intersection of sector and track [5]. This indexed file is the input for proposed Firefly, which is used to find the keyword in the cursive document image in different font case (upper and lower case), font style (bold, italic, underline), the font size (10–24) and a variety of fonts (Vivaldi, Freestyle Script, Brush Script MT, Edwardian Script ITC, Kunstler Script and Rage Italic). Figure 2 represents the 35×35 track and sector representation.

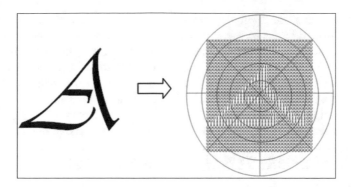

Fig. 2. Tracks and sectors matrix

3.5   Word Spotting Technique

Finding keywords from document images, word spotting technique is used. It identifies the keywords from document images that are more similar to that particular word by using indexed files. In this section the proposed Firefly optimization technique is compared with existing Enhanced Dynamic Time Warping (EDTW) for word spotting. The performance of the proposed work is compared with the existing EDTW technique.

3.5.1   Enhanced Dynamic Time Warping (EDTW)

The Enhanced Dynamic Time Warping (EDTW) is performed based on Dynamic Time Warping algorithm. This technique has calculated the distance between the users specified keywords and find the distance between relevant words from document images. The performance of EDTW based on a DTW word spotting technique. It calculates the optimal alignment ∞ between two points in the relevant keywords from document image [12]. The optimal alignment is utilized to minimize the cumulative distance between aligned samples [13]. The EDTW is find the keyword in partially, which means it finds all similar words from the document images, by using the coefficient function [5].

3.5.2   Firefly Optimization Technique

Firefly optimization is the metaheuristic algorithm which is inspired by swarm intelligence algorithm to resolve the optimization issues especially NP hard problems. Firefly is an insect that yields short and rhythmic flashes, which helps to attract the other fireflies [14]. In this research work, the firefly algorithm is used to find the keywords from cursive document images. This algorithm is inspired by the aspects of the real fireflies. The two important rules of basic firefly algorithm as follows,

1. Fireflies are unisex that, despite their sex, they will attract other fireflies.
2. The desirability is proportionate to their brightness, which depended with the light intensity among the fireflies. The firefly which has the less bright will fly on the way to brighter firefly [14].

In word spotting, the firefly algorithms are used to search the required keyword, which is performed based on the user's query. First, the initial populations of firefly algorithm is generated and calculate the initial light density (similar words). Based on the initial parameters, distance between the particular keyword is calculated by using the coefficient function. Attractiveness is given query word correspondingly to the light intensity seen by another similar words. The updation of light intensity will be taken place and based on the intensity it will move to the next keyword. This will represent the keywords selected from the particular keywords from cursive document images.

Algorithm 1: Proposed Firefly based Word Spotting Technique

Input: Cursive Document Images
Output:Spot the keywords from document images.
Procedure:
Step1: Generate the initial population of firefly x_i where i = 1, 2...n. (n is defined
      as the number of fireflies such as keywords)
Step 2: Estimate the initial light intensity, I is the total intensity of relevant words
Step 3: Define the initial coefficient value= 1, the randomized parameter α= 0.2 and
      the initial attractiveness β=0.1
Step 4: for i= 1 to N and for j= 1 to N
Step 5: if $(x_i < x_j)$
Step 6: Define the keyword is W, s is an index of the keyword and k is initializing
      the index.
Step 7: Calculate the distance between i, j by using $(x_i, y_j) = C = \sum_{s=1}^{k} W_s$
Step 8: Estimate the attractiveness by using $p = Cd(x_i, y_j)$
Step 9: Move the keywords i to j in all D dimensions
Step 10: Update the light intensity
Step 11: Evaluate the new solution
Step 12: End for j; End for i

---

4 Experimental Results

In this section, a number of experiments were implemented to illustrate the efficiency of the proposed Firefly Optimization technique. To perform intensive experimentation, synthetic dataset was used, it consists of scanned and captured document images with different image formats.

4.1 Word Spotting Technique

The proposed Firefly optimization technique is compared with existing Enhanced Dynamic Time Warping (EDTW) technique. Experimental results proved that the Firefly optimization technique found the keywords properly from a cursive document image than EDTW technique.

4.1.1   Accuracy Measure

Standard precision, recall and F-Score measures have been used to calculate the performance of the proposed and the existing techniques. Table 1 shows the performance Measures for EDTW and Firefly based word spotting technique. Figure 3 displays the performance measures for different font types. From the performance analysis we found that the performance of Firefly optimization is better than existing EDTW.

Table 1. Performance measures for word spotting techniques

S. no	Font name	Enhanced Dynamic Time Warping (EDTW)			Firefly optimization		
		Rec.	Pre.	FSc.	Rec.	Pre.	FSc.
1	Freestyle Script	39.87	43.73	52.21	72.51	77.33	86.42
2	Brush Script MT	49.25	54.31	63.38	73.52	78.21	87.53
3	Edwardian Script ITC	32.34	36.11	47.46	69.93	73.93	82.49
4	Kunstler Script	47.04	51.63	62.45	72.34	76.54	85.38
5	Rage Italic	52.34	57.24	65.38	74.82	80.18	89.92

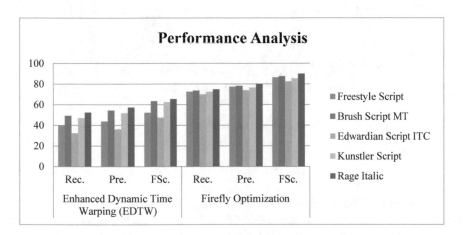

Fig. 3. Performance analysis

4.1.2   Execution Time

Table 2 gives the execution time required for existing and proposed techniques. From Fig. 4, it is found that the Firefly optimization technique has required minimum execution time compared with existing EDTW technique.

Table 2. Execution time (milliseconds)

S. no	Font name	Enhanced Dynamic Time Warping (EDTW) (milliseconds)	Firefly optimization (milliseconds)
1	Freestyle Script	23478	12763
2	Brush Script MT	31097	18673
3	Edwardian Script ITC	19300	9876
4	Kunstler Script	27906	15861
5	Rage Italic	33850	21973

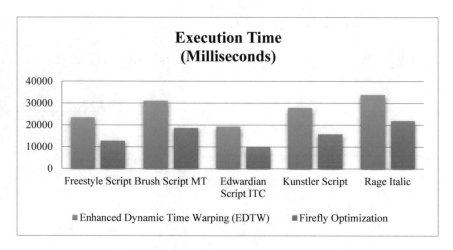

Fig. 4. Execution time

4.1.3   Results of Firefly Based Word Spotting Technique

Firefly optimization results are given from Figs. 5 and 6. Figure 5 represents the result of "shake" keyword and it also finds the word which contains the given keyword, i.e. Shakespeare's and Shakespeare. In the keyword "eco" finds the relevant words such as economics, economies and economists from the cursive fonts in the document images which is shown in Fig. 6.

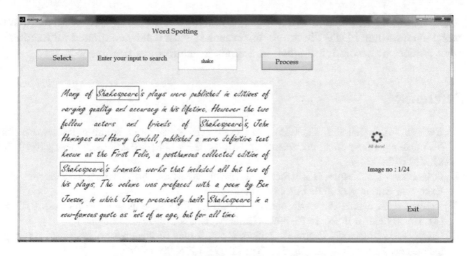

Fig. 5. Search query "shake"

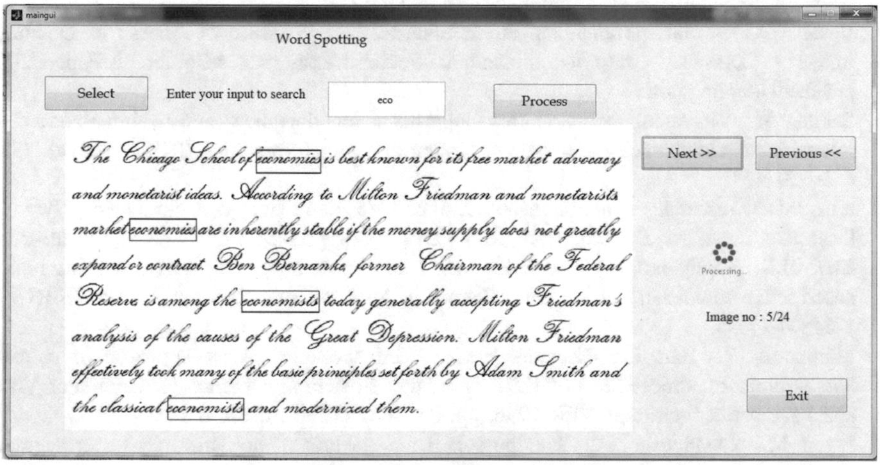

Fig. 6. Search query "eco"

5   Conclusion and Future Work

This research work has proposed a Firefly based word spotting technique for searching keyword from cursive (decorative) document images and it provides optimal solution. It is compared with existing Enhanced Dynamic time warping (EDTW) word spotting technique. The execution time and standard performance measures are calculated for existing and proposed techniques. From the experimental results, it is found that the proposed Firefly optimization technique gives higher accuracy rate and minimum execution time than EDTW. In future, this research work will be continued for handling more number of decorative English font types in the form of document images.

References

1. Kesidis, A.L., Galiotou, E., Gatos, B., Pratikakis, I.: A word spotting framework for historical machine-printed documents. IJDAR **14**, 131–144 (2010). https://doi.org/10.1007/s10032-010-0134-4
2. O'Gorman, L., Kasturi, R.: Document Image Analysis. In: IEEE Computer Society Executive Briefings (2009). ISBN 0-8186-7802-X
3. Vijayarani, S., Sakila, A.: Performance comparison of OCR tools. Int. J. UbiComp (IJU) **6**(3), 19–30 (2015). ISSN: 0975–8992 (Online); (Online); 0976–2213 (Print)
4. Giotis, A.P., Sfikas, G., Gatos, B., Nikou, C.: A survey of document image word spotting techniques. Pattern Recogn. **68**, 310–332 (2017). ISSN 0031-3203
5. Sakila, A., Vijayarani, S.: Content based text information search and retrieval in document images for digital library. J. Digit. Inf. Manag. **16**(3) (2018)
6. Singh, H., Kaur, H.: Content based Image Retrieval using Firefly algorithm and Neural Network. Int. J. Adv. Res. Comput. Sci. **8**(1), (2017)
7. Jawahar, C.V., Balasubramanian, A., Meshesha, M.: Word-Level Access to Document Image Datasets. http://cdn.iiit.ac.in/cdn/cvit.iiit.ac.in/images/ConferencePapers/2004/jawahar04docimg.pdf
8. Sakila, A., Vijayarani, S.: A hybrid approach for document image binarization. In: International Conference on Inventive Computing and Informatics (ICICI) (2017). ISBN 645-650
9. https://www.researchgate.net/publication/261253745_Extraction_of_Line_Word_Character_Segments_Directly_from_Run_Length_Compressed_Printed_Text_Documents
10. Chai, H.Y., Supriyanto, E., Wee, L.K.: MRI brain tumor image segmentation using region-based active contour model. In: Latest Trends in Applied Computational Science. ISBN 978-1-61804-171-5
11. Manmatha, R., Han, C., Riseman, E.M.: Word spotting: a new approach to indexing handwriting. In: Proceedings of IEEE Computer Society Conference on Computer Vision and Pattern Recognition, CVPR 1996, pp. 631–637 (1996). ISBN 0-8186-7259-5
12. Javed, M., Nagabhusha, P., Chaudhuri, B.B.: Extraction of line word character segments directly from run length compressed printed text documents
13. Rath, T.M., Manmath, R.: Word Image Matching Using Dynamic Time Warping, January 2002. http://ciir-publications.cs.umass.edu/pdf/MM-38.pdf
14. Ali, N., Othman, M.A., Husain, M.N., Misran, M.H.: A review of firefly algorithm. ARPN J. Eng. Appl. Sci. **9**(10) (2014). ISSN 1819-6608

A Supervised Machine Learning Approach to Fake News Identification

Anisha Datta and Shukrity Si[✉]

Department of Computer Science and Engineering,
Jalpaiguri Government Engineering College, Jalpaiguri, India
dattaanishadatta@gmail.com, sukriti.si98@gmail.com

Abstract. In the era of digitization, we no longer need to struggle to connect to the outer world. There are so many sources that provide us with news reports of our surroundings, national and international activities. But the problem is there are many sources which are fooling people by spreading fake news about everything. This can be fatal sometimes as it encourages human rage. So to prevent this, the data scientists are eager to detect this kind of sources automatically. We have proposed a new model for this supervised data to predict if the provided news is fake or not. We have used some machine learning algorithms like- Gradient Boosting, Random Forest, Extra tree, XGBoost etc. and among these, GBM gives best result as accuracy of 95%. Then a majority voting classifier is made with these algorithm which gives accuracy of 94.15%.

Keywords: Fake news detection · Supervised machine learning · Confusion matrix · Classification · ROC-AUC curve · Majority voting

1 Introduction

Nowadays people rely on social media for everyday news more than the actual sources. There are many good reasons behind this like-(1) The social media (facebook, twitter, instagram) attracts us more with so many features around, so we can easily get the news from many pages and connected people as well. (2) It helps to share their opinion with so many people. And if we search on Google, there are so many websites we can find to get the news. People tend to limit their search to some of these and not worry about the accuracy of the news. This often leads to misinformation spreading which causes many undesired situations in the society.

According to a Times of India's online article (May 25,2019), About 68% of Indian smartphone users rely on social media to get the news. As their survey says, about 57% of them are concerned about the accuracy of the news and about 35% were reported to have been exposed to fake news that they were sure of. Likewise there are many other countries also who are facing this.

© Springer Nature Switzerland AG 2020
D. J. Hemanth et al. (Eds.): ICICI 2019, LNDECT 38, pp. 197–204, 2020.
https://doi.org/10.1007/978-3-030-34080-3_22

So this is a bit of concern and needs to be prevented. Researches have been going on to make a system which does this automatically. A dataset with 20,800 news-samples related to american politics is used in the work and an ensemble machine learning approach is used for classification here.

2   Related Work

In this present time, news can be spread from any news channel source or newspaper and as well as from social media. And most of the time, the news from social media has no reliable source, so all the news has to be checked to separate the real news from fake news. So, many researches are going on this topic. We studied some papers and will discuss their approaches in this section.

In [1], the model is based on knowledge, style, propagation and credibility. The different types of attribute used in the paper are - Quantity(character count, word count, noun count, verb count, paragraph count etc.), Complexity(average no of clauses per sentence, words per sentence, punctuations per sentence and characters per word), Uncertainty(percentage of modal verbs, certainty terms etc.), Subjectivity(percentage of subjective verbs, report verbs etc.), Non-immediacy(percentage of passive voice, rhetorical questions, number of quotations etc.), Sentiment(percentage of positive words, negative words, the dynamics of emotional states etc.), Diversity(lexical diversity, content word diversity, redundancy), Informality(typographical error ratio), Specificity(temporal ratio, spatial ratio, sensory ratio etc.) and Readability. In [2], Tf-Idf of bi-grams fed into stochastic gradient descent model is used with an accuracy 77.2%. In [3], TF-IDF is used as feature extraction techniques and Linear SVM is used as classifiers with an accuracy of 92%. Knowledge based model and style based models are used in [4] and news context features, social context features are used as features. In [5], Linguistic cue method and Network analysis methods are used along with SVM and Naive Bayes classifiers. In [6], Social Article Fusion (SAF) model is proposed, the model is based on news representation learning and social engagement learning which is evaluated by RNN and LSTM. In [7], Logistic Regression, Naive Bayes and Random Forest are used.

In the paper, GBM, Random Forest, Extra Tree and XGBoost algorithms are used as classifier and TF-IDF is used as feature extraction technique. Then a majority voting classifier is made for the final classification.

3   Preprocessing and Features Analysis

Data preprocessing is a very crucial step before we can train on them with any algorithm. Errors arise if we do not preprocess our data accurately. There can be missing values, inappropriate parts, or many byte codes that can not be decoded easily. So all of these have to be removed before we start training.

Missing values: We got some missing values in our dataset. They were like no character present in that space or written as 'NaN' value. We removed them in python and manually also.

Inappropriate parts: There were some inappropriate parts in the samples like-many commas together, or question mark, exclamation mark etc which are of no use in this classification task. So we manually removed them.

After making the data clean and appropriate for training, we then go for feature analysis part. This dataset contains very few attributes like post id, author name, title of the news, the main content and labels. Among them, we do not count post id as a feature. Author name is not a helpful feature here also. We separated the main content column into a list as this is the perfect feature attribute we could get. This is used with Tf-idf vectorizer.

Tf-idf vectorizer: Term frequency inverse document frequency or Tf-idf vectorizer is used to vectorize words into numeric values as a computer do not understand words directly. We have used this to convert our sentences into a numeric feature matrix with the words as features. This matrix is then passed along with the classifiers.

4 Methodology and Results

We have split the dataset into 80:20 ratio for training and testing respectively. The training part is fitted into the classifiers and then prediction is made on the testing part. We have trained our data with many machine learning algorithms but four of them give better results. They are GBM, ExtraTree, XGB and Random forest.

4.1 Gradient Boosting

Gradient Boosting is an example of greedy algorithm and it is a powerful model for prediction. It can be stated as 'Hypothesis Boosting Problem'. This algorithm consists of loss function, weak learner and additive model. This algorithm works on the belief that a weak learner can be a better learner with the help of several regularization methods.

As base learner, Decision Trees of fixed size are used with Gradient Boosting algorithm. It is a ensemble learning process in which sequential boosting algorithm is implemented. In the most essential respect, expectation of the loss function is reduced in GBM (Gradient Boosting Machine). To reach the goal, the residual calculated from the initial model is calculated and then the weak or base learner is fitted to the residual by the gradient descent algorithm. Thus the model is updated by adding the weighted base learner to the previous model. Finally the target model is obtained by conducting the previous steps with several iterations.

Now let assume that the no. of leaves for each tree is J, the space of the mth tree can be divided into J disjoint subspaces like R_{1m}, R_{2m}, ..., R_{jm} and the predicted value of subspace R_{jm} is the constant b_{jm}. The regression tree can be expressed as-

$$g_m(x_i) = \sum_{j=1}^{J} b_{jm} I(x_i \epsilon R_{Jm})$$

$$I(x_i \epsilon R_{Jm}) = 1, if \, x_i \epsilon R_{Jm}$$
$$= 0, otherwise$$

To minimize the loss function, we use the steepest descent method. We take the approximate solution as-

$$F(x_i) = \sum_{m=0}^{M} f_m(x_i), \text{ M denotes the index of the tree}$$

Here we can write,

$f_m(x_i) = -\rho_m g_m(x_i)$

Where gradient $g_m(x_i)$ is-

$g_m(x_i) = [\dfrac{\partial L(y_i, f(x_i))}{\partial f(x_i)}]$ with $f(x_i) = f_{m-1}(x_i)$

And the multiplier ρ_m is-

$\rho_m = argmin_p \sum_{i=1} nL(y_i, f_{m-1}(x_i) + \rho_m g_m(x_i))$

The updated model is expressed as:

$F_m(x_i) = F_{m-1}(x_i) + \rho_m g_m(x_i)$

—- this is the most important step (i.e. iteration) of this algorithm.

4.2   Random Forest and Extra Tree Classifiers

Random Forest Classifier consists of individual decision trees that operate as an ensemble. In normal trees, every node is split focused on the best split among all variables where in Random Forest, every node is split focused on the best split among a subset of predictors indiscriminately chosen at that node. It actually adds an additional layer of randomness to bagging (Liaw and Wiener [8]). At first, a number of bootstraps have to be drawn from the original data. Then randomly samples of the predictor have to be created and the best split among the variables are chosen (Bagging can be thought of as the special case of random forests obtained). Then new data is predicted by aggregating the prediction of the trees. For this randomness, bias is increased but as it takes the average, the variance is decreased balancing the overall output.

Extra Tree is nothing but Extreme Randomized Trees classifier. Instead of choosing proper thresholds from a random subset of feature matrix like in Random Forest, it uses the entire feature matrix to choose random but proper thresholds from each feature point and the best one is chosen for splitting point. Thus bias is increased a bit but the variance is decreased also than the former to balance the overall output.

4.3   XGBoost

XGBoost is an advanced implementation of Gradient Boosted Decision. It stands for Extreme Gradient Boosting. It was introduced in 2016, so a new and comparatively better model. It is a fast learning method and some features of this algorithm are regularization, handling sparse data, weighted quantile sketch, block structure for parallel learning, cache awareness, out-of-core computing etc.

Unique features of XGBoost

• Regularization: A special feature of XGBoost is to penalize complex models by both L1 and L2 regularization. Regularization is used to prevent overfit of data in the model.
• Handling sparse data: Missing values or data processing steps like one-hot encoding make data sparse. XGBoost provides a sparsity-aware split finding algorithm to tackle many types of sparsity patterns in the data.
• Weighted quantile sketch: Most tree based algorithms search the split points when the data points are of same weights (using quantile sketch algorithm). However, they are not equipped to tackle weighted data. XGBoost has a distributed weighted quantile sketch algorithm to effectively tackle weighted data.

We have used four base-classifier here, they are Gradient Boosting classifier (GBM), Extra Tree Classifier, Random Forest and XGBoost Classifier. With these classifiers a Majority Voting Classifier is created which is our final model for classification. It takes each output prediction from the classifiers as a vote and calculates the majority of these to predict the final output.

The accuracy of the base classifiers are shown in Table 1.

Table 1. Accuracy of the base classifiers

Models	GBM	Extra Tree	Random Forest	XGBoost
Accuracy	95%	93%	94%	94%

In our study, we have observed accuracy, precision, recall and F1 score of the used model and made a comparison between these models based on the values. These values we got by implementing the classifiers - GBM, Extra Tree, Random Forest and XGBoost. The details are shown in the figures of confusion matrix.

The confusion matrix is a way to summarize the performance of a classifier for classification tasks. The square matrix consists of columns and rows that lists the number of instances relative or absolute "True class" vs. "Predicted class" ratios.

For the binary classification, if the two outputs are positive and negative then its confusion matrix will be as given in Figs. 1, 2, 3 and 4.

ROC means Receiver Operating Characteristics and AUC means Area under the Curve. ROC is a probability curve between False positive rate (FPR) and True Positive rate (TPR) and AUC represents degree of measure of separability. AUC defines how much a model is capable of distinguishing between the classes, high AUC means the model is good.

We have made a ROC-AUC curve for all of these 4 classifiers. This is in Fig. 5.

Fig. 1. Confusion matrix of extra tree classifier

Fig. 2. Confusion matrix of XGBoost

Fig. 3. Confusion matrix of Random Forest

Table 2. Confusion matrix of binary classification

Predicted/True	Positive	Negative
Positive	TP	FP
Negative	FN	TN

With applying Majority voting on these four, we have got an accuracy of 94.15% and the comparisons with other models are shown in Table 2.

Table 3. Comparison with other works

Model	Shlok Gilda	Hadeer Ahmed	Our Work
	Gradient Descent Model	Linear SVM	Majority voting
Accuracy	77.2%	92%	94.15%

Fig. 4. Confusion matrix of GBM

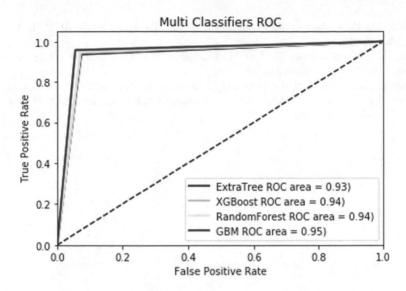

Fig. 5. ROC-AUC curve for all classifiers

Our model performs better than these two models shown in the comparison Table 3, but the work is not tested in the news articles covered all news topics, so the model has to be tested in large dataset.

5   Conclusion

Detecting fake news has become an important work to be done nowadays. We have used a labelled dataset in the work related to american politics. If we use the news articles of all covered topics as dataset and background knowledge and common sense knowledge like ConceptNet for training the data, the results might improve. Ensembling methods like stacking or blending can also be implemented in the work to get better results or deep learning methods like LSTM or CNN can also be used. These can be done in the future.

References

1. Zhou, X., Zafarani, R.: Fake news: a survey of research, detection methods, and opportunities. arXiv:1812.00315v1 [cs.CL], 2 December 2018
2. Gilda, S.: Evaluating machine learning algorithms for fake news detection. In: 2017 IEEE 15th Student Conference on Research and Development (SCOReD) (2017)
3. Ahmed, H., Traore, I., Saad, S.: Detection of online fake news using N-gram analysis and machine learning techniques. In: Conference: International Conference on Intelligent, Secure, and Dependable Systems in Distributed and Cloud Environments, October 2017
4. Shu, K., Sliva, A., Wang, S., Tang, J., Liu, H.: Fake News Detection on Social Media: A Data Mining Perspective

5. Stahl, K.: Fake news detection in social media. Received 20 April 2018; Accepted 15 May 2018
6. Shu, K., Mahudeswaran, D., Liu, H.: FakeNewsTracker: a tool for fake news collection, detection, and visualization
7. Aldwairi, M., Alwahedi, A.: Detecting fake news in social media networks
8. Liaw, A., Wiener, M.: Classification and regression by random forest. R news **2**(3) (2002). ISSN 1609-3631

A Modified Method Using Fusing Weighted Filter for Faded Image Dehazing

Ashwani Vishwakarma$^{(\boxtimes)}$ and M. P. Parsai

Jabalpur Engineering College, Jabalpur, MP, India
ashul994man@gmail.com

Abstract. A novel method for which is an uncommon mix for different accessible strategies likewise it has new methodology so as to picture extending and picture adjustment. In this work concealing picture dehazing procedure reliant on merging a perceptible and a nearby Haze (NIR) pictures for a comparative scene. In outside scenes, picture separation for got pictures is most likely going to be lost by dimness. Close Haze light is less scattered than detectable light thus for its long wavelength; in like manner NIR pictures have unpredictability enough. Our philosophy restores picture separate by merging point for intrigue parts for NIR picture into recognizable picture. To turn away overemphasizing haze free regions by NIR information, we present weighting NIR picture using transmission map. Exploratory results show that our system beats customary techniques.

Keywords: Image dehazing · Profundity data · Picture rebuilding · Generative antagonistic system

1 Introduction

Catching a reasonable picture in cloudiness situations is a fitting issue in picture processing [1]. Impact for applications like as dimness condition assessment or navigational observing have critical job for quality for cloudiness pictures. Taking clear pictures murkiness is intense; for the most part because of fog predominantly because of Color disperse likewise expansion to Color thrown by changing light constriction on different wavelengths [2]. Shading disperse and Color results an obscured subjects and with brought down complexity in all fog pictures. In Fig. 1, demonstrates a model, yellow coral reef (a sort for ocean tree) at right base corner for picture and one yellow fish in right-upper corner are not recognizable because of Color cast; fish and reef in back are misty because of dissipating.

Cloudiness is on the grounds that by many suspended particles like as sand, microscopic fish and minerals that consistently exist in environment. At the point when camera catch pictures reflected light from articles goes to camera, few bit for light meets suspended particles, which ingests few for light and dissipates light. In

© Springer Nature Switzerland AG 2020
D. J. Hemanth et al. (Eds.): ICICI 2019, LNDECT 38, pp. 205–213, 2020.
https://doi.org/10.1007/978-3-030-34080-3_23

conditions which don't have blackbody discharge [3], dispersing regularly extends to numerous dissipating.

Picture dehazing is considered as a troublesome errand since centralization for dinkiness is shifts from spot to put. First researchers use standard strategies for darkness departure. Since single picture can scarcely give much information later researchers attempt to perform dehazing with various pictures. In [1] Narasimhan et al. propose a dehazing method with various pictures for a comparable scene. Significant headway has been made in single picture dinkiness ejection reliant on physical model. An epic darkness ejection strategies by neighborhood separate enhancement for picture subject to Markov Random Field is proposed by Tan [2]. Fattal [3] propose a dehazing system for concealing pictures based o free part assessment. This procedure is repetitive and can't have any kind of effect for dull scale pictures. It have a couple of difficulties to oversee thick darkness pictures. He et al. [4] make diminish channel prior model subject to supposition that in dominant part for non sky locale, something close to one concealing channel has low power at specific pixels. Yu et al. [5] proposes a material science based brisk single picture obscurity removal. It is a novel speedy defogging method subject to a fast complementary isolating methodology. Multifaceted nature for this technique is an immediate limit with regards to amount for data picture pixels.

Fig. 1. Haze and Color cast in haze Images due to Blurry and Bluish Effects

Dimness picture improvement strategies give an approach to improving the article recognizable proof in cloudiness condition. There is part of research begun for the improvement of picture quality, however constrained work has been done in the zone of cloudiness pictures, on the grounds that in murkiness condition picture get obscured because of poor perceivability conditions and impacts like ingestion of light, impression of light, bowing of light, denser medium (multiple times denser than air) and dispersing of light and so on. These are the significant factor which causes the corruption of murkiness pictures.

2 Methodology

The work done in region for dimness picture improvement till now either utilizing mean or middle channels or by utilizing different shading extending strategies or by utilizing balancing, anyway nobody has exhibited a modular which is incorporated for different strategy, Thesis work present a novel system for which is an exceptional combination for different accessible procedures likewise it has new methodology so as to picture extending and picture evening out. Proposed method improves shallow sea optical pictures or recordings utilizing extending cum balancing cum middle channel and furthermore according to wavelength properties. Our key commitments are proposed incorporate a novel division imaging model that repays so as to lessening disparity along proliferation way and a powerful fog scene improvement conspire. Recouped pictures are portrayed by a decreased noised level, better introduction for dull districts, and worldwide complexity where best subtleties and edges are upgraded altogether.

Picture Dehazing Method by Fusing Weighted Near-Haze Image: To begin with, luminance picture and transmission guide are figured from got unquestionable picture. Second, luminance picture and NIR picture are crumbled into gauge and detail pictures using Laplacian pyramid.

The detail NIR pictures are weighted with transmission manual for check distinction overemphasizing in cloudiness free territories. Detail pictures for luminance and NIR picture are stood out and entwined from duplicate new detail pictures for luminance picture. Finally, these gained detail pictures are framed into single luminance picture and united with chrominance information for detectable picture into dehazed concealing picture.

In this paper clear and weighty concealing diminishing prior model is proposed. Using this previous direct model for scene significance for diminish picture is made. With help for coordinated learning model parameters for liner exhibit are found out and relating significance map for diminish picture can fabricated adequately. Using significance map, transmission direct and scene brightness can without much for a stretch restored. This system we can feasibly oust shadiness from a lone picture. In dehazed picture some private data can be concealed and recovered at whatever point essential.

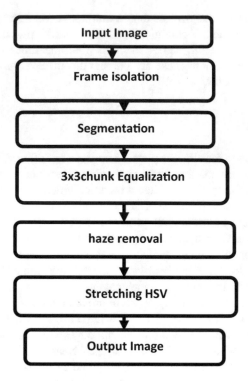

Fig. 2. Proposed design flow processes

Figure 2 above demonstrates the proposed work process outline, the technique can be comprehend by following advances:

Stage 1 input image: here we can include any picture structure our computer, the comparative pictures has been taken by base works, so for reasonable examination it was exceptionally required to take standard pictures.

Stage 2 Frame isolation: red, green and blue piece of the picture gets isolated for the transformation of 3-D shading picture into 2-D pictures for better investigation of the pictures.

Stage 3 Segmentation: here a small segments of original image frames is chucked into 3 × 3 blocks and median of each frame segments is computed.

Stage 4 3 × 3 Chunk Equalization: this part is the key element of our research, this is the method which we have developed for image dehazing, a median base equalization method used instead of probability base equalization on small 3 × 3 chucks of image frames

Stage 5 Haze Removal: this is done using following steps:-

1. Concealing Attenuation Prior: Human personality can without a doubt recognize foggy zone from normal scene without extra information. Splendor and sub-mersion for pixels in dim picture vary distinctly close by obscurity center changes. In dimness free area drenching for scene is high and splendor is

moderate. Dinkiness obsession augments close by movements for significance for scene. We can acknowledge that scene significance is insistently compared with duskiness center and we have

$$d(x) \alpha c(x) \alpha v(x) - s(x), \qquad (1)$$

Where d is significance for scene, c is combination for cloudiness, v is splendor for scene and S is inundation for pixel.

2. Scene Depth Restoration: Straight model can be made using doubt that complexity among brightness and drenching can generally address centralization for mist as seeks after:

$$d(x) = \theta0 + \theta1 \, v(x) + \theta2 \, s(x) + \varepsilon(x), \qquad (2)$$

where $\theta0$, $\theta1$, $\theta2$ are straight coefficients. v and s are brilliance and immersion segment separately. $\varepsilon(x)$ is arbitrary variable speaking to irregular blunder.

3. Getting ready Data Collection: Training data are significant in order to get acquainted with coefficients $\theta0$, $\theta1$, and $\theta2$ accurately. Planning test contain for a dim picture and its relating truth significance map. Significance guide is difficult to get and current significance cameras are not prepared to verify exact significance information. For every obscurity free picture, a discretionary significance map with same size is made.

Ecological Light Estimation: Environmental light is resolved using condition

$$A = I(x), \, x \in \{x | \forall y : d(y) \le d(x)\}, \qquad (3)$$

where An is environmental light and I(x) is force for pixel. To discover barometrical light select pixels with most noteworthy force in cloudy picture among these most brilliant pixel as air light A.

4. Scene Radiance Recovery: As scene profundity d and environmental light An are known scene brilliance can be recuperated utilizing

$$J(x) = (I(X) - A)/(t(x)) + A = (I(x) - An)/e^{(-\beta d(x))} + A, \qquad (4)$$

where J(x) is sans murkiness picture. For maintaining a strategic distance from an excess of clamor, esteem for transmission medium t(x) is confined between 0.1 and 0.9. Scattering coefficient β is considered as consistent in homogenous areas. Figure 2 indicates design graph for proposed framework. From a dim picture input scene brilliance can be recouped utilizing shading weakening earlier model. In dimness free picture some mystery message can insert and recuperate when it is required.

Here least imperative piece for picture is displaced with data bit. Pivot system is associated with recover data from picture.

Stage 6 HSV extending: subsequent to performing RGB extending shade immersion and force change performs and HSV extending done, it will be upgrade the lightning in the picture.

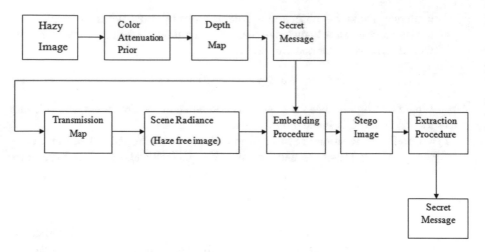

Fig. 3. Proposed architecture for haze removal

3   Results

Figure 4 below shows the GUI developed MATLAB for front end user interface in background of this GUI a MATLAB script runs as per the flow diagram explained in Fig. 3.

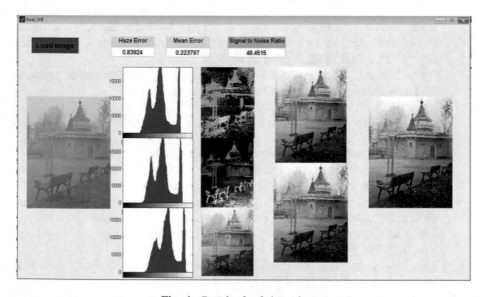

Fig. 4. Results for fad test image

Fig. 5. Input faded test image

Fig. 6. Output dehaze image

Figure 5 above shows the input hazed and fad image and Fig. 6 above show the result output dehazed image. The difference can be clearly observe that the amount of light and contrast in output is more clear then input.

Table 1 below shows the observed results of the proposed simulation for different test images.

Table 1. Observe results for fade image after dehazing

SN	Test image	MSE
1	Mean Square error	0.2237
2	Peak Signal to noise ratio	48.45
3	Haze error	0.83924

The results shows that proposed work simultaneously resolved problem for Color scatter and Color casting also enhanced image contrast and calibrated Color cast and produces high quality haze images or videos. In present paper, proposed work used slide stretching algorithm applies on both RGB and HSI color models in order to enhancing haze images. Main advantage for using two stretching models is because helps to equalize color contrast any type for images and also mention problem for lighting. Proposed approach has produced good results. Quality for images is statistically observed through histograms. Future work will include further evaluation for proposed approach. It may be clearly seen that proposed procedure is best among available procedure with very high PSNR means significantly remove noise and very less MSE hence it has very low error.

4 Conclusion

In this paper a novel shading weakening earlier model dependent on distinction among splendor & immersion segments for pixels inside cloudy picture is proposed. A direct model for scene profundity is made with this shading lessening earlier model. Managed learning model is utilized to learn parameters for model & profundity data recuperated. By utilizing profundity map gotten by proposed strategy, scene brilliance for foggy picture can undoubtedly recuperated. Exploratory outcomes demonstrate that proposed methodology gangs high productivity & extraordinary murkiness evacuation impacts. Proposed approach likewise permits to shroud some mystery message in dehazed picture. It tends to be extremely valuable in numerous security related applications which needs information privacy.

References

1. Kudo, Y., Kubota, A.: Image Dehazing Method by Fusing Weighted Near-Haze Image. 978-1-5386-2615-3/18/2018 IEEE
2. Tan, R.T.: Perceivability in terrible climate from a solitary picture. In: Proceedings of IEEE Conference Computer Vision Example Recognition (CVPR), pp. 1–8, June 2008
3. Fattal, R.: Single picture dehazing. ACM Trans. Graph. **27**(3), 72 (2008)

4. He, J.S., Tang, X.: Single picture cloudiness expulsion utilizing dull channel earlier. IEEE Trans. Example Anal. Mach. Intell. **33**(12), 2341–2353 (2011)
5. Yu, J., Xiao, C., Li, D.: Material science based quick single picture mist expulsion. In: Proceedings of IEEE Tenth International Conference Flag Processing (ICSP), pp. 1048–1052, October 2010
6. Fang, F., Li, F., Yang, X., Shen, C., Zhang, G.: Single picture dehazing and denoising with variational strategy. IEEE (2010)
7. Long, J., Shi, Z., Tang, W.: Quick cloudiness expulsion for a solitary remote sensing image utilizing dull channel earlier. IEEE (2012)
8. Wang, Y., Wu, B.: Improved single picture dehazing utilizing dull channel earlier. In: IEEE International Conference on Intelligent Computing and Intelligent Systems, vol. 2 (2010)
9. Kang, L.-W., Lin, C.-W., Fu, Y.-H.: Programmed single picture based downpour streaks expulsion through picture decay. IEEE Trans. Picture Handling (2012)
10. Tarel, J.-P., et al.: Vision improvement in homogenous and heterogenous mist smart transportation frameworks. Magazine IEEE (2012)
11. Yeh, C.-H., et al.: Efficient image/video dehazing through haze density analysis based on pixel-based dark channel prior. In: International Conference on Information Security & Intelligence Control (2012)

Survey-Iris Recognition Using Machine Learning Technique

Padma Nimbhore[✉] and Pranali Lokhande

School of Computer Engineering and Technology, MIT AOE,
Alandi, Pune, India
{ppnimbhore, pplokhande}@comp.maepune.ac.in

Abstract. In this digital era, Iris identification and detection are most useful and secure to use in banking, a financial section for security as well as it avoids fraud card detection. Iris recognition system gets images of an eyes by CSI scanner, after this, it traces out and senses the iris in the image which is then meant for the feature extraction, training, and matching. In this project, we will make use of two techniques by Iris image extraction for two separate classification method of the machine learning approach. Before feature extraction Normalization and Segmentation is used for the finding out the correct position of iris region in the particular portion of an eye with accuracy. This paper more focuses on machine learning approach to use supervised learning method.

Keywords: Machine learn · Biometrics · Normalization · Classification · Hamming distance

1 Introduction

The digital authentication technology that analyzes and measures human body characteristics either physiological or behavioral for the authentication and security purpose. When the internet has reached its peak and has formed the base for all modern banking systems and business systems, the accurate authentication for accessing our information in online banking sector will be necessary. There is an need for intelligent technology where can make use of security IPIN, OTP, secure signs/images etc. technologies possibly to use for most security purpose in banking sectors now a days [1, 2].

The new technology and method of machine learning are developed. Like an Artificial Intelligence (AI), Decision tree, Decision networks, Self-organizing networks, SVM and many more techniques are developed. The some image feature that is used in this theory is a human iris [2].

Neural networks process any network through their weight and get a single output which we called as binary classification technique in supervised learning of ML. Such networks of ANN, as well as FNN, are known as decision network through which it gets to the optimal solution or reaches to the specific goal which is the complete solution. The family network is one part of this Neural Network defined as the complexity of NN. Decision function is chosen the appropriate weight for neural networks connection where minimum error calculation is done by taking an optimal weight of

© Springer Nature Switzerland AG 2020
D. J. Hemanth et al. (Eds.): ICICI 2019, LNDECT 38, pp. 214–219, 2020.
https://doi.org/10.1007/978-3-030-34080-3_24

neural network architecture. The process of training and matching via multilayer feed-forward NN, feed-forward back propagation NN [1–4].

2 Literature Survey

The author wrote about a machine learning technique of classification and SVM for iris recognition in the paper of Pattern recognition letter in 2016. In order to reflect the research trend, they choose to list relevant work in chronological order. He also studied on Methods based on Fuzzy neural network where he finds some similarity in NN and FNN to calculate the distance via classification technique [2]. Only the difference is to get more flexibility in handling any cluster shape in Minkow Ski distance calculation through which it provides more recognition rate 98.12%. There are three types of SVM technique that are kernel linear, quadratic and polynomial for Iris image(segmented) recognition. Therefore it gives the best performance with the least square and quadratic kernel method to increase the rate of recognition by 98.50% with zero false Acceptance Rate. So nowadays this SVM in support with different classification techniques under ML for supervised and unsupervised method deal with particular recognition and also with applications of iris and eye detection in banking, finance section for security purpose and even nowadays it makes more usefulness to detect card to avoid fraud detection [2–4]. According to the author reference, conditional false reject probability of iris pattern is 109.6 which is generally one in 4 billion. In a single scan of image can analyze more than 200 variable of the iris image as furrows, corona, rings, freckles, etc. The digital scanner deals with the identification of persons different iris pattern extracted from the image eye. The human eye generally consists of the pupil (the inner darkest part), iris (the complete color part) and the Creamish/white part (sclera). Non-concentric areas of eyes are iris and pupil. The accurate Pi/2 area of the internal liner of the iris will not be constant which will be increase or decreases continuously that are depending upon the amount of light incident on the pupil [5, 6].

3 Implementation Concept

It mainly goes through 2 main processes which are registration and authentication. The registration phase goes through 3 basic stages which are image acquisition, normalization, segmentation and feature extraction same for an authentication phase. These phases are described below. The iris recognition system goes (Fig. 1).

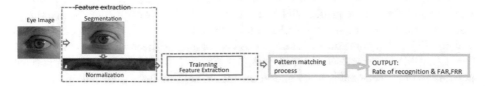

Fig. 1. Implementation workflow

Workflow through the following stages:

- Feature Extraction via 2 stage process
- Training process
- Matching process

3.1 Feature Extraction via 2 Stage Process

The process of image normalization and segmentation process called feature extraction. A clear image normally does not contain much noise that needs to be suppressed. In this stage, the eye image is captured via capture device i.e. CIS 202 3M cogent Iris scanner. Even we can create our own database through this scanner. Completing this step appropriately is of high importance because it goes through the 2 stages of feature extraction using machine learning [5–7].

Machine learning characterizes the main approach of feature extraction via SVM classification. We will do calculation by hamming distance method, it gets an accurate and robust rate of recognition of iris. As SVM hamming distance calculation is second classifier work fails in iris image extraction. So It needs to make use of separate image extraction method for these two classifiers instead of using the same features for both of them because if so it gives low FAR and FRR is not that low. Therefore use SVM on haar wavelet extraction of features method is first and then second feature extraction technique with 1D Gabor wavelets' to get no false.

3.2 Training Process

In this training process, train the capture image using any dataset Like Liam et al. A gaussian dataset which gives 25 to 30 set of five samples per images (150 samples). Here captured images had the number of features gone through 2 stages feature extraction process such that each feature is to train for the process of matching via neural network algorithm like Multi-layer Feedforward NN and Feedforward Back Propogation NN. The process with 2 types of the neural then compared with each other to increase the accuracy of recognition rate by lower down the rejection rate and increase acceptance rate. These two neural networks are Multilayer FNN and Feedforward Back Propogation NN, works with hidden layer had 30 neurons and o/p layer with 10 neurons, weights to be override by backpropagation method [1, 2]. Any distance matrices calculation process to compute the distance between a couple of neurons, this will exploit to upgrade the weight without bias functioning. So the performance of both Multilayered FNN and Feedforward Back Propogation NN will get similar but will be more accurate functioning of Multilayer FNN as compare to Feedforward Back Propogation NN and other Neural Network in time computation. This database of 30 neurons with SVM and Gabor wavelet for extracting iris features which will be deterministic in nature [2]. SVM based architecture gives a very good result of FAR in the closed set and open set situation, wherein open set authorized person use other personal identity and in the closed set, imposter uses authorize person identity. Therefore it will be well protected from attacker like imposters. With a bad rate of FRR, it will increase the rate of recognition and accuracy [12–16].

NN. So for this, we made use of Daugman's iris data set with different signal to noise ratio for given training set [8–11].

3.3 Matching Process

There are 2 types of the neural networks. These two neural networks are Multi-layer FNN and Feedback Propogated NN, where one hidden layer having 30 neurons and an o/p layer with 10 neurons, weights to be override by backpropagation method [1, 2]. Any distance matrices calculation process to compute the distance between a couple of neurons, this will exploit to upgrade the weight without bias functioning. So the performance of both Multilayered FNN and Feedback Feedback Propogated NN will get similar but will be more accurate functioning of Multilayer FNN as compare to Feedback Propogated NN and other Neural Network in time computation. This database of 30 neurons with SVM and Gabor wavelet for extracting iris features which will be deterministic in nature [2]. SVM based architecture gives a very good result of FAR in the closed set and open set situation, wherein open set authorized person use other personal identity and in the closed set, imposter uses authorize person identity. Therefore it will be well protected from attacker like imposters. With a poor rate of FRR, it will increase the rate of recognition and accuracy [12–16].

4 Scanner for Iris Details

Hardware as CIS202 iris scanner, Detects 40 iris images by NIST SAP.

The possibility of change in iris of human eye appears only in cases where the area is affected by damage. Other possibility includes change due to medical operations. İris is considered as security measure in parallel with finger print, only significance of iris is it's human internal organ and has less chance of getting affected throughout the life span of human. Iris itself is unique feature, sensible part of body and appearance is external [1, 2].

CIS 202 Scanner which offers an error-free low-cost solution image catures in within few seconds (Fig. 2).

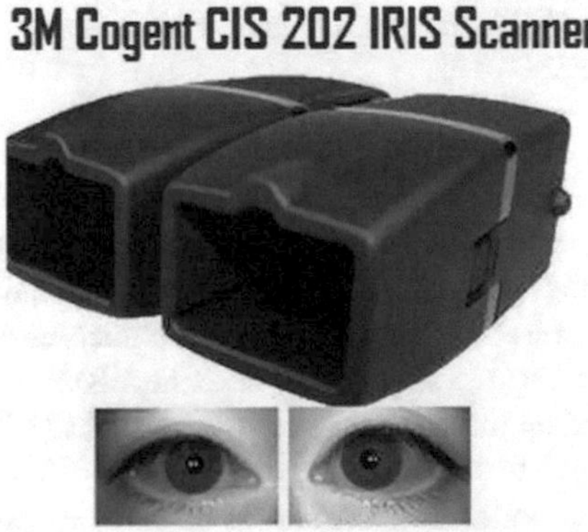

3M Cogent CIS 202 IRIS Scanner

Fig. 2. Iris scanner CIS 203-3M cogent

This scanner has illumination of infrared, USB powered scanner is mostly applicable to catch high-resolution iris pictures, in addition it removes chance of disturbing detection position and controls the spectral image sensitivity. Light capture at the range at some nano meter range and has rating of IP 54, CIS scanner handy design and more ideal for some application such as registry enrollment, applicants object identification, inmate release management etc. and many more [17].

4.1   About a Scanner

We develop this project or recognition system which is able to identify account customer with unique identification of iris. Using this scanner gets best picture quality of iris during project execution. Because of this scanner, it makes to minimize the normalization error and mean square error in segmentation. This will increase the accuracy ratio of iris system. We can use it insecurity, a sensitive business application where security having very much importance. So we can create our own data set in the form of templates or in any latest type database. This scanner image is supportable under the language of Python and Mat-lab. It uses both images of eyes lefe as well as right. Refer Fig. 3 shows the eye captured by the scanner.

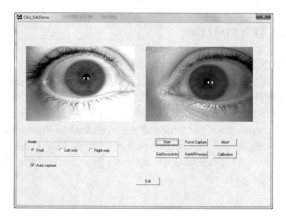

Fig. 3. Image captured by Iris scanner

5   Conclusion

ML technique in this implementation new and is most useful now a day to improve the performance by increasing accuracy in recognition rate and down the robustness characteristic with lower FRR and high false acceptance rate. This is only possible through the machine learning technique two classifier: SVM classifier with hamming distance classifier to calculate the distance between samples (150) with images (30) and One image per 5 samples. In the first stage, SVM classifier will use Haar wavelet feature extraction and at second stage hamming distance classifier will use 1D Gabor wavelets' feature extraction to improve accuracy and non-false acceptance rate.

Machine learning technique with neural network algorithms of Feedforward back propagation network (FBPNN) and Multilayer FNN will also improve the performance of matching training sets. The project is inspired by Machine learning classification with Daugman's algorithm for normalization and segmentation which is the most efficient Iris recognition models that will develop now a day with high performance and maximum rate of recognition.

References

1. Roy, D.A., Soni, U.S.: Iris segmentation using Daughman's method. In: IEEE ICEEOT (2016). ISBN 978-1-4673-9939-5
2. De Marsico, M., Petrosinob, A., Ricciard, S.: Iris recognition through machine learning techniques: a survey. Pattern Recogn. Lett. **82**, 106–115 (2016)
3. Jung, Y., Kim, D., Son, B., Kim, J.: An eye detection method robust to eyeglasses for mobile iris recognition. Expert Syst. Appl. **67**, 178–188 (2016)
4. Daugman, J.: Searching for doppelgangers: assessing the universality of the Iris Code impostors distribution. IEEE IET J. **5**(2) (2016). ISSN 2047 4946
5. Gale, A.G., Salanka, S.S.: Evolution of performance analysis of iris recognition system by using a hybrid method of feature extraction and matching by the hybrid classifier for iris recognition system. In: IEEE ICEEOT (2016). ISBN 978-14673-9939-5
6. Abbdal, S.H., Kadhim, T.A., Abduljabbar, Z.A., Hussien, Z.A., et al.: Ensuring data integrity scheme based on digital signature and iris features in cloud. Indonesian J. Electr. Eng. Comput. Sci. (2016)
7. Nalla, P.R., KumaR, A.: Towards more accurate Iris recognition using cross spectral matching. IEEE (2016). ISBN 1057-7149
8. Nestorovic, N., Prasad, P.W.C., Alsadoon, A., Elchouemi, A.: Extracting unique personal identification number from iris. IEEE (2016). ISBN 978-1-5090-5398-8. School of Computing and Mathematics, Charles Sturt University, Sydney, Australia, Walden University
9. Ali, H., Salami, M.: Iris recognition system using support vector machines. In: Riaz, Z. (ed.) Biometric Systems, Design, and Applications, pp. 169–182. In Tech 2011 (2008)
10. Roy, K., Bhattacharya, P.: Iris recognition with support vector machines. Advances in Biometrics, pp. 486–492. Springer, Heidelberg (2005)
11. Rai, H., Yadav, A.: Iris recognition using combined support vector machine and Hamming distance approach. Expert Syst. Appl. **41**, 588–593 (2014)
12. Patil, S., Gudasalamani, S., Iyer, N.C.: A survey on iris recognition system. In: International Conference on Electrical, Electronics, and Optimization Techniques (ICEEOT) – 2016 (2016). ISBN 978-14673-9939-5
13. Tan, C.-W., Kumar, A.: Towards online iris and periocular recognition under relaxed imaging constraints. IEEE Trans. Image Process. **22**, 3751–3765 (2013). ISSN 1941-0042
14. Ibrahim, A.A., Khalaf, T.A., Ahmed, B.M.: Design and implementation of iris pattern recognition using wireless network system. J. Comput. Commun. (2016). ISSN 2327-5219
15. Chai, T.-Y., Goi, B.M., Tay, Y.H., Nyee, W.J.: A trainable method for iris recognition using random feature points. In: IEEE Conference (2017). ISBN 978-1-5386-4203-0
16. Daugman, J.: New methods in iris recognition. IEEE Trans. **37**(5), 1167–1175 (2007). ISSN 1941-0492
17. https://www.cogentsystems.com

Hybrid Analog and Digital Beamforming for SU-MIMO Frameworks

V. I. Anit[✉] and N. Sabna

Department of ECE, RSET, Ernakulam, India
anitjoseph996@gmail.com, sabnan@rajagiritech.edu.in

Abstract. The analog and digital beamforming performing in hybrid manner that is by superimposing them altogether is the most appropriate scheme for MIMO systems to attain better performance. By utilizing this procedure the framework can lessen the equipment multifaceted nature, power consumption and can increase the spectral efficiency with a few number of RF chains. What's more, interestingly, it's general execution is the same as that of a hybrid beamformer's structure. In this paper the hybrid beamforming structure for mm Wave frameworks with OFDM modulation is performed. That must be accomplished for the instance of single-user multiple-input multiple output framework, the hybrid beamforming architecture were implemented on its transmitter and receiver, and observe the results. That was the proposed idea. For this phase, there is only the transmitter portion of the proposed system is implemented. Expected result is the power spectrum of the transmitted signal. The generation of power spectrum to the hybrid beamforming is based on an analog steering vector designed under certain constraints. Only transmitted portion that is the inverse beamforming is done over here for this phase.

Keywords: Millimeter waves · Beamforming · MIMO-OFDM · Large-scale antenna arrays · Precoding · Steering vector · ULA

1 Introduction

For the implementation of wireless communication systems, millimeter waves play a vital role. The available spectrum is so much congested and so it must have to be think about a serious solution to solve the spectrum scarcity. This motivates to the best remedy that is the use of millimeter waves for wireless communication. Millimeter waves are actually an electromagnetic spectrum radiation situated between microwave and infrared waves and have a very high frequency. It has been proved already that they have the potential to ease the demand for low frequency and allows expansion of frequency depending upon technologies and applications. It offers very large bandwidth, high resolution, low latency, low interference and high security. While considering a communication system the ultimate aim is to increase the data rate with less error probability. Since the wireless scenario is the most common one, it is subjected to noise and other interferences such that the desired data cannot be completely retrieved back from the receiver side. This is mainly because of the poor propagation characteristics of the channel. In order to rectify that, use of large-scale antenna arrays at the transceivers are adequate. For the implementation of

© Springer Nature Switzerland AG 2020
D. J. Hemanth et al. (Eds.): ICICI 2019, LNDECT 38, pp. 220–229, 2020.
https://doi.org/10.1007/978-3-030-34080-3_25

these large scale antenna arrays mm waves are the best choice. And also by using these waves we can embed large number of antennas as a single component, such that it can provide high capacity [1–5]. There exists two issues for designing such a transceiver system using large scale antenna arrays. One of is it consumes too much power. Another thing is that it costs too much. And also it leads to hardware complexity. The transceiver system implementation in conventional fully digital beamforming schemes was impractical because for that it requires one separate RF chain for each antenna element. This results a system with excessive power consumption and high hardware complexity [8–10]. To amend the issues brought about by completely digital beamforming, here it presents hybrid beamforming engineering; where the overall framework comprises of a low dimensional advanced beamformer with low dimension and a simple beamformer with high dimension. The simple beamformer is the analog beamformer and the advanced beamformer is the digital beamformer [11–13]. This inspires us to design a hybrid beamformer with OFDM modulation for the AWGN channel, which uses mm wave frequencies. This paper deals with the case of MIMO system which is worked on the basis of OFDM modulation, which worked on mmwave frequencies, designed for narrowband frequencies [1]. Whole work consists of both transmission and reception of data signals on a MIMO-OFDM system for large scale antenna arrays through an AWGN channel. This paper focuses only on the transmission part of those signals, and it generates the power spectrum of the transmitted signal by using a steering vector based on specific angles and certain design constraints that give a better result than previous ones while considering the case of large scale antenna arrays. The remaining work which consists of the reception of signals and analysis will be published in future works.

1.1 Paper Organization

The rest of this paper is sorted out as pursues. Section 2 presents system model for SU-MIMO. It includes the signal model of the proposed system. Section 3 contains the hybrid beamforming structure for SU–MIMO. And it involves the design of digital precoder, design of analog precoder. Section 4 deals with the simulation results. Conclusions were given in Sect. 5.

2 System Model for SU-MIMO

The overall outline of the proposed system is given in Fig. 1 as a block diagram Here it considers a MIMO-OFDM system with N_t as number of transmitter antennas, N_r as number of receiver antennas which uses N_s data symbols to transmit. The whole process has been performed on the assumption that there is an equal number of subcarriers present at both at the transmitter and receiver side ranging from s[1] to s[k], since for the mmwave framework with highly correlated channels, all the sub channels were commonly acquire low position. The overall system mainly consists of digital precoder, OFDM modulator and an analog precoder at the transmitter side. And an AWGN channel is used here to transmit the data symbols to the receiver. At the receiver side, for decoding there is an analog combiner, OFDM demodulator and digital combiner and thus it is possible to retrieve back the transmitted data symbols.

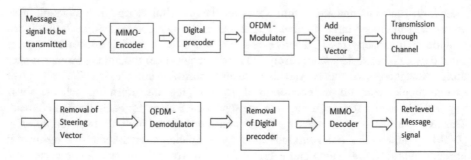

Fig. 1. Block diagram of the proposed system

For the generation of data symbols here it uses random symbols, which were modulated, scrambled and layer mapped to produce data symbols N_s. In the design point of view of the hybrid beamformer, the general beamformer involves advanced beamformer with low dimension and a simple beamformer with high dimension which were executed using simple straightforward parts (Fig. 2).

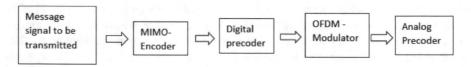

Fig. 2. Block diagram of implementing model

For the implementation side for this phase it has been considered only the transmitter side of the proposed system. At the transmitter it was mainly two sections;

1. Digital precoding section
2. Analog precoding section

These two portions constitute to form the transmitter side. Along with that it is subjected to Inverse Fourier Transform, adding of cyclic chain and has an RF chain mainly used to scale the data symbols to multiple times. RF chain performs both upscaling and downscaling. For our scenario we consider the upscaling process only. And also another function of RF chain is to transmit the data symbols from digital domain to the analog domain. Actually this chain acts as a bridge between the digital precoder and analog precoder. After analog precoding the data symbols has to be transmitted through the channel with the aid of multiple antennas. Up to this portion it was performed on this phase.

2.1 Signal Model of the Hybrid Beamformer

In the proposed structure, the transmitter side comprises of the data symbols that are first precoded digitally on each subcarrier $k = 1, 2 \ldots K$ with the help of a low dimensional digital precoder. Then we perform OFDM modulation. It includes the IFFT's and insertion of cyclic prefix. IFFT helps to transform the data symbols to time domain. Then we insert cyclic prefixes. And we perform analog precoding and thus generates analog precoding matrix is applied to the OFDM modulated data and we transmit the precoded signals. The transmitted signal as,

$$x[k] = V_{RF} V_D[k] s[k]$$

And the received signal after adding the channel matrix at the channel and get transmitted through the channel generates the received signal as,

$$y[k] = H[k] V_{RF} V_D[k] s[k] + z[k]$$

After receiving the signal, the received are initially processed with an analog combiner, cyclic prefix is getting removed, and we find the FFT of the signals, and after passing the signals through a low-dimensional digital combiner we get the transmitted input signal.

3   Design of Hybrid Beamformer

For the design of hybrid beamformer, it has to go through certain steps such as,

1. Design the Digital precoder.
2. Calculate the value of Analog precoding matrix.
3. Find the value of Analog Combiner.
4. Design the Digital Precoder.

And here at first design of analog beamformer has to be done by assuming that the digital beamformer is fixed. And after that for the recently planned simple digital beamformer, at that point, the advanced analog beamformer is structured by exploiting the features of channel precoding. And for the receiver section the analog and digital combining part has to be performed. The implementation of the analog combiner is on the assumption that there is a fixed digital combiner and the design of digital combiner is for that the already designed analog combiner.

3.1   Design of Digital Precoder

The design of digital precoder is on the assumption that analog precoder is already fixed. So for a fixed analog precoder it has been estimated the optimal power allocation of the beamformer circuit. It has been computed the optimum power allocated for the structure by using the quantity of transmitted antennas and number of symbols that were utilized for transmission. By picking the correct number of transmitted antennas and number of symbols it has to make the power as optimized. Then the weights of digital precoder can be easily computed. By using this optimum power allocation technique it has been reduced the complexity of the system, performance has to be improved in a low cost.

3.2   Design of Analog Precoder

Here there it contains the design of the analog precoder for the pre designed digital precoder. And it has been already designed the digital precoder. The analog precoding matrix is designed in such a way that it has been chosen a steering vector which is nothing but a vector that represents the phase or time delays that a plane wave experiences. A wave number based on the frequency of incoming narrowband signal have to be chosen and based on that the steering vector is designed. Based on the vector it has been obtained the value of analog precoding weights.

4   Simulation Results

For performing this hybrid beamforming it contains mainly the generation of data symbols, digital precoding, perform OFDM modulation, perform analog precoding and transmit that signals to the AWGN channel. The data symbols for the digital precoder are derived from random signals as given in Fig. 3. For the generation of data symbols for a MIMO–OFDM system, it has some steps. For that decide the size of transmitter block at first. Then generate the code word from that by applying CRC and turbo coding. And for this simulation there is a transmitter block of size [50 50]. And thus generated two codewords.

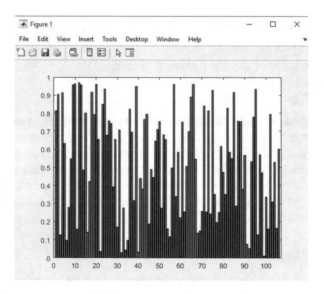

Fig. 3. Plot of input signal

Then the generated codewords have to undergone through scrambling through a scrambling sequence and thus get the scrambled output. In scrambling process up to two codewords were transmitted in a sub outline and for each codeword, bits are mixed with an alternate scrambling data sequence. Then the scrambled sequence gets modulated through a modulator. Here for this simulation there is a "16-QAM" and have done the modulation process for the respective scrambled outputs generated from the two codewords. The complex modulated symbols are then mapped to one or a few layers. It was done by using a layer mapper. The number of layers is not exactly or equivalent to the quantity of receiving antenna ports utilized for transmission of the physical channel. Layermapping is based on the scheme of transmission used. The layer mapped symbols are now given as inputs to the digital precoder in our block diagram. And is processed then as per the algorithm mentioned above. These are the data symbols to the digital precoder for further processing. They were given as Fig. 4.

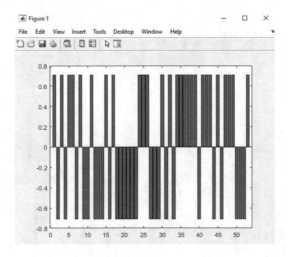

Fig. 4. Plot of data symbols generated

The data symbols were digitally precoded based on the algorithm from [2]. This is given in Fig. 5.

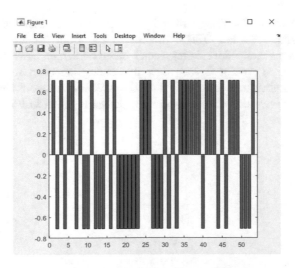

Fig. 5. Plot of digitally precoded symbols

The digitally precoded data symbols were OFDM modulated in order to avoid the inter carrier interference and inter symbol interference. And also since it has to transmit large number of data symbols so in order to avoid the frequency overlapping and also to increase the spectral efficiency. The next step is to perform OFDM modulation. And those OFDM modulated symbols were in Fig. 6.

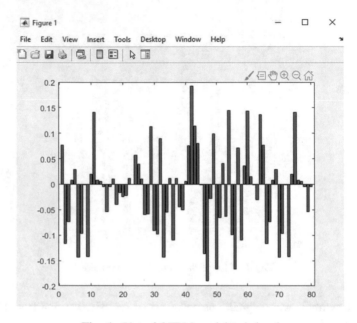

Fig. 6. Plot of OFDM modulated signal

The main work is the design of a hybrid beamformer transmission section by designing a digital precoder, OFDM modulated and from the analog beamforming section it has been generated a steering vector based on certain constraints. And on the basis of that steering vector it was done the analog beamforming to the digitally precoded data symbols and transmits it to the AWGN channel. The corresponding steering vector that we had chosen as per our design is given in Fig. 7.

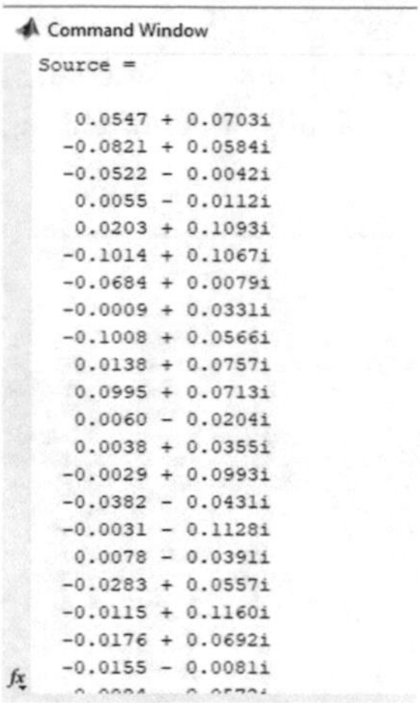

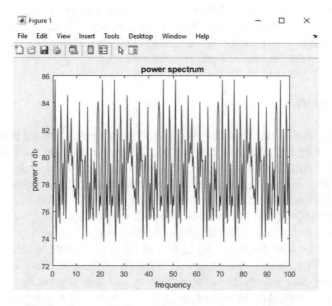

Fig. 7. Matrix of steering vector

The power spectrum of the corresponding signal hybrid beamformed using steering vector at the analog section is given in Fig. 8.

Fig. 8. Plot of power spectrum of analog precoded signal

Then this signal is passed through a ULA with 4 elements which creates an array. Then AWGN noise is added to the analog precoded signal and generated the final transmitted signal. The matrix of finally transmitted signal is given in Fig. 9.

```
>> OUT2

OUT2 =

    0.3747 + 0.2548i    0.3701 + 0.2609i    0.3655 + 0.2668i    0.3607 + 0.2726i
   -0.4225 + 0.1697i   -0.4253 + 0.1632i   -0.4279 + 0.1567i   -0.4304 + 0.1501i
   -0.2889 - 0.0461i   -0.2878 - 0.0514i   -0.2865 - 0.0566i   -0.2852 - 0.0618i
   -0.0548 - 0.1296i   -0.0540 - 0.1310i   -0.0530 - 0.1324i   -0.0521 - 0.1338i
    0.2579 + 0.5823i    0.2492 + 0.5863i    0.2404 + 0.5900i    0.2314 + 0.5936i
   -0.5376 + 0.3792i   -0.5438 + 0.3699i   -0.5498 + 0.3605i   -0.5555 + 0.3509i
   -0.3085 + 0.0082i   -0.3076 + 0.0020i   -0.3066 - 0.0042i   -0.3053 - 0.0104i
   -0.0757 + 0.0431i   -0.0767 + 0.0419i   -0.0776 + 0.0408i   -0.0785 + 0.0396i
   -0.4658 + 0.1756i   -0.4685 + 0.1693i   -0.4710 + 0.1629i   -0.4734 + 0.1565i
   -0.0253 + 0.3933i   -0.0321 + 0.3928i   -0.0389 + 0.3920i   -0.0457 + 0.3911i
    0.4136 + 0.3264i    0.4092 + 0.3338i    0.4046 + 0.3411i    0.3999 + 0.3482i
    0.0415 - 0.2979i    0.0478 - 0.2982i    0.0542 - 0.2982i    0.0605 - 0.2982i
   -0.0345 + 0.1966i   -0.0375 + 0.1959i   -0.0404 + 0.1951i   -0.0433 + 0.1942i
    0.0014 + 0.3351i   -0.0054 + 0.3345i   -0.0122 + 0.3338i   -0.0189 + 0.3328i
   -0.1347 - 0.1292i   -0.1336 - 0.1326i   -0.1323 - 0.1360i   -0.1310 - 0.1394i
    0.0382 - 0.4223i    0.0448 - 0.4221i    0.0513 - 0.4218i    0.0578 - 0.4213i
    0.0509 - 0.1809i    0.0531 - 0.1824i    0.0552 - 0.1839i    0.0575 - 0.1853i
    0.0190 + 0.2099i    0.0152 + 0.2109i    0.0112 + 0.2119i    0.0073 + 0.2128i
   -0.0438 + 0.4458i   -0.0517 + 0.4444i   -0.0597 + 0.4429i   -0.0676 + 0.4411i
   -0.0900 + 0.2666i   -0.0945 + 0.2651i   -0.0990 + 0.2635i   -0.1034 + 0.2617i
```

Fig. 9. Matrix of the transmitted signal

From the simulation results it has been understood that by using OFDM for the case of SU-MIMO, large number of data symbols were transmitted and so the power spectrum shows a better high value. That is the data rate increases. That is as per the input symbols that were transmitted it has been obtained a higher data rate. It was to be clearly indicated by the obtained power spectrum.

5 Conclusion

Hybrid analog and digital beamforming is the most appropriate scheme for MIMO systems. By using this technique the hardware implementation becomes simple, power consumption should be reduced and it can increase the spectral efficiency. And the great thing is that its overall performance is congruent to that of a fully digital beamforming structure. For simulation of this phase we have generated the data symbols that pass through the MIMO-OFDM Communication system as the inputs to the digital precoder. The code word is scrambled, modulated, and layer mapped on here for its generation. Then that is to be used for precoding in digital and analog domain and is to be transmitted. It has been generated the transmitted signal from these symbols. It has to be used for our further processing that is for retrieving the data signals back. Here it was contemplated the hybrid beamforming structure for mmwave frameworks which utilises OFDM modulation. Here it was done only the transmitter section and obtain the result as the power spectrum of the transmitted signal by using a steering vector as per a chosen dimension for ULA and observed the results. And by observing the power spectrum it was found out that the data transmission rate is high and is the same as that of the input symbols that have chosen. So this method is highly recommended for the transmission with the aid of large scale antenna arrays to improve processing gain.

References

1. Sohrabi, F., Yu, W.: Hybrid analog and digital beamforming for mmWave OFDM large-scale antenna arrays. IEEE J. Sel. Areas Commun. **35**, 1432–1443 (2017)
2. Sohrabi, F., Yu, W.: Hybrid digital and analog beamforming design for large-scale antenna arrays. IEEE J. Sel. Topics Sig. Process. **10**, 501–513 (2016)
3. Liang, L., Xu, W., Dong, X.: Low-complexity hybrid precoding in massive multiuser MIMO systems. IEEE Wirel. Commun. Lett. **3**, 653–656 (2014)
4. Kong, L., Han, S., Yang, C.: Wideband hybrid precoder for massive MIMO systems. In: Proceedings of IEEE Global Conference Signal Information Processing (GlobalSIP), December 2015
5. Pi, Z., Khan, F.: An introduction to millimeter-wave mobile broadband systems. IEEE Commun. Mag. **49**, 101–107 (2011)
6. El Ayach, O., Rajagopal, S., Abu-Surra, S., Pi, Z., Heath Jr., R.W.: Spatially sparse precoding in millimeter wave MIMO systems. IEEE Trans. Wirel. Commun. **13**, 1499–1513 (2014)
7. Sohrabi, F., Yu, W.: Hybrid beamforming with finite-resolution phase shifters for large-scale MIMO systems. In: Proceedings of IEEE Workshop Signal Process. Advanced Wireless Communication (SPAWC), Stockholm, Sweden, June 2015
8. Pi, Z.: Optimal transmitter beamforming with per-antenna power constraints. In: IEEE International Conference Communications (ICC), Ottawa, Canada, pp. 3779–3784 (2012)
9. Yu, X., Shen, J.-C., Zhang, J., Letaief, K.: Alternating minimization algorithms for hybrid precoding in millimeter wave MIMO systems. IEEE J. Sel. Topics Sig. Process. **10**(3), 485–500 (2016)
10. Palomar, D.P., Cioffi, J.M., Lagunas, M.A.: Joint Tx-Rx beamforming design for multicarrier MIMO channels: a unified framework for convex optimization. IEEE Trans. Sig. Process. **51**(9), 23812401 (2003)
11. Zhang, X., Molisch, A.F., Kung, S.-Y.: Variable-phase-shift-based RF-baseband codesign for MIMO antenna selection. IEEE Trans. Sig. Process. **53**(11), 40914103 (2005)
12. Cheung, C., Cheng, R.S.: Adaptive modulation in frequency spreading OFDM system with low transmit power spectral density constraint. In: Proceedings of IEEE Wireless Communications Networking Conference (WCNC), Hong Kong, March 2007, pp. 1433–1438 (2007)
13. El Ayach, O., Heath Jr., R.W., Abu-Surra, S., Rajagopal, S., Pi, Z.: The capacity optimality of beam steering in large millimeter wave MIMO systems. In: Proceedings of IEEE Workshop Signal Processing, Advances Wireless Communications (SPAWC), Cesme, Turkey, pp. 100–104, June 2012

Automatic Vehicle Recognition and Multi-object Feature Extraction Based on Machine Learning

E. Esakki Vigneswaran$^{(\boxtimes)}$ and M. Selvaganesh

Sri Ramakrishna Engineering College, Coimbatore, India
{esakkivignesh, selvaganesh.m}@srec.ac.in

Abstract. Over the past few years the influence of surveillance has left a massive impact on the social lives of people. The introduction of controlling the surroundings with the surveillance system would provide zero error and faster reaction time. In this proposed work, we have developed a system that will detect and analyze the traffic signal images. Thousands of traffic signal images are fed into the computer and trained based on the margins of particular classes. A weight file is generated from this training process. YOLO (you only look once) is an algorithm used for training the images and detecting the images. It is a network for object detection. In this project, the objects are vehicles and human beings. The identification of objects is done by searching the location on the image and arrange those objects with its prediction level. Existing methods like R-CNN and its variations, used a pipeline methodolgy to analyze and segment the images in multiple steps. Accuracy and speed of recognition is very slow in existing methodologies because of the individually done component training. Proposed methodology with YOLO is performed by a unique neural network. The output describes the computation and name of the classes from a fed image, which is used as vehicles and human beings.

Keywords: YOLO · Machine learning · OpenCV · R-CNN · CNN

1 Introduction

YOLO (You Only Look Once) is an algorithm which is used for detecting the objects in the input image or video. The main objective of this algorithm is to classify the type of object present in the image and to lay a boundary over it. This method proves to be more optimum than the previous methods like CNN, R-CNN and FR-CNN [1]. A python script is then developed to count the number of items in each class and display them in detail [4]. Using the concept of machine learning with YOLO algorithm we have developed a program to identify type and number of objects fed as input to the computer. Applications like traffic control and stand-alone cars works fine without any manual errors [2].

© Springer Nature Switzerland AG 2020
D. J. Hemanth et al. (Eds.): ICICI 2019, LNDECT 38, pp. 230–237, 2020.
https://doi.org/10.1007/978-3-030-34080-3_26

2 Proposed System

The objective of the proposed system is to detect the type of objects and display the count of each object in detail. An algorithm named YOLO is used for detecting and bounding the objects in the images. The thousands of traffic signal images are given into the algorithm for classification and analyzing the image components. Therefore, algorithms like R-CNN, YOLO have been developed to find these occurrences in a faster way.

2.1 Literature Survey

Shaif Choudhury describes a vehicle detection technique that can be used for traffic surveillance systems. Traffic monotiring system is proposed with Haar based cascade classifier [6]. Rashmika Nawaratne demonstrates the video surveillance system with monitoring, hazard detection and amenity management. Existing methods are lagging from learning from available videos [7]. Nirmal Purohit proposed a technique for identifying and classifying an enemy vehicle in military defense system using HOG and SVM are utilized [3]. Jia-Ping Lin show that the image recognition with YOLO can be applied in many applications of Intelligent Transportation System [5].

2.2 Training Process

Training a dataset is to make sure that the machine recognizes the input provided to the camera at the time of processing. LABELIMG is a Manual image process used for manually defining regions in an image and creating a textual description of those regions. This process is also called as Annotation. The Explicitation for the training process is illustrated in Fig. 1. First, we added a thousand images for the purpose of training. The images are split in the ratio 0:9 for validation and training respectively. The images that are trained has to be finally checked with the images saved for validation.

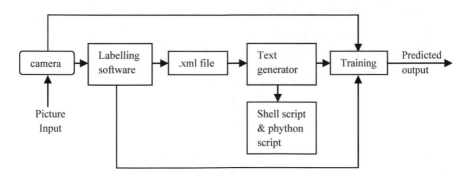

Fig. 1. Explicitation of the training process

In an LABELIMG software, the annotation for each thousand images are done i.e., all these images undergo a process called annotation where all the coordinates of the objects in the images are marked and named according to their class. After Annotation an .xml file is created. It contains the names of the class with their coordinates inside the image. The information about the annotated files are serially arranged according to the time of detection. The .xml files are then converted into text files. A text generator is used for converting the xml files into text files. A python script and a shell script are used for executing the Text generator process. The training process understands only text files so the text generator scripts are added before the process. The main principle of annotation process is to make the system learn to name and detect the objects in the image to the original one. The Schema chart for the training process is illustrated in Fig. 2.

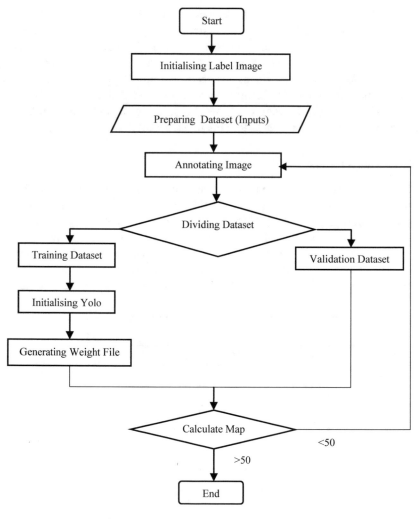

Fig. 2. Schema chart of training process

2.3 Prediction Process

At the time of process, a raw image is captured by the camera and sent to the server as shown in the Fig. 3. The server is setup for the making the process faster. The server contains the program for YOLO algorithm along with our pre-trained models. The pretrained models are nothing but the images that we trained after the validation process at the time of splitting up in the ratio 0:9. The raw image obtained from the camera now compares itself with the pre-trained images using the concept of YOLO. Now the algorithm divides the pictures into N × N grids. The system checks each grid for the number of objects inside it. If there are more than one object in a grid, the grid is then further divided into multiple grids. Once the unique objects are detected the boundaries for them declared.

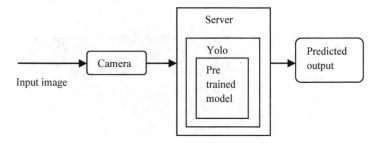

Fig. 3. Explicitation of the prediction process

Now an image with bounded objects and corresponding names in it is produced as the output. According to the code, the number of times each object detected is counted and the count is displayed. The schema chart of prediction process is shown in Fig. 4.

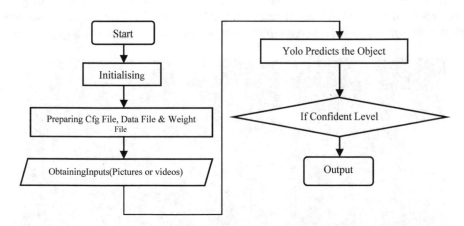

Fig. 4. Schema chart of prediction process

3   Experimental Results

Once the prediction process is completed the predicted output image with objects being bounded and names being named in the image are obtained. The confident level of the object detection is increased. The output in terminal contains the accuracy level and the total count of each objects in number.

Annotation

Annotation is the initial stage of training process. The Fig. 5 shows the objects being marked as named according to their respective classes. A thousand images are trained in such a manner for prediction. For more accurate prediction a further more images are trained in the similar way.

Fig. 5. Annotating the image

Image After Prediction

The process of training should have created a weight file with the details of the similar images. The YOLO compares the present image with the details available in the generated weight files and delivers an accurate predicted output. Now the algorithm will predict the type of object based on the dataset which is fed into the training process.

Fig. 6. Predicted output with bounded class and names (sample 1)

Fig. 7. Predicted output with bounded class and names (sample 2)

The objects can be identified up to 150 mm using the camera. This range varies with respect to the camera. The objects belonging to similar class are marked by a similar colour boundary as shown in the Figs. 6, 7 and 8.

Fig. 8. Predicted output showing numbers of each classes and increased confident level (sample 1)

The predicted output consists of objects being bounded by a box and named correspondingly which is shown in Fig. 9.

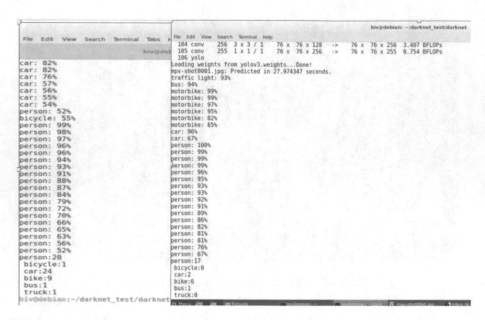

Fig. 9. Predicted output showing numbers of each classes and increased confident level (sample1 & sample 2)

The predicted output consists of objects being bounded by a box and named correspondingly which is shown in Fig. 10.

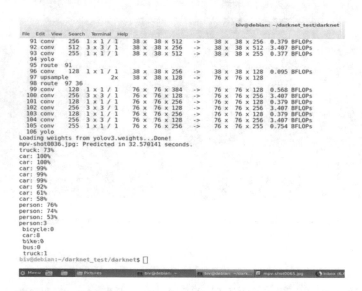

Fig. 10. Predicted output with bounded class and names (sample 3)

4 Conclusion

In this obligation we have determined and classified the type of objects like vehicles and human beings from the obtained images. The usage of YOLO algorithm decreased the process time and produces a more optimised output than the existing methodologies. By fixing a certain area in a captured image, people those who violate traffic rules can also be identified and a penalty can be made in a digital way. The number of vehicles in the image is found out and it can be used for traffic control. YOLO accessess to the whole image by predicting boundaries. We enforces spatial diversity in making predictions. Self driving (unmanned) cars can use this proposed technique for detecting the objects in front of them and driving without collisions. The outgrowth shows that, this process is more efficient and faster than the existing methods.

References

1. Braun, M., Krebs, S., Flohr, F., Gavrila, D.: EuroCity persons: a novel benchmark for person detection in traffic scenes. IEEE Trans. Pattern Anal. Mach. Intell. 9(1), 1–8 (2019)
2. Chen, Z., Huang, X.: Pedestrian detection for autonomous vehicle using multi-spectral cameras. IEEE Trans. Intell. Veh. 11(7), 1–9 (2019)
3. Purohit, N., Israni, D.: Vehicle classification and surveillance using machine learning technique. In: IEEE International Conference on Recent Trends in Electronics, Information & Communication Technology, vol. 20, no. 34, pp. 910–914 (2018)
4. Zhao, M., Zhao, C., Qi, X.: Comparative analysis of several vehicle detection methods in urban traffic scenes. In: IEEE International Conference on Sensing Technology, vol. 7, no. 19, pp. 119–126 (2018)
5. Lin, J.P., Sun, M.T.: A YOLO-based traffic counting system. In: IEEE Conference on Technologies and Applications of Artificial Intelligence, vol. 10, no. 23, pp. 82–85 (2018)
6. Choudhury, S., Chattopadhyay, S.P., Hazra, T.K.: Vehicle detection and counting using haar featurebased classifier. In: IEEE Annual Industrial Automation and Electromechanical Engineering Conference, vol. 17, no. 11, pp. 106–109 (2017)
7. Nawaratne, R., Bandaragoda, T., Adikari, A., Alahakoon, D., De Silva, D., Yu, X.: Incremental knowledge acquisition and selflearning for autonomous video surveillance. In: IEEE Annual Conference of the IEEE Industrial Electronics Society, vol. 10, no. 15, pp. 4790–4795 (2017)
8. Saribas, H., Cevikalp, H., Kahvecioglu, S.: Car detection in images taken from unmanned aerial vehicles. In: IEEE Signal Processing and Communications Applications Conference, vol. 10, no. 8, pp. 840–845 (2017)
9. Lin, C.-Y., Chang, P., Wang, A., Fan, C.-P.: Machine learning and gradient statistics based real-time driver drowsiness detection. In: IEEE International Conference on Consumer Electronics, vol. 4, no. 12, pp. 1801–1807 (2018)
10. Zhigang, Z., Huan, L., Pengcheng, D., Guangbing, Z., Nan, W., Wei-Kun, Z.: Vehicle target detection based on R-FCN. In: IEEE Chinese Control And Decision Conference, vol. 10, no. 19, pp. 5739–5743 (2018)
11. Lee, K.H., Hwang, J.N.: On-road pedestrian tracking across multiple driving recorders. IEEE Trans. Multimedia 17(9), 1429–1438 (2015)
12. Liu, W., Lau, R.W.H., Wang, X., Manocha, D.: Exemplar-amms: recognizing crowd movements from pedestrian trajectories. IEEE Trans. Multimedia 18(12), 2398–2406 (2016)

INTERNET OF THINGS
Its Application Usage and the Problem Yet to Face

Shivam Verma$^{(\boxtimes)}$, Nazish Jahan, and Paresh Rawat

Sagar Institute of Science and Technology,
Gandhinagar, Bhopal 462036, MP, India
Shivamverma4956@gmail.com, nazish9009@gmail.com,
parrawat@gmail.com

Abstract. With the increasing technological growth, technological develop-ments, the demand for digital security system is increased in the last two decades, which is actually required to design a highly secured digital locker in this current scenario. The paper is about to show the application of IOT Technology in the bank locker and the home set up locker technology. The future of Internet of Things (IOT) is already upon us. IoT plays a vital role in various domains like security, communication, Healthcare, Energy and industrial Automation technologies. IoT based smart locker is relatively a new concept, where it focuses on the ease of management. It focuses on easy management and assignment of lockers to users by taking the entire hassling system online using the fundamentals of Internet of Things (IOT) by connecting the entire mechanism to the cloud by means of cloud computing technologies, Although the IOT technology provides a safer, better, smarter solution for the domestic and industrial automation and security service to our data and the operands. For the secure system design we can use the IOT because at provide one more level of security. The IOT was used in locker as a security channel so that our user can operate the Locker by itself with or without his presence for the IOT channel security purpose we use the most common type of device having an AI (Artificial Intelligence) Cloud inside named as ESP8266 (A micro controller contain device that is use to connect any hardware to the internet) it is the launchpad for the whole IOT system. The major contribution of the paper to make the locker safer and to show the IOT technology application in the various domain. The smart locker is very convenient, efficient than the traditional used locker which required the repairing of lock key etc. the analog and digital method is provided to the locker in order to make it more robust in nature.

Keywords: IOT · Launchpad · Smart locker · ESP8266 · Cloud computing · AI cloud

1 Introduction

Internet of Things (IOT) based smart locker develop a thought of remotely connecting and monitoring locker and its locking mechanism through the Internet. While using the con-venience and the efficiency that IOT provides us, new threats also have generated. There are increment in the research of the solution to rid the threats away from the actual data.

© Springer Nature Switzerland AG 2020
D. J. Hemanth et al. (Eds.): ICICI 2019, LNDECT 38, pp. 238–244, 2020.
https://doi.org/10.1007/978-3-030-34080-3_27

The Analog Digital Wi-Fi Interface locker was made up of 4 layer of the security system

- Analog Security
- Digital Security
- IOT security
- Mechanical lock (solenoid lock)

These four layer contain the whole locker security system .the analog part contain the resistor balance channel that should be balance at the particular analog value of order of that we provide to the microcontroller. After successful authentication of the analog channel the locker signal goes to the digital channel that contain a hex pad to enter the correct password to go to the further process of operating the locker, if the digital channel authentication gets successful it sends a data to the mobile ap cloud through which the operating devise is connect than if the user allows the data to be operate then the data returns to the controller and the resultant signal from the IOT channel switch on the solenoid valve so the locker get automatically open by using some motion device. Hence the whole process is a step by step process if anyone get fail to access or the authentication of any of the channel gets fail so the locker set to the initial value. So the 4 layers including the IOT channel gets the locker more smarter, safer, better than the other lockers presented in the market. The locker present in the market is the Bluetooth enabled, by connecting through Bluetooth, Mobile phones cannot operate the locker from a distant location, so the major problem in using the Bluetooth sensor in our locker is the operation of the locker from the far location but in this project we provided the IOT launchpad that is ESP8266 it is enabled with the home or office Wi-Fi network, so using the IOT launchpad instead of using the Bluetooth gives a very wide range of operating the locker.

The block diagram of the processing of the locker is shown in the Fig. 1.

Since many kind of locker presented in the market shown below.

Table 1. Comparison of existing locker systems in the market and features

S. no	Name of the company	Type of technology used
1	Vaultek Electronic locker Manufacturer	Traditional locking, hex pad locking
2	Godrej access electronic safe SEEC (9060)	3–6 digit pin with led and basic mechanism
3	Godrej goldilocks personal safe(2.2) litres	Anti theft alarm with touchscreen
4	Godrej Taurus electronic safe(Ivory)	Auto shut system in case of wrong password

The existing locker systems and their review is briefly described and compared on the URL as: https://www.bestforyourhome.co.in/best-electronic-lockers/. Since all kind of locker present in the market having traditional locking locking and simple basic electronic locking system as we seen in Table 1 so in comparison to this kind of lockers the smart IOT based locker is much reliable, efficient, secured, and cost efficient due to 4 layer of security so that it can be used for the protection of the valuable things in banks as well as home. The comparison between using the other technology supported locker and the IOT supported locker

Table 2. Technological security modules available and there're features

S. no	Comparison point	IOT enable locker	Bluetooth locker	Normal lockers
1	Operating range	Worldwide	10 m	Manual operation
2	Layer of security	Can be have multilayer with IOT	Relatively less than IOT enable locker	Can be have 2–3 layer
3	Speed of processing	Very fast (can be adjusted)	Relatively slow	Fast
4	Cloud computing	Required	Required	Not required
5	Deal with big data	Yes	No	NO
6	Threat of data loss	Low	High	Relatively low
7	Presence of keyholder	Not required can be operate by internet	Required after 10 m (without hopp)	Must required

2   Advantages

- The cost efficient locker can take care of the home locker as well as the Bank locker security in the efficient manner.
- This analog digital Wi-Fi locker system use IOT Launchpad so that it can be operate from anywhere in the world so it is more lenient then to input the key in manual locker system.
- To operate the locker the user just need an smart phone and the IOT launchpad is connected to the Home/Office/Bank secured WI-FI.
- The optional smart phone application takes care of the fact that the user may also wish to control his locker.
- To operate locker system the user need not have data connection enabled in his phone. The system runs ne with the launchpad connected to Wi-Fi at home.
- Since we observed from the Table 2 that the IOT enabled locker is more safer, faster and reliable to the user.

Since the locker needs a power supply and if there is no electricity in the house so the unknown user cant operate it but it has a backup power supply that only the user knows.

Flow Chart Process of the Locker Operation

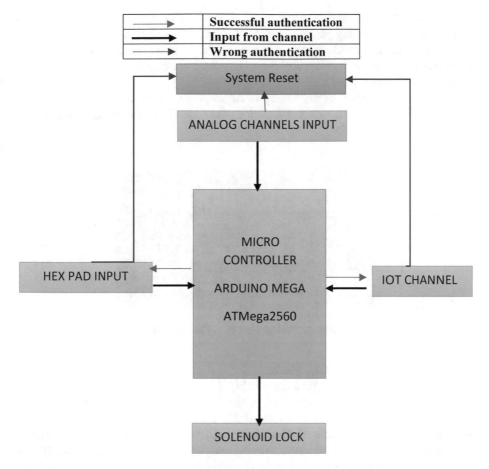

Fig. 1. Block diagram representation of locker operation

2.1 Setup and Equipment

- Arduino mega (AT Mega 2560) microprocessor board.
- ESP8266 Wi-Fi module (AS Launchpad for internet).
- Solenoid lock (12 V DC)
- Motor Driver (to operate the motor)
- Hex pad
- Analog channel (Variable resistor)

Arduino Mega (ATMega2560)
The Arduino is the Open source platform device. The Arduino has many kind of development Board but we used the Arduino Mega Development Board, which has the

ATMega 2560 Microcontroller. The specification of the Arduino Mega is shown below.

- It has the 54 input/output pins and the 15 pins can be used as PWM pins,16 pins can be used for the analog pins and 4 UART (Hardware and Serial port).
- The 16 MHz crystal Oscillator can generate the 16 MHz clock pulse.
- The Arduino Mega development Board has the USB connector, Power jack, ICSP Header and Reset button.3
- The Ac-to-Dc adapter or the battery is use to power the board.
- The ATMega 2560 is compatible with some shield which is designed for the Arduino UNO (Fig. 2).

Fig. 2. Arduino Mega ATMega2560

ESP8266(Launch Pad) WI-FI Module

The **ESP8266** is a low-cost WI-FI microchip having full TCP/IP Stack and micro-controller capability. It is manufactured by the Espressif system Shanghai, China. The first time it comes in the market in August 2014 with the ESP-01 module and it was made by another manufacturer Ai-Thinker. The ESP8266 allows the microcontroller to connect to WI-FI network by using hyes-style commands it builds a simple TCP/IP connections. Since it is designed by the chinese manufacturer so the initially it is use to send the command and receive the data in chinese language but some of the additional external component use and the hardware as well as software was developed in English language. The ESP-32 microcontroller is released after the predecessor device (Fig. 3).

Fig. 3. ESP8266 WI-FI module

The Arduino mega is connected to the 5 V dc power supply adapter, that control the operation of the locker it is like the brain of the locker. The launchpad that is ESP8266 connected to the internet present in the home or office where the locker is to be setup, the Wi-Fi key can be insert in the program. the launchpad is connected with the 5 V DC supply, when the Arduino mega gives an instruction to the esp8266 by its port than the launchpad send the data to the app cloud that is using to operate the locker and the app we used in our project is the BLYNK App, so the data transferred by the Arduino mega is send to the blynk data cloud it is synchronized by the operating device with a security key and token that only the user have, so when any kind of data come on the cloud than the app shows that "AUTHENTICATION SUCCESSFUL" after the successful authentication the user can easily send the data to Arduino to operate the locker or not as shown in Fig. 1.

2.2   Challenges

Since IOT is an internet operated technology so it send and receive so the data needs a particular protocol for the communication so the IOT device uses Lightweight communication protocol also sometimes we have to face the traffic in the data gateway. The some of the problems using the IOT is shown below.

- Routing – the routing is the most important for the communication because it decide the path for the information from the source to destination. The best part can be determine for travelling the information like hopping network, increasing the bandwidth etc.
- Network Security and privacy- the security of the IOT system aims to protect the data from the threat. Threat are of 2 types

  1. internal threat
  2. external threat
  the internal threat stands for the misuse of the data and the external threat stands for the attacking from the outside by cyber attackers and hackers. Since IOT offers the light weighted internet protocol so the threats chances gets increase as well as the probability of the data lose gradually increase.

2.3   Conclusion and Future Scope

Paper proposes the IoT based security model for Digital locker. At the time of project the smart IOT platform has been built to make the locker more safe than before by providing multi-layer security. The IOT Technology is provided as a security layer for the locker. The paper reviews and compares the performance of various existing security system and also highlights their merits and demerits. The basic implementation of proposed IOT based security system was described, and the overall paper is considered as the significant contribution to the field of digital security system.

- The future of this IoT enabled locker is it can be used in the bank so we will implement it for two channel.

- A camera can be mounted in the locker so that if anybody tries to break the locker, it can snap and send the picture of the victim.3
- Apply image processing technique to identify the user so that it may be added as a new security layer.

References

1. Parab, P., Kulkarni, M.: Smart Locker Management System Using IOT
2. Kumar, D.M., Hanumanthappa, D.M., Suresh Kumar, D.T.V., Ojha, M.A.K.: Android based smart door locking system with multi user and multi level functionalities. Int. J. Adv. Res. Comput. Commun. Eng. (IJARCCE) (2016)
3. Basha, S.N., Jilani, D.S.A., Arun, M.S.: An intelligent door system using Raspberry Pi and Amazon web services IOT. Int. J. Eng. Trends Technol. (IJETT) **33**, 84–89 (2016)
4. Srilekha, M.R., Jayakumar, M.R.: A secure screen lock system for android smart phones using accelerometer sensor. IJSTE – Int. J. Sci. Technol. Eng. **10**, 96–100 (2015)
5. Verma, G.K., Tripathi, P.: IIT Allahabad a paper on A digital security system with door locks system using RFID technology (2015)
6. ATmega 328 Pdatasheet. http://www.atmel.com/Images/doc8161.pdf
7. Sayar, A.A., Pawar, D.S.A.: Review of bank locker system using embedded system. Int. J. Adv. Res. Comput. Commun. Eng. (IJARCCE) **5**, 282–285 (2016)
8. Sarp, B., Karalar, T., Kusetogullari, H.: Real time smart door system for home security. Int. J. Sci. Res. Inf. Syst. Eng. (IJSRISE) **1**, 121–123 (2015)
9. Swetha, J.: RFID based automated bank locker system. Int. J. Res. Eng. Technol. **03**(05) (2014). EISSN: 2319-1163 | pISSN: 2321-7308
10. Bramhe, M.V.: SMS based secure mobile banking department. Int. J. Eng. Technol. **3**(6)
11. Chavan, G., Dabke, S., Ghandghe, A., Musale, K.A.: Bank locker security system using Android application. Int. Res. J. Eng. Technol. (IRJET) **02**(01) (2015)
12. Sankar, S., Srinivasan, P.: Internet of Things based digital lock system. J. Comput. Theoret. Nanosci. **15**, 1–6 (2018)
13. Sridhar, S., Smys, S.: Intelligent security framework for IoT devices cryptography based end-to-end security architecture. In: 2017 International Conference on Inventive Systems and Control (ICISC), 19 January 2017, pp. 1–5. IEEE (2017)
14. Oommen, A.P., Rahul, A.P., Pranav, V., Ponni, S., Nadeshan, R.: Int. J. Adv. Res. Electr. Electron. Instrum. Eng. **3**, 7604 (2014)
15. http://Arduino.cc
16. http://blynk.io
17. https://en.wikipedia.org/wiki/ESP8266

Biometric Authentication in Cloud

Kshitij U. Pimple[(⊠)] and Nilima M. Dongre

Department of Information Technology, Ramrao Adik Institute of Technology,
Nerul, Navi Mumbai, India
pimplekshitij99@gmail.com, nilimarj@gmail.com

Abstract. Cloud computing is used to store information and supply computing services and resources to humans available all over the internet. Huge amount of data is being saved in data centers, which could therefore create chance for records leakage and get a right of entry to develop illegitimate customers. Authentication of users therefore helps to make sure and confirm a person's identification earlier than granting him/her get admission into cloud Storage system. The main goal of this method is to authenticate at client-side server, server side, or as third party, and the use of special techniques to maintain the information in a relaxed and secure way. We present a overview on the exceptional authentication mechanisms used in cloud. Our Main focus is on providing biometric authentication for cloud-based structures. In addition, we develop awareness on physical and behavioral based biometric authentication techniques and offer a comparative analysis as which method is better in each instance and thereby provides a better authentication mechanism for cloud.

Keywords: Biometric authentication · Physical biometric · Behavioral biometric · Cloud computing · Authentication schemes

1 Introduction

Cloud computing has prompted the advancement in the innovation field which gives on-request benefits for IT endeavors. All areas are moving to distributed computing in light of cost decrease, accessibility of storage room and because of its practical alternative to modernize the legacy frameworks. Be that as it may, security issues in distributed computing have assumed a noteworthy part in backing off its acknowledgment. Fake users can act like legitimate clients and subsequently get unapproved access to basic data. The clients are not guaranteed of the secrecy and honesty of information put away in cloud. It is very difficult w.r.t the Cloud environment to enable clients to get to their administrations whenever, wherever, and anyplace, free of the systems and devices being utilized. Security assumes an indispensable job in shielding clients from their individual rights yet in multi-domain conditions and administration [20] arranged engineering it is critical to execute multi-domain strategy mix and secure administration creation. The fundamental parts of protection are the absence of client control, potential unapproved optional use, information multiplication and dynamic provisioning. Accordingly [19], significant prerequisite for verifying and accessing is to make it a consistent experience for end-clients. The essential of is straightforward

© Springer Nature Switzerland AG 2020
D. J. Hemanth et al. (Eds.): ICICI 2019, LNDECT 38, pp. 245–254, 2020.
https://doi.org/10.1007/978-3-030-34080-3_28

end-client confirmation and security over diverse system advances. A few protection saving methodologies or the systems for privacy protection and user authentication are mentioned below with a brief writing overview.

1.1 Authentication in Cloud Computing

Authentication is the way toward approving a client on the credentials as indicated by him while enrolling for a unique service. Authentication in distributed computing is a noteworthy worry as just verified individual is permitted to get to an information. The principle goal of this procedure is to confirm at the customer side, server side, or as the outsider utilizing an alternate strategy to keep the information in a safe and safe way ensure against dangers. Verification stop attackers from misusing the content which is put away and remained careful in cloud storage. Confirmation checks whether or not the consumer has a consent keeping in mind the end aim to allow access to the framework. The invalid clients are not permitted in the framework.

1.2 Issues in Cloud Authentication

Cloud professional agencies ask for customers to store their document data in the cloud that is an advantage, cloud provider vendors have accessed this information. There is absence of straightforwardness in the cloud that allows the clients to monitor their personal particular data. Many SLAs have indicated the safety of the critical statistics, anyways, it is difficult for customers to ensure that appropriate guidelines are authorized and actualized. When a consumer chooses to make use of several cloud benefits, the consumer must store his/her password in different clouds. This is a safety issue for the cloud expert organizations. The more cloud advantage the customer is subscript to, the extra duplicate of the client's information will be coursed prompting numerous verification forms.

2   Literature Review

2.1 Authentication Schemes

- **Passwords:** A password is a phrase or series of characters applied for customer verification to demonstrate identity or access to an asset. Programmers at that point can make a dictionary attack or brute force attack to pick up a secret phrase and get right of entry to the framework. Shoulder surfing is a most regular attack wherein client secret key is compromised [17].
- **One-Time Passwords:** A single password or pin (OTP) [18] is a watchword that is big for a single login or swap session, a PC framework or a distinct computer gadget. The otp is legitimate for a unique time-frame so it cannot be mishandled by programmers. It is difficult to predict successor OTP's via an attacker and hash functions that can be used to obtain a value but are hard to decrypt.

- **Two factor Authentication:** Two factor Authentication [18], in any other case called two stage confirmation, is a further layer of protection this is referred to as" multifaceted validation" that requires a secret word and username as well as something that only, and just, that client has on them.

- **Secure Shell (SSH):** It is a convention for cryptographic equipment that benefits securely over an unsecured scheme [14]. It is a convention which offers secure login and secure systems administration. The exceptional software is used by customers for remote login to laptop frameworks. The server keys are each put away on the customer side or probably disperses by means of utilizing a key appropriation convention provides a protected channel in a client-server architecture over an unsecured computer, connecting a SSH client software to a SSH server. Normal apps include remote order line login and Remote load execution; however, SSH can be used to secure any system administration.

- Kerberos is a computer network authentication protocol that operates on the notion of tickets to enable nodes that communicate across a non-secure network to easily demonstrate each others identity [15]. It is a model of the consumer-server and it provides mutual authentication-both the customer and the server check the identity of each other. Kerberos protocol messages are shielded from escape and replay attacks.

- **Biometric authentication:** It is a safety method based mainly on an individual's distinctive biological features to verify that he is who he say he is. Biometric authentication schemes compare the capture of biometric data to the saved genuine information confirmed in database [9]. If the biometric data matches both samples, authentication will be verified. Biometric authentication is normally used to handle access to physical and digital assets such as houses, rooms and computer equipment.

2.2   Biometric Authentication in Cloud

The usage of biometric to cloud depicts that the cloud administrations can be used through an online interface. This interface can either be an internet browser or a versatile application. The essential design of any biometric recognizable proof framework stays same independent of the methodology that is utilized. Biometric framework is easy to setup. This methodology includes moving both the biometric database and the product part to the cloud. A cloud based framework has a few different perspectives, for example, continuous and parallel preparing capacities that make it additionally engaging. The far reaching accessibility of cell phones makes it open for some applications and administrations that depend on portable customers. The current age of biometric frameworks offers numerous new conceivable outcomes for distributed computing security. Receiving biometrics in distributed computing and applications will assist customers with ensuring data security and additionally give a financially savvy security answer for the specialist co-ops.

3 Physical Biometric Systems

Biometric devices comprise five primary parts: Information capture and digitization sensors, biometric template sign processing algorithms record storage units, a matching algorithm comparing fresh templates with earlier recorded and stored templates and a selection method using matching algorithm outcomes to accept or dismiss a new person. Biometric systems are able to define or check an individual's identity.

3.1 Fingerprint

No two people share the same fingerprints in human beings. Therefore, the identity of the guy or girl individual while appearing authentication is very powerful proof-of-function. Sub-characteristics along with crossover, core, Bifurcation, ridge ending, island, delta, pores and so on are used to identify fingerprint patterns.

The author [1, 2] suggested a solution to use a fingerprint recognition machine to photograph the fingertips through the digital camera of the cell phone. The intention to transform the fingertip image acquired with the help of the digicam cell phone, pre-process the photograph, extract ridge structure from it to be as comparable as possible to the ridge structure acquired from the fingerprint sensor. The comparison between the extracted features and characteristics stored in a database is the consequence of the similarity rating(S). The alternative suggested is not only to secure unauthorized access, but also to shield user string input databases from injection assaults.

Some Authors [3] suggested a two-phase automated fingerprint identification scheme. The use of the hardware device is recorded in the off-line stage and the quality of the captured image is enhanced using distinctive algorithms; then important components of the fingerprint are obtained and stored as a template in a database. During the online stage, the person's fingerprint is recorded, improved and the fingerprint characteristics are analyzed under binarization in which the contrast variant is removed; and then the thinning method is used to decrease the ridge width to a single pixel. The current fingerprint is then compared to the template that was stored in the online phase of the database. Although this scheme has been noted to be efficient, owing to the various phase method it can cause slower processing time.

3.2 Iris Identification

Iris patterns have a magnificent and abundant shape and full of complex textures. The iris texture is unique from one person to different and cannot be robbed. Also, we can readily capture iris picture from distance without touching the body. At some point in life, these characteristics are strong. This technique's authentication includes verification by matching the sub-characteristics across the eye pupil. Using idiosyncratic characteristics such as arching ligaments, fiber, freckles, furrows Ridge rings, corona, rifts and so on, an iris pattern of the eye is built. Targeted sub-characteristics include pupils, sclera, pupil region, collarette, radial furrows, crypts, pigment spots and concentrated furrows.

The authors in [4] have proposed a system using the eye image for authentication purpose. An eye picture is acquired live video camera and in addition pre-processed for enhancement using localization, normalization and transformation techniques. To detect the edge map of the picture, Canny edge detection is implemented. The pattern of wrapped iris is transformed into unwrapped model of rectangular iris. This is contrasted to the iris templates already stored by calculating the distance between two templates. The writers used the code of the whole iris. Probability is calculated to the input picture for each stored model. A threshold value is selected and the iris codes with a probability value equal to or greater than that selected.

3.3 Retinal Scanning

Every person has different retina pattern. Hence retina can be used to authenticate and verify an individual. Infrared illumination can be used to distinguish between retina images. We consider outer part of iris, outline of pupil and vessels for retina matching. Vladyslav et al. [6] have proposed a method that can be utilized for other biometric system on Linux operating system. Authors have shown various ways of utilizing different authentication methods. Paper enables us to pick accurate biometric device based on multiple criteria and factors.

3.4 Hand Geometry

To enable verification, this authentication mechanism includes studying hand geometric characteristics such as thickness, palm width, length and finger width at minute level. It may, however, also used as a method for verifying a user but not identifying him. However, study has shown important success when used in conjunction with other authentication characteristics.

3.5 Face Recognition

Authentication of facial recognition is performed through verification of matching a human face. Changing a hair style makes it difficult to recognize an authenticated individual. The fact that human face is dynamic is nature has not been discovered to be robust. The identifying components used to facilitate verification are the characteristics of eyes, eyebrows, nose, chin, mouth, hair etc. The disadvantages that consist of changes in hair fashion, beard and ageing that make it hard to recognize are very powerful, which is taken into account.

In [10], The writers suggested a fresh face identification scheme (FRS) that overcomes all the disadvantages of traditional and other biometric authentication methods and only allows approved clients to access cloud server information or services. The picture is originally caught in which its face is noticeable. In the event of 2d facial recognition, a standard resolution digital camera is required. Additional face detection includes facial identification in the picture captured. The face captured in the digital camera may not be completely perpendicular to the camera and therefore it is necessary to determine and compensate the alignment so it is ready to use the process of recognition. Finally, the extraction of features includes measuring different facial elements and establishing a facial template for matching and identifying purposes (Figs. 1, 2, 3 and 4).

Fig. 1. Fingerprint

Fig. 2. Iris and retinal scanning

Fig. 3. Hand geometry

Fig. 4. Face recognition

4   Behavioral Biometric Systems

Behavioral biometrics is the field of research connected with the measurement of uniquely identifiable and measurable patterns in human activity. The word contrasts with physical biometrics involving inherent human features along with fingerprints or iris patterns. Methods of biometric behavioral verification include keystroke dynamics, gait analysis, voice, features of mouse use, signature analysis and cognitive biometrics.

4.1   DNA Recognition

In [15], a novel biometric template protection algorithm based entirely on DNA encoding was suggested by the writers. First, the biometric aspects of the multimodal template are converted by DNA encoding into a DNA sequence, and a chaotic sequence is generated and converted into a DNA sequence, then the DNA adding operation performs the two DNA sequences, and the sum is converted into decimal numbers so that the encoded template is obtained. The experiment's impacts indicate that the suggested biometric multimodal template protection system does not influence the efficiency of recognition and ensures the safety of the biometrics' multimodal model template.

In [16], the writers have a solid watermarking strategy to conceal DNA sequence data in fingerprint templates for copyright protection based on the discrete transformation of the wavelet. A multi-bit watermark, based mainly on DNA information, is integrated in the fingerprint image's low frequency sub-hand. The DNA decoding is used to obtain the picture signature from the watermarked picture during the reverse phase. The writers used statistical measures to evaluate the suggested method, including mean squared error, peak signal to-noise ratio, and structural similarity index and the experimental findings indicate that the watermarks produced by the suggested strategy are invisible.

4.2 Signature Recognition

Dhagat et al. [11] has mentioned that distance signature can be defined as an authenticity technique for any information. Digital signature ensures that information is sent via authentic user during the transmission of information. Hence digital signature is a technique which is used for trusted data transmission from authentic source. Dhagat et al. [11] has proposed to protect proxy signer. Method proposed by author uses a secure signature in order to distinguish between false and true signer.

4.3 Voice Recognition

In [13], the writers given a unique method for storing individual voice prints, mainly based on the version of Hidden Markov. VoizLock task investigates a distinct manner of using HMM for voice authentication than voice recognition. This voiceprint will then be used for voice authentication, using text-independent speaker recognition techniques in which the system does not depend on speaking a particular text, but only on speaker's speech. More about the user training stage, the article describes how a person's voice print is saved inside the scheme by extracting certain waveform values using HMM. This analyzes, apart from the training stage, the results acquired from the tests conducted covering distinct voice authentication situations.

4.4 Typing Behaviour

Keystroke dynamics is a word indicating the timing detail of the consumer while typing. In [14], writers launched dynamic interval (SADI) sequence alignment algorithms from keystrokes to model-based authentication systems. An interval function is basically the length of each label of attributes and is used in a sequence algorithm alignment to split each attribute into parts. However, the dynamic interval characteristics suggested in this study are comparable to interval characteristics but split each attribute. In a distinctive section number. Dynamic interval characteristics are selected to maximize the comparability of keystroke information similarity measurements. Experimental outcomes at the benchmark dataset of the CMU show that the suggested SADI is similar to other published methods and sometimes outperforms them.

5 Comparison of Physical and Behavioural Authentication Schemes

The physical and behavioral biometrics authentication has been compared with respect to the following properties: Universality, Distinctiveness, Permanence, Collectable, Performance, Acceptability, Uniqueness, Privacy Concept, Safety, Cost,

Popularity, Ease of use. Based on the detailed study, we provide a comparison of the techniques as follows.

5.1   Comparative study of Physical and Behavioral Authentication Techniques

(See Fig. 5).

Technique	Universitality	Distinct	Permanence	Collectable	Performance	Acceptance	Unique	Privacy	Safe	Cost	Popularity	Ease of use
Fingerprint	M	H	H	M	H	M	H	H	M	L	H	H
IRIS	H	H	H	M	H	L	H	H	H	H	H	M
Retinal Scanning	H	H	M	L	H	M	M	L	H	H	M	L
Hand Geometry	M	M	M	H	M	M	M	L	M	M	L	H
Face Recognition	H	H	M	H	L	H	L	H	M	H	H	H

Technique	Universitality	Distinct	Permanence	Collectable	Performance	Acceptance	Unique	Privacy	Safety	Cost	Popularity	Ease of use
DNA Recognition	H	H	H	L	H	H	H	L	H	H	H	L
Voice Recognition	M	L	L	M	L	H	L	M	H	L	H	H
Signature Recognition	L	L	L	H	L	H	L	H	H	M	H	H
Typing Behaviour	L	L	L	M	L	M	L	L	L	M	L	L

Fig. 5. Comparative study of physical and behavioral authentication techniques

5.2   Analysis

From the above comparative analysis, it can be concluded that; Fingerprint and IRIS are best in physical biometrics authentication techniques. This is because of the safety, permanence and distinctiveness properties that these two techniques reflect. Also in Behavioral authentication mechanism, DNA Recognition and Signature Recognition are the best because of their good performance and ease of availability.

6   Conclusion

In this paper, we studied the specific physical and behavioral authentication strategies which can be utilized in cloud. We finished a comparative analysis and observed that Fingerprint and IRIS are best in physical biometrics authentication techniques. Also in Behavioral authentication mechanism, DNA recognition and Signature recognition are the best because of their good performance and ease of availability. From the results obtained, we conclude that a combination of both physical and behavioral authentication in the form of a multi-factor authentication scheme will be best suited in today's cloud storage systems.

References

1. Lohiya, R., John, P., Shah, P.: Survey on mobile forensics. Int. J. Comput. Appl. (2015)
2. Rassan, I.A.L., AlShaher, H.: Securing mobile cloud computing using biometric authentication (SMCBA). In: 2014 International Conference on Computational Science and Computational Intelligence (2014)
3. Venakatesan, N., Rathan Kumar, M.: Finger print authentication for improved cloud security. In: 2016 International Conference on Computational Systems and Information Systems for Sustainable Solutions (2016)
4. Raghava, N.S.: IRIS recognition on hadoop: a biometrics system implementation on cloud computing. In: Proceedings of IEEE CCIS (2011)
5. Sabri, H.M., Elkhameesy, N., Hefny, H.A.: Using Iris Recognition to Secure Medical Images on the Cloud. IEEE. 978-1-4673-9669-1/15/31.00/2015
6. Borodavka, V., Tsuranov, M.: Biometrics: analysis and multi-criterion selection. In: The 9th IEEE International Conference on Dependable Systems, Services and Technologies, DESSERT 2018, Kyiv, Ukraine, 24–27 May 2018 (2018)
7. Bapat, A., Kanhangad, V.: Segmentation of hand from cluttered backgrounds for hand geometry biometrics. IEEE. 978-1-5090-6255-3/17/31.00 2017
8. Varchol, P., Levicky, D., Juhar, J.: Multimodal biometric authentication using speech and hand geometry fusion. In: 2016 15th International Conference on Systems, Signals and Image Processing (2016)
9. Naveen, S., Shihana Fathima, R., Moni, R.S.: Face recognition and authentication using LBP and BSIF. In: 2016 International Conference on Communication Systems and Networks (ComNet), Trivandrum, 21–23 July 2016 (2016)
10. Pawle, A.A., Pawar, V.P.: Face recognition system (FRS) on cloud computing for user authentication. Int. J. Soft Comput. Eng. (IJSCE) 3(4) (2013). ISSN 2231-2307

11. Dhagat, R., Joshi, P.: New approach of user authentication using digital signature. In: 2016 Symposium on Colossal Data Analysis and Networking (CDAN) (2016)
12. Bharadi, V.A., D'Silva, G.M.: Online signature recognition using software as a service (SaaS) model on public cloud. In: 2015 International Conference on Computing Communication Control and Automation (2015)
13. Jayamaha, R.G.M.M., Senadheera, M.R.R., Gamage, T.N.C., Weerasekara, K.D.P.B., Dissanayaka, G.A., Kodagoda, G.N.: VoizLock – human voice authentication system using hidden Markov model. In: ICIAFS 2008. IEEE (2008). 978-1-4244-2900-4/08/25.00 c 2017
14. Ho, J., Kang, D.-K.: Sequence alignment of dynamic intervals for keystroke dynamics based user authentication. In: SCISISIS 2014, Kitakyushu, Japan, 3–6 December 2014
15. Dong, J., Meng, X., Chen, M., Wang, Z.: Template protection based on DNA coding for multimodal biometric recognition. In: 2017 4th International Conference on Systems and Informatics (ICSAI 2017) (2017)
16. Ghany, K.K.A., Hassan, G., Hassanien, A.E., Hefny, H.A., Schaefer, G., Ahad, M.A.R.: A hybrid biometric approach embedding DNA data in fingerprint images. In: 3rd International Conference on Informatics, Electronics, Vision (2014)
17. Babaeizadeh, M., Bakhtiari, M., Mohammed, A.M.: Authentication methods in cloud computing: a survey. Res. J. Appl. Sci. Eng. Technol. 9(8), 655–664 (2015)
18. Meena, S., Syal, R.: Authentication scheme in cloud computing: a review. IEEE. 978-1-5090-3239-6/17/31.00/2017
19. Sridhar, S., Smys, S.: A hybrid multilevel authentication schemeFor private cloud environment. In: 2016 10th International Conference on Intelligent Systems and Control (ISCO), 7 January 2016, pp. 1–5. IEEE (2016)
20. Karthiban, K., Smys, S.: Privacy preserving approaches in cloud computing. In: 2018 2nd International Conference on Inventive Systems and Control (ICISC) 19 January 2018, pp. 462–467. IEEE (2018)

Demosaicing Using MFIT (Matrix Factorization Iterative Tunable) Based on Convolution Neural Network

Shabana Tabassum[1]([✉]) and SanjayKumar C. Gowre[2]

[1] Department of E&CE, KBN College of Engineering, Kalaburagi 585104, India
shabanat99@rediffmail.com
[2] Department of E&CE, BKIT, Bhalki 585328, India
sanjaygowre@gmail.com

Abstract. Demosaicing is introduced for estimation of the other missing color for producing the absolute color image. Enormous research has been done in past to develop the Demosaicing algorithm, however all these methods have shortcomings, hence in this paper, we propose an algorithm called MF-IT (Matrix Factorization Iterative Tunable), this methodology is based on the CNN (Convolution Neural Network), the main aim of this research is to improvise the reconstruction quality of the given image. This is achieved by using the image-block adjustments based transform, which helps in utilizing the image block based transform. One of the main advantage of MFIT is shift invariance, which can be easily obtained without any support of striding image blocks. Henceforth, In order to evaluate the performance of our methodology it is compared against the various state-of-art method. The result analysis shows that proposed methodology simply outperforms the other method.

Keywords: Demosaicing · Image block · Stride · Non-stride

1 Introduction

When the monochrome camera captures an image, CMOS sensor is used for sampling the intensity of light, which is projected on the sensor [1]. Similarly, the colored images are also captured, moreover in the color images the color images are mapped in various parted color bands. For absolute measurement, three different sensors were used i.e. for Red, Green and Blue [2]. Since this approach was very much expensive, CFA was introduced that can capture the color image by using the single sensor. CFA is nothing but Color Filter Array, which samples the single band color at every location [3]. In CFA pattern, red filters and blue filters are available at the horizontal and vertical directions respectively; at remaining locations, the green filters are placed.

CFA demosaicing is technique through which three whole color plane data are created [4]. Moreover, the demosaicing is still in its infancy stage hence there need to extensive research. Moreover, the primal issue with the existing technique regarding the demosaicing is that they assume, the assumption crosses the limit of applicability, hence, it can be said that method are agile. In simple words, the demosaicing can be

© Springer Nature Switzerland AG 2020
D. J. Hemanth et al. (Eds.): ICICI 2019, LNDECT 38, pp. 255–264, 2020.
https://doi.org/10.1007/978-3-030-34080-3_29

defined as the technique of reconstructing the wholethree-color image with the estimation of missing pixel in the each and every plane [5] (Fig. 1).

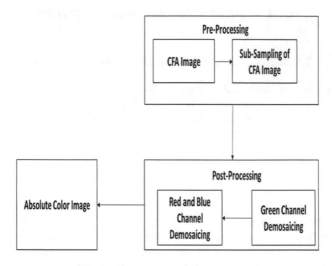

Fig. 1. Typical demosaicing process

The above diagram shows the typical demosaicing process. It mainly contains three distinctive blocks, first block shows the pre-processing and it shows the initial process for data processing. Here the image is sub-sampled into the three different channels, it can follow any pattern which will primary contain the three colors as Red, Green and Blue. Next block contains the Post-Processing, in this the demosaicing initiates with the green color due to the large availability of the color green in given CFA. In the Demosaicing algorithm the first channel, i.e. green channel is made to act as a guide channel later the R& B channel is applied which results in the output as the absolute image.

The demosaicing method is classified into two distinctive categories i.e. dictionary based [6, 7] and interpolation based [8]. Moreover in past various methodologies has been proposed; moreover the prior work has assumed the problem to solved for the Bayer color pattern [9], moreover many camera follows the other pattern of CFA [10]. Many demosaicing methods have been employed and these methods uses the repetitive refinement, for constructing the initial solution sampled channel [11] is used. In paper [12], the demosaicing problem was analyzed in the given frequency domain, Moreover, CFA are decomposed into the 3 components i.e. two chrominance frequencies and one luminance frequencies, later in the [13] design of CFA was done to increase the distance among luminance and chrominance, this achieved the optimal pattern through the exhaustive search. Some of the method which applies the CNN (Convolution Neural Network) has been used. [14] proposed a novel technique for the Bayer demosaicing. Here, the model uses the bilinear interpolation to generate the initial image it has two stage, in first stage the estimation of green and Red/Blue takes place separately, whereas in second stage the three channels ae joined and then estimated.

Here the CNN were designed for the end –to-end model, similarly [15] proposed a particular CNN model for both i.e. demosaicing and images are down-sampled into the comparatively low-resolution feature map. Here the input taken is mosaic image, it is combined with the un-sampled residual, and the last convolution group generates the output.

Motivation and Contribution of Our Research
Demosaicing involves in predicting the missing information, this inevitably generates error, which leads to different visual artifacts in the reconstructed image. Matrix factorization method can be implemented by using striding or non-striding image block. Segregation which consists of striding image block reduces the artifacts, however it is more expensive and the segregation which consists of non-striding part reduces the load of computation but leads to image artifacts. Meanwhile this paper focuses on reconstructing the image with better quality by using the CNN. Here, a methodology named as MFIT (Matrix Factorization Iterative Tunable) is proposed i.e. matrix factorization is implemented with iterative tunable which effectively utilizes the Image-Block based transform with less cost. Later, in order to show the evaluation of algorithm we have compared our methodology with the existing methodology; comparison is done by considering the PSNR and SSIM, which is explained in the later section.

This research is organized in such a way that first section consists of the background research with introduction; later part of the section contains the existing methodology. Second section presents motivation towards this research and the contribution of this particular research third section presents the methodology proposed along with the MFIT methodology, fourth section presents the results and analysis where the evaluation of our proposed methodology is done and the comparative analysis is done with various state of art technique.

2 Proposed Methodology

Demosaicing is nothing but the process of reconstructing the FCI (Full Color Image), in past various methods have been proposed for demosaicing. Moreover in recent days several researcher has focused in solving the various problem arises while reconstructing the image, however the methods which has been proposed have several limitation, one of the major limitation is use of many number of image blocks, this in terms causes the extended load on computation which restricts the model in to real time environment. To overcome this we have proposed the matrix factorization algorithm, which works in iterative.

2.1 Matrix Factorization Method

Lets consider an image by $X \times Y$, then the forward model can be represented by the below equation.

$$U_D = U_oP + G_n \tag{1}$$

Where U_D signifies the under sampled data and n signifies the Gaussian noise, in order to from the matrix factorization, the given underlying image is parted into the two image block i.e. stride and non-stride, the image blocks within the Ω is denoted by $|\Omega|$. In a given set of image blocks, an image is parted into the various ways, this can be done by displacing the partition by various amount of pixels with the every dimensions. Displacement in shifts can be denoted by using N_Ω. These image blocks have the dimensions of $x \times y$. let us consider M_r as the linear operator, which has been, extracted from the given image data. This data corresponds to the rth image block of segregated Ω, this in terms generates the matrix $M_r(p)$. The inner product $\langle E, F \rangle_{D^{xy \times D}} = recon(tr(E^H F))$ then the M_r^* satisfies the below equation.

$$\langle M_r(P), Q \rangle_{D^{xy \times D}} = \langle X, M_r^*(Q) \rangle_2 \tag{2}$$

$P \in D^{XYD}$ and $Q \in D^{XYD}$, from the above equation the linear operator can be defined as $L_\Omega: D^{DXY} \to P$, where $X = D^{|\Omega| \times D \times xy}$, the $L_\Omega P$ component is given by below equation.

$$[L_\Omega P]_r = M_r(P) \tag{3}$$

The inner product is given by the below equation i.e. Eq. 4 and the norm is given as in Eq. 5

$$\langle P, Q \rangle_P = \sum_{q=1}^{|\Omega|} Recon\left(tr\left(P_r Q_r^G\right)\right) \tag{4}$$

Norm is given as where P_r, $Q_r \in D^{xy \times D}$

$$\|P\|_P = \left(\langle P, P \rangle_P\right)^{1/2} \tag{5}$$

Another adjoin $L_\Omega^* : P \to D^{Dxy}$ is said to be the linear operator which satisfies the below equation.

$$\langle L_\Omega P, Q \rangle_p = \langle P, L_\Omega P \rangle_2 \tag{6}$$

For any $Q \in P$ and $P \in D^{dxy}$, this can be defined by the below equation.

$$L_\Omega^* Q = \sum_{r=1}^{|\Omega|} M_r^* Q_r \tag{7}$$

Since all the partition contains only non-stride image blocks which can cover entire image, then we have the following equation, since this problem is considered as the NP hard problem and the matrix is nonconvex function, rank of given matrix is approximated using the Matrix norm

$$P = L_\Omega^*(L_\Omega P)$$
$$Q = L_\Omega(L_\Omega^* Q)$$

(8)

Let there be any matrix Z which belongs to $D^{G_{n_1} \times G_{n_2}}$ as

$$\|\sigma(Z)\|_p = \|Z\|_{S_p}$$

(9)

$\sigma(Z)$ is said to be the singular values of given vector Z, based on primal dual algorithm [28] the image-block based matrix factorization can be defined in terms integrated norm i.e. for any element $P \in p$ is defined as below

$$\|P\|_{1,1} = \sum_{r=1}^{\Omega} \|P_r\|_{S_1}$$

(10)

Similarly the optimization problem can be formulated as in the below equation, λ is said to be the regularized parameter which maintains the trade-off between the constancy of data and matrix factorization. Below equation presents the generalized version to recover the Matrix Factorization image block from the given under-sampled measurement, assuming the segregation Ω of non-striding/

$$\hat{p} = \begin{array}{c} \arg\min \frac{1}{2}\|U_D - U_o P\|_2^2 + \lambda |L_\Omega P|_{1,1} \\ P \in D^{DXY} \end{array}$$

(11)

2.2 Optimization Through Iterative Tunable

We use the sparsity driven [29] to solve the above equation, this reduces the order of surrogate functions. Iterative tunable process is applied.

$$f(P, P_0) = \frac{\alpha}{2}\|P - S\|_2^2 + \lambda\|PL_\Omega\|_{1,1}$$

(12)

Here, $S = P_0 + \frac{1}{\alpha}U_o^H(U_D - U_o P_0)$ and $\alpha \geq \lambda_{max}(U_o^G U_o)$. The implementation of algorithm is done by minimizing the function in the given equation. This can be written as below equation:

$$\|P\|_{1,1} = \begin{array}{c} max\langle\psi, P\rangle_P \\ \psi \in MF_{\infty,\infty} \end{array}$$

(13)

$MF_{\infty,\infty}$ is matrix factorization

$$\check{P} = \begin{array}{c} \arg\min \frac{1}{2}\|P - S\|_2^2 + \frac{\lambda}{\alpha} \begin{array}{c} max \\ \psi \in MF_{\infty,\infty} \end{array} \langle P, L_\Omega^*\psi\rangle_2 \\ P \in D^{DXY} \end{array}$$

(14)

Later our algorithm makes sure that segregated Ω remains constant throughout the iterations and the updation of segregation is done in each segregation, this minimizes the block artifacts. Moreover, each image block is independent from one another.

3   Results and Analysis

When demosaicing is considered, the CFA possesses various advantages, such as reconstruction of images. There have been various ways of finding and improvising the possesses various advantages, such as reconstruction of images. There have been various ways of finding and improvising the demosaicing results through our algorithm we tend to improvise the reconstruction scenario. Hence, in the section implementation details about our algorithm is discussed, later our algorithm is evaluated by performing the comparative analysis with various method along with the existing one. Similarly, the comparison is done based on the SSIM (Correlation), the higher the SSIM value the better the performance.

In this paper, we have tested our model by using the IMAX datasets, here images have been used with their size 500×500, and these are basically cropped from their original size 2310×1814 of high resolution. Moreover, we imply the low-resolution image for evaluation.

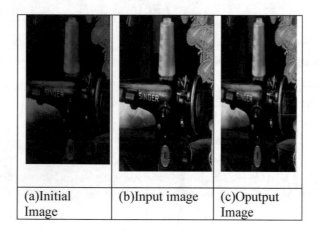

| (a)Initial Image | (b)Input image | (c)Oputput Image |

Fig. 2.   Image processed using MFIT

The above figure shows the different types of images, Fig. 2(a) shows the initial image taken Fig. 2(b) is image that is taken for the input whereas Fig. 2(c) shows the output image.

4 Comparative Analysis

In order to evaluate our algorithm, the comparative analysis has been with the state-of-art technique, comparison is done based on the PSNR or Peak to signal NR (Noise Ratio) and SS (Structural similarity) Index. PSNR is the parameter, which is used for comparing the quality of image compression. When standard algorithm and state of art technique is compared, which is shown in the Table 1, here the IMAX dataset is used. The comparison is done by taking the average of all 18 images and our method performs comparatively better than the other state of art method by possessing the PSNR value of 44.82152. The higher the PSNR value the better the reconstruction of the image. In later part of this section, the comparison is also done based on the SSIM (Structural Similarity) Index. SSIM is the metric, which quantifies the degradation in the image, these degradation might occurred due to the data compression. Our methodology possesses the value of 0.997102 and performs better than the other state-of-art method.

Table 1. Average comparison PSNR and SSIM (co-relation)

Methodology	PSNR	SSIM (co-relation)
DLMMSE [16]	40.110	0.9858
GBTF [17]	40.623	0.9887
LSSC [18]	41.445	0.9936
NAT [19]	37.714	0.9818
OSAP [20]	39.165	0.9900
NN [21]	40.603	0.9925
DJDD [22]	36.927	0.9868
ARI[23]	39.749	0.9905
MLRIwei [24]	36.894	0.9866
FR [25]	37.449	0.9822
JD [26]	36.532	0.9676
Existing (GCBI) [27]	37.621	0.9882
Existing (GBTF) [27]	40.052	0.9913
Proposed	44.82152	0.997102

Table 2 shows the pictorial comparison along with the PSNR value, the PSNR value of proposed methodology is 46.1134 compared to the existing models GCBI possesses 41.58 dB and GBTF possesses 41.57 dB. The comparison shows that proposed method performs better than not only existing methodology but several other state-of-art method as well.

Table 2. Pictorial comparison along with the PSNR value

(a) Original Image	(b) Origin patch	(c) DLMMSE[16] (40.14dB)	(d) GBTF [17] (39.93dB)
(e) LSSC [18] (38.59dB)	(f) NAT [19] (33.79dB)	(g) OSAP [20] (37.35dB)	(h) NN [21] (33.33dB)
(i) DJDD [22] (41.16dB)	(j) ARI [23] (41.27dB)	(k) MLRIwei [24] (34.99dB)	(l) FR [25] (36.36dB)
(m) JD [26] (41.56dB)	(n) Existing(GCBI)[27] (41.58dB)	(o) Existing(GBTF) [27](41.57dB)	(p) Proposed(4 6.113426)

5   Conclusion

We have proposed an algorithm namely MFIT (Matrix Factorization Iterative tunable) methodology for demosaicking. MFIT methodology is based on the image block, by taking the advantage of CNN along with iterative tunable method, we are able to achieve the better reconstructed image. Here first we design the problem definition through MFIT and then through applying the method iteratively optimization is achieved. Moreover MFIT is evaluated by the comparison analysis, when compared with the other method we observe that our method performs better and gives the higher value of PSNR and SSIM, higher value of these two metrics clearly indicates the better image reconstruction. We see that our methodology possesses the average value of 44.8512, whereas existing model has 40.052 and SSIM value is 0.997102 compared to the existing model, which has the SSIM value of 0.9913. This, comes to the conclusion that our methodology reconstructs the image with better quality. However, since this method is still in its infancy stage meanwhile in future image reconstruction can be enhanced by considering the other metrics.

References

1. Gamal, E., Eltoukhy, H.: CMOS image sensors. IEEE Circ. Dev. Mag. **21**(3), 6–20 (2005)
2. Moghavvemi, M., Jamuar, S.S., Gan, E.H., Yap, Y.C.: Design of low cost flexible RGB color sensor. In: 2012 International Conference on Informatics, Electronics & Vision (ICIEV), Dhaka, pp. 1158–1162 (2012)
3. Lukac, R., Plataniotis, K.N.: Color filter arrays: design and performance analysis. IEEE Trans. Consum. Electron. **51**(4), 1260–1267 (2005)
4. Zhang, C., Li, Y., Wang, J., Hao, P.: Universal demosaicing of color filter arrays. IEEE Trans. Image Process. **25**(11), 5173–5186 (2016)
5. Gunturk, B.K., Glotzbach, J., Altunbasak, Y., Schafer, R.W., Mersereau, R.M.: Demosaicing: color filter array interpolation. IEEE Signal Process. Mag. **22**(1), 44–54 (2005)
6. Mairal, J., Elad, M., Sapiro, G.: Sparse representation for color image restoration. IEEE Trans. Image Process. **17**(1), 53–69 (2008)
7. Mairal, J., Bach, F., Ponce, J., Sapiro, G., Zisserman, A.: Non-local sparse models for image restoration. In: Proceedings of IEEE ICCV 2009, pp. 2272–2279 (2009)
8. Monno, Y., Kiku, D., Tanaka, M., Okutomi, M.: Adaptive residual interpolation for color image demosaicing. In: Proceedings of IEEE ICIP 2015, pp. 3861–3865 (2015)
9. Bayer, B.E.: Color imaging array. U.S. Patent 3971065, 20 July 1976
10. Wang, J., Zhang, C., Hao, P.: New color filter arrays of high light sensitivity and high demosaicing performance. In: Proceedings of the 18th IEEE International Conference on Image Processing (ICIP), pp. 3153–3156, September 2011
11. Taubman, D.: Generalized Wiener reconstruction of images from colour sensor data using a scale invariant prior. In: Proceedings of the International Conference on Image Processing (ICIP), vol. 3, pp. 801–804 (2000)
12. Alleysson, D., Susstrunk, S., Hérault, J.: Linear demosaicing inspired by the human visual system. IEEE Trans. Image Process. **14**(4), 439–449 (2005)
13. Hirakawa, K., Wolfe, P.J.: Spatio-spectral color filter array design for optimal image recovery. IEEE Trans. Image Process. **17**(10), 1876–1890 (2008)

14. Gharbi, M., Chaurasia, G., Paris, S., Durand, F.: Deep joint demosaicing and denoising. ACM Trans. Graph. **35**(6), 191:1–191:12 (2016)
15. Tan, R., Zhang, K., Zuo, W., Zhang, L.: Color image demosaicing via deep residual learning. In: Proceedings of IEEE ICME 2017 (2017)
16. Zhang, L., Wu, X.: Color demosaicing via directional linear minimum mean square-error estimation. IEEE Trans. Image Process. **14**(12), 2167–2178 (2005)
17. Pekkucuksen, I., Altunbasak, Y.: Gradient based threshold free color filter array interpolation. In: Proceedings of the International Conference on Image Processing, ICIP, 26–29 September 2010, pp. 137–140 (2010)
18. Mairal, J., Bach, F.R., Ponce, J., Sapiro, G., Zisserman, A.: Nonlocal sparse models for image restoration. In: IEEE 12th International Conference on Computer Vision, ICCV, 27 September–4 October 2009, pp. 2272–2279 (2009)
19. Zhang, L., Wu, X., Buades, A., Li, X.: Color demosaicing by local directional interpolation and nonlocal adaptive thresholding. J. Electron. Imaging **20**(2), 023016 (2011)
20. Lu, Y.M., Karzand, M., Vetterli, M.: Demosaicing by alternating projections: theory and fast one-step implementation. IEEE Trans. Image Process. **19**(8), 2085–2098 (2010)
21. Wang, Y.: A multilayer neural network for image demosaicing. In: IEEE International Conference on Image Processing, ICIP, 27–30 October 2014, pp. 1852–1856 (2014)
22. Duran, J., Buades, A.: A demosaicing algorithm with adaptive interchannel correlation. IPOL J. **5**, 311–327 (2015)
23. Gharbi, M., Chaurasia, G., Paris, S., Durand, F.: Deep joint demosaicing and denoising. ACM Trans. Graph. **35**(6), 191 (2016)
24. Beyond color difference: residual interpolation for color image demosaicing. IEEE Trans. Image Process. **25**(3), 1288–1300 (2016)
25. Wu, J., Timofte, R., Gool, L.J.V.: Demosaicing based on directional difference regression and efficient regression priors. IEEE Trans. Image Process. **25**(8), 3862–3874 (2016)
26. Hua, K., Hidayati, S.C., He, F., Wei, C., Wang, Y.F.: Contextaware joint dictionary learning for color image demosaicing. J. Vis. Commun. Image Represent. **38**, 230–245 (2016)
27. Tan, D.S., Chen, W., Hua, K.: Deep demosaicing: adaptive image demosaicing via multiple deep fully convolutional networks. IEEE Trans. Image Process. **27**(5), 2408–2419 (2018)
28. Luong, H.Q., Goossens, B., Aelterman, J., Pižurica, A., Philips, W.: A primal-dual algorithm for joint demosaicing and deconvolution. In: 2012 19th IEEE International Conference on Image Processing, Orlando, FL, pp. 2801–2804 (2012)
29. Liu, H., Liu, D., Mansour, H., Boufounos, P.T., Waller, L., Kamilov, U.S.: SEAGLE: sparsity-driven image reconstruction under multiple scattering. IEEE Trans. Comput. Imaging **4**(1), 73–86 (2018)

Statistical Analysis for RAMAN Amplifier for 16×10 Gbps DWDM Transmission Systems Having Pre-compensating Fiber

Bhavesh Ahuja$^{(\boxtimes)}$ and M. L. Meena

Department of Electronics, Rajasthan Technical University, Kota, India
Bhaveshahuja5l@gmail.com, madan.meena.ece@gmail.com

Abstract. Optical communication system have undoubtedly revolutionized the satellite communication system. Such networks have become popular not just because of the higher data rate performances but also because of the higher bandwidth allotment and comparatively less attenuation over the long-haul channels. In the proposed optical research the DWDM network with 16-channels is designed for communicating using the Raman Amplifier (RA). The amplifier was studied with a Compensation technique for a 84 km fiber link. The performance evaluated was studied on the basis of bit error rate (BER), Eye-opening, Q-factor (Quality factor).

Keywords: Raman Amplifier · Dense wavelength division multiplexing · Eye-diagram · BER · Dispersion compensation fiber

1 Introduction

Optical amplifiers have become the spine for digital communication system since last few decades. With the rising demands for incrementing the transfer speed and to better the transferring data rate, such amplifiers were envisaged for the communication services requiring such higher bandwidth services. Services such as video conferencing, high definition data exchange or real time virtual reality and more on. The higher bandwidth were therefore attained with help of Optical fiber. The extensive use in the diverse application requiring long-haul transmission of up to hundreds of kilometers, path loss exceeds the available margin. As the use of the repeaters were increased, more conversions takes place such as photon to electron, electron to photon, retiming, electrical amplification, pulse reshaping again adds up to the losses. For system using single wavelength operation and compatible speed, such performances can be handled. But as per the trends, the requirements are in abundance for multiple wavelength system and higher speed.

Thus DWDM method became popular as data carrying ability of the system is increased and thus utilization of optical networks for long optical communication link is attained [1, 2].

D. J. Hemanth et al. (Eds.): ICICI 2019, LNDECT 38, pp. 265–272, 2020.
https://doi.org/10.1007/978-3-030-34080-3_30

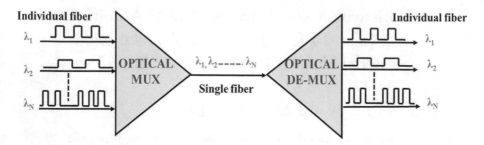

Fig. 1. Building block of basic optical dense wavelength multiplexing system

DWDM is a method of transmission system that uses multiplexer at transmitter division and de-multiplexer at receiver division to communicate to the respective channel through a single fiber as can be seen in Fig. 1.

The spacing between information bearing wavelength is kept at 100 GHz (0.8 nm). In optical transmission system, long distance communications are mainly restricted due to fiber non-linearity, dispersion and attenuation. Higher order mode filters, Dispersion Compensation Filter, DCF (Dispersion compensation fiber) and Fiber Bragg Grating (FBG) are suggested for neutralizing the dispersion [3, 4]. The FBG is the most efficient and cost effective dispersion compensation technique amongst all the dispersion justifying technique [5]. Optical amplifier then became an alternative approach for the loss management, as these amplifiers need not convert the optical signal to electrical and vice-versa. Thus eliminating the optoelectronic repeater for amplification of the signal, and thus help lowering the losses in adequate amount [6, 7].

This research is dedicated to the Raman Amplifier (RA), as it discusses the amplifier for 16-channel network. The transmission power was also optimized and tabularized according to the power budget.

2 Optical Amplifiers

Presently, (RA) Raman Amplifier, EDFA (Erbium doped fiber amplifier), (SOA) Semiconductor Optical Amplifier are generally used in DWDM systems to overcome the attenuation phenomena.

2.1 Raman Amplifier

This Amplifier has an Optical amplification mechanism, in which signal is amplified by Stimulated Raman Scattering (SRS). To achieve stimulated scattering, light is scattered through atoms from a lower to higher wavelength. In SRS, powerful pump beam is propagates through silica fiber for rising gain by transferring the power from shorter wavelength to longer wavelength (known as down shift in frequency or Stokes shift) [8]. Therefore, Raman scattering is a nonlinear interaction between pump wavelength (lower) and signal wavelength (higher) in any optical fiber (both signal and pump wavelength are different). The basic concept of fiber-based Raman amplifier can be seen in Fig. 2.

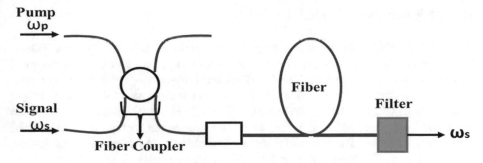

Fig. 2. Basic sketch for fiber-based Raman amplifier

The advantage of RAs are low noise figure and wide gain spectrum (up to 10 nm), it can be changed by varying the pump's wavelength and the number of pumps. Further, Higher Pump power is been required by Raman Amplifier for attaining batter efficiency and expensive high power lasers, for delivering high power into single mode fiber [9]. Two types of RAs are basically used on their design knows as DRA (Distributed Raman amplifier), in which longer transmission fiber (approx. 100 km) is used as medium gain by combining pump wavelength with signal wavelength. Another is Discrete Raman amplifier, in which shorter length of fiber (approx. 20 km) is used for amplification.

3   Dispersion Compensating Fibers

Dispersion compensating fiber uses an efficient way to compensate for positive dispersion in single mode fiber links. Dispersion compensation fiber have negative dispersion of −70 to −90 ps/nm/km and are thus used to compensate the positive dispersion of fibers.

Compensation can be done by three methods depending on the position of the DCF:

- **Pre-Compensation:** DCF is placed before a certain length of conventional SMF.
- **Post-Compensation:** DCF is placed after a certain length of conventional SMF.
- **Mix-Compensation:** DCF having negative dispersion is placed before and after SMF for compensating positive dispersion.

Compensation fibers cause the over-all dispersion of the optical link to be zero over the total fiber distance using the Eq. (1):

$$D_{smf} \times L_{smf} = -D_{dcf} \times L_{dcf} \tag{1}$$

Where, D represents dispersion and L represents length, whereas 'smf' stands for single mode fiber and 'dcf' stands for dispersion compensation fiber.

4   Designed System Model

The 16-channels WDM transmission system is optimized for evaluating the performance of Raman amplifiers (RAs) based on in-line amplifier topology. The designed model is simulated in Optisystem7 environment to investigate how 16-channels amplifier topology perform in WDM transmission environment at the data speed per channel of 10 Gbps. The proposed system is shown in Fig. 3. The proposed system can be divided into three main categories which are; transmitter section, channel (media) and receiver section. The transmitter section contains four basic blocks (data source, pulse generator, CW laser source and Mach-Zehnder modulator). In each channel the transmitter produces a pseudo random bit sequence through a data source at 10 Gbps bit rate from where a binary signal is transmitted to non-return-to-zero pulse generator where binary data is then transformed to electrical data (pulse), on other hand Continuous Wave Laser provides optical signal and the MZM which stands for mach-zehnder modulator is used to modulate the continuous wave optical signal. The Mach-Zehnder modulator receives electrical signal as well as optical signal and induces phase shift between two signals and thus providing an optical signal at output. The insertion loss, chirp factor and extinction ratio of MZ-modulator is selected to be 5 dB, 0.5 and 30 dB, respectively. The frequencies of each channel are selected from 193.1 THz to 194.6 THz as per the recommendation of ITU-T G.694.1.

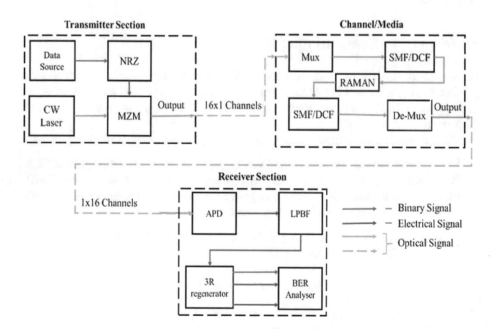

Fig. 3. Blocks of proposed optical system

Table 1. Fiber and simulation parameters with optimized values for the model

Fiber parameters		
Parameter	*SMF*	*DCF*
Length (Km)	70	14
Attenuation (dB/km)	0.2	0.33
Dispersion (nm/ps/km)	16	−80
Differential group delay (ps/km)	0.2	0.2
PMD coefficient	0.5	0.5
Simulation parameters		
Parameters	*Value*	
Carrier frequency of 1st channel	193.1 THz	
Channel spacing	100 GHz	
Data rate	10 Gbps	
Sequence length	16	
Sample/bit	256	

The proposed system uses Continuous Wave Laser (CW) for generating optical lasers of different wavelength and the spacing between channels is kept to be 100 GHz (0.8 nm). Then the incoming signals are combined using multiplexer (DWDM MUX) and transmitted over a fiber channel consisting of DCF, RA and SMF, this section is known as channel/media. The length of DCF and SMF are 14 km and 70 km respectively. Therefore, total transmission length of channel is 84 km with differential group delay, polarization mode dispersion (PMD) coefficient are 0.2 ps/km and 0.5 ps/sqrt (km), respectively. We have used an In-line amplifier topology, in which the amplifier is to be placed between two fibers (either SMF or DCF). The Raman amplifier is introduced in an optical channel with a length of 10 km. The power and frequency of pump laser for RA is to be select on 0.1dBm, and 193.5 THz, respectively and working at wavelength range of 1550 nm. The parameters of RA are such that the affective interaction area is 72 μm^2 and Raman gain peak of 1e−13. Furthermore, the receiver section contains de-multiplexer (DWDM DEMUX) in which the optical signals splits into sixteen channels respectively. The output of de-multiplexers are given to APD (avalanche photo-detector) having an ionization ratio of 0.9 and shot noise distribution select on Gaussian profile and then the signal pass through low pass Bessel filter respectively. Further, data pulse is provided to 3R-regenerator circuit where reshaping, retiming, and re-amplification are to be accomplished. In last the Eye/BER analyzer is used for measuring the BER, Q-factor and eye-diagrams. The component parameters of DCF, SOA and SMF of the designed optical WDM system are displayed in Table 1.

5   Results and Discussion

This article focuses main on analyzing the performance of Raman Optical Amplifiers for 16 × 10 Gbps DWDM transmission systems. For investigating the performance of the proposed network, the length of SMF, the length of DCF, input power, attenuation coefficient and RA parameters are optimized to pact with the environment as given in Table 1.

To investigate the performance of transmission network having 10 Gbps data rate, an optical link of 84 km with justifying standards of BER and Q-factor values are obtained for 16-channels DWDM transmission system using In-line RAs compensation topology. The performance parameters Q-factor and BER were tabularized and shown in Tables 2 and 3 for 16 channels. The Q-factor varies from 47.81 to 32.80, and BER various from 0 to 5.41e−313 respectively as shown in Tables.

Amplifier Raman displayed low output power in comparison to EDFA and SOA, for higher wavelength Raman Amplifiers provides even improved results [10]. The semiconductor optical amplifier is said to have more attenuation when the no. of channels are more and thus affecting the Q-factor and BER as suggested [11, 12] (Fig. 4).

Table 2. *Q-factor* parameters from designed model

Channel frequency (THz)	Raman amplifier Q-factor
193.1	43.449
193.2	41.384
193.3	43.59
193.4	47.810
193.5	44.974
193.6	44.146
193.7	37.0506
193.8	32.807
193.9	39.148
194	34.149
194.1	39.88
194.2	41.88
194.3	39.76
194.4	47.714
194.5	37.79
194.6	34.57

Table 3. *BER* parameters from designed model

Channel frequency (THz)	Raman amplifier BER
193.1	0
193.2	0
193.3	0
193.4	0
193.5	0
193.6	0
193.7	7.652e−301
193.8	1.950e−236
193.9	0
194	5.8196e−256
194.1	0
194.2	0
194.3	0
194.4	0
194.5	5.419e−313
194.6	2.46e−262

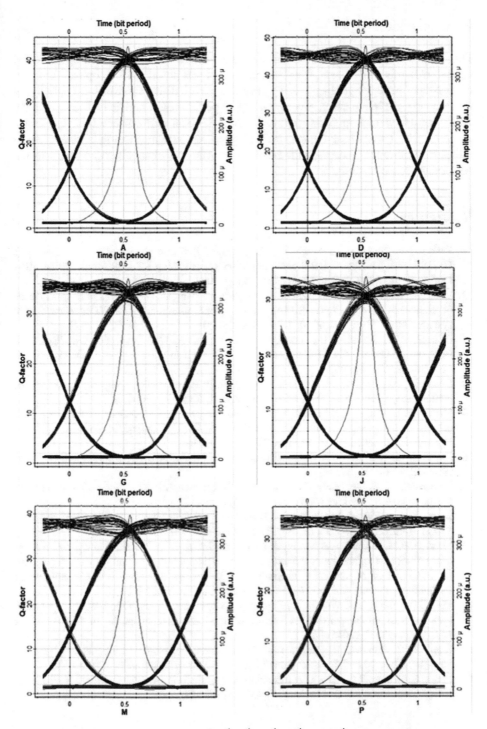

Fig. 4. Eye diagram from 1st, 4th, 7th, 10th, 13th and 16th output channel

6 Conclusion

This Proposed work highlights the performance of designed 16×10 Gbps DWDM optical transmission system. The parameters of proposed 84 km optical link are investigated through eye-diagrams, Quality factor and BER (bit error rate) by varying input power (mW). Further, we evaluate the performance and it is found that the optimized 16-channel amplifier topology gives the best performance (better Q-factor and least BER) with these Parameters. The proposed schemes can thus be applied in future to a complex network with increased number of channels to compensate the attenuation complications.

References

1. Parkash, S., Sharma, A., Singh, H., Singh, H.P.: Performance investigation of 40 Gb/s DWDM over free space optical communication system using RZ modulation format. Adv. Opt. Technol. **2016**, Article ID 4217302, 8 p. (2016)
2. Bobrovs, V., Olonkins, S., Alsevska, A., Gegere, L., Ivanovs, G.: Comparative performance of Raman-SOA and Raman-EDFA hybrid optical amplifiers in DWDM transmission systems. Int. J. Phys. Sci. **8**(39), 1898–1906 (2013)
3. Gnauck, A.H., Garrett, L.D., Danziger, Y., Levy, U., Tur, M.: Dispersion and dispersion-slope compensation of NZDSF over the entire C band using higher-order-mode fibre. Elect. Lett. **36**, 1946–1947 (2000)
4. Sumetsky, M., Eggleton, B.J.: Fiber Bragg gratings for dispersion compensation in optical communication systems. J. Opt. Fiber Commun. **2**, 256–278 (2005)
5. Meena, M.L., Gupta, R.K.: Design and comparative performance evaluation of chirped FBG dispersion compensation with DCF technique for DWDM optical transmission systems. Optik-Int. J. Light Electron Opt. **188**, 212–224 (2019)
6. Agrawal, G.P.: Fiber-Optic Communication Systems, 3rd edn, p. 576. Wiley, New York. ISBN-13 9780471215714
7. Srivastava, R., Singh, Y.N.: Fiber optic loop buffer switch incorporating 3R regeneration. J. Opt. Quantum Electron. **42**(5), 297–311 (2011)
8. Islam, M.N.: Raman Amplifiers for Telecommunications 2: Sub-Systems and Systems, p. 432. Springer, New York (2004). ISBN-13: 9780387406565
9. Mustafa, F.M., Khalaf, A.A.M., Elgeldawy, F.A.: Multi-pumped Raman amplifier for long-haul UW-WDM optical communication systems: gain flatness and bandwidth enhancements. In: Proceedings of the 15th International Conference on Advanced Communication Technology, Pyeong Chang, 27–30 January 2013, pp. 122–127 (2013)
10. Malik, D., Pahwa, K., Wason, A.: Performance optimization of SOA, EDFA, Raman and hybrid optical amplifiers in WDM network with reduced channel spacing of 50 GHz. Optik **127**, 11131–11137 (2016)
11. Kaur, R., Randhawa, R., Kaler, R.S.: Performance evaluation of optical amplifier for 16×10, 32×10 and 64×10 Gbps WDM system. Optik **124**, 693–700 (2013)
12. Singh, S., Kaler, R.S., Singh, A.: Performance evaluation of EDFA, RAMAN and SOA optical amplifier for WDM systems. Optik **124**, 95–101 (2013)

JavaScript Function in Creating Animations

Artur Lipnicki[1(✉)] and Jerzy Drozda Jr.[2]

[1] Faculty of Mathematics and Computer Science,
University of Łódź, Banacha 22, 90-238 Łódź, Poland
`artur.lipnicki@wmii.uni.lodz.pl`
[2] IT Media, Jagiellońska 88, 03-215 Warszawa, Poland
`maltaannon@gmail.com`

Abstract. In many cases, the focus is to create dynamically-driven animations while avoiding excessive use of keyframes. In such a situation, expressions can be utilized. Provided are a number of functions for a certain automation in 2D/3D animation for the effect of counting numerical values and movement in 2D/3D space on the surface of a sphere and ellipsoid. Various pitfalls will be explored regarding direct control of an expression's speed or frequency and potential solutions to overcome the issues. In this paper we present specific solutions as expressions in JavaScript for the purposes of automating animation.

Keywords: Lattice · Key frames · Covering radius · Expressions

1 Introduction

Motion graphics are exciting to watch, but creating them involves quite a commitment to time and detail. The standard animation process most often consists of animations using keyframes. Despite keyframes' flexibility, changing one of them can set off a chain reaction where some, if not all, have to be adjusted to keep the animation consistent. In a situation where a design with an exact path of motion in a specific shape is required using keyframes, the process demands a substantial amount of time because many parameters must be manually set. If adding additional points between specific points in time (accuracy of position in time, correlation with other points, etc.), Expression Language can be utilized. An expression is a compact piece of code (similar to a script). After Effects uses the JavaScript engine when evaluating expressions. The Expression Language menu can be used at anytime to insert methods and attributes into an expression, and the pick whip (@) can be used at any time to insert properties. Some argument descriptions include a number in square brackets. This number indicates the dimension of the expected value or array. Some return value descriptions include a number in square brackets. This number specifies the dimension of the returned value or array. If a specific dimension is not included, the dimension of the returned array depends on the dimension of the input.

© Springer Nature Switzerland AG 2020
D. J. Hemanth et al. (Eds.): ICICI 2019, LNDECT 38, pp. 273–280, 2020.
https://doi.org/10.1007/978-3-030-34080-3_31

In the following chapters, a number of JavaScript function expressions will be reviewed. In the case of these functions, sliders have been added as elements supporting control when editing animations. The last chapter discusses the expressions in the sphere and the ellipsoid in 3D space. These expressions can be combined with the "Store 3D" effect, as well. These functions also work for any other layer or its effect (for three coordinates as an array). Expression functions can still be extended, improved, and supplemented with elements allowing even more animation possibilities while increasing the project's speed and efficiency.

2   Basic Function in Expresssions

2.1   Time Function

The "time" expression is very suitable for objects with perpetual motion. The "time" function allows a user to enter the time value in seconds (timeline run) into the selected parameter. Using time as the rotation parameter tells the object to rotate 1 degree per second. Using the function in the Y axis position parameter as *time*10* moves the object towards the bottom at 10 pixels per second. Because in $2D$ a user has the position x and y, they can decode a simple array (necessary to operate):

[value[0], time * 10]

The necessity of plugging the table results from the fact that the time function works on a one-dimensional structure. Therefore, to achieve a similar effect on both axes, a change should be made:

[time * 20, time * 10]

The above expression will give the speed on the X axis 20 pixels per second, and on the Y axis, 10 pixels per second.

Arrays (multi-dimensional properties, like position parameter) behave similarly to regular matrixes, which means they can be added together, have scalar values added to them, or they can be multiplied by them. This allows for a short and compact solution for changing the first element of the array (in this case, the x position), where the value is the original array to which the time expression is then added to create movement, resulting in horizontal movement:

value + (time * 10)

If dynamic change of the speed of our point (object) is required, a slider can be used. Specifically, one-dimensional or two-dimensional sliders should be used to parameterize the time function. Assume a user is going to use one one-dimensional type slider. Suppose that in the composition there is a solid layer and a "Null 1" object and that control over the speed of the time function in the horizontal and vertical axes is required. Denoted by sx the velocity multiplier along the horizontal axis and by sy the velocity multiplier along the vertical axis, assume that these values are the next two one-dimensional sliders in the "Null 1" object, respectively. The resulting function then shows:

```
sx=thisComp.layer("Null 1").effect("Slider Control")
("Slider");
sy=thisComp.layer("Null 1").effect("Slider Control 2")
("Slider"); [time*sx,time*sy];
```

The "time" function is widely used in traffic dynamics. Almost every move can be supported by this function multiplied by a slider, which will be our indicator. In this example, the main task of our object will be the movement on the elliptical path. We can use the parametric description of the ellipse with given half-axes. The path is given as a half-axis ellipse, respectively a and b, where a, b > 0. Include three silders added to the object type "Null 1". The value of the first slider is defined as the angle of rotation, and the value of the second and third slider as the value on the axes of the ellipse. We obtain

```
ang=time*thisComp.layer("Null 1").effect("Slider Control")
("Slider");
vala=thisComp.layer("Null 1").effect("Slider Control 2")
("Slider");
valb=thisComp.layer("Null 1").effect("Slider Control 3")
("Slider");
nx=vala*Math.cos(ang); ny=valb*Math.sin(ang);
[nx,ny]+value;
```

In the previous notation, the values of "vala" and "valb" indicate the length of the ellipse axis, and "ang", the angle at which our object is located in relation to the center of the ellipse. We can streamline similar code in a different way:

```
vala=thisComp.layer("Null 1").effect("Slider Control 2")
("Slider");
valb=thisComp.layer("Null 1").effect("Slider Control 2")
("Slider");
ang=time*thisComp.layer("Null 1").effect
("Slider Control 3")("Slider");
[(thisComp.width/2), (thisComp.height/2)] +
+[Math.sin(ang)*vala, -Math.cos(ang)*valb]
```

We will use sliders to determine animation, movements in 3D after a three-dimensional sphere (ellipse). Sliders act as controllers (potentiometers) which facilitate access to the most crucial arguments of the code.

2.2 Randomness and Pseudo-randomness

Randomness is effectively a lack of purpose, cause, order, or predictable behavior. A process whose results cannot be accurately predicted, only the description of their distribution can be described as "random". A basic random expression allows a certain value to be drawn from the set range (see also [3,9]). If in the "source" part of the text layer, consider a type function *random()*. In this case a random numbering of (0, 1) in a random (pseudo-random) way is produced.

The expression can be rewritten as $random(0,1)$ With this function, the scope of the draw can be defined as $random(a,b)$ where $a,b \in \mathbb{R}$ and $a < b$. The drawn numbers can also be represented as integer numbers by entering the following functions:

```
value=random(a,b); value=Math.round(value) or
```

```
value=random(a,b);
value=Math.ceil(value)   \\ or value=Math.floor(value)
```

In the last two instructions, approximations from below or above to an integer number are given.

Corollary 1. *Asking for integer-numerical approximations of the table values presents a unique situation. The n - dimensional array of real numbers with an integer array must be described. Taking into account lattice theory, when considering multidimensional arrays and their actual values, approximations with integer values can be asked. Considering a given n - dimensional point with real components, the nearest total point is asked for. In many situations, an error of such approximation can be asked for. In order to define the problem more precisely, following are several concepts. Consider treating $\mathbb{R}^n$ as an n-dimensional Euclidean space with the norm $\|\cdot\|_2$, as the usual norm in L_2. By $d_2(P,Q)$ the corresponding distance of a point P from a subset $Q \subset L_2$ is denoted. The closed unit ball in $\mathbb{R}^n$ will be denoted by B_n. By a lattice L in $\mathbb{R}^n$ means a nonzero finite dimensional discrete additive subgroup of $\mathbb{R}^n$. Given the lattice L, by $\mu(L; \mathbb{R}^n)$ covering radius is defined:*

$$\mu(L; \mathbb{R}^n) := \max_{x \in spanL} d(x, L).$$

In other words, the covering radius of a lattice is the minimal r such that any point in space is within distance at most r from the lattice. To simplify the notation, $\mu(L)$ is used instead of $\mu(L; \mathbb{R}^n)$. To simplify example, it can be written $\mu(\mathbb{Z}) = 1/2 (n = 1)$, $\mu(\mathbb{Z}^2) = \sqrt{2}/2 (n = 2)$. Additional resources can be found in [2,10,11] and more details in [1,5,6,8].

The wiggle function works differently. In this case, no random interval is needed, but only the frequency and amplitude (range from the initial value). More specifically, having a function on the "position" property in a Solid Layer allows the function to be checked $wiggle(a,b)$ which allows the position a times per second within a radius of b from the home point to be drawn (see also [4,7]). Let

```
seedRandom(x,y); wiggle(a,b)
```

where x is a random box and y is logical value (false - randomly draw, true - block the lottery). A true/false logical value allows the possibility to use an expression in order to create a basic molecular system. Consider an animation consisting of a series of randomly appearing elements (as copies of layers containing an expression about the randomness of certain values: position, number, angle, etc.

In this case, a slider is most effective with the value changed in the first argument of the seedRandom function (box selection). The whole lottery procedure can utilize sliders further allowing editing the draw interval or the animation itself (drawing draw ranges). Assume in the "Null 1" type layer a user defines the sliders responsible for the minimum and maximum value for the interval from which they will draw. The resulting function is:

```
seedRandom(x,y);
a=thisComp.layer("Null 1").effect("Slider Control 2")
("Slider"); b=thisComp.layer("Null 1").effect
("Slider Control 2")("Slider"); wiggle(a,b)
```

3 Numbers

A countdown, for example, uses keyframes which points to the appropriate numerical value. The process is straightforward, but time-consuming if edits are required. The following function demonstrates a countdown using sliders:

```
function pad(e, s) { return (1e15 + e + "").slice(-s)};
```

Next, let us set a value that is the number of keyframes for each change in the value of the counter. Therefore:

```
period = thisComp.frameDuration * a;
```

where the a parameter will mean the number of keyframes per change of the counter value, then the number of deductions that have occurred are determined. As a result:

```
i = Math.floor(time/period);
```

Then how many seconds must be reduced for reach "period":

```
dec = b;
```

where the value of parameter b will be a function of the leader giving the above-mentioned value of changes. The remaining number of seconds remaining until the end must be specified. Then:

```
result = thisComp.duration - (i*dec);
```

Values are then specified to stay at 0. Therefore:

```
result = clamp(result, 0, thisComp.duration);
```

then the function can be left in the above version or added to the display format as $mm : ss$. Then:

```
result = Math.floor(result); //rounding down
secs = result % 60; // how many seconds were left in a minute
mins = Math.floor(result/60); //how many min. have been left
pad(mins,2) + ":" + pad(secs,2); // "print" of value
```

Corollary 2. *Consider a text layer with the "source text" function given. Then the countdown process can be implemented using the following function:*

```
function pad(e, s) { return (1e15 + e + "").slice(-s)};
a=thisComp.layer("Null 1").effect("Slider Control")
("Slider") period = thisComp.frameDuration * a;
i = Math.floor(time/period);
b=thisComp.layer("Null 1").effect("Slider Control 2")
("Slider")dec = b;
result = thisComp.duration - (i*dec);
result = clamp(result, 0, thisComp.duration);
result = Math.floor(result);
secs = result; mins = Math.floor(result/60);
pad(mins,2) + ":" + pad(secs,2);
```

where a as the value of slider 1 and b will be a function of the leader giving the above-mentioned value of changes (value of slider 2).

The animation can be started by defining the initial sliders that allow the possibility of editing values, so the text layer is defined with the "source text" field and the "Null 1" type object. The "Null" object assumes that it has four defined sliders: "BeginTime" "EndTime", "StartValue", "EndValue".

The function will therefore take into account the start time of the countdown and the range of its value. The "linear" function can now be utilized for this purpose. This function allows scaling of the range of values of numeric parameters (linear interpolation). The function's form is:

```
linear(t, tMin, tMax, value1, value2)
```

where t - input parameter (scalar), such as time, rotation, etc.

```
tMin - the min value of t, tMax - the max value of t,
value1 - the minimum value from the new range for t,
value2 - the maximum value of the new range for t.
```

A simple countdown can be achieved by the procedure:

```
beginTime, endTime, startVal, endVal = arguments
```

Corollary 3. *Taking into account the above subfunctions, a calculation based on sliders that manage the time of the countdown and the distribution of its value at that time is now obtained:*

```
bt=thisComp.layer("Null 1").effect("BeginTime")("Slider");
et=thisComp.layer("Null 1").effect("EndTime")("Slider");
sv=thisComp.layer("Null 1").effect("StartValue")("Slider");
ev=thisComp.layer("Null 1").effect("EndValue")("Slider");
beginTime = bt;
endTime = et; startVal = sv; endVal = ev;
Math.round(linear(time,beginTime,endTime,startVal,endVal));
```

4   Sphere

Defining the movement of an object after the sphere itself in the After Effects environment is quite difficult. With the help of "Null" layers, a series of sliders can be defined that will help manage the animation as a whole. In the code, the parametric coordinates of the circle (for two-dimensional space) and the sphere (in the case of three-dimensional space) are used. A circular motion is created and the task is to animate the length of the radius and the speed of the movement of the point. Assuming that object "Null" together with two one-dimensional sliders is present.

Corollary 4. *The composition contains a layer of the "solid" type and a "Null 1" layer with the Slider included in it. Then the expression:*

```
rad1=thisComp. layer ("Null   1"). effect ("Radius1")("Slider");
rad2=thisComp. layer ("Null   1"). effect ("Radius2")("Slider");
ang=linear (time ,0 ,30 ,0 ,2∗Math. PI );
nx=rad1∗Math. cos (ang );   ny= rad2∗Math. sin (ang );
[nx , ny]+value
```

defines the motion of the solid type layer ("position" parameter) on the elliptical path in the two-dimensional space.

3D space can also be considered. Consider the function that sets the parameterization of motion on the sphere in three-dimensional space using classic spherical coordinates.

A slider carries the weight of the procedure as a parameter of the frequency of the draw, time of completion, etc. As a result:

Corollary 5. *Assume that the composition contains a layer of the "solid" type (two-dimensional in three-dimensional space) and the "Null" object has one-dimensional sliders:*

```
freq1  =  thisComp. layer ("Null   1"). effect ("Frequency1")
("Slider");
freq2  =  thisComp. layer ("Null   1"). effect ("Frequency2")
("Slider");
rad  =  thisComp. layer ("Null   1"). effect ("Radius3D")
("Slider");
alpha  =  degreesToRadians ( wiggle (freq1 ,   180)[0]);
theta  =  degreesToRadians ( wiggle (freq2 ,   180)[1]);
nx  =  rad ∗  Math. sin (alpha )  ∗  Math. cos (theta );
ny  =  rad ∗  Math. sin (alpha )  ∗  Math. sin (theta );
nz  =  rad ∗  Math. cos (alpha );
point  =  [nx ,  ny ,  nz ]  +  [width ,  height ] ∗  .5;
 toComp (point );
```

performs random movement of solid objects on the edge of the sphere with a given radius (slider).

Corollary 6. *Assume the composition contains a layer of the "solid" type (two-dimensional in three-dimensional space) and the "Null" object is equipped with one-dimensional sliders:*

$freq1 = thisComp.layer("Null \ 1").effect("Frequency1")$
$("Slider");$
$freq2 = thisComp.layer("Null \ 1").effect("Frequency2")$
$("Slider");$
$rad1 = thisComp.layer("Null \ 1").effect("Radius13D")$
$("Slider");$
$rad2 = thisComp.layer("Null \ 1").effect("Radius23D")$
$("Slider");$
$rad3 = thisComp.layer("Null \ 1").effect("Radius33D")$
$("Slider");$
$alpha = degreesToRadians(wiggle(freq1, \ 180)[0]);$
$theta = degreesToRadians(wiggle(freq2, \ 180)[1]);$
$nx = rad1 * Math.sin(alpha) * Math.cos(theta);$
$ny = rad2 * Math.sin(alpha) * Math.sin(theta);$
$nz = rad3 * Math.cos(alpha);$
$point = [nx, \ ny, \ nz] + [width, \ height] * .5;$
$toComp(point);$

performs random movement of solid objects on a three-dimensional ellipsoid.

References

1. Aparicio Bernardo, E.: On some properties of polynomials with integral coefficients and on the approximation of functions in the mean by polynomials with integral coefficients. Izv. Akad. Nauk SSSR. Ser. Mat. **19**, 303–318 (1955)
2. Banaszczyk, W., Lipnicki, A.: On the lattice of polynomials with integer coefficients: the covering radius in $L_p(0,1)$. Annales Polonici Mathematici **115**(2), 123–144 (2015)
3. Chris, J.: After Effects for Designers. Taylor and Francis (2018)
4. Christiansen, M.: Adobe After Effects CC Visual Effects and Compositing Studio Techniques. Adobe Press (2013)
5. Ferguson, L.B.O.: Approximation by Polynomials with Integer Coefficients. American Mathematical Society, Providence (1980)
6. Ferguson, L.B.O.: What can be approximated by polynomials with integer coefficients. Am. Math. Monthly **113**, 403–414 (2006)
7. Geduld, M.: After Effects Expressions. Amazon Digital Services LLC (2013)
8. Kantorowicz, L.V.: Neskol'ko zamecanii o priblizenii k funkciyam posredstvom polinomov celymi koefficientami. Izvestiya Akademii Nauk SSSR Ser. Mat. 1163–1168 (1931)
9. Lefebvre, F.: The Power of Expression, Ebook (2019)
10. Lipnicki, A.: Geometric properties of the lattice of polynomials with integer coefficients (2019, in prepare)
11. Lipnicki, A.: Uniform approximation by polynomials with integer coefficients. Opuscula Math. **36**(4), 489–498 (2016)
12. Trigub, R.M.: Approximation of functions by polynomials with integer coefficients. Izv. Akad. Nauk SSSR Ser. Mat. [Math. USSR-Izw.] 261–280 (1962)

Design and Development of an Internet of Things Based Test and Measuring Instrument

Ayushi Badola[✉]

Manesar, India
ayushi.badola.1306@gmail.com

Abstract. As number of sensors and instruments are increasing exponentially, huge amount of data is generated. This data must be analysed efficiently. Thus, applications to visualize and analyse the data is required. IOT involves devices being embedded with software and sensors to internet, further enabling to collect and exchange data without human interaction.

Designing and developing an Internet of Things based test and measuring instrument is required for better analyzing and measuring data. Data is accessed at real-time using ThingSpeak Cloud and the web application is developed in MEAN stack. Data is gathered from devices example sensors/instruments which are connected to a microcontroller, NodeMCU ESP 8266 to further send data to cloud.

Generated data is stored locally and in cloud, it can be further analyzed and is useful for many industrial applications. Major areas are home automation, electrical appliances, testing of handheld instrument and many more.

Keywords: Internet of things · ThingSpeak · NodeMCU · MEAN stack

1 Introduction

Internet of things includes physical devices, sensors, software and most important network connectivity which allow these environments to collect data. Internet of things is an infrastructure that provides information services by interconnecting physical and virtual devices. The interconnection is based on popular communication technologies.

IOT and cloud together provide a platform to build products that can be virtual, reliable and be connected to physical devices at real-time. Real- time connectivity is crucial with internet of things and its connection to physical devices to overcome portability issues. Keeping data live makes it accessible anywhere. Example a multimeter can measure voltage, current and resistance readings. Major application of a multimeter is to test battery of any device and testing electrical appliances. Thus, by connecting a multimeter to internet via Bluetooth technology will enable it to remotely send data to cloud and further to a web or mobile application.

There are many forms of presentation, one of the most popular are web application, either web or mobile. In the paper to develop an IOT based test and measuring instrument e.g. to measure data from sensors. Web Application can be developed using

D. J. Hemanth et al. (Eds.): ICICI 2019, LNDECT 38, pp. 281–286, 2020.
https://doi.org/10.1007/978-3-030-34080-3_32

MEAN stack or LAMP stack. MEAN stack-based web applications are developed using JavaScript both for user interface and server. One of the most important reasons to use MEAN stack is because mongo dB is built for cloud and with advancement in IOT, cloud is an important factor. AngularJS deals with client-side of the application, nodeJS is used on the server side, expressJS binds both frontend and backend, mongoDB is for storing data. Any instrument or sensor can send data via NodeMCU ESP 8266 microcontroller to the web application. Further, data can be visualized in tabular and graphical form as well. This data can be analysed and thus measurement of instrument is very beneficial in various industries.

2   Literature Review

The detailed analysis of the previous work in field of IOT, cloud and remote monitoring is described in tabular form.

References	Technologies used	Hardware requirement	Software requirement	User centric or research centric	Application
[1]	Light fidelity technology	Pressure sensor, photo diode	Android mobile application	User centric	Analysis rate of patients increased. Applicable in hospitals and industries
[2]	Android, IOT	DHT11, Arduino board, ESP 8266	Android mobile application	User centric	In home automation system
[3]	Cloud technology	None	Analysis of different cloud technologies	Research centric	Review only
[4]	Cloud technology.	Arduino Yun, voltage sensor	ThingSpeak	User centric	In off-grid homes and home automation
[5]	Cloud technology	None	Comparison of different cloud technologies	Research centric	Review paper
[6]	Cloud technology	Sensor	ThingSpeak	User centric	Cloud based sensor and IOT device to monitor data
[7]	Research study	Gateway, Sensor	Cloud technology and GUI	Research centric	Real time device monitoring

(*continued*)

(*continued*)

References	Technologies used	Hardware requirement	Software requirement	User centric or research centric	Application
[8]	Comparison of microcontrollers	Arduino, Raspberry pi	NA	User centric	Creating own Internet of things
[9]	IOT	Arduino microcontroller	ThingSpeak	User centric	Twitter doorbell
[10]	Cloud technology	CC3200 board	MQTT protocol	User centric	Testing and validation of UART
[11]	Sensors	Resistor temperature detector	Database, server	Research centric	Recalibration of sensors for better performance
[12]	Cloud technology	Arduino, Edison	Xively, ThingSpeak	User centric	Data processing and analyzing
[13]	Cloud and IOT	ESP 8266, DHT11, OLED	ThingSpeak	User centric	Weather information prototype

3 Problem Statement and Proposed Architecture

Internet of things is an extension of internet connectivity to physical devices, sensors and on integrating it with cloud technology, real-time data retrieval is possible. Internet of things starts at the sensor level where pressure, temperature, vibration, analytical and other sensors collect data and send this collected information to control and monitoring systems via wired and wireless network. Remote monitoring of data is an important criterion for maximizing the production in any industry With development of modern industry requirement for instrument monitoring is increasing. Thus, measuring parameters at real-time is crucial for better analysis.

Software defined instrumentation is based on a modular architecture is best way to achieve high level of reconfigurability. As more data is collected, there is more need to manage all data, ensuring it is stored in right format. This can be done by choosing correct stack to build applications. MEAN stack allows data to be stored seamlessly and this stack is best suited for cloud based application and IOT. The cloud services will help to get the IOT functionality of remote access. Using ThingSpeak cloud data can be read and written at real-time. The web application interacts with ThingSpeak cloud throught APIs. Further, web application displays all data generated in tabular form. Filtering of data is done based on date and time, thus visualizing data in graphical form.

Block Diagram of the Proposed Architecture

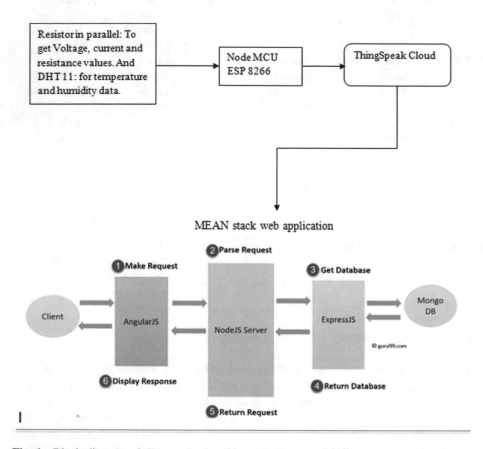

Fig. 1. Block diagram of the proposed architecture (Source of MEAN stack block diagram: https://www.guru99.com/mean-stack-developer.html)

- Resistor in parallel and DHT 11 are connected to Node MCU ESP 8266, and data is sent to ThingSpeak IOT cloud platform (Fig. 1).
- Further, data is fetched from ThingSpeak to the MEAN stack web application. Complete explanation is given in implementation section.

4 Implementation

Step 1: Connect DHT 11 and Resistor in parallel to NodeMCU ESP 8266 board. After connecting using wires, connect board to laptop at port COM 3 or COM 6. Select board NodeMCU 1.0 (ESP 12E Module) and port based on availability.

Step 2: After all connections are made and code is written in Arduino IDE. Check the code for any errors; ensure to have a proper internet connection, as to send data to ThingSpeak proper connection is required.

Step 3: Compile the code and load. After successful loading, data will be read seamlessly from sensor and resistor, and will be sent to ThingSpeak cloud.

Step 4: In ThingSpeak.com, channel has to be created, the channel ID and API keys will be used to read and write data. The channel can be in either private view or public view.

Step 5: The channel will receive all updates at real-time and can be seen anywhere. As it is in cloud storage, thus cloud can be accessed everywhere. In ThingSpeak data can be exported in form JSON, XML and CSV files.

Step 6: Fetching data from ThingSpeak cloud to web application. It is done using the readapikey generated specific to the ThingSpeak channel. Coding is done in javascript to fetch data live from ThingSpeak.

Step 7: The web application is developed using MEAN stack i.e. MongoDb is database to store all data, ExpressJS binds front-end and back-end. AngularJS is used to develop front-end i.e. the user interface of the application. NodeJS is back-end server of the application.

Step 8: It is a multi-page application. In the localhost:4200/list, is the first page of the application, where data is imported from thingspeak at real-time and displayed in tabular form.

Step 9: Further, filtering is done based on start_date, end_date, start_time and end_-time. Thus, all values for that particular time period are displayed as user gives the input. This is part of single page application.

Step 10: After filtering based on date and time, the data can be viewed in graphical representation as well, using Google Charts. Further, on clicking the button user can view graphical form of the filtered data. There is also live data feature added.

5 Conclusion and Future Work

The system has been developed on MEAN stack architecture which enables the flexibility for enhancing the application in iterative manner. The system is designed to overcome the portability issue of a physical device. To monitor data earlier we needed a physical device, using the system data can be monitored remotely and live data values can be fetched anywhere. The system fetches data from DHT 11 and resistor in parallel, using NodeMCU microcontroller. Further data is sent to ThingSpeak cloud which is an IOT platform, this will allow live data rendering. Next, data is fetched to web application from ThingSpeak for further visualization. In the web application data is fetched at real-time and displayed in tabular form. Filtering is done based on range of date and time, those values are displayed. These values are posted to the next pages of application, where graphical representation of data is displayed, which can be further analyzed.

In the paper firstly introduction of the major technologies used is written. Further in literature review, all papers are compared in tabular form. In the next section, there is problem statement and proposed architecture in which precise explanation and block diagram are written. Further implementation explains details of the application.

In future there are other features which will be added to the system, primarily deployment of the web application on cloud, will allow users to access it anywhere. Currently the application is in localhost of the installed system. Further, for better analysis of fetched data, more filters can be added in the graphical representation page of the application.

References

1. Kannusamy, R., Nandakumar, P.: A novel method used to measure the biological parameters via light-fidelity and IoT. Int. J. Eng. Technol. (2017)
2. Zafar, S.: An IoT based real-time environmental monitoring system using arduino and cloud service. ETASR 8(4), 3238–3242 (2018)
3. Ray, P.P.: A survey of IoT cloud platforms. Future Comput. Inform. J. 1, 35–46 (2016)
4. Miron-Alexe, V.: IOT power consumption monitoring system for off grid households (2017)
5. Ganguly, P.: Selecting the right IoT cloud platform. In: International Conference on Internet of Things and Applications (2016)
6. Emeakaroha, V.C., Cafferkey, N., Healy, P., Morrison, J.P.: A cloud-based IoT data gathering and processing platform. In: International Conference on Future Internet of Things and Cloud (2015)
7. Kang, B., Kim, D., Choo, H.: Internet of everything: a large-scale autonomic IoT gateway. IEEE Trans. Multi-scale Comput. Syst. 3(3), 206–214 (2017)
8. Singh, K.J., Kapoor, D.S.: Create your own Internet of Things: a survey of IOT platforms. IEEE Consumer Electron. Mag. 6, 57–68 (2017)
9. Maureira, M.A.G., Oldenhof, D., Teernstra, L.: ThingSpeak – an API and web service for the Internet of Things. Semantics Scholar (2014)
10. Kiran, B.M., Madhusudhan, C., Farhan, M.M., Manzoor, M.B., Kariyappa, B.S.: Testing and validation of UART using cloud based GUI. In: IEEE International Conference on Recent Trends in Electronics Information & Communication Technology, May 2017
11. Sandrić, B., Jurčević, M.: Metrology and Quality Assurance in Internet of Things, May 2018
12. Jang, R., Soh, W.: Design and implementation of data-report service for IoT data analysis. In: CMFE-2015 (2015)
13. Kodali, R.K., Sahu, A.: An IoT based weather information prototype using WeMos. IEEE (2016)

Service Composition in IoT - A Review

Neeti Kashyap[1](✉), A. Charan Kumari[2], and Rita Chhikara[1]

[1] The NorthCap University, Gurgaon, Haryana, India
neeti7777@gmail.com
[2] Dayalbagh Educational Institute, Agra, India

Abstract. Typically, the Internet of Things (IoT) incorporates the provisioning of powerfully adaptable and virtualized things such as cameras, sensors as valuable resources over the Internet. There are different types of services that can be combined to meet user demands. Capturing the user requirements and composing various services from the existing set of services is required in order to facilitate the adaption of an IoT based application. The industry has a strong demand for service composition mechanisms in order to meet the fast and growing demands of the user on the web. There are various techniques and methodologies which can be applied to provide the optimized solution to the problem of Service Composition (SC) in IoT environments. This study involved a survey of existing methodologies that can be addressed to make effective SC in IoT.

Keywords: Service composition · Internet of Things · Optimized · Metaheuristic

1 Introduction

With advancements in Radio Frequency Identification (RFID) Technology, the idea of IoT was introduced in 1999. IoT is the system comprises of physical entities namely things. These things can be implanted with various gadgets to empowers these gadgets to work on data. IoT facilitates these things to be accessible via the world wide web. For example, consider an example of a smart home, a person can switch on or off the air conditioner of his room while sitting in the office from his mobile application. He can also check and manage the energy consumption of his house. This application not only provides comfort but also helps the person in preserving the resources. This brings improved productivity, exactness and monetary advantage.

The things in IoT helps in gathering data which helps in establishing links between various different things. It has been predicted by Gardner that IoT will reach 26 billion units by 2020 [1]. This will affect the information available on the web. This is anticipated to transform the business processes from the scenario of manufacturing a commodity and sales by making the flow of products more visible in real-time. Various businesses use IoT to improve product tracking and optimize manufacturing costs [2]. This paper carefully examines the need and relevance of SC in IoT. It aims to identify various techniques applied to solve this problem and compare in terms of various QoS parameters for the optimization of composed services.

© Springer Nature Switzerland AG 2020
D. J. Hemanth et al. (Eds.): ICICI 2019, LNDECT 38, pp. 287–291, 2020.
https://doi.org/10.1007/978-3-030-34080-3_33

2 Related Work

The overwhelming development and increase in IoT devices led to an immense increase in the number of services. The IoT devices like cameras, sensors provide their functionality via the web and can be remotely accessible via the web. The number of services providing specific functionality is also increasing. The utilization of the best service serving a particular functionality can result in better utilization of resources. Keeping this in mind, composing of services is done by using QoS parameters. Li et al. in 2012 focussed on designing the solution to the service composition problem as a finite state machine (FSM) keeping cost and reliability under consideration. The probabilistic approach has been used by the authors [3]. In 2013, Kim et al. introduced a new service model. The idea is to compose set of services from a set of similar services. This concept was proven with the help of an application to detect a vacant room. This model is based on cloud architecture. As the internet plays a key role in connecting various IoT devices together on a common platform [4]. Chen et al. worked towards the design of a trust which is collected in the form of feedback collected from the devices which have common social interests. They used a filtering approach to combine two types of trusts namely direct trust and indirect trust. They demonstrated the proposed algorithm through a Service Oriented Architecture (SOA) based service composition application in IoT. The social IoT systems comprise of uncensored IoT devices that provide various types of services. Thus this type of environment has a great exposure towards malicious attacks thus there is a need to have a trust management protocol in order to ensure security. To demonstrate the application of this protocol, the trust parameters to minimize the convergence rate [5, 6]. Han et al. explained various challenges associated with smart objects which are part of IoT in research.

As the essence of IoT lies in the efficacy of the interaction among the things or IoT devices which are collaborated and communicate in order to build a smart application in IoT. This paper discussed the importance of service composition along with research challenges along with various applications [7]. One of the challenges in IoT is the aggregation of heterogeneous devices and machines that are being connected to the web, to the people and with other devices as well. The use of IP enabled smart devices functionality is available to different IoT based applications in the forms of device services. This explained the significance of IP technology for IoT devices (Table 1).

Due to the massive generation of data in the IoT application, there is a need to optimize the usage of services for a composition. This can further help in reduction in exchange of data which ultimately can help in decreasing the energy consumption of the network. The movement of data sets exchanged among the services also imposes a load on the network, thus increasing the energy consumption. Baker et al. in 2017 used one of the heuristic search optimization algorithms with Particle Swarm Optimization (PSO). They computed Quality of User Experience(QoUE) using a fuzzy-based inference algorithm. The efficacy of the algorithms has been demonstrated with the help of real-world web services in terms of QoUE and QoS except for the execution cost of the service [8].

Table 1. Summary of various service composition methodologies in IoT

Sno	Year	Author	Problem	QoS parameter	Technique
1	2012	Li, Jin, Li, Zheng and Wei [3]	To address the properties related to SC in IoT	Cost and reliability	Probabilistic approach
2	2013	Im, Kim and Kim [4]	To develop a new service	Feasibility	Modeling based on software component model
3	2014	Chen, Guo and Bao [5]	To implement SC based on trust in IoT	Trust	Adaptive filtering technique
4	2016	Chen, Guo and Bao [6]	Io incorporate trust-based SC	Trust	Developed multi-cloud IoT SC algorithm called (E2C2)
5	2016	Han, Khan, Lee, Crespi, Glitho [7]	To study concepts and research challenges	NA	Defined concept of service composition and its challenges
6	2017	Baker, Asim, Tawfik, Aldawsari and Buyya [8]	To obtain energy-aware SC for IoT applications using multiple clouds	Energy	Uses minimum no. of services in service composition to fulfill requirements of the user
7	2017	Balakrishnan, Sangaiah [9]	To select optimal service composition	Reputation	Optimal service composition selection on the internet of services (IoS)
8	2018	Alsaryrah, Mashal and Chung [10]	Service composition using two objective functions	Service price, latency and execution time and energy	To select optimal service composition using two objective optimizations using PSO
9	2018	Khansari, Sharifian & Motamedi [11]	A multi-objective for service selection and composition in cloud-based IoT	Cost, reliability, delay, and availability	Metaheuristic algorithm
10	2018	Kurdi, Ezzat, Altoaimy, Ahmed and Youcef-Toumi [12]	To compose services to meet the user requirements	QoS parameters	Meta-heuristic search algorithm
11	2018	Yang, Jin and Hao [13]	Service composition in IoT using multi-objective optimization	Energy and execution time	Bio-inspired self-learning co-evolutionary algorithm (BSCA) is used for dynamic multi-objective optimization of IoT services
12	2018	Kashyap, Kumari [14]	Service composition in IoT	Service cost, execution time and reliability	GA and Hyper-heuristic search algorithm

In one of the recent papers by Kansari et al. an approach is shown for service composition in IoT. They considered five QoS parameters namely cost, availability, reliability, latency and response time while aggregating the atomic services. They developed quantum-based genetic algorithms to obtain the optimized solution and helped in improving the composition process [9]. Service composition provides a mechanism to combine various services in order to fulfill the user requirements. Service composition utilizing QoS parameters is an NP-hard problem. However, there are many studies in the recent past years to illustrate the mechanism and methodology of service composition in IoT. Alsaryrah et al. have proposed an optimized service composition technique using bi-objective optimization. The energy consumption has been improved which helped in enhancing the network lifetime while QoS is also maintained. They utilized pulse function to solve the SC problem in IoT by comparing the three algorithms in terms of consumed energy [10]. IoT paradigm in technology which provides connectivity in object to enable connectivity, communication, and collaboration to support various IoT based applications [11].

There are several studies in the past that have addressed service discovery and selection in IoT. Kurdi et al. explained a heuristic search algorithm, multiCuckoo on the service composition problem. They compared the proposed algorithm efficacy and effectiveness by measuring the running time with the benchmark algorithms [12]. Yang et al. in 2018 have explained the multi-objective optimization using bio-inspired self-learning coevolutionary algorithm (BSCA). They composed the proposed algorithm with five popular algorithms [13]. Neeti et al. in 2018 described the usage of hyper-heuristic search algorithm to solve the SC problem in IoT. The algorithm has been compared with the genetic algorithm by varying number of services and candidates for a particular application in IoT [14].

3 Conclusion

This paper addresses the techniques used to address the QoS based SC for various IoT applications. With a tremendous increase in IoT devices, the selection mechanism of a combination of devices has become critical and important. This paper includes various mechanisms and methods which can be employed to provide solution to this problem. This further can help in analyzing and improvement of the composition process. There is a need to identify an efficient method that can help in IoT device monitoring and composing services. Execution cost, Availability, service cost, reliability, and delay are some of the important QoS parameters which help in facilitating the diversity of IoT. Metaheuristic search algorithms have been applied by various authors to address the SC problem efficiently and effectively. Simulations have been used to prove their efficacy. Therefore, service composition problems can be considered as a challenging problem and various optimization techniques can be used to enhance the composing services in IoT.

References

1. http://www.gartner.com/newsroom/id/2684616
2. Gubbi, J., Buyya, R., Marusic, S., Palaniswami, M.: Internet of Things (IoT): a vision, architectural elements, and future directions. Future Gener. Comput. Syst. **29**(7), 1645–1660 (2013)
3. Li, L., Jin, Z., Li, G., Zheng, L., Wei, Q.: Modeling and analyzing the reliability and cost of service composition in the IoT: a probabilistic approach. In: IEEE 19th International Conference on Web Services, Honolulu, HI, pp. 584–591 (2012)
4. Im, J., Kim, S., Kim, D.: IoT mashup as a service: cloud-based mashup service for the Internet of Things. In: 2013 IEEE International Conference on Services Computing, Santa Clara, CA, pp. 462–469 (2013)
5. Chen, I., Guo, J., Bao, F.: Trust management for service composition in SOA-based IoT systems. In: 2014 IEEE Wireless Communications and Networking Conference (WCNC), Istanbul, pp. 3444–3449 (2014)
6. Chen, I., Guo, J., Bao, F.: Trust management for SOA-based IoT and its application to service composition. IEEE Trans. Serv. Comput. **9**(3), 482–495 (2016)
7. Han, S.N., Khan, I., Lee, G.M., Crespi, N., Glitho, R.H.: Service composition for IP smart object using realtime Web protocols: concept and research challenges. Comput. Standards Interfaces **43**, 79–90 (2016)
8. Baker, T., Asim, M., Tawfik, H., Aldawsari, B., Buyya, R.: An energy-aware service composition algorithm for multiple cloud-based IoT applications. J. Netw. Comput. Appl. (2017). http://dx.doi.org/10.1016/j.jnca.2017.03.008
9. Balakrishnan, S.M., Sangaiah, A.K.: Integrated QoUE and QoS approach for optimal service composition selection in internet of services (IoS). Multimed. Tools Appl. **76**(21), 22889–22916 (2017)
10. Alsaryrah, O., Mashal, I., Chung, T.: Bi-objective optimization for energy aware Internet of Things service composition. IEEE Access **6**, 26809–26819 (2018)
11. Khansari, M.E., Sharifian, S., Motamedi, S.A.: Virtual sensor as a service: a new multicriteria QoS-aware cloud service composition for IoT applications. J. Supercomput. **74**(10), 5485–5512 (2018)
12. Kurdi, H., Ezzat, F., Altoaimy, L., Ahmed, S.H., Youcef-Toumi, K.: MultiCuckoo: multi-cloud service composition using a cuckoo-inspired algorithm for the Internet of Things applications. IEEE Access **6**, 56737–56749 (2018)
13. Yang, Z., Jin, Y., Hao, K.: A bio-inspired self-learning coevolutionary dynamic multiobjective optimization algorithm for Internet of Things services. IEEE Trans. Evol. Comput. (2018)
14. Kashyap, N., Kumari, A.C.: Hyper-heuristic approach for service composition in Internet of Things. Electron. Gov. Int. J. **14**(4), 321–339 (2018)

An Energy Efficient Glowworm Swarm Optimized Opportunistic Routing in MANET

D. Sathiya[✉] and S. Sheeja

Department of Computer Science, Karpagam Academy of Higher Education,
Coimbatore, India
sathiyaranjith22@gmail.com, sheejaajize@gmail.com

Abstract. A mobile ad-hoc network (MANET) is a decentralized network where the MN is moved randomly. Every mobile node (MN) in network is joined wirelessly. Owing to random movement of MN in network, energy efficient routing is major issues because the MN has lesser battery power. Energy Efficient Glowworm Swarm Optimization based Opportunistic Routing (EEGWSO-OR) technique is designed to improve routing in MANET. Initially the numbers of glowworms (i.e., mobile nodes) are randomly created. Each glowworm has luminescence quantity (i.e., energy) called luciferin. For opportunistic routing process, MN chooses nearest node to transfer the data packets. The fitness of each mobile is computed for selecting the optimal one with the threshold energy level. This assists to enhance network lifetime (NL). After finding the energy optimized node, the neighboring node is determined using stepwise regression by transmitting the route request and reply messages. Followed by, route path from source to destination is determined with minimum distance. The data packets (DP) are transmitted after the route discovery process, to destination MN without any duplicate transmission. This assists in enhance the data packet delivery and lessens the routing overhead (RO). Experimental analysis is performed with diverse parameters namely energy consumption (EC), NL, data packet delivery ratio (DPDR) and RO with number of MN and DP. The simulation outcome evident that EEGWSO-OR improves the NL, DPDR and lessens EC and RO as evaluated with conventional methods.

Keywords: MANET · Routing · Energy consumption · Glowworm Swarm Optimization · Route discovery · Stepwise regression

1 Introduction

An ad-hoc wireless network is a group of devices linked wirelessly without fixed access point. MANET is a communication network in which all MN communicate with other nodes via wireless connections. The structure of the MANET is shown in Fig. 1.

© Springer Nature Switzerland AG 2020
D. J. Hemanth et al. (Eds.): ICICI 2019, LNDECT 38, pp. 292–310, 2020.
https://doi.org/10.1007/978-3-030-34080-3_34

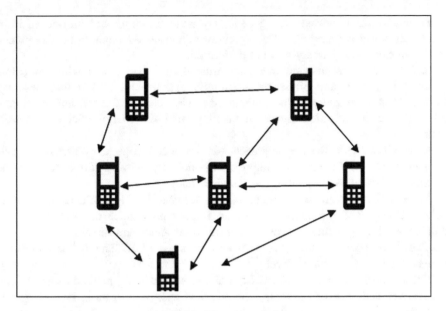

Fig. 1. Mobile ad-hoc network

MANET is a communication network in that nodes are called as mobiles and communicating with others directly. Nodes in the network connect or leave at any time within their communication range. In MANET, the communications between MN are established through the routing process. To transmit the important data, routing is employed to discover path from source to destination. The foremost issue in MANET is to enhance the network operation. Energy efficiency is the major concern in the routing process because the nodes have less power. Therefore, the routing mechanism considers the energy as the main resources for MN communication to improve the NL.

Stable and Energy-Efficient Routing Algorithm Based on Learning Automata Theory (LASEERA) was presented in [1]. But the method failed to minimize the routing overhead.

In [2], an ant colony-based energy control routing protocol (ACECR) was introduced to enhance NL and PDR and lessen the energy of MN. The neighboring node selection was not performed in the routing process.

An Intelligent Energy-aware Efficient Routing protocol was designed in [3] with multi-criteria decision-making approach to enhance DPDR and lessens end to end delay. The approach has high latency in route path finding for large networks.

In [4], an Innovative ACO based Routing Algorithm (ANTALG) was described to enhance the throughput with minimum delay. The algorithm does not enhance the NL since the algorithm failed to select the energy efficient nodes.

A Binary Particle Swarm Optimization algorithm (BPSO) was designed in [5] to enhance energy efficient routing, NL and data delivery. BPSO algorithm consumes superior time for routing DP from source to destination.

To resolve the multicast routing issue, a genetic algorithm was introduced in [6] with lesser power consumption. The algorithm minimizes the delay in routing process but the efficient data delivery was not performed.

In [7], a new bio-inspired integrated trusted routing protocol with ant colony optimization (ACO) and physarum autonomic optimization (PAO) was designed. Among the several route paths, optimal one was selected for efficient routing in MANET. The optimization algorithm failed to find the energy-efficient nodes for routing the DP.

A Tree-based Grid Routing approach was developed in [8] to construct the paths between the nodes. The approach improved the delivery ratio but it does not consider the energy efficient routing and it has a higher delay.

An ant colony optimization algorithm was designed in [9] to enhance the routing, packet delivery, and throughput. However, it does not reduce the RO. In [10], A Reliable and Energy Efficient Protocol depend on distance and remaining energy was introduced to reduce control overhead and improve PDR. The method does not enhance the NL at a required level.

The certain issues are identified from above said existing methods such as high routing overhead, lack of improving the NL, less energy utilization, lack of optimal node selection, failure to select the neighboring node for improving the packet delivery and so on. To conquer above said issues, an EEGWSO-OR technique is introduced.

The major contributions of the proposed energy efficient routing are described as follows,

- EEGWSO-OR technique is introduced to attain the major contributions namely energy efficient node selection and route path discovery. In the first process, the MN (i.e. glowworms) population is initialized with similar luciferin value (i.e. energy level). Due to the movement, the luciferin value of each glowworm is updated. The glowworm with lesser luciferin value moves towards the higher one. Then the position of each glowworm gets updated. After that the fitness is calculated and compared with the threshold level. The node with high fitness compared to the threshold is selected as energy efficient nodes for routing. This assists to enhance NL and lessens the EC.
- Stepwise regression is used for determining the route paths from source to destination by selecting the neighboring nodes with two control message distribution to other MN. After discovery of path, source node (SN) forwards DP along the route path to enhance DP delivery and lessen the RO.

This paper is arranged as follows. Section 2 reviews the related works. In Sect. 3, the issues and challenges are identified from the Sect. 2 are overcome by introducing EEGWSO-OR technique with neat diagram. In Sect. 4, Simulation settings are presented with number of nodes and DP. The simulation results and discussions are presented in Sect. 5. The conclusion of the paper is provided in Sect. 6.

2 Related Works

In [11], an integrated genetic algorithm (GA) and African Buffalo Optimization (ABO) was designed to select the optimal path. The optimization technique enhances NL and lessens the delay. But it has high routing overhead. In order to minimize the RO and packet loss, a reliability factor based routing protocol was designed in [12]. The approach failed to use the efficient optimization technique for selecting the optimal node with less EC to enhance the network lifetime.

K-means cluster formation and firefly cluster head selection based MAC routing protocol was introduced in [13]. But, it failed to lessen delay in data packets flow. In [14], a neighbor coverage-based probabilistic rebroadcasting method was introduced to minimize the overhead. Node energy is not considered to enhance the NL.

With fitness function, an energy efficient multipath routing algorithm was designed in [15] to discover the best route path. The algorithm failed to enhance the EC as well as network lifetime. A novel cross-layer routing approach was introduced in [16] for achieving the better performance and improving the NL. The approach does not minimize the RO and PDR.

A weight-based clustering technique was designed in [17] for routing the data in MANET. Though it lessens the overhead, the NL was not enhanced. In [18], an energy efficient stable multipath routing approach was designed with congestion awareness. The approach obtains high delivery ratio with less overhead in the multipath routing. But the performance of NL and EC remained unsolved.

An ant-based routing algorithm was presented in [19] for minimizing the RO in DP routing but the energy efficient node selection was not performed. To lessen end-to-end delay and enhance the PDR, an Adaptive Protocol for Stable and Energy-Aware Routing was presented in [20]. But, it has superior RO.

The major issues are identified from the above-said literature are addressed by introducing a new technique called EEGWSO-OR technique. The various process of EEGWSO-OR technique is presented in the next section.

3 Methodology

EEGWSO-OR technique is introduced for enhancing the routing process in MANET with minimum energy consumption. The EEGWSO-OR technique comprises three processes namely optimal node selection, route path discovery and DP transmission. In the first process, the optimal node selection is performed by applying an Energy-Efficient Glowworm Swarm Optimization (EEGWSO) technique. The proposed EEGWSO is a nature-inspired optimization algorithm that inspired by the behavior of the lighting worms (i.e., mobile nodes). Compared to existing optimization technique, the glowworm swarm has a high capability for providing the global optimization resulting in improving the accuracy. Therefore, EEGWSO-OR technique uses a glowworm swarm optimization technique rather than the other optimization techniques. After identifying the energy efficient nodes, the route path among the nodes is discovered. At last, opportunistic routing process forward the DP to destination MN. Opportunistic routing is a new model in routing for MANET that chooses node closest to SN to forward the data. The EEGWSO-OR technique designed using the following system model.

3.1   System Model

MANET is arranged in a graphical model '$g(v, e)$' where 'v' denotes a number of MN $mn_i = mn_1, mn_2, mn_3 \ldots mn_n$ distributed over a square area of '$m * m$' within communication range 't_r' and 'e' signify a links among MN in network. Among numerous MN, energy efficient nodes are detected and route data packets $dp_i = dp_1, dp_2, \ldots, dp_n$ from source node s_n and destination node d_n via neighboring node $nn_i = nn_1, nn_2, \ldots, nn_n$. The proposed EEGWSO-OR technique is designed by using the above system model.

3.2   EEGWSO-OR Technique in MANET

EEGWSO-OR technique includes numerous MN moved dynamically in network. Energy efficient MN is chosen for routing DP to improve NL. The diverse process of EEGWSO-OR technique is explained in the architecture diagram.

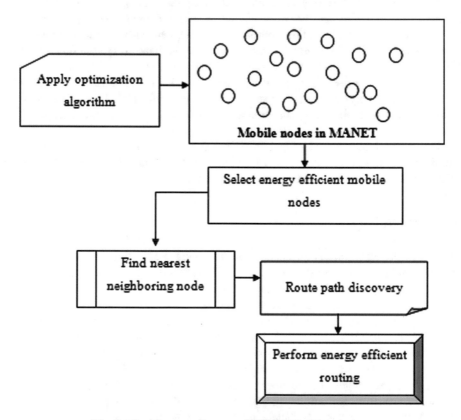

Fig. 2. Architecture diagram of EEGWSO-OR technique

Figure 2 depicts the architecture diagram of EEGWSO-OR technique to improve the routing. MN is distributed arbitrarily in the network. Then, energy efficient glowworm swarm optimization technique chooses the more suitable node among number of MN for routing DP. Optimal MN is chosen depends on the energy level. The node which utilizes the minimum energy for transmitting the DP is selected as optimal for enhancing the network lifetime. After that, the neighboring nodes are determined to construct the route paths. Among the route paths, the path with minimum distance is selected. At last, the DP is transmitted via the path to increase the DPDR and reduce the overhead. The various process of EEGWSO-OR technique is explained in the following subsections.

3.2.1 Energy Efficient Glowworm Swarm Optimization Based Node Selection

In EEGWSO-OR technique, the first process is to select optimal nodes for routing in MANET. In EEGWSO, a swarm of glowworms (i.e. MN) is randomly distributed in the search space depending on the lighting behavior. Each glowworm has luminescence capacity (i.e., energy) called luciferin. The less lighting behaviors of glowworm are attracted to the brighter one. The luciferin intensity (i.e. energy) is associated with the fitness of their current locations. For each iteration, the entire glowworms (i.e. nodes) position is updated, and then the luciferin value also updated due to the dynamic movement of glowworms. Figure 3 depicts the flow process of EEGWSO for optimal node selection. The population of 'n' glowworms (i.e. mobile nodes) and their position are randomly initialized.

The EEGWSO algorithm starts by locating the glowworms arbitrarily in the search space. Initially, the entire glowworms comprise the similar quantity of luciferin (i.e. energy). The energy of MN is computed using below equation,

$$E(mn_i) = p_r * time \tag{1}$$

From (1), $E(mn_i)$ indicate an energy of MN based on the product of power (p_r) and time (t). Energy of MN is evaluated in joule (J), the power is calculated in watt (W), and time is measured in second (s).

After that, the luciferin is updated based on lighting behaviors of luciferin value at glowworm position. At first, all the glowworms in search space contain an equal amount of luciferin value. These values get changed at current positions. It means the energy of each node is changed after transmitting and receiving the DP. Therefore, the luciferin (i.e. energy) is updated with the prior luciferin level of each glowworm. A luciferin amount is depends on computed value of sensed profile. A fraction of luciferin value is subtracted to decay (i.e. energy drops) in luciferin with time. Every glowworm varies luciferin value for objective function value of current location. Luciferin update is formalized as below equations,

$$L_{(t+1)} = (1 - \vartheta) * L_{(t-1)} + \beta E(mn_i(t)) \tag{2}$$

From (2), $L_{(t+1)}$ represents the updated value of luciferin, ϑ denotes a luciferin decay constant $(0 < \vartheta < 1)$. $E(mn_i(t))$ denotes a initial luciferin of the glowworm at time 't'.

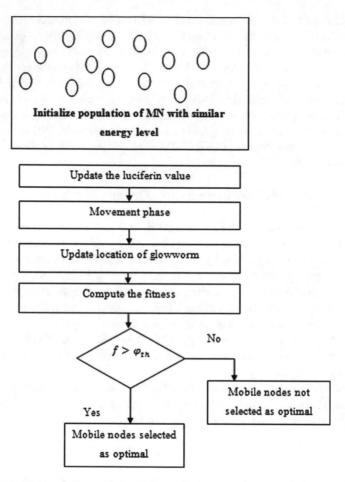

Fig. 3. Flow process of the energy-efficient glowworm swarm optimization based node selection

$L_{(t-1)}$ denotes a prior luciferin level of the each glowworm at time '$t-1$'. β denotes a luciferin improvement constant.

The other process in the EEGWSO-OR technique is the movement phase in which every glowworm with lesser luciferin value moves towards other brighter one with a certain probability. Let us consider the glowworm 'b', the probability of moving to the other brighter one 'a' is computed by Eq. (3),

$$P = \frac{L_b(t) - L_a(t)}{\sum L_n(t) - L_a(t)} \tag{3}$$

From (3), P denotes a probability, $L_b(t)$ denotes a luciferin value of the glowworm 'b' at the time 't'. $L_a(t)$ represents the luciferin value of the glowworm 'a' at the time' t'. $L_n(t)$ denotes a luciferin value of the glowworm 'n' at the time 't. Every glowworms move towards its other having minimum distance and then the luciferin value gets improved. After the movement of the glowworm, the position is updated by using the following mathematical equations,

$$x_{a(t+1)} = x_{a(t)} + \delta\left(\frac{x_b(t) - x_a(t)}{D_{ba}}\right)$$ (4)

From (4), $x_{a(t+1)}$ denotes an updated position of the glowworm 'a' at time $t+1$. $x_{a(t)}$ denotes a position of glowworm 'a' at time 't'. δ denotes a step-size. $x_b(t)$ is the position of the glowworm 'b' at time 't'. D_{ba} denotes a distance between the position of the two glowworms such as 'a' and 'b'. After updating the position of the glowworm, the fitness of each glowworm is computed. Fitness function is an objective function that is exploited to accomplish optimal one from the population. The fitness is computed as follows,

$$f = e_i - e_c$$ (5)

From (5), f indicates a fitness function, e_i signifies a total energy and e_c signifies a consumed energy of each MN. The residual energy is measured as the difference between initial energy and consumed energy of node. Residual energy is remaining energy after the MN. After computing fitness, then the threshold is set to compare the energy level of the each node.

$$y = \begin{cases} f > \varphi_{th}, & mn_i \ selected \ as \ optimal \\ f < \varphi_{th}, & mn_i \ not \ selected \ as \ optimal \end{cases}$$ (6)

From (6), y denotes an output of the energy efficient glowworm optimization technique, φ_{th} represents the threshold energy level, mn_i denotes a mobile node. The MN is elected as optimal node, when the fitness of MN is higher than the threshold energy level. If the MN having less fitness value than the threshold value, then the nodes are not selected as optimal one. As a result, the energy efficient nodes are selected for further processing.

3.2.2 Route Path Discovery

After selecting the energy efficient nodes using optimization technique, the route path discovery is performed for data transmission from SN to the destination. With the assist of neighbouring nodes, the route path from source to destination is identified. The nearest neighboring nodes are determined using stepwise regression. Stepwise regression techniques are used to analyze the MN and are related to the objective function (i.e. minimum distance) and generate a particular outcome together. Stepwise regression technique uses the forward selection process to choose the neighboring node towards the destination node (DN) with two control messages, whose selection provides the most statistically significant improvement in route path discovery.

The two control messages namely RREQ and RREP are used in the nearest neighbor selection to construct the route path. Initially, the SN transmits request messages (RREQ) to all the energy efficient nodes. After receiving the RREQ message, the other MN sends the RREP message back to the source node. EEGWSO-OR technique performs the stepwise regression on the basis of reply message arrival in the source node to find neighbor node with lesser time.

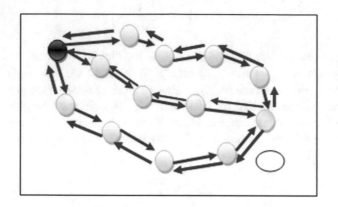

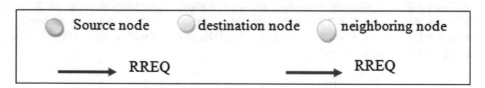

Fig. 4. Stepwise regression based neighbor node selection

As shown in Fig. 4, the neighboring node selection is carried out through the regression analysis. The neighboring node is discovered by the regression analysis through the distance measures and the two control messages.

SN sends the RREQ message to all the nodes,

$$s_n(RREQ) \rightarrow mn_i \rightarrow d_n \tag{7}$$

From (7), s_n denotes a source node, mn_i denotes a mobile node, d_n represents the destination nodes. The symbol '$\rightarrow$' denotes a forward selection of the stepwise regression. After receiving the RREQ message, the RREP message reply from the other MN which is expressed as follows,

$$d_n(RREP) \rightarrow mn_i \rightarrow s_n \tag{8}$$

Based on the RREQ and RREP message distribution, the regression function finds the neighboring node with minimum distance. Let the current coordinate for s_n is (p_1, q_1) and for d_n is (p_2, q_2). The shortest path from source to destination is discovered by measuring the distance among the MN. The distance is measured as follows.

The distance between source and destination is estimated using Euclidean distance.

$$d_{ij} = \sqrt{\sum_{i=1}^{n} (q_i - p_i)^2} \qquad (9)$$

From (9), d_{ij} signify a distance among SN and DN. Hence, the nodes with lesser distance are chosen for routing DP. After finding the route path, SN forwards the DP to the destination. In order to improve the routing process, EEGWSO-OR technique maintenance the route between s_n and d_n. In the route maintenance phase, link failure is quite normal since the nodes are not controlled by any access point. Therefore, the EEGWSO-OR technique tolerate the link failure and then selects other alternative route path with minimum distance. This assists to enhance DP delivery and lessens the RO. The algorithmic process of EEGWSO-OR technique is described as follows.

Input: Number of MN $mn_i = mn_1, mn_2, mn_3 \dots mn_n$, data packets $dp_i = dp_1, dp_2, \dots, dp_n$,

Output : Improved energy efficient routing

Begin

\\ **optimal node selection**

1. Initialize the number of mobile nodes mn_i with similar E

2.   Update the luciferin value $L_{(t+1)}$

3.   Compute the probability P of moving the glowworm

4.   Update the location $x_{a(t+1)}$

5.   **for each** mn_i

6.   compute the fitness f

7.   **if** $(f > \varphi_{th})$ **then**

8.   select the node as optimal

9.   **else**

10.   node does not selected

11.   **end if**

12.   **end for**

\\ **nearest neighbor selection**

13.   For all selected nodes mn_i

14. s_n determine nn

15.   Construct route between s_n and d_n

16. s_n send dp_i to d_n

17.   **end for**

end

Algorithm 1 Energy Efficient Glowworm Swarm Optimization based Opportunistic Routing

Algorithm 1 describes the Energy Efficient Glowworm Swarm Optimization for efficient routing in MANET. The populations of MN are initialized with similar energy level. After that, the luciferin value of each glowworm is updated. Then the probability of lesser luciferin value of the glowworm move towards the higher value is computed. Based on the luciferin value, the location of the glowworm is updated. After that, the fitness of each glowworm is computed and compared with the threshold value. The glowworm with high fitness than the threshold is selected as optimal for energy efficient routing in MANET. This assists to lessen EC and increases the NL. The route paths among the energy efficient nodes are determined and forward the DP along the route path through the neighboring node resulting in improving the packet delivery and minimizing the routing overhead.

4   Simulation Settings

The simulation of EEGWSO-OR technique and existing methods [1] and [2] are executed in NS2.34 simulator. 500 MN are positioned over the square area of A^2 (1500 m * 1500 m). For the simulation purpose, the Random Waypoint mobility is used as node mobility model. The energy efficient MN are chosen for routing number of DP from source to destination. The numbers of DP are ranges from 10 to 100. Simulation time is set as 300 s. DSR protocol is exploited to carry out energy efficient routing. Table 1 portrays the Simulation parameters.

Table 1. Simulation parameters

Simulation parameter	Value
Simulator	NS2 .34
Network area	1500 m * 1500 m
Number of sensor nodes	50, 100, 150, 200, 250, 300, 350, 400, 450, 500
Protocol	DSR
Simulation time	300 s
Mobility model	Random Way Point model
Nodes speed	0–20 m/s
Data packets	10, 20, 30, 40, 50, 60, 70, 80, 90, 100
Number of runs	10

With the above-said simulation parameters, the simulation is performed with 10 different runs. The results of proposed and existing methods are estimated with diverse parameters like EC, NL, DPDR and RO. Results of diverse parameters are discussed in forthcoming section.

5 Results and Discussion

The simulation analysis of EC, NL, DPDR and RO are discussed in this section. EEGWSO-OR technique is evaluated with existing methods namely LASEERA [1] and ACECR [2]. The performances of these three methods are discussed using table values and graphical representation.

5.1 Simulation Results of EC

EC is calculated as amount of energy utilized by MN. The formula for calculating the EC is measured as follows,

$$Energy\ consumption = No.\ of\ mobile\ nodes * Energy\ (single\ mobile\ node) \quad (10)$$

EC of MN is computed in joule (J).

Table 2. Number of mobile nodes versus energy consumption

Number of mobile nodes	Energy consumption (Joule)		
	EEGWSO-OR	LASEERA	ACECR
50	34	38	40
100	39	43	48
150	42	50	54
200	44	52	56
250	50	58	63
300	51	57	69
350	53	58	70
400	56	60	72
450	59	63	68
500	60	65	70

Table 2 describes the results of EC versus a number of MN. Above table values illustrate the EC is significantly reduced using EEGWSO-OR technique when compared to LASEERA [1] and ACECR [2]. The number of MN is ranges from 50 to 500. To show the performance of proposed and existing algorithms, diverse runs are performed.

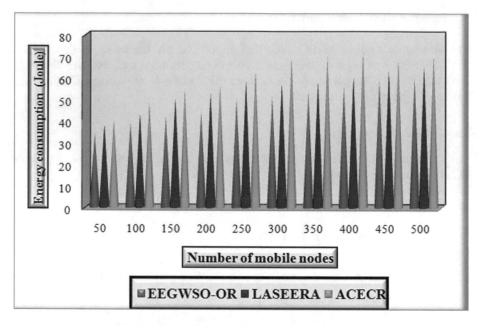

Fig. 5. Performance results of energy consumption

Figure 5 depicts the performance results of EC with three different methods namely EEGWSO-OR technique, LASEERA [1] and ACECR [2]. The above figure clearly illustrates ten different results for EC with the number of MN. From Fig. 5, the EC using EEGWSO-OR is reduced than the conventional algorithm. By selecting the energy efficient nodes, this considerable improvement is attained for DP routing. At first, the entire MN consumes similar energy level. MN is moved independently from one place to another and it is not stable. Owing to instability of MN, the energy of nodes is varied. The EEGWSO-OR technique computes the fitness of the all the nodes. The remaining energy of nodes is measured in the fitness computation from the total energy. Then the fitness is compared with the predefined energy threshold value. The node which has higher residual energy level than the threshold energy level is selected as optimal for routing in MANET. The quality of EEGWSO-OR technique minimizes the EC of the MN.

Let us consider the 50 MN for computing the EC with three different methods EEGWSO-OR technique, LASEERA [1] and ACECR [2]. EC of EEGWSO-OR technique is obtained after the simulation is 34 J. EC of the other two existing LASEERA [1] and ACECR [2] are 38 joules and 40 J respectively. Likewise, the nine remaining runs are performed. Therefore, EEGWSO-OR technique lessens the EC by 10% and 20% as compared to existing [1] and [2] respectively.

5.2 Impact of Network Lifetime

NL is measured as ratio of number of higher energy MN is elected as optimal to total number of nodes in MANET. NL is formalized as follows,

$$network\ lifetime = \frac{Number\ of\ higher\ energy\ nodes\ are\ selected}{n} * 100 \qquad (11)$$

From (11), 'n' symbolizes a total number of MN. NL is evaluated in percentage (%).

Table 3 shows the performance of NL versus a number of MN which is ranges from 50 to 500 with three diverse methods namely EEGWSO-OR technique, LASEERA [1] and ACECR [2]. From Table 3, it is evident that the performance results of NL are improved with energy efficient nodes for routing in MANET. The comparison graphs for three different methods are illustrated in Fig. 6.

Table 3. Tabulation for Network lifetime

Number of mobile nodes	Network lifetime (%)		
	EEGWSO-OR	LASEERA	ACECR
50	82	76	70
100	80	73	68
150	81	75	70
200	79	69	61
250	84	78	73
300	85	77	71
350	89	81	75
400	90	82	78
450	92	84	76
500	94	86	82

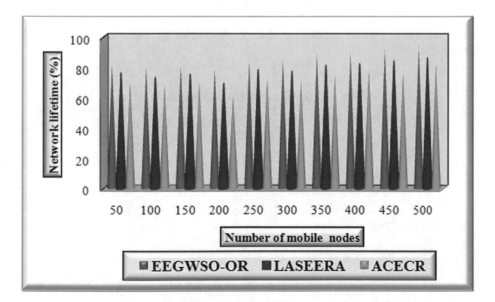

Fig. 6. Performance results of network lifetime

Figure 6 depicts the NL versus number of MN. The above two-dimensional graph illustrates the NL of three different methods of EEGWSO-OR technique, LASEERA [1] and ACECR [2]. From Fig. 6, the NL of EEGWSO-OR technique is enhanced as compared to conventional methods. For data transmission, the EEGWSO-OR technique utilizes energy efficient optimization technique in MANET. The energy efficient nodes are selected using fitness computation. Through computing fitness, the node that has superior remaining energy is identified and chosen as optimal nodes. The nodes with lesser energy are not suitable for performing the more tasks at a longer duration. Therefore, EEGWSO-OR technique selects the node which utilizes the minimum energy for performing the certain tasks. The selection of high energy nodes improves the lifetime of the network.

Ten different simulation results are performed to evaluate the EEGWSO-OR technique with existing methods [1] and [2]. At first, 50 MN are taken as input. Among the MN, 41 energy efficient nodes are selected using EEGWSO-OR technique. But the technique LASEERA [1] and ACECR [2] selects 38 and 35 MN. The above results show that the EEGWSO-OR technique effectively selects the energy optimized nodes than the existing methods. Similarly, the various runs are carried out to obtain the different performance results. EEGWSO-OR technique enhances the NL by 10% and 18% as evaluated with LASEERA [1] and ACECR [2] respectively.

5.3   Impact of Data Packet Delivery Ratio

DPDR is measured as number of DP received at destination to total number of DP sent from SN via neighboring node. The DPDR is mathematically computed as follows,

$$DPDR = \frac{Number\ of\ data\ packets\ received}{Number\ of\ data\ packets\ sent} * 100 \qquad (12)$$

From (12), *DPDR* denotes a data packet delivery ratio and evaluated in percentage (%).

Table 4. Tabulation for data packet delivery ratio

Number of data packets	Data packet delivery ratio (%)		
	EEGWSO-OR	LASEERA	ACECR
10	70	60	50
20	80	75	60
30	83	73	67
40	80	75	70
50	90	82	78
60	88	83	75
70	91	86	76
80	93	88	78
90	90	86	77
100	92	87	83

Table 4 illustrates the results of DPDR with three diverse methods namely EEGWSO-OR technique, LASEERA [1] and ACECR [2]. The evaluation results show that the DPDR is improved using EEGWSO-OR technique than the existing method.

Figure 7 demonstrates the DPDR versus a number of DP which is ranges from 10 to 100. For each instance, the different numbers of DP are taken to compute the data delivery ratio. The DP delivery is improved using EEGWSO-OR technique compared to existing methods. EEGWSO-OR technique performs the route path discovery from source to destination.

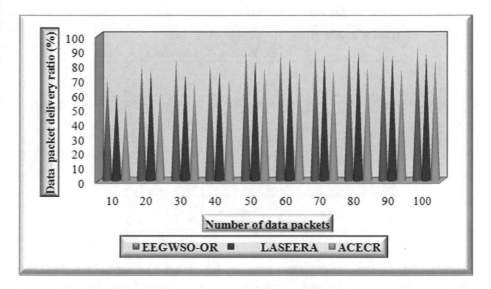

Fig. 7. Performance results of data packet delivery ratio

SN transmits DP to DN via neighboring node. The neighboring node selection is carried out using stepwise regression. In this regression analysis, the two control messages namely request and reply are distributed to all the MN. After getting request message, the nodes broadcast a reply message to source node. Based on the control message distribution, the links between the nodes are established. Then the route path is identified. EEGWSO-OR chooses route path with lesser distance. Once detecting the route path, DP is sent from source to destination. Thus, the DPDR is enhanced.

Let us consider the 10 DP for computing the data delivery ratio. EEGWSO-OR technique delivered 7 DP at the destination end. But the existing methods LASEERA [1] and ACECR [2] delivered 6 and 5 DP respectively. From the above calculation, the DPDR of EEGWSO-OR technique is 70% and the LASEERA [1] and ACECR [2] are 60% and 50% respectively. Similarly, remaining runs are carried out with various data packets. Finally, the obtained results of proposed and existing methods are compared. EEGWSO-OR technique enhances DPDR by 8% and 21% than existing [1] and [2].

5.4   Simulation Results of Routing Overhead

RO is evaluated as amount of time consumed to broadcast DP from source to destination. RO is expressed as follows

$$RO = n * time(routing\ one\ data\ packets) \tag{13}$$

From Eq. (13), *RO* represents routing overhead and evaluated in milliseconds (ms).

Table 5. Tabulation for routing overhead

Number of data packets	Routing overhead (ms)		
	EEGWSO-OR	LASEERA	ACECR
10	17	22	25
20	24	30	36
30	30	33	39
40	32	40	48
50	35	45	55
60	36	42	54
70	35	46	53
80	36	44	55
90	38	43	52
100	41	46	53

Table 5 depicts the results of RO with number of DP. The overhead metric offers efficient routing for transmit the DP from source to destination. The simulation result of the RO is illustrated in Fig. 8.

Figure 8 depicts the RO versus a number of DP. From figure, it is observed that RO of EEGWSO-OR technique attains better performance than [1] and [2]. This is because of EEGWSO-OR technique maintenance the route path whenever the link failure occurs. If the link failure occurs, the EEGWSO-OR technique finds the other alternative route path for packet transmission. This process lessens the RO when sending the DP from source to destination. In addition, the EEGWSO-OR technique also finds the efficient route path by finding the neighboring node for data transmission.

From the above discussion, the RO of EEGWSO-OR technique is considerably minimized by 17% when the comparison performed with the existing LASEERA [1]. Similarly, the overhead in the DP routing also minimized by 31% using EEGWSO-OR technique compared to ACECR [2].

The above results and discussion of various parameters clearly observe that the proposed EEGWSO-OR technique increases the energy efficient routing and NL with minimum overhead.

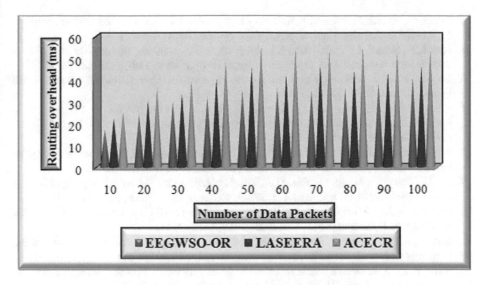

Fig. 8. Performance results of routing overhead

6 Conclusion

In this work, a new technique called, EEGWSO-OR is developed for transmitting the DP with lesser EC and RO in MANET. The EEGWSO-OR technique performs the energy efficient node selection and route path discovery. The energy efficient nodes are selected by applying the efficient optimization technique. It selects an optimal node through the remaining energy. For energy efficient routing, the node with higher residual energy is selected. After finding the energy efficient nodes, the route from source to destination is discovered via selecting the neighboring node. The stepwise regression analysis is performed for finding the neighboring node with two control message distribution. An optimal route is determined that with less distance and consumes less energy with higher PDR. In order to minimize the routing overhead, the route maintenance is carried out whenever the links between the MN are broken. This also improves the DP delivery rate. The simulation is conducted with various performance metrics such as EC, NL, DPDR and RO. The simulation result depicts that EEGWSO-OR technique enhances NL, DPDR with lesser EC and RO than the state-of-art methods.

References

1. Hao, S., Zhang, H., Song, M.: A LASEERA for MANET. J. Commun. Inf. Netw. **3**(2), 52–66 (2018)
2. Zhou, J., Tan, H., Deng, Y., Cui, L., Liu, D.: ACECR for mobile ad hoc networks under different node mobility models. EURASIP J. Wirel. Commun. Netw. **105**, 2–8 (2016)
3. Das, S.K., Tripathi, S.: Intelligent energy-aware efficient routing for MANET. Wirel. Netw. **24**(4), 1139–1159 (2018)

4. Singh, G., Kumar, N., Verma, A.K.: ANTALG: an innovative ACO based routing algorithm for MANETs. J. Netw. Comput. Appl. **45**, 151–167 (2014)
5. Jamali, S., Rezaei, L., Gudakahriz, S.J.: An energy-efficient routing protocol for MANETs: a particle swarm optimization approach. J. Appl. Res. Technol. **11**(6), 803–812 (2013)
6. Lu, T., Zhu, J.: Genetic algorithm for energy-efficient QoS multicast routing. IEEE Commun. Lett. **17**(1), 31–34 (2013)
7. Zhang, M., Yang, M., Wu, Q., Zheng, R., Zhu, J.: Smart perception and autonomic optimization: a novel bio-inspired hybrid routing protocol for MANETs. Future Gener. Comput. Syst. **81**, 505–513 (2018)
8. Al-Maqbali, H., Day, K., Ould-Khaoua, M., Touzene, A., Alzeidi, N.: A new hybrid grid-based routing approach for MANETS. Procedia Technol. **17**, 81–89 (2014)
9. Li, Y., Wang, Z., Wang, Q., Fan, Q., Chen, B.: Reliable ant colony routing algorithm for dual-channel mobile ad hoc networks. Wirel. Commun. Mob. Comput. **2018**, 1–10 (2018)
10. AlQarni, B.H., AlMogren, A.S.: Reliable and energy efficient protocol for MANET multicasting. J. Comput. Netw. Commun. **2016**, 1–13 (2016)
11. Hassan, M.H., Muniyandi, R.C.: An improved hybrid technique for energy and delay routing in mobile ad-hoc networks. Int. J. Appl. Eng. Res. **12**, 134–139 (2017)
12. Khan, S.M., Nilavalan, R., Sallama, A.F.: A novel approach for reliable route discovery in mobile ad-hoc network. Wirel. Pers. Commun. **83**(2), 1519–1529 (2015)
13. Rao, M., Singh, N.: Energy efficient QoS aware hierarchical KF-MAC routing protocol in MANET. Wirel. Pers. Commun. **101**(2), 635–648 (2018)
14. Zhang, X.M., Wang, E.B., Xia, J.J., Sung, D.K.: A neighbor coverage-based probabilistic rebroadcast for reducing RO in mobile ad hoc networks. IEEE Trans. Mob. Comput. **12**(3), 424–433 (2016)
15. Taha, A., Alsaqour, R., Uddin, M., Abdelhaq, M., Saba, T.: Energy efficient multipath routing protocol for MANET using the fitness function. IEEE Access **5**, 10369–10381 (2017)
16. Carvalho, T., Júnior, J.J., Francês, R.: A new cross-layer routing with energy awareness in hybrid mobile ad hoc networks: a fuzzy-based mechanism. Simul. Model. Pract. Theory **63**, 1–22 (2016)
17. Pathak, S., Jain, S.: A novel weight based clustering algorithm for routing in MANET. Wirel. Netw. **22**(8), 2695–2704 (2016)
18. Pratapa Reddy, A., Satyanarayana, N.: Energy-efficient stable multipath routing in MANET. Wirel. Netw. **23**(7), 2083–2091 (2017)
19. Wu, Z.-Y., Song, H.-T.: Ant-based energy-aware disjoint multipath routing algorithm for MANETs. Comput. J. **53**(2), 166–176 (2010)
20. Sarkar, S., Datta, R.: An adaptive protocol for stable and energy-aware routing in MANETs. IETE Tech. Rev. **34**(4), 1–13 (2017)

Image Synthesis Using Machine Learning Techniques

Param Gupta and Shipra Shukla$^{(\boxtimes)}$

Amity School of Engineering and Technology,
Amity University Uttar Pradesh, Noida, UP, India
ershiprashukla88@gmail.com

Abstract. Image synthesis is the generation of realistic images using a computer algorithm. This can be difficult and time-consuming. Image synthesis using machine learning aims to make this process easier and more accessible. The most prominent machine learning model for generating content is known as generative adversarial networks. This paper reviews and evaluates various generative model based on GANs. These various models are evaluated using inception score and Fréchet inception distance. These are common metrics for the evaluation of generative adversarial networks.

Keywords: Image · Generative adversarial networks · Machine learning

1 Introduction

Image synthesis is the creation of images using computer algorithms. An image is a two-dimensional spatial data. Vision evokes a response in the visual perception system. Images aim to capture that response. Images are used to store information in such a way as to evoke the same response as the response evoked by the subject of the image. Creating images be difficult and time-consuming. Technical barriers for creating images can hinder the creative process. Machine learning [1] can be used to make the process of image creation easier.

There are different ways that are available for synthesising images. This paper will classify the various methods and techniques into the following categories based on application

- Random image synthesis
- Paired Image-image translation
- Image synthesis using text
- Unpaired image-image translation.

1.1 Random Image Generation

Random image synthesis is the synthesis of random images of a particular class. The random image synthesis model synthesises random images of a particular class. A random image synthesis generator that is trained using a set of real images of faces will synthesise realistic images of new faces that do not exist and are previously unseen

© Springer Nature Switzerland AG 2020
D. J. Hemanth et al. (Eds.): ICICI 2019, LNDECT 38, pp. 311–318, 2020.
https://doi.org/10.1007/978-3-030-34080-3_35

by the generator. The major limitation of the process is that a large amount training images of the target class is required.

1.2 Paired Image-Image Translation

Paired Image-image translation is used to synthesising an image that belongs to a certain category or set using an image of another category or set when paired images belonging to both sets are available.

1.3 Image Synthesis Using Text

Image synthesis using text is used for synthesising images using a description of the content of the image to be synthesised in the form of text. Image synthesis using text models will produce an image of a particular class that the model is trained for when provided a detailed description of the image of that particular class. For example, a model can be created for synthesising images of birds using a detailed description of the bird. For training images of a particular class along with paired text description must be provided.

1.4 Unpaired Image-Image Translation

Unpaired image-image translation is used when paired data belonging to the input set and the target set do not exist for training. Images of both sets must be used for training but each input image does not require a corresponding target image present in the training dataset.

2 Literature Review

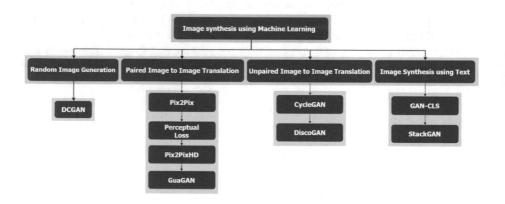

2.1   Random Image Generation

"Deep Convolutional Generative Adversarial Network" or DCGAN is a variation to GANs [2]. It can be used to synthesise random images of a certain class trained using sample images of the target class such as faces, digits, characters, living room, landscapes etc.

In DCGAN the GAN takes in a vector of noise randomly-sampled from a Gaussian distribution re-shaped into a 4-dimensional tensor to be used at the beginning of the convolutional network. Transposed convolutional layers are used to upsample the input and synthesise an image. Batch normalisation layers are also added after the convolutions. Batch normalisation is normalisation with zero mean and variance with learnable parameters.

The generator and the discriminator are trained adversarial using GAN loss and sample images of the target class as the real input to the generator.

2.2   Paired Image-Image Translation

"Pix2Pix" is a model based on the conditional GANs which is used for image-image translation [3]. It can be used to synthesise images conditioned on an input image. The generator is fed images as input and generates the target image. For the loss function, pix2pix uses the Euclidian distance in addition to traditional GAN loss. In pix2pix the U-Net model is used for the generative model. U-Net [4].

The authors dubbed the discriminator they used as PatchGAN. PatchGan tries to discriminate whether each patch of size N x N is generated or real. Since Euclidian distance focuses on high-level features, it produces blurry results. For the discriminator the authors used PatchGAN, a discriminator that focuses on the low-level texture and details.

"Perceptual Loss" is a loss function that uses feature data of pre-trained image classifiers such as VGG to help synthesise images with higher detail [5].

"Pix2PixHD" is a proposed improvement to the Pix2Pix model [6] that has the following improvements.

Coarse-to-fine generator is a proposed improvement to the pix2pix model's generator. The authors have used two generators. The first generator is fed a downsampled image and synthesises an image with the matching downsampled dimension. The second generator, first downsamples the original input to the size of the input of the first generator. The input of the next layer is the sum of the downsampled image and the last layer of the first generator. The second generator then subsequently upsamples the output and synthesises an output with original dimensions.

A multi-level discriminator is also used by the authors. There are three patch-based discriminators (as used in pix2pix) that take input of different sizes. The first discriminator takes the image of the original size and the two subsequent discriminators take images downsampled by a factor of two and four. By downsampling the image, the range of the effective receptive field of the discriminator increases. The first discriminator focuses on higher detail while the other two discriminators check for larger global consistency.

"**GauGAN**" uses a technique called spatially adaptive normalisation for improving image synthesis using semantic segmentation maps [7]. Spatially adaptive normalisation is similar to instance normalisation. In instance normalisation output of a layer is normalised and the multiplied and added to learned parameters across the channels. In Spatially adaptive normalisation the output of layer is normalised and the multiplied and added to spatial parameters obtained from convolution layers with semantic segmentation map as input.

2.3 Image Synthesis Using Text

"**GAN-CLS**" is a conditional GAN used to synthesise images from a text description of the image [8]. It embeds the text using a text embedder and compresses it using a fully connected layer. The compressed text embedding is concatenated to a randomly sampled noise vector and is then fed to de-convolutional layers for the synthesis of the image. The discriminator is fed the real and the synthesised images through convolutional layers. Then the text embedding which is compressed using full connected layer is spatially repeated and concatenated to the output of convolutions along the depth axis. This is followed by more convolutional layers for the discriminator to provide a prediction.

"**StackGAN**" uses two stacked generative adversarial networks for higher resolution images synthesis using a detailed text description [9]. A text embedding is formed from a text description. As the number of dimensions in the text embedding is very high and the amount of data would be very limited, this would cause the latent data manifold to be discontinuous. To solve this problem the authors used a technique called conditional augmentation. For conditional augmentation, an additional conditional variable, used for input, is randomly generated using a gaussian distribution using mean and diagonal covariance matrix of the text embedding. The conditional variable is randomly sampled from the distribution using the re-parameterization trick.

2.4 Unpaired Image-Image Translation

"**Cycle-consistent adversarial network**" or CycleGAN is a GAN which is used for the unpaired image-image translation [10]. It consists of two generator and two discriminators. The first generator converts the input image to target image and the second generator reconstructs the original image. The losses of these two generators is used to train the GAN. This technique is very similar to another technique known as DiscoGAN [11].

3 Methodology

Inception score is a method used for evaluating generative adversarial networks [12]. Inception score measures two criteria that are the quality and the variety of the synthesised images. Inception score is evaluated by first passing an array of synthesised images to a pre-trained image classifier called inception. It outputs a probability distribution of the predicted labels for the images. If the probability distribution is narrow or peaked, that would mean the image is distinct and signify quality. Marginal distribution can be obtained by adding all the probability distributions of the labels. If the marginal distribution is uniform it would signify variety. These two characteristics can be combined by finding the mean of K.L. divergence of marginal distribution to the probability distribution of all images to get a final inception score where a higher value signifies a better result.

$$IS(g) = exp(\mathrm{E}_{s \sim p_g}[D_{kl}(p(l|s)||p(l)])$$

where g is the generative model, s is synthesised output and l is the labels form inception model.

Algorithm: Inception Score

---

    i.   Generate images using the model under evaluation.
    ii.   Feed all generated image to the inception classifier to get logits for every image.
    iii.   Calculate the probability distribution of each image using SoftMax on the logits.
    iv.   Compute the marginal distribution by summing all probability distributions.
    v.   Compute the KL divergence between each probability distribution and the marginal distribution.
    vi.   Find the mean of all the KL divergences to compute the inception score.

Algorithm: Fréchet Inception Distance

---

    i.   Generate images using the model under evaluation
    ii.   Collect the number of real images of the target class of the model.
    iii.   Feed the generated images and the real images to the inception classifier.
    iv.   Extract the activation from the pool_3 layer of the inception classifier.
    v.   Find the mean and variance of the activations of ground truth and synthesised images.
    vi.   Create a Gaussian using the mean and variance.
    vii. Find the Fréchet distance of the Gaussians to calculate the FID.

Fréchet Inception Distance is another method used for the evaluation of images synthesis by generative adversarial networks [13]. Unlike the inception score, Fréchet inception distance evaluates the synthesised images' similarity to real images. It takes the activations from an intermediate layer in the inception model obtained from an array of real and synthesised images and then measure their distance. Lower distance signifies a better result as lower distance means the synthesised images are more similar to the real images.

$$FID(r,s) = ||\mu_r - \mu_s|| + Tr\left(\Sigma r + \Sigma s - (2\Sigma r \Sigma s)^{\frac{1}{2}}\right)$$

Where r is real image and s is synthesised image.

4 Results and Discussion

Images synthesised using machine learning is desired to be as realistic as possible. The synthesised images should also be diverse. A diverse output demonstrates the model has learned synthesise new and original images. The synthesised images should also be similar to real images. Inception score [12] measures both, realism and diversity of images synthesised from a particular model. Fréchet inception distance [13] measures the similarity to ground truth images for images synthesised by a model.

Table 1. Paired image-image translation

	Pix2Pix	Pix2PixHD	GauGAN
Inception score	1.33	1.43	1.57
Fréchet inception distance	366.67	275.34	282.52

Table 1 contains the evaluation of models for paired image-image translation. The models that are evaluated are Pix2Pix [3], Pix2PixHD [6] and GauGAN [7]. Images that are evaluated were synthesised using semantic maps using a model trained on the Cityscapes dataset where the images were paired with a corresponding semantic segmentation map [14].

Table 2. Unpaired image-image translation

	CycleGAN	DiscoGAN
Inception score	1.31	1.54
Fréchet inception distance	345.13	469.17

Pix2Pix [3] utilizes a deep convolutional generative adversarial network with U-Net [4] generator, patch-based discriminator. This model produces the least realistic and diverse output. The synthesised images are also least similar to ground truth images.

Pix2PixHD [6], with Pix2Pix [3] as a baseline, uses multi-level discriminator, coarse-to-fine generator along with feature matching loss to produce more realistic and diverse output. The output is also most similar to the ground truth images

GauGAN [7] uses spatially adaptive normalisation and produces the most realistic and diverse output. Among models for paired image-image translation, GauGAN [7] produces the most realistic and diverse output. Though, Pix2PixHD [6] produces output which is more like real images. Both Pix2PixHD and GauGAN have better inception score and FID. Thus, both are an upgrade over Pix2Pix [3].

Table 2 contains the evaluation of models for unpaired image-image translation. For unpaired image-image translation, CycleGAN [10] and DiscoGAN [11] were evaluated. The images evaluated were synthesised using a semantic map. The models were trained on the cityscapes dataset [14]. During training, the images provided to the model were not paired with a semantic segmentation map [15].

CycleGAN [10] is conditional GAN with cyclic loss for training. DiscoGAN is also a conditional GAN with cyclic loss for training but considers losses of its two discriminators separately.

For models that synthesise image using unpaired training data, DiscoGAN [11] produces results that are more realistic and diverse but images synthesised from CycleGAN [10] are more similar to ground truth images.

Table 3. Image synthesis using text

	GAN-CLS	StackGAN
Inception score	2.18	1.65
Fréchet inception distance	345.63	363.37

Table 3 contains the evaluation of models for image synthesis using text. The models evaluated for image synthesised using text are GAN-CLS [8] and StackGAN [9]. These were trained using the Oxford 102 dataset [16] with images of flowers along with their detailed text description.

GAN-CLS [8] is a conditional generative adversarial network. It produces the most realistic output which is also most similar to the ground truth images. StackGAN [9] is also a conditional generative adversarial network. It utilizes layered GANs to produce an output of higher resolution.

For images synthesised using text, GAN-CLS [8] produces images with that are more realistic and diverse than StackGAN [9]. Images synthesised using GAN-CLS also syntheses images that are more like real images.

5 Conclusions

Several models for image synthesis using machine learning were evaluated. These models were classified into various categories that are Paired image-image translation, Unpaired image-image translation and Image synthesis using text. Models in each of these categories were evaluated and compared. Inception score was used to evaluate the realism and diversity of images synthesised by the model. Fréchet inception distance was used to evaluate the similarity to the ground truth images of the synthesised image by the model. Among models for paired image-image translation, GauGAN produced the most realistic and diverse output and Pix2PixHD produced the output most similar to ground truth images. Among the model for unpaired image-image translation, DiscoGAN produces the most realistic and diverse output and CycleGAN produces the output most similar to ground truth images. For image synthesis using text, GAN-CLS produces the most realistic and diverse output as well as it being more similar to real images.

References

1. Bhatnagar A., Shukla S., Majumdar, N.: Machine learning techniques to reduce error in the internet of things. In: 9th International Conference on Cloud Computing, Data Science & Engineering, (Confluence), pp. 403–408. IEEE, January 2019
2. Radford, A., Metz, L., Chintala, S.: Unsupervised representation learning with deep convolutional generative adversarial networks. arXiv preprint arXiv:1511.06434 (2015)
3. Isola, P., Zhu, J.Y., Zhou, T., Efros, A.A.: Image-to-image translation with conditional adversarial networks. In: Proceedings of the IEEE Conference on Computer Vision and Pattern Recognition, pp. 1125–1134 (2017)
4. Ronneberger, O., Fischer, P., Brox, T.: U-Net: convolutional networks for biomedical image segmentation. In: International Conference on Medical Image Computing and Computer-Assisted Intervention, pp. 234–241. Springer, Cham, October 2015
5. Johnson, J., Alahi, A., Fei-Fei, L.: Perceptual losses for real-time style transfer and super-resolution. In: European Conference on Computer Vision, pp. 694–711. Springer, Cham, October 2016
6. Wang, T.C., Liu, M.Y., Zhu, J.Y., Tao, A., Kautz, J., Catanzaro, B.: High-resolution image synthesis and semantic manipulation with conditional GANs. In: Proceedings of the IEEE Conference on Computer Vision and Pattern Recognition, pp. 8798–8807 (2018)
7. Park, T., Liu, M.Y., Wang, T.C., Zhu, J.Y.: Semantic Image Synthesis with Spatially-Adaptive Normalization. arXiv preprint arXiv:1903.07291 (2019)
8. Reed, S., Akata, Z., Yan, X., Logeswaran, L., Schiele, B., Lee, H.: Generative adversarial text to image synthesis. arXiv preprint arXiv:1605.05396 (2016)
9. Zhang, H., Xu, T., Li, H., Zhang, S., Wang, X., Huang, X., Metaxas, D.N.: StackGAN: text to photo-realistic image synthesis with stacked generative adversarial networks. In: Proceedings of the IEEE International Conference on Computer Vision, pp. 5907–5915 (2017)
10. Zhu, J.Y., Park, T., Isola, P. and Efros, A.A.: Unpaired image-to-image translation using cycle-consistent adversarial networks. In: Proceedings of the IEEE International Conference on Computer Vision, pp. 2223–2232 (2017)
11. Kim, T., Cha, M., Kim, H., Lee, J.K., Kim, J.: Learning to discover cross-domain relations with generative adversarial networks. In: Proceedings of the 34th International Conference on Machine Learning-Volume 70, pp. 1857–1865. JMLR.org, August 2017
12. Salimans, T., Goodfellow, I., Zaremba, W., Cheung, V., Radford, A., Chen, X.: Improved techniques for training GANs. In: Advances in Neural Information Processing Systems, pp. 2234–2242 (2016)
13. Heusel, M., Ramsauer, H., Unterthiner, T., Nessler, B., Hochreiter, S.: GANs trained by a two time-scale update rule converge to a local Nash equilibrium. In: Advances in Neural Information Processing Systems, pp. 6626–6637 (2017)
14. Cordts, M., Omran, M., Ramos, S., Rehfeld, T., Enzweiler, M., Benenson, R., Franke, U., Roth, S., Schiele, B.: The cityscapes dataset for semantic urban scene understanding. In: Proceedings of the IEEE Conference on Computer Vision and Pattern Recognition, pp. 3213–3223 (2016)
15. Majumdar, N., Shukla, S., Bhatnagar, A.: Survey on applications of internet of things using machine learning. In: 9th International Conference on Cloud Computing, Data Science & Engineering, (Confluence), pp. 562–566. IEEE, January 2019
16. Nilsback, M.E., Zisserman, A.: Automated flower classification over a large number of classes. In: 2008 Sixth Indian Conference on Computer Vision, Graphics & Image Processing, pp. 722–729. IEEE, December 2008

A Novel Interference Alignment (IA) Method for Improving the Sum-Rate of the Hetnet Users

D. Prabakar$^{(\boxtimes)}$ and V. Saminadan

Department of Electronics and Communication Engineering,
Pondicherry Engineering College, Puducherry, India
{prabakar.ece,saminadan}@pec.edu

Abstract. Communication between small and macro cells is cooperative to provide seamless access and to evade the issues caused due to open wireless interfaces. The major concern in these kinds of communication is the interference due to common channel exploitation. This article discusses a novel interference alignment (IA) method modeled using multi-objective least-square (MOLS) optimization. In this IA optimization, the received signal is classified using alignment vector boundary using least-square function. This helps to determine data and noise present in the signal. By deploying appropriate alignment vectors, using least-squares, the interference and data signal is classified at the receiver end. The integrated optimization method is efficient in classifying interference by mitigating the slope errors using alignment plots that help to reduce error rate. As the error is mitigated, the degree of freedom of the users is leveraged that improves the sum rate of the network.

Keywords: Interference alignment · HetNets · Lease-square measure ·
Multi-objective optimization · Signal classification

1 Introduction

The development of wireless and mobile communication is enormous in the recent years, by granting interoperability of diverse technologies. Heterogeneous networks (HetNets) comprises of multiple wireless technologies, devices and communication standards that enables universal access to users. HetNets are designed to leverage the performance of wireless communication by scaling and providing flexible interoperable access between the users and radio resources [1]. To improve reliability, based on the user concentration, HetNet is segregated into small and macro cells. A macro cell and its associated base station (BS) serves as the access hub for a group of users in the small cell. Small cell communications are managed by deploying local BS that interacts with the macro BS directly. Macro scale/core network provides licensed spectrum that is shared by the small cell network in slotted time intervals [2]. The communicating users share a common channel due to which the users are defaced due to interference. Interference is caused internally, or from the macro cell, etc. due to which communication reliability and sum rate of the network is degraded. Henceforth, HetNets are

© Springer Nature Switzerland AG 2020
D. J. Hemanth et al. (Eds.): ICICI 2019, LNDECT 38, pp. 319–327, 2020.
https://doi.org/10.1007/978-3-030-34080-3_36

designed to evade interference issues with the consideration of reliability irrespective of the density of the users and transmitters [3]. Some interference suppression methods rely on offloading radio resources from macro to small cells. Obviously, this causes overloaded communication where, every single user is stuck in paused transmission or self-interference in the same channel. The rate of convergence in achieving reliable communication measures in the network is lessened and hence the degree of freedom (DOF) of the users is affected. The offloaded radio resources cannot be handled in the small scale network due to which it becomes congested and drops connections [4].

Interference Alignment (IA) is a prominent technique that is widely adopted for suppressing overloading and congestion issues in a network. More specifically, IA is effective in HetNets due to the varying characteristics of the networks encompassed. IA is augmented as a association with space/time/frequency for processing the received signal from the transmitter. In either of the alignment method, IA derives optimal solutions connected to space/time/frequency at the receiver end. It differentiates the interference and data in the received signal through the optimal solutions [5]. Some formal analysis method such as vector assessment, linear modeling, and non-convergence search space are employed for classifying interference. The fundamental requirement for IA with the different communication users is the cooperative nature and the synchronized access irrespective of the available radio resources. Signal sensing and power estimation are the other requirements in analyzing the signals in the receiver end [6].

A sub-channel optimization and assignment method is designed by the authors in [7] for improving the degree of freedom (DOF) of the users. In this method, K-user IA is considered by proportionating transmitting and receiving antennas. Though the DOF achieved is greater than K/2, the immature constraints result in precoding failures. This issue is addressed in [8] by designing a close-coupled IA system that resolves interference alignment issues exploiting the channel state information (CSI).

The methods designed in [9] also exploit the advantages of CSI along with a quadratic function to improve alignment in small cells. This information carries the details of transmitters and concurrent users irrespective of the size of the network.

To address the issues of complexity in large scale HetNet, the authors in [10] introduced a cluster based IA (CIA). The clusters are used for administering cooperation within the network as a part of synchronized communication. Synchronized communication ensures uninterrupted signal processing within the allocated channel such that interference is suppressed at the receiver using the traditional optimization.

Aggregate cluster based IA is designed in [11] for reducing the signal processing complexity of the communicating devices in a HetNet. This method classifies the network on the basis of observed CSI and noise to mitigate smaller to large interference throughout.

A partial IA scheme is designed by the authors in [12] in which scheduling is a prominent theme for optimization. This method focuses in leveraging the achievable sum rate of the network in a multi-cell environment. The scheduling algorithms differentiate the communication and interference links to ensure the receiver data is noise free. Semi-blind interference alignment (SBIA) method using clustering concept is specifically designed for small cell networks in [13]. This alignment method reduces the length of super-symbol by classifying same slot users.

2 Proposed Method

A spectral effective IA is introduced in this article that exploits multi-objective least-square (MOLS) optimization. Spectral efficiency is achieved through mutual co-existence of multiple small scale devices sharing a common HetNet wireless medium. Different from the methods discussed, MOLS balances IA along with co-existence features. This optimization method follows two-objectives for any received signal analysis namely, less interference and least alignment extraction. The two objectives are separated using a non-linear objective boundary for alignment. In Fig. 1, the considered HetNet model is presented.

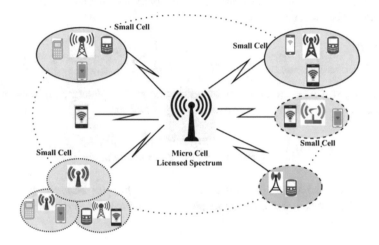

Fig. 1. HetNet with micro and small cell illustration

2.1 System Model

We deem a HetNet comprising of N small cells users that communicate using the unlicensed spectrum of the base network. The network consists of different I/O and employs both uplink and downlink communication. The uplink communication is facilitated using primary users and the downlink communication is pursued by the secondary small scale users. The communicating channels are represented using matrices C_i and D_i. The suffix "i" indicates the ith receiver. Let m denote the antennas used in a channel c communicating with the help of k-transmitters such that $N = 2 \times k - 1$ is an alignment criterion.

2.2 Multi-objective Least-Square Optimization

In the proposed optimization method, the two objectives as mentioned earlier is addressed to retain $N = 2 \times k - 1$ to achieve alignment. This optimization method follows plot boundary for differentiating the received signal and its interference. Let Y_n represent the received signal of the nth user that is expressed as in Eq. (1)

$$Y_n = \sum_{i=1}^{N} c_{in} \, C_i D_i \, d_{in} + \gamma_{in}, \, \forall n \in N \qquad (1)$$

Where, d_{in} is the data representation in the received signal Y_n and γ_{in} is the noise. The received Y_n is decoded for d_{in} and γ_{in} such that $\vec{a}_1$ and $\vec{a}_2$ are the alignment vectors for Y_n classification. In a conventional decoding process, let Y_i represent the mediate data signal segregated from γ_{in} represented as

$$Y_i = \sum_{i=1}^{N} d_{in} \, c_{in} \times \frac{1}{C_i} + d_{in} \frac{1}{D_i} c_{in} + d_{in}(\vec{a}_1 + \vec{a}_2) \qquad (2)$$

The factor $d_{in}(\vec{a}_1 + \vec{a}_2)$ represents noisy data received. Using a co-variance matrix of C_i and D_i, the noise is separated as

$$\gamma_{in} \sum_{i=1}^{N} d_{in}(\vec{a}_1 + \vec{a}_2) + \sum_{i=1}^{N} \sum_{j=1}^{k} [C_i \times \frac{1}{D_i} \times p_k^i \, C_{ij}^N - D_i \times \frac{1}{C_i} \times p_k^i \, C_{ij}^N] \qquad (3)$$

As per the objectives, the definite classification boundary is determined between γ_{in} and Y_i, $Y_i \in Y_n, \forall n \in N$. The two objectives are expressed as

$$d_{in}(\vec{a}_1 + \vec{a}_2) = \begin{cases} \arg \min \|Y_N - Y_i\| \, \forall \, \gamma_{in} \\ \arg \min \gamma_{in} \, \forall \, n \in N \, such \, that \, N = 2k - 1 \end{cases} \qquad (4)$$

By the MOLS optimization, for each Y_i in Y_N, the plot between $(\vec{a}_1 \, \& \, \gamma_{in})$ and $(\vec{a}_2 \, \& \, \gamma_{in})$ pairs are estimated. A sample plot of the above pairs is represented in Fig. 2.

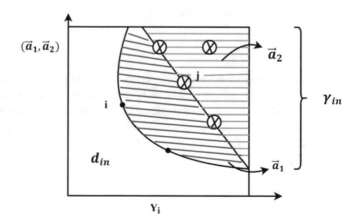

Fig. 2. Plot illustrations for boundary detection

The boundary between d_{in} and $\vec{a}_1$ and d_{in} and $\vec{a}_2$ are estimated $i \in d_{in}$ to classify γ_{in}. In a joint $(\vec{a}_1 + \vec{a}_2)$ evaluation, the target optimization is achieved as $||Y_i - i||, ||Y_i - j||$, if $j = 0$ or $i = 0$. Now, the trade-off in boundary estimation is given by Eq. (5)

$$[i + (e \times j)] = ||Y_i - i||^2 + e \times ||Y_i - j||^2, \ e > 0 \tag{5}$$

Here, e represents the boundary trade off that is estimated as the slope between i and j. The trade-off value is estimated as

$$e = \begin{cases} \sqrt{\frac{c_{in}}{C_i}} \times \frac{1}{d_{in}(\vec{a}_1 + \vec{a}_2)}, & \text{if } j = 0 \\ \sqrt{\frac{c_{in}}{D_i}} \times \frac{1}{d_{in}(\vec{a}_1 + \vec{a}_2)}, & \text{if } i = 0 \end{cases} \tag{6}$$

Now, the sum rate is computed for all $N = 2 \times k - 1$ condition as

$$sum\,rate = \sum_{i=1}^{N} \sum_{j=1}^{k} log_2 \left(1 + \frac{d_{in} c_{in}^k \frac{1}{C_i D_i} \cdot P_k^i}{(d_{in})^k \cdot \gamma_{in}^k} \right) \tag{7}$$

The process of interference alignment is instigated between d_{in} and $\vec{a}_1$ and $\vec{a}_2$ using conventional rank model. The filters are employed at each "i" and "j" (Refer Fig. 2) to segregate the inference with alignment. In the conventional rank model,

$$\left. \begin{array}{l} rank\left(Y_i d_{in}, (CD)^{ii}\right) = \frac{\emptyset^i}{c_{in}} \cdot \vec{a}_1, \ if \ i \neq 0 \\ rank\left(Y_i d_{in}, (CD)^{ii}\right) = \frac{\emptyset^i}{c_{in}} \cdot \vec{a}_2, \ if \ j \neq 0 \\ such\,that\,d_{in}(\vec{a}_1 + \vec{a}_2) = 0, \ \forall \frac{c_{in}}{C_i} \neq 0 \ or \ \frac{c_{in}}{D_i} = 0 \ \& \\ \gamma_{in} \neq 0 \ \& \ N \neq k \end{array} \right\} \tag{8}$$

After the alignment as represented in Eq. (8) for all the rank derived solution $\emptyset^i$, degree of freedom is compute using Eq. (9) as

$$DOF = \max_{d_{in}} \lim_{SNR \to \infty} \frac{c_{in}(SNR, d_{in}, \gamma_{in})}{\log(SNR)} \tag{9}$$

Where, SNR is the signal to noise ratio observed in $Y_i \forall i \in n \in N$, the alignment process is illustrated in Fig. 3.

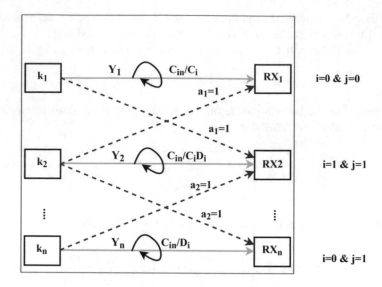

Fig. 3. TX and RX IA

In Fig. 3, k-transmitters (TX) send signals to the receiver where a set of conditions is analyzed. The analyzed set of conditions and the alignment vector requirement is given in Table 1.

Table 1. Signal analysis condition and values

"i"	"j"	a_1	a_2
0	0	1	0
1	1	1	1
0	1	0	1
1	0	1	0

Based on the signal analysis for "i" and "j", the alignment vectors are determined. The alignment vectors follow the rank based optimization as in Eq. (8) to maximize DOF. The implication of one/two alignment vectors relies on the input of "i" and "j". The process is repeated until all $Y_i, i \in N$ is classified for d_{in} and γ_{in} is suppressed.

3   Results and Discussion

The performance of the proposed MOLS based IA is experimented using MATLAB simulations. The network is modeled with 50 small cells consists of uneven number of users. The network is wide spread to 500 m × 500 m area and the users are specified with a transmit power of 10 dBm. In Table 2 a detailed simulation requirement is given

Table 2. Simulation parameter and values

Parameter	Value
Network area	500 m × 500 m
Small cells	50
Path loss model	Rayleigh fading
Bandwidth	180 kHz
Path loss exponent	3.76

For evaluating the consistency of the proposed MOLS-IA, the metrics such as spectral efficiency, DOF and CDF are compared with the existing CIA [10] and SBIA [13].

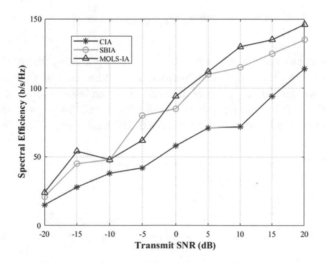

Fig. 4. Spectral efficiency analyses

In Fig. 4, the spectral efficiency with respect to the transmit SNR is compared with the existing and proposed methods. In the proposed method by classifying the interference and the data using boundary based least-square analysis. This helps to retain the DOF the users occupying optimal channel for utilization. This in turn improves the spectral efficiency of the network.

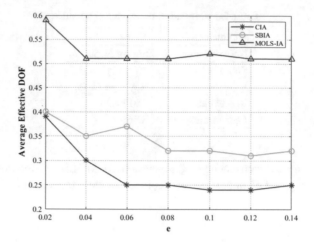

Fig. 5. DOF analyzes

The proposed MOLS-IA achieves better DOF by optimizing the receiving signal in the boundary using proper alignment vectors. The implication of alignment vectors is dependent on the number of k and N received in the end-user device. Another reason for DOF maximization is the early detection of error in received signal. The identified e fixes the number of alignment vectors required for maximizing DOF (Fig. 5).

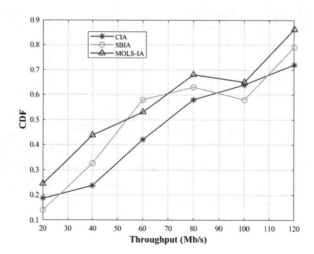

Fig. 6. CDF analysis

In Fig. 6, the CDF with respect to the two alignment vectors are estimated for all the existing and proposed method. In the proposed method, the least-square method achieves better CDF by identifying e and suppressing interference by implying appropriate vectors.

4 Conclusion

In this article a multi-objective least-square optimization for improving the sum rate of the small cells in a HetNet, is proposed. This optimization method achieved IA by differentiating data and interference from the received signal using least-square optimization. The alignment vector requirements are determined using boundary analysis for IA process. Experimental results shows that the proposed optimization based IA retain the sum rate of the small cell users by leveraging the DOF with accordance to the alignment vectors and channel matrix.

References

1. Dehghani, M., Arshad, K., MacKenzie, R.: LTE-advanced radio access enhancements: a survey. Wirel. Pers. Commun. **80**(3), 891–921 (2015)
2. Nakamura, T., Nagata, S., Benjebbour, A., Kishiyama, Y., Hai, T., Xiaodong, S., Ning, Y., Nan, L.: Trends in small cell enhancements in LTE advanced. IEEE Commun. Mag. **51**(2), 98–105 (2013)
3. Lopez-Perez, D., Guvenc, I., de la Roche, G., Kountouris, M., Quek, T.Q.S., Zhang, J.: Enhanced intercell interference coordination challenges in heterogeneous networks. IEEE Wirel. Commun. **18**(3), 22–30 (2011)
4. Acharya, J., Gao, L., Gaur, S.: Heterogeneous Networks in LTE Advanced. Wiley, Hoboken (2014)
5. Jafar, S.A.: Interference Alignment—A New Look at Signal Dimensions in a Communication Network. Now Publishers Inc., Breda (2011)
6. Yang, L., Zhang, W.: Interference Coordination for 5G Cellular Network. Series SpringerBriefs in Electrical and Computer Engineering. Springer, New York (2015)
7. Cadambe, V.R., Jafar, S.A.: Interference alignment and degrees of freedom of the-user interference channel. IEEE Trans. Inf. Theory **54**(8), 3425–3441 (2008)
8. El Ayach, O., Peters, S.W., Heath, R.W.: The practical challenges of interference alignment. IEEE Wirel. Commun. **20**(1), 35–42 (2013)
9. Mungara, R.K., George, G., Lozano, A.: Overhead and spectral efficiency of pilot-assisted interference alignment in time-selective fading channels. IEEE Trans. Wirel. Commun. **13**(9), 4884–4895 (2014)
10. Tresch, R., Guillaud, M.: Clustered interference alignment in large cellular networks. In: 2009 IEEE 20th International Symposium on Personal, Indoor and Mobile Radio Communications, pp. 1024–1028 (2009)
11. Chen, S., Cheng, R.S.: Clustering for interference alignment in multiuser interference network. IEEE Trans. Veh. Technol. **63**(6), 2613–2624 (2014)
12. Zhang, Y., Zhou, Z., Li, B., Gu, C., Shu, R.: Partial interference alignment for downlink multi-cell multi-input-multi-output networks. IET Commun. **9**(6), 836–843 (2015)
13. Kavasoglu, F.C., Huang, Y., Rao, B.D.: Semi-blind interference alignment techniques for small cell networks. IEEE Trans. Sig. Process. **62**(23), 6335–6348 (2014)

Efficient Resource Utilization to Improve Quality of Service (QoS) Using Path Tracing Algorithm in Wireless Sensor Network

N. Tamilarasi[2]([⊠]) and S. G. Santhi[1]

[1] Department of Computer Science and Engineering, FEAT,
Annamalai University, Annamalainagar, India
sgsau2009@gmail.com
[2] Department of Computer and Information Science, Faculty of Science,
Annamalai University, Annamalainagar, India
sjarasi08@gmail.com

Abstract. Wireless Sensor Network consists of a group of independent wireless devices, which is capable of exchanging information with one another without having the knowledge of predefined infrastructure or any centralized node. It functions of WSN depends on the participation of all the nodes in the network. The more nodes involved in the network traffic, the more powerful a WSN acquires. The Quality of Service (QoS) of a routing protocol is constructed successfully only if it knows the bandwidth of a coding host. Nevertheless, it is a challenging issue to identify the coding host and its bandwidth consumption in a WSN. Sending packets from one device to another is done via a chain of intermediate nodes. Detecting routes and forwarding packets consumes local CPU time, memory, network-bandwidth, and energy. We find that the existing, Authenticated Routing for Ad Hoc Network (ARAN) uses Dynamic Source Routing (DSR) Protocol, which has greater performance cost. So we propose a novelty path tracing algorithm using Ad hoc On Demand Distance Vector (AODV) routing protocol for finding the packet droppers in the WSN. The proposed Path Tracing Algorithm (PTA) also detects the Wormhole attack using per hop distance and link frequent appearance count parameters. The performance cost of the proposed method is minimal and outweighed when the security increases. As a result, there is a possibility for a node to delay the packet forwarding and at the same time it utilizes their own resources for data transmission. In the course of broad experimentation we demonstrate that the proposed method detects the Wormhole attacks and reduces the overhead required if the network size increases. Hence it is proved that the QoS is improved when compared to the existing ARAN protocol. The above proposed work is implemented using Network Simulator2 (NS2).

Keywords: ARAN · WSN · PTA · QoS · Wormhole attack · AODV

1 Introduction

Usually a collection of wireless sensor nodes forms a Wireless Sensor Network (WSN) and is capable of transferring information with each other without any fixed infrastructure. In nature WSNs are independent battery powered devices which forms

© Springer Nature Switzerland AG 2020
D. J. Hemanth et al. (Eds.): ICICI 2019, LNDECT 38, pp. 328–337, 2020.
https://doi.org/10.1007/978-3-030-34080-3_37

the networks. Each and every node in the network is allowed to move freely towards any path, and will consequently change its connection to different device during its life time [1, 2]. Ad hoc nodes are wireless in nature and it is vulnerable to various wireless attacks. The major challenge while organizing sensor node is to maintain routing table in order to establish efficient routing. These networks are controlled either by themselves or can be coupled to a bigger network [3]. The autonomous decentralized sensor nodes support multi hop routing and scalability (Fig. 1).

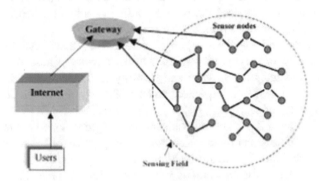

Fig. 1. Wireless sensor network.

Comparatively in WSN, reactive routing protocols perform well when compared to proactive routing protocol. Reactive routing protocols reduces overhead to find Wormhole attacks [4, 5]. As its independent feature, ad hoc is vulnerable to many attacks. Wireless nodes can be fixed and removed at anywhere at any place. Because of their independent nature, we cannot provide full security to the sensor nodes. No centralized node is present in the network, so we believe that all the sensor nodes are cooperative [6]. The main intention of the invader is to break the sensor node cooperativeness, eavesdropping, wormhole attacks, denial of service etc.

To identify, track and show the negative effects of Wormhole nodes, we proposed Path Tracing algorithm. This algorithm detects the Wormhole nodes using per hop distance and link frequent appearance count. To simulate the proposed work in NS2 regarding Quality of Service in WSN and evaluate the results through delay, packet delivery ratio and throughput.

Section 2 deals with literature review and existing system of the project. Section 3 deals with the existing problem. Section 4 deals with module description of the proposed work such as Packet Dropper In WSN, Packet Dropper Prevention And Detection and Estimate Packet Delivery. Section 5 deals with Network Simulator2 Software, performance analysis of existing and proposed mechanism. Finally Sect. 6 concludes with conclusion.

2   Literature Review

Mbowe and Oreku [7] analyzed quality of service in wireless sensor networks based on Reliability, Availability and Serviceability (RSA) parameters rather than the traditional metrics and proved that the proposed mechanism works better than that of existing method.

Parikh, Patel, and Rizvi [8] proposed an improved quality of service method called timestamp Optimization technique. They presented both mathematical and analytical models to describe the proposed mechanism and it considers the sensor nodes that is having only one hop distance from the source to the sink.

Karare, Sonekar and Akanksha [9] proposed a SAFEQ and Watchdog algorithm to improve the quality of service in wireless sensor network. These two collaborative approaches provide Integrity, Privacy and Security to the wireless network and also prevents the networks from unwanted attacks. So the above proposed approach improves the quality of service.

Elakkiya, Santhana Krishnan and Ramaswamy [10] developed a quality of service approach named relative coordinate Rumour Routing protocol based on straight line random walk technique. The main intention of this paper is to calculate the performance metrics under various conditions such as position of a node, network size etc.

3   Problem Definition

The existing Authenticated Routing for Ad hoc Network (ARAN) works on Distance source Routing, which makes many contributions to the secure design of ad hoc routing protocols. In reactive routing protocols, Detection and prevention of Wormhole Attacks using ARAN provides verification, message reliability, and non-repudiation in routing in an ad hoc atmosphere [11, 12]. Here we are implementing packet dropping using DSR routing protocol. Almost all the protocols assume the existence of some routing security that guarantees that the selected route is free of malicious nodes. Although we found that the existing method ARAN needs greater performance cost. Hence the QoS is affected in the existing mechanism. So we propose a PTA for improving the Qos factors in WSN [13].

4   Proposed Mechanism

The proposed Path Tracing algorithm (PTA) will be structured into the following three main phases, which will be explained in the subsequent subsections:

- Packet dropper in WSN
- Packet dropper Detection and Prevention
- Estimation of packet delivery

4.1 Packet Dropper in WSN

One of the crucial attacks in ad hoc network is wormhole attack. Two malicious wormhole nodes are created in the network in out of bound type. These malicious nodes build a tunnel and steal the network data and transmit between each other with high transmission speed.

Wormhole nodes establish a tunnel path and start collecting the wireless information from the environment and promote it to one another via tunnel. It also creates an illusion to other nodes that they are the instant neighbors and get packets from them and dispatch it to another malicious node location in the network. Routing protocols such as on demand and proactive are affected by wormhole attack.

In this paper, AODV Protocol is used to analyze its performance in WSN and thus it is achieved by Path Tracing Algorithm during packet sending and receiving.

4.2 Packet Dropper Detection and Prevention

This section briefly addresses the system model, notations, assumptions and unconsidered proposal. Source routing is used to forward the data packets. Let R = {r1, r2, r3..., rn} be the set of random sensor nodes. They can communicate with each other through radio transmission and the neighboring nodes communicate each other in a bidirectional mode. The distance between the neighboring nodes must be less than the predefined distance 'd'. To increase the access in radio transmission, MAC protocols are used in the wireless nodes [14] (Fig. 2).

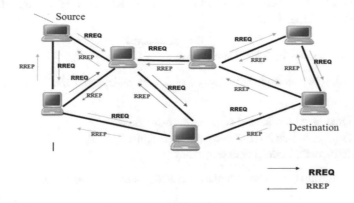

Fig. 2. AODV routing structure.

The network is designed in the manner of lose clock synchronization. Each node in ad hoc atmosphere may or may not be resource restricted. Route Request (RREQ) packets are sent by source node via abrupt neighbors to the destination node [15, 16]. The destination node sends a Route Reply to the source node After receiving the Route Request from the source in the same path. The intermediate routing path information is stored in the routing table cache. Additional information such as prior per hop distance field and time stamp fields are added to the packet header in order to detect the

Wormhole nodes. Malicious Activity in the network is detected by comparing the earlier per hop distance and current per hop distance. If the difference is less than the Maximum Threshold value (M_{th}), then the network is under secure condition otherwise malicious activity is detected. Each and every sensor node in the networks is participated in the routing process to perform this function.

The value for time stamp field is assigned at the time of the first bit of RREQ is sent. Per hop distance field can be changed by intermediary nodes but time stamp field cannot be altered by any other nodes. Whenever an intermediate node obtains RREQ packet, then per hop distance to its instant neighbor is calculated and compared it with the earlier per hop distance in the header value.

After the comparison, it updates per hop distance in the prior per hop distance field of the packet header and forwards RREQ to its neighboring nodes. On obtaining RREQ, the receiver computes per hop distance with its neighbor in the reverse path and it places in the packet header. Every intermediate node forwards one RREP for each RREQ. Every RREP holds each hop distance of all path in which it is related. In addition to per hop distance value, it also holds the time stamp of the time taken between sending and receiving the RREQ and RREP correspondingly between two nodes. The computation of per hop distance of each node is described in the next section.

4.3   Per Hop Estimation

The presence of malicious activity can be known by computing the distance between each hop in a path. We consider that per hop distance is calculated by using the Round Trip Time (RTT). Time taken between the sending of RREQ and receiving of RREP by source node is referred as RTT.

4.4   Variables Used in RTT Calculation

Let us consider two non-packet dropper nodes A and B and the RTT calculated as follows:

  T rep: Time at which RREP is received from B.
  T req: Time at which RREQ is transmitted from A to B.
  IPD: Intermediate node processing delay

  By using formula, we can calculate the RRT between two nodes

$$\Delta T = RTT = T_{rep} - T_{req} - IPD$$

.

4.5   Estimation of Packet Delivery

After detecting packet dropper node, we need to take accurate measurements for packet sending and receiving and it is our responsibility to design an effective defense mechanism for detecting malicious attack. To achieve this, we propose a mechanism to detect the malicious node using per hop distance and link frequent appearance count Parameters using AODV routing protocol [17, 18].

With the estimated value of ΔT, per hop distance between A and B 'DAB' is considered by assuming the routing signals are compared with the speed of light 'v' [19].

$$DAB = (v/2) * \Delta T.$$

Four major operations are needed to detect the packet dropper

1. Calculate current per hop value and compare it with earlier per hop value
2. If the compared result value is larger than the Maximum Threshold rate (M_{Th}), then the packet dropper is detected and it is communicated to all other nodes to make a malicious alert and also count the number of time a link is used in a particular path in order to confirm packet dropper attack.
3. If $DBC - DAB > M_{Th}$ and $F_{Account} > F_{ATh}$ then it is a packet dropper link.
   To make our proposed mechanism energy efficient, Per hop distance is calculated at the time of route discovery. Many routes are identified from the route discovery process. The packet header stores the per hop distance that calculate each path for all the nodes. By comparing the per hop distance between of all nodes in a path, a packet dropper can be detected. If it exceeds a maximum threshold range M_{Th}. Then the node related to that particular path is packet dropper node.
4. For the effective packet dropper detection, we take another parameter called frequent appearance using Path Tracing Approach. If $F_{Account} > F_{ATh}$ then it is a packet dropper link. After the detection of the packet dropper, a node intimates the presence of packet dropper to other nodes in the network. To prevent the packet dropper node participation further, their identities are added to the packet dropper list in each node. So that it is not necessary to calculate per hop distance each time when a path is discovered. Thus our proposed mechanism extends the computation energy by storing the estimated per hop distance in a cache. Hence the Quality of service has been improved in the wireless sensor networks.

4.6   Examination of Frequent Appearance of a Link

Check every link participated in the routing frequently, in order to detect malicious activity. Let Lj represents frequent appearance (F) of a link in a path. The highest number of times Lj takes part in a path is considered as $F_{Account}$.

$$F_{Account} = L_j/\text{Total number of existing links in a path N}$$

If the count of a frequent link in a specific path is larger than the threshold frequent appearance, that path is identified as wormhole attacked path. This count values are monitored and collected in the cache. RTT should be calculated by the node's own clock. This proposed mechanism is easy to implement and reduces requirements overhead.

4.7   Path Tracing Algorithm (PTA)

Following are the Steps for Path Tracing Algorithm to detect Wormhole node.

Step 1: RTT values can be calculated based on the time when the RREQ is submitted and when the RREP is received in a particular path. The RTT computation is done on every node by its individual's clock.

Step 2: calculate per hop distance rate from RTT. In each packet header, the time stamp and the per hop distance value are stored.

Step 3: To identify the Wormhole attack, the above RTT values are stored. Each node in the routing path computes existing per hop distance with its neighbor and compares it with the earlier per hop distance. If the per hop distance is greater than the threshold value go to step 4.

Step 4: Check out the highest count of a link takes part in the particular route. If $F_{Account} > F_{ATh}$, then the path is identified as Wormhole path.

Step 5: The path is declared as Wormhole attacked path and this information is communicated to other nodes to prevent data transmission.

5   Result and Discussions

The proposed Path Tracing algorithm with increased quality of service in wireless sensor network is implemented using Network Simulator2 (NS2). In this simulator, 27 sensor nodes are performed in the region 700 m × 500 m Nam window. The results shows that the proposed path tracing algorithm produces better results in the parameters such as throughput, packet delivery ratio and the delay time.

Simulation Parameters. See Table 1.

Table 1. Simulation parameters

No. of nodes	50, 100, 150, 200, 250
Area size	1000 m × 1000 m
MAC TYPE	MAC/802_11
Propagation	TwoRayGround
Antenna	OmniAntenna
Interference range	550 m
Transmission range	250 m
Simulation time	100 s
Traffic source	CBR
Packet size	1024 bytes
Rate	80 kbps

5.1 Packet Delivery Ratio

PDR is referred as the difference between the quantities of packets sent from source to the amount of packets received at destination. If the number of malicious nodes and its mobility increases, then the PDR decreases gradually.

PDR = Amount of packets received at Destination/Amount of packet sent from Source

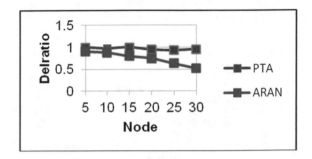

Fig. 3. Nodes vs Packet Delivery Ratio

Figure 3 shows how the Packet Delivery Ratio varies when there is increase in the size of network for the existing and the proposed mechanism. The proposed PTA shows that the PDR slightly varies, when the nodes in the network increases. At the same time, in the existing ARAN, it increases drastically. The above figure proves that Packet Delivery Ratio of PTA is high when compared to ARAN.

5.2 Throughput

If the quantity of malicious node rises, then the throughput will be decreased. The throughput of general DSR is 87% at the node mobility of 10 m/s for 10 malicious nodes and that of AODV is 95%. However the PT algorithm gives 97% of throughput.

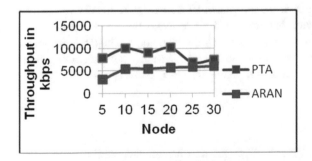

Fig. 4. No. of nodes vs Throughput

Figure 4 represents comparison of throughput between existing ARAN and the proposed PTA and finally the graph depicts that the proposed PTA results in high throughput when compared to existing ARAN.

5.3   Average Delay

The average delay is the delay time that is calculated by measuring the elapsed time between the packet sent and received. The following Fig. 5 shows that there is a drastic change in average delay for the existing system, if we increase the number of nodes. But the proposed system gives better results for the same.

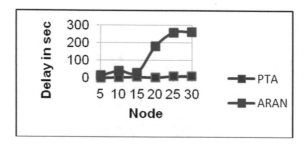

Fig. 5. No. of nodes vs Delay

6   Conclusion

In this reputation-based scheme, built on top of normal AODV secure routing protocol the malicious attack which is one of the network layer attacks (Wormhole). This initiates attacks by creating a tunnel between two or more Wormhole nodes and drops all the packets. To detect and prevent the Wormhole attack, we proposed Path Tracing Algorithm (PTA). The proposed algorithm detects and prevents the Wormhole attack using per hop distance between two nodes and link frequent appearance count parameters. The simulation result clearly proves that our projected algorithm is more efficient in preventing the Wormhole attack with greater throughput and less average delay. The performance analysis of PTA has also reduced overhead and delay. Thus the Quality of service has improved while using the proposed mechanism and it doesn't need any additional hardware for implementation. Thus, the proposed design, proves to be more efficient and more secure than existing secure routing protocol in defending against both malicious and authenticated malicious node.

References

1. Chiang, W., Zilic, Z., Radecka, K., Chenard, J.-S.: Architectures of WSN wireless sensor network nodes. In: ITC International Test Conference, vol. 43, no. 2, pp. 1232–1241 (2004)
2. Yick, J., Mukherjee, B., Ghosal, D.: Wireless sensor network survey. Comput. Netw. **52**(12), 2292–2330 (2008). https://doi.org/10.1016/j.comnet.2008.04.002

3. Tomur, E., Erten, Y.M.: Security and service quality analysis for cluster-based wireless sensor networks. In: Fifth International Conference on Wired/Wireless Internet Communications (WWIC 2007), Coimbra, Portugal, May 2007
4. Sachan, V.K., Imam, S.A., Beg, M.T.: Energy-efficient communication methods in wireless sensor networks: a critical review. Int. J. Comput. Appl. **39**(17), 3548 (2012)
5. Rajesh, T., Kumari, V.S.R.: Design and analysis of an improved AODV routing protocol for wireless sensor networks and OPNET. Int. J. Adv. Res. Electron. Commun. Eng. **3**(10), 1267–1278 (2014)
6. Sivakumar, N., Gunasekaran, G.: The quality of service support for wireless sensor networks. Int. J. Adv. Res. Comput. Sci. Softw. Eng. **3**(1), 297–302 (2013)
7. Mbowe, J.E., Oreku, G.S.: Quality of service in wireless sensor networks. Wirel. Sensor Netw. **6**, 19–26 (2014). http://www.scirp.org/journal/wsn. https://doi.org/10.4236/wsn.2014.62003
8. Parikh, S., Patel, A., Rizvi, S.: Increasing Quality of Service (QoS) in Wireless Sensor Networks (WSN) by using timestamp optimization scheme. In: ASEE 2014 Zone I Conference, 3–5 April 2014, University of Bridgeport, Bridgpeort, CT, USA (2014)
9. Karare, A.R., Sonekar, S.V., Akanksha, K.: Improving the quality of services in wireless sensor network by improving the security. Int. J. Eng. Res. Appl. (IJERA). ISSN 2248-9622 International Conference on Industrial Automation and Computing, ICIAC, 12th & 13th April 2014
10. Elakkiya, A., Santhana, B., Ramaswamy, M.: Performance evaluation of QoS based improved rumour routing scheme for WSN. Int. J. Wirel. Commun. Netw. Technol. http://warse.org/IJWCNT/static/pdf/file/ijwcnt01522016.pdf
11. Xiao, H., Seah, W.K.G., Lo, A., Chua, K.C.: A flexible quality of service model for mobile ad-hoc networks. In: Proceedings of IEEE 51st Vehicular Technology Conference, Tokyo, Japan, May 2000
12. Iyer, R., Kleinrock, L.: QoS control for sensor networks. In: ICC 2003, 11–15 May 2003, vol. 1, pp. 517–521 (2003)
13. Sanli, H.O., Çam, H., Cheng, X.: EQoS: an energy efficient QoS protocol for wireless sensor networks. In: Proceedings of the 2004 Western Simulation Multi Conference (WMC 2004), San Diego, CA, USA, 18–21 January 2004
14. Sharifi, M., Taleghan, M.A., Taherkordi, A.: A middleware layer for QoS support in wireless sensor networks. In: Networking, International Conference on Systems and International Conference on Mobile Communications and Learning Technologies, Maurlius (2006)
15. Lee, S.-K., Koh, J.-G., Jung, C.-R.: An energy-efficient QoS aware routing algorithm for wireless multimedia sensor networks. Int. J. Multimedia Ubiquitous Eng. **9**(2), 245–252 (2014)
16. Perillo, M., Heinzelman, W.: Providing application QoS through intelligent sensor management. In: 1st sensor network protocols and applications workshop (SNPA 2003), Anchorage, 11 May 2003, pp. 93–101 (2003)
17. Wang, J., Xu, J., Xiang, M.: EAQR: an energy-efficient aco based QoS routing algorithm in wireless sensor networks. Chin. J. Electron. **18**(1), 113–116 (2009)
18. Prabha, R., Shivaraj Karki, M.S.H., Venugopal, K.R., Patnaik, L.M.: Quality of service for differentiated traffic using multipath in wireless sensor networks. Int. J. Inven. Eng. Sci. **3**(1), 61–66 (2014)
19. Tiwari, M., Arya, K.V., Choudhari, R., Choudhary, K.S.: Designing intrusion detection to detect wormhole and selective forwarding attack in WSN based on local information. In: Fourth International Conference on Computer Sciences and Convergence Information Technology, ICCIT 2009, pp. 824–828. IEEE (2009)

Advanced Features and Specifications of 5G Access Network

Sudhir K. Routray[1], Abhishek Javali[2(✉)], Laxmi Sharma[3],
Aritri D. Ghosh[2], and T. Ninikrishna[2]

[1] Department of Electrical and Computer Engineering,
AASTU, Addis Ababa, Ethiopia
sudhir.routray@aastu.edu.et
[2] Department of Electronics and Communication Engineering,
CMR Institute of Technology, Bangalore, India
{abhishek.j,aritri.d,ninikrishna.t}@cmrit.ac.in
[3] Department of Telecommunication Engineering,
CMR Institute of Technology, Bangalore, India
laxmi.s@cmrit.ac.in

Abstract. The recently developed fifth generation (5G) of mobile communication has several advanced features. It would be very much different from the previous versions of the mobile generations in several aspects. Based on the performance, 5G will have gigantic data rates, very low end-to-end latency, high spectral efficiency, very high device densities, high energy efficiency, and seamless connectivity with devices and sensors. These advanced features will bring several differences in the 5G architecture. The differences will be both in the core, and access parts of these new networks. While the core will be comprised of high speed optical fibers, the access will have advanced wireless technologies such as massive MIMO, beamforming, energy efficient waveforms, new radio access technologies, and new full duplex channels. In this article, we present the new features and their performance characteristics of 5G access networks. We show the newly proposed standards in place for the 5G new radio (NR). The 5G NR is going to frame the practical aspects of 5G in the wireless form. Based on the 5G NR, the access networks of 5G will have several new initiatives such as new spectrum for communication, too many small cells, massive MIMO antennas, exploiting the space diversities, and advanced signal processing techniques to deal with these complexities. Finally, at the user's end, a high data rate is delivered with very low latency.

Keywords: 5G · 5G access network · 5G NR · 5G specifications · 5G deployment

1 Introduction

The fifth generation (5G) of mobile communication is going to be a very new and advanced form of wireless cellular communication. It promises a large number of advanced features such as very high data rates of the Gb/s order, very low latency compared to the fourth generation (4G), high spectral efficiency, high device densities,

© Springer Nature Switzerland AG 2020
D. J. Hemanth et al. (Eds.): ICICI 2019, LNDECT 38, pp. 338–347, 2020.
https://doi.org/10.1007/978-3-030-34080-3_38

and seamless connectivity with the surrounding IoT networks. 5G is a complete paradigm shift in the current mobile communications landscape. It is very much different from the previous generations in several aspects. It will use all advanced contemporary technologies available in the world. Moreover, 5G networks are going to be very much different from the previous cellular networks. It will be a complex hybrid of both optical and wireless networks. Optical networks will carry the huge core traffic in the core, regional and metro area networks. However, the final hop of the access to the end customers will take place through the wireless networks. Therefore, 5G access network means the final hop wireless part of the whole 5G network. In fact, this is the most challenging part of the 5G networks for the designers. The wireless access part will use several advanced technologies such as massive multi-input multi- output (MIMO) antennas for better quality of services and complex beam forming. For the 5G access networks new standards have been framed by the standardization groups. It is known as 5G new radio (5G NR).

2 Related Work

There are several works already been carried out on the characteristics and the applications of the 5G mobile communications. In [1], a systematic evolution of the cellular communication technologies has been presented. In this work, the evolution from 4G to 5G has been explained using the current standardization practices. It is shown that 5G will be very much different from its previous legacy system 4G. The transition from 4G to 5G will not be a sudden one; rather it will be very smooth through an intermediate version 4.5G. This intermediate version, 4.5G is much better than 4G and has several similarities with the 5G such as the use of massive MIMO and increased spectral efficiency. So, 4.5G is considered as a milestone along the road to 5G. In [2], the recent trends in global communications have been studied. It shows that the mobile communication subscriptions across the world have reached almost a state of saturation because the number of mobile subscribers has exceeded the global population. In [3], new millimeter waves have been proposed for 5G. This is the very first time that millimeter waves will be used in the mobile communications. In [4], optical wireless hybrid networks are proposed for 5G. In 5G, the data rates are very large and other performance parameters too are exceptionally high when compared with the corresponding 4G parameters. It clearly indicates that the bandwidth requirements for the 5G networks are very high. In the core networks only optical fibers can handle this high demands. Similarly, in the regional and sub-regional (or metro) networks too optical fibers are needed. Only the last mile communication is going to be in the wireless form. Some of the key architectures for 5G have been presented in [4]. In 5G, the data rates and other critical performances need a large amount of energy [5]. In order to make 5G sustainable in the long term, it has to be very much energy efficient. This is possible only through the green initiatives proposed in [5]. The energy sent per bit in 4G is of the order of a few µJ. In 5G, it will be in the tens of nJ regime. Similarly, the spectral efficiency too can save the energy significantly. In 5G the spectral efficiency is set to be of the order of 4.5 bits/Hz. Similarly, in every part of the 5G networks the energy saving mechanisms will be deployed. There are several flaws in the current networking practices which are not advisable for the high performance networks. In the practical networks, service segregation or slicing or partition is essential to perform the tasks in a streamlined fashion [6]. This possible in the software defined networking framework. For 5G, software defined networking in implemented though network function virtualization [6]. In fact, it is a kind of softwarization of the networks. In [7], new bandwidths for the emerging communications have been presented. In this work, several new bands have been proposed for emerging communications such as 5G. Millimeter waves and beyond 6 GHz unlicensed bands are now the main choices for 5G. Similarly, IoTs can use the guard bands and in-band frequencies of cellular networks [7]. There are also new proposals for beyond mm waves and terahertz waves. However, the standardization and testing of these bands will take some years. In [8], energy consumption aspects of 5G waveforms have been analyzed using the statistical parameters of the waveforms. Energy consumed by the existing waveforms of 4G and 4.5G are not very much suitable for the 5G applications. Therefore, new waveforms are needed for increased energy efficiency and spectral efficiency. 5G will incorporate a large number of Internet of things (IoT) networks in its ambit [9]. These IoT networks use a large number of sensors and their coverage is normally more than the 5G coverage. Therefore, the energy consumption issues of these networks also pose a challenge for the network architects. In [9], all these energy consumption related issues of IoTs have been addressed. For better control and management of the IoT networks, software defined mechanisms can be extended to the IoT frameworks. In [10], the main challenges and issues of software defined IoTs have been presented. Overall, the control and management of the IoTs get better with the software defined framework. Optical networks are going to play crucial roles in 5G. Mainly in the backhaul and several parts of the fronthaul of 5G, optical networks will be the key drivers. In [11], main trends of optical communication have been presented. In [12], several beyond 5G issues have been addressed. In fact, the beyond 5G scenarios are going to be quite interesting as several limits will be achieved in 5G. However, the main consolidation will happen in the sixth generation (6G) only [12]. Several value added services over the 5G enabled IoT networks have been addressed in [13–15]. These services are very much essential in several applications. In [15], the security aspects of the IoT networks have been addressed. It is suggested that quantum cryptography can be a suitable choice for these networks.

3   A Brief Introduction to 5G Specifications

5G, as the name suggests, will be the new generation of mobile communication which is going to replace the 4G. In fact, the hybrid of 4G and 5G which is known as 4.5G is already available in several countries. It is very much advanced than the 4G and has several features similar to the 5G. However, 5G will be still a very much different technology than 4.5G. The International Telecommunication Union (ITU) has formed the new standard IMT 2020 for 5G. Based on the IMT 2020 standards, ITU has proposed the new specifications for 5G. The data rates in 5G are going to be higher than 4.5G. The average data rates in 5G are expected to be around 3 Gb/s with the peak data rates in the downlink is set at 20 Gb/s. The latency between the end users will be reduced to 1 ms. The energy efficiency is expected to be 100 times better than the 4G. The device densities in 5G will me much higher than the current densities. Spectral efficiency will be increased at least three times more than the best possible case of 4G. Data processing in 5G will be enhanced by at least 10 folds of the current data processing speed. The mobility will also be improved in 5G by almost 50% with respect to 4G. In 5G the handovers will not face many complexities up to the speed of 500 km/h. The main ITU specifications of 5G have been presented in Table 1.

Table 1. ITU proposed specifications of 5G [5].

#	Performance parameters	Typical ranges	Technologies to achieve it
1	Data rate	0.1–20 Gbps (mean: 3 Gbps)	Massive MIMO, MM wave spectrum, Optical core
2	Spectral efficiency	4.5	Full duplex, Massive MIMO (more than 16 antennas)
3	Data processing	10 Mb/s/m^2	Radio access network vitalization, small cells
4	Device density	1 million/km^2	Device to device (D2D), Small cells
5	Mobility	Up to 500 km/h	Heterogeneous network architectures, beam tracking
6	Transmission delay	1 ms	D2D, Content caching close to users
7	Energy consumption	1 μJ per 100 bits	Massive MIMO

Here it is noteworthy that the specifications presented in Table 1 are the preliminary specifications of standardization bodies such as ITU and several others across the world. The peak performances presented in Table 1 are to be achieved only in the suitable laboratory conditions. The real user experiences will be lower than these laboratory values. For instance, the average 3 Gbps data rate will be hardly experienced by the end users in the early days of 5G. It looks very much certain that after several years of operation, the performance will be improved to consolidate around the high values specified in Table 1. Similarly, it is very difficult to bring down the end to end latency to 1 ms in the wireless setup. The laboratory testing are carried over in a very organized setup. Thus the practical communications over the wireless channel will not provide the 1 ms latency in the initial versions of 5G. Initially, in the early years of 5G, it will remain in a few ms order. However, the consolidation of the technologies will improve the performances over the years. The data rates proposed in 5G are really very high for the wireless communication networks. The average data rate of 3 Gbps is the laboratory tested 5G data rates in the static conditions. In the mobile conditions, the data rates would come down to 100 Mbps. This is set as the lower limit of data rates in

5G. The peak data rate of 20 Gbps is the laboratory testing speed under the perfect line of sight conditions. Similarly, the end to end latencies are of the order of several ms. This will be gradually improved with the consolidation of different technologies in the 5G framework. The delays have been reduced significantly using the signal processing techniques. However there are also some new features such as software defined networking which can increase the end to end delays. The spectral efficiency too has an increasing trend. However, the specified values would not be available in the early versions of 5G.

4   An Overview of 5G Network

Broadly, 5G will consist of two parts: the optical part, and the wireless part. The optical part is needed for the high speed data communication in the core, regional and the metro parts of the 5G networks. For the final hop to the mobile devices, the communication has to be wireless. This will be done through the access networks. The access networks will have the base stations which are known as gNB in 5G. The base stations are connected with several small cells aroud it. The final access to the end users normally come from the small cells with massive MIMO antennas which have the abilities to produce proper beams and can also track the beams according to the users movements. In Fig. 1, we show the basic network architacture of 5G separating the core and access parts. The core parts are very well organised and connected with each other though high speed optical fibers. All the gateways are faciliated with the basic core facilities. The robust core is instrumental in the overall succes of 5G. The central gateways are large scale switching facilities which houses several optical corss connects and signal processing facilities. All the gateways including the local gatewats are equiped with the cloud computing facilities. The cloud services are essential for providing the large scale storage needed for the 5G users.

4.1   An Overview of 5G Core Network

The 5G core network is completely an optical network. It has its main switching and server facilities at the nodal hubs. A typical 5G network may have just one or a few central hubs or gateways depending on the size and coverage of the operator. Each central gateway is connected with several regional gateways. The regional gateways segregate the regional traffic and carry them to the corresponding local gateways. Each regional gateway is connected with several local or metro gateways. The local gateways are connected with the 5G base stations which are commonly known as gNB. The gNB communication with the subsequent stages is the function of the 5G access networks.

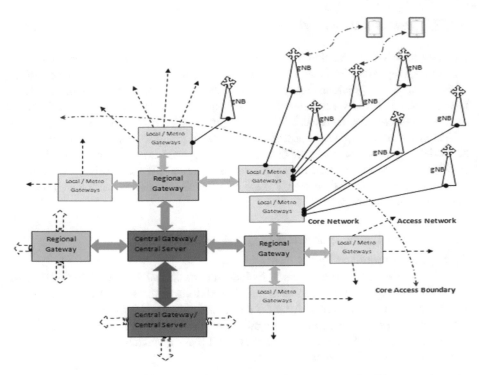

Fig. 1. A typical 5G architecture showing the core and access networks. Here, we show the central gateways, regional gateways, and local gateways of the core network. The core and access networks boundary is shown by red dotted lines. The gNB and the further communication from there onwards is carried out by the access networks.

4.2   An Overview of the 5G Access Network

From the gNBs the access networks communicate with the end devices. As it is last frontier of the 5G network in the service provisioning it is known as access network. Each gNB serves a specific locality. In the urban areas, traffic of each gNB is fragmented in to several small cells. The small cells are massive MIMO antenna holders which are able to exploit the space diversities. These antennas are programmed to follow the target devices though their beams. These beams play key roles in the high data rates and appropriate quality of services. Under a gNB there can be a few small cells to hundreds of small cells depending on the user scenarios. A typical small cell scenario of 5G access network is shown in Fig. 2. It is noteworthy that spectrum around 6 GHz can serve the 5G services without small cells. However, the power radiated in these cases will be more than the small case scenarios.

5   5G New Radio Access Network

Now, we have a clear idea that 5G will not be just another version of LTE similar to the previous legacy Now, we have a clear idea that 5G will not be just another version of LTE similar to the previous legacy. For 5G different new wireless specifications have been proposed. These specifications are different from the existing LTE specifications to a large extent. For instance, LTE so far (up to 4.5G) was using only the below 6 GHz spectrum for its operations. However, that is going to change in the 5G access networks and several bands in the beyond 6 GHz regime have been chosen for 5G NR. In 5G, the spectrum proposed for use ranges from sub 1 GHz to 100 GHz. In addition to the spectrum, there will be scores of changes in 5G NR. In this section we highlight the main principles of 5G NR.

5.1   Different Bands Allocated for 5G NR

5.1.1   Low Frequency Bands

These are in the sub 1 GHz regime. It includes the traditional LTE/UMTS/GSM bands in the 850 MHz to 950 MHz and the further additions of the lower frequencies in the 700 MHz and 600 MHz rages. All these sub 1 GHz frequencies are mainly for the IoT operations which are going to be integral parts of 5G. Several massive machine type communications have been proposed for 5G. All these applications are expected to be carried out in these low frequency bands.

5.1.2   Medium Frequency Bands

The medium frequency bands range from 1 GHz up to 6 GHz. These are the LTE bands and also dedicated for some specific applications. The normal bands in this range are: 3.4 GHz–3.8 GHz, 3.8 GHz–4.2 GHz, and 4.4 GHz–4.9 GHz. These bands are mainly dedicated for enhanced mobile broadband (eMBB), and mission critical applications such as disaster management and rescue operations.

5.1.3   High Frequency Bands

The high frequency bands in 5G NR are normally the beyond 6 GHz bands. These bands are normally in the millimeter (mm) wave ranges. The recently suggested mm wave bands are: 24.25 GHz–27.5 GHz, 27.5 GHz–29.5 GHz, 37 GHz–40 GHz, and 64 GHz–71 GHz. These bands will be used for the main communication of voice, video, and data in 5G NR.

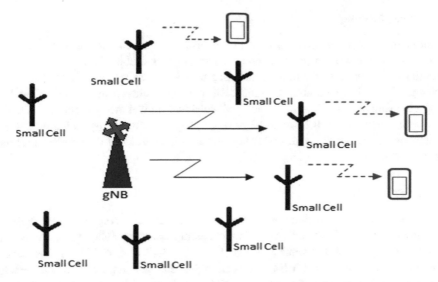

Fig. 2. Access area network of typical 5G architecture showing the gNB and small cells

5.2 OFDM Based Radio Technology

For the 5G NR the LTE access technologies are not suitable. Therefore a new radio access technology (RAT) has been proposed for 5G. This 5G RAT is very much based on orthogonal frequency division multiplexing (OFDM). This is because, OFDM is an advanced broadband access technology and it can provide several advantages over other contemporary technologies. Basically, OFDM is a digital multicarrier modulation technique in which the adjacent carriers are orthogonal to each other. That is how the minor spectral overlap is not a problem in OFDM based RAT. OFDM is very much immune to inter-symbol interference and can also handle the intra-symbol interferences. The OFDM based waveforms are very much scalable and provides a near perfect framework for MIMO spatial multiplexing. This is how the spectral efficiency can be increased in OFDM based 5G NR. OFDM also enhances the windowing and filtering processes required in the 5G receivers. These days the deployment of OFDM is also quite simple with the help of the discrete Fourier transform (DFT) and inverse discrete Fourier transform (IDFT) based signal processing.

5.3 Modes of Operation

5G NR will have two different modes of operation. They are: standalone mode, and non-standalone mode. In the standalone mode, the 5G will not be compatible with the LTE legacy systems such as 4G and 4.5G. In this mode the mm wave spectrum will be used for communication. In the non-standalone mode, 5G and its legacy systems will use several similar technologies and thus they will be compatible with each other. Even the spectrum used by 5G and its LTE legacy systems will be same.

5.4   Beamforming

Beamforming is a signal processing technique used for controlling and customizing the beams of the antennas in practical applications such as mobile communications. It is essential in 5G, because the high frequencies of 5G NR are not suitable for long range coverage. So, a large number of small cells will be deployed in 5G instead of long range covering base stations. Multiple small cells in a limited area are going to interfere with each other. This can be avoided through proper beamforming techniques. These techniques will allow the beams of the antennas to keep track of the devices until they are handed over to another small cell in the vicinity.

5.5   Massive MIMO

MIMO is an essential technology for 5G. It can handle the spatial diversities and provides all the facilities for proper beamforming. Massive MIMO is a large scale deployment of MIMO antennas for multiple users. Massive MIMO has several advantages over the normal MIMO such as higher channel capacity and better energy efficiency. In fact, MIMO uses either two or four antennas for its operations. However, in massive MIMO eight or more antennas are used for transmission and reception of the signals.

5G NR will have full duplex data transmission in both the up and down links. Both frequency division duplex (FDD) and time division duplex (TDD) will be used in 5G access networks. Of course the FDD and TDD will not be symmetrical for the uplink and downlinks. Normally the download data rates are higher than the upload data rates. So, a long interval will be provided for downlink communication than the uplink communication. This is found to be the appropriate solution for the 5G end services as the demand for high data rates in one of the main features of it. Hybridization of both FDD and TDD is required to increase the capacity of the multiple accesses in the last mile networks.

6   Conclusions

In this article, we discussed the main features of the 5G access networks. The 5G wireless parts are designed according to the proposed 5G NR. The standards for 5G NR in IMT 2020 show the effectiveness of the proposed new initiatives. The 5G access networks are going to be very dense compared to the 4G and 4.5G access networks. The 5G NR access network can have either the small cells or the conventional medium and large cells like 4G and 4.5G. The high frequency spectrum in the mm-wave range would force the use of large number of small cells. However, the 6 GHz spectrum provides the options of medium and large cells which may be deployed in the less demanded rural areas. Signal processing is going to play crucial roles in the overall functions of 5G. Massive MIMO and beamforming in 5G will be based on signal processing. Large IoT connectivity is a new dimension in 5G. The real ubiquitous connectivity through IoTs is envisioned in 5G. Hopefully, the 5G access networks will deliver all the proposed features in due course of time.

References

1. Routray, S.K., Sharmila, K.P.: 4.5G: a milestone along the road to 5G. In: Proceedings of IEEE International Conference on Information, Communication and Embedded Systems (ICICES), Chennai, pp. 24–25 (2016)
2. Mohanty, S., Routray, S.K.: CE-driven trends in global communications: strategic sectors for growth and development. IEEE Consum. Electron. Mag. **6**, 61–65 (2017)
3. Rappaport, T.S., et al.: Millimeter wave mobile communications for 5G Cellular: it will work! IEEE Access **1**, 335–349 (2013)
4. Sharma, L., Javali, A., Sarkar, S., Tengshe, R., Jha, M.K., Routray, S.K.: Optical wireless hybrid networks for 5G. In: Optical and Wireless Technologies, pp. 65–71. Springer, Singapore (2020)
5. Routray, S.K., Sharmila, K.P.: Green initiatives in 5G. In: Proceedings of IEEE International Conference on Advances in Electrical, Electronics, Information, Communication and Bio-Informatics (AEEICB), Chennai, pp. 27–28 (2017)
6. Routray, S.K., Sharmila, K.P.: Software defined netwring for 5G. In: Proceedings of 4th IEEE International Conference on Advanced Computing and Communication Systems (ICACCS), Coimbatore, pp. 6–7 (2017)
7. Routray, S.K., Mishra, P., Sarkar, S., Javali, A., Ramnath, S.: Communication bandwidth for emerging networks: trends and prospects. In: Proceedings of IEEE International Conference on Advances in Electrical, Electronics, Information, Communication and Bio-Informatics (AEEICB), Chennai, pp. 27–28 (2017)
8. Routray, S.K., Jha, M.K., Sharma, L., Sarkar, S., Javali, A., Tengshe, R.: Energy consumption aspects of 5G waveforms. In: Proceedings of IEEE International Conference on Wireless Communications, Signal Processing and Networking (WiSPNET), Chennai, pp. 1–5 (2018)
9. Routray, S.K., Sharmila, K.P.: Green initiatives in IoT. In: Proceedings of IEEE International Conference on Advances in Electrical, Electronics, Information, Communication and Bio-Informatics (AEEICB), Chennai, pp. 27–28 (2017)
10. Ninikrishna, T., Sarkar, S., Tengshe, R., Jha, M.K., Sharma, L., Daliya, V.K., Routray, S.K.: Software defined IoT: issues and challenges. In: Proceedings of IEEE International Conference on Computing Methodologies and Communication (ICCMC), Coimbatore, pp. 723–726 (2017)
11. Routray, S.K.: The changing trends of optical communication. IEEE Potentials Mag. **33**, 28–33 (2014)
12. Routray, S.K., Mohanty, S.: Why 6G?: motivation and expectations of next-generation cellular networks. arXiv:1903.04837. https://arxiv.org/ftp/arxiv/papers/1903/1903.04837.pdf
13. Ramnath, S., Javali, A., Narang, B., Mishra, P., Routray, S.K.: IoT based localization and tracking. In: Proceedings of IEEE International Conference on IoT and its Applications, Nagapattinam (2017)
14. Ramnath, S., Javali, A., Narang, B., Mishra, P., Routray, S.K.: An update of location based services. In: Proceedings of IEEE International Conference on IoT and its Applications, Nagapattinam (2017)
15. Routray, S.K., Jha, M.K., Sharma, L., Nymangoudar, R., Javali, A., Sarkar, S.: Quantum cryptography for IoT: a perspective. In: Proceedings of IEEE International Conference on IoT and its Applications, Nagapattinam (2017)
16. 3GPP LTE Release 15, latest version, April 2019. https://www.3gpp.org/release-15. Accessed 11 July 2019

Internet of Things Based Smart Secure Home System

Prema T. Akkasaligar, Sunanda Biradar$^{(\boxtimes)}$, and Rohini Pujari

Department of CSE, BLDEA's V. P. Dr. P. G. Halakatti College of Engineering
and Technology, Vijayapur 586103, Karnataka, India
{cs.akkasaligar, cs.biradar}@bldeacet.ac.in,
rpujari417@gmail.com

Abstract. Smart home system provides more security and safety. Use of IoT technologies make the system easily accessible to the end user. Different hardware/sensor devices maintain more security to the smart home. Any common user can easily handle and operate the system. The proposed home security system has security concerning gas leakage, fire detection, detection of room temperature-humidity and avoiding overflow of water from the overhead tank. The proposed system aims to provide the secured home system by making the accessibility easier to disabled people, children and aged people in particular.

Keywords: Smart-home security system · Sensors · Motor · Raspberry Pi · Telegram

1 Introduction

Physical system is associated to both security and safety of the premises like homes, office, bank, and hospital, etc. Internet of things (IoT) based technologies protect the home system via the internet [1, 2]. Many existing home security systems have limited accessing capabilities at indoor and/or outdoor environments. However, some of the models supporting this feature, have very complex connection along with high initial and maintenance costs. The automated smart homes need many sub-systems to be integrated together with fewer resource requirements. These systems are used to provide security and safety effectively to the living premises of human beings such as home, workplace or other areas that needs protection. The main purpose of this work is to use Raspberry Pi for connecting all sensors and protect the home during security threats and unauthorized functions that occur inside or around the home. Further, detected event is reported to the owner of the home through telegram mobile application messages.

2 Literature Survey

In [3], authors have proposed the IoT based home automation system. In this work, android-mobile device is used to control the home devices. Relay driver circuitry with Raspberry Pi is used in this system. Client user needs to connect to the server using an

© Springer Nature Switzerland AG 2020
D. J. Hemanth et al. (Eds.): ICICI 2019, LNDECT 38, pp. 348–355, 2020.
https://doi.org/10.1007/978-3-030-34080-3_39

android mobile and the internet or GPRS. They have also provided the web user interface to control the home devices. In [4], a single system is designed to remotely function on home appliances and security system. They have used arduino micro-controller and mobile based application is connected with blue tooth technology. Although the system provides multiple functionalities, it has a limitation of distance as the system works on bluetooth technology. The recent developments in Wi-Fi can enable the capability of different devices to connect with each other at any longer distance.

In [5], the authors have implemented home automation using Wi-Fi. This work shows the less involvement of human and maintains a more secure system. Hardware devices implement both monitoring and controlling. They have also developed a mobile application based on Arduino. communication between two components. It is established via Wi-Fi technology as the communication protocol. They have used the PIR sensor, light control, and temperature control sensors to control lights and operates fans automatically. They have used relays and Wi-Fi to integrate all the hardware components.

In [6], authors have proposed IoT based fire accident prevention system. They have used three set of sensors at different places to identify the accurate locality of fire accident. The smoke and temperature sensors are also used for fire detection empowered with alarm. The model is designed using Galileo board.

In [7], the authors have implemented home automation using Wi-Fi. This work shows the less involvement of human and maintain more secured system. Both monitoring and controlling both are implemented by hardware devices. They have also developed a mobile application based on Arduino. Communication between two components is established via Wi-Fi technology as the communication protocol. They have used PIR sensor, light control and temperature control sensors to operate lights and fans automatically. They have used relays and Wi-Fi to integrate all the hardware components.

In [8], the authors have proposed smart home automation by providing more security system through GSM embedded mobile module. In [9] the authors have implemented home automation technique. This paper shows the less involvement of human, it controls the automation system and maintain more security, safety of the system. In [10], the fire alarm and monitoring system are combined with iot technology.

In [11], the authors show a simple application that detects fire, sending an alarm to the authorized person after alerting and controls entire the system. The main goal of the paper [12] is to provide security, safety, and surveillance to home through Wi-Fi. The authors [13] have developed a mobile application based on Arduino communication between two components. It is established via Wi-Fi technology as the communication protocol. They have used relays and Wi-Fi to integrate all IoT components.

3 Methodology

The idea of the project is to overcome the drawbacks faced by current home automation technologies. Figure 1 represents the block diagram of automated smart-secured home/office system. The model consists of microcontroller unit, RS-232 interface, LCD module, crystal oscillator, interfacing circuit, and GSM module. Microcontroller unit is the heart of the controlling system that manages all other modules in the system. GSM module is interfaced with microcontroller using RS-232 module. There is a serial communication between the GSM and the microcontroller unit. All sensors are interfaced to the microcontroller unit using the interfacing circuit. Crystal oscillator circuit provides the necessary clock frequency to run the microcontroller unit. The sensed data is in the analog form which is converted to digital before sending the information to the intended person. The system design can be divided into four modules: temperature and humidity sensor, gas and fire sensor, motion sensor and conductor for water level monitoring. The connectivity of sensor is shown in the Fig. 2.

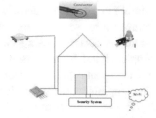

Fig. 1. Architecture diagram of the smart home system.

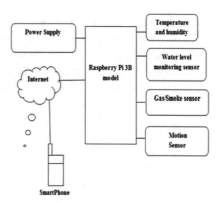

Fig. 2. Block diagram of the smart home system.

Temperature and Humidity Sensor
The DTH11 sensor is very small and fast sensor. It senses humidity and temperature very fastly, which is helpful for measuring temperature and humidity in room. DTH11

sensor contains 8bit microcontroller. The microcontroller converts the analog signal to digital form. It is tiny in size and low-cost. It comes with both three pins and four pins model. DTH11, a 3 pin sensor contains VCC, Signal, GND. DTH11 four pin sensor contains VCC, Signal, No Connection, and GND. DTH11 sensor generates digital output, interface with a microcontroller of Raspberry Pi to get an accurate result. Figure 3 shows DTH11 sensor used to find room temperature and humidity. DTH11 sensor is of high performance and low cost. This sensor has high reliability. It is used for long time stability. The main application of this sensor is monitoring the weather. The temperature ranges from 0–50 °C. DTH11 sensor has humidity range from 20 to 80% with 5% accuracy. DTH11 sensor is better than DTH22 sensor because of the DTH11 reading rate is from 1 Hz or one reading for every second. But, DTH22 reading rate is from 0.5 Hz or one reading for every two seconds.

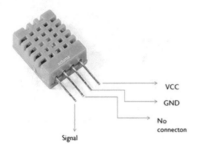
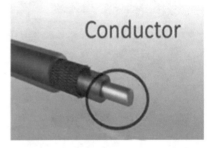

Fig. 3. Temperature - humidity sensor.   **Fig. 4.** Water level monitoring sensor

Water Level Monitoring Sensor
A material that allows electricity to pass through it easily is called a conductor. Metals, especially silver, are good electrical conductors. Most of the metals are good at conducting heat. Thermal conductors are also common in everyday life. Some examples like mercury in thermometers, used in medical field, aluminum in frying pans used in kitchen room, and the iron plate of an electric iron used in iron box.

The Fig. 4 shows the conductor used to find water level monitoring. Three main conductors are used; one for HIGH, LOW and GND. If the water level is HIGH, it indicates the tank is full. If otherwise, the water level is LOW, it indicates the tank is empty. If water level is high motor is turned off automatically. Low level of conductor is used to find water level as low, if water level is low automatically motor is turned on. When motor is turned on, an alert message is sent to the owner of the home.

Gas and Fire Detection Sensor
This sensor covers around 10 m. It is fixed near to gas cylinder or gas related products. It is a low-cost, long life sensor. This sensor can also detect smoke, alcohol, carbon, etc. Gas and fire sensor uses 5 V power supply. It is widely used in industry, hospital, home, etc. The Fig. 5 shows the gas sensor useful for gas detection in home and

industry. It can detect gas and smoke. This sensor is used to keep security and safety always when a person is not present inside the home. When gas leakage is present in the kitchen room, the smell is detected by this sensor and alerts with the alarm and sends message to the owner of the home.

Motion Detection Sensor

PIR sensor is also called as PID passive infrared detector (PID). PIR Sensor is one type of motion detector. PIR sensor detects through infrared light. This sensor consists of pyroelectric sensor. PIR motion sensor is used to control the outdoor and indoor motions of a person in and around the home. This sensor detects the movement of a person near to PIR sensor. The range of this sensor is approximately 5 to 10 m. Using this sensor, we can provide more security and safety to home, office, bank.

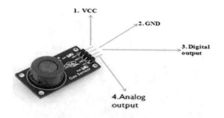

Fig. 5. Gas and fire detection sensor   **Fig. 6.** Motion detection sensor.

When human or animal gets into the range of the sensor, it catches the movement. It means that the movement of humans and animals produce heat energy in the form of infrared radiation. PIR sensors have three pins: (i) VCC Pin (ii) Signal pins and (iii) GND. PIR sensors have to adjust duration and sensitivity of the sensor. The Fig. 6 shows PIR sensor It is very fast in detecting the movement of theft, or a person near to home. It has low cost and of smaller size. This sensor can be hidden inside the wall, or door, so that the theft cannot identify the sensor. The main application of the PIR sensor is a to provide security to the home.

The main benefits of the developed model are:

Low Cost: Initial installation costs are reduced greatly since the cabling is not needed. Wired frameworks need cables, where material and the standard laying of cables are expensive.

Automated system makes use of the concept "possible everywhere and at any time," as the location of the device is exactly known for connecting to the shared network.

Security: One of the essential advantages of the system is security. The proposed system gives security against gas leakage and fire detection.

Monitoring Activities: Monitoring of water level indication results in comfortable living.

4   Results and Analysis

We downloaded and installed telegram android mobile application in mobile phone. Registration of the user mobile number to the application is essential. The android mobile phone is connected to Raspberry Pi. The mobile should be connected to the Wi-Fi network. Raspbian operating system is installed in the Raspberry Pi kit. Further, the motion detection (PIR) sensor, gas leakage-fire detection (MQ3) sensor, temperature-humidity sensor, and conductors for water level monitoring are also connected with the Raspberry Pi. This entire model is connected to the internet. Temperature-humidity (DTH11) sensor is used to find the room temperature and humidity in the room. The Table 1 shows the results of temperature and humidity continuously acquired using DTH11 sensor. The temperature is recorded in degree Celsius and humidity in percentage. These are compared with room temperature shown using mobile location and the temperature given in the newspaper. The Table 1 shows the comparative accuracy of the recorded temperature by DTH11 sensor.

The model is installed within a small space as shown in Fig. 7. The water level monitoring sensor is used to find the level of water in the tank. If the water level is low, motor is turned on automatically to fill the tank with water. If the water level is high, motor is turned off automatically which stops overflow of water from the tank as shown in Fig. 8. The alter message by water level sensor is sent to the owner as shown in the Fig. 9. The gas leakage-fire detection sensor is used to detect the smoke, LPG gas, and alcohol in the home. The motion detection sensor is used to detect the motion of the person. If any unsecured issues occur inside the home, we get alarmed buzzer and a message is sent through telegram application via the Internet as shown in the Fig. 9. Using all these sensors, more safety and security is provided. Compared to other technologies, this model costs less. It detects all unauthorized events within a fraction of a second. The human-friendly resources are used easily. The Fig. 9 shows the sample alert messages, that someone has entered into a home, a water tank is low and gas-fire detection. Thus the owner can prevent these dangerous using proposed model.

Table 1. Records of the temperature – humidity

Days/date	Experimental value	Humidity	Temp. obtained from mobile	News paper data
Tuesday 19/03/2019	31.85 °C	18%	31 °C	32 °C
Wednesday 20/03/2019	32.5 °C	16%	31 °C	32 °C
Thursday 21/03/2019	29.6 °C	17%	30 °C	30 °C
Friday 22/03/2019	31.57 °C	19%	31 °C	32 °C
Saturday 23/03/2019	25.15 °C	17%	25 °C	26 °C

Fig. 7. Home security system.

Fig. 8. Water level monitoring system.

Fig. 9. Display of alert message sent to the android mobile phone.

5 Conclusion

The proposed work uses four sensors namely, temperature-humidity sensor, water level monitoring sensor, fire-gas detection and motion detection sensor. The proposed model records temperature and humidity of the room using DTH11. Moreover it monitors water level of overhead tank of a home using conductor. The MQ3 sensor is used to detect fire and gas, and the PIR sensor detects for person detection. The alert messages are sent to the owner, in case of unusual events. The outcome of this project work is a hardware product that can be used to monitor home/office for any harmful activities. This product increases the comfort and gives security in the home.

References

1. Rajesh, T., Rahul, R., Malligarjun, M., Suvathi, M.: Home automation using smartphone application. Int. J. Adv. Res. Sci. Engg. Tech. **4**(3), 3546–3553 (2017)

2. Akkasaligar, P.T., Shambhavi, T., Soumaya, P.: Review of IOT based health monitoring system. Int. J. Res. Advent Technol. Special issue, 95–99 (2019)
3. Prashant, R., Syed, K., Rashmi, K., Shubum, L.: Raspberry Pi based home automation using Wi-Fi, IoT & Android for live monitoring. Int. J. Comput. Sci. Trends Technol. **5**(2), 363–368 (2017)
4. Isa, E., Sklavo, N.: Smart home automation: GSM security system design & implementation. J. Engg. Sci. Technol. Rev. **10**(3), 170–174 (2017)
5. Khade, N., Dharmik, N., Payal, T., Rutuja, B., Bachhuka, S.: Home Automation. Int. J. Engg. and Mgmt. Res. **7**(1), 390–392 (2016)
6. Kulkarni, B., Aniket, J., Vaibhav, J., Akshaykumar, D.: IoT based home automation using Raspberry PI. Int. J. Innov. Stud. Sci. Eng. Technol. **3**(4), 13–16 (2017)
7. Saumya, T., Shuvabrata, B.: IoT based fire alarm and monitoring system. Int. J. Innov. Adv. Comput. Sci. **6**(9), 304–308 (2017)
8. Vandana, C., Taffazul, I., Shubham, D.: Security issues in home automation. Int. J. Sci. Res. Comput. Sci. Eng. Inf. Technol. **2**(3), 257–261 (2017)
9. Reddy, P.S.N., Reddy, K.T.K., Reddy, P.A.K., Kodanda, G.N.: IoT based home automation using Android application. In: International Conference on Signal Processing, Communication, Power and Embedded System, pp. 285–290 (2016)
10. Himani Singh, D., Nidhi, C., Nishank, S.: Raspberry Pi home automation using Android application. Int. J. Adv. Res. Ideas Innov. Technol. **3**(2), 521–525 (2017)
11. Quadri, S.A.I., Sathish, P.: IoT based home automation and surveillance system. In: International Conference on Intelligent Computing and Control Systems, vol. 7, no. 17, pp. 861–866 (2017)
12. Pampattiwar, K., Lakhani, M., Marar, R., Menon, R.: Home automation using Raspberry Pi controlled via an Android application. Int. J. Curr. Eng. Technol. **7**(3), 962–967 (2017). E-ISSN 2277–4106
13. Aravindhan, R., Ramanathan, M., SanjaiKumar, D., Kishore, R.: Home automation using Wi-Fi interconnection. Int. Res. J. Eng. Technol. **4**(3), 2542–2545 (2017)

Classification of Tweets Using Dictionary and Lexicon Based Approaches

K. Jayamalini[1,2(✉)] and M. Ponnavaikko[3]

[1] Computer Science Engineering, Bharath University, Chennai, India
malini1301@gmail.com
[2] Shree L.R. Tiwari College of Engineering, Mumbai, Maharastra, India
[3] Vinayaka Mission's Research Foundation, AV Campus,
Chennai, India
ponnav@gmail.com

Abstract. Online social media is pervasive in nature. It allows people to use short text messages, images, audios and videos to express their opinions and sentiments about products, events and other people. For example, Twitter is an online social networking and news service where users post and interact with small and short messages, called "tweets". Therefore, nowadays social media become a potential source for business and celebrities to find people's sentiments and opinions about a particular event or product or themselves.

Social media analysis is the process of gathering enormous amount of digital contents generated online from blogs sites and social media networks and examining them to find the insights.

This paper focuses on discovering public opinions and sentiments he on the results of Indian election results declared recently. This Paper also deals Dictionary based Approach and Affective Lexicon based approaches which were used to find the public opinion about election results.

Keywords: Social media data · Opinion Mining (OM) · Sentiment Analysis (SA) · Sentiment Analyzer

1 Introduction

Twitter like Social Media [2, 3] is important for Business and it helps business to produce successful social campaigns, recognise influencers for their brand, product, service & industry, compare key performance metrics and to find strengths, weaknesses of competitors using competitive intelligence, discover the real time trending topics and to keep track the virality of content spreads across the social media and world wide web.

Different Types of inbuilt Social Media Analytical Tools are provided by Social Networks itself on their Dashboards. For example Facebook provides Facebook Insights, Instagram provides Instagram Insights, Twitter [12] provides Twitter analytics, Youtube provides YouTube Analytics and google provides Google Analytics, Other free tools like Followerwonk. ViralWoot, Tailwind, Keyhole and so on are used for analytical purpose. In order to find the user opinion about Indian election results, in this

D. J. Hemanth et al. (Eds.): ICICI 2019, LNDECT 38, pp. 356–364, 2020.
https://doi.org/10.1007/978-3-030-34080-3_40

paper corpus of Twitter data is used. In this paper Dictionary based methods and Affective Lexicon based methods are used to find the public opinion.

2 Overview of Sentiment Analysis System

Sentiment Analyzer [4, 5] is used to analyse the tweet extracted about Indian elections in an enhanced way. The framework of Sentiment Analyser is shown in figure below and it comprises of following modules: Authentication, Tweets Extraction, Pre-processing & Text Cleaning, Feature Extraction and Opinion Analysis. The authentication unit provides the permission to connect to Twitter using Twitter API. The Extraction module connects to Twitter and extracts specific tweets based on given search keyword. The data collected from Twitter contain noisy and irrelevant information. These irrelevant information should be cleaned by preprocessor before it is used by other parts of the system. The preprocessed tweets were categorized into positive, negative, or neutral by the opinion analysis module. The architecture of the opinion analysis system is depicted below in Fig. 1.

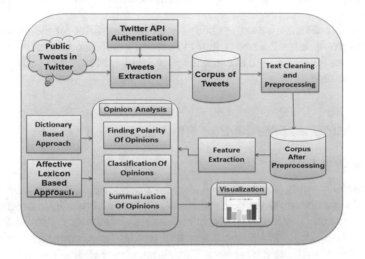

Fig. 1. Opinion analysis system architecture

2.1 Tweets Extraction

Twitter is an incredible social media for social web data analysis. Among all, R offers wide-ranging choices to do many interesting things. To connect to Twitter, a twitter application should be created, which will use the Twitter API to make the connection and provides the authentication using Consumer Key, Consumer Secret, Access token and Access token secret.

2.2   Data Pre-processing and Feature Extraction

The data collected from Twitter contain irrelevant and noisy information [10], which requires a cleanup before further usage. The Pre-processor is used to remove non-English letters, URLs, special characters, punctuation and white spaces and stop words. The extracted tweets contain unwanted attributes which would not add any value to find the opinion. The feature extraction module of the system identifies the most relevant attributes that contain valuable information.

2.3   Opinion Analysis

The opinion analysis is done in three steps as follows:

- Opinion Value Calculation: In this module, the overall value of the sentence is calculated by comparing and assigning each word of the sentence, against a dictionary of positive words and negative words which is predefined and used in the calculation.
- In Opinion Classification step, based on the value assigned to the sentences, the sentences are classified into positive sentences, negative sentences or neutral sentences.
- The Opinion Summarization module is used to summarize all the positive tweets into one group, negative tweets into the second group and neutral tweets into the third group. Also, it is used to count the total number of tweets in each group based on created date.

3   Algorithms and Methods of Implementation

3.1   Raw Data Collection

Tweets are slang words, which are used to express users' opinion or facts about current affairs in Twitter. People tweet personal messages, casual views, links, or anything that fits in 140 character requirements. The tweets consists of retweeted entry as RT, special characters, links, #tags, white spaces and stop words. These component will not add any value while analyzing the contents. These contents are cleaned and made fit for the analysis.

3.2   Preprocessing of Raw Data

The data preprocessing methods often have a important effect on the performance of a supervised Machine Learning algorithms. The steps performed by the preprocessor module are as follows: Case Conversion module converts all the words into either lower case letters or upper case letters in order to make the ascii values of the words "Hello" and "hello" same for further processing. Stop-words Removal removes words like a, an, the, has, have etc. which do not add any value in determining the sentiment value. Punctuation Removal module used to remove the punctuation marks like comma, colon, and full stop often carry no meaning for the text analysis hence they can be removed from input. Stemming usually refers to a process of chopping off the ends of the words to cut derivational affixes. Spelling of the incorrect words can be corrected automatically by the spell checker module.

3.3 Twitter Authentication and Tweets Extraction

R packages "twitteR" and "ROAuth" are used to provide authentication to client application to access Twitter and extract tweets from the user timeline or based on search keyword. Sample tweet about "Election 2019" was extracted and stored.

3.4 Public Opinion Finding

The user opinion hidden in each of the sentence of the corpus is found using following three steps: Finding Opinion Value, Classifying Tweets Using Dictionary [4] based Approach & Affective Lexicon Based Approach [5] and Group wise summarization

- **Finding Opinion Value - Dictionary Based Approach** [6]

  A corpus each of positive words and negative words are maintained. Dictionary based Approach is used to find the value hidden inside each tweet. In this method, for each word (W) in a tweet (t) is compared against each item of the corpus mentioned above. If the word (W) matches the positive corpus the count of 'positive (PosWords)' increases by 1. Similarly, if it matches the negative corpus, the count of 'negative (NegWords)' increases by 1. Finally, the score of the sentence will be calculated. The system creates new corpus called 'Tweet_Scores(TS)' and stores each tweet with its calculated opinion value.

$$TS = \sum_{k=0}^{n} PosWords - \sum_{k=0}^{n} NegWords$$

- **Finding Opinion Value - Affective Lexicon Based Approach**

  The dictionary based approach [7, 8] only categorize the tweets into positive, neutral or negative. But it was not enough to find the exact polarity of a sentence. For example, the "bad" has high rating than the word "worst". The affective lexicon based approach uses a special dictionary with rating of each word to find the rate of positivity and negativity in each tweet. This approach requires another dictionary [9] for managing this task, precisely the dictionary with a rating of words. Sample of words and its affective ratings are shown in Fig. 2 below.

Word	V.Mean.Su	V.SD.Sum	V.Rat.Sum	A.Mean.Su	A.SD.Sum	A.Rat.Sum	D.Mean.Su	D.SD.Sum	[
1 aardvark	6.26	2.21	19	2.41	1.4	22	4.27	1.75	
2 abalone	5.3	1.59	20	2.65	1.9	20	4.95	1.79	
3 abandon	2.84	1.54	19	3.73	2.43	22	3.32	2.5	
4 abandonm	2.63	1.74	19	4.95	2.64	21	2.64	1.81	
5 abbey	5.85	1.69	20	2.2	1.7	20	5	2.02	
6 abdomen	5.43	1.75	21	3.68	2.23	22	5.15	1.94	
7 abdominal	4.48	1.59	23	3.5	1.82	22	5.32	2.11	
8 abduct	2.42	1.61	19	5.9	2.57	20	2.75	2.13	
9 abduction	2.05	1.31	19	5.33	2.2	21	3.02	2.42	
10 abide	5.52	1.75	21	3.26	2.22	23	5.33	2.83	

Fig. 2. Words with affective ratings

This approach uses average rating for calculating tweets polarity based on words affective rating it consists of. For example, if "good" has 4 points and "perfect" has 6 points, the polarity would be calculated as $(4 + 6)/2 = 5$. By doing this way, the influence of several negative words that could have a higher total rating could be avoided.

3.5   Classifying Opinions

The following method is applied on 'Tweet_Scores' Corpus to classify the tweet into positive, negative or neutral.

> *For each Tweet(t) in Tweets Corpus*
> *Calculate Score(t) using (1)*
> *Opinion=null*
> *If score(t) > 0 then*
>   *Opinion="Positive"*
> *Else if score(t) < 0 then*
>   *Opinion="Negative"*
> *Else*
>   *Opinion="Neutral"*

Finally all positive tweets will be grouped and counted under positive head and negative tweets will be grouped under negative head and other tweets are grouped under neutral head.

4   Results Analysis

This section deals with results Analysis. The sample dataset contains around 17000 tweets about public opinion about Indian Elections 2019. Sample Tweet extracted using Twitter API is shown below (Fig. 3):

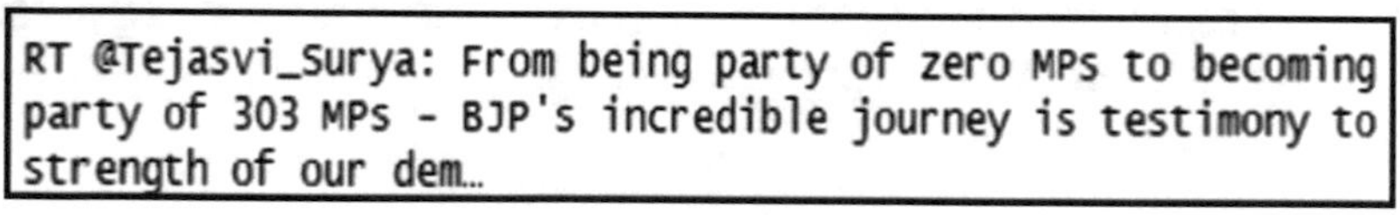

> RT @Tejasvi_Surya: From being party of zero MPs to becoming party of 303 MPs - BJP's incredible journey is testimony to strength of our dem…

Fig. 3. Sample Tweet about Indian Election Results 2019

To find sentiment about each tweets first we need to construct a Term Document Matrix (TDM). In this, the documents are called as rows, the terms are represented as columns and the value in the intersection of rows and columns represent the number of occurrences of the word within a document. Each tweet in the corpus is considered as a document and counts the top occurring words stores them in TDM. Figure below show

TDM for 1500 documents and 4052 terms is constructed i.e. dimension of 1500 X 4052 matrix is constructed with sparsity of 99%. The length of maximal term is 24 (Fig. 4).

```
<<TermDocumentMatrix (terms: 4052, documents: 1500)>>
Non-/sparse entries: 52320/6025680
Sparsity            : 99%
Maximal term length: 24
weighting           : term frequency (tf)
```

Fig. 4. TDM – Term Document Matrix of Indian Election Results 2019

TDM is used to find the Top Frequent Terms of the Election Results Corpus. Figure below shows the occurrence of the terms bjp, modi and great in the documents from 50 to 70 in the corpus (Fig. 5).

	Docs																				
Terms	50	51	52	53	54	55	56	57	58	59	60	61	62	63	64	65	66	67	68	69	70
bjp	1	1	3	0	1	1	0	1	1	1	1	0	1	1	0	0	1	0	1	0	1
modi	2	2	1	1	1	1	2	1	1	1	1	1	1	1	1	1	1	1	1	1	2
great	0	0	0	0	1	0	0	1	0	0	0	0	0	1	0	0	0	0	0	0	0

Fig. 5. Frequent occurrence of terms from the corpus

The figure below shows the top frequent words which occurs more than 500 times in the corpus.

"bjp"	"character"	"description"	"elect"	"heading"	"hour"
"isdst"	"language"	"list"	"mday"	"meta"	"min"
"origin"	"sec"	"sucess"	"wday"	"yday"	"year"
"get"	"great"	"he"	"ji"	"moment"	"narendramodi"
"single"	"we"	"212"	"21"	"even"	"minist"
"take"	"bengal"	"irani"	"kill"	"murder"	"smriti"
"2"	"leader"	"mps"	"12"	"families"	"political"
"media"	"vote"	"1"	"call"	"make"	"amit"
"shah"	"11"	"i"	"will"	"22"	"n"
"paial"	"say"	"so"	"111"	"case"	"win"
"singh"	"up"	"pm"	"member"	"211"	"delhi"

Fig. 6. Top frequent terms of the corpus $\geq$ 500 times

Sentiment Analyzer is used to find and classify opinions and sentiments expressed by public tweets. Dictionary based approach was used to categorize the users' opinions about a particular event, product, or person into positive, negative, or neutral. The figure below shows the classification of Indian Election Results 2019 corpus into three major categories positive, negative and neutral (Table 1).

Table 1. Classification of user opinion

Date	Negative	Neutral	Positive	Grand total
26/05/2019	240	520	572	1332
27/05/2019	320	460	650	1430
28/05/2019	273	745	450	1468
29/05/2019	311	600	604	1515
30/05/2019	413	959	564	1936
31/05/2019	206	348	225	779
Grand total	1763	3632	3065	8460

Visualization [11] of above data is shown below in Fig. 7.

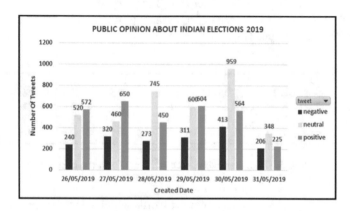

Fig. 7. Classifications of User Opinion about Indian Election Results 2019

But categorize tweets into positive, neutral or negative was not enough to find the exact polarity of a sentence. For example, the "good" has less rating than the word "better". The affective lexicon based approach uses a special dictionary with rating of each word to find the rate of positivity and negativity in each tweet. Figure below shows the rate of positivity of each tweet in the Indian Election Results 2019 corpus (Fig. 8).

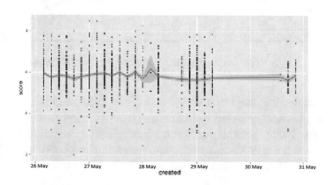

Fig. 8. Classifications of User Opinion about Indian Election Results 2019

Table 2 below shows the overall summary of the results of Opinion Analyzer.

Table 2. Summary of results.

Negative	3758
Neutral	9201
Positive	3858
Total tweets	16817

The graph below in Fig. 9 shows overall result analysis of User Opinion about Indian Elections. The results clearly show that 36% of users expressed positive opinion about Indian Election Results 2019. Majority of the respondents (around 43%) had posted neutral comments about Indian Election Results 2019. Only 21% of users tweeted negatively about Indian Election Results 2019.

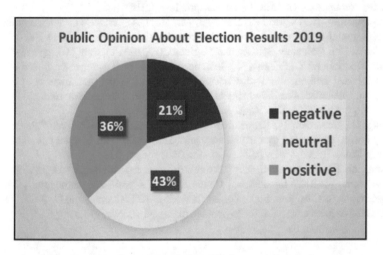

Fig. 9. Overall result analysis of Twitter opinion analyzer

5   Conclusions

This paper explained detailed implementations about two approaches called (i) Dictionary based and (ii) affective Lexicon based for sentiment Analysis by taking election results as an example corpus. Compared to Dictionary based approach affective lexicon based approach gives better results. These approaches help the business to create successful social campaigns, recognize influencers for their brand, service & industry, compare strengths & weaknesses of competitors, discover the real time trending topics i.e. what people are talking about the business and customer opinions & sentiment towards their business.

References

1. Dunđer, I., Horvat, M., Lugović, S.: Word occurrences and emotions in social media: case study on a Twitter corpus, pp. 1284–1287. IEEE (2016)
2. Mishra, P., Rajnish, R., Kumar, P.: Sentiment analysis of Twitter data: case study on digital India. In: 2016 International Conference on Information Technology (InCITe) - The Next Generation IT Summit on the Theme - Internet of Things: Connect your Worlds, Noida, pp. 148–153 (2016)
3. https://twitter.com/login?lang=en. Accessed June 2019
4. Jayamalini, K., Ponnavaikko, M.: Social media mining: analysis of Twitter data to find user opinions about GST. J. Eng. Appl. Sci. **14**(12), 4167–4175 (2019)
5. Jayamalini, K., Ponnavaikko, M.: Enhanced social media metrics analyzer using twitter corpus as an example. Int. J. Innov. Technol. Explor. Eng. (IJITEE) **8**(7), 822–828 (2019)
6. Xu, G., Yu, Z., Yao, H., Li, F., Meng, Y., Wu, X.: Chinese text sentiment analysis based on extended sentiment dictionary. IEEE Access **7**, 43749–43762 (2019)
7. Akter, S., Aziz, M.T.: Sentiment analysis on Facebook group using lexicon based approach. In: 2016 3rd International Conference on Electrical Engineering and Information Communication Technology (ICEEICT), Dhaka, pp. 1–4 (2016)
8. Karamollaoğlu, H., Doğru, İ.A., Dörterler, M., Utku, A., Yıldız, O.: Sentiment analysis on Turkish social media shares through lexicon based approach. In: 2018 3rd International Conference on Computer Science and Engineering (UBMK), Sarajevo, pp. 45–49 (2018)
9. Tiara, Sabariah, M.K., Effendy, V.: Sentiment analysis on Twitter using the combination of lexicon-based and support vector machine for assessing the performance of a television program. In: 2015 3rd International Conference on Information and Communication Technology (ICoICT), Nusa Dua, pp. 386–390 (2015)
10. Cheng, D., Schretlen, P., Kronenfeld, N., Bozowsky, N., Wright, W.: Tile based visual analytics for Twitter big data exploratory analysis. In: 2013 IEEE International Conference on Big Data, Silicon Valley, CA, pp. 2–4 (2013)
11. Jayamalini, K., Ponnavaikko, M.: Research on web data mining concepts techniques and applications. In: 2017 International Conference on Algorithms Methodology Models and Applications in Emerging Technologies (ICAMMAET), Chennai, pp. 1–5 (2017)
12. https://en.wikipedia.org/wiki/Social_media. Accessed June 2019

Internet of Things Based Electronic Voting Machine

Sushil Kumar Singh, Debjani Dey, and Sushanta Bordoloi[✉]

Department of ECE, NIT Mizoram, Aizawl 796012, India
nitsushilkr999@gmail.com, janifeb20@gmail.com,
sushanta.ece@nitmz.ac.in

Abstract. With the advancement in technology, the Internet has become an essential part of our lives. The concept of IoT has caught the attention of industry, society, and academy by enhancing our day to day activities. In this work, the Electronic Voting Machine (EVM) using IoT is presented. As the process of election demands to be fair and transparent, the fingerprint-based EVM can be one of the alternatives. The work focusses on the concept of an online electoral system for election, where the voting information is shared instantly, and mechanism had been put in place to detect a proxy vote. A hardware prototype is developed and to evaluate the performance of the proposed system, field trials were conducted.

Keywords: Electronic Voting Machine · Fingerprint sensor · Twitter · IoT

1 Introduction

Kevin Ashton coined the term Internet of Things (IoT) in the late 90s for supply chain management [1]. In past decades, the notion of it has outperformed encompassing an extensive range of applications such as utilities, healthcare, etc. [2, 3]. With the evolution of technology, the interpretation of 'Things' has changed. The gradual development of the current Internet into a network of interlinked objects helps to harness the information from the environment with interaction of the physical world.

In the context of India, EVM has been a sensitive issue. There has been a lot of concern to address the multifaceted problem such as booth capture, irregularities, tampering, etc. A detailed account can be found in [4]. Though researchers have addressed several issues over the year, real-time data analysis for the EVM is an essential factor [5].

In this paper, we carried out work to eliminate the possibility of booth capture and proxy vote. The significant contribution of the paper are

- Conceptualization of an idea of Electronic Voting Machine (EVM) with an interface to online services to track/detect the real-time activity.
- A prototype of EVM was developed and successfully tested.

The rest of the paper is organized as follows: literature survey is described in Sect. 2, the proposed EVM's details listed in Sect. 3, and the work is concluded in Sect. 4.

D. J. Hemanth et al. (Eds.): ICICI 2019, LNDECT 38, pp. 365–372, 2020.
https://doi.org/10.1007/978-3-030-34080-3_41

2 Prior Artwork

In section, the previous work related to EVM is stated.

Vamsikrishna et al. [6] proposed a system for voting based on the Internet of Things. The data of the voting process is stored in cloud, which is a secure method of data storage using IoT technology. The drawback associated with this system is the use of RFID cards for the process of identification. On displacement of this card, increases the chances of repetitive/double voting which is not desirable in a fair election process.

Akhare et al. [7] has proposed a Mobile based Electronic voting system. The system relies on GSM technology.

Obulesu et al. [8] proposed a system for voting using IoT Technology, incorporating a fingerprint module.

Sudhakar et al. [9] proposed an electronic voting machine based on the bio-metric fingerprint. The use of Ethernet cable for transmission has many disadvantages or limitations like fault in intolerant, difficult troubleshooting, distance limitations as it is limited only for short distance.

3 Design and Implementation

3.1 Proposed System: Block Diagram

The IoT based electronic voting machine basically consist of five main components. These components are NodeMCU, Fingerprint Module, ESP8266 Wi-Fi Module, LCD display and push buttons.

The significant contribution of the paper are

- NodeMCU: It is an open-source IoT platform configured to connect to the Internet. NodeMCU Development board consists of ESP8266 Wi-Fi-enabled chip, which supports serial communication protocols allowing transmission and reception of data. The interface with the fingerprint module helps to read the voter's print and compare it with the existing database. Post-processing of data relevant information is flashed on to LCD.
- Fingerprint Module: It permits the user to scan the fingerprint and transmit it further for necessary processing. The module has a high-security level with an average scanning time of fewer than 0.8 s.
- ESP8266 Wi-Fi Module: It's a Wi-Fi microchip that allows the micro-controllers to connect to a Wi-Fi network. It's a SOC with TCP/IP protocol integrated into it.
- LCD display: It is used for displaying various messages related to the voting process like a voter's ID number, the total number of votes cast.
- Push buttons: The three push button labeled as Party A, Party B, and Party C is used to cast votes for the desired political party.

NodeMCU controls the overall process of the voting system. The buzzer connected to the NodeMCU gives a warning and other signals (refer Fig. 1).

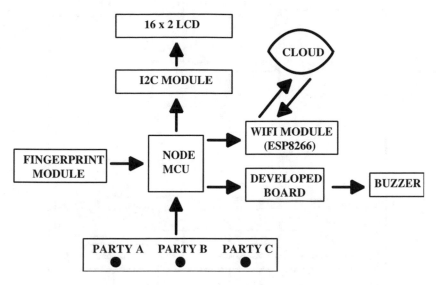

Fig. 1. Block diagram of the proposed system

3.2 Proposed System: Flow Chart

The fingerprint of all voters is stored in the memory of micro-controller. At the time of the voting process, the fingerprint of the voter is matched with [previously] stored database to check the credibility of the concerned voter. The fingerprint module is used to carry out this task with the help of micro-controller. On the verification of the voter, the vote cast is taken into consideration. The overall (incremented) vote is displayed on the LCD. When a person attempts the first time to cast vote, then he/she is allowed to give a vote. However, if the number of the attempt (for a particular voter) is two, the EVM allows the voter to proceed. While this [proxy] vote is not counted, it simply records the party name for which the vote is cast. Also, when the attempt made by a particular voter is three, an alert is sent to the nearest Police station (from the booth) and a News Editor along with his identity. In the end, EVM displays the total valid votes, party that received the most vote, and party name for which the proxy votes were received and information to Election Commission (refer Fig. 2).

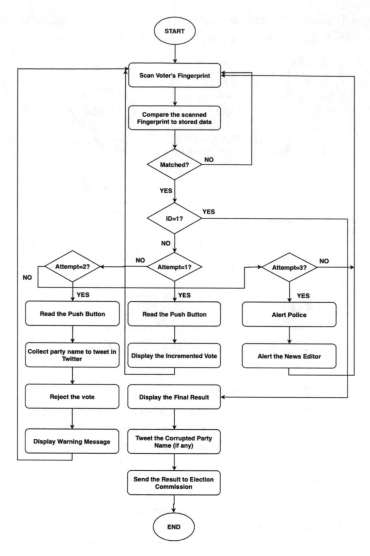

Fig. 2. Flowchart of the proposed system

3.3   Proposed System: Hardware/Prototype

The developed Hardware/Prototype for the EVM is shown in (refer Fig. 3).

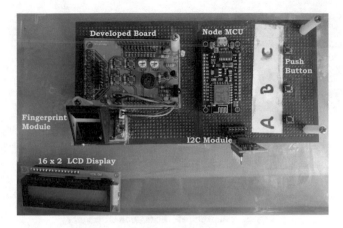

Fig. 3. Snapshot of developed prototype

3.4 Proposed System: Field Trial

Field trial for the prototype was conducted. A database for an eligible voter(s) was created and identification number (ID) was assigned accordingly. The mock voting process was conducted and it was found that the system (EVM) worked perfectly well. The Snapshot of some mock drill is stated in the subsequent section.

3.5 Results and Discussion

In this subsection, the snapshot of mock drill conducted is stated. In Fig. 4, establishment of connection of NodeMCU with WiFi network (IP address assigned) is shown. The verification of voter is shown in Fig. 5. Also, the initial count of the party are indicated as 0 at the start of voting process.

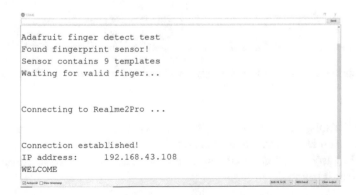

Fig. 4. Snapshot of initial connection of the proposed EVM to cloud

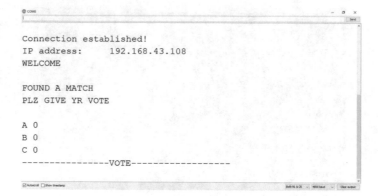

Fig. 5. Snapshot of verification of a voter

Figure 6 shows the message of Alert being sent to nearest Police station and News Editor informing the activity of repetitive attempt (booth capture) by a voter (here ID no. 8).

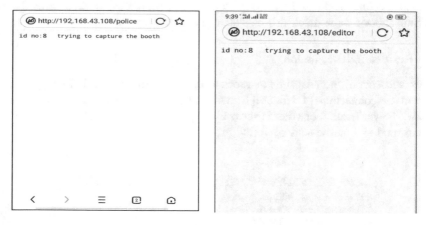

Fig. 6. Snapshot of alert sent to nearest police station and news editor for third attempt by a voter

Figure 7 displays the total vote for the process. The information of total vote cast, winner Party and intimation sent to Election Commission is shown in Fig. 8 (left). The tweet for the party with highest attempt of proxy vote is designated as corrupt party is shown in Fig. 8 (right).

Fig. 7. Snapshot of final count

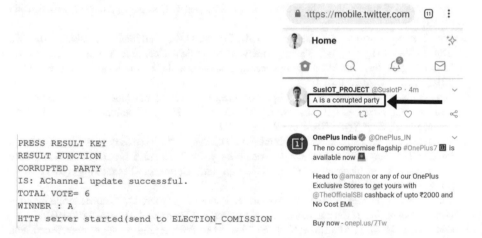

```
PRESS RESULT KEY
RESULT FUNCTION
CORRUPTED PARTY
IS: AChannel update successful.
TOTAL VOTE= 6
WINNER : A
HTTP server started(send to ELECTION_COMISSION
```

Fig. 8. Snapshot of final count being communicated to election commission and information of party receiving proxy vote being tweeted

4 Conclusion and Future Scope

The developed prototype has performed well in the test trail. It takes into account the spot counting of the votes and put in place a mechanism to address the issue of multiple attempts for a suspicious voter(s). The deployment of this prototype in actual use can be a challenging task. Scope of future work includes

- More details about the voter can be incorporated, such as retina scan. It will enhance the security of the election process.
- The database of the eligible voter(s) can be made online (Cloud-based), which will enable a voter to cast his/her vote at any place and addresses the issue of storage of voter's database.

People will come to know about the corrupt political party who tried to capture the booth, and in turn caused a delay in the election process for involving in such practise through their workers.

Acknowledgment. The author/s thankfully acknowledge(s) the support provided by TEQIP-III, NIT Mizoram for carrying out this work. And Sukanta Dey, Research Scholar, Dept. of CSE, IIT Guwahati for his valuable suggestion in preparing manuscript.

References

1. Ashton, K.: That 'Internet of Things' thing. RFID J. **22**(7), 97–114 (2009)
2. Sundmaeker, H., Guillemin, P., Friess, P., Woelfflé, S.: Vision and challenges for realising the internet of things. Clust. Eur. Res. Proj. Internet Things Eur. Commision **3**(3), 34–36 (2010)
3. Gubbi, J., Buyya, R., Marusic, S., Palaniswami, M.: Internet of Things (IoT): a vision, architectural elements, and future directions. Futur. Gener. Comput. Syst. **29**(7), 1645–1660 (2013)
4. Wolchok, S., Wustrow, E., Halderman, J.A., Prasad, H.K., Kankipati, A., Sakhamuri, S.K., Yagati, V., Gonggrijp, R.: Security analysis of India's electronic voting machines. In: Proceedings of the 17th ACM Conference on Computer and Communications Security, 4 Oct 2010, pp. 1–14. ACM (2010)
5. Sen, A., Sen, M., Ambekar, A.: Improved electronic voting machine with real time data analysis. Commun. Appl. Electron. (CAE) **6**(1) (2016). Foundation of Computer Science FCS, New York, USA. ISSN: 2394-4714
6. Vamsikrishna, P., Kumar, S.D., Bommisetty, D., Tyagi, A.: Raspberry pi voting system, a reliable technology for transparency in democracy. In: 2016 IEEE International Conference on Advances in Electronics, Communication and Computer Technology (ICAECCT), pp. 443–449. IEEE (2016)
7. Akhare, A., Gadale, M., Raskar, R., Jaykar, B., Phalke, D., Tiwari, D.: Secure mobile based e-voting system. Int. J. Recent Innov. Trends Comput. Commun. **4**(4), 148–150 (2016)
8. Obulesu, I., Hari, A., Manish, P.N.: Prreethi: Iot based fingerprint voting system. Acad. Sci. Int. J. Innov. Adv. Comput. Sci. **7**(4), 502–505 (2018)
9. Sudhakar, M., Sai, B.D.S.: Biometric system based electronic voting machine using arm9 microcontroller. IOSR J. Electron. Commun. Eng. **10**(1), 57–65 (2015)

A Secured Steganography Technique for Hiding Multiple Images in an Image Using Least Significant Bit Algorithm and Arnold Transformation

Aman Jain[✉]

Jaypee Institute of Information Technology, Noida, Uttar Pradesh, India
jn_aman@yahoo.com

Abstract. Security systems are well-liked in several areas wherever technologies are developing day by day. Using secret writing called encryption and data hiding called Steganography information security are often achieved. A traditionally used technique in Steganography is the LSB technique. Hide info is that the art and science of hidden writing, it's a technique of masking information in an exceedingly different style of data. Additionally, we will send text or a picture disguised by the cover image and change the components of the first to the recipient. A hidden text or image move into the cover image will defend the attacker's original information. In this we proposed a hybrid technique to secure the images in an image. Here Steganography LSB technique is used to hide the images in an image after that used the scrambling Arnold technique to scramble the cover image and improved the security of the hidden image.

Keywords: Cryptography · Steganography · Image steganography · LSB technique · Arnold transformation · Image hiding

1 Introduction

The image of the communication mode employed in the various regions, the median, the analysis space, the negotiation zone, the military zone, etc. The transfer of important pictures is to create a visit from an unsecured Internet network. From that moment, it's necessary to decide a security service so that we can imagine that the person does not have access to important information. The wind impact of the image cubes that additional multimedia system information and protection desires [1]. Cryptography and Steganography are standard and wide used techniques used primarily to method info to encipher or hide its existence, severally [2]. Cryptography may be a tool for methodologyizing the image; It offers the secure method of transmission and get for the image of travel over the net. Security is that the main concern of any system to keep up the integrity, confidentiality and believability of the image. a minimum of cryptography is that the elective methodology, however conjointly the difficulty of proportional severity and gray-scale knowledge is additional [3]. Steganography may be a word derived from the Greek that means "covered writing".

D. J. Hemanth et al. (Eds.): ICICI 2019, LNDECT 38, pp. 373–380, 2020.
https://doi.org/10.1007/978-3-030-34080-3_42

It's a technology that hides data so hidden messages don't seem to be detected. Hide data or image hides the actual fact that a secret message or secret image is shipped. Hide data or images may be a technique that masks the message or image invisibly. An invisible message or image exploitation ways to cover data won't seem suspiciously at the recipient, however an encrypted text message or encrypted image [4]. Data protection exploitation hidden writing and science may be a hidden data technique during a completely different form of data to hide data [5]. Masking info was wide used these days with text, images, audio and video on the duvet. The kind of disguised info is assessed as shown in Fig. 1.

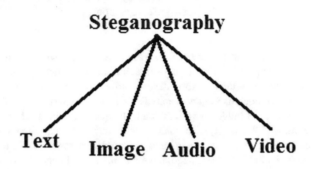

Fig. 1. Categories of steganography [3, 4]

2   Related Work

Afandy et al. discussed the LSB technique is that the simplest methodology and is wide used with info concealment technology. It is depends on combining the key text message bits into the three least important bits of pixels at intervals the quilt image. Hide data depends on the image pixel replacement using LSB technology [6]. Kini et al. present the LSBs of the pixels with the help of image are replaced by the bits of the key image so as to affix the key image within the assistance image. The bits of the key image are processed before encoding [7]. Shankar et al. introduced a method of concealing information that combines data concealment and digital image secret writing. Only real recipients with access to the shared key will retrieve the initial message and cover image. The image is recovered with minimal losses. Arnold Mapping is employed to make sure that the pixels within the image are absolutely merging which the random broadcast step overcomes the limited set period. Thus, secret writing security has been improved [8]. Gulve et al. discuss the pixel value difference methodology, pixel values are calculated between two consecutive pixels. This distinction price is employed to cover the key message. This distinction value is compared to the secret message bits to be converted. If they are unequal, then two serial pixels are adjusted directly that the distinction price will indicate confidential information [9].

3 Least Significant Bit (LSB) Technique

The LSB substitution is that the methodology to easily replace the smallest amount vital bits of the image pixels with the message bits [10]. In the LSB message, the bits are embedded within the least important little bit of the title or cowl image. LSB Steganography are often classified in two ways: its replacement, and also the LSB correspondence was mentioned for the primary time by T. Sharp. The replacement of the LSB Steganography replaces the last little bit of the title image with every bit of the message to be disguised. The second methodology is that of the LSB correspondence, within which every component of the title image is captured primarily in an especially pseudo-random order generated by a secret key. Otherwise, otherwise, if the LSB of the coverage component is that the secret information bit, no changes are created and that they are additional or ablated to the worth of the coverage component. If the key message length contains fewer bits than the quantity of pixels within the title image, pseudo-random exchange changes also are distributed over the image. Therefore there's a modification of one for every bit [11].

4 Arnold Transformation

The Arnold transform is a classical 2D invertible chaotic map defined as [8]:

$$\begin{bmatrix} x\prime \\ y\prime \end{bmatrix} = \begin{bmatrix} 1 & 1 \\ 1 & 2 \end{bmatrix} \begin{bmatrix} x \\ y \end{bmatrix} \, mod \, 1 \tag{1}$$

The inverse transform is defined as:

$$\begin{bmatrix} x \\ y \end{bmatrix} = \begin{bmatrix} 2 & -1 \\ -1 & 1 \end{bmatrix} \begin{bmatrix} x\prime \\ y\prime \end{bmatrix} \, mod \, 1 \tag{2}$$

Arnold's transformation might be a periodic and invertible mapping. In addition, Arnold's transformation applies only to square pictures. The Arnold transform is used to mix digital images and has several applications, particularly digital watermarks [12].

5 Proposed Technique

In the proposed Steganography Technique we hide multiple images in a single image. In this work first we hide the four images one by one in a cover image using the LSB technique. After hiding the all the four images in cover image, the cover image is only shown. Hence all the four hided images are don't shown, only cover image is shown. After hiding the images in cover image we scramble the cover image using the Arnold transformation. Now the scramble image is totally different then the cover image. So now no one can be identify the cover image because it is totally differ then the original image (cover image). Hence the security of the hided images are improved because the images are hided in cover image if cover image is not identify so how can be third person get the hided images.

The proposed technique is done using the following steps:

Step 1: First select the cover image.
Step 2: After selecting the cover image, select the sender image or secret image.
Step 3: Hide the secret images in cover image using the LSB technique.
Step 4: After hiding all four secret images we get the embedded image that is shown same as the cover image.
Step 5: Now we apply the Arnold transformation on this embedded image and we get the scramble image that is totally differ then the original image.

6   Results of Proposed Technique

The results and obtained outputs of the proposed method shown below. First we select the cover image that is shown in the Fig. 2. After selecting the cover image we choose the secrete images that we want to hide in the cover image. In this proposed method we hide the four images that are shown in the Fig. 3. The embedded image is shown in the Fig. 4 and after apply the Arnold transformation we get the scramble image that is shown in the Fig. 5.

Fig. 2. Cover image

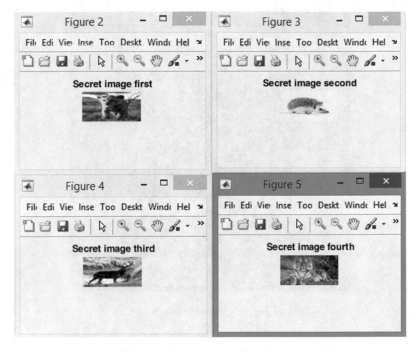

Fig. 3. All four secret images

Fig. 4. Embedded image (cover image with secret image)

Fig. 5. Scramble image

After done the Arnold transformation we get the scramble image that is shown in the Fig. 5 which is totally differ then the original image. If we want to get the original images (cover image and hidden images) then we done the inverse Arnold transformation. After apply the inverse Arnold transformation we get the original images. The original cover image and hidden images are shown in Fig. 6 in the Fig. 7 respectively.

Fig. 6. Original cover image

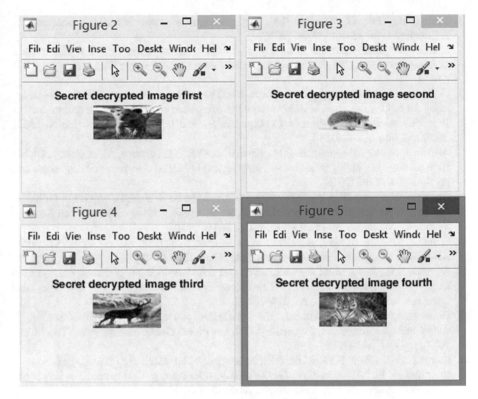

Fig. 7. Hidden or secret images

7 Conclusion

In this paper we showed a secured Steganography method to hide multiple pictures within an image by using LSB technique and Arnold Transformation. In this proposed method we hide the four secret images within an image by using LSB technique. After hiding the secrete images we scramble the embedded image by using the Arnold transformation. After applying the Arnold transformation we get the scramble image, which is totally different from the original image. So the unauthorized person doesn't get the any information about the cover image or secret images. Hence we secure the secret images within an image by using the LSB technique and Arnold transformation.

References

1. Hole, K.R., Gulhane, V.S., Shellokar, N.D.: Application of genetic algorithm for image enhancement and segmentation. Int. J. Adv. Res. Comput. Eng. Technol. (IJARCET) **2**(4), 1342–1346 (2013)
2. Pillai, B., Mounika, M., Rao, P.J., Sriram, P.: Image steganography method using k-means clustering and encryption techniques. In: IEEE International Conference on Advances in Computing, Communications and Informatics (ICACCI), pp. 1206–1211 (2016)

3. Madhu, B., Holi, G., Murthy, S.K.: An overview of image security techniques. Int. J. Comput. Appl. **154**(6), 37–46 (2016)
4. Menon, N., Vaithiyanathan: A survey on image steganography. In: IEEE International Conference on Technological Advancements in Power and Energy (TAP Energy), pp. 1–5 (2017)
5. Gedkhaw, E., Soodtoetong, N., Ketcham, M.: The performance of cover image steganography for hidden information within image file using least significant bit algorithm. In: The 18th International Symposium on Communications and Information Technologies (ISCIT 2018), pp. 504–508 (2018)
6. Al-Afandy, K.A., EL-Rabaie, E.-S.M., Faragallah, O.S., ELmhalawy, A., El-Banby, Gh.M.: High security data hiding using image cropping and LSB least significant bit steganography, pp. 400–404. IEEE (2016)
7. Gopalakrishna Kini, N., Kini, V.G., Gautam: A secured steganography algorithm for hiding an image in an image. In: Integrated Intelligent Computing, Communication and Security, Studies in Computational Intelligence, pp. 539–546 (2019)
8. Shankar, S.S., Rengarajan, A.: Data hiding in encrypted images using arnold transform. ICTACT J. Image and Video Process. **07**(01), 1339–1344 (2016)
9. Avinash, K.G., Joshi, M.S.: A secured five pixel pair differencing algorithm for compressed image steganography. In: IEEE Third International Conference on Computer and Communication Technology, pp. 278–282 (2012)
10. Dumitrescu, S., Wu, X., Memon, N.: On steganalysis of random LSB embedding in continuous-tone images. In: Proceedings of International Conference on Image Processing, vol. 3, pp. 641–644 (2002)
11. Ramaraju, P.V., Raju, N., Krishna, P.R.: Image encryption after hiding (IEAH) technique for color images. In: International conference on Signal Processing, Communication, Power and Embedded System (SCOPES)-2016, pp. 1202–1207 (2016)
12. Zhang, X., Zhu, G., Wang, W., Wang, M., Ma, S.: Period law of discrete two-dimensional Arnold transformation. In: IEEE Fifth International Conference on Frontier of Computer Science and Technology, pp. 265–269 (2010)

Optimum Energy Based Routing Protocols for Wireless Sensor Network: A Short Review and Analysis

Ganesh J. and V. R. Sarma Dhulipala$^{(\boxtimes)}$

Department of Physics, University College of Engineering, Anna University,
BIT Campus, Tiruchirappalli, Tamil Nadu, India
dvrsarma@aubit.edu.in

Abstract. Resource limitation such as power, reliability, storage capacity and computation in Wireless Sensor Network (WSN) which throw a challenge in the recent advancement in sensor network research. In WSN, Sensor Nodes (SN)s are generally operated by battery and hence it is hard to change or recharge the battery in unattended environments. These may be overcome by the design of proper routing strategies. This research presents a brief survey of routing protocols in WSN based on the parametric analysis.

Keywords: Wireless Sensor Networks · Routing protocols · Energy efficiency · Analysis

1 Introduction

Collection of many of SNs of random topologies in distributed environment is known as WSN and used for sense and monitor the environment [1]. WSN has recently come up as a premier technology due to its usefulness in wildlife monitoring, military, telemedicine and healthcare application [2]. The SN sends gathered data via transmitter, to directly or multi-hop transmission to sink or base station. SNs are generally powered by battery and the lifespan of the SN is relatively low due to continuous measuring and monitoring the environment. While the design a routing protocol in WSN, low utilization of the sensor energy is an important issue.

Routing in WSN is very crucial because of numerous design issues with prominence on energy efficiency. The traditional ad hoc routing protocols are not suitable for WSN.

The factors that affect the protocol design [3] are as follows

1.1 Node Deployment

The deployment of sensor node is either deterministic or self organizing. In first method, the SNs are physically installed and transmitted the data by using known route. In self organizing method, the SN is deployed dynamically in a distributed manner.

D. J. Hemanth et al. (Eds.): ICICI 2019, LNDECT 38, pp. 381–390, 2020.
https://doi.org/10.1007/978-3-030-34080-3_43

1.2 Link/Node Heterogeneity

Traditionally, every node in a network is homogeneous in nature. Addition of heterogeneity increases multiple issues associated to data routing due to different configuration of SNs.

1.3 Energy Consumption

The battery is a major power source for SN. The lifetime of WSN is wholly depends on energy level of a SN. The transmission of a single bit takes more energy compared to computation and reception of the bit.

1.4 Network Dynamics

SNs are stationary for many of the applications. For specific application, the mobility of SNs are necessary to gather the sensed data & difficult to utilize the minimum sensor energy.

1.5 Fault Tolerance

Few sensor nodes in a WSN may block or fail due to insufficient energy and these failures should not degrade the performance of sensor networks. Fault tolerance is a capacity to support sensor network performance inspite of the node failure.

1.6 Data Aggregation

Some sensor nodes can produce the redundant data, in order to reduce the sending rate, similar kind of data from the various sources of sensor node can be aggregated. Data aggregation is the collection of sensing information from multiple nodes by using some aggregation function such as sum, average, min and max to eliminate the redundant data. By using data aggregation techniques is a way to decrease the energy utilization of SN and traffic in the network.

Some more factors such as data delivery model, coverage area, hardware constraints, scalability and QoS also to be considered in distributed environment.

2 Related Work

In this section, [4] has been stated the performance of two important protocols in data centric protocols such as Directed Diffusion (DD) and Sensor Protocol Information via Negotiation (SPIN) protocol [5]. Low Energy Adaptive Clustering Hierarchy (LEACH) is a cluster based and Power Efficient Gathering in Sensor Information System (PEGASIS) is a chain based protocol but these two protocols are belongs to hierarchical based protocol. Data Aggregation (DA) is available in LEACH but not in PEGASIS. Due to clustering, LEACH became energy efficient [9]. PEGASIS helps to avoid the cluster formation and form a single data to transmit to the sink. Geographical Adaptive Fidelity (GAF) [11] concentrated the group of SN are formed into virtual grid based on location and make it some node are in active in grid. Some nodes are in sleep in grid for extending the lifespan of the network. M-GEAR [12] forms a group of SNs are into logical region and reduce the distance between the nodes for enhance the network lifetime.

3 Routing Protocol Classification

Routing protocols in WSN can be categorized into three major groups such as data centric routing, hierarchical routing and location based routing. In WSN routing protocols, energy conservation is the important key factor and the routing application has a strong relationship.

3.1 Data Centric Routing Protocol

Each node in WSNs, it is very hard to allocate the ID for SN due to massive deployment of SN and dynamic topology. Due to lack of this global knowledge of the sensor networks, it is not easy to pick a particular portion of SNs to be queried. In this protocol, the sink broadcasts queries to a particular region and till data from the sensor node of a particular region. DD and SPIN are belongs to this category.

DD

It [6] is a type of data centric routing protocol; choose an optimal path based on data content. During the data is transmitted by an origin node, the data can be stored and altered by intermediary node, which is start interests based on data that were lastly stored. Before transmission to the destination, the intermediate node aggregates the data sent by the source node. More interestingly, interests, aggregation and data dissemination occur in a localized manner with the help the neighboring nodes in it. There have a several key features such as data naming, interests, gradients, data transmission and reinforcement. The base station sends their willingness to all active tasks to specify a minimum data rate. A gradient individualize a rate of data along with event direction and pick a specific path from its collection. In this protocol, data collecting from multiple origin nodes are put together and to avoid the redundant data, minimizing the number of data transmissions, it saves the node energy and extends the lifespan of the networks.

SPIN

This protocol [7] is type of negotiation based and data aggregation protocols. Adv: data advertisement message that carries the metadata (information about the sensor data). Req: when a node has interested to receive the actual sensor data, it sends the request message. Data: a node consists of the actual sensor data along with meta data header.

Here sensor node obtains a new data; it transmits the meta-data to its neighbor before actual transmission. The nearby node collects the meta-data; it verified the collected meta-data with already stored meta-data. If newer one and a node are really wants that data, then it broadcasts the response as a request message to sender. The sender sends the actual data to its interested neighbors. There are two important key features of this protocol (i) node had its own actual sensor data, then aggregates the own data and its neighbor node data. (ii) Each and every node may not respond the request message to this protocol. This protocol eliminates the three deficiencies of classical flooding protocols, namely implosion, overlap and resource blindness. The implosion and overlap problems can resolved by using negotiation and resource blindness problem can eliminated by using resource adaptation.

3.2 Hierarchical Routing Protocol

A group of SN is formed into a cluster. Each cluster, one node which is having superior than other node can acts as a Cluster Head (CH). The data from the Non-cluster head (NCH) nodes should send to CH. The data from its member nodes are collects and aggregates the collected data. LEACH and PEGASIS belongs to hierarchical routing protocol.

LEACH

This protocol [8] can be operating several rounds and two phases such that set up and steady state phase per round.

During the first phase, the formation of cluster and selects the CH for each cluster. This phase consists of three steps further that are namely advertised, cluster setup and TDMA schedule preparation. It uses randomization to spin the role of a CH and enhance the network lifetime. In steady state phase, CH aggregates all the data from its member and forwards the aggregated data forwards to the sink through directly or multi-hop communication.

Each node has selected itself CH and also informing that it becomes a member of the cluster under the cluster head. During this phase, an only receiver of all CHs becomes remains on, the message of all the NCH should received by the CH. CH changes periodically in order to balance the energy conservation of SNs. This random selection can be achieved through selecting a number randomly between 0 and 1. Any node in a cluster to become a CH if the RN is lower than the threshold $T(n)$:

$$T(n) = \begin{cases} \frac{p}{1-p*\left(r\,mod\frac{1}{p}\right)}, & if\ n \in G \\ 0, & otherwise \end{cases} \tag{1}$$

In Eq. (1), where

p = the probability to become a CH,
r = the present round &
G = the collection of SNs which are not become a CH till $\frac{1}{p}$ rounds.

The CH prepared the TDMA schedule based on the number of nodes in the cluster. Each starts its transmission during its own allocated time schedule. Data transmission has begun only when the clusters are formed by its corresponding CH and TDMA is fixed.

PEGASIS

Each node has known about entire knowledge of the networks. The Eqs. (2) and (3) are used to calculate the cost for transmitting and receiving for a distance d and n bits length of message.

Transmitting

$$E_{Tx}(n, d) = E_{Tx-ele}(n) + E_{ele} * n + \epsilon_{amp} * n * d^2 \qquad (2)$$

Receiving

$$E_{Rx}(n) = E_{elec} * n \qquad (3)$$

Equation (2), E_{Tx-ele} is energy dissipates to run the transmitter circuitry, Eq. (3), E_{Rx-ele} is energy dissipates to run receiver circuitry, ϵ_{amp} is energy for transmitter amplifier. It is chain based protocol and an enhancement of the LEACH. Each node in PEGASIS [10] is communicated only with their direct neighbor node to send the data. It cannot send directly to sink node. Unlike LEACH algorithm, this protocol forms chains of sensor nodes using greedy approach. Node which is chosen from the chain is eligible to transmit the data to sink. Each SN is collected the data from its neighbor node and data fusion (to form a single packet from one or more data packets from various node) packet, forward to its own neighbors and eventually sent to sink. Nearer nodes are grouped into a chain and further develop a route to sink.

The data fusion saves the data transmission for fuse the data; it minimizes the energy utilization of node and prolongs the lifespan of network. There is no need for cluster formation. It selects the optimal path to the sink. The main demerit is not applicable to real world applications, particularly continuous monitoring application.

3.3 Location Based Routing Protocol

Each node has known its location information calculated by using localization systems. The protocol belongs to this category worked based on location of the SN. M-GEAR and Geographical Adaptive Fidelity (GAF) are belonging to this category.

Geographical Adaptive Fidelity (GAF)
It [11] is a geo-location based and energy aware routing protocol. The SN are equally formed a virtual grid. The appropriate size of the grid is essential as it directly have an effect on the network connectivity. If the grid size is outsized then it is hard to connect the entire network by activating just one node per grid. Each node in the grid uses its precise location information for routing. Finding of location, it may use the technique like GPS, RSSI, etc. nodes consumes energy for transmitting, receiving and also during the idle state few amount of energy depleted but it is lesser than transmitting and receiving.

In this protocol, there are three different states used such that discovery, activity and sleep state. All nodes need not to stay in active state. Only one sensor node makes it awake and transmits the data to the sink while some nodes are in active and remaining node is in the sleep state for decrease the energy depletion. The SN in which contains the highest residual energy is in active state. All nodes start with discovery state and then it goes either active or sleep state. Instead of all sensor nodes in a grid, the active node should take responsible for observing and transmitting the data to the base station.

M-GEAR

SNs have produced large of amount of data and these data transmit to the base station for further processing. It takes much energy for SN and degrades the network lifetime also. Therefore, it decreased the number of transmissions to improve the lifetime of networks. In this protocol [12], DA is a process of combining the huge amount of data into a small set of useful data. Sometimes DA is also called as data fusion. In addition to CH, they introduce a Gateway Node (GN) can place in the center of the network. The main role of GN is to gather data from nearby CHs, aggregates this data & sending to base station. With the help of gateway node, the energy conservation of SN reduced and enhances the network lifetime. The sink spreads a HELLO packet to all of its SNs. The SN sends a response back to the sink and the sinks calculates the distance among SN and store the location details into the data table.

During the setup phase, the whole network region divides into four logical regions. Nodes those who are directly communicate with the base station are grouped into first region. SNs that are closer to GNs are equally divided into two regions and namely this is called as non-clustered regions. The SNs which are far from GN are belongs to fourth region. In CH selection phase, CHs are chosen in each region independently. These CH selection process is like as LEACH. In Scheduling phase, when clusters are formed, then CH prepare a TDMA slots to their SN for the transmission. The entire member node transmits the data to CH when their allotted time schedule. In steady state phase, the entire member node transmits the data to CH. The CH aggregate the data from its member and this data further transmit to GN. The GN again aggregates receive the sensed information from CHs and sent this data to sink directly.

4 Performance Evaluation

In this paper, the performance of various protocols in each category of WSNs calculated and analyzed.

Table 1 shows the simulation parameters for SPIN and DD. The message sizes are equally taken for both SPIN and DD protocols.

Table 1. Network parameter for SPIN and DD protocol.

Parameters	Value for SPIN	Value for DD
Network size	100×100 m^2	100×100 m^2
Number of nodes	100	100
Transmit cost	600 mW	660 mW
Receive cost	200 mW	395 mW
Message size	4000 bits	4000 bits
Metadata size	48 bits	100 bits
Initial energy	10 Joules	10 Joules

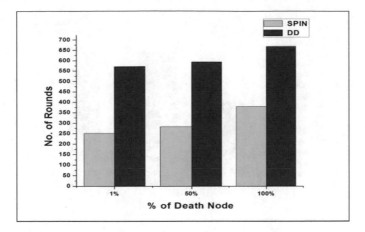

Fig. 1. Number of rounds and percentage of death node for DD and SPIN.

Figure 1 shows the performance comparison of DD and SPIN protocol. The percentage of death node is compared and results are tabulated in Table 2. The first node dies at 251th rounds in SPIN and 572nd rounds in DD. When compare with SPIN protocol, the DD is performs better and this can be clearly understood from Fig. 1.

Table 2. Number of node dies at 1%, 50% and 100% for SPIN and DD diffusion protocol

Protocol/Percentage of node died	% of death node		
	1%	50%	100%
SPIN	251	283	379
DD	572	594	669

Based on simulation parameters, the evaluation of the LEACH, PEGASIS, M-GEAR and GAF shown in Table 3. Hundred SNs are deployed in 100×100 m^2 terrain dimensions, and base station for the clustering is located at 50×150 m^2.

Table 3. Networks parameters

Parameters	Value
Network size	100×100 m^2
Number of nodes	100
Sink node location	(50,150)
E_{elec} (E_{Tx} and E_{Rx})	50 nj/b
E_{DA}	5 pj/b
E_{AMP}	0.0013 pJ/b/m^4
E_{fs}	10 pj/b/m^2
Message size	2000 bits
Initial energy	0.5 J

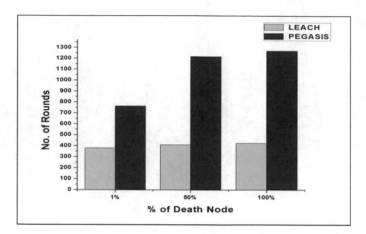

Fig. 2. Number of rounds and percentage of death node for LEACH and PEGASIS.

Figure 2 Shows the rounds until 1%, 50% and 100% node dies a 100X100 m^2 network. Table 4 exhibit the performance comparison of the LEACH and PEGASIS protocol with the same network parameters. In this case, clearly performance of PEGASIS better than LEACH protocol.

Table 4. Number of node die at 1%, 50% and 100% for LEACH and PEGASIS protocol

Protocol	% of node died		
	1%	50%	100%
LEACH	379	408	422
PEGASIS	762	1216	1270

Figure 3 Shows the results of M-GEAR and GAF. Table 5 exhibits the performance of M-GEAR is better than GAF.

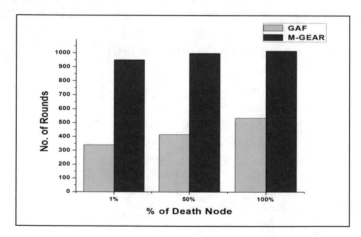

Fig. 3. Number of rounds and percentage of death node for M-GEAR and GAF

Table 5. Number of node die at 1%, 50% and 100% for GAF and M-GEAR protocol

Protocol	% of death node		
	1%	50%	100%
GAF	340	414	534
M-GEAR	951	998	1015

5 Conclusion

In WSN, basically there are three classifications in routing protocol such as data centric, hierarchical and location based protocols. Out of which, we implemented two routing protocols in each category in MATLAB. We found that DD, PEGASIS and M-GEAR perform well by utilizing low energy comparing SPIN, LEACH and GAF respectively. Hence, it is found that prolong the lifespan of each category (DD, PEGASIS and M-GEAR).

Acknowledgement. The author V.R. Sarma Dhulipala wishes to acknowledge the fund received by Department of Science and Technology (DST), Govt. of India, Sanction no. **DST/CERI/MI/SG/2017/080 (AU)(G)**.

References

1. Akyildiz, I.F., Su, W., Sankarasubramaniam, Y., Cayirci, E.: Wireless sensor networks: a survey. Comput. Netw. **38**(4), 393–422 (2002)
2. Ruan, Z., Sun, X.: An efficient data delivery mechanism by exploring multipath routes for wireless sensor networks. Inf. Technol. J. **10**(7), 1359–1366 (2011)
3. Patil, M., Biradar, R.C.: A survey on routing protocols in wireless sensor networks. In: 2012 18th IEEE International Conference on Networks (ICON), pp. 86–91. IEEE (2012)
4. Jain, V., Khan, N.A.: Simulation analysis of directed diffusion and SPIN routing protocol in wireless sensor network. In: 2014 Conference on IT in Business, Industry and Government (CSIBIG), pp. 1–6. IEEE (2014)
5. Angurala, M.: A comparative study between leach and pegasis—a review. In: 2016 3rd International Conference on Computing for Sustainable Global Development (INDIACom), pp. 3271–3274. IEEE (2016)
6. Intanagonwiwat, C., Govindan, R., Estrin, D.: Directed diffusion: a scalable and robust communication paradigm for sensor networks. In: Proceedings of the 6th Annual International Conference on Mobile Computing and Networking, pp. 56–67. ACM (2000)
7. Heinzelman, W.R., Kulik, J., Balakrishnan, H.: Adaptive protocols for information dissemination in wireless sensor networks. In: Proceedings of the 5th Annual ACM/IEEE International Conference on Mobile Computing and Networking, pp. 174–185. ACM (1999)
8. Heinzelman, W.R., Chandrakasan, A., Balakrishnan, H.: Energy-efficient communication protocol for wireless microsensor networks. In: Proceedings of the 33rd Annual Hawaii International Conference on System Sciences, p. 10. IEEE (2000)
9. Nandhini, M., Dhulipala, V.S.: Energy-efficient target tracking algorithms in wireless sensor networks: an overview. IJCST **3**(1), 66–71 (2012)

10. Lindsey, S., Raghavendra, C.S.: PEGASIS: power-efficient gathering in sensor information systems. In: Aerospace Conference Proceedings, vol. 3, p. 3. IEEE (2002)
11. Xu, Y., Heidemann, J., Estrin, D.: Geography-informed energy conservation for ad hoc routing. In: Proceedings of the 7th Annual International Conference on Mobile Computing and Networking, pp. 70–84. ACM (2001)
12. Nadeem, Q., Rasheed, M.B., Javaid, N., Khan, Z.A., Maqsood, Y., Din, A.: M-GEAR: gateway-based energy-aware multi-hop routing protocol for WSNs. In: 2013 Eighth International Conference on Broadband and Wireless Computing, Communication and Applications (BWCCA), pp. 164–169. IEEE (2013)

Critical Analysis of Internet of Things Application Development Platforms

Shubham Agarwal, Ria Dash, Ashutosh Behura,
and Santosh Kr. Pani[(⊠)]

School of Computer Engineering,
KIIT Deemed to be University, Bhubaneswar, India
shubham.agarwal7930@gmail.com, dashria1997@gmail.com,
{ashutoshfcs, spanifcs}@kiit.ac.in

Abstract. The world is tending towards automation and there is relentless competition in the global market. Internet of Things (IoT) and Android OS have become buzzwords due to their vast technological diaspora and rich spectrum of sophisticated features. In future, mobile platforms will go on to play an important role in the IoT environment, particularly where the data needs to be gathered around moving sensors or human resources where as fixed systems can be better served by low-cost and dedicated sensors. In order to fuel such software-led innovations, various companies like Google and Microsoft Inc. have come forth with different operating systems such as Android Things and Windows 10 IoT Core. In this paper, we aim at presenting Android Things as a developing environment for IoT applications. We also compare Android Things with Windows 10 IoT Core and juxtapose their features, performances and constraints.

General Terms: Android · Android Things · Internet of Things · Raspberry Pi · Windows 10 IoT Core

Keywords: Arduino · Embedded systems · IoT development boards · Raspbian OS · Node MCU · Banana Pi · Windows

1 Introduction

Internet of Things (IoT) is an interconnection of things embedded with software, sensors, electronics, actuators and connectivity to internet. When IoT is supplemented with actuators and sensors, the technology results into a more general classification, that is, a cyber-physical system which incorporates other technologies such as smart grids, virtual power plants, automated homes, intelligent vehicles and smart townships [3].

Connectivity enables the things to garner and exchange data, creating a wide array of opportunities for a comprehensive integration of the real world into computer-oriented systems resulting in higher efficiency, economic benefits and reduced human exertion. Low computational IoT devices like Arduino, Econotag, IoT-LAB M3 nodes, OpenMote nodes, TelosB motes, Zolertia Z1, etc. run on traditional Operating Systems as they are immensely resource-constrained. In many respects, the smartphones were

© Springer Nature Switzerland AG 2020
D. J. Hemanth et al. (Eds.): ICICI 2019, LNDECT 38, pp. 391–400, 2020.
https://doi.org/10.1007/978-3-030-34080-3_44

the first widely available IoT devices. Connected via WiFi or cellular networks, each handset use internal sensors to provide relevant data to the user. From GPS coordinates for navigation to motion sensors for screen orientation or fitness tracking, smartphones have been acquiring data via Cloud platforms for quite a long time.

The main objective of this paper is to present an in depth analysis of the IoT development environments. We aim at presenting Android Things as a developing environment for IoT applications. We also compare Android Things with Windows 10 IoT Core and juxtapose their features, performances and constraints.

The IoT system Architecture and application development environment is presented in the next Section. Comparative analysis of the two recent IoT development that is Android Things and Windows 10 IoT core is presented next in Sect. 3. Review of related work is presented in the next section. Next section contains the conclusion of the paper.

2   IoT Architecture and Development Platform

2.1   IoT Architecture

An IoT Application has an underlying architecture. Even though there is no generic reference architecture available for IoT application development, following components are part of almost every IoT application as shown in Fig. 1.

Things: It represents an entity on embedded system with a unique identifier and the ability to exchange data over a network. It can contain operating system, communication protocols and modules from analogue and signal communication. They are generally capable of network connectivity and power management.

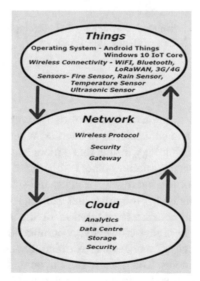

Fig. 1. Simulation of an IoT architecture

Network: It is the communication infrastructure with an operating system, communication and security protocols. A gateway with several modules are a part of this infrastructure.

Cloud: Cloud platform is crafted to contain and process data related to IoT. It is built to deal with massive volumes of data generated by devices, websites, sensors, applications, customers and partners to manage actions for real-time feedback.

2.2 IoT Development Platform

In order to develop several IoT applications, following mentioned boards are popularly used:

Raspberry Pi: Raspberry Pi is used to develop IoT applications and is useful for the processing the algorithm or data at the board. An operating system called Raspbian is used by encoding it with C++, C or Java codes. Raspberry Pi is comparatively costlier and hence, not feasible enough to be used in small-scale projects or projects that do not require processing at IoT board.

Arduino: Arduino is used to develop projects that do not need processing power at IoT board. It is embedded with a combination of C, C++ and Java programs and is further deployed to develop several IoT applications. It is cheaper and hence, easily accessible for small-scale projects.

Omega 2: Omega 2 supports programming languages such as C++, C, Python, NodeJS, Ruby, Rust, PHP, Pearl and GoLang. It have on-board USB-to-Serial chip that permits serial connectivity to its command line terminal via the Micro-USB port. This serial terminal is implemented to access the boot-loader.

There are many constraints in developing an IoT Application using traditional method [4, 5] such as:

- Limited Application Program Interface (APIs) are available. Third party applications needed to be developed for accomplishing simple tasks like connecting via cellular networks or mobile phones.
- It does not integrate with the system and thus demands a separate User Interface to be developed.
- Connection to cloud is quite tedious.

There are several IoT development boards that support Android but they need driver programs to operate. Third party intervention is required due to absence of direct APIs. An analysis of some such popular boards that supports Android is presented in Table 1.

Table 1. Analysis of boards which supports Android Operating System

Name	Features	Strengths	Limitations
Banana Pi	• Allwinner A2 SoC, • 1 GB DDR3 SDRAM • Dual core upto 1.2 GHz AXP209 PMU	• Open source single board computer • Can run Android 4.2, Ubuntu, Debian, Fedora etc.	• Absence of community support for prototyping tools
Beagle board	• 512 MB integrated DDRAM • Integrated power management • 2 × 32-bit 200-MHz programmable real-time units • ARM cortex-M3	• Supports different operating systems like Linux, Symbian, Minix, OpenBSD, FreeBSD, RISC OS and others • Low cost Linux computer with tremendous expansibility	• Cannot be used during complex multimedia and Linux based projects
Odroid Xu4	• Exynos 5422 Octa big.LITTLEARM • Cortex-A15 @ 2.0 GHz quad-core and Cortex-A7 quad-core CPUs	• Can be used in desktop PCs and as micro-controllers • Amazing data transfer speeds	• Requires purchasing of expensive power adapter

2.2.1   Android Things

The reasons why users consider it to be far a more convenient and preferable development platform are cited below:

- The User Interface and back-end are developed together like any mobile application or Website.
- Data communication is easier and convenient as compared to other development environments.
- It provides over the air updates which make it easier for developers to push updates.
- Supports a rich set of APIs.
- It provides a complete product with back end and user Interface.
- It is easy and convenient to connect Android Things Application to cloud.

2.2.2   Windows 10 IoT Core

Windows 10 IoT Core is a version of Windows 10 optimized for smaller devices having or not having a display and runs on both ARM and x86/x64 devices. The Windows IoT Core documentation provides details on connecting, updating, managing and securing devices. It provides a unique full-screen universal windows app for the users. The system displays the interface of the unique full-screen universal windows app at a time. It is the version of the Windows 10 OS version of Microsoft that has been optimized for diminutive devices. The primary features of Windows 10 IoT Core include:

- Booting the device through UEFI, required feature in all images,
- Adds Unified Write Filter (UWF) to protect physical storage media from data writes.
- Provides a means of setting up the device's WiFi connection if no other WiFi profile was configured. It places the WiFi adapter into a Soft-AP mode so that phones or other devices can connect to it.
- Prevents the device from going to sleep due to inactivity.

3   Comparison Between Android Things and Windows 10 IoT Core

We present below a head to head comparison between the two most modern IoT application development environment that is Android Things and Windows 10 IoT core.

3.1   Development Platforms

Android Things applications are developed on Android Studio where as Windows 10 IoT core applications are developed in Microsoft Visual Studio 2017.

3.1.1   Android Studio

Android Studio is the official Integrated Development Environment (IDE) for Google's Android OS [1, 2]. On 16th May, 2013 the Android Studio was launched at the Google I/O conference. In May 2013, it was in early access preview stage beginning from version 0.1. The first stable build was released on December 2014.

PROS:

- Generalized to create any type of program
- Once written, it can run anywhere (uses JVM)
- Has an upgraded standard
- Huge amount of libraries and APIs available

CONS:

- Usually require internet connection to sync projects

3.1.2   Windows 10 IoT Core

Microsoft Visual Studio is an integrated development environment (IDE) from Microsoft Inc. It is used to build computer programs, web apps, websites, web services and mobile apps. Visual Studio uses Microsoft software development platforms. Visual Studio supports 36 different programming languages and allows the debugger and code editor to support (to varying degrees) nearly any sort of programming language, given that a language-specific service exists.

PROS:

- Specially designed for performance, high speed and gaming experiences
- Stable
- Easy Debugging
- Pointers are provided for manual memory management

CONS:

- Prone to memory leaks when pointers are manhandled
- Not specialized for Android
- Relatively small library set

A subtle comparison between both the platforms is demonstrated in Table 2.

Table 2. Comparison between android studio and visual studio 2017

Attribute	Android studio	Visual studio 2017	Remarks
Developer	Google, Jetbrain	Microsoft Incorporation	
Operating system	Microsoft Windows 7/8/10 Mac OS X 10.10 (Yosemite) or higher, up to 10.13 (macOS High Sierra) GNOME or KDE desktop	Windows 10 version 1507 or higher: Education, Professional, Enterprise and Home Datacenter Windows 8.1: Core, Enterprise and Professional and Windows Server 2016: Standard	Android Studio is available for various operating systems whereas Visual Studio is attainable by only Windows OS
Size	500 MB for IDE + 1.5 GB for SDK	500 MB for IDE + 23 GB for SDK	Size of Visual Studio is larger
Language	Java, Kotlin, XML	Visual Basic .NET, CLI/C++, C#, F#, Typescript, JavaScript, XML, CSS, HTML, and XSLT	Visual Studio supports a larger number of languages
License	Freeware + Source Code	Freemium	
Package manager	APK	NuGet Package Manager	

3.2   Supportive Boards

Android Things and Windows 10 IoT core support varieties of development boards. A detailed listing of the supported boards are shown in Tables 3 and 4.

Android Things -

Table 3. Features of android things w.r.t. supporting boards

Name	RAM	Wi-fi	Bluetooth	Ethernet	Board remark
NXP i.MX8M	1 GB/2 GB	Yes	Yes	Yes	Production platform
Qualcomm SDA212	1 GB	Yes	Yes	No	Production platform
Qualcomm SDA624	2 GB	Yes	Yes	No	Production platform
MediaTek MT8516	512 MB	Yes	Yes	Yes	Production platform
NXP Pico i.MX7D	512 MB	Yes	Yes	Yes	Development platform
Raspberry Pi 3 Model B	1 GB	Yes	Yes	Yes	Development platform

Windows 10 IoT Core -

Table 4. Features of Windows 10 IoT Core w.r.t. supporting boards

Board	RAM	Wifi	Bluetooth	Ethernet
AAEON UP SQUARED	2 GB/4 GB/8 GB LPDDR4	No	No	Yes
DRAGONBOARD 410c	1 GB	Yes	Yes	No
MINNOWBOARD TURBOT	2 GB	No	No	Yes (dual)
RASPBERRY PI 2	1 GB	No	No	Yes
RASPBERRY PI 3 MODEL B	1 GB	Yes	Yes	Yes
INTEL JOULE	4 GB	Yes	Yes	No

3.3 Deployment of IoT Applications

Android Things - Programming (Java/Kotlin)/UI Designing (XML) of Android Things is developed in Android Studio. Applications are deployed in hardware by using the IP address of the board and implementing Android Debug Bridge.

Windows 10 IoT Core - Programming (C#)/UI Designing (HTML/CSS) of Windows 10 IoT core is developed in Visual Studio 2017. Applications are deployed in hardware using IP address of the Windows IoT Core device which can be setup using IoT Core dashboard. Debugging is done through Visual Studio Remote debugger. The Windows IoT Portal in Windows 10 IoT Core Platform can be utilized to manage and configure the device.

3.4 Over the Air (OTA) Updates

Android Things - It provides the OTA Update feature usable by the Android Things Console which makes the process of device updation simpler for large scale industrial products. This feature enables to update large number of devices without causing physical disturbance to the system. To push a software update, the appropriate update channel is to be determined. There are four standard channels for each and every product as:

- Canary (canary-channel)
- Beta (beta-channel)
- Development (dev-channel)
- Stable (stable-channel)

The OTA Updating Procedure is shown in Fig. 2.

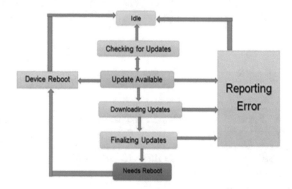

Fig. 2. Over the air update process

Windows 10 IoT Core - OEMs and enterprise customers provide updates for Windows 10 IoT core by Using Microsoft Store, Device Update Center, Azure IoT Device Management or OMA-DM. OEM and enterprise customers using Windows 10 IoT Core can take advantage of device management configuration service providers that allow some control over the process of device updation. Device Management Policy can be set either by using Windows Imaging and Configuration Designer tool or a mobile device management service. Azure IoT Device Management is a highly scalable management solution available on Windows 10 IoT Core.

3.5   Security

IoT devices available in market do not receive software updates and are vulnerable to both cyber breaches, such as encryption, password security and general lack of granular user access permissions.

Android Things - Android Things provides system image, updates and fixes so that the user can focus on creating compelling products. One can push these system updates and app updates to the device using the Android Things Console, enhancing both security and scalability.

Windows 10 IoT Core - Microsoft has introduced strong enterprise grade security features that can be incorporated on smaller, resource constrained class of IoT devices. Given the sensitive nature of such altercations, IoT devices should not behave like "open devices" but "locked-down" ones, similar to mobile phones where access to firmware is usually restricted. The Unified Extensible Firmware Interface (UEFI) standard implements a security feature called Secure Boot that allows a device to only boot trusted software(s). Microsoft featured its enterprise-grade Bit Locker Drive Encryption technology in Windows 10 IoT Core. BitLocker ensures that the stored data on a device remains encrypted, even if the device is tampered with while the OS is not running.

The aforementioned observations are summarized in Table 5.

Table 5. Summarized comparison between android things and windows 10 IoT core

Sl. No.	Windows 10 IoT Core	Android things
1.	Library is small, hence third party libraries are involved	It is generalised to create any type of program
2.	Intel Joule is most preferred in terms of boards and RAM	Raspberry Pi 3 Model B is the most popular board in Android Things
3.	The programming (C#)/UI Designing (HTML/CSS) is done in Visual Studio 2017. We deploy our app in hardware using IP address of the Windows IoT Core device that can be setup using IoT Core dashboard	The programming (Java/Kotlin)/UI Designing (XML) is executed in Android Studio. By implementing Android Debug Bridge, the application is deployed in the hardware
4.	It is an extremely secure operating system, with enterprise grade security features	The system is secure too, but a little less than Windows 10 IoT Core. It provides OTA updates, fixes and system images

4 Related Works

Patil et al. [1] compared Android and Windows based on several attributes such as Developer, OS Family, CPU Architecture supported, Environment (IDE) etc. and explained the results of analysis of various operating systems available which makes it convenient for a manufacturer to choose the most compatible one.

On August 6, 2018, CHI Software [6] came up with a model that enables Touchless Navigation with Android Things. The device senses the user's gestures such as up, down, left and right and then navigates through the content. i..e newsfeed, weather forecast.

Spivey [7] showed a method to build own Security System in Android Things using Raspberry Pi 3 Model B, generic PIR Motion Sensor, USB-A to Micro-USB Cable, Jumper wires and Wia.

On October 16, 2017, Daniele Bonaldo [8] presented an Android Things World Clock using Raspberry Pi 3 Model B, Arduino Nano R3, NeoPixel strip, Google Android Things and Google firebase. The clock allows the user to READ time, control it via nearby API or through voice commands.

5 Conclusion

In this paper, We present a critical analysis of various IoT development platforms. Android Things and Windows 10 IoT Core are compared and contrasted based on various attributes and distinctions. The security, deployment, accessibility, architecture, platforms, updates and pros and cons are discussed in depth thereby fleshing out the details of various aspects associated with both the Developing Platform. This paper will be helpful to the developers and researchers in the field of Internet of Things in selecting the right platform and technology to explore the solutions for encountered problems in IoT. We conclude that it is very hard to predict as both the platforms are aimed at improving the way we look at technology in the near future. While Android Things has an upper hand over the Windows 10 IoT core as Android is a leading and popular operating system with huge number of Android Developers and a lot of APIs and SDK as compared to Windows mobile operating system.

Funding. The study was funded by Kalinga Institute of Industrial Technology, Bhubaneswar, Odisha.

Conflict of Interest. The authors declare that they have no conflict of interest.

References

1. Patil, M.B., Joshi, R.K., Chopade, N.: Comparative study of android and windows operating system
2. Goyal, V., Bhatheja, A., Ahuja, D.: Survey paper on android vs. Ios
3. Mandula, K., et al.: Mobile based home automation using Internet of Things (IoT). In: 2015 International Conference on Control, Instrumentation, Communication and Computational Technologies (ICCICCT). IEEE (2015)
4. Banana Pi Specification. http://www.lemaker.org/product-bananapi-specification.html
5. Banana Raspberry Pi. http://www.myassignmenthelp.net/banana-raspberry-pi
6. Beagle Board. https://beagleboard.org/
7. Comparing Beaglebone Black and Raspberry Pi. https://www.dummies.com/computers/beaglebone/comparing-beaglebone-black-and-raspberry-pi/
8. Google Android Things. https://developer.android.com/things/get-started/

Build Optimization Using Jenkins

M. N. Rakshith[(✉)] and N. Shivaprasad

Department of E&C, Sri Jayachamarajendra College of Engineering,
JSS Science and Technology University, Mysuru, India
rmn2907@gmail.com, nagshivu@gmail.com

Abstract. With the advances in technology there are various software models and tools which have been developed as a platform for the validation and testing the framework. With increase in these platforms the developers face huge challenges in the process of developing a new software for their specific product thus Continuous Integration (CI) comes into the picture. CI is a practice which improves the efficiency and lessens the work complexity by integrating their work in a baseline frequently. One such tool that is widely being used for such practices is Jenkins. As a client server model, is used to trigger the build whenever a user check-in into the repository. Jenkins allows to perform this implementation with the use of numerous plug-ins. This work mainly aims at building the gap between the systematic literature survey and proposes a method which acts as an optimal way of solution for the reduced stop and wait time involved in the CI, with the view of optimization of build time. This method optimizes the serving time, code quality and code coverage capability for a developer.

Keywords: Continuous integration · Build optimization · Build time · Jenkins

1 Introduction

With an accelerated growth towards the automated world, the demand for a high-quality software from the organization is also beaming high. This leads to increased competition among the software industries to develop features accomplishing efficient resource utilization and deliver projects with reduced build time. One among the tool that allow an industry to achieve is jenkins which is an open source CI tool [5]. Cloud computing with this view has gained a wider interest in the research area for optimal performance of resource allocation. In conventional software development the resources used by the developers are very expensive which in turn increases the gross resource utilized by the R&D (Research and Development) developers. To provide a measuring platform with rapid visualization for maintaining the monitoring system, CI techniques evolved.

2 Concepts on Continuous Integration and Jenkins

2.1 Continuous Integration

With the increase in the software development requirements, the need for developers to develop the codes satisfying all the scenarios plays a vital role. In this process the code

D. J. Hemanth et al. (Eds.): ICICI 2019, LNDECT 38, pp. 401–409, 2020.
https://doi.org/10.1007/978-3-030-34080-3_45

needs to be enhanced and updated periodically [3]. Achieving this flexibility becomes a tedious task when it has to be performed manually, thus CI came into the picture. The ideology behind the CI is to provide developers a platform that can integrate the code into a shared repository such as git hub. The robustness of this ideology are monitored by providing check ins as and when the code is updated, which is verified by an automatic build process. Using this developer can identify the errors and debug them quickly [9].

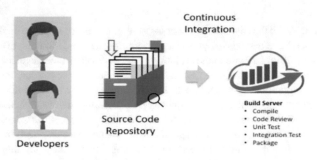

Fig. 1. Continuous integration

Advantages of this CI are

- Providing enhanced communication by increasing the visibility.
- Locate and debug issues in the test bud itself.
- Providing a solid platform.
- Reduces the integration problems which allows developer to deliver software rapidly.

The CI server performs the unit tests and integration tests on the system built and releases deployable artifacts for testing along with assigning a label to each of the updated version as represented in Fig. 1.

2.2   Jenkins

Traceability is the main aspect of developers developing a software which provides flexibility of backtracking to the root cause. Jenkins is a CI tool that allows developers to implement a realistic and complete case study to create an agile environment. This further supports different orchestrate strategies and visualization tools. Plugins are used to perform the CI which acts as a boon for this software. Development life cycle processes of all kinds, including build, document, test, package, stage, deploy, static analysis will be integrated using jenkins.

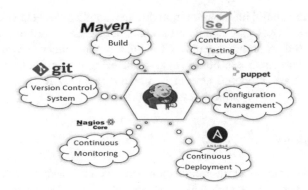

Fig. 2. Jenkins integrating various DevOps stages

Jenkins is mainly chosen for this work because, apart from other competitive tools [11], it supports a number of plugins, integrated to any development and operations (DevOps) stages as shown in Fig. 2, where jenkins is integrated with git, maven, selenium, etc. Some of the plugins that are widely used for this work are

Build Status. This plugin gives an information regarding how many builds are being done at a regular interval of time, be it daily or weekly along with the information about queue and wait time for each build. This enables user to know about the serving time required for each build [13].

Multiple SCM. This provides the user check in and out multiple source control tools that might be one or more repository which overcomes the drawback of SCM that provides only one source control tool at a time [13].

Parameterized Trigger. This plugin provides users to have their input as a variable and use it in run time. This is the most used plugin in dynamic environments where users have lots of options and user-defined values to be used in the build which may keep changing [13].

Perforce. This provides a rapid and seamless path to extract projects into pipeline stages within Jenkins from a perforce plugin. In this context perforce behaves like a repository. This plugin helps to manage the user's CI processes improvising the quality and traceability [13].

2.3 Literature Survey

CI has become a DevOps practice used by different communities as CI allows all the team to deploy rapidly changing hardware and software resources [16], related works on our research has been previously done using CI jenkins, build optimization as an important part of CI. In [3], regarding the build, the projects that use CI are more effective at merging requests, they researched their work on several GitHub projects and provided answers for CI usage, costs and benefits. Performance analysis of CI builds for GitHub projects were carried out specifically focusing on JAVA programming language [14]. In [7], the usage of CI has been researched and have acknowledged that, build information from different sources can help developers. In [8, 14], the CI usage in software quality information and scalable test execution using CI has been discussed. A primitive part of CI is built system, either way researchers tried to improve the performance of builds [6] and enhancing the dependency retrieval [10]. So, researchers have proposed several optimizing methods for builds [4] and also discussed efficient way of running test cases [1].

3   Methodology

This section briefs out about the formulation and proposed design for the optimization of build time. From our research build time can be optimized using full build, parallel build and incremental builds. We discuss the need for CI and how to enhance the build quality and optimize build time using open source tool Jenkins and by making efficient use of available Jenkins plugins. The design formulated in the Fig. 4 gives an overview of the methodology involved in performing the automated continuous integration for the slave node.

3.1   Jenkins Architecture

Jenkins supports master-slave architecture, which is also called Jenkins Distributed Builds. This provides user flexibility of running various environments like LINUX, Windows, Ubuntu, etc., also these distributable builds are capable of running same test cases in parallel on numerous environments. Such a build helps to improve the distributed approach and achieve the desired results instantly. The master node would be responsible for scheduling the jobs, monitoring the slaves, dispatching builds to the slaves for execution [12]. List of slaves configured are shown in Fig. 3.

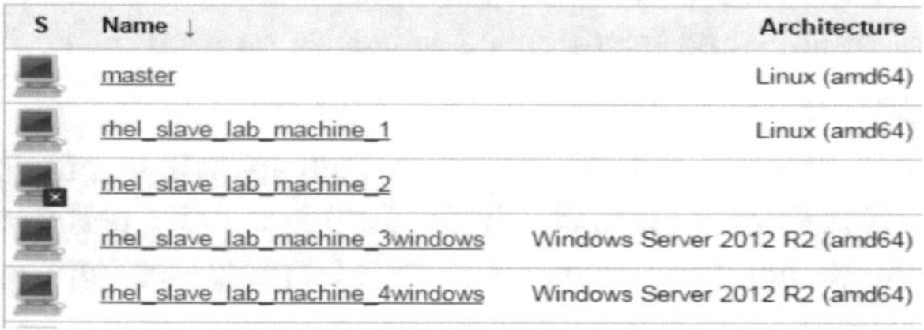

S	Name ↓	Architecture
	master	Linux (amd64)
	rhel_slave_lab_machine_1	Linux (amd64)
	rhel_slave_lab_machine_2	
	rhel_slave_lab_machine_3windows	Windows Server 2012 R2 (amd64)
	rhel_slave_lab_machine_4windows	Windows Server 2012 R2 (amd64)

Fig. 3. Supervising slave nodes

The slave node monitored by master node will be responsible for the execution of parallel builds. The user can configure the master to run a particular build on any slave or can schedule a build for a particular time. Each slave as a number of executors where one slave can run multiple jobs in it.

3.2   Implementation of Jenkins Jobs

Before jumping into the main approach, let us see the initial steps required for setting up a jenkins job, jenkins server will be installed on master node with the use of manage node, different OS slaves will be created based on platforms, one slave agent can run multiple tasks with the help of multiple executors plugin, when all the slave agents are up and running, master will choose a slave agent where a job can run in an OS definitive platform. Our approach is using maven project, which deals with POM (Project Object Model) file which drastically reduces the configurations. To manage source codes jenkins supports Git, Perforce, Mercurial, Multiple SCM's (source code management) etc. Jenkins allows user to pass parameter to build using parametrized build plugin.

Build steps are included in the scripts, since it is a maven build mentioning the POM file, goals like clean and install is the key part for build. Based on the obtained results from the executed jobs, concerned teams are notified and if the build fails particular person who broke the build will be notified. Above procedure are successfully executed with the help of our scripts, configuration files and with the efficient use of available jenkin plugins. Advancing to the optimization part, this process can be completely automated, but our main focus is to approach a way to reduce the build time.

Algorithm 1 Build Script for optimization
___
Input: Repository URL, POM File
Output: Build Result
  1: **if** (Code check-in happens) **then**
  2:      Jenkins server selects a slave platform
  3:      Notify changes from checked in repository
  4: **end if**
  5: **if** (Repo server not running on slave) **then**
  6:      Install required modules on slave machine
  7: **else**
  8:      Workflow Start
  9: **end if**
10: Build and Test the source code
11: **if** (Test cases Fail) **then**
12:      Fix Bug and Build again
13: **else**
14:      Generate report
15: **end if**
16: Email notifications to concerned teams for failed tests.
17: Build will be terminated

As shown in Fig. 4, whenever build is triggered to run the test cases, we need an environment to test these things. From the POM file the required files and software's are copied to the slave agent and run the test cases, this operation is full build, where every time a new job is created full build happens, so whenever a developer check-in build will be triggered. Let us take a use case whenever a build is triggered a clean install happens in other words full build, whenever there is full build it may take more than 3 h to build. To make it efficient the approach is to use incremental build. So, whenever there is a check-in in the same module, we can keep a check, so we can build only that module in other words incrementally building jenkins tasks, the algorithm 1 for build optimization is shown below.

Detailed flow of running jenkins tasks is shown and steps to it are listed below.

1. Initially the process begins with code check-in, if there is no code check-in, tasks will be in the hold state.
2. After successful code check in, Jenkins server selects the appropriate slave platform for which the workflow could be started along with the note on efficient utilization of resource.

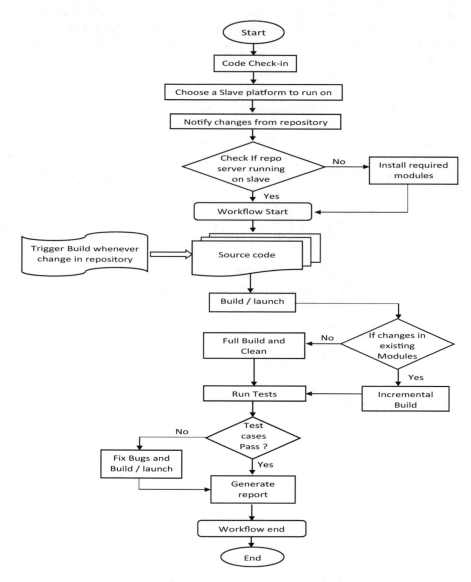

Fig. 4. Proposed methodology

3. It is very important to update the changes in the repository regarding the selected slave and the code check-in status.
4. Check whether the repository sever is running on slave platform, if installed execute workflow or else install the repo server and required modules and then proceed to the workflow.
5. Build/Launch and run the Test source code on the slave machine.
6. If the test cases fail, fix the bugs and again reinitiate the build, if the build is successful generate a report.
7. The report consists of all the information about failed and passed test cases, who initiated the build, who broke the build.
8. So, if there is a code check-in in the same module, first the system checks whether there are any required modules to be installed in the system, if not incremental build will be initiated, it is at this stage the build is being optimized with the help our scripts.
9. Step 5, 6 &7 will be repeated again.
10. Terminate the workflow and the build on successful completion of the pipeline.

4 Results and Discussions

The outcome of build optimization and the time expended for list of jobs are shown in the graph, in Fig. 5. From the graph we can get to know about the time duration of number of builds, total build time and different states of builds, yellow coloured part represents the unstable build and the red coloured part depicts the failed scenarios whereas the black coloured line displays total collective build time on each time range. Logs will be available in the build console and a detailed description of all the builds or a particular build will be available. From Fig. 5, in build No. 30, Full build takes around 2 h to build, progressively the build time is getting reduced, these shows our methodology is fully functional. For example, in Fig. 6, build No. 37 took approximately 2 h to build, while build No. 42 which took approximately 1 h, this is due to incremental build of certain modules. So, whenever there is a check-in, perform full build and if there are only few changes in the builded module incremental build is a paramount option.

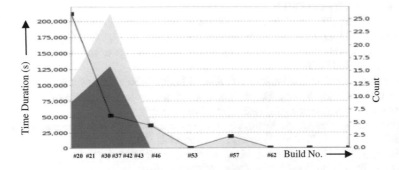

Fig. 5. Build time trend

Also, parallelly building jobs in slaves will save time. There are number of CI tools in the market, since jenkins is an open source tool and availability of efficient plugins makes it as a cost-efficient solution to all CI related problems.

Status	Job name	#	Duration
Unstables	E2E-RunPublicVsims2_11	#43	1 hr 42 min
Unstables	E2E-RunPublicVsims2_11	#42	1 hr 13 min
Unstables	E2E-RunPublicVsims2_11	#37	2 hr 3 min
Unstables	E2E-RunPublicVsims2_11	#36	6 min 40 sec
Unstables	E2E-RunPublicVsims2_11	#31	2 hr 41 min
Unstables	E2E-RunPublicVsims2_11	#30	2 hr 54 min
Unstables	E2E-RunPublicVsims2_11	#21	5 min 42 sec
Unstables	E2E-RunPublicVsims2_11	#20	5 min 40 sec

Fig. 6. Build duration

5   Conclusion and Future Scope

CI is the imperative part for any Software industry. Integration of tasks on a regular basis will optimize developers time, CI helps developers to focus more on key issues and better software quality. Jenkins afford a superior result for these tasks and it's a chosen one for CI automation. These paper gives an idea about how build time is optimized using Jenkins. The detailed steps and procedures for implementing are discussed and an overview of plugins used for Jenkins are discussed in detail in this paper. Finally, these tasks can be optimized much more in the future, allowing to build/rebuild using any repository and test whenever there is a check-in, the entire process can be automated in future.

References

1. Askarunisa, K., Punitha, A.J., Abirami, A.M.: Black box test case prioritization techniques for semantic based composite web services using OWL-S. In: Proceedings of the IEEE-International Conference on Recent Trends in Information Technology, ICRTIT (2011)
2. Fitzgerald, B., Stol, K.-J.: Continuous software engineering: a roadmap and agenda. J. Syst. Softw. **123**, 176–189 (2017)
3. Vasilescu, B., Yu, Y., Wang, H., Devanbu, P., Filkov, V.: Quality and productivity outcomes relating to continuous integration in GitHub. In: FSE (2015)
4. Celik, A., Knaust, A., Milicevic, A., Gligoric, M.: Build system with lazy retrieval for Java projects. In: FSE (2016)
5. Smart, J.F.: Jenkins: The Definitive Guide. O'Really Media, Sebastopol (2011)
6. Laukkanen, E., Paasivaara, M., Arvonen, T.: Stakeholder perceptions of the adoption of continuous integration: a case study. In: AGILE (2015)
7. Beller, M., Gousios, G., Zaidman, A.: Oops, my tests broke the build: an analysis of travis ci builds with github. Technical report, PeerJ Preprints (2016)
8. Brandtner, M., Giger, E., Gall, H.C.: Supporting continuous integration by mashing-up software quality information. In: CSMR-WCRE (2014)

 9. Leppanen, M., Makinen, S., Pagels, M., Eloranta, V.-P., Itkonen, J., Mantyla, M.V., Mannisto, T.: The highways and country roads to continuous deployment. IEEE Softw. **32** (2), 64–72 (2015)
10. Miller, A.: A hundred days of continuous integration. In: AGILE (2008)
11. Seth, N., Khare, R.: ACI (automated continuous integration) using jenkins: key for successful embedded software development. In: 2015 2nd International Conference on Recent Advances in Engineering & Computational Sciences (RAECS), pp. 1–6 (2015)
12. Online Resource. https://jenkins.io/doc/
13. Online Resource. https://wiki.jenkins.io/display/JENKINS/Plugins
14. Gopularam, P., Yogeesha, C.B., Periasamy, P.: Highly scalable model for tests execution in cloud environments. In: 18th Annual International Conference on Advanced Computing and Communications (ADCOM) (2012)
15. Erdweg, S., Lichter, M., Weiel, M.: A sound and optimal incremental build system with dynamic dependencies. In: OOPSLA (2015)
16. Sampedro, Z., Holt, A., Hauser, T.: Continuous integration and delivery for HPC. In: PEARC 2018, 22–26 July 2018, Pittsburgh, PA, USA (2018)

A Detailed Analysis of Data Security Issues in Android Devices

Ditipriya Dutta, Shubham Agarwal, Ria Dash,
and Bhaswati Sahoo$^{(\boxtimes)}$

School of Computer Engineering, KIIT - Deemed to be University,
Bhubaneswar, India
ditipriyadutta0705@gmail.com,
shubham.agarwal7930@gmail.com, dashria1997@gmail.com,
bhaswati.sahoofcs@kiit.ac.in

Abstract. People believe that any sufficiently advanced technology is indistinguishable from magic. The ability to analyze and act on data is increasingly gaining importance in this ever-expanding world. Data forms an integral part of the identity of an individual or an organization. However, as data is getting digitized and more information is being shared online via various social media platforms and apps that are specifically designed to cater to the needs of users, data privacy is becoming a matter of extreme concern. Protecting data is of paramount importance because data in the hands of nefarious individuals or organizations can pose a massive threat to both the digital and real worlds alike. Data protection is not just about protecting one's personal data, it also includes the protection of fundamental rights and freedom of the people associated with it. Every organization is in charge of personal data associated with either the organization itself or to the clients they serve. Protecting the identity of their customers is a very huge task in itself. Illegal extraction of private and personal data by malpractices is taking place every other second and as digitally-enhanced citizens, recognizing the deliberate attempts at data phishing should be our key concern. Personal data can be used to influence and control our decisions and somewhat mould our behaviour and hence, we, as an individual, have a lot at stake when it comes to protecting confidential information. In this paper, we will discuss the various data security issues in Android devices, until the launch of Android Nougat 7.0, and how data extraction without prior information is a threat to the security, integrity, and confidentiality of the users.

Keywords: Android · APK decompiling · Data breach · GPS · Media files · Mobile devices · Permissions · Phishing · Privacy · Security

1 Introduction

To quote Douglas Adams, "First we thought the PC was a calculator. Then we found out how to turn numbers into letters with ASCII—and we thought it was a typewriter. Then we discovered graphics, and we thought it was a television. With the World Wide Web, we've realized it's a brochure." In the 21st century, technology has become ever-invasive, with the challenges in design getting even more discernible. Throughout

© Springer Nature Switzerland AG 2020
D. J. Hemanth et al. (Eds.): ICICI 2019, LNDECT 38, pp. 410–417, 2020.
https://doi.org/10.1007/978-3-030-34080-3_46

history technology has been the impetus for change. It has enabled us as humans to make strides our ancestors could only dream of. Digital revolution has redesigned the conceptions of time and distance. It has generated a wealth of data that is available at the stroke of a key. Technology is the most substantial need in today's world where people need almost everything at their doorstep alongside strengthening human interaction across the globe. Every other person next door possesses not one but multiple smart-phones, and each smart-phone provides a feature to install applications that make life potentially unproblematic. But in this prosaic life, people have become so much engrossed that they frequently forget about being vigilant regarding most of the applications that are in use, thereby compromising with their own security. This paper talks about the various incidents related to data breach in Android devices that have taken place in the recent past (uptil Android Nougat 7.0), how all of them were detected and how data was eventually leaked. Personalized devices contain a lot of confidential information, be it one's phone number, other people's phone numbers, private photos, videos, messages, email addresses, residential addresses and even bank details. Once installed, some of these applications start covertly accessing and extricating personal information from the listed devices. The most disturbing part is that the apps won't work unless you authorize them to access certain basic features of your smart-phone. Thus, this paper is going to elaborate on how these applications manage to access people's data in the first place. Section 1 deals with the Introduction, Sect. 2 talks about the Related Works, citing apps which have been accused of data breach, Sect. 3 discusses the various permissions required by android apps to gather data, Sect. 4 talks about how data is congregated via Android apps, Sect. 5 discusses Android APK decompiling and the paper is concluded in Sect. 6.

2 Related Works

2.1 Truecaller

Truecaller works on the basis of give-and-take policy. If one wants to know about the unknown calls that one gets, one has to part with his own phone-book details. The application, after being installed, demands for permission to access the phone's contact list. After the access is granted, the information is publicly reinforced by a massive number of clients who have downloaded the application on their advanced cellphones. As a component of the end client assertion, the application requests that the client enable access to his address book/contacts. This information is then transferred by the application to the organization's servers. In the wake of experiencing a few information coordinating/refining calculations and algorithms, this information is made accessible to all clients to seek upon. More or less, all your contacts including telephone numbers, contact names and email IDs will be transferred to a protected Truecaller server. These are categorized into records and each number will dole out a Truecaller ID. When somebody looks through the number utilizing Truecaller number inquiry, the guest ID related with that number will be displayed [5].

2.2   Facebook

The Facebook – Cambridge Analytica data scam was a major governmental shakedown in mid-2018 when it was revealed that Cambridge Analytica had reaped individual information off numerous profiles without the users' consent and utilized it for political impetus. It was a watershed minute for the mass. The act of such information being stolen had hastened a massive drop in Facebook's stock prices which called for uncompromising direction of information use by various technical organisations. The unlawful gathering of personal data by Cambridge Analytica was first reported in December 2015 by Harry Davies, a reviewer for The Guardian. He detailed that Cambridge Analytica worked for Ted Cruz, the Senator of United States, by using information gathered from a large number of individuals' Facebook accounts without their authorization [4].

2.3   Google Timeline

Google has added a new feature which entails it to track people's location, by default, even after the GPS location tracker is turned off. Earlier, they claimed that turning off the location tracker would no longer allow the application to accurately keep a track of all the places that the user visits. However, recently it was discovered that Google keeps a track of a user's location via Google maps and location history but just doesn't add it to the Google maps timeline [6]. This signifies that Google Timeline has a very meticulous idea of an individual's location. For instance, if you visit a building which has two restaurants in the same floor, Google Timeline would know which restaurant you are in and would keep an accurate record of it [5] (Fig. 1).

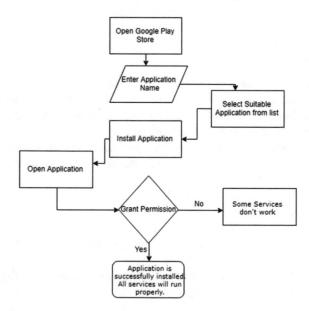

Fig. 1. The step-by-step procedure to install applications in an Android device from Google Playstore.

3 Android Permissions

The principal inducement of permissions is to ensure and protect the privacy of an Android client. Android applications must insist authorization to access classified client data (such as message inbox and contacts), as well as few framework highlights (such as GPS and camera). Android permissions are primarily divided into two sections:

3.1 Normal Permissions

If an app enlists Normal Permissions in its token(s), it speaks about authorizations that do not present much peril to the user's security or the device's activities. In that case, the system automatically concedes those permissions to the applications.

3.2 Dangerous Permissions

In contrast to typical authorizations, i.e., Normal Permissions, if an application documents precarious consents in its manifest, it would require endorsement of the client. Hazardous consents are those which can possibly influence the client's protection or the gadget's working. In such cases, the client needs to unequivocally agree to acknowledge those permissions. Presently the manner in which the framework asks about these authorizations rely upon the performance of the framework [1]. For instance, in android 5.1.1 versions and below, the system requests the user to grant every possible dangerous permission during installation. However, in versions 6.0 and above, the system does no such thing during installation. It asks the user to grant the permission during run-time.

3.2.1 GPS Permission

Few applications track our location after gaining access to the devices' GPS (Global Positioning System) trackers. These applications access the GPS tracker in the background and keep track of all the places that the phone has been. This is done once the user grants the permission to the app for accessing the GPS. However, sometimes, the GPS is accessed without the knowledge of the users.

3.2.2 Contact Permission

Android applications are also capable of accessing phone-book details for certain functionaries. Applications like Truecaller, Ola and other such names access the phone-book details because the functionality of the applications depend on it. However, sometimes, while accessing such data, the server extracts other personal and confidential information along with it, for instance, the application may send the contact book to another server without prior knowledge of the user.

3.2.3 Gallery Permission

One must have noticed that some applications like effect cameras and photo-editing apps ask for permission to access your gallery or camera. Every user's gallery contains classified information and media files. There have been instances, where certain applications which were granted permission to access the gallery, also extracted personal photos and videos of the user.

4 Transmission of Data via Android Apps

The Android platform has been deployed over a wide range of electronic gadgets, especially, mobile phones, conveying sensitive or non-sensitive information to a different set of devices, irrespective of manufacturer and carrier. Modern digital privacy measures separate accumulation and examination, with collection of data ideally only happening once and the consequent analysis depends upon legitimate collection [3].

4.1 Permission to Access Inbuilt Features

When a user installs an application on an Android device, the app asks for permissions like accessing the GPS location, Contacts, Gallery, message inbox, etc. The app-builders provide a "Terms and Conditions" section, usually in Google Playstore, that contain details of permissions and data used by the application. Since they know that maximum users do not read all the clauses before agreeing to the Terms and Conditions, they grant every permission unknowingly and without reading them thoroughly as they do not know about the consequences that would follow. This negligence on part of the user and smart strategy and malicious intent of the app-builder further leads to data leakage and poses a threat to his security.

4.2 Gathering Private Data

After the user grants the permission to an application to access some of his private data, the latter can start accessing sensitive information in the background without the knowledge of its user. The application can then send the data to the server without the approval and knowledge of the user. This data is then either sold or used by companies or techies with vindictive intent to provide services or breach the privacy of the user.

4.3 Breach of Personal Data Through SMS

Certain apps ask you for your mobile number to generate a One Time Password (OTP). These apps send you their OTPs for verification via SMS. The application then devours the SMS from the phone and displays the OTP automatically in the screen. In order to do so, the app seeks our permission to read our messages, which we allow without knowing the consequences. But there might be situations where an app will access our SMS in the background without letting us know. It might so happen that when crucial private information is exchanged via SMS, for instance, bank balance, card details and social media passwords, the app will be able to transmit and store such information in its own server as well. Money can be laundered from the respective bank accounts of the users, huge scams will eventually be probable and security and integrity of the user will be compromised.

4.4 Tracking User's Location Without Consent

Many Android applications provide location based services, owing to which, the application requests for the permission of using Location. After granting the permission, the application can access the Live location of the user. This may be used by certain companies for providing the services like cabs, courier delivery, etc. but on the contrary, the same can be used to monitor and track the location activities of the user. This may intimidate the user's private life or trigger any mis-happening. Reduced security is the primary downside of cellphone tracking. Location statistics collected by various firms can provide a phenomenally intrusive glimpse into the private lives of the cellphone users. However, tracking information remains accessible for phone providers and extremity services [2].

4.5 Collecting Data via Keylogger or Keyboard Capturing

A keylogger (keystroke logging) is a reconnaissance programming that once introduced on a framework, has the capacity to record every keystroke made on that framework. The record is spared in a log document, usually encrypted. This log may contain sensitive information like password or pin and can be sent to the company that manufactures the application, thereby threatening the user's security again. The action of recording the keys struck on the keypad console is done regularly in a clandestine manner, with the goal that the users dealing with the console are unaware of their tasks being monitored. Data will then be recovered by the individual who is working on the logging program [7].

4.6 Accessing Personal Media Files

A lot of applications demand an access of our media files and we allow them to. These apps get all our photos, videos and documents and uploads them to their respective databases. For example, if user X allows app Y to access its gallery, app Y can transfer all of X's photos from his phone to its own. The same goes for important files and documents, too. All of this happens without prior information to the user.

4.7 Using User's Search History in One App to Display Advertisements in the Other

Let us assume a situation where a user views certain products of his choice on an e-commerce website. When he exits from the application and opens another one, he might as well see advertisements that show you similar products or the same ones that he viewed some time back. This signifies that the company backs your search history data up and sells it other such companies which, in turn, display advertisements containing products you viewed in the former application. Thus, the company uses the browsing history of the user to create a log out of it for showing advertisements and endorsing its own products. Companies like Google AdSense are the ones that permit distributors in the Google Network of Content Sites to serve programmed content, video, pictures or interactive media advertisements to target both the crowd and the site content [9].

5 Android APK Decompiling

The source code of any Android application forms the most vital part of app designing. Android APK decompiling is a reverse-engineering tool that enables individuals to decode resource files and monitor and analyse the code accordingly. Once the source code is decompiled, it cannot be recompiled to its original form. This tool is extremely hazardous for amateurs, beginners or students who design various applications without proper security measures/tools, which industrial designers are usually furnished with. This turns out to be one of the most critical issues in the arena of Android App Development [8]. The server credentials and inbuilt algorithms of most apps are under grievous threat.

6 Results and Conclusion

The security of an individual is solely dependent on the user's discretion. Some applications try to seek our permission to gain access to our private data. It is completely on us as to what information we choose to provide or withhold. But even the largest IT companies of the world use our data and instrumentalize the information that we provide to them, to enhance their own services. It is time for android users to start taking app permissions more seriously. Thus, when a user installs and uses such apps and the apps seek permission to access our phone book, GPS location, gallery, camera and so on, he must be careful in identifying the data that can be used against him in the near future. There have been multiple instances in which app permissions have been misused by application developers and an individual's privacy has been compromised. To create transparency, Android Operating Systems must generate and display notifications to the user while the app is reading or accessing sensitive information. As mentioned earlier, some applications usually send an OTP to the user and the user has the option of copying the OTP directly to the app portal. Now, when the app should be reading the SMS with the OTP in it, the user must receive a notification that the app is trying to read data from the device. This way the user will know as to which application is trying to access which particular feature on a smart phone. The Android OS in particular should create a more secure and compatible architecture of APIs that enable the permission of using features like location, camera, contacts, and so on. In case a third-party application wants to access user data, it must first get itself listed at the portal the OS would provide. The app will then get a key to be able to access user records. The OS and its associated algorithms must analyse and decide whether the application is secure enough. Besides, the User Agreement expects the individuals to read each and every clause before agreeing to the Terms and Conditions, but the length of it does not comply with the user. Hence, shorter versions the User Agreement must be displayed during run-time for the user to have a brief idea about the kind of user information an application needs to function properly. The safety of the user is in his own hands, and he can always be selective towards providing personal information to any application.

Funding. The study was funded by Kalinga Institute of Industrial Technology, Bhubaneswar, Odisha.

Conflict of Interest. The authors declare that they have no conflict of interest.

References

1. Shabtai, A., Fledel, Y., Kanonov, U., Elovici, Y., Dolev, S., Glezer, C.: Google android: a comprehensive security assessment. IEEE Secur. Priv. **2**, 35–44 (2010)
2. Gibler, C., et al.: AndroidLeaks: automatically detecting potential privacy leaks in android applications on a large scale. In: International Conference on Trust and Trustworthy Computing. Springer, Heidelberg (2012)
3. Vidas, T., Zhang, C., Christin, N.: Toward a general collection methodology for Android devices. Digit. Invest. **8**, S14–S24 (2011)
4. Davies, H.: Ted Cruz campaign using firm that harvested data on millions of unwitting Facebook users. The Guardian, 11 December 2015. Archived from the original on 16 February 2016
5. http://www.techgide.com/truecaller-works-truecaller-app-works-truecaller-get-data/
6. https://www.ethow.com/2015/10/how-Truecaller-works-is-what-you-might-not-like.html
7. https://www.webopedia.com/TERM/K/keylogger.html
8. https://www.apkdecompilers.com/
9. https://www.greenbot.com/article/2954979/google-apps/what-you-need-to-know-about-your-location-history-timeline.html

Enhanced Routing Protocol for Internet of Things Using Dynamic Minimum Rank Hysteresis Objective Function

C. Tamizhselvan[✉] and V. Vijayalakshmi[✉]

Department of Electronics and Communication Engineering,
Pondicherry Engineering College, Puducherry, India
{cptamilselvan,vvijizai}@pec.edu

Abstract. Internet of Things (IoT) is a prominent development in wireless communication technology that encompasses different networks and applications to provide seamless service for end-user requests. The heterogeneous combination of the devices in the network requires optimal neighbor selection for request processing and data management. In this paper Dynamic Minimum Rank Hysteresis Objective Function (DMRHOF) is proposed to enhance the routing reliability between the devices in the IoT environment. This function relies on Rank, Received Signal Strength and Expected Transmission Count metrics of the devices for identifying the reliable routes to the service provider. This objective function is unanimous and operates over all the devices throughout the routing path. Neighbor selection relies on decision-making process of the objective function that is executed in a step-by-step manner for mitigating routing issues. This objective function-based routing method improves network packet delivery ratio and retains convergence time with less delay.

Keywords: ETX · IoT · Objective function · RSS · Weight analysis

1 Introduction

Internet of Things (IoT) is a collection of different intelligent communicating devices that employ heterogeneous wireless technologies for providing un-interrupted services for end-users. The concept of IoT is to visualize information from diverse environments in digital manner for network level assimilation [1]. Starting from short range communication standards to the advanced long-term evolution standards, the end-user is provisioned with different services in a pervasive manner. User request for access, retrieval and information sharing is granted from different distributed resources that are interconnected through wireless mediums. This development in communication technology grants optimized access and in-time response for the increasing user density and request processing rate [2, 3].

IoT is a combination of both infrastructure and infrastructure-less service architectures that are coherent with multiple sophisticated features of the users. The features of the users such as mobility, interoperability, cooperative sharing, scalability, etc. are satisfied by the IoT devices using appropriate routing procedures. The devices for communication purpose, discover multiple links to reach the service provider (such as

© Springer Nature Switzerland AG 2020
D. J. Hemanth et al. (Eds.): ICICI 2019, LNDECT 38, pp. 418–424, 2020.
https://doi.org/10.1007/978-3-030-34080-3_47

cloud) [4]. But the necessity is the device should discover the optimized route. In this process, the routes are discovered using intermediate gateways and other service providers. Routing made should be precise in determining the reliability of the access and service response for the devices. Routing protocols designed for infrastructure and infrastructure less networks cannot be directly exploited for IoT. The induced protocols and neighbor selection process is amended as per the requirements of the end-user and the type of communication devices. In particular, the routing process needs to satisfy the requirements from user, device and service provider perspective [5, 6].

Depth-First forwarding (DFF) is designed by Yi et al. [7] for preventing failures in data delivery in IoT. This method exploits time stamp for delay-less data delivery through routing paths formed in a search-based ordering process.

Cognitive machine-to-machine Routing Protocol for Low-Power and Lossy (RPL) Networks is proposed in [8] for constructive routing feasibilities in IoT. This protocol assimilates the layer functions and access controls along with the cognitive radio protocols. These protocols coexist with the RPL for improving network throughput and reducing delay.

Routing by Energy and Link quality (REL) [9] is introduced for detecting and mitigating weak links issues in IoT neighbor selection. This routing process is reliable for low power IoT devices where routing decisions rely on available energy and hop count for evading the lossy links.

A fuzzy based Routing Protocol for Low-Power and Lossy Networks (F-RPL) is introduced in [10] for improving the reliability of neighbor discovery. This routing protocol adapts dynamically to the metrics and functions aimed at the time of neighbor selection. Fuzzy decision-making proceeds over the routing metric for achieving an energy efficient and improved packet delivery solution.

Light-weight On-demand Ad hoc Distance vector routing for next generation (LOADng) is designed in [11] for enhancing the performance of low power IoT networks. This protocol focuses in improving the autonomous and dynamic connections between the devices. Therefore, the delivery ratio is improved with less transmission loss.

Medeiros et al. [12] introduced an integer linear programming (ILP) based multi-objective routing aware of mixed traffic (MAXI) for mesh network that supports IoT. The process of neighbor selection in this method relies on the application demands of the mesh network nodes.

2 Proposed Dynamic Minimum Rank Hysteresis Objective Function (DMRHOF)

The proposed routing method enhances the choice of optimal IoT device selection in a distributed communication environment. The routing method is designed using dynamic objective function that relies on the rank factor. As IoT is a collection ranging from smart sensing to interactive machines, neighbor selection and routing is an inherited property from the wireless networks. Routing and optimization is necessary to improve the quality of service and in-time response for the communicating devices. The process of routing is preceded using different communication metrics such as: rank, received signal strength (RSS) and estimated transmission count (ETX). In a conventional routing process, neighbors are selected by evaluating the hop count between the user and the service provider. The shortest distant neighbors and path are opted for relaying requests. Different from this procedure, the proposed routing methodology accounts the fore-mentioned metrics for neighbor selection and request forwarding. The process of neighbor selection is performed as a step-by-step decision process on the basis of the above metrics. Let us consider an IoT environment consisting of $\{d_1, d_2, \ldots, d_n\} \in D$ devices. The devices that are present within the radio range of the neighbors share a weighted direct link. If $\{l_1, l_2, \ldots, l_n\} \in L$ is the set of links between the devices, then $\{w_1, w_2, \ldots, w_n\} \in W$ denotes the link between them. The weight is computed using Eq. (1) as

$$W(L) = \frac{r_s}{r_h} \tag{1}$$

where, r_s and r_h are the requests serviced and handled using the link L.

An illustration of the IoT environment is represented in Fig. 1.

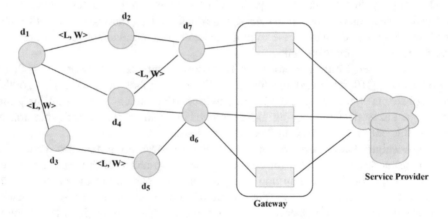

Fig. 1. IoT environment illustration

The communicating user employs a suitable wireless routing protocol for discovering the routes to the service provider gateway. Through proper exchange of control messages, the routes are identified by the user. The routing layer in the user device protocol stack verifies for the weight of the link in each path and the maximum weighted path is esteemed with high rank. The rank R is expressed using Eq. (2)

$$\left. \begin{array}{l} Rank(L) = \arg max\{W(l_1), W(l_2), \ldots, W(l_n)\}, \forall l \in L \\ such\, that,\ W(l_n) = \sum W(L), L\, of\, all\, D \end{array} \right\} \tag{2}$$

The rank estimation is the first step of decision making, wherein, the neighbor with estimated rank is opted for relaying. After the transmission process, the Received Signal Strength (RSS) of the neighbor is verified for further verification. As the communication between the devices occupies wireless medium, it is necessary to estimate the RSS that ensures reliable data transmission. Let *dist* be the physical distance separating the transmission and receiver, and then the RSS is computed using Eq. (3)

$$RSS = 10log_{10}P_t + 10log_{10}\partial - 10log_{10}\left(\frac{dist}{dist^*}\right)^{\beta} + \gamma \tag{3}$$

Here, P_t is the transmit power of the sender, ∂ is the antenna constant, $dist^*$ is the reference distance value, β is the path loss factor and γ is the channel fading factor. The communication reliability between the devices is estimated on the basis of signal strength. The signal strength variation fluctuates r_s and r_h. In the second level of decision making, the RSS is accounted along with the rank where in the devices satisfying both the constraints are selected for data transmission.

Expelling routing paths based on the rank and RSS at the initial state of neighbor discovery is profitable. Contrarily, discarding the path between intermediate nodes in the network degrades the quality of transmission and service response. Therefore, a balanced decision system is required to optimize the rate of request processing in this method. To strengthen the process of optimization, ETX metric between the devices is augmented in determining a reliable neighbor. Equation (4) is used to compute the ETX metric between two devices d_1 and d_2.

$$ETX = \sum_{i=1}^{k} i * r_l^{k-1}(1 - r_l) \tag{4}$$

Where, k and r_l represents the transmission and request loss respectively. In Fig. 2, the process of destination information object (DIO) message validation in accordance with the estimated metrics is illustrated.

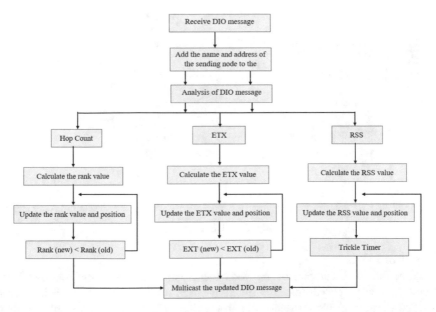

Fig. 2. DIO message validations

The received messages are then validated for the ETX between the devices and then a common reliable factor $rf(D)$ is estimated using Eq. (5).

$$rf(D) = Rank(L) + [(\Delta_1 * RSS) + (\Delta_2 * ETX)] \tag{5}$$

Where, Δ_1 and the Δ_2 are the adjustable values for RSS and ETX respectively and $(\Delta_1 + \Delta_2) = 1$. These adjustable weights are balanced with the $Rank(L)$ for in-range neighbors such that the $\max\{rf(D)\}$ neighbor is selected as the next neighbor of the path. The routing process is then followed by optimizing $Rank(L)$ such that $W(l_n) = \sum W(L)$, L of all D until the service provider is reached.

3   Performance Analysis

The evaluation of the proposed dynamic minimum rank hysteresis objective function (DMRHOF) is assessed using contiki cooja 2.7 open source IoT simulator. The IoT environment is modeled with 60 devices accessing a cloud resource through request messages of size 64-Kb. An average of 1 Mb/s bandwidth is allocated for the accessing wireless channels for the devices in the network. The maximum request that is processed in a unit time is 10. The time-out for the IoT response is 240 ms. The consistency of the proposed method is verified through a comparative analysis with the existing method such as F-RPL [10] and Loadng-IoT [11]. For a comparative analysis, packet delivery ratio (PDR), response delay and convergence time are considered.

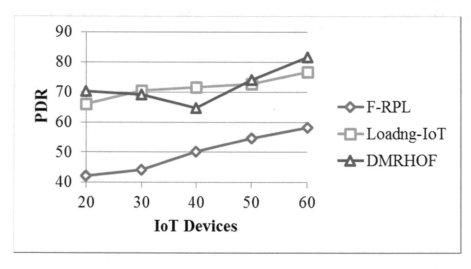

Fig. 3. PDR analyses

Analysis of PDR
In Fig. 3, the PDR comparison between the existing and proposed method is performed. In The proposed objective function-based routing method, the neighbors are classified through a step-by-step decision-making process. The un-reliable neighbors

are expelled during initialization as well as intermediate transmission. This is performed to maximize the rate of request transmission and processing. The rank factor estimated helps to retain the neighbors on the basis of request delivery and response, increasing the PDR of DMRHOF.

Analysis of Response Delay

The communicating user devices select route neighbors by evaluating three different metrics of the intermediate devices in a joint manner. This helps to prevent unnecessary link drop outs and delayed response from the service provider. The path nodes retain uniformity throughout the communication links between the all the path devices using the same objective function. The objective function retains the reliability of the devices by suppressing additional delay in request response (Refer Fig. 4).

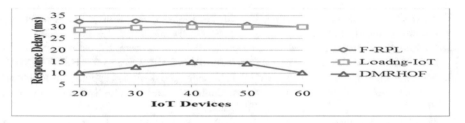

Fig. 4. Response delay analyses

Convergence Time Analysis

The convergence time analysis for the existing and proposed methods is presented in Fig. 5. The convergence time of the proposed objective function is prolonged due to the consideration of three different metrics that are interlinked with each other. The rate of earlier convergence is prevented in this optimization as it is unanimously followed throughout the communication path. This retains the available solution space for the devices in identifying routes consistently.

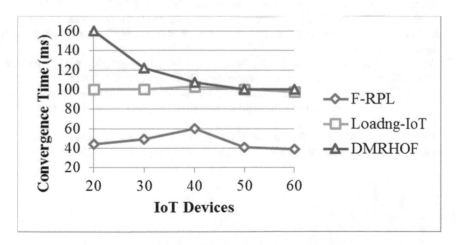

Fig. 5. Analysis of convergence time

4 Conclusion

In this proposal, a novel Dynamic Minimum Rank Hysteresis Objective Function (DMRHOF) is presented to improve the routing decisions in an IoT environment. Routing decisions are facilitated using tri-metric objective function for reliable neighbor discovery mitigating route issues. This process is seamless in a steady state for data management and request processing in the service provider end. Experimental results verify the consistency of the proposed method by improving PDR and convergence time, with less response time.

References

1. Musaddiq, A., Zikria, Y.B., Hahm, O., Yu, H., Bashir, A.K., Kim, S.W.: A survey on resource management in IoT operating systems. IEEE Access **6**, 8459–8482 (2018)
2. Mubeen, S., Nikolaidis, P., Didic, A., Pei-Breivold, H., Sandström, K., Behnam, M.: Delay mitigation in offloaded cloud controllers in industrial IoT. IEEE Access **5**, 4418–4430 (2017)
3. Kumar, R.P., Smys, S.: A novel report on architecture, protocols and applications in Internet of Things (IoT). In: 2018 2nd International Conference on Inventive Systems and Control (ICISC), 19 January 2018, pp. 1156–1161. IEEE (2018)
4. Mubeen, S., Asadollah, S.A., Papadopoulos, A.V., Ashjaei, M., Pei-Breivold, H., Behnam, M.: Management of service level agreements for cloud services in IoT: a systematic mapping study. IEEE Access **6**, 30184–30207 (2017)
5. Kumar, R.P., Smys, S.: Analysis of dynamic topology wireless sensor networks for the Internet of Things (IoT). Int. J. Innov. Eng. Technol. (IJIET) **8**, 35–41 (2017)
6. Al-Fuqaha, A., Guizani, M., Mohammadi, M., Aledhari, M., Ayyash, M.: Internet of things: a survey on enabling technologies, protocols, and applications. IEEE Commun. Surv. Tutor. **17**(4), 2347–2376 (2015)
7. Yi, J., Clausen, T., Herberg, U.: Depth-first forwarding for unreliable networks: extensions and applications. IEEE Internet Things J. **2**(3), 199–209 (2015)
8. Aijaz, A., Aghvami, A.H.: Cognitive machine-to-machine communications for Internet-of-Things: a protocol stack perspective. IEEE Internet Things J. **2**(2), 103–112 (2015)
9. Machado, K., Rosário, D., Cerqueira, E., Loureiro, A., Neto, A., de Souza, J.: A routing protocol based on energy and link quality for internet of things applications. Sensors **13**(2), 1942–1964 (2013)
10. Araújo, H., Rodrigues, J., Rabelo, R., Sousa, N., Sobral, J.: A proposal for IoT dynamic routes selection based on contextual information. Sensors **18**(2), 353 (2018)
11. Sobral, J., Rodrigues, J., Rabelo, R., Saleem, K., Furtado, V.: LOADng-IoT: an enhanced routing protocol for internet of things applications over low power networks. Sensors **19**(1), 150 (2019)
12. Medeiros, V.N., Silvestre, B., Borges, V.C.: Multi-objective routing aware of mixed IoT traffic for low-cost wireless Backhauls. J. Internet Serv. Appl. **10**(1), 9 (2019)

Different Approaches in Sarcasm Detection: A Survey

Rupali Amit Bagate[1,2(✉)] and R. Suguna[1]

[1] Vel Tech Rangarajan Dr. Sagunthala R&D Institute of Science
and Technology, Avadi, Chennai, India
rupali.bagate@gmail.com
[2] Army Institute of Technology, Dighi Hills, Pune, India
drsuguna@veltech.edu.in

Abstract. Sarcasm is an unwelcome impact or a linguistic circumstance to express histrionic and bitterly opinions. In sarcasm single word in a sentence can flip the polarity of positive or negative statement totally. Therefore sarcasm occurs when there is an imbalance between text and context. This paper surveys different approaches and datasets for sarcasm detection. Different approaches surveyed are statistical approach, rule based approach, classification approach and deep learning approach. It also gives insight to different methodologies used in past for sarcasm detection. After surveying we found deep learning is generating a good result as compare to other approaches.

Keywords: Sentiment analysis · Sarcasm detection · Machine learning · Deep learning

1 Introduction

Sentiment analysis is field of study to analyze and extract the sentiment or opinion of people toward product or topic mentioned in text, facial expression, speech or music. In natural language processing, big data mining and machine learning sentiment analysis is one of the research area. Researchers use opinion mining in place of sentiment analysis. Sentiment analysis (SA) identifies sentiments in text and analyzes its polarity as neutral, positive or negative. SA identifies a state of mind of a person from his emotions expressed in text [1]. This field focuses to obtain opinions, sentiments, emotions based on observations of one's actions that are involved in written text, music, speech, utterance etc.

As per the survey sentiment analysis can be carried out at four respects such as document, sentence, aspect, and lexicon level. Document level considers a full document for sentiment analysis for e.g. blog of specific topic. As name suggests sentence level takes sentence into consideration i.e. paragraph has many sentences. In sentence majority of the polarity decides the person's sentiment towards the topic. Aspect level and lexicon level consider a word from sentence for sentiment analysis. Figure 1 shows a visual taxonomy of sentiment analysis as described above. One step ahead challenge in sentiment analysis is detecting sarcasm. Therefore, sarcasm is one of the prominent

© Springer Nature Switzerland AG 2020
D. J. Hemanth et al. (Eds.): ICICI 2019, LNDECT 38, pp. 425–433, 2020.
https://doi.org/10.1007/978-3-030-34080-3_48

research areas of Sentiment analysis, which can be analyzed on all sentiment analysis (SA) levels.

As Defined in Cambridge Dictionary [2] Sarcasm is usage of words that clearly signify the converse of what someone utters to hurt someone's feelings or to censure something in an amusing way. As per Oxford Dictionary [3] to convey the mock or disrespect sarcastic words have been used. Sometimes Sarcasm has a positive implied surface sentiment with negative sentence (for example 'Visiting my project guide is much exciting!!!'). or negative entail sentiment (for example 'what a bad act, terrible anyway') or no surface sentiment (for Example 'I am feeling sick as dog!').

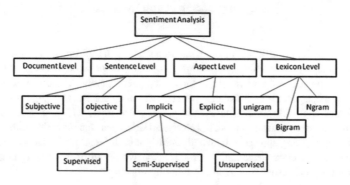

Fig. 1. Taxonomy of sentiment analysis

Sarcasm is a special kind of sentiment with negative, positive or no surface sentence [4]. Sarcasm often expressed with facial expression, textually, utterance while communicating etc. Sarcasm is a very salient aspect in Social media analysis. Data in Social Media is huge and unstructured. Also the absence of face to face contact or the context is very challenging to analyze, be it the mockery or satire in the sentences or words. In this paper we surveyed and compared different techniques to analyze sarcastic sentiments from given a text.

This article surveys state of art in sentiment analysis for sarcasm. Section 3 describes Linguistic presentation of sarcasm. In Sect. 4, various datasets are discussed. Different Approaches are discussed in Sect. 5. The article is concluded in Sect. 6.

2   State of Art

Many authors studied and summarized the different features and methods of text datasets for sarcasm detection. Generally, sentiments are classified into four categories as discussed in Sect. 1. Prominently this review is focusing on Sarcasm detection using lexicon level and implicit aspect level methods. Aspect level SA is categorized into two parts, explicit and implicit level. Implicit aspect extraction can be done in supervised, semi supervised and unsupervised way (refer Fig. 1). For sarcasm identification various mechanism used such as rule Based, deep learning, bootstrapping, statistical etc. [4, 5].

Table 1 shows the detailed survey of sarcasm detection. Tayal et al. [6] used Semi supervised approach to analyze the political sarcastic tweets and polarity detection for a given dataset. The author described process of sentence pre-processing, sentence detection, polarity detection and pos tagging. Filatova [7] used document level sarcasm detection. Sentence level (SA) is not producing accurate result compare to document level. They performed qualitative and quantitative analysis on text by considering document and text utterance [7]. Davidov [8] used semi- supervised approach to study sarcasm detection on text. They used single word feature, n-gram features, pattern features and punctuations features to analyze the text for sarcasm identification. They have used hash-tags utility to identify the sarcasm. Tungthamthiti et al. [9] suggested a novel method of sentiment analysis on concept level considering common sense knowledge with coherent identification of subject relevance. For this it uses machine learning classification. Lunando et al. [10] have considered a sentiment score in SentiWordNet for sentiment analysis for sarcasm identification. Bharti et al. [11] used parsing-based lexicon generation algorithm (PBLGA) to identify Sarcasm based on the interjection word that takes place in the sentence. The blend of two methods is also studied and differentiated with the technique/approach to detect sarcasm. Rajadesingan [12] developed a SCUBA model (sarcasm using behavioral modeling) to detect sarcasm by analyzing users past and current tweets. Riloff et al. [13] presented novel bootstrapping algorithm which automates learning and list out positive idioms and negative idioms from sarcastic tweets. Barbieri et al. [14] and Giachanou et al. [15] worked on short text like micro blogging. Instead of pattern of words as a feature, they worked on set of seven features comprises of frequency of words, written and spoken style, intensity of words, structure of sentences, sentiments, synonyms and ambiguity. Fersini et al. [16] proposed a Bayesian Model Averaging (BMA) combines multiple classifiers as per their fidelity and their minimal probability predictions to better classification of sarcastic and non sarcastic statements, as well as ironic and non ironic text. Roberto González-Ibáñez et al. [17] proposed sarcasm categorization by lexical and pragmatic facet which detects positive negative idioms expressed in Tweets. They compared a performance of automatic sarcasm detection & human classification technique using different studies. Authors have used unigram, dictionary based using support vector machine and logistic regression classifier to classify sarcasm. Bouazizi et al. [18] worked on a pattern based approach for sarcasm detection. They considered four set of features (Punctuation, pattern, syntactic & semantic) sarcasm. Authors classified tweets as a sarcastic and non-sarcastic and suggested method that achieves an accuracy of 83.1% along with precision as 91.1%. Lukin et al. [19] used bootstrapping method, which trains the classifier to distinguish subjective dialogues into sarcasm & nastiness. It uses pattern-based classifier build on syntactic structure which attains a high precision and recall. Davidov [8] used semi supervised sarcasm identification algorithm (SASI) technique on two different domains, Amazon and Twitter. They discussed the difference between two domains dataset. SASI algorithm uses structured dataset of amazon and sarcasm hashtag for twitter analysis. So, from above we can summarize different methods used for sarcasm detection by different authors. Next section explains the linguistic representation of sarcasm.

Table 1. Summary of different methods of sarcasm detection

Author/Year	Domain	Method	Features
Davidov et al. [8]	Twitter Amazon product Review	Semi supervised sarcasm identification algorithm (SASI)	Pattern based and punctuation based
Lukin and Walker [19]	Online dialogue	Bootstrapping Method	Pattern based
Bouazizi and Ohtsuki [18]	Twitter	Pos tagger and SVM	Punctuation based, pattern based, syntactic & semantic [5]
Gonzalez-Ibanez et al. [17]	Twitter	Unigram & Dictionary Based using SVM & logistic regression classifier [5]	Lexical & pragmatic Feature
Fersini et al. [16]	Microblog	Bayesian Model Averaging	Pragmatic piece & part of speech Tags
Barbieri et al. [14]	Twitter	Decision tree	Frequency of words, written and spoken style, intensity of words, structure, sentiments of context, synonyms, ambiguity
Liu et al. [20]	English & Chinese social Media	Multi-strategy ensemble learning approach [5, 20]	Punctuation symbol, lexical syntactic Feature (English) Rhetorical, homophony, construction feature (Chinese)
Riloff et al. [13]	Twitter	Bootstrapping Algorithm along with SVM	Context consideration from Syntactic Structure
Rajadesingan et al. [12]	Twitter	SCUBA framework	Identify user's past tweets, behavioral Modeling
Bharti et al. [11]	Twitter	Parsing-based lexicon generation algorithm (PBLGA) Multiple presence of the interjection words	Negative sentence and positive situation Sentence start with interjection
Lunando et al. [10]	Indonesian social media	Machine learning (Naïve bayes, Maximum Entropy, Support Vector Machine)	Negativity information and the number of interjection words
Tungthamthiti et al. [9]	Twitter	Support vector machine	Concept level & Common sense knowledge
Davidov et al. [21]	Twitter and amazon	Semi Supervised sarcasm identification	Hash tags, 1 gram feature, n-gram features, pattern features and punctuation features [8]
Filatova [7]	Amazon product reviews	Low star rating	Document and a text utterance
Tayal et al. [6]	Twitter	Supervised approach	Punctuation marks, adjective and verb

3   Linguistic Presentation of Sarcasm

Sarcasm means where literal meaning & interpretation of words are contrary to each other [22]. Sarcasm is a form of irony. Irony is a metaphor which is a discrepancy between what actually occurs and what is anticipated. Irony is a rhetorical device where the words are intentionally used to show a meaning in spite of the desired one. Irony is often fallacious for sarcasm. Sarcasm is a form of lingual irony. Sarcasm is wilful insulator induce harm. When someone says, "Oh, great" your extraordinary food skills made my stomach full. In above sentence you don't actually mean that the incident is positive. It induces a negative surface. Here, word 'great' ironically shows a higher negative entail, even after word is positive. Three types of ironies are there such as Verbal, Situational and Dramatic. Verbal irony is utilization of locution to convey something different what a person writes. The main attribute of verbal irony is, that is availed by a speaker or writer intentionally. It comes in text where a person sight to be understood as meaning something unlike to what his words actually mean. Situational irony is disparity between what is expected to happen and what actually happen. Statement states the situational irony, "The fire station burns down while the firemen are out on a call." Dramatic irony occurs when the audience has already a perception of situation, what is happening than a character. "In a drama, the person walks into a scene and the audience already knows the suspect is in the house".

Types of Sarcasm: There are four varieties of sarcasm [23]. Propositional, Embedded, Like-prefixed and Illocutionary. Propositional Sarcasm has implicit sentiments hidden inside a statement. Understanding a context of sentence is very important to interpret a sentiment of sentence. For e.g. "food was amazing!!!". Here sentence can be interpreted as non-sarcastic if circumstances are not known. Embedded Sarcasm where inappropriate words are embedded in sentence for e.g." Because he's been such a fine friend, I've struck him off my list". Word struck is embedded incongruity in sentence. Illocutionary sarcasm comprises of non textual clues in sentences. It contrast the sincere utterance of text. It comprises of writers attitude such as facial expression while uttering a sentence. For e.g. rolling one's eye while saying "That's how assignment should be done!". Like-prefixed sarcasm is implicit refusal of argument being said. For e.g. " you are like your friend".

Tuple Representation: [4] sarcastic sentences are represented as collection of six tuples. <S, H, C, u, P, P'> where S is Speaker, H is listener, C is Context, u is utterance, P is Literal Proposition and P' is Intended Proposition. Above Tuple conveys Speaker S utters a context C in Literal Proposition p in such a way that Hearer H hears intended Proposition P'. For e.g. Customer says "your room service was fantastic!!!". Here manager knows that their service was worst, therefore sarcasm is understood. Below is the representation of six Tuple representations for above statement.

S: Customer **H:** Manager **u:** Your room service was fantastic **P:** Really room service was good **P':** Room service was horrible.

4 Datasets

This section explains different datasets considered for sarcasm detection. Datasets are split up into short, long and dialogues. Example of short text is tweets or reviews of products on Amazon. Long text can be a blog contents and dialogues are chat communication happens online for e.g. in telecom sector. Short text contains mostly one sarcastic utterance, whereas long text contains sarcastic and non sarcastic sentences together in context. Table 2. shows a work done on different datasets by researchers.

Table 2. Summary of datasets used in various fields

Dataset type	Related work
Short text	[6, 8, 9, 11, 12, 14, 16–18]
Long text	[7, 10]
Dialogues	[19, 24, 25]

5   Approaches

This segment describes different methods for sarcasm detection in sentiment analysis. As per the literature survey and study Rule based, statistical based, classification technique [5], deep learning based [26] approached are described below.

Rule Based: Rule Based approach identifies sarcasm using certain evidence occurs in context in terms of rules. These rules are mostly consisting of hashtags # sarcasm #annoyed etc. For example: "Her performance was amazing!!!" #sarcastic. Rajadesingan et al. [12] used this method to identify the sarcasm in text. They have collected 40,000 tweets to with positive and negative sentiments. They filtered sentences with less than 10 frequencies using bigram and trigram features and calculated sentiment score for sentences.

Statistical Based: [20] Sarcasm detection using statistical Approach differs in various features and learning algorithm. They used pattern-based features which points out the existence of distinctive patterns taken out from a huge sarcasm dataset. Davidov et al. [8] considered precise match, partial overlap and no match techniques to classify the pattern base feature. Authors categorized words into HFWs (High Frequency Words) and CWs (Content Words) resulted from corpus frequency more or less than FH [21]. Liu et al. [20] Considers an english and Chinese sarcastic datasets. Authors have not used explicit feature to identify the sarcasm. They have not considered the imbalance between non-sarcastic and sarcastic samples. They have not considered explicit features to identify sarcasm and also ignores the imbalance between non-sarcastic and sarcastic samples in real world applications. They implemented a novel method MSELA to solve the imbalancing problem. They considered Rhetorical, homophony, construction feature (in Chinese) and Punctuation symbol, lexical syntactic Feature (in English) for sarcasm detection [20].

Classification Technique: Several classification techniques are classified as supervised, unsupervised and semi supervised. SVM, naïve bayes, maximum entropy uses supervised techniques. Tayal et al. [6] used supervised approach to detect sarcasm using different features such as punctuation marks, adjective and verb from context. Davidov et al. [8] used Semi supervised approach for sarcasm detection. They have considered pattern based and punctuation based features for SASI [5] algorithm. Turney [27] used unsupervised approach to identify semantic orientation. author used collection of pattern of tags to identify semantic orientation using PMI (Point wise Mutual Information) information retrieval method.

Deep Learning: Deep Learning is becoming a promising area in natural language processing as well as sarcasm detection. [22] Identified subtle form of incongruity from the depth of sarcasm. They used four types of word embedding feature to collect the context disparity when sentiment words are missing. They experimented four kinds of word embedding: LSA, GloVe, Dependency-based and Word2Vec. Only current feature alone is not sufficient to calculate the performance. Inclusion of past feature improves the performance more. Ghosh et al. [28] and Joshi et al. [24] used a blend of a Convolution Neural Network and Recurrent Neural Network (Long Short-Term Memory) followed by a Deep Neural Network yielding F Score of 0.92%. They compared their techniques with recursive SVM. Which show an improvement for the deep learning architecture. [26] Identified sarcasm using pre-trained convolution neural network (CNN) by extracting sentiments(S), emotions (E), personality features (P) along with baseline features (B). Authors applied CNN and CNN-SVM method on three different datasets, balanced (dataset 1), imbalanced (dataset 2) and test dataset (dataset 3). Table 3 shows the summary of how deep learning generates better results as compared to other techniques.

Table 3. [26] Experimental results of 5 cross validation on datasets

B	S	E	P	Dataset1 F1 score %		Dataset2 F1 score %		Dataset3 F1 score %	
				CNN	CNN, SVM	CNN	CNN, SVM	CNN	CNN, SVM
+				95.04	97.60	89.32	92.32	88.00	92.20
	+	+	+		90.70		90.90		84.43
+	+	+	+	95.30	97.71	89.73	94.80	88.51	93.30

6   Conclusion

Sarcasm detection in social media, micro blogs, ecommerce or dialogues provides vital perception for current trends, discussions, views and real time happening on various domains. Sarcasm detection has emerged remarkably in recent years as upcoming research area. This paper surveys different approaches and datasets used for sarcasm detection. As observed, most of authors worked on explicit sentiments of given sentence. Very less work covers the implicit aspect detection from sentence for sentiment analysis for sarcasm. Most of authors used supervised and semi-supervised approach for sarcasm detection. Future direction to work in sarcasm is to work on implicit aspect from sentence taking previous and further context into consideration while doing sarcasm detection. As per survey, very few authors worked on drawing out of contextual Information for sarcasm detection. Deep Learning is one of the best method as per survey to achieve good results. Therefore we concluded Deep Learning is one of the interesting & promising area to work on sarcasm detection.

References

1. Abdi, A., Shamsuddin, S.M., Aliguliyev, R.M.: QMOS: query-based multi-documents opinion-oriented summarization. Inf. Process. Manag. **54**(2), 318–338 (2018)
2. https://dictionary.cambridge.org
3. https://en.oxforddictionaries.com
4. Bhattacharyya, P., Carman, M.J., Joshi, A.: Automatic sarcasm detection: a survey. ACM Comput. Surv. (CSUR) **50**(5), 22 (2017). Article No. 73
5. Chandankhede, C., Chaudhari, P.: Literature survey of sarcasm detection. In: 2017 International Conference on Wireless Communications, Signal Processing and Networking (WiSPNET), 22 March 2017
6. Yadav, S., Gupta, K., Rajput, B., Kumari, K., Tayal, D.: Polarity detection of sarcastic political tweets (2014)
7. Filatova, E.: Irony and sarcasm: corpus generation and analysis using crowdsourcing (2012)
8. Tsur, O., Rappoport, A., Davidov, D.: Semi-supervised recognition of sarcastic sentences in Twitter and Amazon (2010)
9. Tungthamthiti, P., Kiyoaki, S., Mohd, M.: Recognition of sarcasms in tweets based on concept level sentiment analysis and supervised learning approaches (2014)
10. Purwarianti, A., Lunando, E.: Indonesian social media sentiment analysis with sarcasm detection (2013)
11. Babu, K.S., Jena, S.K., Bharti, S.K.: Parsing-based sarcasm sentiment recognition in Twitter data (2015)
12. Zafarani, R., Liu, H., Rajadesingan, A.: Sarcasm detection on Twitter: a behavioral modeling approach (2015)
13. Riloff, E., Qadir, A., Surve, P., De Silva, L., Gilbert, N., Huang, R.: Sarcasm as contrast between a positive sentiment and negative situation (2013)
14. Barbieri, F., Saggion, H., Ronzano, F.: Modelling sarcasm in Twitter, a novel approach (2014)
15. Giachanou, A., Crestani, F.: Like it or not: a survey of Twitter sentiment analysis methods. ACM Comput. Surv. **49**(2), 28 (2016)
16. Pozzi, F.A., Messina, E., Fersini, E.: Detecting irony and sarcasm in microblogs: the role of expressive signals and ensemble classifiers (2015)
17. González-Ibáñez, R., Muresan, S., Wacholder, N.: Identifying sarcasm in Twitter: a closer look, vol. 2. Association for Computational Linguistics (2011)
18. Bouazizi, M., Ohtsuki, T.: A pattern-based approach for sarcasm detection on Twitter (2016)
19. Walker, M., Lukin, S.: Really? Well. Apparently bootstrapping improves the performance of sarcasm and nastiness classifiers for online dialogue (2013)
20. Liu, P., Chen, W., Ou, G., Wang, T., Yang, D., Lei, K.: Sarcasm detection in social media based on imbalanced classification. Springer, Cham (2014)
21. Rappoport, A., Davidov, D.: Efficient unsupervised discovery of word categories using symmetric patterns and high frequency words (2006)
22. Grice, H.P.: Logic and conversation. In: Speech Acts, vol. 3 (1975)
23. Camp, E.: Sarcasm, pretense, and the semantics/pragmatics distinction. Noûs **4**, 587–634 (2012)
24. Tripathi, V., Bhattacharyya, P., Carman, M., Joshi, A.: Harnessing sequence labeling for sarcasm detection in dialogue from TV series 'Friends' (2016)
25. Tepperman, J., Traum, D., Narayanan, S.: "Yeah Right": sarcasm recognition for spoken dialogue systems (2006)

26. Cambria, E., Hazarika, D., Vij, P., Poria, S.: A deeper look into sarcastic tweets using deep convolutional neural networks, Osaka, Japan (2016)
27. Turney, P.D.: Thumbs up or thumbs down?: semantic orientation applied to unsupervised classification of reviews (2002)
28. Ghosh, A., Veale, T.: Fracking sarcasm using neural network (2016)

Comparative Analysis of Deep Neural Networks for Crack Image Classification

Sheerin Sitara Noor Mohamed$^{(\boxtimes)}$ and Kavitha Srinivasan

Department of Computer Science and Engineering,
SSN College of Engineering, Chennai, India
sheerinsitara1615@cse.ssn.edu.in, kavithas@ssn.edu.in

Abstract. Deep Learning (DL) is widely used in different types of classification problems in real time. DL models can be constructed in two ways namely, Convolutional Neural Network (CNN) and using pre-trained models such as VGG16, VGG19 and Inception ResNet V2. In this paper, an automatic crack image classification approach is proposed and implemented using CNN and pre–trained models. For validation, three types of datasets having both enhanced and without enhanced crack images are used and the result of classification are analysed using appropriate quantitative metrics such as accuracy, precision and recall. From the results, it has been inferred that the proposed CNN, derived the highest accuracy of 99% for dataset 1 whereas Inception ResNet V2 model, derived the highest accuracy of 87% and 94% for dataset 2 and dataset 3 respectively.

Keywords: Crack classification · Convolutional Neural Network · Pre-trained models · Transfer learning approach · Deep Learning

1 Introduction

Deep Learning (DL) model is composed of input layer, multiple hidden layers and output layer. It aims at learning features from lower level to higher level in classification. DL can be constructed in two ways: A Convolutional Neural Network (CNN) approach by defining appropriate hyper parameters like batch size, epochs, number of convolutional layers, type of filters, normalisation, fully connected and softmax layer and, Transfer Learning (TL) approach in which pre-trained model is used to extract features and classification is done using supervised machine learning algorithms.

The CNN model comprises of two or more convolutional layers followed by fully connected layers. Initial layer learns the basic features namely edge and corner. The middle layer recognises part of an image and the last layer identifies the object for classification. CNN requires minimum level of pre-processing when compared to other supervised machine learning algorithms.

Pre-trained model is a defined model which has been trained on large dataset to solve classification problems. Some of the pre-trained models are AlexNet, Visual Geometry Group Network (VGGNet), Inception, Residual Network (ResNet) and Inception Residual Network (Inception ResNet). Pre-trained models are used in three

© Springer Nature Switzerland AG 2020
D. J. Hemanth et al. (Eds.): ICICI 2019, LNDECT 38, pp. 434–443, 2020.
https://doi.org/10.1007/978-3-030-34080-3_49

different ways, depends on the appropriate applications. They are: (i) Training the entire model, (ii) Training some layers and leave the other layers as frozen, (iii) Freeze the convolutional base (feed the classifier with output of the convolutional base) as transfer learning approach.

The CNN and pre-trained models have wide range of applications in image classification like land classification, disease classification, animal classification, crack classification, etc. In this paper, CNN is designed for automatic crack image classification, and the results are compared with chosen pre-trained models with appropriate quantitative metrics. In addition, results are analysed with state-of-art research works and inferences are discussed.

The remaining part of this paper spans across the following subsections. In Sect. 2, Literature survey related to machine learning, CNN and pre-trained models are explained. In Sect. 3, the design of proposed CNN model, chosen pre-trained models with implementation are described. A brief summary about results and discussions of both models for three datasets are given in Sect. 4 and conclusion is given at the end.

2   Literature Survey

The research work on crack image detection and classification has been carried out using image processing, machine learning and deep learning techniques. Basically, image processing technique processes the image and extracts significant features from an image region. Some of the image pre-processing techniques are wiener filter, sobel filter and wavelet transform [1], morphological operation, least square method and Otsu's method [2]. From the pre processed image, features are extracted and classified using machine learning techniques. The features extracted for crack classification are mean, intensity, pattern, texture, gradient, area, width and height of crack region.

Machine Learning techniques are applied to classify or predict the type of crack image. Some of the techniques includes Support Vector Machine (SVM), Decision tree, Adaboost, Multi Level Perceptron (MLP) and Random Forest, each technique has some limitation. The kernel function selection is not entirely solved in SVM [1]. Adaboost is not suitable for noisy data and outlier. Further decision tree leads to over fitting, when more features are used in classification [3]. Finally, MLP has the capability to classify unknown pattern by learning from the known pattern but it requires more number of iterations [4]. To address these issues, DL techniques are emerging in the current period.

Deep Learning (DL) performs mathematical manipulation to convert the input into output by linear or non-linear relation. In deep learning approach, subsamples of each image are generated and learns the features deeply. Some of the DL techniques are Convolutional Neural Network (CNN), Recurrent Neural Network and Recursive Neural Network. CNN gives better accuracy for image classification while increasing the number of layers and also gives poor result when the training and test images are collected from different locations [5]. Recurrent Neural Network allows variable sized images for training and testing [6]. Recursive Neural Networks are able to learn deep structured information but it is not widely accepted because of its complex information processing characteristics [7].

Pre-trained models are a kind of Deep Learning model, which has been trained on large dataset (ImageNet dataset - 1000 classes) for classification problems. The most widely used pre-trained models includes AlexNet, Visual Geometry Group Network (VGGNet), Inception, Residual Network (ResNet) and Inception Residual Network (Inception ResNet). AlexNet is composed of 5 convolutional layers followed by 3 connected layers and uses 3 × 3 kernel sized filter. In addition, it uses dropout layer after each fully-connected layer to reduce over fitting. The error rate is considerably reduced using Rectified Linear Unit (ReLU) instead of sigmoid function [8]. The different versions of VGGNet are VGG11, VGG13, VGG16 and VGG19 and it differs by number of weighted layers (11, 13, 16 and 19 respectively). Each VGGNet contains convolutional layer followed by 3 fully connected layers and it uses multiple stacked smaller sized kernel instead of larger sized kernel. This structure increases the depth of the network and enables to learn more complex features at a lower cost and the error rate is reduced to 7.3%. In other words, 11 × 11 and 5 × 5 kernel-sized filter is replaced by multiple 3 × 3 kernel-sized filter one after another. In VGG-D, multiple and different kernel filters are used to extract more complex features [9] where as selecting the right kernel size for the convolution operation is difficult because of the huge variation in the distribution of data. Also, deep network structure leads to over fitting problem and it is addressed in Inception model using wider network instead of deeper network and the error rate is reduced to 6.67% [10]. ResNet (VGG with residual connection) is either two layers or three layers in structure. When the depth increases, accuracy is saturated and hence using minimum number of layers ResNet attains an error rate of 3.37% [11]. Inception ResNet is developed by Third Research Institute of Ministry of Public Security, but the basic models such as Inception and ResNet are developed by different corporations namely Google and Microsoft. These two models are combined, since the ResNet performance is high and the Inception has wide collection of models. Inception ResNet is able to achieve higher accuracy at lower epoch, where the error rate is 2.29% only. These pre-trained models are widely used in image classification [12] to reduce the training time and also to find the suitable model. [13], used VGGNet for crack image classification and achieved an accuracy of 92.27%. The dataset includes 3500 images with and without cracks captured using Unmanned Aerial Vehicle (UAV).

From the above discussion, it has been inferred that each pre-trained model has its own pros and cons. The accuracy of VGGNet is proportional to the inclusion of additional layers. In Inception model, the wider network is efficient than the deeper network irrespective of filter size. ResNet is similar to VGGNet, except it is recursive. Inception ResNet is a hybrid model developed from Inception and ResNet model by combining the characteristics of both the models.

3   Proposed Design

An automatic crack image classification approach is proposed and implemented using CNN and chosen pre-trained models. The classification results are analysed with three types of datasets. Datasets are divided into training and test set with 80% and 20% respectively. The CNN model and pre-trained models are generated from the training images. The crack type is predicted for the test set using the models and the results are compared and analysed. The overall workflow is shown in Fig. 1.

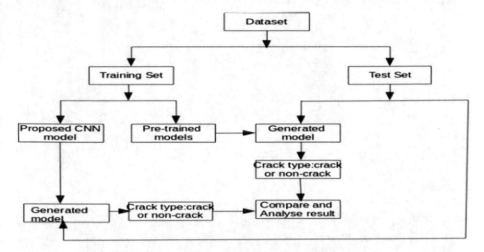

Fig. 1. Workflow diagram of CNN and pre-trained models

3.1 Convolutional Neural Network (CNN)

The proposed CNN model consists of four convolutional layers (marked as CONV), four max pooling layers (marked as MAX), two fully connected layers (marked as FC), followed by one softmax layer (marked as SOFTMAX) at the network's end. It uses dropout layer in between the fully connected layers. The design and detailed architecture of CNN is represented in Figs. 2 and 3.

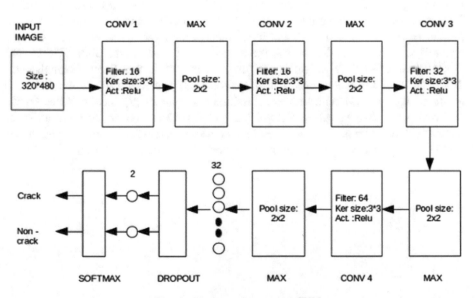

Fig. 2. Design of proposed CNN

```
Using TensorFlow backend.

Layer (type)                     Output Shape              Param #
=================================================================
conv2d_1 (Conv2D)                (None, 318, 478, 16)      448

max_pooling2d_1 (MaxPooling2     (None, 159, 239, 16)      0

conv2d_2 (Conv2D)                (None, 157, 237, 16)      2320

max_pooling2d_2 (MaxPooling2     (None, 78, 118, 16)       0

conv2d_3 (Conv2D)                (None, 76, 116, 32)       4640

max_pooling2d_3 (MaxPooling2     (None, 38, 58, 32)        0

conv2d_4 (Conv2D)                (None, 36, 56, 64)        18496

max_pooling2d_4 (MaxPooling2     (None, 18, 28, 64)        0

flatten_1 (Flatten)              (None, 32256)             0

dense_1 (Dense)                  (None, 32)                1032224

dropout_1 (Dropout)              (None, 32)                0

dense_2 (Dense)                  (None, 2)                 66
=================================================================
Total params: 1,058,194
Trainable params: 1,058,194
Non-trainable params: 0
```

Fig. 3. Architecture of proposed CNN model

3.2   Pre-trained Model

The existing pre-trained models are VGG16, VGG19, Inception, ResNet and Inception ResNet. Among these pre-trained models, VGG16, VGG19 and Inception ResNet V2 are chosen in our research work and its corresponding hyper parameters are given in Fig. 4. The reason behind choosing these pre-trained models are: Both VGG16 and VGG19 have 16 and 19 weight layers respectively and it uses Batch Normalisation (BN) instead of Local Response Normalisation (LRU). Additionally, three 1x1 convolutional layer is used to increase the classification accuracy [9]. Inception ResNet V2 performs better than Inception ResNet V1 and it achieves better accuracy at lower epochs [12]. In addition, Inception ResNet has the characteristics of both Inception and ResNet such as wider network, kernel filter and hyper parameters [10, 11].

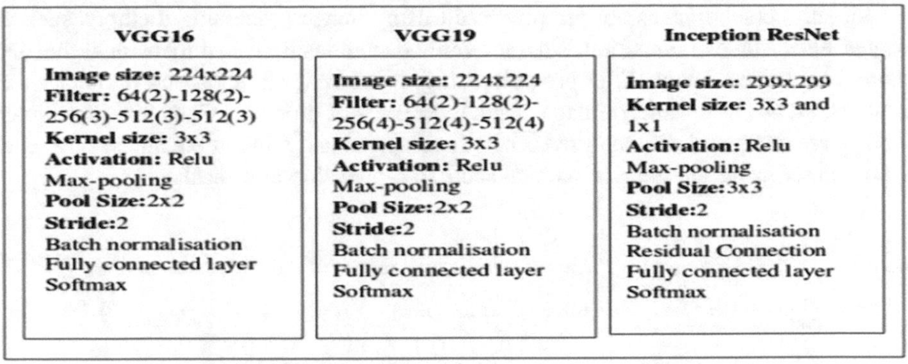

VGG16	VGG19	Inception ResNet
Image size: 224x224 **Filter:** 64(2)-128(2)-256(3)-512(3)-512(3) **Kernel size:** 3x3 **Activation:** Relu Max-pooling **Pool Size:** 2x2 **Stride:** 2 Batch normalisation Fully connected layer Softmax	**Image size:** 224x224 **Filter:** 64(2)-128(2)-256(4)-512(4)-512(4) **Kernel size:** 3x3 **Activation:** Relu Max-pooling **Pool Size:** 2x2 **Stride:** 2 Batch normalisation Fully connected layer Softmax	**Image size:** 299x299 **Kernel size:** 3x3 and 1x1 **Activation:** Relu Max-pooling **Pool Size:** 3x3 **Stride:** 2 Batch normalisation Residual Connection Fully connected layer Softmax

Fig. 4. Chosen pre-trained models and its hyper parameters

4 Results and Discussion

The proposed CNN model and chosen pre-trained models for crack image classification are validated using three different types of dataset. The datasets used are named as dataset 1, dataset 2 and dataset 3 for ease of explanation. Dataset 1 contains 40000 images (20000 crack images and 20000 non-crack images), captured using low cost sensor devices where the crack region is prominent. Dataset 2 has 2947 images (2356 crack images and 591 non-crack images), collected from different locations. Some of the images of this dataset are of low resolution, contrast and/or contains shadow. Dataset 3 comprises 545 images (437 crack images and 108 non-crack images), collected from DIGITLab and Google [14, 15] and also acquired from our campus through mobile camera. The images of each dataset are divided into training set and test set as given in Table 1.

Table 1. Dataset description

Dataset	Crack type	Training set (no. of images)	Test set (no. of images)	Total images
Dataset 3 (proposed dataset) [16, 17]	Crack	245	60	305
	Non-crack	192	48	240
	Total	437	108	545
Dataset 1 [14]	Crack	16000	4000	20000
	Non-crack	16000	4000	20000
	Total	32000	8000	40000
Dataset 2 [15]	Crack	1306	327	1633
	Non-crack	1050	264	1314
	Total	2356	591	2947

Initially, crack images are pre-processed using image processing methods such as wiener filter, wavelet transform, singular value decomposition, K-dimensional tree and morphological operation. After pre-processing, input image is improved in quality and contrast, named as enhanced image. Then the results of proposed CNN and pre-trained models are compared and analysed for both enhanced and without enhanced images of each dataset using the metrics such as accuracy, precision and recall.

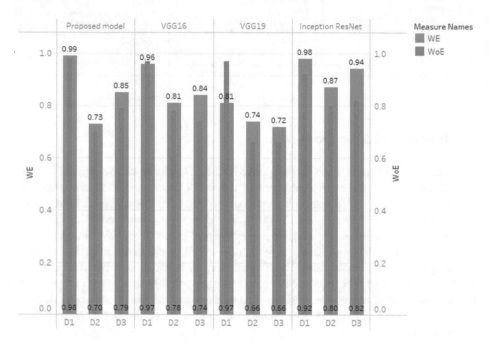

Fig. 5. Accuracy measure of proposed CNN and pre-trained models vs. datasets

Dataset 1 attains improved accuracy, precision and recall for proposed CNN model, since the crack regions are prominent in the input image for enhanced images. For dataset 2 and dataset 3, Inception ResNet model has resulted improved accuracy, precision and recall for enhanced images as shown in Figs. 5, 6, and 7. In the Figures, WE represent With Enhancement and WoE represent With out Enhancement of the crack image.

From the above results, it has been inferred that proposed CNN and Inception ResNet model performs better than VGGNet for enhanced image set. The proposed CNN model resulted with 1%, 2% and 1% improved accuracy, precision and recall than Inception ResNet for dataset 1 of enhanced images as shown in Figs. 5, 6 and 7. The Inception ResNet model resulted with 14% and 9% of improved accuracy than the proposed CNN for dataset 2 and dataset 3 respectively as shown in Fig. 5. For precision, Inception ResNet model resulted 4% and 10% improvement than the proposed CNN for dataset 2 and dataset 3 respectively as shown in Fig. 6. Finally for recall, Inception ResNet model attains 7% and 13% improvement than the proposed CNN for dataset 2 and dataset 3 as shown in Fig. 7.

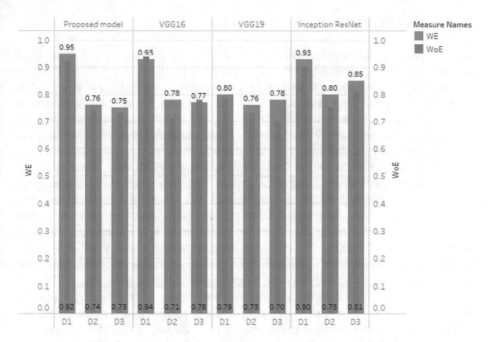

Fig. 6. Precision measure of proposed CNN and pre-trained models vs. datasets

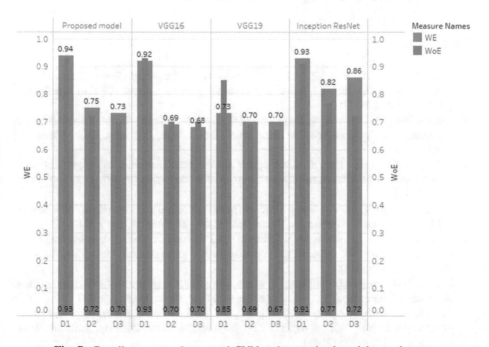

Fig. 7. Recall measure of proposed CNN and pre-trained models vs. datasets

5 Conclusion and Future Work

The Comparative analysis of automatic crack image classification using Convolutional Neural Network (CNN) and pre-trained models for three datasets with and without enhancement is proposed and analysed using appropriate quantitative metrics. The proposed CNN and Inception ResNet model performs better than VGGNet for enhanced images. The proposed CNN model resulted with highest accuracy, precision and recall of 99%, 95% and 94% for dataset 1 whereas Inception ResNet resulted with significantly high value for dataset 2 and dataset 3. The accuracy, precision and recall are 87%, 80% and 82% for dataset 2 and 94%, 85% and 86% for dataset 3 respectively.

The dataset with large number of images and prominent crack region gives improved result for both proposed CNN model and Inception ResNet model. Irrespective of dataset size, enhanced image set gives better accuracy than the without enhancement image set when there is a wide collection of low contrast and resolution images. The overall inferences are: deeper network performs well for limited depth, wider network is preferable to overcome the vanishing gradient problem, batch normalisation is better than local response normalisation in neural networks.

In general, the pre trained models obtains better accuracy for ImageNet dataset but fails for other type of datasets acquired in real time using normal camera with low resolution and contrast and, the datasets may be small in size. In the future research, the performance of classification for real time images can be further improved by identifying suitable image processing and learning methods.

References

1. Chen, Y., Mei, T., Wang, X., Li, F.: A bridge crack image detection and classification method based on climbing robot. In: Proceeding 35th Chinese Control Conference, pp. 4037–4042 (2016)
2. Sheerin Sitara, N., Kavitha, S., Raghuraman, G.: Review and analysis of crack detection and classification techniques based on crack types. Int. J. Appl. Eng. Res. **13**(8), 6056–6062 (2018)
3. Sheng, P., Chen, L., Tian, J.: Learning-based road crack detection using gradient boost decision tree. In: IEEE Conference on Industrial Electronics and Applications, pp. 1228–1232 (2018)
4. Salari, E., Ouyang, D.: An image-based pavement distress detection and classification. In: IEEE International Conference on Electro/Information Technology, pp. 1–6 (2012)
5. da Silva, W.R.L., de Lucena, D.S.: Concrete crack detection based on deep learning image classification. In: 18th International Conference on Experimental Mechanics, Brussels, Belgium, pp. 1–6 (2018)
6. Zhang, D.-Q.: Image recognition using scale recurrent neural networks. In: Computer Vision and Pattern Recognition, pp. 1–5. https://arxiv.org/abs/1803.09218v1 (2018)
7. Chinea, A.: Understanding the principles of recursive neural networks: a generative approach to tackle model complexity. In: International Conference on Artificial Neural Networks, pp. 952–963 (2009)
8. Krizhevsky, A., Sutskever, I., Hinton, G.E.: ImageNet classification with deep convolutional neural networks. In: Neural Information Processing Systems, Tahoe, Nevada, pp. 1–9 (2012)

9. Simonyan, K., Zisserman, A.: Very deep convolutional neural networks for large scale image recognition. In: International Conference on Learning Representations, Banff, Canada, pp. 1–14 (2014)
10. Szegedy, C., Liu, W., Jia, Y., Sermanet, P., Reed, S., Anguelov, D., Erhan, D., Vanhoucke, V., Rabinovich, A.: Going deeper with convolutions. In: Conference on Computer Vision and Pattern Recognition, Columbus, Ohio, pp. 1–9 (2014)
11. He, K., Zhang, X., Ren, S., Sun, J.: Deep residual learning for image recognition. In: ImageNet Large Scale Visual Recognition Challenge, pp. 1–12 (2016)
12. Szegedy, C., Ioffe, S., Vanhoucke, V., Alemi, A.A.: Inception-v4, inception-ResNet and the impact of residual connections on learning. In: Proceedings of the Thirty-First Association for the Advancement of Artificial Intelligence Conference on Artificial Intelligence, pp. 4278–4284 (2017)
13. Özgenel, C.F., Gönenç Sorguç, A.: Performance comparison of pretrained convolutional neural networks on crack detection in buildings. In: 35th International Symposium on Automation and Robotics in Construction, pp. 1–8 (2018)
14. Zhang, L., Yang, F., Zhang, Y.D., Zhu, Y.J.: Road crack detection using deep convolutional neural network. In: IEEE International Conference on Image Processing, pp. 3708–3712 (2016). Dataset
15. Maguire, M., Dorafshan, S., Thomas, R.J.: SDNET2018: a concrete crack image dataset for machine learning applications (2018). Browse all Datasets. Paper 48. https://digitalcommons.usu.edu/all_datasets/48. https://doi.org/10.15142/t3td19. Accessed Feb 2019. Dataset
16. Dataset: http://lasir.umkc.edu/cdid/. Accessed Aug 2017
17. Dataset: https://www.google.co.in/search?q=pavement+crack&safe=off&rlz=1C1GGGEen US410US410&espv=2&biw=1067&bih=517&tbm=isch&tbo=u&source–univ&sa=X&ei= WSz5VNb4C8YggTJ6IH4Ag&ved=0CB0QsAQ&dpr=1.5&gwsrd = cr#gwsrd = cr&imgrc =/. Accessed Sept 2017

Comparative Study Between a Conventional Square Patch and a Perturbed Square Patch for 5G Application

Ribhu Abhusan Panda[1]([✉]), Tapas Ranjan Barik[2],
R. Lakshmi Kanta Rao[2], Satyawan Pradhan[2], and Debasis Mishra[1]

[1] Department of Electronics and Telecommunication Engineering,
V.S.S. University of Technology, Burla, Odisha, India
`ribhupanda@gmail.com, debasisuce@gmail.com`
[2] Department of Electronics and Instrumentation Engineering,
GIET University, Gunupur, Odisha, India
`bariktapas2016@gmail.com, rrlakshmikanta@gmail.com,`
`satyawanp772@gmail.com`

Abstract. In this tabloid, an exclusive patch has been designed with a perturbation of the conventional square patch. Taking the important parameters like S_{11}, Antenna Gain, Directivity etc. into account a comparison has been made between the two designed antennas. The antenna has been designed for 5G communication which includes the frequency range from 26 GHz to 28 GHz. Simulation has been carried out by the HFSS. Return loss ($S_{11} < 10$ dB) plot and antenna gain is considered as the prominent parameters to determine which one is better for 5G application between conventional square patch and the perturbed square patch. The perturbation is done by implementing a circular cut from one side of the conventional patch. For the substrate FR4 epoxy material has been chosen for its wide availability.

Keywords: Perturbed square patch · 5G · S_{11} · Antenna gain · Circular cut · FR4 epoxy

1 Introduction

The modification of the conventional rectangular or square shaped patches began long ago but recent advances in these modifications have given a new approach to design the microstrip patch antenna [1–5]. Many elements have been added to boost the antenna gain and directivity [6–9]. A novel array structure has been proposed for X-Band application [10–12]. Inimitable shaped patches have been advanced in the year 2017 for different applications [13, 14]. Here a conventional square patch is designed and it has been altered in such a way that the resultant structure will be novel one. A comparison has been made between both the designed patches for 5G application which includes the 26 GHz to 28 GHz frequency bands.

© Springer Nature Switzerland AG 2020
D. J. Hemanth et al. (Eds.): ICICI 2019, LNDECT 38, pp. 444–449, 2020.
https://doi.org/10.1007/978-3-030-34080-3_50

2 Antenna Design with Specific Design Parameters

The conventional square shape has been designed with dimensions 40 mm × 40 mm × 1.6 mm for the substrate. The dimension of the ground plane is 40 mm × 40 mm × 0.01 mm. Regular square patch has been designed with dimensions 21.42 mm × 21.42 mm × 0.01 mm. The perturbed structure has the substrate and the ground plane with same dimension. The only difference the shape of the patch. The perturbed patch is designed by subtracting a semicircle having a radius same as that of the half of the length of the square patch. The prosed design is revealed in the Fig. 1. Figures 2 and 3 illustrates the designs of conventional square and perturbed square patch. In Fig. 4 the complete design of the perturbed patch has been shown (Table 1).

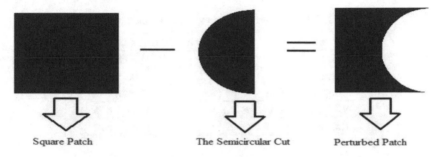

Square Patch   The Semicircular Cut   Perturbed Patch

Fig. 1. Pictorial representation of the methodology to design the perturbed patch

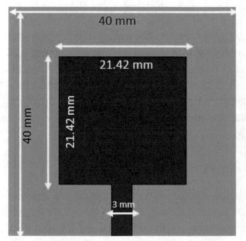

Fig. 2. Design of square patch

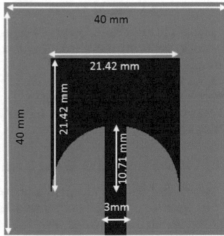

Fig. 3. Design of perturbed patch

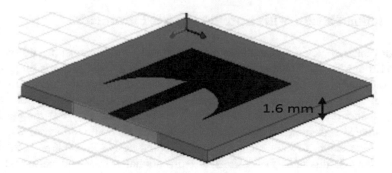

Fig. 4. Perturbed patch antenna with specified substrate height

Table 1. Specific design parameters for the antenna design

Parameters	Symbol used	Value
Frequency	f	28 GHz
Wavelength	λ	10.71 mm
Width of the feed	wf	3 mm
Width of the square patch	w	21.42 mm
Radius of the cut	r	10.71 mm

3 Simulation Result

3.1 Return Loss and VSWR

Emphasis has been made on the S-Parameter plot to know the resonant frequency and return loss of both the structures. For the conventional patch the resonant frequency and return loss is 27.5 GHz and −25.74 dB respectively where as for the perturbed patch the resonant frequency and return loss is 27.6 GHz and −30.96 dB. This comparison is shown in Fig. 5. In similar manner the Standing wave ratio measured in terms of Voltage for the square patch is 1.1089 and for the perturbed patch is 1.06 that are shown in Fig. 6.

3.2 Enhancement of Gain and Directivity after the Perturbation

The principal purpose of perturbation with the square patch to increase the antenna gain. From the observation of the simulation result it is palpable that the improvement of the antenna gain and directivity can be achieved by the alteration of the shape of conventional square patch which resulted a Plano Concave shape (Figs. 7 and 8) (Table 2).

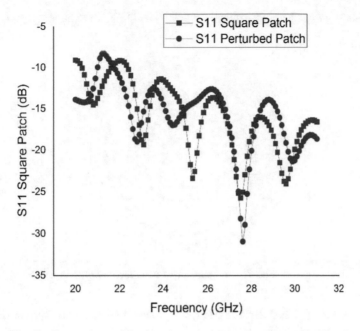

Fig. 5. Comparison of S$_{11}$ between conventional and perturbed patch

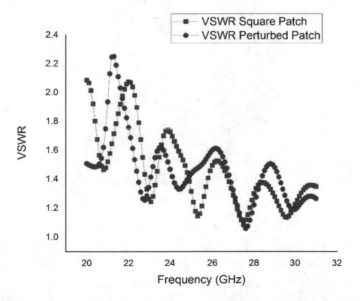

Fig. 6. Comparison of VSWR between conventional and perturbed patch

PeakDirectivity of Perturbed Patch ━━━ PeakDirectivity of Square Patch

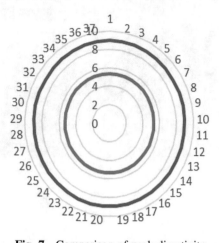

Fig. 7. Comparison of peak directivity

PeakGain of Perturbed Patch ━━━ PeakGain of Square Patch

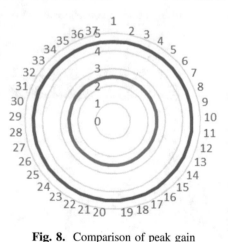

Fig. 8. Comparison of peak gain

Table 2. Parameters from the simulation results

Parameters	Resonant frequency (GHz)	VSWR	S11 dB	Directivity dB	Gain dB	Radiation efficiency
Conventional square patch	27.5	1.11	−25.74	5.34	2.51	47%
Perturbed square patch	27.6	1.06	−30.96	9.007	4.52	50.68%

4   Conclusion

After an exclusive perturbation to the conventional square patch the directivity has been increased by a value of 3.667 dB and the antenna gain has been increased by a value of 2.51 dB. The proposed perturbed square patch resonating at a frequency 27.6 GHz with return loss −30.96 dB which includes the frequency of 5G communication. So the projected antenna can be used efficiently for 5G.

References

1. Panda, R.A., Mishra, D., Panda, H.: Biconcave lens structured patch antenna with circular slot for Ku Band application. Lecture Notes in Electrical Engineering, vol. 434, pp. 73–83. Springer, Singapore (2018)
2. Panda, R.A., Dash, P., Mandi, K., Mishra, D.: Gain enhancement of a biconvex patch antenna using metallic rings for 5G application. In: 6th International Conference on Signal Processing and Integrated Networks (SPIN), pp. 840–844 (2019)
3. Panda, R.A., Panda, M., Nayak, S., Das, N., Mishra, D.: Gain enhancement using complimentary split ring resonator on biconcave patch for 5G application. In: International Conference on Sustainable Computing in Science, Technology & Management, SUSCOM-2019, pp. 994–1000 (2019)
4. Panda, R.A., Mishra, S.N., Mishra, D.: Perturbed elliptical patch antenna design for 50 GHZ application. Lecture Notes in Electrical Engineering, vol. 372, pp. 507–518. Springer, India (2016)
5. Panda, R.A., Mishra, D., Panda, H.: Biconvex patch antenna with circular slot for 10 GHz application. In: SCOPES 2016, pp. 1927–1930. IEEE (2016)
6. Attia, H., Yousefi, L.: High-gain patch antennas loaded with high characteristic impedance superstrates. In: IEEE Antennas and Wireless Propagation Letters, vol. 10, pp. 858–861, August 2011
7. Rivera-Albino, A., Balanis, C.A.: Gain enhancement in microstrip patch antennas using hybrid substrates. In: IEEE Antennas and Wireless Propagation Letters, vol. 12, pp. 476–479, April 2013
8. Kumar, A., Kumar, M.: Gain enhancement in a novel square microstrip patch antenna using metallic rings. In: International Conference on Recent Advances and Innovations in Engineering, ICRAIE-2014, Jaipur, pp. 1–4 (2014)
9. Ghosh, A., Kumar, V., Sen, G., Das, S.: Gain enhancement of triple-band patch antenna by using triple-band artificial magnetic conductor. IET Microw. Antennas Propag. 12(8), 1400–1406 (2018)
10. Panda, R.A., Panda, M., Nayak, P.K., Mishra, D.: Log periodic implementation of butterfly shaped patch antenna with gain enhancement technique for X-band application. In: System Reliability, Quality Control, Safety, Maintenance and Management, ICICCT-2019, pp. 20–28. Springer, Singapore (2020)
11. Panda, R.A., Mishra, D.: Log periodic implementation of star shaped patch antenna for multiband application using HFSS. Int. J. Eng. Tech. 3(6), 222–224 (2017)
12. Panda, R.A., Mishra, D.: Modified circular patch and its log periodic implementation for Ku Band application. Int. J. Innov. Technol. Explor. Eng. 8(8), 1474–1477 (2019)
13. Panda, R.A.: Multiple line feed perturbed patch antenna design for 28 GHz 5G application. J. Eng. Technol. Innov. Res. 4(9), 219–221 (2017)
14. Panda, R.A., Mishra, D.: Reshaped patch antenna design for 60 GHz WPAN application with end-fire radiation. Int. J. Mod. Electron. Commun. Eng. 5(6), 5–8 (2017)

Insecticide Spraying Using Quadcopter

Ashwini G. Mule[✉] and R. P. Chaudhari

Electronics and Telecommunication Department,
Government College of Engineering Aurangabad (MS), Aurangabad, India
mule.ashwini85@gmail.com, rpchaudhari@yahoo.com

Abstract. India is second highest populated country in the world. India has almost 13 billion populations. In other side the area under farming degrading day to day. So, now to fulfill the need of food we have to increase the food production. While achieving the high production rate farmer needs to accept advance technology in their farms. But, Indian farmers are not rich to buy costly equipment for spraying pesticides or insecticides in their farm. To solve this problem, we are presenting low cost insecticide quadcopter. Use of pesticides plays important role in agricultural fields for a better crop yielding. To increase crop production, speed of every process related to farming should be increased. Considering climate conditions, environmental situations, manual spraying of pesticides may consume more time. Our proof of concept model will be low cost and ideal for Indian farmer who have low surface area for farming.

Keywords: Unmanned Aerial Vehicle · ESC · Atmega 2560 pesticide mechanism

1 Introduction

As the world population is increasing, demands for basic needs are also increasing at an unprecedented rate. Human have three basic needs such as, food, cloth, and shelter. Among these three needs food and cloth are totally dependent on agriculture so by fulfilling this need, the agricultural production rate should be considerably increased. But, in agriculture field automation development is much poor and every farmer cannot afford it. In our country 70% peoples are farmers and this farmer does not have a large cultivation area due to small pieces of farm most of the agriculture work like sowing, ploughing, harvesting and spraying takes places manually.

Among this process spraying fertilizer are frequently used to kill insects and increase the growth of crops. According, to the World Health Organization (WHO) report, more than 1 million cases of pesticide are being recorded and among them near about 1 lakhs people are suffering from health problem. These fertilizers affect on human nervous system and lead to directly affect on body. This pesticides chemical leads so many health issues like asthma, hormone disruption and having carcinogenic agent which may cause cancer.

Considering this issue Unmanned Aerial Vehicle (UAV) is the only option to save the farmer life and health. Also, considering the Indian framer we developed the low cost UAV for spraying pesticide and fertilizer.

D. J. Hemanth et al. (Eds.): ICICI 2019, LNDECT 38, pp. 450–456, 2020.
https://doi.org/10.1007/978-3-030-34080-3_51

The quadcopter is an identical type of aerial vehicle which can be used for various applications. It is a multi-rotor device and which has vertical take-off and landing vehicle.

It comes under the category of Unmanned Aerial Vehicle (UAV). Quadcopter is also classified as rotor craft as it is made from four rotors.

It shows an identical control system which allows a balanced flight so it can eliminate in stability problems.

In quadcopter, each rotor plays very important role in direction and balanced of vehicles as well as lifting of vehicles. It is different from traditional helicopter which shows particular tasks to every rotor it has. That is, task lift or directional control.

To achieve high productivity, agriculture requires spraying of pesticides for protection of yield. Fertilizers and chemicals are frequently required to kill insects. Pesticides sprayed by human beings can be improper in amount which is desired for crops. So it may affect the nervous system or fewer amounts can make production decrement.

So, remote controlled unmanned aerial vehicle is preferred to spray the pesticides.

Quadcopter is chosen for this purpose because of two reasons like it has high stability and great lifting power. The Quadcopter is easier to control while helicopter model or other similar vehicles.

2 Related Work

Meivel et al. [1] demonstrate the quadcopter using flight controller this ad pilot control all operations which commanded by user. Author used X cross configuration in quadcopter configuration. This quadcopter net payload was 4 kg to navigate that UAV GPS system was used. Sprayer controlled by RF transmitter remote which is operated by user.

Prasad Reddy et al. [2] suggest the system UAV which controlled by radio controller and spray kept continuously on which west lots of fertilizer.

Gilles and Billing [3] use a heavy (100 kg) UAV i.e. small petroleum aircraft (model RMAX, Yamaha motor) and it was controlled by radio link, 60 MV dual joystick. This UAV does not have any provision for autonomous navigation is not suitable for small farming area. It also has high manufacturing cost.

Huang et al. [4] developed a system where they used two stroke gasoline engines and in other model they used battery powered. This autonomous flight controlled system receive command from ground station via wireless telemetry system freedom IMU, 3 axis magnetometer, GPS, proprietary radio receiver and Linux based flight computer. In this system message send from base station to API.

Patel et al. [5] demonstrates the Quadcopter for agriculture surveillance in which the author used an infrared camera for taking image. In this image temperature sensing of plants by infrared thermography is done. Then using this image disease are detected and for that specific area fertilizer are spray over it.

Kedari et al. [6] developed a Quadcopter which is used for pesticides spraying and this quadcopter is handle by an android application.

Koriahalli et al. [7] explains a system automatically controlled drone, where quadcopter is made up with flight controlled board, GPS, and wireless transmitter was used for communication. Spraying pump is controlled from ground station by sending a signal to Arduino.

Chavan et al. [8] developed a Quadcopter using At mega 328 where ESC, magnetometer, gyroscope and water pump is connected and controlled signal is send using RF technology.

Sadhana et al. [9] done with Quadcopter system using Arduino Uno in which gyroscope and accelerometer is used for balancing the quadcopter. For spraying pesticide radio transmitter is used.

Gheorghita and Vintu et al. [10] present two types of approach for mathematical modeling of quadcopter i.e. kinematics and dynamic. First is based on equation of classical mechanics and other one is design from Denavit – Hartenberg. In that dynamic quadcopter was discussed. The model was simulated in math works. Simulation were used to tune the parameter of PID controller that ensure the response to step input reference.

Kamarth and Hereford et al. [12] design swarm of autonomous drone that can fly and hover with no input from user. They used crazy file 2.0 as the based platform then added sensor and made software revision for it operated autonomously. It overcomes unexpected issue such as noisy sensor data that disrupted control loop, weight imbalance. They were able to fly the single PAQ (programmable autonomous quadcopter) and make some measurement of speed and battery life.

3   Hardware Description

3.1   ESC

Electronics speed controller is regulating and control the speed of electric circuit. ESC follows a speed obtain from throttle level, joystick and manually input signal. Esc generate three high frequency signal with different controllable phase continually to keep the motor turning. Esc have source of lots of current can draw the lots if power of the motor. It has 3 sets of wire, one wire will plug into main battery of quadcopter, second wire will have typical servo wire that plugs into receiver throttle channel and third wire is used for powering the wire.

3.2   BLDC

Brushless motor is outer runner motor. It is specially made for quadcopter and multirotor. It is 1400 kV motor. Feature of the motor is 3.2 mm hardened still shaft, dual ball bearing and has 3.5 mm gold spring male connector is already attached and include three female connectors for speed control. It provides high performance super power and brilliant efficiency i.e. 80%. 30 A electronics speed controller can be used to drive motor.

3.3   Radio Receiver

It receives 2.4 GHZ signal from transmitter. It has 10 channels which are independent to receive the signal from transmitter and send it to the controller to further processing. It has low current consumption excellent receiver selectivity and blocking performance. It has current consumption less than 40 mA and works on 5 V power supply.

3.4   MPU 6050

MPU 6050 sensor has combination of accelerometer and gyroscope in single chip. It has 16 bit analog to digital conversion hardware channel to each other. It catches x, y, and z channels at the same time. MPU6050 sensor combine 3 axis accelerometer and 3 axis gyroscope on same silicon chip. It is together with on board digital motion processor which process 6 axis motion fusion algorithm. I2C bus is used to interface with controller.

3.5   LIPO Battery

Lithium polymer batteries are the preferred power sources for most electric modellers. It offers high energy storage ratio, weight ratio and high discharge rate. LIPO battery is in a single cell 2.2 V to in a pack in 10 cells is connected in series 37 V. For Quadcopter 3 cells are connected in series as one parallel which give us 12 V supply. It is 4 stages fully automatic charging process controlled by MCU. It has 100% full load burn-in test. It has high efficiency, long life and high reliability.

4   System Development

In the quadcopter term quad means four that means four motor are used in this system. Mainly there are two type of configuration type are available in quadcopter.in that one is '+ type configuration' and another is 'x type configuration'. In this system we used "x type configuration" [1]. The Quadcopter module turns on by which is remotely located far away. The system mainly divided into two parts i.e. transmitter and receiver. The main objective of the system is to spray insecticide in 100 m^2 area with the help of quadcopter and it is activated by module which is in placed in user location. Advantage of this system is, quadcopter is activated by user with help of module and wireless communication is done between transmitter and receiver [2].

Transmitter section:

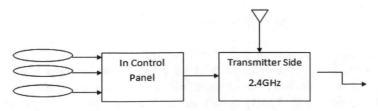

Fig. 1. Block diagram of transmitter section

Figure 1 shows the block diagram of transmitter section. In the transmitter section user have a remote to control the system which generated a signal according to movement of switch on the remote [3]. In the input side potentiometer is varies between 0 to 5 kΩ. It gives signal to the ADC i.e. analog to digital converter. It has 10-bit resolution. As per input given to the ADC the value is convert into 1 byte [4]. Transmitter transmits the signals in the form of channels. No. of channels are transmitted through the transmitter. Transmitter section consist of control panel in which mechanical switches, joystick is including. Transmission rate of the transmitter is 2.4 GHz.

Receiver section:

Figure 2 shows the block diagram of receiver section. Receiver consist of RF 2.4 GHz, BLDC, GPS, Compass, ATMEGA 2560 and water spray unit. It receives the signal through 2.4 GHz RF signal in the form of string. In that string the channel is separated by 5 bytes. The channel are pitch, roll, throttle, yaw, pump. An inertial measurement unit recognizes the difference in pitch, roll and yaw used by gyroscope. The IMU unit consist of gyroscope and accelerometer sensor [5, 6].

Transmitter control assist the measurement sensor like gyroscope and accelerometer as per input send signal to the electronics speed controller [7]. GPS is used to show the location and compass sensor shows the direction. It hold the position x, y axis [10]. Sensor give the signal to the controller [8]. It controls the system. Controller gives 20 mA current and single phase PWM signal to the electronics speed controller. In this system we lift 2 kg weight hence ESC gives 30 A current and 3 phase PWM signal to the motor. ESC give the power to control the speed of motor.

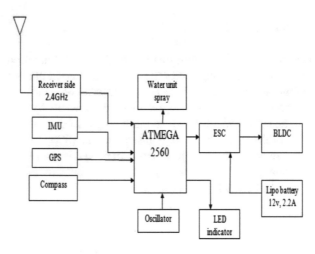

Fig. 2. Block diagram of receiver section

For the pesticide spraying mechanism we use pesticide tank of capacity 300–400 mg. 12 V dc water pump single inlet 1.5 A battery switch pipes fitted to T-split and single nozzles [9]. When the brushless motor is turned on, pesticides through the pipe with the help of the battery spread through the nozzle on define particular agricultural area [11, 12].

5   Result

Insecticide spraying using Quadcopter provides optimal results by covering major area for spraying of pesticides. This is used to avoid human contact with pesticides with high content of chemicals. This can lead to reduction in spraying time. To increase quantity pesticides, we need to increase capacity of tank. With height of nearly five feet we can able to spray pesticides. Following Table 1 gives an idea of results parameters (Figs. 3 and 4):

Fig. 3. Output image

Table 1. Result parameter

Sr. no.	Parameters	Results
1	Height of Quadcopter	5 feet
2	Direction of landing	Vertical
3	Motor load	12 V
4	Area covered by Quadcopter	10 m in square area
5	Weight lifting capacity	2 kg
6	Average flight time	5 to 10 min

Fig. 4. Flying Quadcopter

6   Conclusion

This system provides automatic way of spraying pesticides with lesser computational time. This new approach gives optimal result when compared with the conventional model. Proposed system can provide high performance analysis by attending spraying of pesticide. Human body can be protected from highly harmful pesticide. This is feasible in all environment. This can reduce problem of finding workers in remote locations. This can also encounter attacks by insects on crops which might be helpful for reduction of pollution in agricultural atmosphere. In coming days, we can make a system ecofriendly by implementing system with solar energy.

References

1. Melvel, S., Maguteeswaran, R.: Quadcopter UAV based fertilizer and pesticide spraying system. IARJS **1**(1), 8–12 (2016)
2. Prasad Reddy, P.V., Sudhakar Reddy, K., Vijayarani Reddy, N.: Design and development of drone for agricultural application. IJLEMR **2**, 50–57 (2017)
3. Giles, D.K., Billing, R.C.: Deployment and performance of a UAV for crop spraying. AIDIC **44**, 307–312 (2015)
4. Huang, Y., Hoffmann, W.C., Lan, Y.: Development of a spray system for an unmanned aerial vehicle platform. Appl. Eng. Agric. **25**(6), 803–809 (2009)
5. Patel, P.N., Patel, M., Faldu, R., Dave, Y.R.: Quadcopter for agricultural surveillance. Adv. Electron. Electr. Eng. **3**(4), 427–432 (2013)
6. Kedari, S., Lohagaonkar, P.: Quadcopter a smarter way of pesticide spraying. IJIR **2**, 1257–1260 (2016)
7. Koriahalli, K.B., Jturi, N.: An automatically controlled drone based aerial pesticide sprayer
8. Chavan, P.S., Jagtap, K., Mane, P.P.: Agriculture drone for spraying fertilizer and pesticide. IJRTI (2) (2017)
9. Sadhana, B., Naik, G.: Development of quadcopter based pesticide spraying mechanism for agriculture application. IJIREEICE (5) (2015)
10. Gheorghita, D., Vintu, L., Mirea, L., Braescu, C.: "Quadcopter control system" modelling and implementation. In: 2015 19th (ICSTCC), October 2014–2015
11. Gaponov, I., Razinkova, A.: Quadcopter design and implementation as a multidisciplinary engineering course. IEEE, August 2012
12. Kamarth, L., Hereford, J.: Development of autonomous quadcopter. IEEE (2017)
13. Haque, Md.R., Muhammad, M., Swarnkar, D.: Autonomous quadcopter for product home delivery. In: ICEEICT. IEEE (2014)

Sentiment Analysis for Movies Prediction Using Machine Leaning Techniques

Manisha Jadon[1], Ila Sharma[1(✉)], and Arvind K. Sharma[2]

[1] Department of CSE, R.N. Modi College of Engineering, Kota, India
mjadon09@gmail.com, sharma.ila8@gmail.com
[2] Computer Science & Engineering, University of Kota, Kota, India
drarvindkumarsharma@gmail.com

Abstract. The main intention of this paper is to determine the sentimental analysis for movies prediction using machine learning techniques. A novel methodology along with necessary algorithms is presented here. The input dataset obtained from IMDb is chosen for prediction of box office collection. Firstly preprocessing is performed to remove the redundant features such as stop words, punctuations, numbers, whitespace, lowercase conversion, word stemming, sparse term removal etc.. Preprocessing technique involves two steps namely tokenization and stop word removal. The preprocessed text is sent for transformation process. This formation involves calculating the values of TF and IDF. These are performed for estimating the sentiment analysis. Afterwards, fuzzy clustering is performed that helps in getting both positive and negative outcomes on basis of different movies datasets. In this way, a classification model is made utilizing SVM classifier for foreseeing the pattern of the box office incomes from the reviews of movies.

Keywords: Machine learning · Sentiment analysis · Movie reviews · IMDb · SVM classifier

1 Introduction

Nowadays, due to the rapidly growing of e-commerce, vast comments in the online for the services and the products are created. Text mining is utilized to separate significant data from huge measure of information. A key segment is used to interface together the extricated information to outline new substances or new speculations to be researched further by increasingly standard technique for experimentation. There are numerous difficulties in Sentiments investigation. The essential is that an opinion word that is contemplated to be positive in one situation is additionally considered negative in another situation. The subsequent challenge is that people don't persistently all out sentiments inside similar methods. The standard content processing relies upon the very reality that little varieties between two things of content don't change the which means impressively. Sentiment investigation searches out words that show assessment and realizes the association between textual reviews and furthermore the outcomes of these reviews [1].

D. J. Hemanth et al. (Eds.): ICICI 2019, LNDECT 38, pp. 457–465, 2020.
https://doi.org/10.1007/978-3-030-34080-3_52

The proposed framework demonstrates that a rearranged version of the sentiment-aware autoregressive model can create generally good exactness for foreseeing the movies deal utilizing on online reviews. It uses document level sentiment investigation that comprises of TF and IDF. In sentiment process, real tasks recorded are subjectivity and sentiment grouping, sentiment lexicon generation, sentiment spam identification and nature of reviews [2].

The remainder of the paper is sorted out as follows: Sect. 2 presents related works. Section 3 clarifies the sentiment analysis for films forecast. Section 4 shows performance examination of this proposed system. Section 5 talks about the Review results and anlysis. Section 6 finishes up the paper while at the last references are given.

2   Related Works

Here, related works of the sentiment analysis area is presented. A few of them are reveiwed.

In 2016 the author proffers the SVM classifier that utilizes the sentiment analyses based on the affective feature [3] and [4] and [5] presents the tasks of sentiment analyses survey utilizing the techniques of the machine learning and the engagement of the machine learning techniques in the sentiment classification of the twitters respectively.

In Trupthi et al. [6], investigated AI approaches with various component determination plans, to distinguish the most ideal methodology and found that the characterization utilizing high data highlights, brought about more precision than Bigram Collocation. They likewise recommended that there was a degree for development utilizing mixture systems with different arrangement calculations and the sentiment analyses using the hybrid approach was proffered by the orestes apple [7] in the year 2016. In Brindha et al. [8], displayed an overview on various arrangement systems (NB, KNN, SVM, DT, and Regression). They found that practically all grouping procedures were fit to the qualities of content information. They inferred that further investigation on grouping advancement could get the improved nature of content outcomes and precise information alongside limited getting to time. The sentiment analysis employing the ML applying the syntax features was proposed by Zou et al. [9], the sentiment classification [10] employing the algorithm k-nearest-neighbor utilizing genetic algorithm that is based on the natural process of selecting the optimal solution. [12] presents the twitter based semantic analyses for the sentiment.

3   Sentiment Analysis for Movies Prediction

A notion of the means for TF-IDF normally utilized in the characterization of Sentiments methods. Part of speech (POS) model in which an archive is referred to as a vector, whose passages compare to individual terms of the Vocabulary. POS form information should be a huge pointer of sentiment articulation. The work on subjectivity recognition uncovers a high connection between's the nearness of adjectives and sentence subjectivity [1]. The evaluation results demonstrate that utilizing just adjectives as features bringing up much more performance execution than utilizing a frequency words with the resembalance. As it prompts to compare closeness dependent on capabilities which happen at all documents and having negligible distinctions.

3.1 Text Processing

Pre-processing systems are isolated into two subcategories which are examined as under-

- **Tokenization:** Textual information includes blocks of characters called tokens. The documents are isolated as tokens and utilized for further preparing.
- **Pre-processing:** Pre-processing stage includes primary activities which assit in changing the information for usability before the actual estimation task can be completed. So as to exhibit the impact of pre-processing on the classification models, the analysis includes recording results by considering the total pre-handling step in one methodology and eliminating this phase in the subsequent methodology. In this manner two cases are considered: one for non pre-processed and other one for pre-processing training information.

 Pre-processing stage includes the accompanying undertakings:

- Number Removal
- Punctuation Removal
- Stop Words Removal
- Whitespace Removal
- Lowercase Conversion
- Word Stemming
- Sparse Term Removal.

3.2 Algorithm for Preprocessing

The algorithm for Preprocessing is as follows:

Data → Input data for the system.
Preprocess → Data was preprocessed by using two different processes called tokenizing and stop word removal.
Tokenizing → Data comprises into block of characters.
Input Reviews are separated as tokens and start preprocess.
Stop Words → some of the frequently used stop words in English "a", "of" etc.
Those words are removed from the reviews. Preprocess completed.

4 Performance Investigation

The performance investigation of this proposed approach relies upon different significant parameters, viz Precision, Recall, F1-measure, and Accuracy. Which are examined and explained here.

Review ascertains the precision or the affectability of a classifier. It is a bigger review implies as less false negatives when less review implies as progressively false negatives.

4.1 Precision

Precision estimates the exactness of the classifier. It is bigger precision implies as low false positives, when the lower precision implies high false positives. It could be computed as-

$$\text{PRECİSİON} = TP \, / \, (TP + FP) \tag{1}$$

Also, this is an easy way to improve the Precision is to lesser the Recall.

4.2 Recall

Recall ascertains the affectability or sensitivity of a classifier. It is a bigger Recall implies as less false negatives when less Recall implies as progressively false negatives. It could be computed as-

$$\text{RECALL} = TP/(TP + FN) \tag{2}$$

It improves the Recall may often the Precision decrease because it gets hard to precise.

4.3 F1-Measure

It is the Precision and Recall metric is merged to produce the unique metric is known as F1-measure. It is the weighted harmonic mean of the evaluation metric as Precision and Recall.

$$\text{F1} - \text{Measure} = 2 \times (\text{Precision} \times \text{Recall})/(\text{Precision} + \text{Recall}) \tag{3}$$

4.4 Accuracy

It is an effective classifier for evaluating the metrics. Two more useful metrics are Precision and Recall, it also used to the statistical measures.

$$\text{Accuracy} = (TP + TN)/(TP + TN + FP + FN) \tag{4}$$

The accuracy of the propagation is true results between the total numbers of cases is examined. Accuracy is a usual calculation for the classification evaluation.

5 Movies Review Results and Analysis

Movies reviews are utilized in e-commerce to give the customers an opportunity of comment and rate on the products. The review of movies in sentiment analysis are preprocessed by the searching process of several predefined reviews. This movies reviews have numerous dataset, which comprises positive and negative reviews. The positive sentiment analysis exhibits the good outcomes and the negative sentiment analysis displays the poor outcomes.

5.1 Comparison of Positive Reviews

Table 1 exhibits the comparison of positive reviews for the different techniques and the different parameters like Precision, Recall, F1-measure, and Accuracy. The Table 1 shows the process of positive reviews when the number of approaches compared.

Table 1. Comparison analysis of positive reviews

Approach	Precision	Recall	F1-measure	Accuracy
	Positive	Positive	Positive	Positive
Semantic based	79.7	77.5	78.6	78.6
SVM based	76.3	70.2	73.2	74.7
Hybrid approach (label features only) [16]	78.5	70.1	74.1	75.6
Hybrid approach (all features)	81.2	78.2	79.7	80.6
Proposed approach	88.6	84.6	83.4	92.6

Following Fig. 1 shows the graphical representation of the comparison of positive reviews. This specifies the varies methods like semantic-based, SVM based, hybrid techniques of label features, hybrid techniques of all features, and proposed techniques. The positive values of the parameters as precision, recall, F1-measure, and accuracy measured the positive reviews. From this, existing methods are compared by the proposed method. It gives the accuracy of positive reviews as high than the traditional method by the values of 0–100.

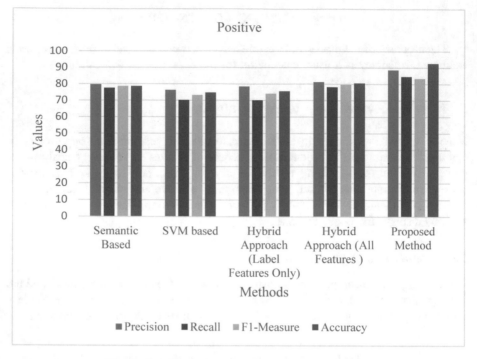

Fig. 1. Perspective on comparison of positive reviews

5.2 Comparison of Negative Reviews

Table 2 reveals the comparison of negative reviews for the different techniques and the different parameters like precision, recall, F1-measure, and accuracy. The Table 2 shows the process of negative reviews when the number of approaches compared. In the semantic-based the precision, recall, F1-measures and accuracy of the values as the 68.2, 72.4, 70.2, and 72.8 as the negative reviews.

In the SVM based approaches is compared by the negative reviews of the parameters as the values (76.3, 70.2, 73.2, and 74.7). In the label features of hybrid techniques has the comparison of the negative reviews by precision, recall, F1-measures and accuracy as (74.5, 66.8, 70.4, and 73.4). In the overall features based hybrid approaches by the negative reviews of such parameters consists of various average values as 79.6, 81.6, 80.1, and 77.5. In the proposed method, the values are given as highest to compare the existing methods.

Table 2. Comparison analysis of negative reviews

Approach	Precision	Recall	F1-measure	Accuracy
	Negative	Negative	Negative	Negative
Semantic based	68.2	72.4	70.2	72.8
SVM based	74.5	66.8	70.4	73.4
Hybrid approach (label features only)	75.8	72.8	74.3	71.9
Hybrid approach (all features)	79.6	81.6	80.1	77.5
Proposed method	80.2	82.1	81.2	90.7

Figure 2 Perspective on Comparison of Negative Reviews.

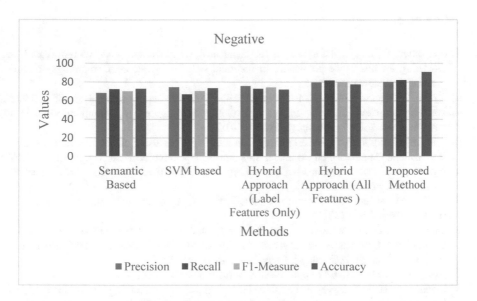

Fig. 2. Comparison of negative reviews

Figure 2 shows the methods and the average values. From the parameters of accuracy, precision, recall, and f1-measure calculate the negative reviews by the proposed techniques. Also, the existing methods are compared by the present method it shows the accuracy as more than the existing methods.

The graphical representation provided depicts the accuracy and authors, methods of the reviews. This represents the existing methods of the SVM based which varies in the accordance of the number of accuracy values and authors provided. The maximum values shows the accurateness on the existing methods and authors, which improves the accuracy one by one of the reviews. The earliest movies of the reviews show the accurate values.

6 Conclusion

The aim of this research to present a novel sentimental framework with the use of SVM is presented with the reviews of different movies by its negative and positive ratings and rankings. Moreover, the SVM classifier is utilized to enhance the performance of the proposed model. Also, a different semantic analysis characteristics concerning POS tagging, removal of stop words are employed for the attainment of reviews which could be both positive and negative. Therefore, the obtained results shows the efficiency of proposed classifier techniques on comparing other classifier approaches. From this it gives evident that our model attains very good level of accuracy that could more effective on comparing the other existing models.

References

1. Nagamma, P., et. al.: An improved sentiment analysis of online movie reviews based on clustering for box-office prediction. In: International Conference on Computing, Communication & Automation (2015)
2. Tyagi, E., Sharma, A.K.: An intelligent framework for sentiment analysis of text and emotions-a review. In: 2017 International Conference on Energy, Communication, Data Analytics and Soft Computing (ICECDS) (2017)
3. Luo, F., et al.: Affective-feature-based sentiment analysis using SVM classifier. In: 20th International Conference on Computer Supported Cooperative Work in Design. IEEE (2016)
4. Aydoan, E., et al.: A comprehensive survey for sentiment analysis tasks using machine learning techniques. IEEE (2016)
5. Qasem, M., et al.: Twitter sentiment classification using machine learning techniques for stock markets. IEEE (2015)
6. Trupthi, M., et al.: Improved feature extraction and classification-sentiment analysis. In: International Conference on Advances in Human Machine Interaction, HMI-2016. R.L. Jalappa Institute of Technology, Bangalore, 3–5 March 2016
7. Apple, O., et al.: A hybrid approach to sentiment analysis. IEEE (2016)
8. Brindha, S., et al.: A survey on classification techniques for text mining. In: 3rd International Conference on Advanced Computing and Communication Systems, ICACCS-2016, Coimbatore, India, 22–23 January 2016
9. Zou, H., et al.: Sentiment classification using machine learning techniques with syntax features. In: International Conference on Computational Science and Computational Intelligence. IEEE (2015)
10. Kalaivani, P., et al.: An improved K-nearest-neighbor algorithm using genetic algorithm for sentiment classification. In: International Conference on Circuit, Power and Computing Technologies (ICCPCT). IEEE (2014)
11. Karamibekr, M., et al.: A structure for opinion in social domains. IEEE (2013)
12. Saif, H., et al.: Semantic sentiment analysis of Twitter. In: 11th International Semantic Web Conference, ISWC-2012 (2012)
13. Internet Movie Data Base. https://www.imdb.com
14. Tyagi, E., Sharma, A.K.: Sentiment analysis of product reviews using support vector machine learning algorithm. Indian J. Sci. Technol. **10**(35), 1–9 (2017)

15. Kumari, U., Sharma, A.K., Soni, D.: Sentiment analysis of smart phone product review using SVM classification technique. In: IEEE-2017 International Conference on Energy, Communication, Data Analytics & Soft Computing (ICECDS) (2018)
16. Zhao, K., Jin, Y.: A hybrid method for sentiment classification in Chinese movie reviews based on sentiment labels. In: International Conference on Asian Language Processing (IALP) (2015)

An Automated System for Driver Drowsiness Monitoring Using Machine Learning

Suvarna Nandyal and P. J. Sushma[✉]

Department of Computer Science and Engineering,
Poojya Doddappa Appa College of Engineering, (Affiliated to VTU Belagavi,
and Approved by AICTE), Kalaburagi, India
suvarna_nandyal@yahoo.com, jujaresushma@gmail.com

Abstract. Now a days the road accidents are increasing, the primary cause for these accidents is the drowsy driving which leads to death. This is the reason that the detection of driver drowsiness and its indication is foremost in real world. Most of the methods used are vehicle based or Physiological based. Some methods are intrusive and distract the driver, some require expensive sensors manually. But in today's era real time driver's drowsiness detection system is very much essential. Hence, the proposed system is developed. In this work a webcam records the video where the driver's face is detected. Facial landmarks are pointed on the recognized face and the Mouth Opening Ratio (MOR), Eye Aspect Ratio (EAR), and head bending values are calculated, subjective to these values drowsiness is detected based on threshold, and an alarm is given. As the drowsiness is the stage where the driver is unmindful of persons walking on the road, so the pedestrian is detected to avoid any calamity and potholes are identified to avoid the sudden changes in the driving speed which is caused by drowsiness. From the observation, it is found that proposed system works well with 98% of accuracy. If the user is slightly bend for head and mouth then accuracy is less.

Keywords: Drowsiness · Facial landmarks · Eye aspect ratio · Mouth opening ratio · Head bending values

1 Introduction

Drowsiness the of driver is a state in which the person is neither fully alert nor completely in sleep mode it is in between state. This drowsiness may cause inactiveness in the driver which is an unfavourable condition. The large number of deaths occur due to drowsiness related road accidents and detection of driver drowsiness contributes to the decrease in the number of deaths occurring in traffic accident. In this context, it is extremely important to use technologies to implement the systems those are able to continuously monitor drivers and measure their level of alertness during driving. The detection of drowsy driver is classified into three types: behavior based, vehicle based, and physiological based. In behavioral based method, blinking and closing of eye, head bending, yawn, etc. are analyzed to detect drowsiness. In vehicle based method, a number of standards like movement of steering wheel, accelerator, speed of vehicle, deviations

© Springer Nature Switzerland AG 2020
D. J. Hemanth et al. (Eds.): ICICI 2019, LNDECT 38, pp. 466–476, 2020.
https://doi.org/10.1007/978-3-030-34080-3_53

from position of lane etc. are monitored. It is considered as driver drowsiness when there is detection of any unusual change in these values. In physiological based method, the signals like Electrocardiogram (ECG), heartbeat, Electro-oculogram (EOG), pulse rate, Electroencephalogram (EEG), are monitored and from these drowsiness is detected.

In the growing nations like India, with advancement in the technology of transportation and uplift of vehicle numbers on the roads, there is increase in road accidents. Hence the proposed work is to give alertness by means of voice alert system or alarm to the driver when he is inattentive to drive and shift from normal alert mode to non-alert. In this system real time data is collected by video camera, this data gives information about driving condition of the driver which acts as input to controller. Based on the input appropriate measures are taken by the controller to alert the driver.

2 Related Work

Horng and Chen has proposed a fatigue detection system for driver. The face of the driver is located, and using edge detection the eye regions are located. The obtained eye images are converted to the HSI model to distinguish eyeball pixels to determine the closing and opening of eyes for judging driver fatigue [11, 13]. Tian and Qin proposed a system which uses the skin-color matching & vertical projection in the daytime and it also uses the intensity adjusting & vertical projection at night to detect and track the face region, and then validates it by geometric symmetry of face, secondly apply the horizontal projection to notice the eye position and certify it by some features of face [8, 15]. Hong and Qin proposed a system to detect driver's drowsiness and state of eye. Firstly, the Haar-like feature detection is used for face region detection. Secondly, to get the eye region a horizontal projection and geometrical position are identified for face detection and eye position detection [12]. Ashardi and Mellor proposed a drowsiness detection model by using the input paramerters such as steering wheel angle and the distance between the lane [1].

Bhowmick and Chidanand proposed method called Otsu thresholding where the facial landmarks are located for Eye localization. Morphology operations and k-means is used for accurate eye segmentation. Then to extract the position of the eye, features of shape are calculated and these are trained using SVM [5, 14]. Picot and Charbonnier proposed drowsiness detection system of driver. A channel of electroencephalographic (EEG) is used for brain activity monitoring. Visual activity is observed through blink detection. Features of blinking are taken out from an electro-oculographic (EOG) channel. These features are merged to produce an EOG-featured detector of drowsiness [2, 6]. Alshaqaqi and Baquhaizel has proposed ADAS (Advance Driver Assistance System) module. It uses an algorithm to analyze, locate and track the drivers eyes and face to compute PERCLOS (percentage of eye closure), a scientifically supported course of drowsiness associated with slow eye closing [4, 10]. Ahmad and Borole proposed system is implemented using cascade object identifier from vision toolbox of Mat lab, which detects face, eyes, nose and mouth from the image which is captured from web camera [9]. Sengupta and Dasgupta proposed a scheme for estimate the levels of alertness of an particular person using concurrent asset of multimodal physiological signals and combining the data into a single standard for quantification of alertness [3, 7].

3 Proposed Methodology

Figure 1 shows proposed methodology of the overall working of the system. Methodology is the approach for the analysis of the new methods used for the study. It comprises of system methods and theoretical analysis with that system. It consists of drowsiness monitoring system connected to raspberry pi and camera. This system monitors main three parameters such as the status of opening and closing of eye, yawning, head bending. And also it has features like pedestrian detection and pothole detection.

Fig. 1. Schematic image of the system

The system including these devices work together and transfer all the information to the raspberry pi which processes the data and automatically set the appropriate course of action. The functions of each device working are as follows:

Raspberry Pi: This controls the whole system to work properly and take the prescribed action such as detecting face from the video and giving alarm, data processing, analysis.

Web Camera: The camera records the video of the driver and detects the face of driver and then it detects the eye region, mouth region, head bending. Also it detects the human movements to give the pedestrian detection, also the pot hole detection is done using camera.

Speakers: The speakers give the alarm as voice messages to the driver to be alert if the driver is detected drowsy based on eye aspect ratio, mouth opening ratio, and head bending. And also it gives alarm when it detects pedestrian.

USB: USB is the standardized connection between computer peripherals and personal computers, these together communicate to supply power. Here it connects the raspberry pi to PC to get the power supply. It has widely replaced interfaces such as parallel ports, and serial ports.

Ethernet Cable: In the wired network an Ethernet cable is the most used type. The Ethernet cable connects devices within a LAN (local area network) like PCs, routers. Here, it connects the personal computer to raspberry pi for sharing files and internet access.

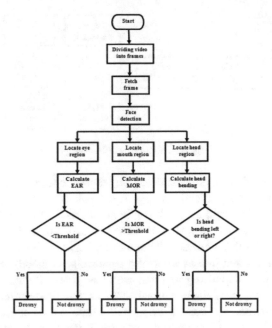

Fig. 2. Flow diagram of system

The Fig. 2 shows the flow diagram of the proposed system where it elaborates each stages of the system. In this the images are taken from the video and face is detected, from this specific regions like eyes, mouth and head are marked on the face. To calculate these specific regions some threshold values are given. The threshold value of EAR is 0.3 and for MOR threshold is 0.6 for head bending the z-axis threshold is 15. Depending upon these calculated values the drowsiness is detected.

3.1 Drowsiness Detection

The proposed system block diagram is given. The video is recorded using a webcam. The camera will be positioned in front of the driver which captures the image of the face. The frames are extracted from the video to obtain images. Face is detected in the frames using haar cascade. After detection of the face, facial landmarks are marked like positions of eye, head, and mouth. From the facial landmarks, eye aspect ratio, mouth opening ratio and position of the head are estimated and the drowsiness of the driver is detected using these features and machine learning approach. If the driver is found drowsy and is detected, an alarm will be sent to the driver to alert him/her. The details of each block are discussed as follows in Fig. 3.

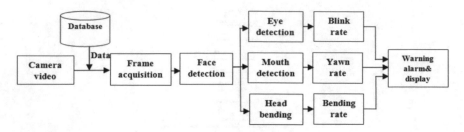

Fig. 3. Proposed methodology

The Fig. 3 shows the eye position detection where the frames are acquired from the video later on face is detected and eye region is located. Then eye position is tracked continuously during this stage the blink rate is monitored if it is found for more frames then is considered as not the normal blink then it gives alarm to the driver. Similarly as depicted in figure after face detection the mouth region is landmarked where the mouth opening of person is taken into account. If a person opens mouth for more number of frames then it is considered as yawn and alarm is given. Further the head position detection, initially the head is in normal state as the driver tilts the head due to drowsiness its head bending rate is evaluated. These values are considered to find whether the bending normal or due to drowsiness. If it found as drowsy alert is given.

The working stages of system is as follows:

Data Acquisition: In this the video is captured using webcam and each frames are extracted and are processed for face detection. Using image processing techniques on these images driver face is detected using webcam with periodic eye blinking, eye closing, yawning and head bending.

Face Detection: First the face of driver is detected from the frames. In this Haar cascade is used. In Haar Cascades a function is trained from a large number of positive and negative images. The extracted images from that can be used to differentiate between one image to other image. At a location of target the Haar feature takes adjacent regions which are rectangular, the pixel intensities sums up in all region and the difference between these sums is calculated. Most of the features calculated are irrelevant. In the detection phase, the target size window is incorporated over the input image and Haar featured values which means the value at rectangled region are calculated for each subsection. The difference is compared to threshold which is learned value, that separates objects of samples of eye, mouth, and head from non-objects (Fig. 4).

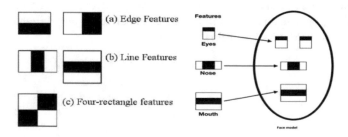

Fig. 4. Haar features

The classifiers at each stage label the region either positive or negative given by the sliding window of the current location. Positive regions indicate that an object was found and non-objects are indicated by negative regions. The region is passed by the classifier to the next stage if the label is positive, the label is given as negative if the classification of this region is found complete, and the window is slided to the next location by detector. At the final stage when it is classified the region as positive. Using detector the object is found at the current window location.

Facial Landmark Marking: In this we need to find different facial features like the regions of the eyes, mouth and head. Initially the face images are normalized in order to reduce the effect of distance from the camera, varying image resolution and non-uniform illumination. After image normalization the Face Landmark Detection algorithm is used. This technique utilize simple feature i.e. pixel intensities differences to approximate the landmark positions. By a cascade of regressor the estimated positions are refined with an iterative process. The regressors produce a new estimate from the previous one, trying to reduce the alignment error of the estimated points at each iteration. The algorithm is fast, it takes about few milliseconds to detect a set of landmarks on a given face (Fig. 5) (Table 1).

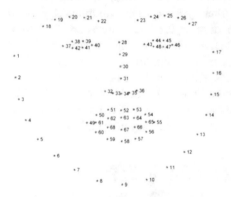

Table 1. Facial landmark points

Parts	Landmark points
Right eye	37–42
Left eye	43–48
Mouth	49–68

Fig. 5. The facial landmark points

Feature Extraction
The features are as described below.

Eye aspect ratio (EAR): From the regions of eye which are located, the ratio of height and width of the eye is given by:

$$EAR = [(P_{38} - P_{42}) - (P_{39} - P_{41})]/2(P_{40} - P_{37})$$

Where P_i represents point marked as i in facial landmark and $(p_i - p_j)$ the distance between points marked as i and j. Hence when the eyes are completely open, EAR value is high and when the EAR value goes down towards zero the eyes are closed. Thus, monotonically decreasing EAR values indicate closing eyes and it is zero for completely closed eyes (eye blink). Accordingly, due to drowsiness the driver blinks eyes and EAR values indicate the drowsiness of the driver.

Mouth opening ratio (MOR): To detect yawning during drowsiness the MOR is calculated as:

$$MOR = [(P_{51} - P_{59})(P_{52} - P_{58})(P_{53} - P_{57})]/(P_{55} - P_{49})$$

When mouth opens due to yawning the MOR value increases rapidly and remains at high for a while due to yawn (indicating that the mouth is open) and again decreases rapidly towards zero. One of the characteristics of drowsiness is yawn.

Head Bending: Driver's head tilts towards either side (left or right) due to drowsiness with respect to axis. With this the driver drowsiness can be detected.

3.2 Pedestrian Detection

The Fig. 6 illustrates pedestrian detection which is done using mobile net SSD (Single-Shot Detector). This is very important part as driver is drowsy the persons walking on the lanes of road may not be visible hence pedestrian detection is done. It uses Mobile net for classification and recognition and the SSD is a framework to realize the multibox detector. The combination of both can do human detection. A SSD is to predict the class probabilities and bounding boxes. The model takes the input an image which passes through with different sizes of filter by multiple convolutional layers, to predict the bounding boxes on detection of pedestrian the feature maps are used from convolutional layers at different position of the network. Bounding boxes are produced by extra feature layers of 3 × 3 filters of a specific convolutional layer are used. Each box has the parameters they are: the coordinates of the center, the height and the width. Only the relevant bounding boxes are considered in the final mode of SSD model. The subpart of selecting the boxes is done by Hard Negative Mining. The boxes are ordered and the top is selected depending on the ratio between the negative and the positive.

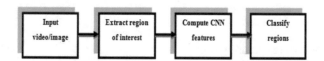

Fig. 6. Block diagram of pedestrian detection

3.3 Pothole Detection

The Fig. 7 shows the pothole detection. In this the image is captured from the camera and is converted to its HSV (hue, saturation, value) channels as it finds the maximum variance from the image. The performance of HSV, the region of interest can be selected. With specific colour channel the detection of pothole is done. The average colour model which does not match with complete roads is not included. The area of the road is taken and put under analysis for the detection of potholes. From the image the recognition of the regions differing in properties such as contrast, brightness is considered and the potholes with different textures compared to road area mainly from dark areas and then pothole is detected.

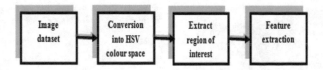

Fig. 7. Block diagram of pothole detection

4 Results and Discussion

The Fig. 8 shows the output of eye detection where the eye rate is given continuously when the change in rate is found which is less then the threshold 0.3 then it gives alarm. Similarly the Fig. 9 shows the yawn rate values, here if the values are increased more then the threshold 0.6 for more number of times then MOR rate increases which is indication as drowsy.

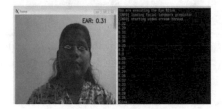

Fig. 8. Eye aspect ratio and blink detection

Fig. 9. Mouth opening ratio and yawning

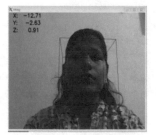

Fig. 10. Head position detection

Fig. 11. Pedestrian detection

Figure 10 shows the head position detection, when the driver tilts due to drowsiness then alert is given and in Fig. 11 shows the pedestrian detection to detect persons walking on the road lanes when the drowsy driver unable to identify the presence of persons. The Fig. 12 shows the pothole detection on the roads for safe driving.

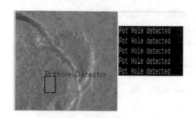

Fig. 12. Pothole detection

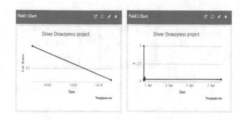

Fig. 13. Threshold value of EAR and EAR values

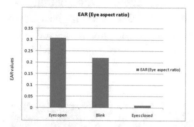

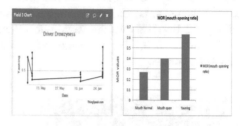

Fig. 14. EAR values at different values of eye position

Fig. 15. MOR values at different mouth opening ratio

The Fig. 13 shows the EAR values at different time stamps and Fig. 14 gives the analysis of eye state values at different conditions such as eyes open, close and blink. The Fig. 15 gives the analysis of variations in MOR values as in normal, mouth open state and yawn.

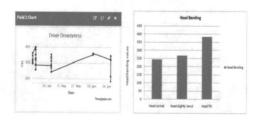

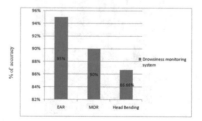

Fig. 16. Head bending values at different positions of head

Fig. 17. % of accuracy of EAR, MOR, and Head bending value

The Fig. 16 is the analysis of position of head when drowsy and normal state. And the Fig. 17 gives the percentage of accuracy of all the parameters considered for drowsiness detection such as EAR, MOR and head bending (Table 2).

The % of accuracy and error rate is calculated as:

$$\% \text{ of Accuracy} = \frac{\text{Total no.of images recognized correct}}{\text{Total no.of images}} * 100$$

$$\% \text{ of Error} = \frac{\text{Total no.of images recognized wrong}}{\text{Total no.of images}} * 100$$

Table 2. % of accuracy and % of error in the system

Features	% of accuracy	% of error
EAR	95%	5%
MOR	90%	10%
Head bending	86.66%	13.34%

5 Conclusion

A real time driver drowsiness monitoring system has been proposed based on visual behavior and machine learning. Here, features like eye aspect ratio, mouth opening ratio and head bending values are computed from the video. The implemented system works accurately with the data generated by taking images from video. Where the data is continuous stream which includes the EAR, MOR and head bending values. When driver is inattentive to drive on roads the system includes the features like pedestrian detection which is useful to recognize the persons walking on road lanes to avoid accidents. Then Pothole detection is done to avoid the sudden change in the driving speed due to drowsiness. Each parameter value analysis and accuracy calculation is done by considering the data streamed.

References

1. Abas, A., Mellor, J., Chen, X.: Non-intrusive drowsiness detection by employing Support Vector Machine. In: 2014 20th International Conference on Automation and Computing (ICAC), Bedfordshire, UK, pp. 188–193 (2014)
2. Picot, A., Charbonnier, S.: On-line detection of drowsiness using brain and visual information. IEEE Trans. Syst. Man Cybern.—Part A: Syst. Hum. **42**(3), 764–775 (2012)
3. Sengupta, A., Dasgupta, A., Chaudhuri, A., George, A., Routray, A., Guha, R.: A multimodal system for assessing alertness levels due to cognitive loading. IEEE Trans. Neural Syst. Rehabil. Eng. **25**(7), 1037–1046 (2017)
4. Alshaqaqi, B., Baquhaizel, A.S., Ouis, M.E.A., Bouumehed, M., Ouamri, A., Keche, M.: Driver drowsiness detection system. In: IEEE International Workshop on Systems, Signal Processing and Their Applications (2013)
5. Bhowmick, B., Chidanand, K.S.: Detection and classification of eye state in IR camera for driver drowsiness detection. In: IEEE International Conference on Signal and Image Processing Applications (2009)

6. Belakhdar, I., Kaaniche, W.: Detecting driver drowsiness based on single electroencephalography channel. In: International Multi Conference on Systems, Signals and Devices (2016)
7. Chui, K.T., Tsang, K.F., Chi, H.R., Ling, B.W.K., Wu, C.K.: An accurate ECG based transportation safety drowsiness detection scheme. IEEE Trans. Ind. Inform. **12**(4), 1438–1452 (2016)
8. Dalal, N., Triggs, B.: Histograms of oriented gradients for human detection. In: IEEE Conference on CVPR (2005)
9. Ahmad, R., Borole, J.N.: Drowsy driver identification using eye blink detection. Int. J. Comput. Sci. Inf. Technol. **6**(1), 274 (2015)
10. Vitabile, S., De Paola, A.: Bright pupil detection in an embedded, real-time drowsiness monitoring system. In: 24th IEEE International Conference on Advanced Information Networking and Applications (2010)
11. Singh, S., Papanikolopoulos, N.P.: Monitoring driver fatigue using facial analysis techniques. In: IEEE Conference on Intelligent Transportation System, pp. 314–318, October 1999
12. Hong, T., Qin, H.: An improved real time eye state identification system in driver drowsiness detection. In: IEEE International Conference on Control and Automation, Guangzhou, China, 30 May–1 June 2007
13. Horng, W.B., Chen, C.Y., Chang, Y., Fan, C.H.: Driver fatigue detection based on eye tracking and dynamic template matching. In: IEEE International Conference on Networking, Sensing and Control, Taipei, Taiwan, 21–23 March 2004
14. Ou, W.L., Shih, M.H., Chang, C.W., Yu, X.H., Fan, C.P.: Intelligent video-based drowsy driver detection system under various illuminations and embedded software implementation. In: 2015 International Conference on Consumer Electronics, Taiwan (2015)
15. Tian, Z., Qin, H.: Real-time driver's eye state detection. In: IEEE International Conference on Vehicular Electronics and Safety, 14–16 October 2005, pp. 285–289 (2005)

A Comparative Review of Recent Data Mining Techniques for Prediction of Cardiovascular Disease from Electronic Health Records

M. Sivakami[1] and P. Prabhu[2(✉)]

[1] Department of Computer Applications, Alagappa University,
Karaikudi, Tamilnadu, India
sivakamiurcw@gmail.com
[2] Directorate of Distance Education, Department of Computer Applications,
Alagappa University, Karaikudi, Tamilnadu, India
prabhup@alagappauniversity.ac.in, pprabhu70@gmail.com

Abstract. Cardiovascular disease is one of the key diseases spreading all over the world. Conventional method of curing ailment generates voluminous patient data which are left uncared during and after treatment. The collected data can be mined systematically using various tools & techniques and the wealth of information obtained could systematically assist the clinicians' in decision making. Literature has cited a plenty of work concerning the stated problems using various data mining techniques are reviewed here critically. The reported accuracy level clearly revealed that the arrival of distinct decision needs more accuracy for better decision making support for doctors. Hence the mining of electronic health records using hybrid model which combines benefits of various algorithms for pre-processing and modeling may provide more insight to discussed problem with improved scalability, speed and accuracy.

Keywords: Data mining · Heart disease · Prediction · Decision making · Modeling · Hybrid techniques

1 Introduction

Health care systems are generating peta bytes of data which are to be reclaimed when required. Data mining is a systematic assignment of abstracting valuable information present in a data [1]. The mining concepts enabled hospital data is useful to identify specific patterns which could render decision-making for diagnosis and treatment planning. The predictive model enabled hospitals information systems will reduce subjectivity and time for decision making between healthcare providers and beneficiaries [2]. The hidden wealth of information present in a database is retrieved and meaningfully transformed into facts are primary goals of the data mining task. Development of smart devices and cloud computing methodologies and analytical tools benefits mankind in exploring useful information from a dataset. The ascertained facts are used properly to enrich the excellency of clinical support system. This paper reviews the utility of data mining techniques for prediction of heart disease with different data mining tools. The following chapter and sections will define the

© Springer Nature Switzerland AG 2020
D. J. Hemanth et al. (Eds.): ICICI 2019, LNDECT 38, pp. 477–484, 2020.
https://doi.org/10.1007/978-3-030-34080-3_54

mentioned attributes of data mining and signify its importance in the field of health care industries. The Sect. 2 discuss about the cardiovascular disease. The review Sect. 3 comprises ensuing sections such as various learning techniques to understand the accuracy level for heart disease prediction. Secondly, based on the clinical health records, the review concentrates on the machine learning approaches which were employed to identify the different types of heart disease. Improved efficacy in predicting heart disease via hybrid models is suggested in third section. The fourth section of review is limited to precise prediction of coronary heart disease using various mining methods. The extent of cardiac risk accessed by mining algorithm using electronic health records are conceded in section five. Lastly the review, reports the various accuracy level on comparison of individual algorithm with others. The Sect. 4 deals with the inference from the comparative review related to cardiovascular disease prediction. The Sect. 5 discuss about the conclusion and future work.

2   Cardiovascular (Heart) Disease

Cardiovascular Disease (CVD) is a condition of narrowing or blocking blood vessels leading to a heart attack, heart failure, heart valve disease, heart muscle disease, chest pain (angina), abnormal heart rhythm and/or stroke. Generally heart related disease is induced due to physical idleness, irregular diet and detrimental usage of alcohol and tobacco. The CVD are classified as angina, arrhythmia, coronary artery disease, cerebrovascular disease (stroke), heart attack, and rheumatic heart disease [3]. To access the same, data collected from patients are classified as structured data such as patients' history and unstructured data such as diagnosis report, images and ECG structures etc. Thus the medical data are found to be quite complex and conclusion obtaining from such a data is a herculean task. Database technology, machine learning and statistical analysis are employed to unveil the facts present in the health record via supervised and unsupervised learning strategies [4]. Depending on the modeling objectives the said data mining techniques will cater the needs. The commonly employed modeling objectives are classification and prediction. Classification and prediction are the modeling objectives routinely employed to analyze the data sets such as Naive Bayes, Decision Tree, neural network, bagging algorithm, and support vector machine [5].

3   Related Works

This section discuss about literature review of decisive support system by various data mining techniques for clinical decision making.

3.1   Data Mining Techniques

Different data mining techniques were applied to identify CVD earlier with its merits and demerits but still needs improvements. Amin et al. [6] studied a method using 13 attributes in seven classification techniques viz., k-NN, Support Vector Machine (SVM), Naive Bayes, Logistic Regression(LR), Vote (a hybrid technique with Naïve Bayes,

Logistic Regression) and Neural Network Decision Tree. The study concluded that the good accuracy levels were reached in vote model and the extent of severity of heart disease was also analyzed. Emre et al. [7] employed different supervised learning approach were used to analyze the health records. The study revealed that the appreciable results were obtained using Classification and Regression Tree algorithm and climatic conditions has influences over the cardiac arrest. Domadiya et al. [8] analyzed the problem of privacy preserving in the medical data set and found out that vertically partitioned health care sets were effectively scrutinized using association rule mining. The time and rule restricted parameters enhanced machine learning was done by association rule to predict heart disease as accounted by Ordonez et al. [9]. The study concluded that the records were mined with minimum support of two transactions. Support Vector Machine classifier enabled automatic diagnosis procedure was developed by Dolatabadi et al. [10] to assess the conditions of coronary artery disease by comparing normal and affected patients heart rate variability and electrocardiogram data sets. Clear aortic stenosis and clear mitral regurgitation were clearly differentiated by rule based method of decision trees on heart sound data sets was effectively done by Pavlopoulos et al. [11]. Heart rhythm risk rate with therapeutic assistance was developed by Karaolis et al. [12] using association rule based apriori algorithm. Statistical classification of genders based on coronary heart disease was done by Nahar et al. [13] using association rule mining and explored clinical factors causing heart disease upon males and females.

3.2 Types of Heart Disease

The following authors identified the acute coronary syndrome and Coronary Artery Heart Disease along with risk level prediction using risk scoring methods. Huang et al. [14] developed a learning approach from Electronic Health Record (EHR) to address the clinical risk factors for Acute Coronary Syndrome (ACS) in comparison with small set of traditional ACS risk scoring methodologies. The studied method identified handful of ACS patients with different risk levels. El-Bialy et al. [15] integrated the results of the system learned outcomes on various attributes of diverse datasets for accessing the coronary artery disease and developed a learning method for accurate diagnosis.

3.3 Hybrid Techniques

Manogaran et al. [16] suggested that high sensitivity 98% high specificity up to 99% was achieved using multiple kernel learning with adaptive neuro-fuzzy inference system on KEGG Metabolic Reaction Network dataset for heart disease diagnosis and obtained results were compared with deep learning methods. Arabasadi et al. [17] learned a hybrid method with neural network and effectively detected coronary artery disease without the aid of angiography. Polat et al. [18] presented a hybrid approach using artificial immune recognition system (AIRS) with association rule and fuzzy logic techniques and formulated a predictive analysis for identification of CVD and the proposed hybrid model's specificity was tested using AIRS system. Relief and Rough Set (RFRS) method based hybrid classification system produces accurate results on medical data set was investigated by Liu et al. [19]. Prabhu et al. [20, 21] proposed models using frequent sets for decisive business transactions. K-groups based features are used to group the users of similar type. Business intelligence is obtained to recommend top N itemsets in e-commerce domain. Table 1. Display the comparison of hybrid prediction models.

Table 1. Comparison of hybrid prediction models

Author	Year of publication	Type of disease	Tool	Techniques	Accuracy
Manogaran et al. [18]	2017	Heart Disease diagnosis	Adaptive Neuro-Fuzzy Inference System (ANFIS) Classifier	Multiple Kernel Learning (MKL)	98%
Arabasadi et al. [19]	2017	Coronary artery disease (CAD)	—	Genetic algorithm with Neural network	93.85%
Polat et al. [20]	2007	Heart disease and Hepatitis disease	Artificial Immume Recognition System (AIRS) classifier	C4.5 decision tree algorithm	92.59%
Liu et al. [21]	2016	Cardio Vascular Disease (CVD)	Ensemble classifier	Relief and Rough Set (RFRS) method	92.59%

3.4 Prediction of Cardiovascular Diseases

Tayefia et al. [22] utilized traditional risk factors with clinical biomarkers in the mining algorithm for accurate, specific and sensitive prediction of coronary heart disease. The result suggested that the added biomarker as attribute assist the task in the machine learning process. Classification of heart failure based on its types was attempted over the public of Ontario by Austin et al. [23] and found out that the tree-based classification offers better than the conventional classification technique. Using the heart perfusion measurements and risk factors as attributes Ordonez [24] mined by association rules on patient's four arteries data set and found out that the prescribed attributes of mining reduce the number of rules. One-hot encoding and word vector modeling in neural network platform was attempted by Jin et al. [25] to predict heart failure incidences to support diagnostic systems. Chahal et al. [26] have discussed the importance of protein coding gene to assess the cause of congenital heart disease with diagnosis attributes. The techniques utilize genes sequenced identification in these regions with non-coding variant data sets. Tsipouras et al. [27] suggested an interpretation technique with efficiency and consistency for the finding of coronary artery disease by machine learning methods. Rémy et al. [28] devised a data mining model to find out a physician for effective diagnosis of heart diseases. Frequent itemset instructed association mining rule was employed by Ilayaraja et al. [29] to access the extent of risk level during cardiac disorder. The routine clinical attributes were employed in classification techniques for predicting the heart diseases and encouraging results were obtained by Princy et al. [30].

3.5   Risk Level Prediction

Anooj [31] used a fuzzy logic assisted health record mining for cardio vascular disease prediction and reported that the process of diagnosis was easy with fuzzy learning system. Rare association rule assisted machine learning method was utilized on data sets of cardiovascular disease, hepatitis and breast cancer and the obtained the results of mining forecast the memory usage with execution time as reported by Borah et al. [32]. Karaolis et al. [33] devised a mining rule for assessing risk factors involved during coronary heart disease. Pu et al. [34] added a new genetic marker in prediction model to assess the risk factor. The study concluded the importance of adding genetic markers in cardiovascular disease (CVD) diagnosis.

3.6   Comparative Analysis

Ordonez et al. [35] worked with constrained association rules using inter related attributes for heart disease identification. The studied patient attributes on learning revealed the healthy arteries and diseased arteries. Ali et al. [36] developed a diagnostic tool based on DNN classification. The mining model proved its efficiency. Kurt et al. [37] analyzed various classification techniques as indicated in Table 2 and compared their corresponding performances as coronary artery disease (CAD) predictive tool.

4   Inference from the Review

The review summarized that various preprocessing techniques such as sampling, feature selection (Filter and Wrapper), dimension reduction (using Principle Component Analysis, Facto Analysis, Forward feature construction, Backward feature elimination) and noise removal are used in conventional methods. Mining of electronic health record using routine algorithms predicts the CVD with varying accuracy ranging from 57.8% to 98%. Further there is need to improve the accuracy for better decision making. On comparison of various data mining techniques, support vector machine reported the greater accuracy in predicting the heart diseases. On seeing various heart diseases, data mining techniques are applied to detect Acute Coronary Syndrome (ACS) and coronary artery disease. Hybrid techniques of data mining report high accuracy (high sensivity 98% high specificity up to 99%) when comparing with the other methods in the literature. The tools/software used are Mat lab, R statistical programming, Python, Java. Weka, Rapid miner and Map reduce are used for making effective decision support models.

5   Conclusion and Future Work

Prediction of cardiovascular diseases using mining techniques aids clinicians for making substantiate decision for treatment. Risk level prediction find out the various risk level of cardiac patients using sensitivity, specificity and accuracy. The literature study on the said papers clearly revealed that the data mining of health care reports using various algorithms are suggestive of predicting cardiovascular disease with various accuracies. However in general, use of conventional algorithm for the stated task will forecast the problem in minimal level as compared to hybrid model. Hybrid models may provide solutions to various issues such as scalability, sparsity, speed and accuracy. Hence in the succeeding works, the design and development of hybrid models in combination with many algorithmic features such as optimization and parallelization using hadoop environment will mine the ECR accurately with better in lights to the clinicians for serving the society. The results of various prediction algorithms may be tested with synthetic and real-world datasets.

Acknowledgements. This research work was carried out with the financially support of RUSA-Phase 2.0 grant sanctioned vide Letter No. F24-51/2014-U, Policy (TNMulti-Gen) Dept. of Edn. Govt of India, Dt.09.10.2018 at Alagappa University, Karaikudi, Tamilnadu, India.

References

1. Frawley, W.J., Shapiro, G.P., Matheus, C.J.: Knowledge discovery in databases: an overview. AI Mag. **13**, 57–70 (1992)
2. Bushinak, H., AbdelGaber, S., AlSharif, F.K.: Recognizing the electronic medical record data from unstructured medical data using visual text mining techniques. Int. J. Comput. Sci. Inf. Secur. (IJCSIS) **9**, 25–35 (2011)
3. Srinivas, K.: Analysis of coronary heart disease and prediction of heart attack in coal mining regions using data mining techniques. In: IEEE International Conference on Computer Science and Education (ICCSE), pp. 1344–1349 (2010)
4. Thuraisingham, B.: A primer for understanding and applying data mining. IT Prof. **2**, 28–31 (2000)
5. Han, J., Kamber, M., Pei, J.: Data Mining Concepts and Techniques. Morgan Kaufmann Publishers, Burlington (2006)
6. Amin, M.S., Chiam, Y.K., Varathan, K.D.: Identification of significant features and data mining techniques in predicting heart disease. Telemat. Inform. **36**, 82–93 (2019)
7. Emre, I.E., Erol, N., Ayhan, Y.I., Ozkan, Y.I., Erol, C.I.: The analysis of the effects of acute rheumatic fever in childhood on cardiac disease with data mining. Int. J. Med. Inf. **123**, 68–75 (2018)
8. Domadiya, N., Rao, U.P.: Privacy preserving distributed association rule mining approach on vertically partitioned healthcare data. Proc. Comput. Sci. **148**, 303–312 (2019)
9. Ordonez, C., Omiecinski, E., de Braal, L, Santana, C.A., Ezquerra, N., Taboada, J.A., Cooke, D., Krawczynska, E., Garcia, E.V.: Mining constrained association rules to predict heart disease. In: IEEE International Conference on Data Mining (ICDM) (2001)
10. Dolatabadi, A.Z., Khadem, S.E.Z., Asl, B.M.: Automated diagnosis of coronary artery disease (CAD) patients using optimized SVM. Comput. Methods Programs Biomed. **138**, 117–126 (2017)
11. Pavlopoulos, S.A., Stasis, ACh., Loukis, E.N.: A decision tree – based method for the differential diagnosis of Aortic Stenosis from Mitral Regurgitation using heart sounds. Biomed. Eng. Online **3**, 21 (2004)
12. Karaolis, M. Moutiris, J.A., Papaconstantinou, L., Pattichis, C.S.: association rule analysis for the assessment of the risk of coronary heart events. In: Conference of the IEEE EMBS Minneapolis, Minnesota, USA, pp. 6238–6241 (2009)

13. Nahar, J., Imam, T., Tickle, K., Chen, Y.P.: Association rule mining to detect factors which contribute to heart disease in males and females. Expert Syst. Appl. **4**, 1086–1093 (2013)
14. Huang, Z.X., Dong, W., Duan, H., Liu, J.: A regularized deep learning approach for clinical risk prediction of acute coronary syndrome using electronic health records. IEEE Trans. Bio-Med. Eng. **65**, 956–968 (2017)
15. El-Bialy, R., Salamay, M.A., Karam, O.H., Khalifa, M.E.: Feature analysis of coronary artery heart disease data sets. Proc. Comput. Sci. **65**, 459–468 (2015)
16. Manogaran, G., Varatharajan, R., Priyan, M.K.: Hybrid recommendation system for heart disease diagnosis based on multiple kernel learning with adaptive neuro-fuzzy inference system. Multimed. Tools Appl. **77**, 4379–4399 (2018)
17. Arabasadi, Z., Alizadehsani, R., Roshanzamir, M., Moosaei, H., Yarifard, A.A.: Computer aided decision making for heart disease detection using hybrid neural network - genetic algorithm. Comput. Methods Programs Biomed. **141**, 19–26 (2017)
18. Polat, K., Güneş, S.: A hybrid approach to medical decision support systems: combining feature selection, fuzzy weighted pre-processing and AIRS. Comput. Methods Programs Biomed. **88**, 164–174 (2007)
19. Liu, X., Wang, X., Su, Q., Zhang, M., Zhu, Y., Wang, Q., Hindawi, Q.: A hybrid classification system for heart disease diagnosis based on the RFRS method. Comput. Math. Methods Med. **2017**, 11 (2017)
20. Prabhu, P., Anbazhagan, N.: Improving business intelligence based on frequent itemsets using k-means clustering algorithm. In: Meghanathan, N., Nagamalai, D., Rajasekaran, S. (eds.) Networks and Communications (NetCom2013). Lecture Notes in Electrical Engineering, vol. 284, pp. 243–254. Springer, Cham (2014)
21. Prabhu, P., Anbazhagan, N.: FI-FCM algorithm for business intelligence. In: Prasath, R., Kathirvalavakumar, T. (eds.) Mining Intelligence and Knowledge Exploration. Lecture Notes in Computer Science, vol. 8284, pp. 518–528. Springer, Cham (2013)
22. Tayefi, M., Tajfard, M., Saffar, S., Hanachi, P., Amirabadizadeh, A.R., Esmaeily, H., Taghipour, A., Ferns, G.A., Moohebati, M., Mobarhan, M.G.: Hs-CRP is strongly associated with coronary heart disease (CHD): a data mining approach using decision tree algorithm. Comput. Methods Programs Biomed. **141**, 105–109 (2017)
23. Austin, P.C., Tu, J.V., Ho, J.E., Levy, D., Lee, D.S.: Using methods from the data mining and machine learning literature for disease classification and prediction: a case study examining classification of heart failure sub-types. J. Clin. Epidemiol. **66**(4), 398–407 (2013)
24. Ordonez, C.: Association rule discovery with the train and test approach for heart disease prediction. IEEE Trans. Inf. Technol. Biomed. **10**, 334–343 (2006)
25. Jin, B., Che, C., Zhang, Z.L.S., Yin, X., Wei, X.: Predicting the risk of heart failure with EHR sequential data modeling. IEEE Access **6**, 9256–9261 (2018)
26. Chahala, B.G., Tyagic, D.S., Ramialisona, B.M.: Navigating the non-coding genome in heart development and Congenital Heart Disease. Differentiation **107**, 11–23 (2019)
27. Tsipouras, M.G., Exarchos, T.P., Fotiadis, D.I., Kotsia, A.P., Vakalis, K.V., Naka, K.K., Michalis, L.K.: Automated diagnosis of coronary artery disease based on data mining and fuzzy modeling. IEEE Trans. Inf. Technol. Biomed. **12**, 447–458 (2008)
28. Rémy, N.M., Martial, T.T., Clementin, T.D.: The prediction of good physicians for prospective diagnosis using data mining. Inf. Med. Unlocked **12**, 120–127 (2018)
29. Ilayaraja, M., Meyyappan, T.: Efficient data mining method to predict the risk of heart diseases through frequent itemsets. Proc. Comput. Sci. **70**, 586–592 (2015)
30. Princy, R.T., Thomas, J.: Human heart disease prediction system using data mining techniques. In: International Conference on Circuit, Power and Computing Technologies ICCPCT, p. 9 (2016)

31. Borah, A., Nath, B.: Identifying risk factors for adverse diseases using dynamic rare association rule mining. Expert Syst. Appl. **113**, 233–263 (2018)
32. Anooj, P.K.: Clinical decision support system: risk level prediction of heart disease using weighted fuzzy rules. J. King Saud Univ. – Comput. Inf. Sci. **24**, 27–40 (2012)
33. Karaolis, M.A., Moutiris, J.A., Hadjipanayi, D., Pattichis, C.S.: Assessment of the risk factors of coronary heart events based on data mining with decision trees. IEEE Trans. Inf. Technol. Biomed. **14**, 559–566 (2010)
34. Pu, L.N., Zhao, Z., Zhang, Y.T.: Investigation on cardiovascular risk prediction using genetic information. IEEE Trans. Inf. Technol. Biomed. **16**, 795–808 (2012)
35. Ordonez, C.: Comparing association rules and decision trees for disease prediction. In: ACM, Conference: Proceedings of the International Workshop on Healthcare Information and Knowledge Management, HIKM (2006)
36. Ali, L., Rahman, A., Khan, A., Zhou, M., Javeed, A., Khan, J.A.: An automated diagnostic system for heart disease prediction based on χ^2 statistical model and optimally configured deep neural network. IEEE Access **7**, 34938–34945 (2019)
37. Kurt, I., Ture, M., Kurum, A.T.: Comparing performances of logistic regression, classification and regression tree, and neural networks for predicting coronary artery disease. Expert Syst. Appl. **34**, 366–374 (2008)

Machine Learning Based Optimal Data Classification Model for Heart Disease Prediction

R. Bhuvaneeswari[1(✉)], P. Sudhakar[1], and R. P. Narmadha[2]

[1] Department of Computer Science and Engineering, Annamalai University, Chidambaram, India
anand.anandrajendran@gmail.com, kar.sudha@gmail.com
[2] Department of CSE, Sri Shakthi Institute of Engineering and Technology, Coimbatore, India
nammul4@gmail.com

Abstract. Heart disease (HD) is a greatest reason for high death rate among the number of inhabitants present on the planet. Identification of HD is viewed as a significant subject in the area of medical data examination. The measure of data in the medicinal field is tremendous. Data mining transforms the huge accumulation of actual medical data into useful data for making decisions. In this paper, prediction models were created by utilizing the ML technique called J48 classifier. In order to enhance the results further, a correlation based feature selection (CFS) model is applied to perform the feature selection process. Test results demonstrate that the HD prediction model shows excellent results over the compared methods.

Keywords: CFS · J48 · Feature selection · Prediction

1 Introduction

In last decade, Heart disease (HD) is assumed as a main source of death all over the globe. Recently, World Health Organization (WHO) has evaluated that a total of 17.7 million people have died because of HD [1]. The death ratio of 1:10 happens by the HD and a greater number of people died annually from HD compared to some other reasons. While forecasting the HD and give cautioning earlier, death rate can be reduced. The utilization of data mining carries another measurement to predict HD. Different data mining systems are utilized for distinguishing and filtering helpful data from the medical dataset with insignificant client sources of info and endeavors [2]. Over the previous decade, specialists investigated different approaches to actualize data mining in healthcare so as to accomplish an exact forecast of HD.

The regulations of data mining take it to a great extent on the procedures utilized and the features adapted. The medicinal dataset in the healthcare organization has many irregularities and holds unwanted information. It is more diligently to utilize data mining strategies with no earlier and suitable arrangements. As indicated [4], redundant and inconsistent data in the actual dataset influence the estimate result of the

© Springer Nature Switzerland AG 2020
D. J. Hemanth et al. (Eds.): ICICI 2019, LNDECT 38, pp. 485–491, 2020.
https://doi.org/10.1007/978-3-030-34080-3_55

algorithms. While applying ML (ML) models to its maximum capacity, a compelling readiness is expected to reprocess the datasets. Besides, undesirable features can diminish the exhibition of data mining methods also [3]. Consequently, alongside data planning, an appropriate feature selection strategy is expected to accomplish effective HD prediction utilizing critical features and data mining procedures. Despite the fact, it is evident that feature selection is treated as significant process as the selection of reasonable features will assist the classification process [7] in an efficient manner. As per [5], it is desired to determine the HD to have higher precision yet it is difficult to accomplish it. Moreover, a mix of noteworthy features helps to enhance the precision level. This demonstrates a broad investigation to recognize noteworthy features is important to accomplish that objective. The exhibition of data mining procedures utilized in forecasting HD is incredibly decreased without a decent mix of key features and additionally the inappropriate utilization of the ML algorithm [6]. In this manner, it is crucial to recognize the best blend of critical features that works unimaginably well with the best performing calculation. This examination centers on finding the data mining methods with critical features that will perform well in anticipating heart disease. Be that as it may, it is difficult to recognize the best possible method and select the noteworthy features.

Data mining transforms the huge accumulation of actual medicinal data into useful data for making decisions [8]. In this paper, prediction models were created utilizing the ML technique called J48 classifier. To further enhance the results, a correlation-based feature selection (CFS) model is applied to perform the feature selection process. Test results demonstrate that the HD prediction model shows excellent results over the compared methods.

The remainder of the study is planned as pursues. Section 2 elaborates the presented work and Sect. 3 validates the presented work. And, Sect. 4 concludes the work.

2   Proposed Work

The presented HD prediction model involves two main stages namely Correlation-based feature selection (CFS) and J48 classification which are discussed below. İn the first stage, the number of features present in the input data will be selected by the use of CFS method. Then, the extracted features will be classified by the use of J48 classifier model to predict the presence of HD properly.

2.1   CFS

CFS provides ranking to the features depending upon a heuristic evaluation function which relies on correlations. The function validates the subset composed of feature vector that is integrated with the class labels; however, it is not dependent on one another. CFS model considers that the unwanted features exhibit lay correlation with the class and hence, it is eliminated by CFS. At the same time, excessive features will be investigated and it shows high correlation with one or many features. The way of assessing a group of features is represented below.

$$M_S = \frac{\overline{lt_{cf}}}{\sqrt{l + l(l-1)\overline{t_{ff}}}} \tag{1}$$

where M_S is the validation of a subset comprising S features, n of a subset of consisting of features, $\overline{t_{cf}}$ is the average correlation value among the feature and class label, $\overline{t_{ff}}$ is the average correlation value among two features.

C4.5 is a descendant of ID3 devised by Ross Quinlan and is implemented in WEKA as J48 by the use of Java. Every individual one uses a greedy and a top-down method to construct decision trees (DT). It is utilized to classify data where fresh data will undergo labeling based on the predefined observations (training data set). The induction of DT starts with training dataset that undergo partitioning at every node leads to small portions; hence it follows a recursive divide and conquers process. Additionally, the data comprises collection of objects and a collection of attributes are also sent.

For each tuple in the dataset, it is linked to a class label that determines the choice of belonging to a specific class or not. Division is again carried out when the tuple comes under various classes. The dataset partitioning makes use of heuristic which selects a feature which is optimal. It is called as a metric of attribute selection. It is accountable for the kind of branching which takes place on a node. Gini index, IG is few instances that partitions a node to a binary or multiway, correspondingly.

C4.5 makes use of gain ratio as the attribute selection metric that has the benefit compared to IG employed in the predecessor ID3. As ID3 generates a n-array branch trees when the attribute on which partitions is the data with unique values, and hence it could not employ for classification. The attribute selection should take place as follows:

$$IG(N,A) = Entropy(N) - \sum_{values(A)} \frac{|N_i|}{|N|} Entropy(N) \tag{2}$$

where N is collection of instances at that specific node and $|N|$ is its cardinality, N_i is the subset of N where attribute A holds a value of i, and entropy of the set N is calculated as:

$$entropy(N) = \sum_{i=1}^{No.ofclassses} -P_i \log 2P_i \tag{3}$$

Where Pi is proportion of instances in N which has the ith class value as output attributes.

In every node, IG is determined and the maximum IG is chosen for further processes. In case of completing the procedure of node discovery, fresh branches are included to every value attained by test attribute below the corresponding nodes. During the training process, every node's instances are provided to the branch along with the integrated test attribute value, and in addition this subset of training instances is employed iteratively for creating new nodes. When there is zero modification in the outcome, then leaf generation takes place to terminate the recursion of nodes and assignment of output attribute takes place in the identical class value. Extensions were made to the basic ID3 algorithm to

- It works only on continuous attributes,
- It deals with the missing values
- It prevents the problem of overfitting.

The presented model is tested using a set of different test cases.

3 Performance Validation

A detailed simulation is carried out to investigate the performance of the introduced HD model. The following subsection offers the details of the dataset used, measures and the results analysis.

The details of the applied Statlog HD dataset are given in Table 1. As shown, it is evident a set of 270 instances are presented with the underlying features of 13. In addition, a set of two classes namely presence and absence of HD is present in the dataset. From the total of 270 instances, a collection of samples of around 44.50% comes under the present category. In addition, a collection of samples of around 55.50% comes under the absent category.

Table 1. Dataset details

Description	Dataset-1
Number of instances	270
Number of features Data sources	13
Number of class	2
Percentage of present samples	44.50%
Percentage of absent samples	55.50%
Data sources	[9]

The set of 13 attributes present in the dataset are described clearly in Table 2. In addition, Fig. 1 illustrates the frequency distribution of the attributes present in the applied dataset. For analysis purposes, a set of measures used for classification performance is given below.

- Precision
- Recall
- Accuracy
- F-Score and
- Kappa

Table 2. Comparison of confusion matrix derived by different models

Experts	Proposed		J48		Random tree		RBFNetwork	
	P	A	P	A	P	A	P	A
P	91	29	88	32	89	31	97	23
A	22	128	31	119	33	117	20	130

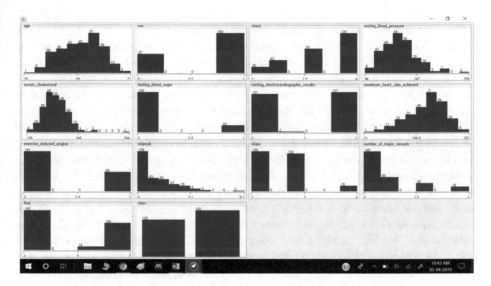

Fig. 1. Frequency distribution of heart disease dataset for all attributes

The values of these measures should be maximum for indicating the effective classification performance.

Table 2 shows the confusion matrix derived for the results of the applied HD prediction model. From the table, it is clear that the presented HD model classifies a total of 91 instances correctly under present category and a total of 128 instances under absent category. Similarly, the J48 classifier model classifies a total of 88 instances correctly under present category and a total of 119 instances under absent category. Likewise, the RT classifier model classifies a total of 89 instances correctly under present category and a total of 117 instances under absent category. At the end, the RBF model classifies a total of 97 instances correctly under present category and a total of 130 instances under absent category.

Next, the classifier results acquired by different models against the applied identical dataset are shown in Table 3. The values are presented in terms of precision, recall, accuracy, F-score and kappa. For comparison actions, a set of three classification models such as J48, RT and RBF are employed.

A detailed comparative results are made in terms of various measures. From the above table, it is apparent the presented CFS with J48 model shows effective outcome with a maximum precision of 81.11. At the same instant, the RBF model shows competitive outcome with the precision value of 80.33. In addition, the RF shows

Table 3. Performance evaluation of different classifier on HD dataset

Classifier	Precision	Recall	Accuracy	F-Score	Kappa
CFS + J48	81.11	81.11	81.11	81.00	61.52
J48	73.33	73.95	76.66	73.64	52.71
RT	74.16	72.95	76.29	73.55	52.08
RBF	80.33	80.91	80.96	80.81	60.13

better outcome over J48, it is still inefficient over compared ones with the precision value of 74.16. at last, the J48 shows least classifier results with a minimum precision value of 73.33. Likewise, it is obvious the presented CFS with J48 model shows effective outcome with a maximum recall of 81.11. At the same instant, the RBF model shows competitive outcome with the recall value of 80.91. In addition, the RF shows better outcome over J48, it is still inefficient over compared ones with the recall value of 73.95. At last, the J48 shows least classifier results with a minimum recall value of 72.95. Similarly, the table values show effective outcome with a maximum accuracy of 81.11. At the same instant, the RBF model shows competitive outcome with the accuracy value of 80.96. In addition, the J48 shows better outcome over RT, it is still inefficient over compared ones with the accuracy value of 76.66. At last, the RT shows least classifier results with a minimum accuracy value of 76.29. At the same time, the table values exhibit efficient performance with a maximum F-score of 81.00. At the same instant, the RBF model shows competitive outcome with the F-score value of 80.81. In addition, the J48 shows better outcome over RT, it is still inefficient over compared ones with the F-score value of 73.64. At last, the RT shows least classifier results with a minimum F-score value of 73.55. Finally, a kappa value analysis is made and the table values exhibit efficient performance with a maximum kappa value of 61.52. At the same instant, the RBF model shows competitive outcome with the kappa value of 60.13. In addition, the J48 shows better outcome over RT, it is still inefficient over compared ones with the F-score value of 52.71. At last, the RT shows least classifier results with a minimum kappa value of 52.08.

A comparison with recently proposed methods is also made as shown in Table 4. From this table, it is clearly noticed that maximum results is attained by the presented model over the existing models. The presented model shows effective classification with the precision of 81.11, recall of 81.11, accuracy of 81.11 and F-score of 80.81 and kappa value of 61.52 respectively.

Table 4. Comparison with recent methods for applied dataset in terms of accuracy

Classifiers	Accuracy
CFS + J48	**81.11**
J48	76.66
RT	76.29
RF ensemble classifiers	80.49
Pruned J48 DT	73.79
Extreme learning machine	80.00

4 Conclusion

This paper has presented a new feature selection based classification model by the use of CFS and J48 classifier on HD prediction. İnitially, the CFS model is applied to extract the features and then the J48 classifies the data produced from the output of CFS model. The experimental analysis of the presented model is carried out on the benchmark HD dataset under several aspects. The presented model shows effective classification with the precision of 81.11, recall of 81.11, accuracy of 81.11 and F-score of 80.81 and kappa value of 61.52 respectively.

References

1. World Health Organization (WHO): Cardiovascular diseases (CVDs) – Key Facts (2017). http://www.who.int/news-room/fact-sheets/detail/cardiovascular-diseases-(cvds)
2. Srinivas, K., Rao, G.R., Govardhan, A.: Analysis of coronary heart disease and prediction of heart attack in coal mining regions using data mining techniques. In: Paper presented at the 5th International Conference on Computer Science and Education (ICCSE), Hefei, pp. 1344–1349 (2010)
3. Paul, A.K., Shill, P.C., Rabin, M.R.I., Akhand, M.A.H.: Genetic algorithm based fuzzy decision support system for the diagnosis of heart disease. (ICIEV). In: 5th International Conference on Informatics, Electronics and Vision, pp. 145–150. IEEE (2016)
4. Kavitha, R., Kannan, E.: An efficient framework for heart disease classification using feature extraction and feature selection technique in data mining. In: International Conference on Emerging Trends in Engineering, Technology and Science (ICETETS), pp. 1–5 (2016)
5. Shouman, M., Turner, T., Stocker, R.: Integrating clustering with different data mining techniques in the diagnosis of heart disease. J. Comput. Sci. Eng. **20** (1) (2013)
6. Dey, A., Singh, J., Singh, N.: Analysis of supervised machine learning algorithms for heart disease prediction with reduced number of attributes using principal component analysis. Analysis **140**(2), 27–31 (2016)
7. Anooj, P.K.: Clinical decision support system: risk level prediction of heart disease using weighted fuzzy rules. J. King Saud Inf. Univ.-Comput. Sci. **24**(1), 27–40 (2012)
8. Liu, X., Wang, X., Su, Q., Zhang, M., Zhu, Y., Wang, Q., Wang, Q.: A hybrid classification system for heart disease diagnosis based on the RFRS method. Comput. Math. Methods Med. **2017**, 11 (2017)
9. http://archive.ics.uci.edu/ml/datasets/statlog+(heart)

Deep Neural Network Based Classifier Model for Lung Cancer Diagnosis and Prediction System in Healthcare Informatics

D. Jayaraj[1]($\boxtimes$) and S. Sathiamoorthy[2]

[1] Department of Computer Science and Engineering,
Annamalai University, Chidambaram, India
jayarajvnr@gmail.com
[2] Tamil Virtual Academy, Chennai, India
ks_sathia@yahoo.com

Abstract. Lung cancer is a most important deadly disease which results to mortality of people because of the cells growth in unmanageable way. This problem leads to increased significance among physicians as well as academicians to develop efficient diagnosis models. Therefore, a novel method for automated identification of lung nodule becomes essential and it forms the motivation of this study. This paper presents a new deep learning classification model for lung cancer diagnosis. The presented model involves four main steps namely preprocessing, feature extraction, segmentation and classification. A particle swarm optimization (PSO) algorithm is sued for segmentation and deep neural network (DNN) is applied for classification. The presented PSO-DNN model is tested against a set of sample lung images and the results verified the goodness of the projected model on all the applied images.

Keywords: CT images · Classifier · Deep learning · Lung cancer · Segmentation

1 Introduction

Lung cancer is an important reason for the increased global mortality rate [1]. It occurs due to unmanageable growth of cells. When it is untreated, it will spread to the whole body and leads to loss of life. The main factor for causing cancer is because of smoking and next reason is seamless explosion to pollution, harmful gas, heredity and second-hand smoking [2]. The symptoms are chest pain due because of long-lasting cough, weight loss, fatigue, difficult to breath to a continuous cough that extends to weight loss, weakness, shortness of breath [3]. Different recent methods are developed to treat cancer through surgeries and diverse therapies which are hurting [4]. Therefore, for avoid this problem, different image processing models acts as an important part where it makes the diagnosis process simple and increases the survival rate. At the same time, the automated lung cancer diagnosis model is also not easier [5]. The CT images are gathered from the LIDC database and unwanted noises are discarded by the use of pre-processing level and the cancer cells are recognized and classified.

© Springer Nature Switzerland AG 2020
D. J. Hemanth et al. (Eds.): ICICI 2019, LNDECT 38, pp. 492–499, 2020.
https://doi.org/10.1007/978-3-030-34080-3_56

Segmentation techniques under classification into various ways and this study concentrated on the optimization algorithms. [6] introduced a fuzzy C-Means (FCM) method which enables the allocation of individual data to many clustering through the assignment of a membership value of every data pattern to every cluster. The allocated membership value lies in the range of [0, 1] for every data from every cluster will be incremented to 1. However, the system suffers from the limitation that the earlier representation of cluster and iteration count is high. [7] presented a K-means clustering model to segment the image through the partition of the objects to K groups by determining the mean compared to the calculation of the distance among every point from the cluster mean. The main drawback of this method is firstly it is difficult to predict the k-value. For different initial partitions, different results are produced, it does not produce good results for different size and different density clusters and finally, it doesn't work well with global clusters.

[8] introduced a Mean Shift Segmentation technique which defines a widow around it and then computes the mean of the data point. This technique operates in the way of representing the window and computes the mean of the data point. It follows the kernel method to calculate the density of the gradient point. The limitation lies in the dependency of the output on window size and processing is high. In addition, it does not work well for more number of features. [9] presented an Expectation Maximization technique which repetitively determines the maximum likelihood estimation of the missing or hidden data. It has the limitation of slower convergence. Here, a novel method for automated identification of lung nodule of computer tomography (CT) becomes essential and it forms the motivation of this study. Recently, deep learning models becomes popular and find its applicability in various domains. In this paper, the deep learning model is applied for lung cancer classification.

This paper presents a new deep learning classification model for lung cancer diagnosis. The presented model involves four main steps namely preprocessing, feature extraction, segmentation and classification. A particle swarm optimization (PSO) algorithm is sued for segmentation and deep neural network (DNN) is applied for classification. The presented PSO-DNN model is tested against a set of sample CT ling images and the results verified the goodness of the projected model on all the applied CT images.

The upcoming sections are arranged here. PSO-DNN model is explained in Sect. 2. Next, results are validated in Sect. 3 and concluded in Sect. 4.

2   Proposed Work

Figure 1 illustrates the overall process involved in the PSO-DNN model. Initially, the images are gathered from the LDIC dataset. Then, the pre-processing of the images takes place to improvise the quality of the image to attain better results. Then, PSO based segmentation of the applied lung CT images takes place. Next, the useful features get extracted from the segmentation image. Finally, the DNN model is applied for classifying the images as normal or abnormal ones from the feature extracted image. These four steps involved in the presented model are provided below.

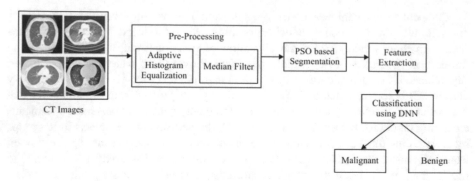

Fig. 1. Overall process of the proposed method

- **Pre-processing:** This step involves two sub-processes namely image enhancing and noise removal. To carry out these processes, adaptive Histogram Equalization and Median filter techniques are applied.
- **Segmentation:** This step makes use of PSO algorithm to identify the boundary to recognize the useful data.
- **Features extraction:** This step finds useful to extract the features such as perimeter, area and eccentricity. They are applied as the training features for building a classification model.
- **Classification:** This step utilizes DNN model for classifying the applied images as malignant or benign. The presented HD prediction model involves two main stages namely Correlation-based feature selection (CFS) and J48 classification which are discussed below. İn the first stage, the number of features present in the input data will be selected by the use of CFS method. Then, the extracted features will be classified by the use of J48 classifier model to predict the presence of HD properly.

2.1 Preprocessing

A set of two sub-processes namely image enhancing and noise removal involved in this stage. Initially, the adaptive histogram equalization technique is applied due to the nature of its efficiency and less computation. It improves the image and its visual effected through the enlargement of the grayscale distribution of the images based on the probability distribution function. It eliminates the limitations of the traditional methods by the automated identification and adaptability to the grayscale value of the image.

At the same, a median filtering technique is applied to reduce noise, specifically impulsive as well as salt and pepper noise. It minimizes the noise appeared randomly and it retains the edge details of an image. It undergoes the replacement of the pixels with the median of every pixel of its neighborhoods (ω) as given in Eq. (1):

$$y(m, n) = \text{median}(x(i, j), (i, j)\varepsilon\omega) \tag{1}$$

2.2   Segmentation

The process of image segmentation and clustering are equivalent. A better choice of the center is essential for the clustering technique. Since k-means and other clustering models are mainly based on the starting clusters and it has the nature of local convergence. Here, segmentation of the lung cancer images takes place using PSO algorithm which utilizes a global search technique for optimizing the choice of initial clusters which can be utilized for clustering. The objective function is defined in a way that it maximizes the inter-class distance among the clusters and reduces the intra-class distance inside a cluster.

$$C_i = \underset{min}{\text{Arg}} \sum_{i=1}^{K} \|C(i) - M_i\|^2 \tag{2}$$

For minimizing the C_i, PSO algorithm is applied. An assumption is made that the initial clusters are set at the beginning of iteration. It leads to the replacement of M_i by I as represented as follows.

$$C_i = \underset{min}{\text{Arg}} \sum_{i=1}^{K} \sum_{j=1}^{MN} \|C(i) - I(j)\|^2 \tag{3}$$

The objective function defined in Eq. (4) is applied for the minimization. The number of particles is selected as the image size and three equal distant centre values are selected to allocate the values to the initial swarm. u^t as shown in Eq. (5).

$$C_i = u^i \sum_{m=1}^{M} I(m)\partial(m) \tag{4}$$

The particle positions get updated as given below.

$$C_{i+1} = C_i + \frac{V_i}{Scalar} \tag{5}$$

2.3   Feature Extraction

Once the image segmentation gets completed, it is applied to extract features. It is an essential process that provides much description related to the image. The substantially extracted features are listed below:

- **Area:** It indicates the aggregation of the pixel values in the image which are generally represented by 1 in the provided binary image A = n (1).
- **Perimeter:** It defines the total number of pixels presented at the edges of the provided objects.
- **Eccentricity:** It computes the distance from major axis to the foci of ellipse.

2.4   DNN Based Classification

A DNN comes under a kind of feed-forward network consisting of consecutive pairs of convolutional (conv), max-pooling (MP) and diverse fully connected (FC) layers. The original intensity of pixels present tin the input images are provided through the hierarchical feature extractor. The generated feature vector undergoes classification via the FC layers. Every weight undergoes optimization by the reduction of the misclassification error on the applied training data. Every conv layer carries out a 2D conv of its input maps using a rectangular filter. The filter is employed to each probable positions of the input map. When the earlier layer has many maps, then, the activation layer of the respective conv are totaled and given to the nonlinear activation function. The DNN has multiple connections, weights and non-linearities. Next to the diverse pair of conv and MP layers, single FC alone integrates to the outcome of the feature vector. The output layer is generally a FC layer with a single neuron under individual classes undergo activation using a softmax function, therefore, verifying ensuring that every neuron's output activation is understandable as the possibility of specific input belongs to the class. The output layer properly indicates the class label of every input and classification is attained.

Training process
By the utilization of ground truth data, every pixel present in the training images are trained under mitosis or non-mitosis. Next, a training set is constructed where every instance will be mapped to a square window of RGB values which undergo sampling from the raw image to the class of middle pixel. When a window is placed slightly outside the boundary of an image, the pixels which are missed are managed by a mirroring method.

3   Performance Validation

3.1   Dataset Used

A series of experiments are made to ensure the goodness of the presented PSO-DNN model on the diagnosis of the lung cancer, a benchmark dataset from Lung Image Database Consortium (LIDC) dataset is utilized [10]. The LIDC dataset holds numerous lung cancer screening CT images which could be applied to develop the automated diagnosis model for lung cancer. It has a total of 1018 images with the dimensions of 512*512 pixels. In general, it is not easier to process the images in DICOM form, therefore, they are converted to the JPEG Grayscale image through MicroDicom.

3.2   Results Analysis

For verifying the superior performance of the presented PSO-DNN model, a qualitative analysis is initially made. Figure 2 shows the sample test images from LIDC dataset is applied. Next, Fig. 3 shows the outcome of the sub processes involved in the PSO-DNN model. Figure 3a shows the outcome of the Adaptive Histogram Equalization technique which is an enhanced form of the applied test image. Then, Fig. 3c shows the outcome of the median filtering technique and Fig. 3d illustrates the outcome of the Segmented Nodule using PSO algorithm.

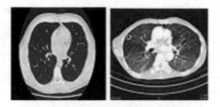

Fig. 2. Sample test images

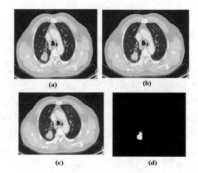

Fig. 3. (a) Pre-processed image using median filtering (b) Segmented nodule using PSO

Table 1 provides a comparative investigation of the results attained by various lung cancer diagnosis approaches with respect to accuracy, sensitivity and specificity. The table values indicated that the presented PSO-DNN model exhibits superior classification performance by attaining the maximum accuracy of 91.87%. Nevertheless, the FCM + PBN model exhibited manageable performance by achieving a slightly lower accuracy value of 90.00%.

Table 1. Classifier results analysis of diverse methods on lung CT image dataset

Classifiers	Accuracy	Sensitivity	Specificity
Proposed	91.87	91.26	92.98
FCM + BPN	90.00	90.87	92.00
SVM	86.60	79.87	74.89
MLP	82.87	76.59	70.54
RBFNetwork	83.89	90.35	46.67
kNN	89.00	87.50	86.45

Simultaneously, the kNN model tried to handle the classification process by obtaining a moderate accuracy value of 89%. On the other hand, the ML classifier model called SVM shows slightly better performance over the MLP and RBF classifiers by acquiring the accuracy value of 86.60%. However, it does not show its effective performance over the presented PSO-DNN model. At the same time, the existing MLP and RBF classifiers showed its inefficiency on the applied test images by achieving minimum accuracy values of 82.87% and 83.89% respectively.

From the above-mentioned figure and table, it is evident that the maximum classifier results are attained by the presented PSO-DNN method. It is completely verified from the maximum accuracy, sensitivity and specificity values of 91.87, 91.26 and 92.98%. Therefore, the PSO-DNN model can be applied as an automated diagnosis tool for lung cancer.

4 Conclusion

In this paper, a new deep learning based classification model for lung cancer diagnosis is presented. Initially, the images are gathered from the LDIC dataset. Then, the pre-processing of the images takes place to improvise the quality of the image to attain better results. Then, PSO based segmentation of the applied lung CT images takes place. Next, the useful features get extracted from the segmentation image. Finally, the DNN model is applied for classifying the images as normal or abnormal ones from the feature extracted image. Benchmark dataset from LIDC is utilized for validation. From the above-mentioned figures and tables, it is evident that the maximum classifier results are attained by the presented PSO-DNN method. It is completely verified from the maximum accuracy, sensitivity and specificity values of 91.87, 91.26 and 92.98%. Therefore, the PSO-DNN model can be applied as an automated diagnosis tool for lung cancer.

References

1. World Health Organization: Description of the Global Burden of NCDs, Their Risk Factors and Determinants, Burden: Mortality, Morbidity and Risk Factors, pp. 9–32. World Health Organization, Switzerland (2011)
2. Siegel, R., Naishadham, D., Jemal, A.: Cancer statistics. CA Cancer J. Clin. **63**, 11–30 (2013)
3. Richard, D.: Lung cancer and other causes of death in relation to smoking. BMJ **2**, 1071–1081 (1956)
4. Ruth, L.K., Jeff, D., Declan, W.: Symptoms of lung cancer. Palliat. Med. **6**, 309–315 (1992)
5. Mark, S.W., Denise, M.Z., Edwin, B.F.: Depressive symptoms after lung cancer surgery: their relation to coping style and social support. Psychol. Oncol. **15**, 684–693 (2005)
6. Jalal Deen, K., Ganesan, R., Merline, A.: Fuzzy-C-means clustering based segmentation and CNN-classification for accurate segmentation of lung nodules. Asian Pac. J. Cancer Prev. **18**, 1869–1874 (2017)
7. Akhilesh Kumar, Y., Divya, T., Sonali, A.: Clustering of lung cancer data using foggy K-means. In: IEEE 2013 International Conference on Recent Trends in Information Technology (ICRTIT), pp. 13: 13–18 (2013)

8. Ying, W., Rui, L., Jin, Z.Y.: An algorithm for segmentation of lung ROI by mean-shift clustering combined with multiscale HESSIAN matrix dot filtering. J. Central South Univ. **19**(12), 3500–3509 (2012)
9. Qian, Y., Weng, G.: Lung nodule segmentation using EM algorithm. In: IEEE 2014 Sixth International Conference on Intelligent Human-Machine Systems and Cybernetics, pp. 14: 20–25 (2014). 24
10. Armato, I., Samuel McLennan, G., McNitt-Gray, F.R., Michael, Charles, Reeves, Anthony, P., Clarke, L.: Data From LIDC-IDRI. The Cancer Imaging Archive (2015). http://doi.org/10.7937/K9/TCIA.2015.LO9QL9SX

Forecast Solar Energy Using Artificial Neural Networks

Sukhampreet Kaur Dhillon$^{(\boxtimes)}$, Charu Madhu, Daljeet Kaur, and Sarvjit Singh

UIET, Panjab University, Chandigarh, India
sukhamd.24@gmail.com

Abstract. Energy harvested from the natural resources such as solar energy is highly intermittent. However, its future values can be predicted with reasonable accuracy. In this paper, a solar energy forecasting model based on Artificial Neural Network is proposed. Intensity of solar radiations is predicted 24 h ahead based on pressure, dew point, temperature, relative humidity, wind speed, zenith angle and historical values of solar intensity. The dataset of 4 months is used from National Solar Radiation Database. Simulations were performed to analyse the impact of each input variable on results. Results indicate the efficiency of the model both in terms of accuracy and computational costs.

Keywords: Artificial neural networks · Solar energy · Forecasting · Data analytics · Machine learning

1 Introduction

If the time duration between the time of prediction and occurrence of event is large NWP and satellite based prediction methods can be used, however, for inter-hour or intra-hour predictions, these methods offer very less accuracy as compared to Artificial Neural Network (ANN) [1]. A large amount of research has been done in this field using different inputs and models [2]. However, the inputs fed directly to the ANN do not deliver much information to the network and hence the models are less accurate [3]. In this paper, feature extraction and addition of an input that is derived from the original inputs and has high correlation with the solar irradiance was incorporated into the system. Furthermore, most of the papers deal with prediction of daily average GHI or monthly average GHI, while most of the applications require hourly forecast [4].

V. Sampath Kumar et al. [5] proposed feed forward neural network model forecast solar radiation in Botswana. This dataset of 15 years was used to obtain the solar constants which were further fed to the feed forward neural network while clearness index was output of the model which is ratio of global solar radiation at the place and extraterrestrial radiation (depends only on the latitude of the place). The model proved to be quite efficient with coefficient of regression 0.99.

Morris Brenna et al. [3] proposed feed forward neural network model trained with Levernberg-Marquardt algorithm for solar irradiance forecasting. The results revealed that neural network with 13 neurons in input layer as well as hidden layer which was

D. J. Hemanth et al. (Eds.): ICICI 2019, LNDECT 38, pp. 500–507, 2020.
https://doi.org/10.1007/978-3-030-34080-3_57

given six inputs such as solar radiations 24 h earlier, day of the year and forecasted ambient temperature is optimum in terms of computational costs and accuracy.

Ahmed Tealab et al. [6] compared various models which were based on ANNs and predicted the future values of non-linear time series. It was concluded that there were just 17 studies from 2006 to 2016 which mentioned the mathematical specifications of their models partly or entirely. Out of these 17, there were only three studies which designed new neural network models; others used autoregressive process of neural networks.

Khalid Anwar et al. [7] designed a feed forward neural network model to forecast monthly mean solar radiations for the states of Andhra Pradesh and Telangana state. Nine inputs: mean temperature, latitude, longitude, mean monthly meteorological parameters, month of the year, altitude, mean sunshine duration, mean wind speed, mean relative humidity were given as inputs to neural network. Results revealed that the network with nine neurons in hidden layer shows maximum accuracy with MAPE of 2.16.

Rachit Srivastava et al. [8] performed six day ahead prediction of average daily solar intensity using Model Averaged Neural Network. Nine parameters including time, maximum temperature, average temperature, minimum temperature, wind, rain, atmospheric pressure, azimuth and dew point were fed to ANNs. Average of monthly RMSE for 12 months was observed be lesser for the proposed model with value of 204.52 W/m^2.

Jabar H. Yousif et al. [2] compared various machine learning techniques for forecasting PV/T solar energy production. The results indicated ANN models are the more efficient in predicting Global Solar Radiation. The study covered 40 research papers from last 10 years and presented the accuracy of each of the models along with other specifications such as location of dataset and input variables.

2 Data Set

The dataset used in this study is derived from National Solar Radiation Database (NSRDB) measured at a place called Patseni in state of Uttar Pradesh, India. Dataset from 2001 till date can be accessed from the NREL website [9]. For this study 4 months dataset is used, as shown in Fig. 1, which is split into three parts for training, validation and testing. The dataset from 1st of January 2001 to 28th of February 2001 is used for training while the data for the month of March and April is used for validation and testing respectively. The dataset provides information about 6 weather parameters apart from the solar irradiance values namely: pressure, temperature, wind speed, relative humidity, dew point and zenith angle.

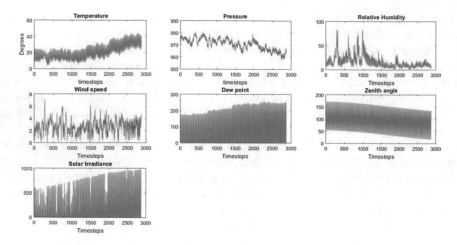

Fig. 1. Data set

3 Artificial Neural Networks

Artificial Neural Networks consist of numerous nodes which are interconnected with each connection carrying fixed or variable weight. When a signal is passed from one node to another, it is multiplied with the weight associated with the connection. There are various neural network architectures as discussed below:

3.1 Feed Forward Neural Network

Feed Forward Neural Network (FFNN) is the most basic type of neural networks. In FFNN, information only goes in forward direction only and no loops are formed. They have least memory requirements among all the neural networks [10].

3.2 Back Propagation Neural Network

Back Propagation Neural Network (BPNN) is prominently used in literature for regression and classification purpose. In BPNN weights are adjusted to reduce the differences between the actual and the desired outputs of network. Memory requirement of BPNN is lesser than that of Deep Neural Network however they are too large to be fulfilled by Wireless Sensor Networks.

3.3 Elman Neural Network

Elman Neural Network (ENN) is basically a 4 layered FFNN except that it has one additional layer namely the context layer which has same activation function as in the hidden layer neurons.

3.4 Model Averaged Neural Network

Multiple FFNNs are created and then the outputs of all the networks are averaged to produce the desired results [6]. In order to design a Model Averaged neural network, multiple networks are designed using different number of neurons and then outputs of all the nets are averaged out.

3.5 NARX Neural Network

Nonlinear Auto-Regressive Exogenous (NARX) neural network is recurrent dynamic network that predicts the future values of a time series based on the present and past values of the series as well as present and past values of some driving series [11]. Driving series is the one which influences the focussed series.

3.6 Recurrent Neural Network

Unlike other neural networks, Recurrent Neural Network (RNN) can efficiently represent and encode hidden states as the output depends on random number of previous inputs. They can model complex non-linear relationship among variables. RNNs remember all the previously processed information so they have large memory requirements.

There are three types of training algorithms to train a neural network:

Gradient Descent Algorithms. These are popular training algorithms that update weights and biases corresponding to the negative gradient of performance index.

Conjugate Gradient Algorithms. In these algorithms, adjustment is not done along steepest gradient descent of performance index rather along some conjugate direction and produce faster convergence than gradient descent algorithms.

Quasi-Newton Algorithms. Newton's methods give better results than other two algorithms. The first step in Newton's method is to calculate the Hessian matrix. However, calculating Hessian matrix involves lot of computational cost so Levenberg-Marquardt algorithm [12] is generally used owing to its low computational costs and higher efficiency for lesser number of neurons.

4 Proposed Model

The model proposed in this study for forecasting solar irradiance is shown in Fig. 2. The weather information related to temperature, pressure etc. is collected by sensors and three parameters from the input variables namely temperature, pressure and relative humidity are used for reference signal generation. The dataset is filtered using *low pass filter*. All sets of filtered data are cross correlated with GHI to find the optimum filter length for each variable. Then the features of the data are extracted using *Independent Component Analysis* algorithm and these features are fed to the FFNN trained by *Levenberg-Marquardt* algorithm.

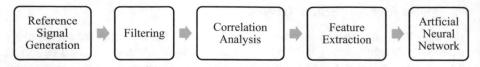

Fig. 2. Solar energy forecasting model

5   Results

The model was designed using MATLAB version R2015a. In order to decide the combination of inputs to be fed to the network, the resulting Root Mean Square Error (RMSE) from each of the combinations are tabulated in Table 1. For the first combination, all the input variables are fed to the network and then 6 out of 7 variables are fed to the network. Furthermore, the results are analysed by feeding 5 to 1 inputs, removing the least significant variable of the remaining inputs each time.

Table 1. Comparison of Error obtained from different input combinations

Combination number	Pressure	Temperature	Relative humidity	Wind speed	Dew point	Zenith angle	Reference signal	RMSE
1								**57.59**
2	×							62.3
3		×						67.28
4			×					63.68
5				×				67.9
6					×			68.5
7						×		58.39
8							×	**77.23**
9	×					×		73.26
10	×		×			×		76.25
11	×	×	×			×		**57.93**
12	×	×	×	×		×		64.68
13	×	×	×	×	×	×		64.06

The results reveal that removing reference signal from the inputs effects the results to the maximum extent which implies reference signal is the most significant variable. Moreover, by adding reference signal to list of inputs, pressure, relative humidity and temperature can be removed from the inputs combination as this would affect the accuracy of the model to very little extent as evident from the error generated by feeding combination number 11 to the network. Figure 3 shows the results with and without reference signal.

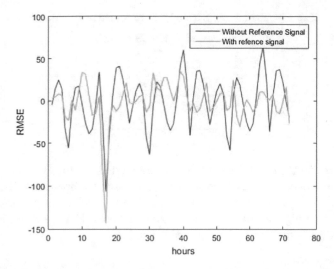

Fig. 3. Comparison of results with (*green*) and without reference signal (*red*)

Another challenge is to choose a suitable architecture for ANN that is to decide the number of hidden layers, neurons count in hidden layers and window size. A number of experiments were performed to take this decision and results are shown in Table 2.

Table 2. Comparison of feed forward neural networks with different configurations

Hidden layers	Neurons count in hidden layers	Memory occupied (kBytes)	RMSE (W/m^2)
3	[24 24 24]	28	52.15
3	[11 11 11]	12	70.3
2	**[24 24]**	**20**	**56.61**
2	[11 11]	8	60.92
1	[24]	8	74.97
1	[11]	5.2	82.39

As the results indicate, ANN with 2 hidden layers and 24 neurons in each of the hidden layer shows good performance. Various combinations of transfer functions were implemented and results revealed that transfer function for the hidden layers should be '*logsig*' and that for the output layer should be '*purelin*'. Further, the window size is varied from 1 to 5 for the network. The results as shown in Table 3 reveal that network with window size 3 shows best performance.

Table 3. Root Mean Square Error for different window sizes

Window size	RMSE (W/m^2)
1	72.05
2	71.66
3	**56.61**
4	73.11
5	121.89

Table 4 shows the comparison of forecasting models in terms of memory occupied by the neural network and resulting error in prediction. Comparison of results reveals that Feed Forward Neural Network is most efficient network owing to its low memory requirements and high accuracy. Recurrent Neural Networks outperform it in terms of accuracy; however they occupy large memory space.

Table 4. Comparison with other Neural Network Models

Model no.	Approach	RMSE (W/m^2) with Pre-processing	Memory requirements (kBytes)
1	Back propagation Network	71.47	212
2	**Feed Forward Neural Network**	**56.61**	**20**
3	NARX	59.63	216
4	Elman back propagation	66.95	212
5	Recurrent Neural Networks	49.17	228
6	Model Averaged Neural Networks	57.82	56

6 Conclusion

In this paper, solar energy forecasting is performed 24 h ahead using Feed Forward neural network using '*trainlm*' as training function. The number of input variables was reduced from 6 to 3 by generating a reference variable which was equivalent to three input variables and network trained using reference variable resulted in better performance. The results indicated a network with 2 hidden layers, 24 neurons in each layer and window size 3 shows best performance. The model was compared with five other models and it outperformed them if optimum trade off is made between accuracy and computational costs with Root Mean Square Error of 56.61 W/m^2 and memory requirements of 20 kBytes.

References

1. Yan, X., Dhaker, A., Bruno, F.: Solar radiation forecasting using artificial neural network for local power reserve. In: International Conference on Electrical Sciences and Technologies in Maghreb, Tunisia, pp. 1–6. IEEE (2014)
2. Yousif, J.H., Kazem, H.A., Alattar, N.N., Elhassan, I.E.: A comparison study based on artificial neural network for assessing PV/T solar energy production. Case Stud. Thermal Eng. 13, 10040 (2019)
3. Brenna, M., Federica, F., Longo, M., Zaninelli, D.: Solar radiation and load power consumption forecasting using neural network. In: 6th International Conference on Clean Electrical Power, Italy, pp. 726–731. IEEE (2017)
4. Tengyue, Z., Shouying, L., Feng, Q., Chen, Y.: Energy-efficient control with harvesting predictions for solar-powered wireless sensor networks. Sensors 16(1), 53 (2016). Brunelli, D. (ed.), MDPI, Basel
5. Kumar, V.S., Prasad, J., Narasimhan, V.L., Ravi, S.: Application of artificial neural networks for prediction of solar radiation for Botswana. In: International Conference on Energy, Communication, Data Analytics and Soft Computing, India, pp. 3493–3501. IEEE (2017)
6. Tealab, A.: Time series forecasting using artificial neural networks methodologies: a systematic review. Future Comput. Inform. J. 3(2), 334–340 (2018)
7. Anwar, K., Deshmukh, S.: Use of artificial neural networks for prediction of solar energy potential in southern states of India. In: 2nd International Conference on Green Energy and Applications, Singapore. IEEE (2018)
8. Srivastava, R., Tiwari, A.N., Giri, V.K.: Forecasting of solar radiation in India using various ANN models. In: 5th IEEE Uttar Pradesh Section International Conference on Electrical, Electronics and Computer Engineering (UPCON), India. IEEE (2018)
9. National Solar Radiation Database. https://rredc.nrel.gov/solar/old_data/nsrdb/
10. Bhaskar, K., Singh, S.N.: AWNN-assisted wind power forecasting using feed-forward neural network. IEEE Trans. Sustain. Energy 3(2), 306–315 (2012)
11. Alzahrani, A., Kimball, J.W., Dagli, C.: Predicting solar irradiance using time series neural networks. Proc. Comput. Sci. 36, 623–628 (2014)
12. Mellit, A., Pavan, A.M.: A 24-h forecast of solar irradiance using artificial neural network: application for performance prediction of a grid-connected PV plant at trieste, Italy. Solar Energy 84(5), 807–821 (2010)

Deep Learning Based Approaches
for Recommendation Systems

Balaji Balasubramanian$^{(\boxtimes)}$, Pranshu Diwan$^{(\boxtimes)}$, and Deepali Vora$^{(\boxtimes)}$

Department of Information Technology, Vidyalankar Institute of Technology,
Mumbai, India
balajib26@gmail.com, pranshusdiwan@gmail.com,
deepali.vora@vit.edu.in

Abstract. Recommendation systems are one of the most widely used Machine Learning algorithms in the industry. Deep learning, a branch of machine learning, is popularly used in fields like Computer Vision, Natural Language Processing etc. Recommender systems have started widely using Deep Learning for generation of recommendations. This paper studies different deep learning methods for the recommendation system highlighting the important aspects of design and implementation.

Keywords: Recommendation systems · Matrix factorization · Collaborative filtering · Deep Learning · Recurrent Neural Network · Convolutional Neural Network

1 Introduction

Recommendation systems have become a part of a consumer's daily life. They help give the user personalized recommendations based on their preferences. Recommendation systems have a huge impact on how a user interacts with the content – may it be viewing or purchasing. Recommendation systems are implemented in majority of applications, from content streaming websites like Netflix or Spotify to shopping websites like Amazon. The successful implementation of recommendation systems has helped several companies to monetize the results and better understand their customers. Recently Deep Learning has witnessed a huge surge in popularity. It has mainly been used to deal with image, text and audio data. It can be seen in many applications such as Self Driving Cars [7], Language Translation [8], Speech Recognition [9] etc. The reason for this is the data explosion due to internet, better hardware for complex computation like GPU, and better algorithms for learning complex ideas. In this paper, different ways of applying Deep Learning to build Recommendation Systems are studied.

2 Input Data Types

The quality of a recommendation system depends on the type of data we use as an input. Depending on the type of data available, one can define two types of systems:

© Springer Nature Switzerland AG 2020
D. J. Hemanth et al. (Eds.): ICICI 2019, LNDECT 38, pp. 508–520, 2020.
https://doi.org/10.1007/978-3-030-34080-3_58

2.1 Recommendation Systems Having Explict Data

Explicit data means data which has explicit action by the user, such as a rating. Explicit data is a very strong indicator of an user's preference.

2.2 Recommendation Systems Having Implicit Data

Implicit data is data consumed by the user without explicitly rating it. It may include various user actions like the number of times a song is played, or the types of movies watched, the active time of users, the click action, or how long they were on the page, and so on. While the user does not explicitly interact with the content, it is possible to achieve confidence in user preferences based on their implicit behavior. For example, a user listening to a song 20 times has higher confidence (rating) than a song the user has listened to only thrice. The drawback is that implicit data contains a lot of noise and many times the data can be irrelevant.

3 Types of Recommendation Systems

3.1 Collaborative Filtering Based

This type of recommendation system makes use of collaborative filtering methods to find user patterns and preferences and recommends items to a new user based on another user's similarity. The major advantage of collaborative filtering is that it does not take into consideration the content of the item. This is especially useful if finding similarity between items is difficult. Deep representation learning can be performed using a hierarchical Bayesian model called collaborative deep learning [4].

Collaborative filtering involves filtering of information by using recommendations of other users. Based on the behavior of users, explicit user profiles are created. Collaborative filtering analyses the interdependencies between users and their behavior to generate new user-item preferences [1].

A collaborative filtering framework is broadly divided into 3 steps [2]:

Matrix factorization (data reduction): This involves factorizing a large matrix (sparse matrix) into smaller but denser matrices which represent the original matrix [3]. These smaller denser matrices contain hidden or latent features, which are used to make recommendations. The product of these denser matrices is equal to the sparse matrix.

Factorization Machine [10] is a generalization of Matrix Factorization and it combines more than two input features. It is described as:

$$y(x) = w_o + \sum_{i=1}^{n} w_i x_i + \sum_{i=1}^{n} \sum_{j=i+1}^{n} \langle v_i, v_j \rangle x_i x_j$$

w_o is bias
w_i is the weight for i-th variable

In $\langle v_i, v_j \rangle$, two input features are described by k latent factors and their interaction is computed. This is pairwise for all the input features.

Factorization Machines is a widely used method because it can deal with highly sparse data and is also computationally efficient.

Collaborative Filtering (CF) method: The algorithm calculates similarities between users and items are recommended based on the user's nearest neighbors [2].

Calculating similarities: In order to calculate nearest neighbors, similarities are needed. Similarities can be calculated in several ways, like cosine similarity or Jaccard similarity.

For generating good recommendations, the user-item matrix sparsity should be lower than 99.5%. A number higher will have a prediction accuracy low. This will result in low quality recommendations [5]. However, at times there's a large amount of data and the sparsity of the user-item matrix can be very high. In such cases, a good accuracy score for recommendations needs to be achieved. Using association rules and mining techniques, similar patterns across users can be found, even for a very sparse matrix [5]. To reduce the size of data, clustering analysis techniques [5] are used.

Alternating Least Squares (ALS)
Alternating least squares is an optimization function used exclusively for implicit type of data. ALS is an iterative process where in each iteration, the goal is to come closer to a factorized representation of our original data [1].

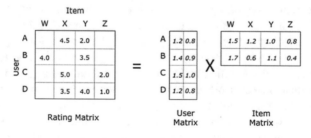

Fig. 1. Matrix factorization [6]

Consider the matrix in Fig. 1, where $M = U \times V$. Using ALS, the values in U and V are found. By randomly assigning values and iterating using least squares values that fit closest to M are found. ALS uses the same approach, but instead iteratively alternates between V and optimizing U and vice versa. This is done for each iteration until a value closest to M is reached [3].

Efficiency issue in ALS: Inverting a matrix is considered an expensive operation and for a Mx M matrix the time complexity is $O(M^3)$. Updating one latent vector has a higher time complexity, and this is ineffective for large scale data. To speed up this process we use uniform weighting [3].

3.2 Content Based Recommendation Systems

In content-based recommendation system, the similarity between the items is calculated. The user is recommended items which are most similar to the items the user has interacted with.

3.3 Hybrid Recommendation Systems

Hybrid recommendation systems take into consideration both collaborative filtering methods to generate similar users, and content-based recommendations to generate similar items. Hybrid system ensembles the output of both systems to give a common output. Hybrid systems are generally preferred as they have the advantage of both collaborative filtering and content-based systems. Major organizations like Netflix use a hybrid recommendation system.

4 Introduction to Deep Learning

Hybrid Neural Networks are algorithms inspired by the human brain. Yann et al.'s [11] LeNet was one of the first successful implementations of a Neural Network. It used Convolutional Neural Network (CNN) to understand Handwritten characters. AlexNet by Alex et al. [12] was the major breakthrough for Deep Learning. It won the Imagenet Competition in 2012 [13] in which it performed with a high accuracy on a dataset that contained 1.2 million images.

4.1 Convolutional Neural Network

Figure 2 shows a basic Convolutional Neural Network (CNN). It is mainly used to understand images. It has also been used to understand audio in a few applications. It has a hierarchical learning structure. The initial layers learn about the simple features like edges and curves in the image. Those are combined in the middle layers to learn more complex features like eyes, nose etc. In the final layers the model learns very complex features like face of cat, dog etc. In the above diagram the CNN contains Convolutional layers, Sub-sampling layers and Fully connected layers. Convolutional layers understand the image to learn about it features like edges, curves, eyes, nose etc. The Sub-Sampling layers reduce the number of features of the model to simplify the learning process, remove noise and save computational resources. The fully connected layers combine all the features learnt by the previous layers to provide the output.

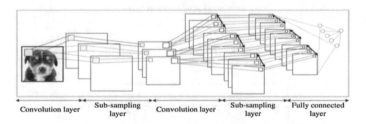

Fig. 2. Basic Convolutional Neural Network (CNN) [14]

4.2   Recurrent Neural Network

Recurrent Neural Network was introduced by Michael I Jordan [15] in 1986. It is used to understand natural language and has been used in applications such as Sentiment Analysis [16], language translation [17] etc.

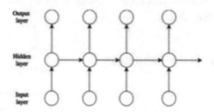

Fig. 3. Basic Recurrent Neural Network [14]

Figure 3 shows Recurrent Neural Network (RNN). It can remember sequential data. A layer at time 't' receives both the input data and output of layer at time 't−1'. Hence the layer can keep track of data and remember a sequence. But the RNN suffers from vanishing gradient problem in which it can forget information when the sequence is too long. To deal with this LSTM [18] and GRU [19] where introduced.

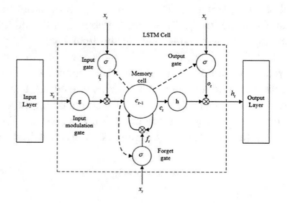

Fig. 4. Structure of LSTM [20]

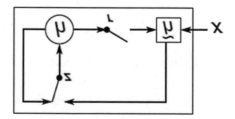

Fig. 5. Structure of GRU [20]

Figure 4 shows the structure of LSTM and Fig. 5 shows the structure of GRU. The main improvement over simple RNN is that these networks can control the information flow and forget information that is not required. For example, if the goal is to understand if the text review is positive or negative, it is required to focus on a few words that describe the sentiment and forget everything else.

5 Deep Learning Based Approaches for Recommendation Systems

5.1 Deep Content-Based Music Recommendation

Collaborative filtering is a popular approach for music recommendation, but it suffers from cold start problem and it cannot recommend new music because there is no user data available. Recommending music is a tough problem because music has a big variety of style and genre. Oord et al. [21] uses a Convolutional Neural Network (CNN) to understand the audio signal in music. Music audio signals are very low level and hierarchical learning approach of Deep Learning can be used to understand high level features such as genre, mood and lyrics of music.

The raw audio signals are not directly passed on to CNN. An intermediate time-frequency representation is produced which is then passed onto the CNN which can learn complex features in the music. Million Song dataset [22] was used to train the CNN. The CNN is used to predict latent factors from audio signal which contains important features that describe the music. Ground truth latent factors for users and items are used to train the CNN and are produced by Weighted Matrix Factorization [23]. Authors also found that CNN outperforms bag of words approach [32].

5.2 Deep Neural Networks for YouTube Recommendations

Covington et al. [25] recommends a two-stage approach for recommending YouTube videos: - Candidate Generation step and Ranking step. In Candidate Generation step, small number of videos which are relevant to the user are selected from the huge YouTube database to speed up the recommendation process. User history and contextual data are used during this step. In Ranking step, the videos selected in the previous step are ranked (Fig. 6).

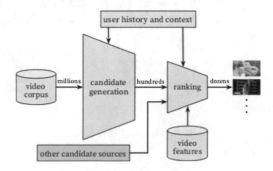

Fig. 6. 2-step process for recommending YouTube videos [25]

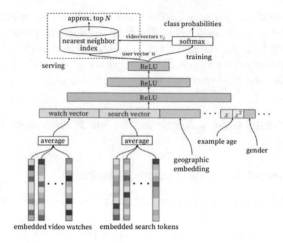

Fig. 7. Candidate Generation Step [25]

Candidate Generation Step - There are millions of videos and a particular user has only seen a few of those, so the dataset is very sparse. Neural Networks cannot deal with sparse data effectively so embeddings are used to create dense representation of this data which can then be passed onto the Neural Network. The embeddings are trained on watched video data, search history, geographic location of user, age, gender etc. and separate embeddings are created for each of these input features. Any number of features can be and they are then concatenated and passed onto Fully Connected layer with Rectified Linear Units (ReLU) activation function. Output layer contains Softmax activation. The output of the SoftMax layer is then passed to a nearest neighbor algorithm which generates hundreds of videos that are relevant to the user (Fig. 7).

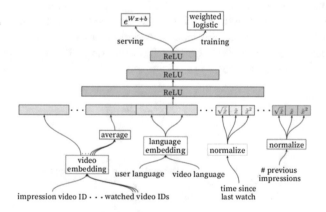

Fig. 8. Ranking step [25]

Ranking Step - In this step, the few hundred videos selected in the previous step are ranked. The Neural Network architecture is similar to that of the previous step, only difference is that in the final step every videos is scored using logistic regression after which they are ranked and presented to the user. Hundreds of features such as video impressions, interaction of the user with the topic of the video and the channel that uploaded the video are used to train the model. The number of features used to train the model is far more than the previous step to get a more accurate answer and the number of videos is just a few hundred and it won't be computationally expensive to use complex data (Fig. 8).

5.3 Session Based Recommendations with Recurrent Neural Networks

Hidesi et al. [26] used Deep Learning for session-based recommendation. Gated Recurrent Units (GRU) was used to keep track of long session of user interaction data which can be used to understand the mood of the user and make recommendations. GRU was preferred to LSTM because it provides similar accuracy while consuming less computational resources. Fig m shows the Neural Network architecture that was used. Input to the GRU contains a sequence of past events in the session and the model predicts the future events. The input data is 1-of-N encoded. 1-of-N encoding contains a vector whose length is equal to the number of items and each item has a place in the vector. For each item, the place in the vector corresponding to the item is one and rest is zero. This is then passed to embedding layers which converts sparse input to a dense output and is then passed to GRU layer. There are multiple GRU layers and all the layers receive input from the previous GRU layer and also the input embedding because it is found to provide better accuracy (Fig. 9).

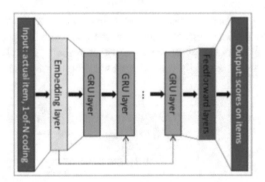

Fig. 9. GRU based architecture for Session-based Recommendations [26]

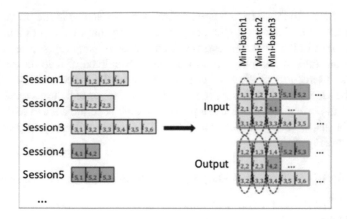

Fig. 10. Session-Parallel mini batches [26]

The authors came up with Session-Parallel mini batch to feed the input to the Neural Network. Each session is of different length, some sessions may contain a handful of event whereas some sessions may contain hundreds of events.

So we pass first event of X sessions and then the second event and so on. Ranking loss function is used to train the Neural Network (Fig. 10).

5.4   Wide and Deep Learning for Recommendation Systems

Cheng et al. [27] combined Deep Learning (deep) and linear models (wide). Linear models memorize the frequently co-occurring pairs and Neural Network can learn complex features and have good generalization power and it is jointly trained. The linear model uses a simple function $y = w^T x + b$ where $x[x_1, x_2,..]$ is a vector of features and w is the model parameters and b is the bias. The Deep Learning model has a Dense Embedding followed by dense layer of neurons with ReLU activation units. The wide and deep part are jointly trained by backpropagating the gradients (Fig. 11).

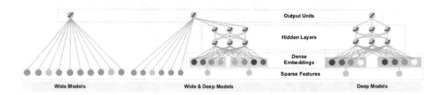

Fig. 11. Wide and Deep Learning model [27]

5.5   Wide Content2Vec: Specializing Joint Vector Representations of Product Images and Text for Product Recommendations

Content2Vec by Nedelec et al. [28] combines the features learnt from multiple data sources such as image, text, user interaction etc. The model is flexible, and we can easily plugin other data sources. Figure a. shows the model. In the first stage there are multiple content-specific data modules that are trained on a specific data source such as image, text, etc. The embedding for the image uses AlexNet [12]. Pretrained AlexNet is used which was trained on the ImageNet dataset [13] and the features are transferred to another task. Word2Vec was used for text embedding module. The Word2Vec was trained on the entire product catalog and it contains a good representation of the important features of the product and Text CNN [29] was used on top of it for the task for text classification. In addition to this, embeddings are used for product purchase sequence, product metadata etc.

In the second stage there is an overall embedding that combines several content-specific modules into a unified embedding that describes the product using all its data sources. Authors introduce Pairwise Residual Unit which is inspired by the original Residual unit [24]. It can preserve the features learnt by the Content-specific Embedding modules and make incremental improvements on the contents learnt by the previous stage (Fig. 12).

In the third stage there is a pair embedding that computes similarity scores between the different products, which is then used to make product recommendations.

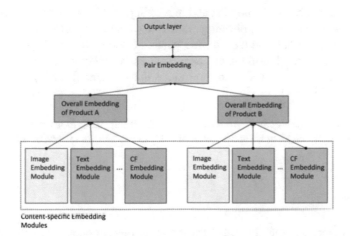

Fig. 12. Combining multiple content-specific Embedding modules [28]

5.6 Wide DeepFM: A Factorization-Machine Used Neural Network for CTR Prediction

DeepFM [30] is an improvement over Wide and Deep Model [27]. It uses the same input and embedding for wide and deep portion of the model and it doesn't require any feature engineering. The wide part is a Factorization Machine [10] that models the order-2 feature interaction. Deep portion of the model contains a dense neural network. ReLU activation function is found to work better than tanh and sigmoid. Dropout [31] is used in which a small number of neurons are randomly set to zero. Each neuron has a probability (for example 0.4) of being set to zero. This forces the Neural Network to represent a complex function with a smaller number of neurons and helps deal with overfitting. The final output layer combines the output from the wide and deep part. The wide and deep part is jointly trained and can learn complementary features (Fig. 13).

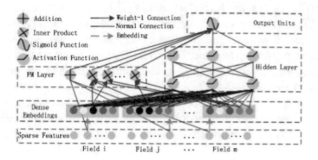

Fig. 13. Wide and deep architecture for DeepFM [30]

6   Conclusion

Deep Learning has become a popular and interesting field of study in machine learning. It is widely used in many applications and Deep Learning methods for Recommendation System have been studied in this paper. It complements the traditional Recommendation System algorithms by learning from complex data sources like image, text and audio. There is a need for better models that can combine all these data sources and build a multimodal algorithm.

Acknowledgement. The authors would like to thank all our anonymous critics for their feedback on this paper, along with the faculty of Vidyalankar Institute of Technology for their unconditional support.

References

1. Hu, Y., Koren, Y., Volinsky, C.: Collaborative filtering for implicit feedback datasets. In: IEEE International Conference on Data Mining (ICDM 2008), pp. 263–272 (2008)
2. Geuens, S., Coussement, K., De Bock, K.: A framework for configuring collaborative filtering-based recommendations derived from purchase data. Eur. J. Oper. Res. (2017). https://doi.org/10.1016/j.ejor.2017.07.005
3. He, X., Zhang, H., Kan, M.-Y., Chua, T.-S.: Fast matrix factorization for online recommendation with implicit feedback. In: SIGIR, pp. 549–558 (2016)
4. Wang, H., Wang, N., Yeung, D.-Y.: Collaborative deep learning for recommender systems. In: KDD, pp. 1235–1244 (2015)
5. Najafabadi, M.K., Mahrin, M.N., Chuprat, S., Sarkan, H.M.: Improving the accuracy of collaborative filtering recommendations using clustering and association rules mining on implicit data. Comput. in Hum. Behav. **67**(Supplement C), 113–128 (2017)
6. Seo, J.D.: Matrix Factorization Techniques for Recommender Systems. Towards Data Science, 22 November 2018. https://towardsdatascience.com/paper-summary-matrix-factorization-techniques-for-recommender-systems-82d1a7ace74

7. Fridman, L., Brown, D.E., Glazer, M., Angell, W., Dodd, S., Jenik, B., Terwilliger, J., Kindelsberger, J., Ding, L., Seaman, S., et al.: Mit autonomous vehicle technology study: Large-scale deep learning based analysis of driver behavior and interaction with automation, arXiv preprint arXiv:1711.06976 (2017)

8. Bahdanau, D., Cho, K., Bengio, Y.: Neural machine translation by jointly learning to align and translate. CoRR abs/1409.0473 (2014). http://arxiv.org/abs/1409.0473

9. Amodei, D., Anubhai, R., Battenberg, E., Case, C., Casper, J., Catanzaro, B., Chen, J., Chrzanowski, M., Coates, A., Diamos, G., et al.: Deep speech 2: end-to-end speech recognition in English and Mandarin, pp. 173–182 (2016)

10. Rendle, S.: Factorization machines. In: ICDM, pp. 995–1000 (2010)

11. LeCun, Y., Bottou, L., Bengio, Y., Haffner, P.: Gradient-based learning applied to document recognition. Proc. IEEE **86**(11), 2278–2324 (1998)

12. Krizhevsky, A., Sutskever, I., Hinton, G.E.: Imagenet classification with deep convolutional neural networks. In: NIPS (2012)

13. Berg, A., Deng, J., Fei-Fei, L.L.: Large scale visual recognition challenge 2010. www.image-net.org/challenges. (2010)

14. Pouyanfar, S., Sadiq, S., Yan, Y., Tian, H., Tao, Y., Reyes, M.P., Iyengar, S.S.: A survey on deep learning: algorithms, techniques, and applications. ACM Comput. Surv. (CSUR) **51**(5), 92 (2018)

15. Jordan, M.I.: Serial order: a parallel distributed processing approach. Adv. Psychol. **121** (1986), 471–495 (1986)

16. Socher, R., Perelygin, A., Wu, J.Y., Chuang, J., Christopher, D.M., Ng, A.Y., Potts, C.: Recursive deep models for semantic compositionality over a sentiment treebank. In: Conference on Empirical Methods in Natural Language Processing. Citeseer, Association for Computational Linguistics, pp. 1631 1642 (2013)

17. Bahdanau, D., Cho, K., Bengio, Y.: Neural machine translation by jointly learning to align and translate. CoRR abs/1409.0473 (2014). http://arxiv.org/abs/1409.0473

18. Hochreiter, S., Schmidhuber, J.: Long short-term memory. Neural Comput. **9**(8), 1735–1780 (1997)

19. Cho, K., van Merrienboer, B., Gülçehre, Ç., Bahdanau, D., Bougares, F., Schwenk, H., Bengio, Y.: Learning phrase representations using RNN encoder-decoder for statistical machine translation. In: The Conference on Empirical Methods in Natural Language Processing, pp. 1724–1734 (2014)

20. Fu, R., Zhang, Z., Li, L.: Using lstm and gru neural network methods for traffic flow prediction. In: 2016 31st Youth Academic Annual Conference of Chinese Association of Automation (YAC), pp. 324–328, November 2016

21. Oord, A.V.D., Dieleman, S., Schrauwen, B.: Deep content-based music recommendation. In: NIPS, pp. 2643 − 2651 (2013)

22. Bertin-Mahieux, T., Ellis, D. P.W., Whitman, B., Lamere, P.: The million song dataset. In: Proceedings of the 11th International Conference on Music Information Retrieval (ISMIR) (2011)

23. Hu, Y., Koren, Y., Volinsky, C.: Collaborative filtering for implicit feedback datasets. In: Proceedings of the 2008 Eighth IEEE International Conference on Data Mining (2008)

24. He, K., Zhang, X., Ren, S., Sun, J.: Deep residual learning for image recognition (2015). arXiv:1512.03385

25. Covington, P.; Adams, J., Sargin, E.: Deep neural networks for youtube recommendations. In: Proceedings of the 10th ACM Conference on Recommender Systems, ACM, pp. 191–198 (2016)

26. Hidasi, F.B., Karatzoglou, A., Baltrunas, L., Tikk, D.: Session-based recommendations with recurrent neural networks. arXiv preprint arXiv:1511.06939 (2015)

27. Cheng, H.-T., Koc, L., Harmsen, J., et al.: Wide & deep learning for recommender systems. In: 1st Workshop on Deep Learning for RecSys, pp. 7–10 (2016)
28. Nedelec, T., Smirnova, E., Vasile, F.: Specializing joint representations for the task of product recommendation, arXiv preprint arXiv:1706.07625 (2017)
29. Kim, A.Y.: Convolutional neural networks for sentence classification. arXiv preprint arXiv: 1408.5882. (2014)
30. Guo, I.H., Tang, R. et al.: DeepFM: a factorization-machine based neural network for CTR prediction. In Proceedings of the 26th International Joint Conference on Artificial Intelligence, pp. 1725–1731 (2017)
31. Hinton, G.E., Srivastava, N., Krizhevsky, A., Sutskever, I., Salakhutdinov, R.R.: Improving neural networks by preventing co- adaptation of feature detectors. arXiv preprint arXiv:1207. 0580 (2012)
32. Zhang, Y., Jin, R., Zhou, Z.-H.: Understanding bag-of-words model: a statistical framework. Int. J. Mach. Learn. Cybern. 1(1–4), 43–52 (2010)

Analysis of Liver Disease and HCC Inducing Factors Using Machine Learning Algorithms

Vyshali J. Gogi$^{(\boxtimes)}$ and M. N. Vijayalakshmi

Department of MCA, RV College of Engineering, Bengaluru 560059,
Karnataka, India
{vyshali.j.gogi,mnviju74}@gmail.com

Abstract. The process of identifying patterns in huge datasets comprising methods such as machine learning, statistics, and database system can be considered for data mining. It is a multidisciplinary field in computer science and it excerpts knowledge from the massive data set and converts it into comprehensible format. The Medical environment is rich in information but weak in knowledge. Medical systems contain wealth of data which require a dominant analysis tool for determining concealed association and drift in data. The health care condition that comprehends to liver disorder is termed as Liver disease. Liver disorder leads to abrupt health status like Hepatocellular Carcinoma (HCC) that precisely governs the working of liver and intern affecting other organs in the body. Machine learning techniques can be used to get the result of a test with indistinguishable degree of accuracy. Data mining classification techniques like Decision Tree, Support Vector Machine Fine Gaussian and Linear Discriminant algorithms are applied. Laboratory parameters of the patients are used as the dataset. Data contains features that can establish a rigorous model using Classification technique. Linear Discriminant algorithm showed the highest prediction accuracy 95.8% and ROC is 0.93.

Keywords: Liver Function Test (LFT) · Data mining · Liver disease · Hepatocellular Carcinoma (HCC) · Healthcare · Linear Discriminant · Support Vector Machine (SVM) · Decision tree

1   Introduction

The derivation of concealed information from massive data stores is performed by potential mechanism called Data Mining. The data mining strength in healthcare domain implements the precise use of data and analytics there by classifying the inefficiencies and analyze the best techniques to augment care and reduce the cost. The complexity of healthcare system necessitates a proper technology adoption for mining the huge hidden information [15].

Hepatocellular Carcinoma (HCC) is a kind of liver disease and one of the deadliest disease which is often diagnosed at advanced stage. HCC is a medical condition where tumor will be formed in the liver which is malignant. Cancer contributes to the maximum death rate and cancerous liver being most prominent amongst them. Every year people suffering from cancer all over the world exceed 700,000 [2]. Liver cancer

© Springer Nature Switzerland AG 2020
D. J. Hemanth et al. (Eds.): ICICI 2019, LNDECT 38, pp. 521–529, 2020.
https://doi.org/10.1007/978-3-030-34080-3_59

contributes to over 600,000 deaths every year. The symptoms that are diagnosed in clinical trials are used to predict the disease. The patient is advised to undergo Liver Function Test (LFT) if predicted with some liver disorder [5]. This paper helps in foresee the presence of liver disease using different classification algorithms [14]. The dataset considered consists of lab reports of 574 patients who are advised with LFT. The algorithm accuracy is analyzed on these datasets. Patients diagnosed of Liver disease can further be considered for diagnosis of HCC. The Mechanism which identifies patterns among huge datasets using process of statistics, machine learning and database systems is data mining [6].

Data mining comprises of four phases namely data preparation phase, data analysis and classification phase, knowledge acquisition phase and prognosis phase. The initial phase includes assorting the information from the available source. This information is from the laboratory test report of the patients who are predicted of having liver disorder. The dataset is checked for any missing values. The dataset is analyzed in the next phase to check which algorithm can be applied to obtain optimal results. The last phase includes classification of data by application of proper algorithm on the dataset and formatting a training model. In the last phase data is tested against the trained model to get proper predictive values. The proposed paper makes use of laboratory parameters of individuals after taking LFT [11].

Classification algorithms like Decision trees, SVM and Linear Discriminant are considered to diagnose the disease. The model that yields the maximum efficiency is considered. Once the patient is predicted with the liver disease, he is further advised for imaging tests to determine the existence of tumor lesions and their stages.

2   Literature Survey

Liver disorder is a life threatening disorders with high mortality rate every year. Prognosis of the liver disorder includes various steps starting with the routine urine and blood tests. Based on the symptoms observed patient is advised for LFT.

Several surveys have been performed to effectively diagnose the liver disorder using various machine learning algorithms [13]. A framework was proposed to classify liver fibrosis considering images obtained from ultrasound. Deep features were learnt via transfer learning and Visual Geometry Group Network (VGGNet). Fully connected network (FCNet) classifier was trained for predicting normal, early stage and late stage fibrosis liver status [1]. Machine learning algorithms were used to diagnose Chronic Liver Disorder. Two methodologies for predicting the disease were explored. The molecular biology approach deciphers the human anatomy which is often affected by diet, age and ethnicity. Another approach was a chemical approach which is an effective method of prediction [2]. LFT is performed for patients depending predictions made from clinical trials. Decision Template and Demster-Shafer classifier fusion techniques along with data from LFT contributed in diagnosing jaundice. The classifier accuracy increased for Multi-Layer Perceptron (MLP) classifier and Support Vector Machine (SVM) classifier [3]. One requires a proper knowledge about the LFT dataset to extract valid information from the report which helps in improving the investigation and better management of the disease. A paper was presented which gave clear vision to the generalists about extracting greater information from the test [12]. Cirrhosis is a type of liver disease. The data of the patients with primary biliary cirrhosis was considered for experimentation.

Machine learning algorithm PCA-KNN (Principal Component Analysis-K- Nearest Neighbor) approach was implemented for improvement of the diagnostic process and reduction of need for invasive procedures [4]. Liver cancer stages was predicted depending on extent to which liver is damaged. A KNN based intelligent computational method was presented for detecting the impact of liver damage at four different stages. The efficiency of classification was compared with diagonal quadratic, quadratic, linear, diagonal linear discriminant analysis classifiers along with classification and regression trees [5].

The datasets are heterogeneous in nature. The attributes in the dataset were removed in order to determine the dataset accuracy. The focus was to identify the best classification algorithm [6].

Machine learning (ML) application has made huge significance in cancer research. A review discussed about machine learning concepts and their application in prognosis or prediction of cancer [9]. Predictive models developed used supervised ML methods and classification algorithms which aimed at predicting valid disease outcomes. Heterogeneous data is assimilated along with implementation of different techniques for selecting and classifying features, contribute in developing models for cancer domain [7]. Blood test results contributes for diagnosis of liver disorder. Artificial Immune Systems (AIS) can be employed for this purpose. Computer simulation was used for comparing the efficiency of AIS algorithm with Artificial Neural Network (ANN) and other classification methods. AIS and ANN surpass the traditional methods [8]. A cancer prediction system was modeled which could provide which provides warning to the patients thereby reducing the cost and saving time. Validation of the system is done by comparing the predicted results with the previously available case study of the patient [10].

3 Methodology

Classification algorithms like SVM, Linear Discriminant and Decision tree is implemented in the current work.

Support Vector Machine (SVM) is a potential method followed by regression, classification and general pattern recognition methods. SVM requires no prior knowledge to be added which results in efficient generalizations and throughput [10]. The algorithm is good classifier as it intents in predicting the best classifier and provides classification function that differentiates training data into two classes.

Linear Discriminant analysis (LDA) dimensionality reduction method is a preprocessing step in classification of patterns and machine learning. It reduces the dimensions and removes the redundant and dependent features. LDA does the process of transformation from higher to lower dimension space [5].

Classification systems with multiple covariant can be utilized to build decision tree in data mining. Prediction algorithms are developed for target variables using decision tree. It can handle massive problematic datasets and does not require complicated parametric structure [16]. The Sample dataset is categorized into training and validation dataset whenever the dataset is huge.

The proposed work comprises of following phases:

3.1   Collection and Pre-processing of Data

Patients who are predicted of liver disorder through the initial clinical tests are considered for LFT. These parameters help in prediction of liver disease. Laboratory report of 584 patients who are advised LFT are taken as dataset. The pre-processing is carried out by cleansing, integrating, eliminating redundancy and transforming the data. The unwanted and repeated data is reduced in the dataset in cleansing stage. Structured and unstructured data are integrated. Redundancy is removed by eliminating the unwanted data and retaining only essential attributes in the dataset. The data is then transformed into scaled values and made to fit within minimal range.

Data pre-processing also include detection of missing values in the dataset. The missing values are handled effectively to avert imprecise assumption with regard to the data. The missing values are detected in the dataset.

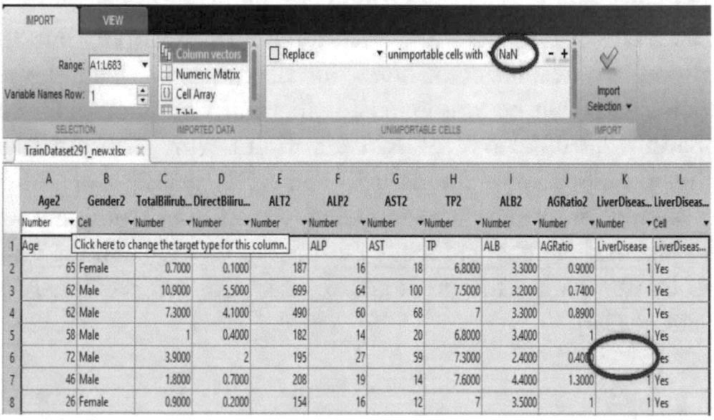

Fig. 1. Missing values in the dataset.

Figure 1 gives missing values in dataset.

3.2   Dataset Preparation and Classification

Data extracted from source is organized as collection of data or individual analytical data. Each attribute of data serves as variable and every instance has individual characterization. Liver disease is predicted using LFT dataset. The LFT dataset is created by adapting the methods of data collection and pre-processing. The LFT dataset helps us in diagnosing the disorder based on its parameters. In this proposed work LFT dataset with 11 attributes are considered for obtaining accurate results. Each attribute should lie between the specified threshold ranges. If the attribute values deviated from the specified threshold, the liver disorder is predicted.

Classification is a process in data mining that comprises of identification of problem statement. The characteristics of the liver disease among the patients are diagnosed through the observation and best performing algorithm is predicted based on the output of MATLAB's statistical output. Attributes considered in the dataset are: Age, Gender, Total Bilirubin, Direct Bilirubin, ALT, ALP, AST, TP, ALB, AGRatio, Liver Disease prediction (Table 1).

Table 1. Liver Function Test; mg/dL - milligrams per deciliter; U/L units per liter; g/dL - grams per deciliter; g/L - grams per liter.

Attributes	Min. value	Max. value
Total Bilirubin	0.1 mg/dL	1.2 mg/dL
Direct Bilirubin	0.1 mg/dL	1.2 mg/dL
ALT	7 U/L	55 U/L
ALP	45 U/L	115 U/L
AST	8 U/L	48 U/L
TP	6.3 g/dL	7.9 g/dL
ALB	3 g/dL	5.0 g/dL
AGRatio	23	35 g/L

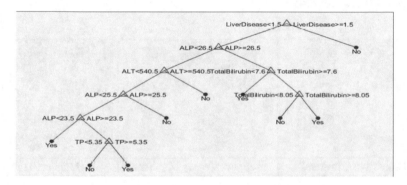

Fig. 2. Tree structure for LFT dataset

The decision tree shows the all the parameter of LFT dataset. Root node contains the prediction attribute. The leaf nodes contain the parameters and the values got during the LFT for the patient (Fig. 2).

4   Results

Decision tree, SVM and Linear Discriminant algorithms are implemented on the dataset. The results are explained with the help of diagrams and graphs obtained from applying machine learning algorithms. Linear Discriminant algorithm gave 95.8% accuracy, while the SVM and decision tree gave 82.7 and 94.9 accuracy respectively.

The disease is predicted depending on predictors plotted as graph considering various features. The Scatter plot graph, ROC curve and confusion matrix for Decision Tree, SVM and Linear Discriminant Algorithm as below.

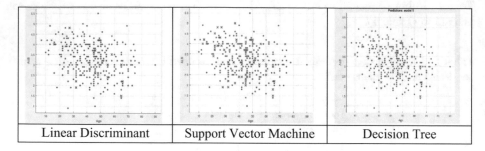

| Linear Discriminant | Support Vector Machine | Decision Tree |

Fig. 3. Scatter plot for linear discriminant, SVM and decision tree algorithm

Figure 3 depicts that classification accuracy attained from Linear Discriminant Model is 95.8%. The red dot shows disease is predicted "Yes" and blue dot indicates "No". The state in which the condition is detected when it is present is called true positive rate and the state in which it is not detected even when it is present is called false positive rate (Fig. 4).

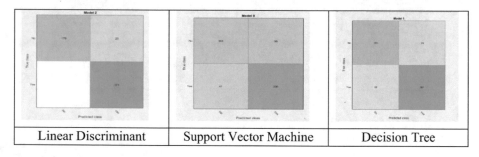

| Linear Discriminant | Support Vector Machine | Decision Tree |

Fig. 4 Confusion Matrix obtained with respect to each algorithm

The following table gives the performance of algorithms with respect to accuracy and other factors (Table 2).

Table 2. Performance of algorithms on the dataset

Algorithm	Accuracy	True Positives	True negatives	False positives	False negatives	Misclassification rate
Linear discriminant	95.8%	371	178	24	0	0.04
SVM	82.7%	338	144	58	41	017
Decision tree	94.9%	361	180	19	10	0.06

The sensitivity is when the result is actually yes and when it is predicted yes, specificity is when actually no and predicted no. The dataset contains many false positives which means that the patients actually does not have disease but its predicted yes which may cost money and time for patients. The false negatives predicts no when the patient actually have the disease which may cost patient life.

Linear Discriminant gave the maximum accuracy, zero false negatives and low misclassification rate compared to SVM and Decision tree.

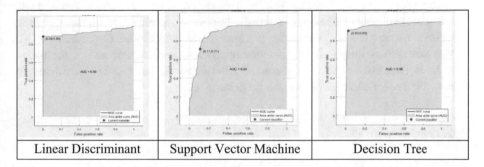

| Linear Discriminant | Support Vector Machine | Decision Tree |

Fig. 5 ROC curve for each algorithm

Figure 5 shows the ROC (Receiver Operating Characteristic). The true positive and false negative rates for different possible cut points have been plotted. The Area under Curve (AUC) observed is 0.93 which is almost equal to 1 (Table 3).

Table 3. TPR, FPR, sensitivity and specificity for each algorithm

Algorithm	True positive rate (TPR)	False positive rate (FPR)	Sensitivity	Specificity (1-FPR)	AUC
Linear discriminant	1	0.11	1	0.89	0.93
SVM	0.89	0.28	0.89	0.71	0.88
Decision tree	0.97	0.09	0.97	0.91	0.96

The above table shows the ROC curve obtained for each algorithm. The ROC (Receiver Operating Characteristic) Curve plotted shows the tradeoffs between sensitivity and specificity. Linear Discriminant gave the curve closest to the top left corner which indicates better performance than SVM and Decision tree.

The Area under Curve (AUC) for linear discriminant is 0.93 (which is almost equal to 1) which indicates that the predictions are better compared to other algorithms.

5 Conclusion and Future Work

Healthcare data mining is the most challenging area with structured, unstructured formats of data. This paper makes use of the lab test reports of the patients who has undergone Liver Function Test. MATLAB2016 is used and classification algorithms Linear Discriminant, SVM and Decision trees are applied. Linear Discriminant gave high accuracy of 95.8%. Various predictors are tested by plotting graph that determined the existence of disorder in liver. Confusion matrix gave the true positive and false positive rates of the classifier. The AUC was almost equal to 1. Patients diagnosed with liver disease can be considered for further diagnosis such as ultrasound imaging, biopsy and tumor marker test to predict the presence of Hepatocellular Carcinoma (HCC). Further research is proposed for considering the tumor characteristics of the patient once he is diagnosed with liver disorder. Also large dataset can be considered for training the model and algorithms can be determined.

References

1. Meng, D., Zhang, L., Cao, G., Cao, W., Zhang, G., Hu, B.: Liver fibrosis classification based on transfer learning and FCNet for ultrasound images. IEEE Access **5**, 2169–3536 (2017)
2. Sontakke, S., Lohokare, J., Dani, R.: Diagnosis of liver diseases using machine learning. In: 2017 International Conference on Emerging Trends & Innovation in ICT (ICEI) Pune Institute of Computer Technology, Pune, India, 3–5 February 2017
3. Saha, S., Saha, S., Bhattacharya, P.P., Subhash, N.: Classifier Fusion for Liver Function Test Based Indian Jaundice Classification. In: 2015 International Conference on Man and Machine Interfacing (MAMI). IEEE (2015). 978-1-5090-0225-2/15
4. Singh, A., Pandey, B.: Classification of primary biliary cirrhosis using hybridization of dimensionality reduction and machine learning methods. In: 2016 International Conference on Inventive Computation Technologies (ICICT) (2016). 978- 1-5090-1285-5, 7823232
5. Singh, A., Pandey, B.: An euclidean distance based KNN computational method for assessing degree of liver damage. In: 2016 International Conference on Inventive Computation Technologies (ICICT) (2016). 7823222
6. Karthick, R., Malathi, A.: Preprocessing of various data sets using different classification algorithms for evolutionary programming. Int. J. Sci. Res. (IJSR) **4**(4), 2730–2733 (2015)
7. Dixon, S., Yu, X.H.: Liver disorder detection based on artificial immune systems. In: 2015 11th International Conference on Natural Computation (ICNC) (2015). 978-1-4673-7679-2/15/$31.00
8. Song, Y.Y., Lu, Y.: Decision tree methods: applications for classification and prediction. Shanghai Arch. Psychiatry **27**(2), 130 (2015)
9. Arutchelvan, K., Periyasamy, R.: Cancer prediction system using data mining techniques. Int. Res. J. Eng. Technol. (IRJET) **02**(08), 1179–1183 (2015)
10. Bahramirad, S., Mustapha, A., Eshraghi, M.: Classification of Liver Disease Diagnosis: A Comparative Study (2013). ISBN: 978-1-4673-5256-7/13/$31.00
11. Hall, P., Cash, J.: What is the real function of the liver 'function' tests? Ulster Med. J. **81**(1), 30–36 (2012)
12. McLernon, D.J., Dillon, J.F., Sullivan, F.M., Roderick, P., Rosenberg, W.M., Ryder, S.D., Donnan, P.T.: The utility of liver function tests for mortality prediction within one year in

primary care using the algorithm for liver function investigations (ALFI). PLoS ONE 7(12), e50965 (2012)

13. Kurt, I., Ture, M., Kurum, A.T.: Comparing performances of logistic regression, classification and regression tree, and neural networks for predicting coronary artery disease. 0957-4174/$ - see front matter 2006 Elsevier Ltd (2008)

14. Dreiseitla, S., Ohno-Machado, L.: Logistic regression and artificial neural network classification models: a methodology review. 1532-0464/02/$ - see front matter 2003 Elsevier Science (USA) (2002)

15. Kim, S., Jung, S., Park, Y., Lee, J., Park, J.: Effective liver cancer diagnosis method based on machine learning algorithm. In: 2014 7th International Conference on Biomedical Engineering and Informatics (2014). 978-1-4799-5838-2 2014

Accident Detection and Prevention Using Cloud Database and GPS

U. S. Pavitha[(✉)] and G. N. Veena

Ramaiah Institute of Technology, Bangalore, India
{pavitha,Veenagn}@msrit.edu

Abstract. One of the most commonly used modes of transport is the roadways. The roadways in our country extend from one end to another. When travelling from one place to another by road, accidents can occur due to many reasons like loss of concentration, natural calamities a many more. Many people lose their lives because they are not able to get medical help in time. In this project, we propose a system to avoid accidents and also if an accident occurs get help to the place of accident as soon as possible. In this project, we detect the oncoming vehicles towards us and if the distance is less than the threshold, we alert the driver about it. If an accident occurs, we retrieve the location of the vehicle. This location is then stored in the database. This data is then used to find the location of the nearest hospital and a communication with the locality of the vehicle is sent to them. Thus medical help can reach the point of accident as soon as possible. Also messages will be sent to their family members to inform them about the accident.

Keywords: Keil μvision · GPS · GSM · SQL

1 Introduction

According to the survey made by Times of India in the year 2016 nearly one lakh persons were killed in road mishaps in India. Unfortunately, due to delay in ambulance facility about 30% of deaths have occured. Golden Aid increases the danger of saving a life by preventing traffic jams excessive delay in a crisis. The shortest path is shown to the accident place when the ambulance approaches the traffic lights and the main server monitors the location of the cars and switches the traffic signal. To achieve your goal in less time and without any individual interference.

2 Related Works

In this document, cars are provided with an accident identification and prevention technique that will offer a higher opportunity to decrease the crashes that take place on the highways very day. In the event of an accident, the system will detect its place and

© Springer Nature Switzerland AG 2020
D. J. Hemanth et al. (Eds.): ICICI 2019, LNDECT 38, pp. 530–537, 2020.
https://doi.org/10.1007/978-3-030-34080-3_60

inform those individuals who can take immediate action instinctively. A system based in Arduino that uses a global positioning system and global systems [2].

Authors have suggested and performed in this document by linking to the car with separate sensors, which notifies registered participants at any moment of an incident. The GPS sends the registered user geographic location and the tilt sensor is used to identify the collision triggered by tilting. If there is a serious car vibration owing to road conditions, the vibration sensor will detect the accident [3].

An unpredicted event that causes many lives to be lost is called an accident. Collision can happen because of the driver's quick driving, drunk driving or no adequate driving understanding, bad road conditions, etc. In many cases we may not be able to discover the location of the accident because we do not understand where the accident will occur. GSM is used to transmit location messages [4].

3 Proposed Work

The Arm7 based vehicle tracking system is as shown in Fig. 1. It has two partitions: the hardware and software description. Arm7 board with microcontroller, LCD, GPS and GSM shield are the hardware specification. Arm7 software and Google map issused are the software description.

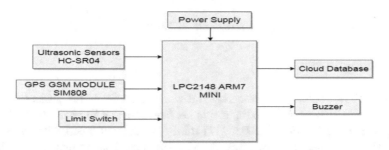

Fig. 1. Block Diagram

Arm7's cell phone based car monitoring system allows monitoring of the vehicle's present place. Figure 2 Displays the flowchart of the vehicle tracking system based on Arm7.

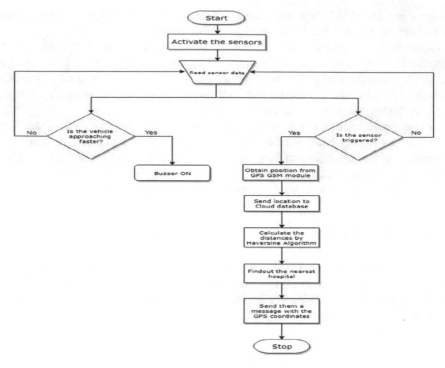

Fig. 2. Flow Chart

4 Implementation

The Fig. 3 shown below is representing the working model of the project. The entire project can be allocated into two fragments i.e. hardware and software. In the hardware part, we are using an ARM7 TDMI [11] mini development board, ultrasound sensor HC SR04, Buzzer and limit switches. ARM 7 [8] is used as the main processing unit for the project. All the sensors are coordinated using ARM7. The language used for programming is embedded C. Kiel uVision is used for programming the IC. Flash magic is used for uploading the data on the chip. The ultrasound sensor is used for measuring the distance of the nearest vehicle. SIM 900 module is used for obtaining the GPS spot of the motor vehicle and also sending this data. Limit switches are used for detection of accident.

The entire process first begins by initializing various sensors that are connected to ARM7. All the sensors i.e. ultrasound sensors, GPS and GSM [5, 12] module are initialized. After initialization of sensors, the distance of the next vehicle is found using the ultrasound sensor. A proper threshold has been set for detecting the proper range of the means of transportation. Uncertainty the vehicle is very close to the operator's vehicle, the buzzer will sound, making the user aware that his vehicle is close to the next vehicle. Behind the ultrasound sensor, there is a limit switch.

Fig. 3. Working model

In the event of an incident, the ultrasound sensor will affect the limit switch and the circuit will send a signal to ARM7 to initialize the automotive location retrieval process. The ARM7 will then communicate with the module GPS-GSM [10] to obtain the automotive GPS coordinates. The ARM7-SIM 900 interaction is performed with the use of AT commands.

The GPS co-ordinates that are obtained from the module are in raw format and require further modification for determination of exact location. As soon as an accident is detected, the location of the accident is sent as a message to the GSM module connected to the laptop. An SMS is used because even in the areas with least signal, an SMS can be sent, This SMS contains the location of the vehicle with the prefix of a particular string, $GPMRC. This particular string is searched for in the message received at the receiving end. Any other messages are discarded.

A SIM800 module is used as an interface for communication. Python 3 is used for creating the software part of the project. pySerial module of python is used for interfacing the SIM800 module. Whenever a message is received on the module, it is checked using a particular AT command, These AT commands are generally used for executing a certain number of queries using the SIM module. First the connection is setup to the module using the COM port of the laptop. The function serial. PySerial is used for creating this connection. The data will be communicated between the laptop using this COM port.

As soon as a message is received on the SIM module, it reads the number of the message. This number is used to then store the message in memory for further processing. The message that is stored contains the location in a raw format and hence it has to processed. After the proper latitude and longitude values are obtained, we use the reverse geolocator module to discover the approximate spot of the accident. For this module we pass the latitude and longitude as a tuple and the function prints back the approximate location at that given latitude and longitude. For example, if the given coordinates for this function are (13.030878, 77.565043), then the value returned by this function will be Bangalore.

This module is mainly used to decrease the area that is covered and also helps to narrow down the search for hospitals. After searching for the approximate location of the accident, we use Google maps API to find the list of hospitals is in this area. It is a free API and can be used readily. We need to generate and API key for using this service. We used the previously obtained location and search the area for hospitals. The search query is executed and the result is obtained in json format. This response contains the name and location of the hospital. We receive the list of 20 nearest hospitals. From this list to find the nearest one, we use Google's distance matrix API. This API gives us the approximate distance between the two points and also the estimated time it will take to reach the point. A cloud [7] database is created to store the data of the customer and also the mobile number of the hospital to which the message has to be sent. We have used Google cloud database for storing this data online.

The database is accessed using the MySQL. connector module of python. First an instance of the database is run. This instance is then used for accessing the data that is stored in the database. As soon as we find the nearest hospital, the database is accessed for finding the number of the hospital that is to be contacted and also the contact number of the customer to which the emergency message has to be transmitted to. After obtaining these numbers, we use the SIM module that is connected to the laptop to send message to the required numbers.

After retrieving this data from the cloud database, we collect this data and send messages using the SIM module. Messages that are sent contain the location of the accident. These messages are sent to the nearest hospital and also the contact person of thecustomer. The main advantage of using SMS for transmission of data is that it doesn't require a good signal quality to send a message. A message can be easily sent from the GSM module in areas of low coverage and hence help can reach the point as soon as possible.

5   Results

(i) The geographic coordinates of places are calculated using Google geocoding API. The process of converting addresses into geographic coordinates, which can be used to mark position on map is called Geocoding. To use the Google Maps Geocoding API, one must need an API key, which can be get from here.

The modules required for this are as follows:

(1) Importrequests
(2) Importjson.

(ii) Reverse Geocoding to get location on a map using geographic coordinates Reverse geocoding is the process of finding a place or a location address from a given pair of geographic coordinates (latitude and longitude).

The modules required for this are as follows:

(1) Reverse geocoder: It is a python library for offline reverse geocoding.
(2) Print: A module which helps to "pretty-print" any arbitrary python data structure.

```
import reverse_geocoder as rg
import pprint

def reverseGeocode(coordinates):
    result = rg.search(coordinates)
    # result is a list containing ordered dictionary.
    x=result[0]
    y=x['name']
    pprint.pprint(y)
    return y

# Driver function
if __name__ == "__main__":
    # Coorinates tuple.Can contain more than one pair.
    coordinates = (13.030878, 77.565043)
    reverseGeocode(coordinates)
```

(iii) pySerial

The module termed "serial" spontaneously selects the suitable backend. The backend is provided for Python operating on Windows, OSX, Linux, BSD (perhaps any POSIX compliant system) and Iron Python, this encapsulates the access for the serial port.

(iv) Using google distance matrix API, Google Map calculates distance and duration between two places. Distance Matrix API provides travel distance and time taken to reach destination. To use this API, one must need the API key. The modules required for this are as follows: (1) Importrequests (2) Importjson.

```
{
    "destination_addresses" : [ "New York, NY, USA" ],
    "origin_addresses" : [ "Washington, DC, USA" ],
    "rows" : [
        {
            "elements" : [
                {
                    "distance" : {
                        "text" : "225 mi",
                        "value" : 361715
                    },
                    "duration" : {
                        "text" : "3 hours 49 mins",
                        "value" : 13725
                    },
                    "status" : "OK"
                }
            ]
        }
    ],
    "status" : "OK"
}
```

(v) Get a collection of search query locations using Google Places API Google Places API Web Service to enable users to search for location data on a multitude of classes, such as institutions, prominent points of interest, geographic locations, etc. One can search by closeness or a text string for locations. For instance, "Delhi Hotels" or "Oshawa Shoe Shops". The service has a list of locations that match the text string and any location bias that has been set in reaction. To use this service, the use of rmust requires an API key. The modules required for this are as follows : (1) Importrequests (2) Importjson.

6 Technology

(i)   MDK Microcontroller Development Kit
      For Arm Cortex-M based microcontroller devices, the overall software development environment is Keil MDK.
(ii) Arm Development Studio
      The most dedicated software development tool-chain for the architecture is intended specifically for Arm processors Development Studio.
(iii) Philips Utility Tools
      Phillips Semiconductor offers two distinct tools for uploading firmware to the microprocessor. Their Windows XP/Vista program is called "LPC2000 Flash Utility," which is called their Windows 7 Flash Magic utility. The LPC2000 Flash ISP Utility enables you to download programs from the NXP (Philips) LPC2000 devices to the on-chip Flash boot loader.
(iv) PyCharm IDE
      In computer programming, it is possible to use PyCharm, an integrated development environment (IDE), specifically for the Python language. It promotes code analysis, a graphical debugger, an embedded unit tester, integration with version control (VCSes) schemes, and Django supports web development.
(v)   Python
      It is a language of interpretation, high-level, general-purpose programming. It is dynamically typed and garbage-collected that supports various programming paradigms, including procedural, object-oriented, and functional programming. Because of its extensive standard library, it is defined as a language "including batteries."
(vi) Google CloudSQL
      Cloud SQL is used to handle relational databases on Google Cloud Platform. It is a fully managed database service which makes setting up, maintaining and managing easy.

7 Conclusion

The scheduled technique will ensure car, driver and passenger security and security. Using GPS and GSM technology, the Arm7 placed vehicle tracing scheme was created and effectively tested to trace the precise position of a moving or stationary car in real time. A cell phone was used to reveal the vehicle's place on the Google map. The system provides for better service and cost-effectiveness.

References

1. Rahman, M., Mou, J., Tara, K., Sarker, M.: Real time Google map and arduino based vehicle tracking system. In: 2nd International Conference on Electrical, Computer & Telecommunication Engineering (ICECTE) (2016)
2. Mahamud, S., Monsur, M., Zishan, S.R.: An arduino based accident prevention and identification system for vehicles. In: IEEE Region 10 Humanitarian Technology Conference (HTC) (2017)
3. Kumar, S., Soumyalatha, N., Hegde, S.G.: IoT approach to save life using GPS for the traveller during accident. In: IEEE International Conference on Power, Control, Signal and Instrumentation Engineering (ICPCSI) (2017)
4. Berade, P., Patil, K., Tawate, P., Ghewari, M.U.: Intelligent Accident Identification and Prevention System Using GPS and GSM Modem. Int. Res. J. Eng. Technol. (IRJET) (2018)
5. Maurya, K., Singh, M., Jain, N.: Real time vehicle tracking system using GSM and GPS technology- an anti-theft tracking system. Int. J. Electron. Comput. Sci. Eng. ISSN 2277-1956/V1N3- 1103-1107
6. Song, K.T., Yang, C.C.: Front vehicle tracking using scene analysis. In: Proceedings of the IEEE International Conference on Mechatronics & Automation. National Chiao Tung University, Taiwan (2005)
7. Alexe, A., Ezhilarasie, R.: Cloud computing based vehicle tracking information systems. IJCST 2(1), March 2011. ISSN: 2229 - 4333 (Print) | ISSN: 0976 - 8491 (Online)
8. Ramya, V., Palaniappan, B., Karthick, K.: Embedded controller for vehicle in-front obstacle detection and cabin safety alert system. IJCSIT 4(2), 117–131 (2012)
9. Chen, H., Chiang, Y., Chang, F., Wang, H.: Toward real-time precise point positioning: differential GPS based on IGS ultra rapid product. In: SICE Annual Conference, The Grand Hotel, Taipei, Taiwan August, pp. 18–21 (2010)
10. Asaad, M., Al-Hindawi, J., Talib, I.: Experimentally evaluation of GPS/GSM based system design. J. Electron. Syst. 2(2), 67 (2012)
11. Bhumkar, S.P., Deotare, V.V., Babar, R.V.: Intelligent car system for accident prevention using ARM-71. Int. J. Emerg. Technol. Adv. Eng. 2(4), 56–78 (2012)
12. Khan, A., Mishra, R.: GPS – GSM based tracking system. Int. J. Eng. Trends Technol. 3(2), 161–169 (2012)

Contemplation of Computational Methods and Techniques to Predict COPD

Shaila H. Koppad[1]([⊠]), S. Anupama Kumar[1], and K. N. Mohan Rao[2]

[1] Department of MCA, RV College of Engineering, Bengaluru 560 059, India
Shaila.koppad@gmail.com, kumaranu.0506@gmail.com
[2] Department of Pulmonary Medicine, Rajarajeshwari Medical College,
Bangalore 560 074, India
kotnur.rao@gmail.com

Abstract. Recent developments in healthcare industry gives an insight on the increase in the various diseases and its complications across the world. Some diseases can be diagonised and prevented in the early stage itself while others are not. The development in science and technology helps the people all over world to get their disease diagonised and treated for betterment of life. One chronic lung disorder that is geographically spread among people of all ages is Chronic Obstrchuctive Pulmonary Disease (COPD). This paper gives an insight on the health care issues related to COPD and how researchers handle the issues using statistical and computational technologies. This paper brings in the state of art research works carried over by various researchers regarding the cause and effect of the disease. The findings of the researchers in understanding the challenges in predicting the disease and handling the patients using statistical and computational tools is tabulated. This papers gives an insight into the shortcomings of the statistical techniques and the advantages of using computational techniques over it. The application of big data and analytics in health care industry to predict and diagonise COPD is convened.

Keywords: Chronic Obstructive Pulmonary Disease (COPD) · Global Initiative for Chronic Obstructive Lung Disease (GOLD) · Logistic Regression (LR) · Generalized Estimating Equations (GEE) · Electronic Health Record (EHR)

1 Introduction

COPD is one of the lung disease which affects people across the globe and right now fourth driving reason for death on the planet [1]. The risk factors for COPD are air pollution, biomass fuel, occupational exposer, genetic abnormalities, abnormal lung development, accelerated aging and major is tobacco smoking [2]. It is difficult to identify the disease through its symptoms because asthma also has the identical symptoms. The basic test to diagnose the COPD and differentiate it from asthma is spirometry test. There are different spirometry devices which are used in different hospitals and data obtained is available in diversified formats because of different spirometry devices. The result of the test is interpreted by the doctor to diagnose the disease in the current model.

© Springer Nature Switzerland AG 2020
D. J. Hemanth et al. (Eds.): ICICI 2019, LNDECT 38, pp. 538–545, 2020.
https://doi.org/10.1007/978-3-030-34080-3_61

The diversity of data is one of the major challenges in identifying the disease and a lot of researchers are involved in finding solution to integrate the various parameters and attributes that be used to predict the prevalence of the disease. This paper gives a comparative analysis of the existing statistical and computational techniques available globally to comprehend the problems related to COPD. Finally a proposed solution is discussed which can be built to integrate and analyses the datasets in future.

This paper is dealt with as takes after: Sect. 2 survey to the related work with respect to the predominance of COPD worldwide, Sect. 3 explains the factors affecting the patients and also summary of statistical and computational methods used on COPD patients. Section 4 provides the overall analysis of existing techniques. Finally, Sect. 5 provides a proposed solution using big data analytics and machine learning followed by acknowledgment and references.

2   Background Work

Researchers all over the world try to find means to identify a common methodology to predict the prevalence of the disease so as to control the disease rate. The following statistics data is released by WHO every year discussing about the morality and causes of the death rate worldwide.

The Fig. 1 discusses about the top 10 global causes of deaths in the year 2000 among them COPD stands at 4th position as non-communicable disease Fig. 2 shows COPD stands at 3rd position causing more deaths worldwide for the year 2016 [3]. The reasons of increase in the disease may be air pollution, bio mass, occupation exposure, life style habits etc. Some of these parameters are measurable and can be directly analysed. Some of the parameters are patient specific like the life style of the patient which includes whether the patient is a smoker, his age, health condition etc (Fig. 3).

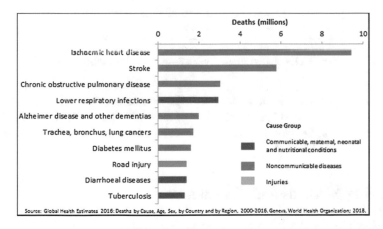

Fig. 1. Causes of deaths in 2016 [3]

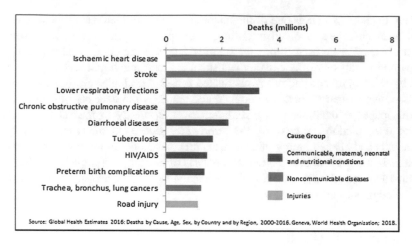

Fig. 2. Causes of deaths in 2000 [3]

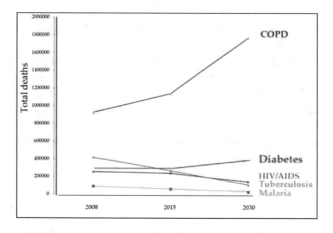

Fig. 3. Evaluated mortality because of various ailments [4]

By this comparison it is clear that the COPD is an increasing disease year on year and predicted to be the first leading cause of death worldwide by 2030 [4] if actions are not taken to prevent the globoal warming, air pollution, occupational stress and change the food habits and life style.

3 Existing Methods and Techniques

This section gives a detailed analysis about the factors that are responsible for the disease, the existing methods and techniques used by the researchers to identify and predict the prevalence of COPD.

The GOLD characterizes COPD as wind stream impediment which is irreversible and is dynamic. It is related with an anomalous incendiary reaction of the lungs to breathed in harmful particles or gases. The reason for the disease is multiple factors and depends on each person's immune power. This section gives an insight into the various symptoms that may cause COPD.

The prediction and the prevalence of COPD research was conducted in different countries but the parameters considered in each research work were varying. Some of the attributes considered were sex ratio, mean age, wheezing, fever, smoking history, breathlessness, BMI of the patients. The spirometry test was the basic test to diagnose COPD and statistical tools were used to compare and analyze these results [5] but it has not yielded as an efficient prediction due to various reasons.

In Canada [6] the impediment was two to six times higher than estimates based on self-reports of having been diagnosed with COPD by a health professional. The researchers also concluded that the severity of the disease increases with the increase in age and the related factors. In case of Netherlands, the researchers have considered the GOLD standards including the FEV/FVC symptoms and concluded the risk factors of COPD is more for male when compared to female patients.

In [7] related the presence of asthma as a major concern for the prevalence of COPD in the Italian public. Regression techniques were applied over the data and concluded that Asthma and COPD coincide in a considerable extent of the population. The Asthma COPD Overlap Symptoms (ACOS) cover disorder speaks to an imperative clinical phenotype that merits more restorative consideration and further research.

Frederik van Gemert and et al. [8] conducted spirometry analysis for people with mean age 45 and exposed them to biomass spoke. The prevalence of COPD was high for the people aged 30–39 (both men and women) and the presence of COPD was more for men than women. The analysis of the test was carried out using statistical tools.

The authors in [9] has discussed about the usage of different medical devices like inhalers, MDIs, DPIs, and nebulizers, dry powder inhalers etc., by the COPD patients of different age groups and discussed about the increase in the severity of the disease when the elderly people are not able to use the devices.

In [10, 11] the researchers have discussed about the maintenance of the medical records and its efficiency in predicting the prevalence of COPD. The medical records based on the clinical information about the patients are analyzed. Both the researchers have used different indicators and different methods to analyze the records.

Hilary Pinnock et al. [12] have applied machine learning algorithms to predict the admission of COPD patients in the health center using their baseline data from the randomized controlled trial and data extracted from their electronic health record. The machine learning algorithm was found AUC = 0.71 when compared to the normal counting method.

Table 1. Details of Computational and Statistical Techniques used on COPD Patients

Reference	Year	Techniques	Purpose	Parameters	Results
Computational techniques					
[13]	2017	AMNN, BN, ID3 and C4.5	Diagnosing asthma and COPD	Age group, Sex, Oxygen saturation, Pulse, Cough, Breath shortness, FEV1, FVC, TIFF, PEF, EF25, EF50, MEF75, MEF2575, Inhaler, Smoker, Wheeze, Spit, Chest pain, Patient name, Asthma, COPD	80.3%
[14]	2016	Decision Tree J48	Prediction COPD	Identity Number , age, Gender, Patient Type, Cough, Expectoration, Dyspnea grade, Wheeze, Fever, Hemoptysis, Packs NOYs, PY, Height, Weight, BMI, Pulse, HB, Chest Xray, FEV1, Type	ROC area is 1
[15]	2012	Artificial Neural Network	Identification of COPD disease	Age (years), Weight (kg), Height (cm), FEV1 (L), FEV1 (% pred), FEF/FVC (%), FEV1/FVC (%)	AUC (0.95)
[16]	2011	Decision Tree	To find severity and prediction of stable COPD	sex, age, dyspnoea, FEV1 % pred and previous hospital admissions prognosis	AUC (0.74)
Statistical techniques					
[17]	2018	Multinomial Logistic Regression	Severity of COPD disease	Age, Gender, Cough, Fever, Wheezing, Dyspnea, Smoking, Body Mass Index (BMI) , Hemoglobin, FEV1	
[18]	2017	Statistical Analysis System	Effectiveness of COPD Drugs	Not explained	
[19]	2015	Logistic regression	Forecasting hospitalization in the clinic	Age, Sex, Active Smoker(%), BMI, FEV1% predicted, mMRC dyspnea scale, Exacerbations, Hospital admission previous year, ADO index, GOLD risk groups : A, B, C, D	AUC (0.988)
[20]	2014	LR and (GEE)	Hospital Readmission Prediction	Alcohol, Drug, LAMA, Flu, Oxygen, ER, HOS, HOSC	0.974
[21]	2008	Negative binomial mode	Analysis of exacerbation rates in COPD Patients	Not explained	

The above Table 1 provides compilation of the various statistical and computational techniques used by researchers in recent years over COPD patients. The purpose of their research and the outcome of the research using the techniques is also given in the table.

4 Collation of the Existing Techniques

From the survey it is clearly understood that:

- The symptoms used to predict the prevalence of COPD is not standardized among the countries
- Smoking and other clinical data are considered as an important risk factors and identified mean age and sex ratio as one of the major concern for the prevalence of the disease
- Multiple parameters and different criteria's are involved in Diagnosis of COPD
- Statistical analysis has been used by the researchers and it has not yielded an efficient prediction due to various reasons, but the accuracy of their results were not measure
- Clinical history of the patient is not sustained
- Computational techniques were implemented using different parameters unanalyzed for accuracy, but the parameters used for the research work was varying
- Electronic Health Records are not considered in the research works because of the variety and volume.

From the above summary it is understood that since the number of patients are increasing, heterogeneous data is available which cannot be handled by statistical technique. Hence there is a need to implement a computational technique and efficient algorithms to provide better health care solutions.

5 Conclusion

This research paper provides clear insight that data of COPD patients is maintained as Electronic Health Record. Medical exams conducted are in different formats which might be structured, unstructured or semi-structured [22–24]. Electronic health records are substantial and complex that they are troublesome to deal with conventional programming and are not able to be effectively dealt with customary administration techniques. The human services industry truly has created a lot of information, driven by record keeping, consistence and administrative prerequisites, and patient care. The above said statistical and computational techniques are not sufficient to handle these versatile variety and huge volume of data.

Big data in healthcare is one of the promising application in the current scenario [25]. The research work has to be carried for development of organized structure to find out the presence/absence of COPD disease among the patients using the reports and symptoms collected from them. The generated data has to be structured and stored in the necessary format, clinical and operational data should be configured for analyzing resource utilization.

References

1. WorldHealthOrganization. http://www.WorldHealthOrganization.int/respiratory/copd/burden/en
2. Eisner, M.D., Anthonisen, N., Coultas, D., Kuenzli, N., Perez-Padilla, R., Postma, D., Romieu, I., Silverman, E.K., Balmes, J.R.: An official American Thoracic Society public policy statement: novel risk factors and the global burden of chronic obstructive pulmonary disease. Am. J. Respir. Crit. Care Med. **182**(5), 693–718 (2010)
3. WorldHealthOrganization. http://www.who.int/news-room/fact-sheets/detail/the-top-10-causes-of-death, 24 May 2018
4. The Global burden of Disease, WHO October 2008. www.who.int/healthinfo/global_burden_disease/projections/en/index.html. Accessed 22 Dec 2011
5. Koppad, S., Kumar, S.A.: Investigating COPD using big data analytics. In: International Conference on Advanced Computing and Communication Systems, ICACCS-2017, Coimbatore, India, 06–07 Jan 2017, pp 1984–1987. IEEE (2017). ISBN: No . 978-1-5090-4558-7
6. Evans, J., Chen, Y., Camp, P.G., Bowie, D.M., McRae, L.: Estimating the prevalence of COPD in Canada: reported diagnosis versus measured airflow obstruction. Health Rep. **25**(3), 3–11 (2014)
7. de Marco, R., Pesce, G., Marcon, A., Accordini, S., Antonicelli, L., et al.: The coexistence of asthma and chronic obstructive pulmonary disease (COPD): prevalence and risk factors in young, middle-aged and elderly people from the general population. PLoS ONE **8**(5): e62985 (2013). https://doi.org/10.1371/journal.pone.0062985
8. van Gemert, F., Kirenga, B., Chavannes, N., et al.: Prevalence of chronic obstructive pulmonary disease and associated risk factors in Uganda (FRESH AIR Uganda): a prospective cross-sectional observational study, vol 3, January 2015. www.thelancet.com/lancetgh
9. e Taffet, G., Donohue, J.F., Altman, P.R.: Considerations for managing chronic obstructive pulmonary disease in the elderly, Article in Clinical Interventions in Aging, January 2014
10. Duenk, R.G., Verhagen, S.C., Janssen, M.A.E., Dekhuijzen, R.P.N.R., et al.: Consistency of medical record reporting of a set of indicators for proactive palliative care in patients with chronic obstructive pulmonary disease: a pilot study. Chronic Respir. Dis. **14**(1), 63–71 (2017)
11. E Himes, B., Dai, Y., Kohane, I.S., Weiss, S.T., Ramoni, M.F.: Prediction of COPD in asthama patients using electronic medical records. J. Am. Med. Inform. Assoc. **16**, 371–379 (2009)
12. Pinnock, H., Agakov, F., Orchard, P., Agakova, A., Paterson, M., McCloughan, L., Burton, C., Anderson, S., McKinstry, B.: European Respiratory Journal **46**: PA3858 (2015). https://doi.org/10.1183/13993003.congress-2015.pa3858
13. Spathis, D., Vlamos, P.: Diagnosing asthma and chronic obstructive pulmonary disease with machine learning, 18 August 2017
14. Koppad, S.H., Kumar, A.: Application of big data analytics in healthcare system to predict COPD. In: 2016 International Conference on Circuit, Power and Computing Technologies (ICCPCT). IEEE (2016)

15. Amaral, J.L., Lopes, A.J., Jansen, J.M., Faria, A.C., Melo, P.L.: Machine learning algorithms and forced oscillation measurements applied to the automatic identification of chronic obstructive pulmonary disease. Comput. Methods Programs Biomed. **105**(3), 183–193 (2012). https://doi.org/10.1016/j.cmpb.2011.09.009

16. Esteban, C., Arostegui, I., Moraza, J., Aburto, M., Quintana, J.M., Pérez-Izquierdo, J., Aizpiri, S., Capelastegui, A.: A development of a decision tree to assess the severity and prognosis of stable COPD. Eur. Respir. J. **38**(6), 1294–1300 (2011). https://doi.org/10.1183/09031936.00189010

17. Shaila Koppad, H., Anupama Kumar, S.: Efficacy of knowledge mining and machine learning techniques in healthcare industry. In: Anouncia, S.M., Wiil, U. (eds.) Knowledge Computing and Its Applications. Springer, Singapore (2018). ISBN978-981-10-6679-5

18. Shepherd, S., McGaugh, M.: Analyzing the effectiveness of COPD drugs through statistical tests and sentiment analysis Indra kiran Chowdavarapu, Oklahoma State University, OSU CHSI, Oklahoma State University Paper 1031-2017

19. Abascal-Bolado, B., Novotny, P.J., Sloan, J.A., Karpman, C., Dulohery, M.M., Benzo, R.P.: Forecasting COPD hospitalization in the clinic: optimizing the chronic respiratory questionnaire. Int. J. Chronic Obstructive Pulm. Dis. **10**, 2295 (2015)

20. Zeng, L., Neogi, S., Rogers, J., Seidensticker, S., Clark, C., Sonstein, L., Trevino, R., Sharma, G.: Proceedings of the 2014 International Conference on Industrial Engineering and Operations Management Bali, Indonesia, 7–9 January 2014

21. Keene, O.N., Calverley, P.M.A., Jones, P.W., Vestbo, J., Anderson, J.A.: Statistical analysis of exacerbation rates in COPD: TRISTAN and ISOLDE revisited. Eur. Respir. J. **32**(1), 17–24 (2008)

22. Futrell, K.: MT(ASCP), Structured Data: Essential for Healthcare Analytics & Interoperability, October 2013

23. Big data — Changing the way businesses compete and operate, Insights on governance, risk and compliance, April 2014

24. Deep Analytics, Unstructured data: A big deal in big data. http://www.digitalreasoning.com/resources/Holistic-nalytics.pdf

25. Liang, Y., Kelemen, A.: Big data science and its applications in health and medical research: challenges and opportunities. J Biom. Biostat. **7**, 307 (2016). https://doi.org/10.4172/2155-6180.1000307

Analysis on Preprocessing Techniques for Offline Handwritten Recognition

Krupashankari S. Sandyal[1]([✉]) and Y. C. Kiran[2]

[1] Department of Information Science and Engineering,
Dayananda Sagar College of Engineering, Bangalore 560 078, India
krupasandyal@gmail.com
[2] Department of Computer Science & Engineering,
BNM Institute of Technology, Bangalore 560 070, India
kiranchandrappa@gmail.com

Abstract. Analysis of document images for information recovery has been exceptionally prominent in the later past. Wide collection of information, which has been generally put away on paper, is as of now being changed over into electronic form for superior storage and intelligent processing. This needs processing of documents by utilizing different processing strategies. Pre-processing techniques are advantageous in the recognition of document images as well as medical images. The most significant goal of image recognition system is to perform operations on images to find patterns and to retrieve information from the images for feature extraction process. In this paper, we emphasize on various techniques for pre-processing an image that aids in the further process of Image recognition.

Keywords: Image · Filter · Preprocess · Techniques · Morphological operations

1 Introduction

Image processing could be an efficient approach where the digitized picture is balanced in order to improve its quality. Particular sorts of areas in image processing exist based on the kind of input image, a number of them are satellite image handling, medical image handling, document image handling, etc. Document image processing is the method of analyzing the documents and the arrangement of auxiliary data that can be helpful for the long-run utilize. There are generally 2 ways of approaches in document image handling they are, Online and offline approach. Offline handwritten recognition is commonly treated as a moving undertaking when stood out from online handwriting recognition [1]. The nature of the image varies based on different grounds such as poor contrast image, noised image, and obscured image. But some of the images have serious noise issue because of environmental variations, capturing conditions, etc. Subsequently, image processing should be done for the effectual improvement of the quality of images. Pre-processing is fulfilled by distinctive filtering strategies like a max filter, min filter, median filter, mean filter and wiener filter, which are applied to discard the commotion from an image [3].

© Springer Nature Switzerland AG 2020
D. J. Hemanth et al. (Eds.): ICICI 2019, LNDECT 38, pp. 546–553, 2020.
https://doi.org/10.1007/978-3-030-34080-3_62

This paper is prepared and depicted as follows with five diverse main headings. Sect. 1 presents the Introduction, Sect. 2 portrays the background and research work, Sect. 3 aims to talk about the image processing techniques, Sect. 4 talks about experimentation outcome and Sect. 5 reports conclusions finally with Reference.

2   Related Work

Otsu [4] which is one of the global binarization techniques, is more capable for scanned document images. Background deviations lead to advanced techniques such as Niblack [5] and Sauvola [6], which are turning points in document image binarization [7]. Skew estimation is utilized as a preprocessing step in computerized image analysis frameworks. A skew rectified page is more pleasant for visualization [8]. Kaur et al. [2] analyzed two existing algorithms, they are Fast Fourier transform and nearest neighbor to estimate the text skew angle in a document image. There are diverse sorts of noises that distort images, like salt and pepper, Gaussian and speckle noise. Jaiswal et al. [9] contributed his work on denoising of severe noises like impulse and Gaussian noise. PSNR (peak signal to noise ratio) as well as MSE (mean square error) are the two error metrics used to measure the quality of image compression. Results of PSNR and MSE are calculated for analysis on image quality. Saxena et al. [10] has presented a survey on image denoising techniques also referred as pre-processing that helps in the further process of image recognition. Diverse sorts of noises and filters for denoising images are depicted in this paper. Image having salt and pepper noise can be enhanced by applying the filtering approach, it has been proved to be the best. Meenal et al. [11] analyzed different traditional image denoising methods. This paper examines various sorts of noises and results of filters applied to denoise the pictures.

3   Image Pre-processing Techniques

Document image recognition framework is broadly classified into four subproblems: Image acquisition, pre-processing, feature extraction and lastly recognition.

3.1   Pre-processing

There are different series of steps related to the strategy of pre-processing, here we are going be analyzing a number of of them like, binarization, skew identification and correction, normalization, denoising and thinning.

3.1.1   Binarization
Binarization is a technique or process that helps in modification of an image of up to 256 gray levels to a black and white image.

3.1.2   Skew Identification and Rectification
Digitalization of documents emerges as a bridge over the gap of past and present technologies. Sometimes because of not feeding pages properly into the scanner causes

skewness of these image pages. Skew identification and Skew rectification to deskew text in images are widely used in the pre-processing stage, which influences the performance of an recognition system significantly. Numerous techniques have been proposed as alternatives for skew angle detection of document images, some of them are Fast Fourier transformation, K-Nearest Neighbour, Hough Transformation and Cross-Correlation (Fig. 1).

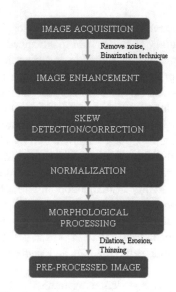

Fig. 1. Flow chart for Pre-processing

3.1.3   Normalization

The Normalization procedure is highly influential in case of changing the range of pixel intensity values. In this process, the intensity value of the pixel is reestablished to the range of [0, 1]. The different dimension images are changed over into fixed dimensions.

3.1.4   Denoising

Image denoising is the basic challenge within the field of image processing where the essential objective is to discover the original image by overcoming the noise from an undermined image. Filters are of more advantageous in case of suppressing either the high frequencies or the low frequencies within the image, that helps in updating and recognizing edges within the image.

3.2   Different Image Filtering Techniques

Image filtering is more beneficial for numerous applications like smoothing, sharpening, denoising and edge detection. A filter is characterized by a small scale matrix called as kernel, which is portrayed as a small array applied to each pixel and its neighbors inside an image [12]. Filters are broadly classified into two sorts they are, Linear Filter and Non-linear filter. Here we consider only Median filter, Min filter, Max filter, Wiener and Averaging filters. Gaussian noise and impulse noise are introduced as noise and filters are applied as shown for comparison in Fig. 3, 4, 5, 7 and 9.

Linear filters have a characteristic that their yield could be a linear association of their input. A few of the linear filters are Weiner filter and Mean filter. A non-linear filter is one that cannot be done with convolution or Fourier multiplication. A few of the useful Non-linear filters are Median filter, Min filter and Max filter.

Median Filter: A median filter is one of the non-linear filter, which is performed by taking the dimension of all of the vectors inside a mask and after that sorting according to the magnitudes or dimensions [13]. Figure 3 depicts how to expel impulse noise from an image using a median filter (Fig. 2).

 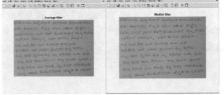

Fig. 2. Introduce salt and pepper noise   **Fig. 3.** Applying average and median filter

Max filter: The maximum filter is referred as the biggest of all pixels inside a neighborhood locale of an image. Maximum filter erodes shapes on the image. It is usually applied to an image to expel negative outlier noise conjointly utilized to discover the brightest points in an image.

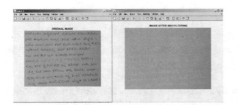 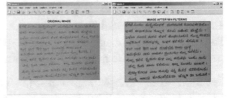

Fig. 4. Apply max filter on Gaussian noise   **Fig. 5.** Apply min filter on Gaussian noise

Min filter: The minimum filter is portrayed as the least of all pixels inside an adjacent locale of an image. The minimum filter broadens object boundaries, it is ordinarily applied to an image to get rid of positive outlier noise and moreover utilized to discover the darkest centers in an image.

Mean Filter: Average or mean filtering is a straightforward and simple technique to smoothen the images by lessening the amount of intensity variation between the neighboring pixels. It is a basic spatial filter which is utilized to reestablish every single pixel esteem in an image with the acquired mean estimation of its neighbors, including itself (Fig. 6).

Fig. 6. Simulate blur and noise   **Fig. 7.** Apply average filter on Gaussian noise

Wiener filter: The Wiener filter is amazing in terms of the mean square error [14, 15] because it diminishes the overall mean square error in the process of inverse filtering and noise smoothing. This filter is valuable in dispensing with the additive noise and inverts the obscuring at the same time from an image (Fig. 8).

Fig. 8. Simulate blur and noise   **Fig. 9.** Apply Weiner filter on Gaussian image

3.2.1 Morphological Processing

To attain excellent image quality the morphological operations like dilation and erosion are generally utilized in conjunction with image processing.

Erosion: In the morphological disintegration, it discards pixels on the object boundaries. It is mainly utilized for shrinking features and expelling bridges, branches and little projections.

Dilation: Enlargement could be a morphological method that is fundamentally utilized by growing features for filling gaps and holes of certain shape and size by structural element.

Filling: Because of the inconsistency in the binary conversion recognized by the optimal thresholding, a set of background regions are totally inside the foreground regions, it is known as a gap inside the foreground objects. Consequently, we make use of image filling to fill such gaps within the input binary image.

Thinning: It is proficient in discarding specific foreground pixels from binary images which results in single pixel width image to identify the handwritten character easily (Figs. 10, 11 and 12).

Fig. 10. Morphological operation: Erosion

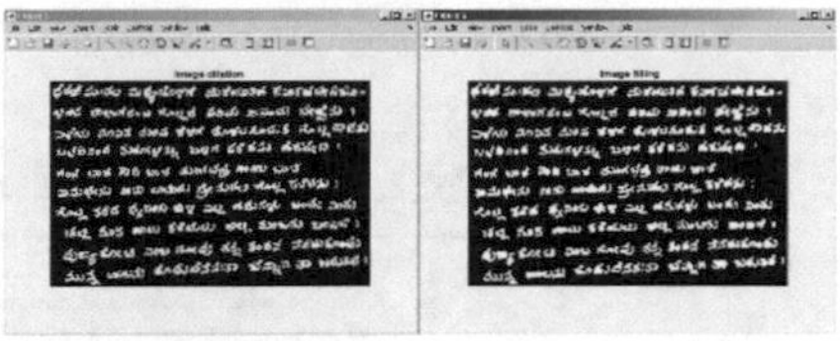

Fig. 11. Dilation and filling

Fig. 12. Image thinning

4   Results and Discussions

There are primarily two error measurements which are competent of differentiating image compression quality, namely Mean Square Error (MSE) and Peak Signal to Noise Ratio (PSNR). The cumulative squared error between the compressed and the original image is represented by using MSE and PSNR. The MSE represents the cumulative squared error between the compressed and the original image, whereas PSNR represents a measure of the peak error. The reduced MSE valuation shows a small mistake. The elevated value of PSNR shows the better quality of the rebuilt image.

The original input images are corrupted for experimental purposes by simulated Gaussian and impulse noise. Average, wiener, median, min and max filters were applied to get denoise images. We calculate PSNR values for performance evaluation of the above-mentioned filters. PSNR is utilized for the outcomes shown in Tables 1 and 2 as a performance evaluation parameter for distinct denoising methods. The most noteworthy value in PSNR is regarded to be the highest [16]. For the determination of an ideal filter for an image, the high PSNR value is considered.

Table 1. PSNR values for images with salt and pepper noise

Salt and pepper noise	Types of filters				
	Average filter	Max filter	Min filter	Median filter	Weiner filter
PSNR	22.727	19.988	22.822	22.900	22.776

Table 2. PSNR values for images with Gaussian noise

Gaussian Noise	Types of Filters				
	Average filter	Max filter	Min filter	Median filter	Weiner filter
PSNR	18.096	19.356	19.371	19.926	19.988

5   Conclusion

This paper discusses the implementation details of various pre-processing techniques required for document image processing. The techniques described in this paper are used to improve the quality of document images and the different filters discussed are used in order to eliminate the blur and noise. Moreover, performance evaluation is also analysed using PSNR method. According to the performance evaluation as per Tables 1 and 2, median filter is best suited for denoising images with salt and pepper noise and wiener filter gives better results for denoising images with Gaussian noise. This paper is beneficial for researchers who are working in the area of image processing.

References

1. Plamondon, R., Srihari, S.N.: On-line and off-line handwriting recognition: a comprehensive survey. IEEE Trans. Pattern Anal. Mach. Intell. **22**(1), 63–84 (2000)
2. Kaur, R., Kumar, S.A.S.: A hybrid approach to detect and correct a skew in scanned document images using fast fourier transform and nearest neighbor algorithm. Int. J. Adv. Electron. Comput. Sci. **3**(5), 1–6 (2016). Issn:2393-2835
3. Kanagalakshmi, K., Chandra, E.: Performance evaluation of filters in noise removal of fingerprint image. In: 3rd International Conference of Electronic Computing Technology, vol.1, pp. 117–121 (2011)
4. Otsu, N.: A threshold selection method from gray-level histograms. IEEE Trans. Syst. Man Cybern. **9**(1), 62–66 (1979)
5. Niblack, W.: An Introduction to Digital Image Processing. Prentice-Hall, Upper Saddle River (1986)
6. Sauvola, J., Pietikäinen, M.: Adaptive document image binarization. Pattern Recog. **33**, 225–236 (2000)
7. Howe, N.: A Laplacian energy for document binarization. In: Proceedings of the International Conference on Document Analysis and Recognition (ICDAR), pp. 6–10 (2011)

8. Lins, R.D., Avila, B.T.: A new algorithm for skew detection in images of documents. In: Image Analysis and Recognition, LNCS, vol. 3212, pp. 234–240. Springer, Berlin (2004)
9. Jaiswal, A., Upadhyay, J., Somkuwar, A.: Image denoising and quality measurements by using filtering and wavelet-based techniques. Int. J. Electron. Commun. (AEU) Science Direct **68**(8), 699–705 (2014)
10. Saxena, C., Kourav, D.: Noises and image denoising techniques: a brief survey. Int. J. Emerg. Technol. Adv. Eng. **4**(3), 878–885 (2014)
11. Jain, M., Sharma, S., Sairam, R.M.: Effect of Blur and noise on image denoising based on PDE. Int. J. Adv. Comput. Res. (IJACR) **3**(8), 236–241 (2013)
12. http://idlastro.gsfc.nasa.gov/idl_html_help/Filtering_an_Imagea.html
13. Kumar, S., Kumar, P., Gupta, M., Nagawat, A.K.: Performance comparison of median and wiener filter in image de-noising. Int. J. Comput. Appl. (0975–8887) **12**(4), 27–31 (2010)
14. Gonzalez, R.C., Woods, R.E.: Digital Image Processing, 3rd edn. Prentice Hall, Upper Saddle River (2008)
15. Guan, L., Ward, R.K.: Restoration of randomly blurred images by the Wiener filter. IEEE Trans. Acoust. Speech Signal Process. **37**(4), 589–592 (1989)
16. https://in.mathworks.com/help/vision/ref/psnr.html

Modeling Transport Pattern Using Artificial Intelligence

L. V. Arun Shlain$^{(\boxtimes)}$, A. Gayathri, and G. Aiswarya

Bannari Amman Institute of Technology, Sathyamangalam, India
{arunshalinlv,gayathri.it17,
aishwarya.it17}@bitsathy.ac.in

Abstract. The rapid pace of advancements in Artificial Intelligence (AI) is delivering unprecedented opportunities to upgrade and enhance the performance of various technologies and organizations, mainly in the transport area. AI needs a good understanding of the information and connection among AI on one hand, and transportation system characteristics and factors, on the other hand, to use the AI successfully. In addition, it is promising for every transport specialists to determine which is the best approach or method to utilize these advances to predict a quick improvement in diminishing congestion in the transport system. And also for making travel time progressively solid which is very useful to their customers and improve their financial status and productivity of their fundamental resources. Artificial Intelligence systems which mainly involves addressing the transportation issues mostly in traffic the board, traffic safety, problems during travelling in public transportation, urban portability. It is too difficult to handle in the transportation system when the transport issues become a challenge to model and anticipate the transport patterns. In this paper, we have a perfect view of the Artificial intelligence that will help the future transport scenarios thereby making it easy for people to travel using smart cars. We have also discussed about the different ways by which a road, rail track, airway can be modernized using artificial intelligence and its impact on people's life style.

Keywords: Self-driving · Drone air taxi · Traffic pattern · Automatic vehicle location (ALV)

1 Introduction

Artificial Intelligence (AI) is a growing technology which creates a machine that having the brain and those brains are work like human brains. It can recognize the speech, learn like a human, plan something and it also solves the problems on its own like a human being. AI can play a vital role in upcoming digital world. In this, we have the advancement of transportation in terms of AI. Artificial Intelligence means automate the things so in future, transportation in all domain like roadway, railway and airway all should automate by using AI. It will create a tremendous help for the users in the following automation areas

- Sense and predict the traffic
- Calculate the shortest root
- peer-to-peer ridesharing

© Springer Nature Switzerland AG 2020
D. J. Hemanth et al. (Eds.): ICICI 2019, LNDECT 38, pp. 554–560, 2020.
https://doi.org/10.1007/978-3-030-34080-3_63

- All vehicle are automated by using Self-driving.

[1, 2] focus on the aspects of artificial intelligence in traffic management. [3, 4] explains how Artificial intelligence may impact in transport Current and future developments.

1.1 AI in Public Transports

AI plays an important role in many sectors specifically it is well developed in the transport sectors. Thus, it could bring a major impact on our transportation system and also develop a rapid change in public transports. If the traffic lights are fitted with it will minimize the workload for people and also for traffic inspector like if the people are not approaching from one direction, they don't need to wait unnecessarily to wait for the signal to go green and for the inspector they don't want to get to that place and clear the traffic by themselves in rainy days and in hot summer days. There will be a decreasing in the accidents once it learns the traffic rules and it can also be used to access our route with GPS and eventually it can go to our desired destinations. So, if a person is not prepared a file or something else, he can do it while driving because he has AI interfaced intelligent vehicle. In recent days, AI in transport has grown a lot in each and every mode of transport. It is mainly focussing in the area of autonomous vehicles which is a driverless vehicle. The main problem in the transportation area and the most challenging one is capacity issues, pollution which does not affect the environment. In this current century, many other technologies are implemented in many sectors daily life and some of the greatest technologies are still in the developing stage. Recently a new vehicle was invented which is like the car and one or two persons can travel on a road which is similar to the rope car.

2 Future Works in AI

2.1 Smart Cars

Current vehicles are additionally furnished with a wide scope of detecting capacities by using GPS. This technique can be used in personal cars in the year of 2001. GPS helps drivers while giving huge scale data to innovation organizations and urban areas about transportation designs. In 2025 an average automobile can have more than 70 sensors including gyroscopes, accelerometers, ambient light sensors, and moisture sensors. Self-driving car idea was found in the year of 2000 and many companies produced the self-driving cars in the year 2015. The complete automatic cars were not given to the market that will be definitely come under normal usage in the year 2025.

Self-driven cars and trucks have been of high enthusiasm for upcoming years. In the business and transport sectors, Uber and Elon Musk have created and newly evolved the self-driving trucks to lessen the total number of accidents which happens mainly on highways and also to increase efficiency. First, Uber announced a driverless car, trucks and many other vehicles by implementing Artificial Intelligence in the transportation field. The Daimler trucks have created an 18-wheeler semi-self-ruling truck with an autopilot framework is the one automated truck which was implemented successfully. The costs of labours in the transport area will persistently diminish with an increase in the use of AI and also giving higher benefits to other industry players. The issue of long driving a car and other vehicle and ceasing for a break will no longer be a major concern when it completely becomes a robotized fleets.

2.2   Traffic Patterns

In those days traffic plays a major role in the day to day life. Traffic congestion in the United States meets the cost of $50 billion per year. In the year of 2025 by using AI we can Introduce the smart traffic congestion controller can be introduced it will allow streamlined traffic patterns and a significant reduction in congestion. In future Artificial Intelligence can be used to resolving, control and optimize the problems. Prediction and detection of traffic accident and condition by converting traffic sensors into 'intelligent' agents using cameras.

In 2025 there are AI's ability to make sense of large amounts of information and change the way that the vehicle moves around our cities.

Transport is mainly concerned only when influenced by traffic flow. Traffic issues by vehicles in the US costs around $50 billion per year. AI can bring a solution to those issues, this information is adjusted for traffic the board by means of AI, it will allow the streamlined traffic designs which result in a huge decrease in transportation problems. Several systems or algorithms to overcome the issues by Artificial Intelligence are:

- Smarter traffic light algorithm.
- Real-time tracking system.

This system can also be applied to public transportation for the best booking and routing.

2.3   Artificial Intelligence in Railways

In future we the train will have the self-driving and the efficiency of the time will be enhanced by using AI. We have introduced the sensor between the gateway. The sensed data can be feed into the AI and which aggregate there on the edge gateway, enabling onboard real-time decision-making.

2.4   Smart Highways

Can you imagine a road that talk to you!!!

It's a very interesting as it manages the traffic signals and it automatically adapts the weather conditions. Virtual lines will replace white yellow lines in future. In night times it will be glow thereby alerting the road user about the curves and edges of the roads.

2.5   Electronic License Plate

We will have the new form of an electronic license plate that is the digital display license plate. This is the identification sign of a vehicle that emits wireless signal is used for tracking and digital monitoring systems. The electronic tag has a mind-boggling plan that works straightforwardly with the vehicular PC framework and can speak with outside sources that may remotely get to it for the guideline of vehicle use.

2.6   Autonomous Delivery Trucks

Nowadays many departmental stores are having door delivery. For this purpose, we need lots of employees. For reducing the cost of employee we have the automatic delivery trucks. These trucks will collect the products and it will be delivered to the customer site without any delay.

2.7   Drone Air Taxis: Commuting by Air

In future, we will have the flying taxis. we can easily book in online by using an app like uber, red and Ola. This is creating a wonderful way to avoid traffic jams and accident in streets and highways. It should be cost-effective in future. It should bring the people to the third generation and also it is pollution free, less fuel, less noise and economically comfortable to users. It has an extra feature of having the self-flying and also may have very safety flying.

2.8   Tickets Are Bio-Transport

Suburbanites in Sweden have been trying microchip embeds as tickets, and the innovation could spare individuals a gigantic measure of cash in fines for charge avoidance.

2.9   Virtual Tracks

Trains are running by using virtual tracks that will more useful to avoid the accident and we don't have the specific railway station and tracks that will be congested.

3   Artificial Intelligence for Next-Gen Transportation

Mainly buses play a significant role in the public transportation part than other sectors. The numerous scientists have been led to make the transport (bus) journeys and destination more safety and reliability for customers. Artificial Neural Networks (ANNs) are efficient to use to decrease the waiting time of passengers by predicting transport landing times by themselves. Application in transport sectors has also been stretched out to implement the design of automated transports. IBus which was presented with a strong dual mode architecture is one of the features in AI. And another feature in the AI progressions of demand responsive public transportation is the flexible on-demand transport administration which works under flexible schedule and courses.

An Automatic Vehicle Location (ALV) framework has been acquainted with enhances the operational efficiency of public transport and also manages operational control and upgrade by the very large quality of public transport administration.

4 Benefits of AI in Vehicles

Self-healing: Vehicles can perceive the errors with themselves and fix it without any external help.

Self-socializing: The total capacity of a vehicle to socialize with the encompassing infrastructure, other vehicles and people in common language.

Self-learning: The vehicle uses its own practices and learns the features by themselves, driver, inhabitants, and the surroundings for a better environment.

Self-driving: The vehicle has the capacity to drive automatically by learning itself, with some automated impediment in a controlled condition.

The future vision for insightful urban mobility is smarter, basic leadership dependent on real-time information in transportation, and in system advancement by increasing the efficient utilization of infrastructure. Significantly, it encourages a better, safe, secure and more advantageous transportation system which makes an insightful availability in a system to achieve a manageable, accordant, compatible, feasible and environmentally friendly network. Recently, the world was at the cusp of introducing autonomous vehicles (AV), that is, the vehicles fit for moving or driving without the help and guidance of a human driver.

Artificial Intelligence is achievable in structuring and also enhancing an ideal system for the network, finds a better action plans for the public transport authorities, improving other planning plans for traffic flag, developing and advancing courses for individual drivers. AI is also connected for system episode identification, allowing abnormalities during flights and information can be collected from the streets for image processors/video successions. Besides, AI has also been developed to use in traffic request forecast, climate condition expectations, also predict the future traffic state for the executives and control purposes and to lighten clog and a fast basic leadership during road accidents. It can also help the transport specialists to choose what decision during an incident i.e., road accident or serious climate conditions and possibility on the measure of cash required for support and rehabilitation.

5 AI with Machine Learning and Deep Learning

The major part of AI in a machine and deep learning concepts involves artificial intelligence which plays a very huge role in many fields including transportation sectors. Machine learning means the in which only the machines that follow the instructions and guidance which are given by the humans that are available as data. Humans are training the particular machines that are assigned for any work to do and also trains the machines to learn the data model and predict high and better accuracy. Deep leaning is a machine learning technique that learns and also gains the features and tasks directly from the data. The data can be anything including images, sound or text. It can also perform big data and computers with high performance. Some of the examples are surveillance, many social media, real-time data and applications, email spam and so on.

6 AI in Robotics

AI also plays a major role in robotics. It is like a software robot is a type of computer program in which operations can do automatically and complete the specific tasks. It helps in navigation, manipulation of particular objects, identifying location and mapping etc. These are the tasks that can be added to humanoid robots to increase their performance. Recently a drone which is called as the self-driving air vehicle is developed for multiple purposes like transferring the blood in forest areas and emergency areas from nearest location and a box with blood is developed and it is maintained at a particular temperature. Currently, the impact of AI is developed all over the world especially in robotics to reduce the human intervention.

7 Drawbacks

There are certain drawbacks in this process like autonomous vehicles of aero planes due to this increase in autonomous driving, ships like are really challenging because the weather condition report in certain areas may go wrong because of the changing climate conditions, and also the implementation of this AI among may face a serious trust issue in many sectors. Because every system is autonomous it provides access to every individual and there may be a chance to hack by the hackers.

So, to prevent from this kind of incidents most of the AI developers have to work hard in so that it will overcome all of these drawbacks and public can use it without any fear and they must have lot of knowledge to use modern technologies and also about AI.

8 Conclusion

By the year 2025, we have more than 10 million self-driving vehicles and more than 250 million smart cars in roads. Elon Musk and Stephen Hawking found that in future AI will predict the correct root than the human. All the transportation is being processed only with Artificial Intelligence Technology. It plays a major role to reduce the human inconvenience and wastage by the way of transportation. In the year 2025, we will not able to control the traffic without using AI. This will lead to a revolution ther by making the lives of people even more easier.

References

1. Stencl, M., Lendel, V.: Application of selected artificial intelligence methods in terms of transport and intelligent transport systems. Transp. Eng. **40**(1), 11–16 (2012)
2. Sustekova, D., Knutelska, M.: How is the artificial intelligence used in applications for traffic management. In: Scientific Proceedings XXIII International Scientific-Technical Conference ISSN 1310-3946
3. Abduljabbar, R., et al.: Applications of Artificial Intelligence in Transport: An Overview. Sustainability **11**, 189 (2019) https://doi.org/10.3390/su11010189

4. Niestadt, M., et al.: Artificial intelligence in transport Current and future developments, opportunities and challenges. European Parliamentary Research Service
5. Liu, J., et al.: Artificial intelligence in the 21st century special section on human-centered smart systems and technologies. IEEE Access
6. Bagloee, A., Tavana, M., Asadi, M., Oliver, T.: Autonomous vehicles: challenges, opportunities, and future implications for transportation policies. J. Mod. Trans. **24**(4), 284–303 (2016)
7. Machin, M., et al.: On the use of artificial intelligence techniques in intelligent transportation systems. In: IEEE (2018)
8. Li, J., et al.: Survey on Artificial Intelligence for Vehicles. Automot. Innov. **1**(1), 2–14 (2018). https://doi.org/10.1007/s42154-018-0009-9
9. Elkosantini, S., Darmoul, S.: Intelligent public transportation systems: a review of architectures and enabling technologies. In: ICALT 2013 (2013)
10. Miles, J.C., Walker, A.J.: The potential application of artificial intelligence in transport. IEEE **153**(3), 183–198 (2006)
11. Matthew, N.O.S., Adebowale, E.S., Sarhan, M.M.: Smart Transportation: A Primer. Int. J. Adv. Res. Comput. Sci. Softw. Eng. **7**(3), 6–7 (2017)
12. Sherly, J., Somasundareswari, D.: Internet of things based smart transportation systems. Int. Res. J. Eng.Technol. (IRJET) 02(07)

CP-ABE Based Mobile Cloud Computing Application for Secure Data Sharing

Prakash H. Unki[✉], Suvarna L. Kattimani, and B. G. Kirankumar

Department of Computer Science and Engineering,
BLDEA's V.P. Dr. P.G. Halakatti College of Engineering and Technology,
Visvesvaraya Technological University, Vijayapur,
Belagavi 586103, Karnataka, India
prakashhunki@gmail.com, suvarnaky1977@gmail.com,
bgkiran16@gmail.com

Abstract. Mobile cloud computing becomes more popular in recent years because of the availability of the internet and smartphones. Maintaining security in mobile clouds is the most challenging issue because mobile devices are having limited computational resources and most of the operations are done through the internet. Therefore, it is necessary to safeguard data access from unauthorized access. In this paper Ciphertext based Attribute-Based Encryption is implemented. It offers better data security and privacy using efficient access control and privilege management. We have measured the performance of Key Policy Based-ABE and Cipher Text Policy Based-ABE, results show that Cipher Text Policy-Based ABE takes less time to perform encryption, key generation, decryption and offers better security compared to KP-ABE.

Keywords: KP-ABE · CP-ABE · Mobile cloud computing · Access policy

1 Introduction

Nowadays, the mobile cloud applications are very popular because of globally available internet and advancement in smartphone technology. Mobile cloud is a new computing technology delivers unlimited storage and other computational resources for their clients based on their needs or payment on the go access mode. Some of the applications and services are freely available in the public cloud, where users can have access to these applications up to some time limit or storage limit. In the public cloud, it is very difficult to trust data because of malicious users. Due to this it led to security and privacy attacks. Hence to handle these issues some of the security algorithms need to be adopted in mobile clouds. There are several security standards available like Key Policy-Attribute Based Encryption, Cipher Text Policy Attribute Based Encryption, Homomorphic algorithm, etc. but these security standards are developed for High computational devices and cloud servers. To apply these in the mobile cloud require little modifications because the mobile devices are the lack of storage and computing power capabilities. In such cases, it is required to utilize fewer resources on the mobile side. And heavy computational operations should be outsourced to the proxy server. It effectively reduces the computational overhead at the mobile cloud. KP-ABE and

© Springer Nature Switzerland AG 2020
D. J. Hemanth et al. (Eds.): ICICI 2019, LNDECT 38, pp. 561–568, 2020.
https://doi.org/10.1007/978-3-030-34080-3_64

CP-ABE standards suitable for mobile cloud computing, but CP-ABE offer more benefits compare to KP-ABE in terms of security, less time taken for encryption and key generation and decryption, overall it reduces the communication cost and provides better access control facility. In this paper, we developed the prototype CP-ABE Based Mobile cloud computing application for secure data sharing.

2 Related Work

The concept of cloud hosted key management in management in mobile cloud applications is discussed in [2]. This scheme highlights the process of generating and distributing the keys in encrypted format. The fully homomorphic encryption implementation in cloud increases the level of security during data sharing. Solutions of the disclosure attacks can be solved using matrix mapping of symmetric keys explained in [3]. The need of one to many sharing algorithms like attribute-based encryption in cloud computing. And its different types are explained in [4]. Big Data in cloud computing offers large amount of data is warehoused. It involves large number of users are join and leave the cloud services. Key generation and distribution are one of the challenges. For this solution is re-encryption of keys and data update is discussed in [5]. Cloud servers are incorporated security policies for maintain privacy and security in cloud. But these security algorithms are designed and developed for high computational devices. Whenever cloud services are accessed from mobile devices it takes lot time for encryption and key generation. To solve this different framework is proposed i.e. light weight data sharing scheme for portable devices discussed in [6]. Cloud data outsourcing is a very popular trend for start-up companies to maintain the data storage at online. cloud data sharing offers great number of data access aids to its users. possible reasons to adopt ABE in cloud and its implementation details are discussed in [7]. Mobile cloud environment includes data owner and user and trusted authority auditor. Maintaining the privacy of data and security is issue in this environment. Solution for this issue is half decryption and implementation of this scheme is discussed in [8]. In multi cloud, health record are accessed by malicious users. To protect this concept of key policy encryption scheme is proposed and discussed in [9]. In the mobile cloud environment increased file transfer taken place it increases the communication cost. To solve this problem aggregate key forward security key encryption and re-encryption standard is proposed in [10].

3 Existing System

KP-ABE Scheme has four function Setup, encryption, key-Generation, Decryption

(a) Setup (R_n, P_k, M_k): Setup function uses uniform random numbers Rn to produces the master key and public key.

(b) Encryption (Msg, $U_{Attributes}$, $Public_{key}$): Data Authority performs Encryption, it generates the ciphertext using the user attributes and Original message M, public key.

(c) Key Generation ($U_{Attributes}$, $Public_{Key}$, $Master_{Key}$): trusted authority creates secret keys using user attributes and public key and master key.

(d) Decryption (C_T, SK): User performs the decryption using Cipher text and secret key of the user.

Significant problems of KP-ABE algorithm [1] in cloud computing has several restrictions to adopt, in this scheme First limitation is, actual data owner has no control to choose who can be able to decrypt the data. Second, it increases communication overhead when more attributes are added to the control tree. Its access policy is not suitable for mobile cloud devices.

4 Proposed System

Sahai and Waters [1] implemented the Attribute Based Encryption which is initially produced from the Identity Based Encryption (IBE). IBE is adopted in broadcasting systems. Basically, ABE is classified into two schemes Key policy based-ABE and Cipher Text Based-ABE. In KP-ABE encryption algorithm access policy is enclosed with data user key. In CP-ABE algorithm [6] access control procedure is enclosed with encrypted message. CP-ABE is commonly used in real time cloud application because it has better control over access control policy and data owner has an option design his own access policy based on number of attributes. only authorized user whose attributes match with access control procedure can decrypt the original data.

The Cipher Text Policy-ABE Algorithm comprises of four functions they are explained in the following sections.

1. **Setup-** It accepts security parameters k and produces master key (MK) and public key (PK). Mixture of master key and public keys are used to generate Secret key.
2. **Encrypt-** This function performs encryption using public key, message, access control tree to generate Cipher Text (CT).
3. **Key Generation-** It accepts input of User Attributes and Master key (MK) to produce Secret Key (SK).
4. **Decrypt-** This function accepts Secret key (SK) and Cipher text (CT) to produce decrypted data.

4.1 Overview

The CP-ABE Based Mobile cloud computing application for secure data sharing framework is shown in Fig 1. Data owner (DO): Data owner encrypt data then upload the data onto mobile cloud storage. DO govern the access strategy and its attributes.

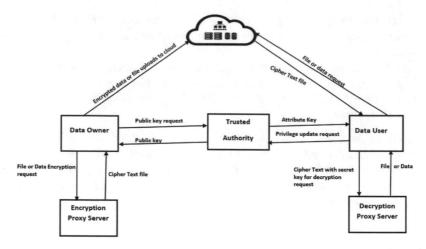

Fig. 1. CP-ABE Based Mobile cloud computing for secure data sharing framework

Data Users (DU): Data User is used to retrieve data from the mobile cloud environments.

Trusted Authority (TA): TA is performing the role of key generation and distribution.

Overall operations of this framework will be explained in the following paragraph.

DO register with cloud and Trusted Authority cloud provides encrypted storage facility. And Trusted authority TA will generate the public key for data owner. DO states the policy that decides how access the data and encryption is outsourced to Encryption proxy server (EPS) to decrease the computational load at mobile cloud. EPS returns Encrypted data to data owner for uploading on to cloud. Trusted authority generates the keys and sends to data users and data owner. Data user request the decrypted file from the cloud. If the access privileges match then request sent to cloud for access. Else request will be sent to DO, meanwhile data owner has option to revoke the users from accessing the data. DO checks the user authenticity by verifying the access policy and his credentials if it found correct secret key will be shared through email-id. Users decode the data or file with help of the secret key shared by data owner.

5   Proposed System Performance Analysis

In this section, we inspect the efficiency of proposed system CP-ABE Based Mobile cloud application for secure data sharing.

5.1   Experimental Settings

The test is performed on intel core i3 processor with 3.2 GHz CPU, 4 GB RAM, Windows 10 operating system and Prototype Application is developed in JAVA, JSP using NetBeans IDE, for data storage My SQL data base was used and for performance testing we used Apache JMeter application. Performance measure is evaluated by comparing the KP-ABE and CP-ABE schemes w.r.t security, time efficiency of Encryption and Key-Generation, Decryption.

5.2 Performance Analysis

5.2.1 Encryption Time

The analysis of encryption time with varying file size. The increase in file size gradually increases the encryption time. In existing KP-ABE, the encryption time is more for maximum file sizes. But, the proposed CP-ABE offered the minimum time requirement as shown in Table 1(a). The graphical illustration of comparison between proposed CP-ABE and KP-ABE is depicted in Fig. 2. The encryption time for maximum file size (100 MB) the encryption time will be 93 ms and 65 ms respectively. The encryption time for CP-ABE is 65 ms which is less compared to KP-ABE Method.

Table 1. KP-ABE VS CP-ABE performance analysis (a) Encryption Time (b) Decryption Time (c) Key-Generation Time (d) Security ratio (e) Compative Analysis of KP-ABE and CP-ABE

Algorithm	File Size(MB)	Time (ms)
KP-ABE	100	93
CP-ABE	100	65

(a)

Algorithm	Attributes	Time (secs)
KP-ABE	100	0.348
CP-ABE	100	0.251

(b)

Algorithms	Time (ms)
KP-ABE	80
CP-ABE	55

(c)

Algorithm	Security Performance in (%)
KP-ABE	60%
CP-ABE	92.91%

(d)

Parameters	KP-ABE	CP-ABE	Analysis
Encryption Time(ms)	93	65	-28
Decryption Time(sec)	0.348	0.251	-0.09
Key-Generation Time(ms)	80	55	-25
Security Performance in(%)	60	92.91	32.91

(e)

566   P. H. Unki et al.

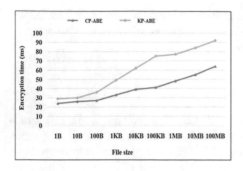

Fig. 2. Encryption time analysis

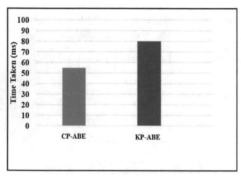

Fig. 4. Key generation time analysis

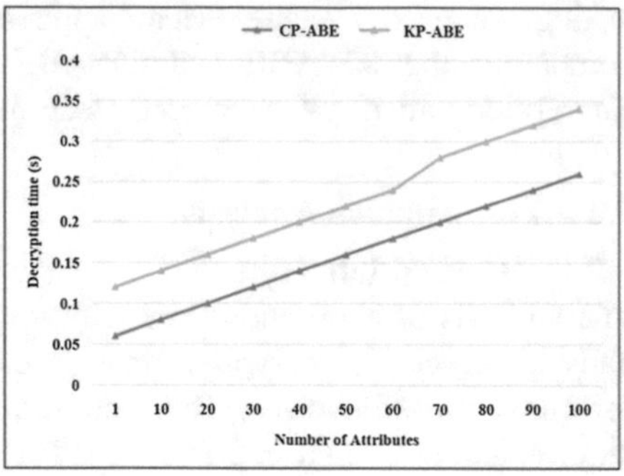

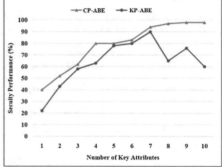

Fig. 3. Decryption time analysis

Fig. 5. Security performance analysis

5.2.2 Decryption Time

The proposed CP-ABE offered the minimum time compared to KP-ABE Algorithm as shown in Table 1(b). Figure 3 shows the visual representation of decryption time analysis with the number of attributes. The proposed CP-ABE consumes 0.251 s for and for the maximum attributes (100) the CP-ABE and KP-ABE offers 0.251 and 0.348 s respectively. Hence, the proposed CP-ABE offers 90 ms lesser decryption time than the KP-ABE for minimum and maximum attributes respectively.

5.2.3 Key Generation Time

The proposed CP-ABE offered the minimum time compared to KP-ABE Algorithm as shown in Table 1(c). The comparative analysis is graphically presented in the following Fig. 4. The time required for key generation in KP-ABE method and proposed CP-ABE are 80 and 55-mile seconds respectively. From the comparative analysis, the proposed CP-ABE offered 25 ms reduction compared to existing methods respectively.

5.2.4 Security Analysis

The proposed CP-ABE offered the better amount of security compared KP-ABE Algorithm as shown in Table 1(d). The graphical representation of comparative analysis between the security performance and minimum number of attributes is illustrated in Fig. 5. The proposed CP-ABE offered the better security performance compared to the existing methodologies for the maximum number of attributes. For the maximum attribute of 14, the security performance for CP-ABE is 92.91 which is maximum compared to KP-ABE method.

6 Conclusion

In this paper, we have implemented CP-ABE Based Mobile Cloud Application for secure data sharing. Our formulated results optimize time and security progressively. The performance of proposed CP-ABE is improved in terms of encryption time, decryption time, key generation time and security ratio respectively. The resultant values are −28, 0.09, −25, 32.91. Both KP-ABE and CP-ABE Results are shown quantitively and qualitatively by state-of-the-art algorithms. Our proposed methodology increases the time efficiency in mobile cloud and offers better secure data sharing. In the future it would be interesting to apply the method for the increased file size and incorporates the mobile cloud domain knowledge.

References

1. Bethencourt, J., Sahai, A., Waters, B.: Ciphertext-policy attribute based encryption. In: IEEE Symposium Security and Privacy, Oakland, CA (2007)
2. Tysowski, P.K., Hasan, M.A.: Cloud-hosted key sharing towards secure and scalable mobile applications in clouds. In: 2013 International Conference on Computing, Networking and Communications (ICNC), San Diego, CA (2013)
3. Gupta, C.P., Sharma, I.:A fully homomorphic encryption scheme with symmetric keys with application to private data processing in clouds. In: 2013 Fourth International Conference on the Network of the Future (NoF), Pohang (2013)
4. Nimje, A.R., et al.: Attribute-based encryption techniques in cloud computing security: an overview. Int. J. Comput. Trends Technol. **4**(3), 419–422 (2013). ISSN: 2231-2803
5. Yasumura, Y., Imabayashi, H., Yamana, H.: Attribute-based proxy re-encryption method for revocation in cloud storage: reduction of communication cost at re-encryption. In: 2018 IEEE 3rd International Conference on Big Data Analysis (ICBDA), Shanghai, pp. 312–318 (2018)
6. Jin, Y., Tian, C., He, H., Wang, F.: A secure and lightweight data access control scheme for mobile cloud computing. In: 2015 IEEE Fifth International Conference on Big Data and Cloud Computing, Dalian, pp. 172–179 (2015)
7. Alansari, S.F., et al.: Privacy-preserving access control in cloud federations. In: 2017 IEEE 10th International Conference on Cloud Computing (CLOUD), Honolulu, CA, pp. 757–760 (2017)
8. Anmin, F., et al.: A new privacy-aware public auditing scheme for cloud data sharing with group users. IEEE **99**, 1 (2017)

9. Joshi, M., Joshi, K., Finin, T.: Attribute based encryption for secure access to cloud based EHR systems. In: 2018 IEEE 11th International Conference on Cloud Computing (CLOUD), San Francisco, CA, pp. 932–935 (2018)
10. Chu, C.K., et al.: Key- aggregate cryptosystem for scalable data sharing in cloud storage. IEEE Trans. Parallel Distrib. Syst. **25**(2), 468–477 (2014)

A Multicast Transmission Routing Protocol for Low Power Lossy Network Based IoT Ecosystem

D. R. Ganesh[1(✉)], Kiran Kumari Patil[2], and L. Suresh[3]

[1] Reva University, Bangalore, India
ganesh1990.mtech@gmail.com
[2] School of Computing and IT, Reva University, Bangalore, India
kiran_b_patil@rediffmail.com
[3] CI - TECH, Bangalore, India
suriakls@gmail.com

Abstract. Internet of Things (IoT) emerges as one of the highest potential technologies for NextGen communication, which has exploited IPv6 routing protocol for Low Power Lossy Networks (LLNs), often called RPL to make optimal routing decision. However, exponential increase in QoS demands, delay-resilient and energy-efficient transmission in wireless communication has triggered researchers to achieve better solution. Multicast transmission has been found as a robust technique for timely and energy-efficient communication; however no significant effort is made towards exploiting its efficacy over LLNs, especially under different network conditions. With this drive, in this paper a vigorous multicast transmission protocol with dual objectives (RSSI and ETX) based parent node forwarding and selection decisions has been developed. The key robustness of the proposed model is its ability to perform multicast transmission over dynamic network condition while preserving high packet delivery ratio, low delay and packet loss ratio and better energy efficiency. Contiki - Cooja based simulation has confirmed robustness of the proposed multicast routing protocol, which recommends it to be used for LLNs- IoT communication purposes.

Keywords: Multicast transmission · Internet of Things · Low-Power lossy networks · RPL routing

1 Introduction

IoT Machine-to-Machine (M2M) communication demands event driven, scalable, and real time communication, the predominant focus is made on achieving a routing model which minimizes the latency, data drop, etc. Many efforts have been made to achieve it, but realizing fault-tolerant communication under dynamic network condition has been explored a few. Noticeably, being constrained and dynamic topology M2M communication especially under mobility conditions undergoes fault condition caused due to immature link-outage, energy depletion, node death etc. In addition, conventional WSN protocols are not suitable to cope up with IPv6 standard interface with internet to

© Springer Nature Switzerland AG 2020
D. J. Hemanth et al. (Eds.): ICICI 2019, LNDECT 38, pp. 569–582, 2020.
https://doi.org/10.1007/978-3-030-34080-3_65

enable major IoT applications. To alleviate such problems IPv6 Routing Protocol for LLNs, often called RPL has been proposed to support QoS centric M2M and/or IoT communications. Undeniably, RPL has been found robust to meet QoS centric communication over LLNs; however enabling energy efficiency and mobility feature have always been the challenge for academia-industries. Typically, a link outage during topological changes forces node to make 4–5 retransmissions scheduled at the interval of 10 ms and thus causes significantly high delay that eventually violates QoS provision and cause energy consumption. On the other hand, transmission using multiple path and/or multi-hops too imposes significantly large energy consumption thus reducing life-time of the sensor nodes. To alleviate such issues encompassing delay and energy exhaustion, multicast transmission has been found a potential solution. Multicast routing can be an effective approach to ensure reliable transmission as well as time and energy efficient routing. This as a result could achieve QoS provision for RPL based LLN communication. Multicast routing can be stated as a process where the same message is transmitting to the different receivers at the same time within the transmitter's radio range. There are numerous efforts made to enable multicast protocol, such as tree based, mesh based. However these have been especially designed to cope up with the Mobile Ad-hoc Networks (MANET); which are not optimal for LLNs. Topology based protocols need periodic flooding of control messages to manage the underlying overlay structure up-to-date, thus causes the early depletion of energy. In such cases multicasting can be a potential approach to support group communication as compared to unicasting, which seems a significant value addition to the RPL based LLNs. To preserve the battery resources and bandwidth routing protocols must minimize the number and size of control messages it transmits. Many existing multicast protocols in WSN have large control overhead, COM generates multicast tree structures using greedy algorithm to reduce the control head. Tree based multicasting has also been found effective to reduce signaling overheads, which can be vital for RPL based LLN. A few efforts like On Demand Multicast Routing Protocol (ODMRP) which has been especially designed for Ad-hoc networks with dynamic moving hosts has been found good for Ad-hoc network. Similarly, the use of Scalable position-based multicast (SPBM), Hierarchical Differential Destination Multicast (HDDM) protocols has been identified as potential to support scalable communication. Though, these approaches have performed satisfactory, their efficacy has remained unexplored for RPL assisted LLN. On the other hand, majority of these approaches avoid use link information such as Received Signal Strength Indicator (RSSI) or control packets such as Expected number of Control packets (ETX) to make routing decision or acyclic (destination oriented acyclic graph) tree formation. This as a result could achieve higher efficiency in terms of energy and delay as well as reliability. This can be a potential approach to meet major QoS centric communication demand over IoT using LLNs. In our previous work [15] we introduced RPL with mobility to achieve QoS centric and fault-tolerant communication; however it could not exploit multicast transmission, which could have achieved significant gain with respect to delay and energy-efficiency. It can be considered as the key driving force behind this research. In this paper, the predominant emphasis has been made on applying multicasting approach to transmit data timely with minimum signaling overheads and energy exhaustion. Additionally, we have applied ETX and RSSI information to constitute

DODAG formation for best forwarding path selection for multicast transmission across LLNs. Noticeably; we have used the dual objective based forwarding path selection method as proposed in [15] to ensure reliable data transmission.

2 Related Work

Gebhardt et al. [1] developed a QoS multicast routing protocol (QMRP) for partially mobile wireless TDMA networks for predicting delays, and provide guaranteed communication bandwidth. Afifi et al. [2] suggested power allocation with a wireless multi-cast aware routing for virtual network embedding to describe various applications using overlay graphs which includes functional blocks with predefined interconnections having feedback loops.

Aiming at reduction of energy consumption and enhanced network lifetime, Chang et al. [3] developed an energy-efficient transport protocol where every source sensor computes the multicast costs of numerous multicast trees among them and destination sinks, then trees having minimum control overhead are selected. Furthermore, Ajibesin et al. [4] developed the input-oriented variable return to scale (VRS) envelopment with slacks models for energy efficiency in Ad-hoc wireless multicast networks. Lu et al. [5] implemented a Neighbor Knowledge-Based Multicast (NKBM) scheme for dense wireless mesh network to improve the performance. Leão et al. [6] presented geographic multicast routing protocol using IoT-LAB infrastructure as a platform for real life experiments. Tsimbalo et al. [7] aimed to characterize the performance of lossy multicast network by focusing on general problems, considering arbitrary field size and number of destination nodes, and a realistic channel. Lee et al. [8] implemented a Sink-Initiated Geographic Multicast (SIGM) protocol for mobile sinks in wireless sensor networks to reduce location updates from sinks to a source, and achieve fast multicast tree construction and data delivery. Similarly, Sanchez et al. [9] proposed a Geographic Multicast Routing (GMR) protocol for WSNs to efficiently deliver multicast data messages to multiple destinations. Experimental observations demonstrated reduced computation time. Tomaszewski et al. [10] accentuated on packet routing and frame length optimization by considering multi-hop wireless mesh networks that serve multicast periodic packet traffic. In addition, to alleviate the issues of frame length minimization authors developed an integer programming optimization model and the associated near-optimal algorithms. Xie et al. [11] developed multisource, multi-destination, and multi-relay coding schemes by considering networks with multiple source destination pairs involving multicast where there are various nodes that can act as potential reliable nodes. Conti et al. [12] developed a consistent and secure multicast routing protocol for IoT systems. The primary motivation behind REMI is to empower effective communication in low-power and lossy networks by guaranteeing that a message will be gotten by the entirety of its planned goals, regardless of the system size and the presence of nodes which are misbehaving, it utilizes a group based routing approach that triggers a faster multicast dispersal of messages inside the system. Smys et al. [13] concentrated on mobility management in adhoc network utilizing power based routing and they have given a framework for mobility management with transmission power and speed of the backbone node. Kumar et al. [14] gave an overview of the IoT with highlighting on enabling Architecture, protocols, and application issues.

3 RPL: An Overview

RPL routing protocol is particularly designed for data transmission or communication over LLNs, which makes this appropriate for IoT systems. RPL routing protocol works on the basis of Directed Acyclic Graph (DAGs) paradigm, in this it uses three network parameters; they are DIO, DAO and DIS. RPL uses these attributes or control messages to generate DODAG, this could be identified as best forwarding node for sending data packets. DAG indicates the RPLs default tree structure over LLN. However, as per the topological variations and communication demands this undergoes various modifications. DODAG is the basic topological attribute of RPL. RPL has numerous DODAGs, and they are defined as following attributes.

- DODAG ID: This indicates the DODAG root.
- RPL Instance ID: This ID identifies autonomous DODAGs set of IDs which can be used for particular conditions.
- *DODAG Version Number:* This augments various events such as DODAG formation.
- *Rank:* It is position of a node in combination with root node in DODAG, in which rank is provided to each node which increases the downstream direction of DAG and vice versa.

Generally, to achieve data transmission, link information, and node discovery RPL employs different control messages. The control messages are as below:

- *RPL Control Messages*
  They are also a type of ICMPv6 (RFC 2463), primarily there are four types of control messages:
  - DIO: DODAG root invokes the DIO to form new DAG, the recently formed structures of DODAG allows the multicast transmission. DODAG represents the network information which helps the nodes to display the RPL instance discoveries; construction and maintenance of DODAG parent.
  - DAO: To discover the nodes visited towards the "bottom to top", it sends the reverse route data. In general, to update the route tables containing the information of related children DOA transmits the control messages by each node of the network except the DODAG root.
  - DIS: DIS recognizes the neighboring nodes in the connected DODAGs.
  - DAO - ACK: DOA recipient transmits the control messages as reply in the form of unicast packets.

Overview of RPL is shown in Fig. 1.

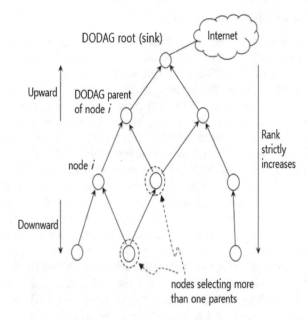

Fig. 1. RPL overview

In this process, every nodes rank data is sent through DIO in a way such that the nodes situated at far place from the root would acquire highest ranks than the near ones. This characteristic plays a very important role in looping problem, also removes the fault delay. RPL uses a factor named OF to create DODAG for enabling data transmission, this OF factor decides the suitability of nodes to become parent nodes to form forwarding path. Objective Function is selected in a way such that it must assure QoS delivery and timely data delivery. Generally, the default Objective Function is commonly known as OF0, and ETX is also applied for evaluating the suitability of a node to form path. In this, DIO messages works to build bottom up route, at the same time to form bottom path. DAO is unicast by child nodes to their parent nodes. However, the DIO message frequency affects the signaling and computation costs. To avoid unwanted DIO messages RPL uses the trickle timer based scheduling. Trickle timer supports the early network convergence, and also decreases the number of additional control messages. Moreover, the trickle timer plays an important role in optimizing computational efficiency of a network. When any topological modifications occur trickle timer will be reset followed by node discovery where the rank matrix of every node is achieved according to their suitability to become parent node is inspected to ensure the reliable data transmission.

4 Our Contribution

Since, the proposed multicast routing protocol is developed especially for RPL routing purposes to be used in LLN conditions, Sect. 3 discussed the basic functions and allied control messages of the RPL. This section briefs the proposed multicast routing protocol. Some of the key design principle and its implementation are given as follows:

Step-1 Clustering or Cluster Formation

In our proposed routing model, once deploying the nodes across the network we execute clustering within DODAG tree of the RPL protocol. Here, we have generated clusters virtually on the DODAG tree, where each constituted cluster is considered as an individual DODAG with cluster head (CH). Noticeably, in this approach CH functions as the root mote of the cluster formed. In our proposed model, we have applied RSSI and ETX information to perform CH selection. Unlike conventional approaches such as REMI, where classical OF0 is used as objective function to perform CH selection, in this paper a function with dual objectives, RSSI as well as ETX. Here, we used DIO messages to form DODAG which itself is a cluster. The obtained DAO control packet as ACK contains very important information such as RSSI and ETX, which are used to perform rank estimation of the node. For each cluster we obtain the RSSI and ETX information for the connected nodes and a node-rank table is obtained. Noticeably, in the proposed proactive node-table, nodes are ranked or shorted in the decreasing order of rank matrix, which is nothing else but the (high) RSSI and low (ETX). Realizing the significance of high link quality, the use of RSSI which is highly related to PDR is considered. On the other hand, RPL being control signals based routing approach employs multiple messages to form DODAG and make further decision. Such control signals and allied messages overhead get accumulated in case of large scale or multihop transmission. To avoid it, maintaining lower ETX can be vital. Considering this fact, in our proposed model both RSSI as well as ETX has been used as objective function, where this information are used to estimate node rank for each connected node in the cluster. The obtained node rank table can also be called as the Parent-table, where nodes are arranged in the decreasing order of respective node rank value. In our proposed routing model, the node with the highest rank is considered as CH. Thus, identifying the CH for each cluster, its DODAG is formed. In DODAG only the children nodes of the root can serve as CH and hence in our deployed network the number of CHs will be equal to that number of root mote children. Here, each node in cluster contains a unique cluster ID (CiD), which is different from the mote ID (MiD). The common cluster ID (i.e., CiD) is shared by all motes available in that cluster.

Step-2 DIO Extraction and Network Information Retrieval

In our proposed routing approach, the routing model follows RPL mode of operation that enables each mote to store significant information including IDs, link information, etc. Predominantly, each mote in the cluster gathers information about the neighbor motes and respective CiD. This information helps communicating to the connected nodes or motes (CNs) about the network condition and managing DODAG topology which is achieved through message transmission by participating motes. As stated in above section, obtaining periodic DIO message as the information received from one hop mote it exploits mote's CiD. Similarly, the mote ID as well as CiD are collected by each mote where i signifies i-th cluster. Here, motes with respective ID, MiD are connected for each cluster ID, CiD.

Step-3 Loop-Avoidance or Duplicate Avoidance

One of the common issues in RPL is the duplicate unicast messages that often come into existence due to repeated unicast transmission to form DODAG. To avoid such duplicate unicast messages in our proposed model we exploit source address and sequence values of a message, which are obtained through DIO. Once identifying the same source ID, the receiving motes discards such duplicate messages. This as a result reduces signaling overheads, bandwidth exhaustion as well as energy consumption.

Step-4 Multi Directional Forwarding

Being a multicast transmission, multicast packets are forwarded in different directions that comprise neighbor nodes or the connected motes, in downwards (intended children), upward (preferred parent) motes. In addition, the multicast messages are transmitted towards the neighbor motes, i.e., those mote(s) which is (are) the member of other cluster, simultaneously. This approach reduces end to end delay as well as hop count that enhances overall transmission reliability.

- *Q-FRPML assisted Multicast Routing Protocol*

Being a cluster based multicast routing protocol, our proposed RPL routing protocol employs multicast message forwarding mechanism to distribute messages across the LLN comprising multiple nodes or motes operating independently. In the proposed routing protocol, there are two key phases. In the first phase of implementation we perform multiple objective functions (RSSI and ETX) based DODAG formation and Clustering. On contrary, in the subsequent (second) phase considering the fact that a node can be a root or non-root mote, mote transmits multicast message within the network by considering cluster of the motes (CiD) as their destination. Noticeably, the first phase is performed before making any transmission between the source and destination node. Utilizing the inter node distance information (i.e., the distance between source and sink), the transmission mode can be decided such as multi-hop or one-hop transmission. Being multicast transmission, the source mote transmits packet in three directions, upwards, downward and towards neighbors belonging to other cluster to transmit multicast packet in DODAG tree. In case the source mote contains multiple neighbors which are part of same cluster then packet is transmitted to only one mote because in a cluster if any node contains a packet these packets will be disseminated in the entire cluster. In our proposed multicast routing protocol, we have performed communication in two modes:

1. DODAG Parent (root node) as the source mote of multicast packets,
2. Non-root mote as the source of multicast packets.

A snippet of these transmission models is given a follows:

A. *DODAG Parent (root node) as the source of multicast packets.*

For DODAG parent or root node as multicast-source, in our proposed routing protocol the data or packet is transmitted to all interested children nodes in the cluster. To perform it, at first parent node or mote verifies its routing table to identify connected children motes for the multicast address, which is then followed by multicasting of the packets to the children motes towards the downward direction. In the existing REMI protocol, in case parent node doesn't find any children more interested to collect the message or data, it drops the entire packet; however such act might adversely affect overall performance and reliability. On contrary, in our proposed model, parent node transmits towards neighbor motes (i.e., towards the motes connected to the other clusters) that ensures that the intended data or message reaches to its destination. Noticeably, in our proposed multicast routing approach in case for a received packet if the total number of interested children is lesser than the 50% of total number of children or connected motes, the parent node or root sets a flag named Inter-Cluster-Flag (ICF) as 1 (i.e., ICF = 1) in IPV6 header before transmitting packets to interested children, which is then followed by packet forwarding towards the interested children nodes. This approach seems viable and significant, especially for an IoT ecosystem or network comprising large number of interested children. Noticeably, unlike REMI protocol where authors have applied Optimized Forwarding Mechanism (OFM), we have followed the approach proposed in Q-FRPML [15] to form fault-resilient best forwarding path. The detailed discussion of the employed best Q-FRPML can be found in [15]. During multicast transmission process, adding or appending ICF to the header, when parent mote (node) forwards packets all clusters receive these packets simultaneously. This is because the CHs are the children of the root mote. Receiving packet with ICF = 1, it is not forwarded to the neighboring cluster, as it would have already received. Root accepts the multicast packet to forward it to the respective destinations. To achieve it, root node stores the information pertaining to the received packet along with respective MiD from which it was received. This information is saved in the table known as Discard Duplicate Forwarding (DDF), which is armored with a timer that updates entry continuously. Here, the root node gathers MiD for all the replicas of this packet till reset of timer. Noticeably, prime motive behind usage of timer is to reduce replication broadcasting by the root across the various clusters. Once timer resets, the main node checks MiD of CHs which are allied with the packet in its DDF table. Checking the MiD of the clusters it avoids transmitting or forwarding that packet to those motes as these clusters have already processed the packet in upwards direction.

B. *Non-root mote as the source of multicast packets.*

As in RPL non-root motes, also called connected nodes or motes intend to transmit as source node, it transmits the packet by means of unicast or Point-to-Point transmission manner in upward direction (i.e., towards root node or preferred parent node). On contrary, in case of any interested children node, it transmits packet towards downward. In addition, it transmits (downwards) the packet to the neighbors belonging to the different clusters as well. Once a mote gets an IPv6 packet with multicast address as the destination, it verifies DDF before forwarding it to destination. This method avoids duplicate packet transmission significantly. In case mote finds that MiD in DDF where it matches the source address and allied sequence number of packet under transmission, it marks that packet as duplicate-packet and discards it from further processing or forwarding. In order to reduce duplicate packets in the LLN, all connected motes including the parent node or root avoids forwarding packet to same mode/ cluster from where it was received. This approach significantly reduces overheads, resource exhaustion, and delay as well as energy consumption. The overall data transmission is accomplished in three distinct ways. These are

- *When packet is receive from its parent node*
- *When packet is received from children or neighbor node*
- *When a non-root node or source mote transmits packets to the other motes.*

A snippet of these transmission schemes is given as follows:

- *When packet is receive from its parent node*:

Phase-1   Mote verifies packet header for its associated ICF flag status. In proposed approach, a mote can only be allowed to forward the packet to neighbour motes which belongs to other clusters, when ICF = 0.

Phase-2   In case of ICF = 1, the connected mote exploits their routing table to find interested children who are identified and mentioned for that particular multicast address with received packet, and identifying the (interested) mote it executes Q-FRPML to transmit or forward the packets.

Phase-3   In case a (receiving) mote itself is a part of multicast group (as address mentioned in the received packet), it transmits packet to the network stack for further processing, else drops the packet.

- *When packet is received from children or neighbor node*

Phase-4   Mote forwards packet to the parent node using Q-FRPML protocol to ensure that the packet reaches to the parent node timely with high throughput.

Phase-5   Mote performs earlier phases (1–3).

- *When a non-root node or source mote transmits packets to the other motes*

Phase-6   Mote performs Phase-1, Phase-2 and Phase-4 to complete packet forwarding or transmission.

5   Results and Discussions

Considering the ever rising significance of IoT ecosystems and allied M2M communication requirements, in this research the predominant emphasis was made on exploiting efficacy of RPL routing protocol with Q-FRPML routing to perform multicast transmission over LLNs. Undeniably, Q-FRPML routing protocol exhibited satisfactory in our previous research [15]; however its efficacy could not be examined in previous work. On contrary, enabling multicast transmission with Q-FRPML can be of paramount significance to achieve an optimal routing protocol with minimum delay, high reliability and minimum energy consumption. This as a result could be vital for LLNs, a resource constrained network. Unlike REMI protocol [12] that doesn't deal with optimal parent node selection, which is can be vital for RPL routing under LLN conditions, we have exploited efficiency of Q - FRPML by applying ETX, RSSI as objective function to identify best (say, preferred) selection of parent node. In REMI authors seem to be applying basic objective function of OF0 which cannot ensure reliable transmission under noisy condition and mobility, which is common in the contemporary wireless networks. On contrary, the proposed model exhibits Q - FRPML routing decision while accommodating multicast transmission and therefore achieves higher reliability as well as minimum energy consumption and delay. Noticeably, Q-FRPML applies both RSSI, ETX as objective function for deciding parent node and thus ensures that the transmission occurs with healthy forwarding path with negligible or zero link-outage probability and computational overheads (along with low latency, redundant packets and energy exhaustion). Furthermore, since our multicast transmission model exploits efficacy of the Q-FRPML, it can be well suited for mobile WSN based IoT or mobile-RPL purposes. The complete proposed protocol was developed utilizing a Simulation platform known as Contiki- Cooja. It was developed and simulated over Instant Contiki - 2.7 on Ubunto 14.4 version OS. To preserve backward compatibility with native RPL, as recommended by IETF, our proposed multicast transmission protocol is applied in similar to standard RPL with IEEE 802.15.4 protocol stack. The simulation environment considered in this paper is given in Table 1. The simulation environment is given as follows:

Table 1. Simulation conditions

Simulation Conditions	
OS	Ubunto 14.4, Instant Contiki - 2.7
Simulator	Cooja
Radio	Unit Disk Graph Medium (Distance Loss)
Motes	Tmote (SkyMotes)
No. of motes during simulation	10, 20, 30, 40 and 50 motes
Radio range	100 m
Network	Routing (RPL) Adaption (6LoWPAN)
Data Link layer	CSMA-CA
MAC	Standard IEEE 802.15.4 MAC
Physical	Standard IEEE 802.15.4 PHY
Protocol	Q-FRPML assisted multicast transmission

To observe performance of proposed multicast transmission protocol, we have obtained performance variables such as PDR, PLR, Delay and Energy Consumption. Performance obtained with our proposed multicast routing protocol has been compared with a recent work [12], REMI which applies conventional default objective function to perform parent node selection and doesn't consider any fault condition during packet forwarding from source to destination. Considering the fact that with increase in network size or motes, the protocol might show different or varied performance and therefore we examined relative performance with the different node density conditions. In other words, to assess performance different network size with varying node or mote densities (10, 20, 30, 40 and 50 motes) was considered. Figures 2 and 3 present the PDR and PLR performance, respectively. Observing the results it can easily be observed that proposed protocol performs batter when compared to existing REMI protocol [12]. This can be primarily due to the robustness of Q-FRPML based multicasting that exploits multiple objective functions (RSSI & ETX) to decide parent node and forwarding decision. Being fault resilient protocol, our proposed model ensures minimum or negligible link-outage probability and allied packet drop. This as a result has exhibited high PDR and low PLR as compared to REMI. Noticeably, REMI doesn't use any kind of link-outage resiliency and hence might undergo packet loss (Fig. 3), and hence low PDR (Fig. 2).

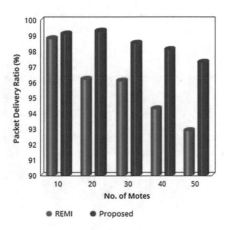

Fig. 2. PDR performance

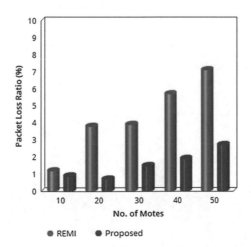

Fig. 3. PLR performance

As stated above the robustness of our proposed multicast routing protocol ensures that the PDR remains significant high or satisfactory than the classical OF0 based REMI. Consequently, our proposed routing protocol maintains high PDR and hence low retransmission probability.

In addition, in case of any fault it employs backup path as discussed in [15], and therefore the time to be consumed during node discovery and parent node selection is reduced significantly. On the other hand, multicast enables our approach achieving lower latency than the native Q-FRPML. This makes our proposed multicast routing model efficient in terms of delay. Figure 4 shows that our proposed protocol presents lower delay than REMI.

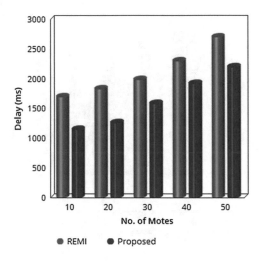

Fig. 4. Delay performance

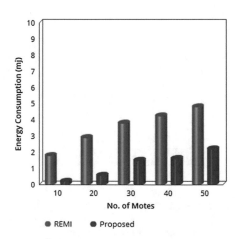

Fig. 5. Energy consumption

As stated being low retransmission scenario, our proposed routing approach saves energy, and hence it is more energy-efficient than the REMI protocol (Fig. 5). Thus, observing overall performance, it can be found that the proposed multicast protocol can be a potential solution towards time, reliability and energy-efficient communication over LLNs in IoT ecosystems.

6 Conclusion

In this paper a RPL based multicast routing protocol was developed for LLN communication, which is a common network condition for IoT ecosystem. Unlike conventional approaches, in this paper multilevel efforts were made where at first dual objective functions RSSI and ETX were used for optimal parent node selection in clustering based LLN. In addition, a multicast transmission scheduling was developed that ensures that the data or packet reaches its destination reliability, even at low communication or end-to-end delay. The use of Q-FRPML as best forwarding path formation technique and parent selection strengthened proposed multicast transmission approach to achieve high packet delivery ratio, low packet loss ratio, low delay and high energy efficiency. The Contiki based simulation and performance comparison with the state-of-art recent work REMI too affirmed that the proposed multicast protocol performs better to meet contemporary QoS centric communication over LLNs.

References

1. Gebhardt, J., Gotzhein, R., Igel, A., Kramer, C.: QoS multicast routing in partially mobile wireless TDMA networks.In: 2015 IEEE Global Communications Conference, San Diego, CA, pp. 1–7 (2015)
2. Afifi, H., Karl, H.: Power allocation with a wireless multi-cast aware routing for virtual network embedding. In: 2019 16th IEEE Annual Consumer Communications & Networking Conference (CCNC), Las Vegas, NV, USA, pp. 1–4, (2019)
3. Chang, W., Hou, Y., Chen, M.: A multicast routing scheme for many-to-many applications in WSNs. In: 2015 IEEE 12th International Conference on Networking, Sensing and Control, Taipei, pp. 99–104 (2015)
4. Ajibesin, A.A., Ventura, N., Murgu, A., Chan, H.A.: Data envelopment analysis with slacks model for energy efficient multicast over coded packet wireless networks. IET Sci. Meas. Technol. 8(6), 408–419 (2014)
5. Lu, D., Dong, S., Jiao, D.: NKBM: a neighbor knowledge-based multicast scheme for dense wireless mesh networks. In: 2015 34th Chinese Control Conference (CCC), Hangzhou, pp. 6501–6506 (2014)
6. Leão, L., Felea, V., Guyennet, H.: Performance of geographic multicast approach on a real-life platform. In: 2019 Wireless Days (WD), Manchester, United Kingdom, pp. 1–8 (2019)
7. Tsimbalo, E., Tassi, A., Piechocki, R.J.: Reliability of multicast under random linear network coding. IEEE Trans. Commun. 66(6), 2547–2559 (2018)
8. Lee, E., Park, S., Lee, J., Kim, S.: Geographic multicast protocol for mobile sinks in wireless sensor networks. IEEE Commun. Letters 15(12), 1320–1322 (2011)
9. Sanchez, J.A., Ruiz, P.M., Liu, J., Stojmenovic, I.: Bandwidth-efficient geographic multicast routing protocol for wireless sensor networks. IEEE Sens. J. 7(5), 627–636 (2007)

10. Tomaszewski, A., Pióro, M.: Packet routing and frame length optimization in wireless mesh networks with multicast communications. In: 2016 17th International Telecommunications Network Strategy and Planning Symposium, Montreal, QC, pp. 1–6 (2016)
11. Xie, L., Kumar, P.R.: Multisource, multidestination, multirelay wireless networks. IEEE Trans. Inf. Theory **53**(10), 3586–3595 (2007)
12. Conti, M., Kaliyar, P., Lal, C.: REMI: a reliable and secure multicast routing protocol for iot networks. In: ARES 2017, Reggio Calabria, Italy (2017)
13. Smys, S., Bala, G.J., Jennifer, S.: Mobility management in wireless networks using power aware routing. In: 2010 International Conference on Intelligent and Advanced Systems, pp. 1–5. IEEE, 15 June 2010
14. Kumar, R.P., Smys, S.: A novel report on architecture, protocols and applications in Internet of Things (IoT). In: 2018 2nd International Conference on Inventive Systems and Control (ICISC), pp. 1156–1161. IEEE, 19 January 2018
15. Ganesh, D.R., Patil, K.K., Suresh, L.: Q-FRPML: QoS-centric fault-resilient routing protocol for mobile-WSN based low power lossy networks. Wirel. Pers. Commun. **105**(267), 267–292 (2018)

Classification of Autism Gene Expression Data Using Deep Learning

Noura Samy[1,2(✉)], Radwa Fathalla[1,2], Nahla A. Belal[1,2],
and Osama Badawy[1,2]

[1] Arab Academy for Science and Technology and Maritime Transport,
Alexandria, Egypt
nou.moh91@gmail.com, {radwa_fathalla, nahlabelal,
obadawy}@aast.edu
[2] Collage of Computing and Information Technology, Alexandria, Egypt

Abstract. Gene expression data is used in the prediction of many diseases. Autism spectrum disorder (ASD) is among those diseases, where information on gene expression for selecting and classifying genes are evaluated. The difficulty of selection and identification of the ASD genes remains a major setback in the gene expression analysis of ASD. The objective of this paper is to develop a classification model for ASD subjects. The paper employs: Deep Belief Network (DBN) based on the Gaussian Restricted Boltzmann machine (GRBM). Restricted Boltzmann machine (RBM) is considered a popular graphical model that constructs a latent representation of raw data fed at its input nodes. The model is based on its learning algorithm, namely, contrastive divergence, and information gain (IG) is used as the criterion for gene selection. Our proposed model proves that it can deal with gene expression values efficiently and achieved improvements over classical classification methods. The results show that that the most discriminative genes can be selected and identified with its gene expression values. We report an increase of 8% over the highest achieving algorithm on a standard dataset in terms of accuracy.

Keywords: Restricted Boltzmann machine · İnformation gain · Feature analysis · Gene expression · Autism · Deep learning

1 Introduction

Autism spectrum disorders (ASD) (also known as Autism) among individuals are commonly found in societies. Autism refers to a group of developmental brain disorders. Autism involves a range of disabilities and impairment symptoms of different levels. ASD are related to brain growth. It affects how people recognize and deal with others, causing problems in interaction and social communication. The disorder also includes specific patterns and frequent behavior. Autism has a prevalence ratio of 1:4 among males to females. Genetic composition and environment may play a role in the ASD. Researchers assert that autism has genetic causes, which are the main cause of the disease. Autism comes in different degrees. An individual can be mildly impaired by the symptoms or severely disabled [1]. A class of disorders distinguishes autism

© Springer Nature Switzerland AG 2020
D. J. Hemanth et al. (Eds.): ICICI 2019, LNDECT 38, pp. 583–596, 2020.
https://doi.org/10.1007/978-3-030-34080-3_66

degree. These disorders are called Pervasive Developmental Disorders (PDDs). The PDDs includes: Autistic Disorder (classic autism: Fragile X syndrome), Asperser's Disorder (Asperser syndrome), Pervasive Developmental Disorder not Otherwise Specified (PDD-NOS), Rett's Disorder (Rett syndrome) and Childhood Disintegrative Disorder (CDD) [1]. It is difficult to explain the disease through mutations that appear, or through rare interactions that are multi-genes. Autism genetic variables or DNA does not change, but are inherited, affecting the so-called gene expression, such as Fragile X syndrome. There is no way to prevent autism spectrum disorder. However, early diagnosis can improve behavior and skills [1]. Microarray technology allowed gene expression data to be highly available. Microarray allows the gathering of data, which determine the expression pattern of thousands of genes [2]. The profile pattern of mRNA might exhibit the genes or the environmental factors, which will cause the disorder. The larger number of features with small sample size is the cause of high dimensionality problems. This is the focal problem in classifying gene expression data [2]. Consequently, the gene selection can distinguish differentially expressed genes, or eliminate unrelated genes. There are two main types of gene selection: filter methods and wrapper methods. The first type of gene selection, filter methods, are efficient because it respects computational time and works without using classifiers [3]. Hence, filtering methods are selected in examining the high-dimensional data of microarray datasets. Information gain is used in our method to filter genes. The filter method calculates the degree of gene correlation of each gene to the target group through the internal characteristics of the data set [4]. There are two types of machine learning algorithms: supervised learning and unsupervised learning. Supervised machine learning involves algorithms that use input variables to forecast a target classification (i.e., the dependent variable). Supervised machine learning algorithms are either categorical or continuous. Among the supervised learning algorithms are: Support Vector Machines (SVM), K-Nearest Neighbors (KNN), Bayesian Network (BN), Multilayer Perceptron's (MLP or NN), and Decision Trees (DT). The second type of machine learning algorithms, unsupervised learning, is also referred to as descriptive learning. Unsupervised learning is training through input data without any known output [5]. Examples of Unsupervised techniques include: Clustering and Self-Organizing Maps (SOM). Unsupervised techniques analyze the relationships among diverse genes. Lately, Deep learning methods have gained huge popularity due to their high performance. The hierarchical structure of deep neural network conduct the nonlinear transfer of the input data. During the analysis of gene expression, there are frequent challenges in the selection and identification of the most relevant genes to autism. This struggle is due to variations associated with the experiments. This struggle could also be due to the existence of alterations in the genes. In an autistic case, the variance may be associated to the existence of alterations in many genes. There are further difficulties found when there are small samples (in the range of hundreds) versus the large samples (in the range of tens of thousands) [6]. In this paper, a Deep Belief Network (DBN) based on Gaussian–Bernoulli Restricted Boltzmann Machine (GBRBM) is used as a classifier that employs deep learning for autism disease classification. The IG filter is used as a gene selector to remove irrelevant genes, and to select the most relevant genes. DBN is a stacking of RBMs. The conditional probabilities are considered as the input of the RBM. The conditional probabilities take into consideration the real values normalized to a range from [0, 1]. The output is the per class probability. The Contrastive Divergence (CD) method is used for training the model. The proposed model is tested using a gene expression dataset downloaded from NCBI for fragile X syndrome and it is proved effective in binary classification problems that contain gene expression values.

The current paper contains the following sections: Sect. 2 presents the related works regarding the research problem; Sects. 3 illustrates the proposed model and the data used; Sect. 4 explains the experiments and the emerged results; and Sect. 5 concludes and explores future directions.

2 Related Work

This section presents existing studies and work related to feature selection and classification of autism disorder. It reviews several works done in different methods of machine learning that were applied on microarray gene expression data. Scientists can observe gene expression through microarray technology on a genomic scale. Microarray technology has increased the possibility of classification and diagnosis on the gene expression level.

Heinsfeld et al. [7], identify ASD with the application of deep learning algorithms, a brain-based disorder. They indicated in their study that ASD reflect social deficits and repetitive behaviors. They objectively identified ASD participants, using data regarding functional brain imaging. They explained that the brain distinguished ASD from typically developing controls. The results achieved an accuracy of 70%.

Hameed et al. [6] research used different statistical filters and a wrapper-based geometric binary particle swarm optimization-support vector machine (GBPSO-SVM) algorithm. In their study, they investigated the gene expression analysis of ASD. They explained that there is always difficulty in the process of seeking the genes relevant to autism.

Tomasz Latkowski and Stanislaw Osowski [8] study focused on an application of data mining. This application mainly recognized the significant genes and its sequences in gene expression microarray datasets. Their research used several analyses: the Fisher discriminant analysis, relief algorithm method, two sample t-test, Kolmogorov–Smirnov test, Kruskal Wallis test, stepwise regression method, feature correlation with a class and SVM recursive feature elimination. These tests were used to assess the autism data. Accordingly, their research classification accuracy increased 78% after integration.

Tang et al. [9] used the dimensionality reduction in assessing the autism data. This assessment is a popular technique to delete noise and redundant features. Results specified that the dimensionality reduction techniques categorized data into: feature extraction and feature selection.

In another study [10], Nur Amalina Rupawon and Zuraini Ali Shah applied several methods for data mining to select the informative genes of autism in gene microarray. Their research used a two-stage selection, genetic algorithm hybrid with three different classifiers, namely: "K-nearest Neighbor (KNN), Support Vector Machine (SVM) and

Linear Discriminant Analysis (LDA)". Their proposed model reported an 8% increase over the results obtained in [8].

Tanaka et al. [11] suggested an inference of the RBM. Their research studied the input of the RBM (random binary variable). It also studied the straightforward derivation. They showed the development from the model of stacked RBM to the model DBN-DNN and proposed an improved DNN model. They also showed that the conventional inference was insignificant. Thus, the suggested inference is sensible than RBM conventional algorithm.

Gupta et al. [12] projected the classification of gene expression data. In their study they used a Gaussian Restricted Boltzmann Machine (RBM). RBM is considered a machine learning model. This model focuses on the Neural Networks. In their research, RBM is applied on a binary classification problem. This aided in identifying if certain individuals are affected by lung adenocarcinoma. Thus, the analyses main focus was on the gene expression values and Random Forest used as a gene selector.

Hyde et al. [13] research reviewed various ASD literature, giving scholars a proposed method for identifying and describing supervised machine-learning trends. Their insights acted as empirical evidence used to fill in the academic gap found in literature. The emerged empirical evidence increased the body of mining ASD data, relating to clinically, computationally, and statistically sound approaches.

Gao et al. [14] used several analyses for the data in their study. These analyses provided sufficient accuracy for classifying cancer, using the whole set of genes as data. In the beginning of the research, the IG was used to filter redundant and irrelevant genes. Next, the SVM was used to remove the noise in the datasets. Finally, IG-SVM was conducted to indicate and select the informative genes. Results of the IG-SVM showed a 90.32% classification accuracy for colon cancer.

In [15], a novel approach was used for classifying and analyzing the cancer detection. The study mainly focused on the determination of cancer and its subtypes. The classification was performed on selected features, derived from both (1) Particle Swarm Optimization algorithm and (2) Ant colony Optimization algorithm. The study used breast cancer gene microarray datasets.

In another study [16], Andrey Bondarenko et al. compared between DBNs and RBMs classification performance against other accepted classifiers. Some of the classifiers that the study compared were: SVMs and Random Forest Trees. The study used several datasets: UCI datasets, and a proprietary document classification dataset, which was single mid-sized. In conclusion to their study, the existing approaches allowed RBMs and DBNs to cope with high dimensional data. RBMs allowed the performance of training on unlabeled data.

Smolander et al. [17] focused on the arrangement of patients with inflammatory bowel disease and breast cancer. Their analyses focused on high-dimensional gene expression data. They investigated classifiers that integrated deep belief networks and vector machines. By combining classifiers, it aided to solve high dimensionality problem in genomic data. The research studied a computational diagnostics task. The results of their study were able to introduce guidelines for the complex usage of DBN. Their study showed how DBN could be used to classify gene expression data from complex diseases.

Koziol et al. [18] research study used Boltzmann machines for the classification problem regarding the diagnosis of hepatocellular carcinoma. Their study used two methods for the classification problem: "logistic regression and restricted Boltzmann machines (RBMs)". The analysis that was used in this study was the 10-fold cross-validation for the determination of operating characteristics of the classifiers. Results of the study indicated that: "RBMs typically had greater sensitivities, but smaller specificities, than logistic regression with the same input variables".

Jiang et al. [19] study identified the vital genes that affect disease progression. The study sought to: (1) identify hierarchical structures, and (2) apprehend differential analysis of gene expression datasets. The research focused and used the restricted Boltzmann machine. The study conducted the investigation by using Huntington's disease mice at different time periods. In conclusion, the results showed that SRBM-II outperformed other traditional methods.

Ray et al. [21] established a granular self-organizing map (GSOM). In the study, they combined a fuzzy rough set with the SOM. They explained that during the progress of the GSOM, weights of the neighborhood neurons and the winning neuron are updated. This update was caused by a modified learning procedure. Results of the study showed that GSOM have an effect on the clustering samples and the development fuzzy rough feature selection.

In the current study, the proposed model is inspired from [12]. This model used Gaussian restricted Boltzmann machine to solve the classification problem from cancer gene expression dataset. However, in the current study, we applied an RBM based model, which is devised by Tanaka [11]. The inference for RBM will be used to classify autisms' gene expression. This study also incorporates information gain into the proposed model to be used as a gene selector.

3 Proposed Model

In this paper, a deep learning method for classification of gene expression data is presented. Prior studies focused on feature selection on both supervised and unsupervised learning. We decided to focus on the problem of supervised learning (classification) in autism, where the class labels are known beforehand. In order to exclude irrelevant genes from the given gene expression, we applied the information gain procedure to generate dependency weight vector. The weight vector denotes the correlation of the gene and ASD. The main outcome of this phase is choosing the most relevant genes to autism. In this work, we sort these genes' weight vector in a descending order. Based on a manually preset threshold, we select only the most relevant genes. The updated data records will be used during the classification steps in the next phase.

The main contribution of this study is the identification of the process of classifying gene expression. In the first phase, this study uses small sample data size with large number of genes (features) that manifests severe high dimensionality. High dimensionality is the main problem because its increase the computational complexity, increase the risk of over fitting and the sparsity of the data will grow.

According to studies, there is a huge amount of high-dimensional datasets. In addition, feature selection has drawn great interest in the field of machine learning. Former studies addressed various approaches, such as: clustering, regression and classification, in both supervised and unsupervised ways. Generally, the field is challenged by 3 difficulties: "Class ambiguity- Boundary complexity- Sample sparsity". Today, high-dimensional, data of small sample size are common in various fields; they include: "genetic microarrays previously presented, chemometrics, medical imaging, text recognition, face recognition, finance, and so on". The features of these problems hinder the execution of a reliable and efficient classification system. Thus, feature selection is significant to avoid this problem [22]. The feature selection is used: (1) to identify different expressed genes, (2) to choose the appropriate features, and (3) to remove the irrelevant genes (not harming the remaining genes). The outcome of this phase is then passed to a classification module that uses a DBN based on GBRBM.

A Gaussian RBM contains normalized real values on visibles between the ranges of [0:1]. In this study, we used a Deep belief network-Deep Neural Network (DBN-DNN) hybrid architecture. First, stand-alone RBMs are pre-trained, with seen data by the Contrastive Divergence (CD) training algorithm. Stacking the trained RBMs forms a DBN. The outputs of the trained RBM on the hidden nodes are supplied as the training data on the input nodes of the RBM of the next layer. Then, the DBN is unfolded into a DNN by adding a topmost layer with nodes corresponding to classes. The final move is fine-tuning the architecture parameters by Back-Propagation (BP) algorithm of the error on the topmost layer. The following block diagram explains our model (Fig. 1).

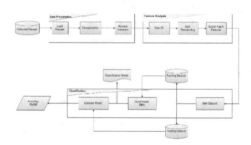

Fig. 1. Block diagram of our proposed model.

3.1 Data Preparation

The data records were subjected to a normalization process yielding the values in the ranged from 0 to 1.

3.2 Feature Analysis

In this phase, the dataset is ready to be handled by feature selection. The study uses information gain (IG). Filter techniques (IG) contain several advantages. They are able to scale to very high-dimensional datasets. They are simple and fast to use. They are independent of the classification algorithm. The IG method assesses the applicability of features by observing the intrinsic properties of the data. Thus, the IG method is applied as a filter in gene selection. It helps rank the genes based on entropy. Entropy is an information theory measure [4]. It can be viewed as the expectation of how useful the information in a message, which is represented by IG value. Filter techniques (IG) relies on the degree of the entropy, which reflects the amount of information this attribute contributes to the data set. IG value is calculated for each feature. It aids to decide whether this feature is to be chosen, or not. The gene contribution with more information yields a high IG value [20]. Thus, IG values are sorted in descending order, and a cut-off point is applied. The study assumes that dataset have M instances. M = {1, 2, 3... m} with x classes. Calculate the entropy of the dataset using Eq. (1) [4]:

$$\text{Entropy}(M) = -\sum_{i=1}^{x} P(C_i, M) * \log P(C_i, M) \tag{1}$$

Where P (C_i, N) represents C_i, the ratio of and M where C_i, i = 1, 2, .., x are set of instances that belong to the i the class. The entropy of the dataset from gene y calculated using Eq. (2) [4]:

$$\text{Entropy}_y(M) = \sum_{j-1}^{n} \frac{|M_J|}{M} * \text{Entropy}(M_j) \tag{2}$$

If y is gene that has distinct valued L = {$L_1, L_2......L_n$} and letting $M_j \in M|y = L_n$

Calculate IG value of gene u using Eq. (3) [4]:

$$IG(u) = \text{entropy}(M) - \text{entropy}_U(M) \tag{3}$$

Because gene expression values are within different ranges for different, we need to normalize the calculated IG equation using the following Eq. (4).

$$X = (Y_i - \min(Y))/(\max(Y) - \min(Y)) \tag{4}$$

Where Y = ($Y_1......Y_n$) and X_i is the i the normalized data point. Sort descending according to normalized IG value calculated by Eq. (4). Select top of k mean (manually) that we only accept the most important genes. The result is transferred for use during classification phase in the next phase.

3.3   Classification

In this phase, we utilize the emerged filtered dataset originated from the previous phase of feature selection. The top *k* features are the new data. Applying a novel inference for RBM model constitutes the main building block of our proposed DBN. It is able to portray the proposed inference and reflects on RBM the probabilistic properties. Nevertheless, the exact calculation of the proposed inference is intractable. Consequently, the closed form approximation is carried out. Training, DBN, and CD training applied with the proposed inference, which is devised by Tanaka [11]. Figure 2 shows the architecture of the RBM. It is composed of a visible layer. The normalized gene records are fed into the machine through this visible layer. Weights and biases are used to deduce the transformation function which maps these input genes to higher level features on the hidden layer. The visible and hidden layer nodes are fully connected with undirected edges. However, there are no intra connections among nodes of each layer, according to the independence assumption. A RBM has m visible units, the input data, and n hidden units, the features [11]. The energy of the whole configuration is calculated according to Eq. (5) [11], which gives the joint probability distribution of visible and hidden nodes.

$$E\left(\vec{v}, \vec{h}\right) = \sum_{i=1}^{m} a_i v_i - \sum_{j=1}^{n} b_j h_j - \sum_{i=1}^{m} \sum_{j=1}^{n} v_i h_j w_{ij} \tag{5}$$

Where and indicate the states of the visible and hidden nodes; are the biases, and is stands for the weight assigned to each interconnectivity between the visible and hidden nodes. The weights and biases form the model parameters that are fine-tuned throughout the training. These parameters are initialized completely random and updated in the learning epochs. The state of the neurons at the hidden layer is obtained by Eq. (6) [1].

$$P\left(h_j = 1 | \vec{v}\right) = \text{sig}\left(b_j + \sum_i v_i w_{ij}\right) \tag{6}$$

Equation (6) indicates the probability that hidden layer node will be set to 1. It is a stochastic process in which the utilized activation function is commonly the sigmoid given by Eq. (7) [11].

$$\text{sig}(x) = \frac{1}{1 - e^{-x}} \tag{7}$$

The RBM learns by the Contrastive Divergence (CD) algorithm. It is a biphasic process. Firstly, the hidden states are updated as shown above. Secondly, in the "reconstruction" phase, the states of the visible nodes are determined as follows by Eq. (8) [11].

$$P\left(v_i = 1 | \vec{h}\right) = \text{sig}\left(a_i + \sum_j h_j w_{ij}\right) \tag{8}$$

This step is denoted as. The nodes on the hidden layer is recalculated based on the reconstructed values on the visible layer in the second phase. Last step of the training epoch, is updating the model parameters according to the following Eqs. (9, 10, 11) [11].

$$\Delta w_{ij} = \varepsilon < v_i h_j >_0 - < v_i h_j >_1 \tag{9}$$

$$\Delta a_i = \varepsilon \left(v_i^0 - v_i^1\right) \tag{10}$$

$$\Delta b_j = \varepsilon \left(h_j^0 - h_j^1\right) \tag{11}$$

Where ε is the learning rate. These independently learning RBMS piled above each other to form a DBN. The more layers of RBMs, the deeper the DBN and the more abstract the features resulting at the upper levels. The process is sequential. Each layer maps the input data to values on the hidden layer. These values are passed to the upper layer as input on the visible. This is iteratively repeated until reaching the upper most layer. The DBN is a directed generative model is calculating by Eq. (12) [11].

$$P(x, h^1, h^2, \ldots h^l) = P(x|h^1)P(h^1|h^2)\ldots P(h^{l-2}|h^{l-1})P(h^l, h^{l-1}) \tag{12}$$

Conditional layers P () are factorized conditional distributions for which the computation of probability. Once the neural network parameters for every layer, the weights W and therefore the biases b are initialized by RBMs. The DBN learning is termed fine-tuning. Fine-tuning uses the class-label information of the training data set that was omitted in pre-training. The main goal of this research is to able to generalize to new unseen samples. So, a back-propagate was needed in the final layer of the derivatives. This is calculated by Eqs. (13, 14, and 15) [11].

$$\frac{\partial L}{\partial W_{ij}} = \delta_j \frac{\partial \mu_j}{\partial W_{ij}} + T_j \frac{\partial p_j^2}{\partial W_{ij}} \tag{13}$$

$$\delta_j = \frac{\partial L}{\partial \mu_j} = \frac{\partial L}{\partial O_j} \frac{\partial o_j}{\partial n_j} \frac{\partial n_j}{\partial \mu_j} \tag{14}$$

$$T_j = \frac{\partial L}{\partial P_j^2} = \frac{\partial L}{\partial O_j} \frac{\partial O_j}{\partial n_j} \frac{\partial n_j}{\partial p_j^2} \tag{15}$$

Where j is the output nodes, is the input node L -loss function, -the output of the j - the node and -the input to the activation function of j -the node. The activation function input for the algorithm is given by Eq. (16) [11].

$$n = \frac{\mu}{\sqrt{1 + \rho^2 \pi / 8}} \tag{16}$$

The derivatives of the hidden layer for the proposed algorithm is calculated by Eqs. (17, 18, 19, 20) [11].

$$\frac{\partial L}{\partial w_{ij}} = \delta_j \frac{\partial \mu_j}{\partial w_{ij}} + T_j \frac{\partial \rho^2}{\partial w_{ij}}, \tag{17}$$

$$\delta_j = \alpha_j \frac{\partial n_j}{\partial \mu_j}, \tag{18}$$

$$T_j = \alpha_j \frac{\theta n_j}{\theta \rho_j^2},$$ (19)

$$\alpha_j = \left[\sum_k \left\{\delta_k \frac{\theta \mu k}{\theta o_j} + T_k \frac{\theta \rho_k^2}{\theta o_j}\right\}\right] \frac{\theta o_j}{\theta n_j},$$ (20)

Where the node of the previous layer identifies with k, and are the back-propagated errors from the upper layer (Fig. 3).

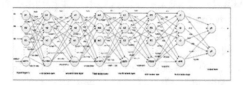

Fig. 2. Proposed RBM model.

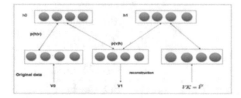

Fig. 3. Data and reconstruction in the CD training [11].

4   Experiment Results and Discussion

This section explains the data used for experimentation and the results obtained, along with analysis of the achieved results. This study uses MATLAB for implementation on a machine with Intel Processor, 2.8 GHz and 64-bit architecture and on Windows 10 Professional.

4.1   Experiment Dataset

The datasets used in this study are available on NCBI [23]. In this experiment, we used two datasets. The first dataset refers to the fragile X syndrome gene expression dataset (GEO: GSE7329). The first dataset contains 30 samples (observations) with 43,931 genes (features) in addition to the ground truth class. Classes divided to 15 autistics and 15 non autistic.

The second dataset refers to the peripheral blood lymphocytes (GEO: GSE25507). The second dataset contains 146 samples and 54,613 genes and is used in previous studies [6, 8]. The ground truth class is also given. Classes divided to 69 non autistic and 77 autistics.

4.2   Performance Evaluation

The proposed model evaluated through Accuracy and RMSE. Using Eqs. (21, 22) [11] to evaluate RMSE of the test and training data.

$$\text{RMSE} = \sqrt{\frac{1}{MN} \sum_{K=1}^{N} \|y_k - f(X_k)\|_2^2} \tag{21}$$

M is the number of output classes, N is the number of data, the input is k, the ground truth output and the inference of the trained DBN is represented by y_k.

$$\text{Accuracy} = \frac{Tp}{N} \tag{22}$$

True Positive (TP) refer to the positive tuples which were classified correctly.

4.3   Experiment

The experiments focused on an inference for RBM, which is applied to build a DBN. And, it uses IG for dimensionality reduction. Tanaka inspired this approach [11]. The data are not binary data. Succeeding, before dimensionality reduction, different architectures are built, used to obtain the highest accuracy with the best architecture (Table 1). Here, the architectures are given in order.

- A_1: Five-hidden-layers with [43931-4096-1024-512-128-32-2] nodes.
- A_2: Six-hidden layers with [43931-4096-2048-1024-512-128-32-2] nodes (Fig. 2).
- A_3: Four – hidden- layers with [43931-1024-512-128-32-2] nodes.

The training hyper parameters used in DBN are as follows: total number of iterations = 100, mini-batch data = 10, Learning Step Size = 0.01. Then, it uses IG for dimensionality reduction. 10000 genes (features) are selected from the dataset. The second architecture (A_2) achieves higher accuracy than the others before using feature selection.

As for the second dataset, we apply the 6-layered architecture [54,613-4096-2048-1024-512-128-32-2], similar to A_2, shown (Fig. 2). After dimensionality reduction, the dataset reduces from 54,613 to 10000 with the same weight and bias parameters. Classical classifiers are compared against this architecture in Tables 2 and 3.

Table 1. Comparison of different classifiers on dataset-1.

	Dataset-1 before dimensionality reduction				Dataset-1 after dimensionality reduction			
	Decision Tree	K-NN	Naïve Bayes	DBN	Decision Tree	K-NN	Naïve Bayes	DBN/IG hybrid
Accuracy	90%	70%	80%	98.77%	53.3%	83.3%	86.67%	98.64%
RMSE	–	–	–	0.667	–	–	–	0.667

Table 2. Comparison of different classifiers on dataset-2.

	Dataset-2 before dimensionality reduction			Dataset-2 after dimensionality reduction
	(GBPSO-SVM) algorithm [6]	Relief algorithm SVM [8]	DBN	DBN/IG hybrid
Accuracy	92.1%	78%	98.66%	98.62%
RMSE	–	–	0.667	0.499

Table 3. Comparison time on dataset 1&2.

	Dataset-1 dimensionality reduction		Dataset-2 dimensionality reduction	
	Before	After	Before	After
	Time	Time	Time	Time
Accuracy & RMSE	2:30 h	50 min	3:10 h	1 h

4.4   Result and Discussion

The experiment is implemented on matlab2018A and on a machine that having CPU 2.8 GHz and 64 bit. The research evaluated the algorithm before using feature selection. This allows the research to obtain the best accuracy with the best architecture. Table 1 illustrates the comparison of dataset-1 before and after reduction in classical classifiers, included in our experiments. Also, the table compares between proposed model before using feature selection (represented in the column called DBN) and after feature selection (represented in the column DBN/IG). The research obtains that the highest accuracies using DBNs on both datasets. When the IG is used, the result decreases contrary to expectations. Further, the time decreases approximately to its third table. This outcome saves computational power and reduces memory consumption. The shortcut visible nodes (from 43,931 to 10000) lead to decrease equations and calculations, while preserving the performance measures. When the proposed model is compared to other classical classifiers (such as Decision Tree, K-Nearest Neighbor (K-NN) and Naïve Bayes), or related research [6, 8] the results shows that our proposed model surpasses all the other algorithms by a large margin (as shown in Tables 1 and 2).

5 Conclusion

In this study, the newly proposed (IG/DBN) hybrid model is applied to the classification of autistic gene expressions based on RBMs. This research supports the early diagnosis for ASDs. It explains that ASDs is significant and it is considered as a clinical best practice. When ASDs is spotted early, it leads to early intervention. Nevertheless, it remains a challenge to diagnose before the age of 3 yearrs. For example, there are many cases in where normal developmental variance are overlapped with ASDs symptoms [24]. RBMs have been shown to be a highly performing classification tool in areas of image analysis, video sequence, and motion capture [12]. The utilized datasets consist of real-valued genes. The competitive performance, which is reported in these applications, inspired us to investigate its use in the analysis of gene expression values. We developed a novel framework for RBMs based on the Tanaka's approach [11]. This study also uses IG as a filter to reduce dataset dimensionality. In this study, the research ranked features, and selected the top k, reducing the size of the feature vector. This greatly reduced the model complexity in terms of the density of parameters to learn. Thus, it elevated computation efficiency of the proposed framework. In all cases, with/without dimensionality reduction, the presented classification tool achieved unprecedented accuracy figures, beating the state-of-the-art on both datasets. This proved that stacked RBMs forming a DBN learn the data model in a more accurate and efficient manner, overcoming the bottleneck of the disproportion between the size of samples and features.

In future work, modern GPUs should be used instead of CPUs for increasing the speed of the computation. According to research, modern GPUs can be 2500 times faster than CPUs. This will allow stacking more RBM layers, in hope of achieving even higher accuracies and ultimately supporting real time processing.

References

1. Gordon, J.A.: A parent's guide to autism spectrum disorder: national institute of mental health, USA, pp. 1–27 (2018)
2. Pushpa, M., Swarnamageshwari, M.: Review on feature selection of gene expression data for autism classification: international journal of innovative research in science. Eng. Technol. **5**, 3166–3170 (2016)
3. Saengsiri1, P., Na, S., Wichian, U., Meesad, P., Herwig U.: Integrating Feature Selection Methods for Gene Selection: Semantic Scholar, pp. 1–10 (2015)
4. Lai, C.M., Yeh, W.C., Chang, C.Y.: Gene selection using information gain and improved simplified swarm optimization: Neurocomputing, pp. 1–32 (2016)
5. Han, J., Kamber, M., Pei, J.: Data Mining Concepts and Techniques, 3rd edn, pp. 1–740. Elsevier, Hoboken (2012)

6. Hameed, S.S., Hassan, R., Muhammad, F.F.: Selection and classification of gene expression in autism disorder: use of a combination of statistical filters and a GBPSOSVM algorithm. PLoS ONE **12**(11), 1–25 (2017). e0187371

7. Heinsfelda, A.S., Francob, A.R., Craddockf, R.C., Buchweitzb, A., Meneguzzia, F.: Identification of autism spectrum disorder using deep learning and the ABIDE dataset, pp. 16–23, Elsevier (2017)

8. Latkowski, T., Osowski, S.: Data mining for feature selection in gene expression autism data, pp. 864–872. Elsevier (2015)

9. Tang, J., Alelyani, S., Liu, H.: Feature selection for classification: a review. In: Data Classification: Algorithms and Applications, pp. 37–64 (2014)

10. Rupawon, N.A., Shah, Z.A.: Selection of Informative Gene on Autism Using Statistical and Machine Learning Methods. In: UTM Computing Proceedings Innovations in Computing Technology and Applications vol. 6, pp. 1–8 (2016)

11. Tanaka, M., Okutomi, M.: A novel inference of a restricted boltzmann machine. In: IEEE 22nd International Conference on Pattern Recognition; Tokyo, pp. 1–6 (2014)

12. Gupta, J., Pradhan, I., Ghosh, A.: Classification of gene expression data using gaussian restricted boltzmann machine (GRBM). Int. J. Recent Innov. Trends Comput. Commun. (IJRITCC) **5**(6), 56–61 (2017)

13. Hyde, K.K., Novack, M.N., LaHaye, N., Parlett-Pelleriti, C., Anden, R., Dixon, D.R., Linstead, E.: Applications of supervised machine learning in autism spectrum disorder research: a review. Rev. J. Autism Develop. Disord **6**, 1–19 (2019)

14. Gao, L., Ye, M., Lu, X., Huang, D.: Hybrid method based on information gain and support vector machine for gene selection in cancer classification. Genomics Proteomics Bioinf. **15**(6), 389–395 (2017)

15. KajaNisha, R., Sheik Abdullah, A.: Classification of cancer microarray data with feature selection using swarm intelligence techniques. Acta Sci. Med. Sci. **3**(7), 82–87 (2019)

16. Bondarenko, A., Borisov, A.R.: Technical university: research on the classification ability of deep belief networks on small and medium datasets. Inf. Technol. Manage. Sci. **6**(1), 60–65 (2013)

17. Smolander, J., Dehmer, M., Sterib, F.E.: Comparing deep belief networks with support vector machines for classifying gene expression data from complex disorder. In: Open Bio, pp. 1–26 (2017)

18. Kozio, J.A., Tan, E.M., Dai, L., Ren, P., Zhang, J.Y.: Restricted boltzmann machines for classification of hepatocellular carcinoma. Computat. Biol. J. **2014**, 1–5 (2014)

19. Jiang, X., Zhang, H., Duan, F., Quan, X.: Identify Huntington's disease associated genes based on restricted Boltzmann machine with RNA-seq data. BMC Bioinformatics, pp. 1–13 (2017)

20. Shaltout, N.A., El-Hefnawi, M., Rafea, A., Moustafa, A.: Information gain as a feature selection method for the efficient classification of influenza based on viral hosts. In: Proceedings of the World, London, vol. I, pp. 1–7 (2014)

21. Ray, S.S., Ganivada, A., Pal, S.K.: A granular self-organizing map for clustering and gene selection in microarray data. In: IEEE, pp. 1–17 (2015)

22. Bolón-Canedo, V., Sánchez Maroño, N., Alonso-Betanzos, A.: Feature Selection for High-Dimensional Data, Artificial Intelligence: Foundations, Theory, and Algorithms, pp. 1–163. Springer (2015)

23. National Center for Biotechnology Information. http://www.ncbi.nlm.nih.gov

24. Chuthapisith, J., Ruangdaraganon, N.: Early detection of autism spectrum disorders. In: Autism Spectrum Disorders: The Role of Genetics in Diagnosis and Treatment, Stephen Deutsch, IntechOpen, 1 August 2011. https://doi.org/10.5772/17482

A Lightweight Selective Video Encryption Using Motion Vectors and IPMs for H.264 Codec

Rohit S. Malladar[1(✉)] and R. Sanjeev Kunte[2]

[1] Department of CS&E, Jain I.T, Davangere, India
powerohit@gmail.com
[2] Department of CS&E, J.N.N College of Engineering, Shivamogga, India
sanjeevkunte@gmail.com

Abstract. Selective video encryption is a technique of selectively choosing a part of the video bit stream for encryption. Such kind of encryption can impart good security and computational efficiency in the H.264 bit stream. In this paper, we propose a selective video encryption technique which uses the motion vectors and chrominance layer of each macro block in I frames for encryption. These two are important elements in the bit stream whose encryption shall reflect on the present GoP (Group of Pictures). The proposed technique is implemented in the JM 19.0 reference software under high profile condition. Different YUV sequences are used for the investigational evaluation of the proposed technique. Results are discussed according to the PSNR, bit rate, etc.

Keywords: Encryption · Selective video encryption · Motion vectors · Group of Pictures

1 Introduction

The latest international coding standard on video is the H.264, representing the state of the art in the level of compression in video, suiting to the internet friendly features. Series of applications such as Video on demand, Pay per view [1], video conference and so on is supported by H.264. Impressive level of compression provided by the H.264 codecs [2] makes it possible for any requirement to be fulfilled which needs encoding of video in less time. Continuous development shall result in H.264 dependent products to emerge in the market and hence craving for H.264 security shall also increase. Video information security techniques were based on authentication control in the earlier era instead of encrypting the video content itself. Hence piracy and illegal copying of video was not difficult. The encryption scheme which focuses on using the present key features in video can increase its security. Compression is the main factor which is to be taken care of and compression efficiency acts as a distinguished factor for H.264 video coding standard for a huge variety of applications.

© Springer Nature Switzerland AG 2020
D. J. Hemanth et al. (Eds.): ICICI 2019, LNDECT 38, pp. 597–606, 2020.
https://doi.org/10.1007/978-3-030-34080-3_67

The contribution of our proposed scheme in the field of selective video encryption are:

- Sign bit encryption of motion vectors in the luma channel.
- Encryption of Intra Prediction Modes (IPMs) in the I frame.

The proposed work shows a method of encrypting the motion vectors calculated for the macro blocks and the encryption of chrominance layers during the phase of intra prediction. Sign bit encryption technique is assumed to be a suitable technique considering the computational complications. The presented work is arranged in the paper as described next. In Sect. 2, a background on the selective video encryption method using the motion vectors and IPMs is presented, which shall brief on previous work done on the selected technique. Section 3 presents the theoretical presentation of the proposed technique. The experimental setup for the mentioned work is given in Sect. 4. Section 5 presents the detailed analysis of the results using different parameters like PSNR, and time taken applicable to different YUV sequences. In the end, concluding remarks are presented.

2 Background of Selective Video Encryption

Selective video encryption plays a vital role in minimizing the computation complexity of the entire encryption method. Different regions of a frame like a face or the entire body or any object under motion may be selected for the encryption as Region of Interest (ROI) based on the applications. I frame, DC coefficients, selected macro blocks were used for selective video encryption in the past. In this section, the encryption techniques which concentrated on the motion vectors and the Intra Prediction Mode (IPM) are presented.

Role of motion vector in disclosing the useful information of video was presented by Liu et al. [3]. A two-step encryption scheme is described, where a random iterator is XORed with the motion vectors and the resultants are scrambled to spatially distribute the vectors in the bit stream. Macro Block (MB) starting position in a slice was relocated along with motion vectors by Kwon et al. [4]. No change in the size of bit stream was observed as the only mode of encryption used was permutation. Index based chaotic sequence was used to encrypt the motion vectors by Batham et al. [5]. The real value sequence generated was converted into integer sequence using a key to encrypt the video frames which resulted in a hybrid video codec.

Redundancy of the values in the motion vector were identified and represented in terms of an ordered pair (value, runs), where the 'value' represented the actual value of the motion vector whereas the 'run' represented the frequency of appearance of the value. Further the 'value' was also encrypted using the knapsack algorithm before embedding into the bit stream by Mishra et al. [6]. I Luminance Transform Coefficients (LTC) were used for encryption by Varalaxmi et al. [7]. Considering the sensitivity of human eye for the difference in light luminance, LTCs are encrypted using RC4 stream cipher. Pseudorandom stream of bits were generated by RC4 in combination with the bit wise exclusive XOR.

Data available for the Intra Prediction Mode (IPM) was completely encrypted by Jiang et al. [8] using the chaotic pseudo random sequences in a two step method. In the first step of encryption, H.264 encoding standard uses Exp-Golomb for encoding of IPMs and chaotic sequence is induced into the IPM modes. In the second step, security is added by XORing the resultant data. IPM encryption along with the encryption of other sensitive data such as motion vector difference was proposed by Li et al. [9]. The bits of information in the Scalable Video Coding (SVC) were encrypted by the Leak EXtraction (LEX) algorithm. The round transformation of AES was taken care by this technique. Equivalent key agility was provided by LEX and it was also observed that block performance of short messages was very similar to AES.

Intra prediction mode in H.264 was scrambled by Ahn et al. [10] in which the intra blocks were scrambled which made the inter blocks to remain un-changed. Using the given specific key, the pseudo random sequence was generated for scrambling. Neighboring blocks availability was checked before the application of technique to the prediction modes.

In this section, the related work on selective video encryption is presented starting from the encryption of I frames, DC coefficients, etc. and then the encryption of motion vectors followed by encryption of intra prediction modes. In the next section we propose a novel encryption method which also acts on motion vectors but selectively choosing one of the three components of the input YUV file.

3   Proposed Method

In this work, selective encryption of the H.264 is realized in two different phases. In the first phase, motion vectors in the P-frames are chosen for the encryption. The high profile is chosen since the proposed work targets configurations of H.264 video for the Video on Demand (VoD), Pay Per View (PPV) and web streaming.

Encryption of the motion vectors found in the P-frame affect the temporal redundancy in the GoP. In this method, sign bit of the motion vectors are changed in order to encrypt the key parts in the luma component only. Macro block positions are deeply affected by the sign change of the motion vectors, which misplaces them in the upcoming frames and hence the visual degradation of the video can be noticed.

The theory of luma component lies in the fact that it is derived from the weighted average of the three color components i.e., red, green and blue. The proposed technique exploits the fact that our human eye shows high sensitivity to the certain wavelengths of light affecting the perceived brightness of the color.

$$Y' = 0.299R + 0.587G + 0.114B \tag{1}$$

The luma component can be derived from the Eq. 1. In short, YUV color space allows for the reduced bandwidth for chroma components without perceptual distortion.

Motion vectors can be calculated by matching the blocks between the frames in a search area. The search area can be altered as per the needs of the codec and the best match along the different directions is taken as the displaced block. Distinctive matching criteria may be the Mean Square Error (MSE) or the Mean Absolute Difference (MAD).

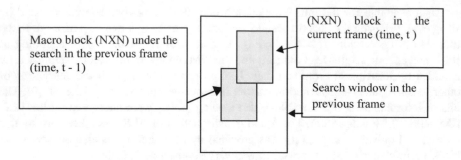

Fig. 1. Macro block movement between the frames

A "motion vector" is calculated by finding the correspondence between macro block at time t, and macro block at time t − 1, where t is the frame index in a video signal which is as shown in Fig. 1. Suppose, 'mv' is a pointer variable of motion vector, then sign bit encryption can be shown as:

$$mv - > mv\_x * = -1; \qquad (2)$$

$$mv - > mv\_y * = -1; \qquad (3)$$

where, mv_x is the movement of the block along X axis
mv_y is the movement of the block along Y axis.

As shown in Eqs. 2 and 3, the sign bit of motion vectors are flipped, but only on the luma component of the YUV sequence. Selecting a region of interest is an important part in the encryption failing which the technique may have less impact on the perception of the video. The Fig. 2 attempts to show the impact of the sign bit encryption with the motion vectors movement captured using the stream analyzer CodecVisa [11].

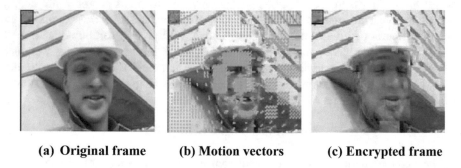

(a) Original frame　　**(b) Motion vectors**　　**(c) Encrypted frame**

Fig. 2. Encrypted frame with its motion vectors

After the motion vector encryption, the Intra Prediction Modes (IPMs) are considered for inducing more security. In the second phase of selective video encryption, the predicted residual values during the intra prediction stage are encrypted. Spatial redundancies is reduced by intra prediction by exploiting 'with in the frame' correlation between the neighboring blocks in a given frame. 4×4 or 16×16 is the macro block size set within the picture and each macro block comprises of luma and chroma components. I frame is coded without any reference to the data outside the existing frame. Prediction error is computed by the Sum of Absolute Differences (SAD) along the predefined directions. Direction in which the smallest error is recorded, that is chosen as the Intra Prediction Mode (IPM). The possible directions are as shown in Fig. 3.

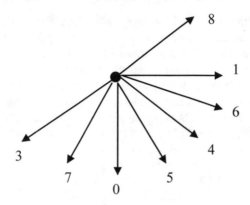

Fig. 3. Direction of nine intra prediction modes in H.264

Formerly encoded and restructured blocks are subtracted from the present block, results in the intra prediction mode. This resultant block 'P' is constructed for every 4×4 or 16×16 macro bock. The difference of the current block and intra block output is considered as the 'residual' block. This residual block is selected for the encryption as shown in the Fig. 4.

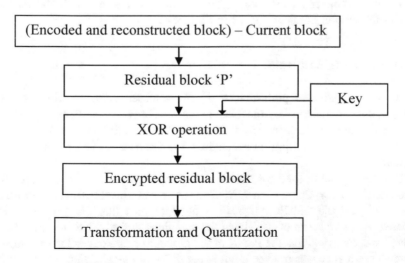

Fig. 4. Intra Prediction Modes (IPM) encryption

Key given by the user is used to XOR these residual values in order to encrypt the data as shown in Eq. 4.

$$C = P \oplus K \tag{4}$$

Where, P is the predicted residual value (plain text)
        K is the key given by the user
        C is the cipher text which is the encrypted residual value.

Since the encryption is carried out on I frame, the effect of the change in the data shall reflect on the adjacent frames and hence the encryption applied on the selected part of the frame is continued in the current Group of Picture (GoP).

In this section, we have demonstrated the novel selective video encryption technique which actively selects the luma component for the sign bit flipping technique and further, encrypts the residual values of IPM. The experimental setup and results of the above operation on selectively choosing only one component is recorded and discussed in the next two sections.

4 Experimental Setup

The proposed selective video encryption scheme is carried out using the JM 19.0 [12]. It is the reference software which is a complete implementation of the H.264 codecs as specified by ITU. Only the P-frames are considered for the encryption and the luma component is separately operated for the calculation of motion vectors. The software package contains a Visual Studio .NET workspace with libraries for encoding and decoding a YUV sequence. The technique is implemented on a 2.13 GHz, i3 processor with a RAM of 4 GB on a windows machine.

Encoder is run with a frame rate of 30.0 fps with a high profile selection. The quantization parameter for the I and P Slice of the frame is set to 28 and max search range for the motion prediction is set to 32 blocks.

5 Results and Analysis

In order to evaluate the performance of the proposed selective video encryption technique, the parameters like time taken to encode the sequence, Peak Signal to Noise Ratio (PSNR) along with the bit rate are considered. Analysis based on the quantization parameter is carried out using three standard YUV sequences i.e., foreman, carphone and akiyo.

FM is the acronym used for Foreman, CP for Carphone, AK for Akiyo, BG for bridge-close, CG for Coast Guard, MA for Miss America sequences. Further, Orig refers to original file and Encr represents the encrypted file. The proposed work is compared with Varalaxmi et al. [7] and the variations in the PSNR are discussed with results. It was observed that the PSNR of the proposed technique is lesser compared to it. Foreman sequence was used as input under different Quantization Parameter levels.

5.1 Time Taken to Encode the Sequence

Every kind of a frame can be set separately using the encoder file and in this proposed technique, Quantization Parameter (QP) is varied for every of the three YUV sequences and the corresponding encoding time is recorded. Table 1 shows encoding time variations with change in QP.

Table 1. Encoding time taken by YUV sequences with varied QPs in seconds.

Seq	QP = 28		QP = 34		QP = 40	
	Orig	Encr	Orig	Encr	Orig	Encr
FM	3.67	3.73	3.52	3.45	3.67	3.73
CP	3.53	3.38	3.40	3.42	3.53	3.38
AK	2.83	2.73	2.71	2.67	2.83	2.73

Table 1 shows the tradeoffs between the quality of encryption and time taken to encode the sequence is not high.

5.2 Bit Rate

Bit rate is defined as the number of bits processed in a second. The bit rate is expected to be constant irrespective of the parameters varied in the bit stream. The proposed technique induces a considerable change in the bit stream and still the bit rate (calculated in bits/second) remains considerably unchanged.

Table 2. Bit rates of P-frames with varied QP.

Seq	QP = 28		QP = 34	
FM	332	339.52	286	285.2
CP	303.6	312.8	266.64	264.57
AK	193.32	197.48	191.28	191.2

The technique described here works on the P-frame of the GOP, the bit rate doesn't vary much and hence it is in the support of the technique proposed and through the Table 2, we present that the format compliance is maintained.

5.3 Peak Signal to Noise Ratio (PSNR)

PSNR is used as the performance metric to evaluate the compressed frame and it measures the peak error between the original and compressed frames. Higher the PSNR, better is the quality of the frame and hence in the field of encryption it is expected to have a lesser value which depicts the low quality of the image. Here the PSNR of the YUV sequences is calculated at different Quantization Parameters (QP).

The effect of the proposed technique in terms of PSNR is shown in the Table 3 and follows the discussion of the same. Primary goal of any encryption scheme shall be to encrypt the payload in considerably less time. The same has been the goal of the proposed technique in this paper. The table depicts the decrease in the PSNR value at a good rate compared to the original.

Table 3. PSNR according to the different QP.

Sequence	QP = 18		QP = 24		QP = 30		QP = 36	
	Orig	Enc	Orig	Enc	Orig	Enc	Orig	Enc
BG	42.01	8.14	38.61	8.59	36.42	8.7	34.93	8.65
CP	45.21	8.84	42.24	9.41	39.27	10.59	36.87	10.75
CG	47.11	7.89	44.21	8.02	41.45	8.19	38.86	8.56
FM	45.46	6.63	42.34	7.26	39.81	7.31	37.40	7.37
MA	44.96	11.17	42.31	11.21	39.33	11.72	37.02	11.75

It can be noticed from the above table that the increase in the QP results in a better image quality. Considering the PSNR values at different QP levels, the proposed method is computationally efficient to encrypt a video in real time and hence may find its use in Video on Demand, Pay Per View, etc.

5.4   Encrypted Images

Visual degradation of the video at right parts may render the entire video meaningless and may create a curiosity in the viewers' mind to see the original video. The visual result of the proposed work which encrypts the motion vector and the IPMs is shown in Fig. 5. The disturbed portion of the image is the result of the motion vector encryption.

(a) Original Frame   (b) Encrypted frame

Fig. 5. Original and encrypted sample frames in the YUV sequences.

The encrypted frame shows the blocks displaced due to the motion vector encryption and the IPM encryption results in the change of residual values which further decreases the PSNR value of the frame.

5.5 Comparison

The two phase encryption scheme proposed in this work is compared with the work of similar method based on H.264 bit stream being affected by the chaos insertion, targeting the motion vector signs. The foreman sequence is taken as the input and quantization parameters are set to 28 and 36 to calculate the PSNR. The recorded values are pictured in the Fig. 6 shown below.

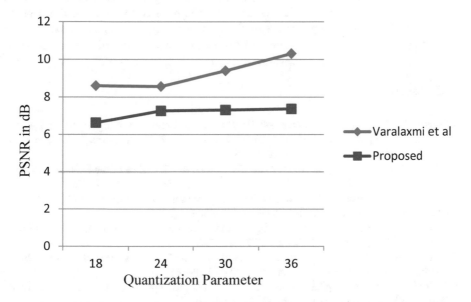

Fig. 6. Comparison of proposed work with Varalaxmi et al. [7].

PSNR is expected to decrease in comparison with the original frame after encryption. The above graph shows that our approach has more effect on video encryption. The motion vector encryption by flipping the sign bit, results in lesser PSNR compared to the encryption of Luminance Transform Coefficients (LTC) by RC4. Thus the selection of the critical part of the sequence plays a vital role in selective video encryption.

In this section, it is evident from the performance metrics used that the proposed selective video encryption technique is suitable for the online streaming of the videos by the content providers as per the demand. In the next section, we conclude our proposed work in the field of selective video encryption.

6 Conclusion

A novel selective video encryption technique is proposed in this paper, which encrypts the motion vectors by flipping its sign bit in the luma component of the YUV raw video file and further uses the XOR encryption to encrypt the IPMs. The implemented work is a lightweight scheme of combined techniques of motion vector encryption and IPMs presented so far. Many video applications demand the security for the content that is delivered to their customers and the proposed work reflects the same. Analysis of the proposed technique started with measuring the time taken for the encryption and the encryption overhead was found to be 0.10 to 0.20 s. Bit rate remained nearly constant compared to the actual codec and the PSNR of the encrypted frames dropped down to a very lesser value in support of the proposed encryption technique. Further, the work was compared with a similar technique to show the proposed technique had a stronger impact. As per the results presented in the previous section, it is observed that the proposed scheme serves the purpose and can be used under the real time constraints.

References

1. Posch, D., Hellwagner, H., Schartner, P.: On-demand video streaming based on dynamic adaptive encrypted content chunks. In: Proceedings in 21st International Conference on Network Protocols (ICNP), pp. 1–10 (2013)
2. Wiegand, T., Sullivan, G.J., Bjøntegaard, G., Luthra, A.: Overview of the H.264/AVC video coding standard. IEEE Trans. Circuits Syst. Video Technol. **13**, 560–576 (2003)
3. Liu, Z., Li, X.: Motion vector encryption using multimedia streaming. In: Proceedings in Tenth International Multimedia Modelling Conference (2004)
4. Kwon, S.G., Choi, W.I., Jeon, B.: Digital video scrambling using motion vector and slice relocation. In: LNCS, vol. 3656, pp. 207–214 (2005)
5. Batham, S., Acharya, A.K., Yadav, V.K., Paul, R.: A new video encryption algorithm based on indexed based chaotic sequence, pp. 139–143 (2014)
6. Mishra, M.K., Mukhopadhyay, S., Biswas, G.P.: Encryption of motion vector based compressed video data, vol. 2, pp. 349–357 (2016)
7. Saranya, P., Varalaxmi, H.M.: H.264 based selective video encryption for mobile applications. Int. J. Comput. Appl. **17**, 21–25 (2011)
8. Jiang, J., Liu, Y., Su, Z., Zhang, G., Xing, S.: An improved selective encryption for H.264 video based on intra prediction mode scrambling. J. Multimedia **5**(5), 464–472 (2010)
9. Li, C., Yuan, C., Zhong, Y.: Layered encryption for scalable video coding. In: 2nd International Congress on Image Signal Processing, pp. 1–4, October 2009
10. Ahn, J., Shim, H.J., Jeon, B., Choi, I.: Digital video scrambling method using IPM. In: LNCS, vol. 3333, pp. 386–393 (2004)
11. http://codecian.com/downloads for CodecVisa4.38
12. http://iphome.hhi.de/suehring/tml/ for JM 19.1 software

A Study on Physical Unclonable Functions Based Security for Internet of Things Applications

Swati Kulkarni[1(⊠)], R. M. Vani[2], and P. V. Hunagund[1]

[1] Department of Applied Electronics,
Gulbarga University, Gulbarga 585106, India
swatikulkarni494@gmail.com,
prabhakar_hungund@yahoo.co.in
[2] University Science Instrumentation Centre, Gulbarga University,
Gulbarga 585106, India
vanirml2@rediffmail.com

Abstract. Wireless communication trends are getting changed day by day. Internet of Things (IoT) is the latest running trend of wireless communication. IoT applications are very demanding in present and many more will come in the near future. IoT applications are running in insure medium i.e. Internet. Security and Privacy are the major challenges in IoT implementation. In conventional cryptography, one dedicated secret key is stored in volatile or non-volatile memory. Which has been used by the processor for every encryption or decryption Even though system is turned off the secret key is remain present in memory and can be access by attacker using side channel attacks. Thus, the traditional approaches to key storage are not favoured. Physically Unclonable function (PuF) can resolve such issues. Instead of using the same dedicated secrete key PuF can generate unique bits for every operation. Here we are presenting a brief survey of security challenges opportunities of PuF for IoT applications.

Keywords: Cryptography · Hardware security · PuF · Arbiter PuF · Ring Oscillator PuF · IoT

1 Introduction

Present days no one can imagine life without internet because Internet has an enormous impact on daily activities of human beings. Education, Business, Government, Humanity and Society etc. All the sectors are connected to internet. Internet of things is new and favourable concept because internet itself is powerful and useful creation. IoT consists of sensor networks along with intelligence resources. All the personal information comes on the internet and there are maximum chances that data may be altered by an unauthorized person. To prevent the unauthorized access of data one should hide the data from unauthorized person. Art of hiding the data is called Cryptography. The expanded version of an embedded systems is the Internet-of-things (IoT). The attackers are targeting to theft the information and do the tempering, sometimes destructive to get

© Springer Nature Switzerland AG 2020
D. J. Hemanth et al. (Eds.): ICICI 2019, LNDECT 38, pp. 607–614, 2020.
https://doi.org/10.1007/978-3-030-34080-3_68

the control on such systems so data security is important. The main overheads associated with data securities are area and power consumptions [1, 3]. The secret key is an
important element in cryptography. Traditional cryptographical methods which are
used to secure secret information in systems for large infrastructures are not suitable for
embedded and IoT systems. Embedded and IoT systems are more susceptible to
invasive attacks because many times they are deployed to remote areas. Like a desktop
system, IoT and embedded systems doesn't have enough amount of memories, powerful CPUs and supporting device drivers and huge infrastructure. Security for IoT
applications is the main focus of research for many researchers and numerous applications are being introduced every year [2]. If someone has stolen that secrete key then
the cryptographic system will fail. Unauthorized people are doing side-channel attacks
to get a secret key. In case of a side-channel attack, the attacker tries to get the
information about power consumption and electromagnetic radiation of memory
devices using timing and power analysis to find out the content stored in memory.
Many cryptographic algorithms are vulnerable to side-channel attacks [18].

Physical unclonable functions (PUFs) are a capable state-of-the-art primitive that
are used for authentication and secret key storage without the requirement of secure
EEPROMs and other expensive hardware. Basically, PuF works on the principle of
process variation. When manufacture is manufacturing Integrated circuit i.e. IC of same
types it won't be possible to prepare exactly the same ICs. Even though in nano-meter
technology also can have process variations. This process variation gets introduces in
various steps of IC fabrication such as epitaxial growth, masking, etching, doping, ion
implantation, etc. IC fabricated by same manufacturer using same process, same
material, same wafer and same die will vary from IC to IC [5]. Biometrics fingerprint is
a unique identification of any person. Similarly, PuF produces a unique fingerprint of
device. Process variations causes propagation delays in logic gates, change in threshold
voltage and temperature and aging of ICs etc. Which should not be acceptable but still
it is not under control. So PuF has to take an advantage of process variations. Process
variations are responsible to generate random delays on electrical paths. A unique PuF
response for each individual device can be confirmed by calculating entropy of delays
[5]. Uniqueness and unpredictability are the properties of PuF (Fig. 1).

n-bit Challenge (Ci)   PUF   **n-bit Response (Ri)**

Fig. 1. Challenge-response of PuF (Source: Securing communication devices via PUFs [4])

More specifically, a PUF is a disordered physical system that can be operated by
external input which is called as called "challenges" Ci. Depending on the type of PUF
number of challenges, can be applied. The output of PUF is nothing but response Ri. Ci
and Ri are combinedly called challenge-response pairs (CRPs) of the PUF. PUFs are
designed in such a way that the response (Ri) depends upon the not only applied
challenge but physical randomness which was produced due manufacturing process
variation of the device. In another way we can say that for two devices of the same type

if we applied the same challenges then the response of two PuF will be different. Uncolorability and unpredictability are the properties of PuF [5].

Depending upon how many numbers of challenges are applied to PuF, it is categorized into 1. Strong PuF and 2. Weak PuF [4].

2 Basic Types of PuF

2.1 Strong PuF

Strong PuF is having a greater number of CRP behaviour from the physical randomness present in the PUF. Response are depending on many physical components, and a very large number of possible challenges that can be applied to the PUF. As strong PuF is having more CRP, for the designer also it would be very difficult to find out all challenge-response pairs. Ideally, strong PuF designs having more challenges but practically for designer, it would be difficult to find out all CRPs. Even though strong PuF gives better performance than weak PuF still strong PuF is used in limited application because of its complexity [5]. First PuF designed was introduced by Pappu et al. in [7]. The paper terms the device a "physical one-way function," but the functionality is identical to that of a strong PUF. That was optical PuF.

2.2 Weak PuF

Compare to strong PuF weak PuF arc having less and fixed challenges so it's response should be kept confidential. They are mainly used to derive a secret key in cryptographic algorithms and are the least vulnerable to modelling attacks. One of the earliest Weak PUFs was a design proposed in 2000 by Lofstrom et al. [5, 6]. Practically weak PuF designs are widely used.

There are different ways, how PuF has can be designed. Such as (a) Arbiter PuF (b) RO PuF (c) Butterfly PuF (d) SRAM based PuF (e) Reconfigurable PuF (f) Duty cycle based PuF (g) Loop PuF (h) Resistive RAM (RRAM) based PuF (i) Current starved Oscillator based PuF (j) Inverter based PuF (k) Multiplexer based PuF and (l) RS Latch based PuF etc. anyone can go for hybrid based PuF also. Every design has its own pros and cons [14–17].

Arbiter PuF is nothing but delay-based PuF and Ring Oscillator PuF (RO PuF) are widely used design structures of PuF.

3 Basic Design of PuF

3.1 Arbiter PuF

Arbiter PuF is also called a delay based PuF. The design consists of Multiplexers and D-Flipflops. Multiplexers are arranged in such a way that they form two parallel paths [8]. Same input bit is applied to first multiplexer of both paths, upper and lower. An output of the upper multiplexer is connected to first input of lower multiplexer and output of lower multiplexer is connected to second input of upper multiplexer.

Similarly rest of multiplexers are cross-coupled to each other as shown in Fig. 2. These two paths having different delays because of internal variations that's why the same input will take different time to reach the output. Here input applied to select line of multiplexers are nothing but Challenges and output of D- flipflop is nothing but the response of PuF. Whichever path will be faster that will reach to D-Flipflop, PuF response will be available. Every different challenge, PuF response is different, and will generate unique fingerprint of device. Many researchers have already proved that, Arbiter PUF circuit is simple, attackers can try to construct a precise timing model and study the parameters from different challenge response pairs [6].

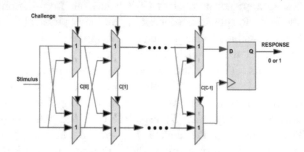

Fig. 2. Arbiter PuF design (Source: Physical unclonable functions and applications [8])

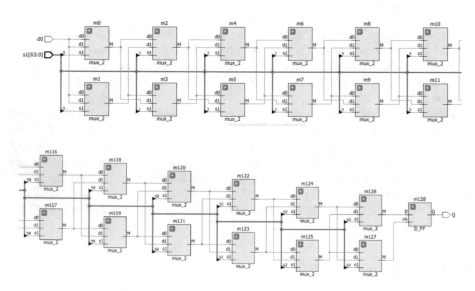

Fig. 3. Schematic of Arbiter PuF implementation

Figure 3 shows Schematic of Arbiter PuF implementation. As arbiter PuF consist of multiple stages of switching elements ate connected as chain. In this implementation 128 multiplexers are cascaded and created two delay paths. Challenges are applied to select lines of multiplexers. Output of multiplexers are fed to D-Flipflop one of the outputs of multiplexer are given to Data input of D-Flipflop and another multiplexer is connected as clock input of D-FF. Output of D flipflop is nothing but response bit or 1-bit PuF output.

3.2   Ring Oscillator Based PuF

It is another simple and widely used PuF design. As the name indicate the main component of RO PuF is Ring Oscillators. Multiple paths of ring oscillators are used, each ring oscillators oscillate with slightly different frequencies because of process variations. The output of all ring oscillators is given to both the multiplexers and challenges are applied to the select lines of multiplexers. The output of both the multiplexers is fed to two different counters as their clock signals. As soon as counter overflows, the counting is stopped and one bit of the output of PuF is calculated, for this calculation comparator is used. Comparator can compare which counter caused the overflow [2, 6, 9] (Fig. 4).

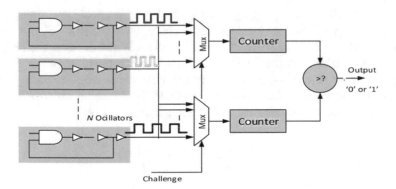

Fig. 4. RO PUF design (Source: Design and implementation of a group-based RO PUF [12])

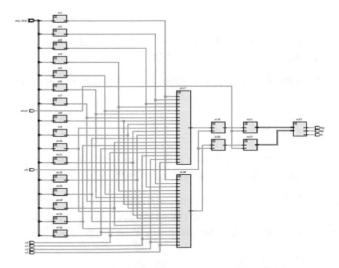

Fig. 5. Schematic of implementation RO PuF

In both type of design PuF gives you a single bit as an output bit. In case of a cryptography algorithm, consider Advance Encryption Standard (AES) minimum Key size requirement is of 128 bits. If you need to produce a 128-bit response in parallel then you have to use 128 parallel connections of PUF design but if you want it serially then you can provide 128 different challenges to achieve a 128-bit response.

Figure 5 is schematic of implementation of Ring Oscillator. This is simplest RO PuF. In this implementation 75 Not gates and 16 NAND gates are used to construct of Ring oscillator. The challenges are applied to select lines of two 16:1 multiplexer. Input lines of both multiplexers are connected to the same ROs and select lines are connected to random signal generator to produce randomness in the implementation. Output of multiplexers are fed to D-Flipflop to introduce random delay in PuF. D-FF is connected to two 4-bit counters. As soon as one of the counter overflows, counting is stopped and output is compared by 4-bit comparator. Comparator gives one bit of response bit i.e. nothing but 1bit output of PuF. In order to get 'n' response bits same circuit should implement 'n' times [12–14].

PuF test results are analysed by either intra-chip or inter-chip performance of the device. Let's assume that you have applied a challenge to the ROPUF, then you will get a single bit response. Similarly, if you have will apply 128 different challenges then you will get 128-bit responses. Let's say the set of 128-bit responses is R and the corresponding set of challenges is C. This measurement is repeated during different environmental conditions. Then you can evaluate the hamming distance, the corresponding hamming distance termed as intra-chip hamming distance.
example:

R1 = set of 128 bit at 25° for Challenge C for a device D
R2 = set of 128 bit at 35° for Challenge C for a device D

Then the intra-chip hamming distance is the evaluated between R1 and R2.
Similarly, if you repeat the experiment for different devices then the obtained hamming distance is termed as inter-chip hamming distance.
example:

R1 = set of 128 bit at 25° for Challenge C for a device D1
R2 = set of 128 bit at 25° for Challenge C for a device D2

Then the inter-chip hamming distance is the evaluated between R1 and R2.

4   Properties of PuF

The PuF design can be used practically only when it passes 3 basic parameters tests. We can also call them as quality measures of a PuF:

1. **Uniqueness** - This is very important parameter of any PuF design. PuF always gives you unique response for different challenges. It can be calculated by finding the hamming distance between the output/response bits [10–12].

2. **Reliability** - It can be measure how stable PuF gives response. Irrespective of any environmental condition or temperature variations PuF always gives same response bit for corresponding challenge. It can be checked using error correction methods [10].
3. **Security** - PuF should be able to prevent any attacks [12].

5 Conclusion

Goal of Cryptography is dealing with data integrity, trust management, privacy and identity management. These goals can be achieved using PuF. There are many ways PuF implementations. PuF is very easy and promising way to achieve Security. PUF applications are very comprehensive. Research is going on how PuF can be used in smart cards also in IoT application as memory and power requirements are very low in these fields. PUFs seem to be an elegant solution for hardware security and device authentication where the required randomness property is obtained from process variation. PUFs have also been used in consumer devices for low-cost authentication purposes.

References

1. Aidasani, L.K., Bhadkamkar, H., Kashyap, A.K.: IoT: the kernel of smart cities. In: Third International Conference on Science Technology Engineering & Management (ICON-STEM), pp. 8–11 (2017)
2. Pocklassery, G., Kajuruli, V.K., Plusquellic, J., Saqib, F.: Physical unclonable functions and dynamic partial reconfiguration for security in resource-constrained embedded systems. In: IEEE International Symposium on Hardware Oriented Security and Trust (HOST), pp. 116–121 (2017)
3. Kulkarni, S., Vani, R.M., Hunagund, P.V.: Review on IoT based case study: applications and challenges. In: Hemanth, J., Fernando, X., Lafata, P., Baig, Z. (eds.) International Conference on Intelligent Data Communication Technologies and Internet of Things (ICICI), ICICI 2018. Lecture Notes on Data Engineering and Communications Technologies, vol 26. Springer, Cham (2019)
4. Sklavos, N.: Securing communication devices via physical unclonable functions (PUFs). In: Reimer, H., Pohlmann, N., Schneider, W. (eds.) Securing Electronic Business Processes, ISSE 2013. Springer Vieweg, Wiesbaden (2013)
5. Joshi, S., Mohanty, S.P., Kougianos, E.: Everything you wanted to know about PUFs. IEEE Potentials **36**(6), 38–46 (2017). https://doi.org/10.1109/mpot.2015.2490261
6. Rührmair, U., Holcomb, D.E.: PUFs at a glance. In: Proceedings of the Conference on Design, Automation & Test in Europe, pp. 347:1–347:6 (2014)
7. Pappu, R.S., Recht, B., Taylor, J., Gershenfeld, N.: Physical one-way functions. Sci. J. **297** (5589), 2026–2030 (2002)
8. Tajik, S., Dietz, E., Frohmann, S., Seifert, J.-P., Nedospasov, D., Helfmeier, C., Boit, C., Dittrich, H.: Physical characterization of arbiter PUFs. Springer, Heidelberg (2014)
9. Herder, C., (Mandel) Yu, M.-D., Koushanfar, F., Devadas, S.: Physical unclonable functions and applications: a tutorial. Proc. IEEE **102**, 1126–1141 (2014)

10. Suh, G.E., Devadas, S.: Physical unclonable functions for device authentication and secret key generation. In: 44th ACM/IEEE Design Automation Conference, San Diego, CA (2007)
11. Gehrer, S., Leger, S., Sigl, G.: Aging effects on ring-oscillator-based physical unclonable functions on FPGAs. In: 2015 International 8th Conference on ReConFigurable Computing and FPGAs (ReConFig), Mexico City (2015)
12. Yin, C.E., Qu, G., Zhou, Q.: Design and implementation of a group-based RO PUF. In: Automation & Test in Europe Conference & Exhibition (DATE), Grenoble, France (2013)
13. Bhargava, M., Mai, K.: An efficient reliable PUF-based cryptographic key generator in 65 nm CMOS. In: 2014 Design, Automation & Test in Europe Conference & Exhibition, Dresden (2014)
14. Anandakumar, N.N., Hashmi, M.S., Sanadhya, S.K.: Compact implementations of FPGA-based PUFs with enhanced performance. In: 30th International Conference on VLSI Design and 2017 16th International Conference on Embedded Systems (VLSID), pp. 161–166 (2017)
15. Katzenbeisser, S., Kocabaş, Ü., Rožić, V., Sadeghi, A.-R., Verbauwhede, I., Wachsmann, C.: PUFs: myth, fact or busted? A security evaluation of physically unclonable functions (PUFs) cast in silicon. In: International Association for Cryptologic Research, pp. 283–301 (2012)
16. Aman, M.N., Chua, K.C., Sikder, B.: Physical unclonable functions for IoT security. ACM. https://doi.org/10.1145/1235
17. Guajardo, J., Kumar, S.S., Schrijen, G.-J., Tuyls, P.: FPGA intrinsic PUFs and their use for IP protection. In: Paillier, P., Verbauwhede, I. (eds.) CHES 2007. LNCS, vol. 4727, pp. 63–80. Springer, Heidelberg (2007)
18. Mughal, M.A., Luo, X., Mahmood, Z., Ullah, A.: Physical Unclonable Function based authentication scheme for smart devices in Internet of Things. In: 2018 IEEE International Conference on Smart Internet of Things, pp. 160–165. IEEE (2018)
19. Johnson, A.P., Chakroborty, R.S., Mukhopadhyay, D.: A PUF-enabled secure architecture for FPGA-based IoT applications. IEEE Trans. Multi-scale Comput. Syst. 1(2), 110–122 (2015)

A Data Security Scheme Using Image Steganography

Karthika C. Nair$^{(\boxtimes)}$ and T. K. Ratheesh

Department of IT, GEC Idukki, Idukki, Kerala, India
karthika93karthik@gmail.com,
ratheeshtk@gecidukki.ac.in

Abstract. This paper proposes a method for protecting data from unauthorized access with the use of both stenographic and cryptographic mechanisms. The system uses block pixel-value differencing (BPVD) and Advanced Encryption Standard (AES) for data protection. The original image is partitioned into 4×4 blocks and the data gets traversed in all the three directions and RGB channel. Based on the hiding capacity of each direction, the proposed method choose a color channel for embedding the encrypted data. AES method is used for encrypting the text data to be embedded. In this new hybrid method the concealing capacity of the image is improved as each direction obtain maximum data. As the data is encrypted, the security of the system is improved much. Using color images the concealing capacity of PVD method is identified and direction based analysis is provided in it.

Keywords: AES · Cryptography · PVD · Steganography

1 Introduction

Data can be made secured during its transmission through internet using various type of mechanisms or schemes. Cryptography is one such scheme for protecting data from unauthorized access. This mechanism converts the original data into unreadable format so that there is no modification possible in the data content during the communication process. But the attacker can detect the existence of the data. Steganography is another scheme used in security areas for hiding data. Image steganography is an important steganography scheme to protect data, where it is concealed inside an image. Digital images are mostly used for hiding the data because of their high frequency. Cover image is the image used to conceal data [3] and stego image is the image with the hidden data. Steganalysis is the process of breaking steganography [4].

Combination of both steganography and cryptography technique is known as stegocrypto which will add security for the data in two dimensions. Here we propose a scheme that combines block pixel value differencing (BPVD), a new method used in steganography to hide data on an image with AES algorithm for securing the data. The image is split into set of blocks and the pixel difference is used for embed data on the image. The data is encrypted with the help of AES before embedding it into the image. To give more security we take a challenge as occur high embedding capacity through

D. J. Hemanth et al. (Eds.): ICICI 2019, LNDECT 38, pp. 615–622, 2020.
https://doi.org/10.1007/978-3-030-34080-3_69

multiple directions. As both of the technologies consist of strongholds, the combination of cryptography and stenography will enhance the security of the data.

2 Related Works

Steganography can be applied for securing data with different technique as pixel value differencing, least significant bit based steganography, cosine transform methods. Some of the efficient methods used for protecting data in steganographic and cryptographic levels combined or not available in literature is described below.

Hameed et al. [5] proposed an adaptive directional pixel value differencing with three direction and which choose adaptively. This system proposed a PVD based technique which divide the image into set of blocks and hide data inside each block. The technique provide high hiding capacity and get maximum PSNR.

Luo, Huang, Huang [7] proposed a PVD based method to enhance the security of data. This method uses an approach of rotation in anticlockwise. Before that the image is divided into 2 square and the square is rotated into 90 and 180°. Consecutive pixels are selected then and is used for embedding the data. The embedded data which make a difference value and it is similar to the original data using PVD.

Khodaei and Faez [8] proposed an LSB and PVD based system to protect information. This system takes the image and split into 3 parts viz. Left, center and right. Center part is used for hiding the data first followed by left block and finally the right block. Then calculate the embedded bits in the base and other pixels. This method also calculate the difference of upper and lower level ranges. Then it achieve the number of bits embedded in the pixel blocks.

Mandal and Das [9] proposed a technique using PVD to protect data on color images. The system separates the image, pixels are extracted one by one and calculate the difference between these pixels. Then it eliminate an overflow problem of each component pixel. Furthermore for providing more security, different pixel components uses different number of bits in it.

Thinn et al. [10] proposed a method for securing data using AES. A symmetric algorithm used for encrypt data. A second key is used for data encryption and an extra method used to reduce the round of subByte operation.

Yanga et al. [11] introduce a PVD method for hiding data in color images. In this paper the edge area is used for hiding the data hence obtain maximum capacity. It always takes two pair of pixels and hence the time complexity is reduced and a pixel value shifting scheme is used for embed data.

3 Proposed Method

The proposed scheme uses Cryptography and Steganography mechanisms where Cryptography method is used for conceal the content of message by encrypting it and steganographic method conceal the existence of message itself. The stego image containing the encrypted data is then exchanged. The proposed system uses a steganography algorithms called pixel value differencing (PVD) and a cryptographic algorithm called advanced encryption standard (AES). The image for hiding data could either grayscale or RGB both of which help to hide data without loss. The proposed system use RGB image for hiding data. The architecture of proposed system shown in Fig. 1. The system consist of four modules viz. Encryption, Encoding, Decoding and Decryption. Both the encryption and encoding encodes the message at sender side and the reverse processes decoding and decryption decodes the message at receiver side.

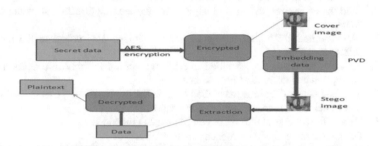

Fig. 1. Architecture of the proposed stegocrypto system

Since the system uses both cryptographic and steganographic mechanisms it imparts multiple level security for the data.

3.1 Encryption and Embedding Process

Image pre-processing is done at first in the cover image. Two filtering processes are applied i.e. sharpening and smoothing. A high pass filter is used for sharpening process and after sharpening the brightness of the image. After sharpening process, the image demarcation is performed which spreads the edge areas with brightness or darkness variation. The low pass filter is used for smoothening process, and is used to cut down the interference or the disparity between two images.

The image is split into 4 × 4 parts and separate it into red, green and blue channel. Pixel value is a simple number which represent the brightness of the image. It vary from 0 to 255. First we extract two consecutive pixel from each block with the help of PVD. After the extraction the difference between the two pixels are calculated. Figure 2 shows the block of image.

P(i,j)		P(i,j+1)	
P(i+1,j)		P(i+1,j+1)	

Fig. 2. Block of image

Choose first block from red channel and calculate the difference and embed the data inside that. Then choose the green channel and follow the same process. The same is then done for blue channel. The whole process is repeated until all the encrypted data got embedded on the stego image. The algorithm of the whole process is described in Fig. 3.

---

1. Split the color image into three channels as R, G ,B. Then divide each channel into number of blocks and assign each direction.
2. Find the difference as: fig 3. 3. shows the embedding direction structure of RGB.
3. PVD method for hiding the data inside an image ,and calculate the new difference value.
4. An encryption process is applied as AES encryption. In this two process the encryption and decryption process are bring.
5. The data embedding bring as embed data in the order of RGB, first embed in R channel then green and finally blue channel..
6. Repeat the process at the end of complete data embedded.
7. Calculated the embedding capacity in each direction.

Fig. 3. Algorithm for encryption and embedding

After calculating the difference value it is checked against values in the range table which is shown in Table 1. The table consist of both range and width values. For example if the range is 0–7 then 3 bit data embed in it, as $2^3 = 8$.

Table 1. Range table

Range	0–7	8–15	16–31	32–63	64–127	128–255
Width	8	8	16	32	64	128

3.2   Decoding and Decryption Process

Decoding and decryption are the reverse process of embedding and encryption. Here the saved data is extracted from the same image and find the difference value in each block. The pixel block is first extracted from red channel and then from green channel and finally from blue channel. The process is repeated and completes the decoding process. After decoding the data from image it is given for decryption process in which the encrypted data is decrypted using AES. The algorithm for decryption and decoding is given in Fig. 4.

---

1. Obtain the pixel value extraction from the stego image.
2. Break the encrypted image using decryption process of AES algorithm.
3. Find new difference from the embedded image.
4. Find the optical range values for each direction, it be achieved from the embedding capacity of each direction.
5. Convert each difference values into corresponding binary values, then obtain the extracted figure.

Fig. 4. Algorithm for decryption and decoding

4 Performance Analysis

The evaluation of the new system and its comparison study with the existing system is performed using different parameters such as mean square error, peak signal to noise ratio and histogram analysis which is detailed in the following sections. Figure 5 shows the cover images used for protecting data which are baboon, peppers and airplane. The performance evaluation of proposed system of BPVD with AES data hiding method is carried out by conducting several experiments for PSNR calculation and histogram analysis. The evaluation shows that the proposed system outperform the existing system.

Fig. 5. Cover image

4.1 Optical Tone of the Stego Image

The optical tone of image identified with PSNR and the images are shown below. The PSNR is defined as a measure of torture in the stego image with decibel (db). The PSNR value can be determined as:

$$PSNR = 20 \log_{20} \left(R^2 / MSE \right)$$

Where MSE is the mean square error which is defined as:

$$MSE = 1/mn \sum_{i=0}^{i=m-1} \sum_{j=0}^{j=n-1} \frac{(Aij - Bij)^2}{256 \times 256}$$

Where A is the cover image pixel and B is the stego image pixel, m n are the size of the image.

Table 2 shows the PSNR values of the images of the proposed system. The proposed system has a maximum PSNR value as 68.60 for airplane image, which means that the system attain less distortion and less MSE value. The PSNR values of 59.31 and 66.34 for the cover images baboon and pepper are also high values as compared to existing systems. The system returns maximum PSNR value with less MSE value compared to the existing systems. So, it is clear that the system is highly secure and the data is hidden completely inside the image.

620   K. C. Nair and T. K. Ratheesh

Table 2. PSNR of proposed method

Image	MSE	PSNR
Airplane	0.01	68.6082845 db
Baboon	0.08	59.3102219 db
Pepper	0.02	66.3441289 db

4.2   Comparison with DCT Methods

The proposed method is compared with DCT based existing method with the same cover images. The DCT system consist of maximum PSNR value as 24.12 with the pepper image. The low value shows that the system is not secure completely because of the high distortion. The proposed method overcome these problem as it uses the block based PVD method. Comparing to DCT method the visual quality is high with the proposed method with high PSNR. Hence it achieve better security. Table 3 shows the PSNR of DCT method.

Table 3. DCT of proposed method

Image	MSE	PSNR
Airplane	255	24.0984455 db
Baboon	251.62	24.1107659 db
Pepper	254.92	24.7899289 db

4.3   Histogram Analysis

Histogram analysis is performed for both cover and stego image and the result are shown in the Figs. 6 and 7. It plots the relationship between the number of pixel occurrence or intensity of the image versus the number of pixels. The similarity of histograms of plain and stego images means that the stego image could hide the data without any distortion. The dissimilarity shows the occurrence of some error which in turn means that the data can easily be hacked. It is seen that after applying BPVD method the histograms of plain and stego images are similar. The proposed system creates a low torture of pixels when stego image is generated hence the hidden data is secure against attacks. Also the system allows hiding more number of bits than that possible with existing system.

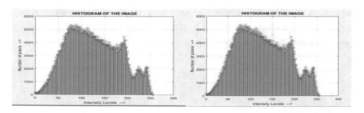

Fig. 6. Cover and stego image of Baboon

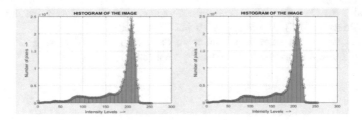

Fig. 7. Cover and stego image of airplane

5 Conclusion

The authors have proposed a new hybrid system for hiding encrypted data on images for the secure transmission of data by combining BPVD and AES methods. The cover images are RGB image and the system supports images in png, jpeg or bmp formats. The proposed method outperforms the existing method by adaptively selecting the embedding direction and the embedding of data bits using each block. Also the data is encrypted using AES before the embedding process. Hence it provides multiple level security to the data. Set of experiments are performed to analyse the performance of the system. Experiment results shows that the PSNR value of the proposed method is much high, when compared with the existing systems even with a lesser MSE values. The histogram analysis showed that the original and stego images generate histograms of much similarity. The experimental results shows that the proposed BPVD system outperforms the existing system in terms of data security.

References

1. Chang, K.C., Chang, C.P., Huang, P.S., Tu, T.M.: A novel image steganographic method using tri-way pixel-value differencing. J. Multimed. **3**(2), 37–44 (2008)
2. Cheddad, A., Condell, J., Curran, K., Mc Kevitt, P.: Digital image steganography: survey and analysis of current methods. Signal Process. **90**(3), 727–752 (2010)
3. Chen, J.: A PVD-based data hiding method with histogram preserving using pixel pair matching. Signal Process. Image Commun. **29**(3), 375–384 (2014)
4. Borges, P.V.K., Mayer, J., Izquierdo, E.: Robust and transparent color modulation for text datahiding. IEEE Trans. Multimed. **10**(8), 1479–1489 (2008)
5. Chang, C.C., Hsiao, J.Y., Chan, C.S.: Finding optimal LSB substitution in image for hiding by dynamic programming strategy. Pattern Recogn. **36**(7), 1583–1595 (2003)
6. Provos, N., Honeyman, P.: Hide and seek: an introduction to steganography. IEEE Secur. Priv. **1**(3), 32–44 (2003)
7. Swain, G.: Adaptive pixel value differencing steganography using both vertical and horizontal edges (2015). https://doi.org/10.1007/s11042-015-2937-2
8. Luo, W., Huang, F., Huang, J.: A more secure steganography based on adaptive pixel-value differencing scheme (2010). https://doi.org/10.1007/s11042-009-0440-3
9. Khodaei, M., Faez, K.: New adaptive steganographic method using least-significant-bit substitution and pixel-value differencing (2011). https://doi.org/10.1049/iet-ipr.2011.0059

10. Mandal, J.K., Das, D.: Colour image steganography based on pixel value differencing in spatial domain (2012). https://doi.org/10.5121/ijist.2012
11. Thinn, A.A., Thwin, M.M.S.: Modification of AES algorithm by using second key and modified SubBytes operation for text encryption (2016)

EYE: Rethinking Electrical Distribution Network with Internet of Things

Ayush Bhardwaj[✉], Vishal Singh, Umesh Kumawat, and S. P. Singh

Indian Institute of Technology, Roorkee, India
abhardwaj@ee.iitr.ac.in

Abstract. Electrical distribution network is a complex network of electrical components (distribution transformers, circuit breakers etc.) working in harmony to ensure delivery of electricity from generating end to end-consumers. The current methods involves manual monitoring for low kVA transformers with limited manpower and in turn fails to guarantee reliability, availability and quality of service in a distribution system. Considering the existing scenario of rural areas, there are no circuit breakers installed for low kVA transformers which eventually lead to catastrophic failures due to thermal stress, overloading etc. and the restoration process is expensive in terms of manpower, capital and time which cause distress to the end users. In this paper we propose "EYE", a system which upgrades the conventional electrical distribution network to a smart IOT enabled network with minimal cost expenditures. The proposed system utilizes the heterogeneous distribution of distribution transformers to form an underlying WANET in the RF band of 433–473 MHz along with environmentally resilient sensors. We quantify the results by deploying the proposed system in one of the rural parts of India (Roorkee) with actual distribution network.

Keywords: IOT · Distribution network · WANET · Over-current · Mesh topology

1 Introduction

IoT (Internet of Things) [3] has lead to a transition in the ways, end-users control and monitor consumer electronic devices and the next generation of IoT or Industry 4.0 is transforming the operation of current industries, hence making them more efficient. IoT has also lead to emergence of newer domains of research for example network slicing in 5G, data analytics in various sectors such as healthcare, waste management etc. Despite of all the benefits of IoT, it has not yet penetrated the crucial government sectors such as AC power distribution and transmission which needs immediate attention as it forms the base for operation of all consumer electronics and industries.

Electricity supports the livelihood of people and as well as technological advances in all the nations across the globe. Reliability, availability of electricity is crucial factor for determining the quality of living in a geographic part of the world. In developing nations, for example like India, Indonesia etc. massive power outages [2] are observed due to failure of distribution, transmission, increased loads etc. The power outage due

© Springer Nature Switzerland AG 2020
D. J. Hemanth et al. (Eds.): ICICI 2019, LNDECT 38, pp. 623–630, 2020.
https://doi.org/10.1007/978-3-030-34080-3_70

to failure of distribution network affects million of people in India itself due to lack of expensive circuit breaking capabilities in low kVA transformers. To avert such outages, a preventive method could be employed which would allow early maintenance of electrical instruments. In [1], authors have utilized GSM modules only for communication but such approach places heavy dependency on a central network operator and uses a centralized approach with all the nodes pushing to cloud and each node with its own GSM connection which is quite expensive as compared to forming clusters.

In this paper we propose "EYE", which allows electrical engineers to perform preventive maintenance to avoid power outage with the help of WANET utilizing the RF band of 433–473 MHz. Our design is motivated by the following questions: *How to upgrade the existing low power distribution network without any upgrades to any heavy electrical components? How to benefit from the geographical heterogeneous distribution of transformers? How to make the monitoring decentralized for improved reliability?*

Our key contributions are:

- EYE leverages the heterogeneous distribution of transformers and distance upper bounds between two transformers due to power loss limitation to form a mesh topology and clusters.
- EYE uses slave-master scheme to make the system cost-effective by employing different hardware configurations for efficient information routing.
- EYE supports multiple interfaces and alert system to allow electrical engineers to be the first responders.
- We test EYE with actual distribution transformer locations and new nodes can be readily added with minimal changes in existing network.

2   Background

2.1   WANET

Wireless Ad-hoc network is a decentralized network with a large number of nodes connected with each other using certain wireless technologies. It doesn't rely on pre-existing topology. WANET utilizes multiple routing schemes [10] for forwarding data in the network.

2.2   Faults

Fault in an electrical ecosystem can be defined as Failure which ultimately lead to diversion of current through its expected path or abrupt changes in current magnitude. Faults are generally caused by some uncertain accidents, excessive internal and external stresses etc. The fault impedance being very low value leads to a flow of enormous amount of current and causing a permanent damage. A transformer is generally damaged the following type of faults [4]:

- **Over Currents due to overloads and external short circuits.** The short-circuit may occurs in multiple phases of an electrical power system. The level of fault current is always very fatal and can cause permanent damage to the electrical equipment. The resistive thermal dissipation of copper in fault scenarios goes high up to some intolerable values. This increasing copper loss causes internal thermal stress.

- **Winding Faults.** The winding faults [5] in transformers are categorized into earth faults or inter-turn faults. Phase to phase winding faults in a transformers are rare. The phase faults in an electrical transformer may be caused by bushing flash over as well as faults in tap changer equipment.
- **Incipient Faults.** They are internal faults that do not pose immediate threat to the transformers. If ignored these faults could lead to series of fatal faults. The incipient faults are categorized into inter-lamination short circuit caused by insulation failure between core lamination, decrease in the oil level due to oil leakages, blockage of oil flow paths. All these faults [5] lead to excessive thermal stress.

These fault causes abnormal variations in various distribution transformer parameters like winding and temperatures, load current, oil level etc. so these variations can be taken into account for detection of fault. The developed system incorporates detection of such faults.

3   EYE Hardware Components

Each EYE node installed at the distribution transformer has an embedded processing capability and a on-board storage capacity along with the required communication protocols for the extraction and distribution of data such as winding temperature, oil temperature and line current from the distribution transformer.

3.1   Communication Interface

Our system involves a hybrid combination of long range radio communication system as well as mobile communication system. Each of the above communication means add to the enhanced capabilities of the monitoring system. An overview of the communication systems has been described below:

- **Long Range Radio Communication:** The nodes uses HC-12 [6] which utilizes working RF band of 433.4–473.0 MHz. With a maximum transmitting power of 100 mW (20 dBm), the receiving sensitivity is −117 dBm at a baud rate of 5000 bps in the air. Baud rates are decided based on different distance values. When distance is 1000 m the module supports FU3 mode at 4800 bps serial speed in open space, whereas when the distance is increased to 1800 m in it supports FU4 mode at reduced baud rate and volume of data. The limited length of electrical lines between two distribution transformers due to power loss limitations allows us to transmit data with the above mentioned transmission modes. It supports UART Bus interface.

- **Mobile Communication System:** SIM900A [9] supports communication in 900 MHz band dual band GSM/GPRS module. It can communicate with on-site computing unit via AT commands (GSM 07.07, 07.05 and SIMCOM enhanced AT Commands). It supports low power consumption when it is sleep mode i.e., 1.5 mA. It supports UART Bus interface.

3.2 Sensing Units

All the on-site nodes for EYE, are installed with long lived and environmentally resilient sensors. Sensing capabilities of each node is supported by the given sensors:

LM75: They [7] are used for monitoring the oil and winding temperatures in order to prevent them from thermal stress. LM75 is a precision integrated circuit temperature sensor with an output voltage linearly proportional to centigrade temperature. It utilizes a low Operating Supply Current 250 µA (typ), 1 mA (MAX) so it has a very low self-heating capability. It has a temperature rating of −55 °C to 125 °C which makes it suitable for high temperature spikes. LM75 utilizes the I2C bus interface.

ACS712: It [8] was employed for measuring the load current for 3 phases as well as the neutral current for unbalanced load systems. A current transformer was employed for stepping down the load current. The device consists of a precise, low-offset, linear Hall circuit and a 1.2 m internal conductor resistance. It requires a 5.0 V, single supply operation and provides an output sensitivity of 66 to 185 mV/A. ACS712 utilizes I2C bus interface. The output of ACS712 is assumed to be linear. Consider the equation given below:

$$y = mx + b \tag{1}$$

where in Eq. 1 'b' is the zero offset. Considering the output to be 'y' the sensors are calibrated for slope 'm' and offset 'b'. In order to determine the zero offset a few sensor readings are taken and their average deviation from 512 is calculated.

$$V_{result} = \left(\frac{Max - Min}{2}\right) * \left(\frac{V}{1024}\right) \tag{2}$$

where in Eq. 2, *Max* is the maximum sensor reading and Min is the minimum sample reading and V is the supply voltage, typical value of which is 5 V.

3.3 Power Supply

Power supply unit of each node provides a desired dc voltage to drive all the utilities of our system. Initially, the 230 V mains voltage is stepped to a 12 V level followed by an ACDC conversion using a rectifier. Output of the rectifier still contains some ripples despite of its conversion to a DC signal and to remove the ripples and obtain smoothed DC, power filter circuits are used. 12 V DC is rated down to 5 V using a positive voltage regulator chip. Figure 1 describes the basic functional units of the power supply unit. Our system could also be supported as a battery operated node.

Fig. 1. Functional diagram of power supply unit.

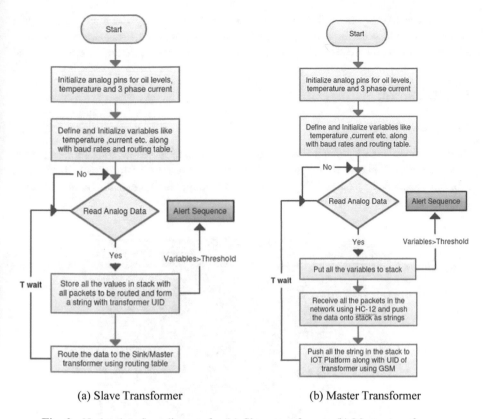

(a) Slave Transformer   (b) Master Transformer

Fig. 2. Nodes data flow diagram for (a) Slave transformer, (b) Master transformer.

3.4 Embedded System

The embedded system is based on Atmega328. It has 12 digital input/output pins, 8 analog inputs, 1 UART (hardware serial port), a 16 MHz crystal oscillator. The on-board system houses a flash memory that provides 32 KB storage space and 2 KB of SRAM. This system requires a 5 V operating voltage.

4 EYE Data Flow Scheme

In the proposed system, the master and slave analogy has been used as the master transformer is responsible for pushing data to online utility server whereas all the slave transformer relays data to the master or sink transformer.

Figure 2(b) describes the flow process for the master/sink transformer. All the data stored in the stack of main transformer i.e. data of the slave transformers is pushed to the IOT platform along with UID for further analysis and monitoring. The flow process for the slave transformers is shown in Fig. 2(a).

All the sensor nodes stack the data they receive for relaying and then push the data to the next node according to the routing table along with its own parameter values. Each packet has an alert flag that is set in case the value crosses certain tolerance levels.

5 Evaluation

5.1 Network Topology

EYE has been tested for a Mesh Topology (CW1SN) network for multi hop data routing. Figure 3 shows the actual position of distribution transformers in radial distribution network in Roorkee Region obtained through the global positioning system, which was our test environment.

Fig. 3. Distribution network in the geographical location of Roorkee, India.

5.2 Results

The experimental results obtained from deploying EYE in Roorkee's distribution network is summarised in this section. The currents of high magnitudes cant be fed directly to current sensing ICs, so it needs to be scaled down. For this purpose, current transformers (150/5 A) are utilized to achieve the desired values. The transformer secondary line current after being scaled down is fed to the system.

To create warning conditions, we fed the current to one of our nodes in laboratory using potentiometers. The reference current is 3.5 A.

By employing potentiometer phase voltage has been altered. If the current crosses the reference value, the system will issue the warning message to the concerned first responder. An alert message as shown in Fig. 4(a) are obtained during the operation.

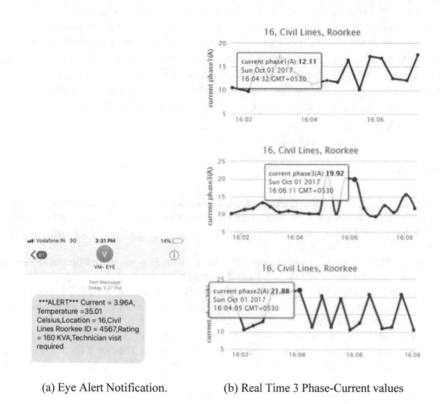

(a) Eye Alert Notification.   (b) Real Time 3 Phase-Current values

Fig. 4. Output of EYE.

Figure 4(b) shows real time current data values obtained from the EYE network for one of the slave transformers with time stamps and precise location with following name plating:

kVA rating: 100 kVA, Primary Voltage: 11 kV, Secondary Voltage: 415 V Secondary Current: 140 A, Vector Group: Dyn11
Maximum rise in Oil temperature: 40 oC, Impedance (%): 4%.

6   Conclusion

This paper proposes "EYE", which allows preventive maintenance of distribution transformers by deploying a WANET with minimal expenditures and leverages the heterogeneous distribution of distribution transformers and distance limitation between two nearby transformers.

7 Future Works

The system could be upgraded by making it more intelligent and efficient by adding more enhanced capabilities such as (1) Zero cross detection for power factor calculations, checking phase balance, detection of intrusion currents in case of failures, (2) Load Forecasting could be performed for smarter distribution network and dynamic consumption rates, (3) A more compact and portable system with greater data acquisition speeds and processing capabilities. We would like to thanks (1) Department of Electrical Engineering Indian Institute of Technology, Roorkee, Uttarakhand, India, (2) Circle Office, Uttarakhand Power Corporation Ltd. (UPCL), Roorkee, Uttarakhand, India for supporting our work.

References

1. Al-Ali, A.R., Khaliq, A., Arshad, M.: GSM-based distribution transformer monitoring system. In: Proceedings of the 12th IEEE Mediterranean Electrotechnical Conference (IEEE Cat. No. 04CH37521), Dubrovnik, Croatia, vol. 3, pp. 999–1002 (2004). https://doi.org/10.1109/MELCON.2004.1348222
2. Abrar, Md.M.: Power cut off and power blackout in india a major threat-an overview. Int. J. Adv. Res. Technol. 5(7), 8–15 (2016). ISSN 2278-7763
3. Čolaković, A., Hadžialić, M.: Internet of Things (IoT): a review of enabling technologies, challenges, and open research issues. Comput. Netw. **144**, 17–39 (2018). https://doi.org/10.1016/j.comnet.2018.07.017, ISSN 1389-1286
4. Sacotte, M., Wild, J.: Transformer faults and protection systems. In: 14th International Conference and Exhibition on Electricity Distribution. Part 1. Contributions (IEEE Conf. Publ. No. 438), Birmingham, UK, vol. 1, pp. 10/1–10/5 (1997). https://doi.org/10.1049/cp:19970450
5. Bengtsson, C.: Status and trends in transformer monitoring. IEEE Trans. Power Deliv. **11**(3), 1379–1384 (1996)
6. Elecrow: HC-12 Wireless Serial Port Communication Module. https://www.elecrow.com/download/HC-12.pdf. Accessed 25 July 2019
7. Texas Instruments: LM75 Precision Centigrade Temperature Sensors Data Sheet (2016). http://www.ti.com/lit/ds/symlink/lm35.pdf. Accessed 25 July 2019
8. Allegro Microsystems: Fully Integrated, Hall Effect-Based Linear Current Sensor IC with 2.1 kVRMS Isolation and a Low-Resistance Current Conductor ACS712 Data Sheet (2017). www.allegromicro.com/media/Files/Datasheets/ACS712-Datasheet.ashx. Accessed 25 July 2019
9. SimCom: SIM 900 Hardware Design (2017). https://simcom.ee/documents/SIM900/SIM900_Hardware%20Design_V2.05.pdf. Accessed 25 July 2019
10. Zanjireh, M.M., Shahrabi, A., Larijani, H.: ANCH: a new clustering algorithm for wireless sensor networks. In: 27th International Conference on Advanced Information Networking and Applications Workshops, WAINA 2013 (2013). https://doi.org/10.1109/waina.2013.24

A Comparative Study of Thresholding Based Defect Detection Techniques

Yasir Aslam$^{(\boxtimes)}$ and N. Santhi

Department of Electronics and Communication Engineering,
Noorul Islam Centre for Higher Education, Kumaracoil, India
yasiaslam@gmail.com

Abstract. This paper presents a comprehensive literature analysis of defect detection techniques. There are several techniques for image segmentation developed by the analyst in turn to make images lustrous and simple to evaluate. In digital image processing, the thresholding is a renowned technique for the segmentation of images. Defect detection is currently a practicable field for improvising performance as well as retaining the products quality. The extensive applications of common thresholding methods such as adaptive thresholding, Otsu thresholding and seven other thresholding techniques have been discussed. In this paper, a comparative study of detection of defect or crack using various thresholding techniques is conferred.

Keywords: Thresholding · Segmentation · Defect detection · Image processing

1 Introduction

The simplest method of image segmentation is thresholding which can be used to create binary images from a gray scale image, also from color images the binary images are produced by segmentation. The procedure of segmentation involves constituting each pixel in the image source into two or more classes. Several binary images are obtained, if there are more than two classes than the usual result. In image processing, the image are split into smaller segments by thresholding, with a minimum of one gray scale value defining the boundary. In thresholding process the pixel intensities reduced, that is values lower than the certain value is taken as non-defect region and higher pixel values are taken as defect region. Also, whenever the background illumination is uneven the method of global thresholding is not appropriate. Classically, in adaptive thresholding the input is a color image or gray scale [1] and merely obtained binary image as outputs characterizing segmentation. For each pixel, a threshold value is calculated in the image input and is set to background value if the pixel value falls below threshold, if not it takes on the foreground value. For different local areas, different threshold values are used in adaptive thresholding.

In recent years, nondestructive testing (NDT) methods [2] have been commonly adopted for the defect detection in materials. The vision based defects detection gives generous scenario of computerization and legitimate analysis. The study put forward a practice meant for identification and segmentation of defects using thresholding. Presently, there are several researchers toward a group of works in the area of competing

© Springer Nature Switzerland AG 2020
D. J. Hemanth et al. (Eds.): ICICI 2019, LNDECT 38, pp. 631–637, 2020.
https://doi.org/10.1007/978-3-030-34080-3_71

for developing defect and scratch detection. The CAD method in image processing [3] for detection of solder defects, process is practical for defects on the surface of weld joint. The wavelet transform [4] for the analysis of images to spot the casting defects seized from X-ray inspection. For analysis the images used exhibited distinctive castings defects which are consequent from X-ray inspection equipment.

The Particle Swarm Optimization (PSO) [5] together with adaptive thresholding method is used for image defect detection, thereby searching the global optimum result with a very fast convergence speed across the whole search space. The use of the statistical features and Gabor filter [6] features routinely categorize the surface defects of product within glaze and crumbling. Besides, used the Gabor filter aspect for inquiring the fragile cracks present inside unrefined steel block through deprecating the energy cost function parting defected regions as well as the defect-free criteria.

2 Survey of Thresholding Techniques

Zhang [7] introduced a crack detection algorithm based on histogram-based approach. By preprocessing, feature extraction, and image segmentation have obtained about the crack image information. Aminzadeh et al. [8] introduced an approach for selection of threshold that facilitate for selecting the threshold value in intensity range present in the boundary of defects and by means of examining background histogram modes as defective regions and background. The background region histogram, values of threshold are routinely chosen by the background histogram mode extents and for some standard images of surface defects the method proved to be practical. The parameters for algorithm implementation are the percentage of histogram and threshold of standard deviation. Lee [9] presented the edge-based adaptive thresholding technique for deinterlacing defining correct edge path that resolves the performance of interpolation. The process identified direction and edge information of binary image with the basis of locally adaptive thresholding, in addition utilizes window size of adaptive appropriately for detecting the edge slope for well-organized calculation. The thresholding problem in image defect detection and segmentation is considered as an optimization problem and can be solved with the help of combining the adaptive thresholding with the Cuckoo Search [10] and Firefly Optimization [11].

Truong et al. [12] introduced the Otsu method variants developed for defect identification. The Otsu's method is regarded as one of the commonly referenced techniques of thresholding which produces satisfactory results as objects within image is outsized and duly unlike as of background. Otsu's method is the basis for reasonably lot of applications in instinctive defect identification. It resolves the optimal threshold value thereby maximizes the foreground and background between-class variances. The entropy of image is the numerical extent of uncertainty that signifies the image characteristics.

Liu et al. [13] introduced the approach to identify the image bugholes through analyzing concrete surface, also signifies analysis factor of surface bugholes. The shooting area is impaired by the shooting distance and image recognition quality.

A partition method was utilised for validating the detection accuracy with concrete surface. In Otsu threshold segmentation, gray image and contrast improvement take out bugholes characteristics inside concrete surface, whereas threshold values for shape discriminative factor differentiates bugholes as well as cracks on surfaces. Xie et al. [14] proposed an algorithm for image enhancement which is based on the shear-let transform for the surface defect detection of magnetic tile. The shearlet transform is in particular efficient for the identification of the discontinuities on edges and other abnormalities. The image scale parameter, shear direction and translation are three-parameter functions of shearlet transform. This method is discrete that confides with Laplacian pyramid shared through numerous shearing filters. The algorithm used improves the image contrast stuck between regular area and defected area. By the advanced distinctiveness of the shear-let transform process, an enrichment algorithm projected for finding the defected area.

Karlekar [15] introduces four kernels of two dimensional inhering low pass filter and high pass filter. The wavelet transform, fraction of wavelet image that decomposed confers the information about the fabric texture pattern. The fabric image defect identification is composed up to two levels. The wavelet transform functions much enhanced to line defects detection as such upright, flat and transverse line defects. Vijaykumar et al. [16] proposed an inventive practice using Gabor transform approach for detecting the defect on the surface of rail heads. Consequently for the defect identification, it is necessary imminent to take out the background rails from and auxiliary improve the image for thresholding. The Gabor transform utilizes the algorithm of Binary Image Based Rail Extraction (BIBRE) for background rails surface image extraction. The corresponding precision and recall values results were 89.9% and 89.7% shows the detection performance. The extricated rails improved toward achieving the consistent background using the method of undeviating enhancement. This approach is utilized for improvising the difference in brightness between backgrounds and their objects of an image. The Gabor filters is used by the improved rail image for the rails defects exposure. The Gabor filters method expands variance in energy amongst the defect less surface and defected surface. The result of thresholding depends on the defects energy and imperfections are recognized from the thresholded image.

Pham et al. [17] proposed segmentation with split and merge technique of the images for expediting the detection on fruit peel defects. Firstly the technique segments unique image in color space by k-means clustering technique. The tiny region is subsequently filtered out as a result of assimilation toward the nearby regions. The Region Adjacent Graph (RAG) was constructed depending on the regions obtained from preceding stage to make available the process of merging. The regions are iteratively combined or merged for segmenting the defects depending upon technique of minimum spanning tree. Ahmed et al. [18] introduced a method of fuzzy classification, pixels to segment an image consisting of two steps. Initially the group of pixels divided among specified non-disjoint class number, containing each images with three classes. At initial part, it is necessary to keep useful information of the longest possible, whereabout the pixels composing the crack till the decision making time. Thereafter, consequently define the assorted image regions, whereas it is carried out by selecting pixel of each and assigning toward the class having the maximum membership degree. Through this way for an optimal image the technique is adopted for finding the threshold of histogram with one mode and the method achieves 82% detection rate.

Lu et al. [19] introduced a method which is based on choosing image segmentation threshold with the help of gray level threshold with two iteration selection, region containing defects of constituent image surface is segmented. Depending upon this, the major defects present in the image are identified and classified with the help of the morphological operation, obtaining the feature matching and feature of the defect area, it achieves above 97% defect rate. Dinh et al. [20] proposed a self-regulating peak detection algorithm on segmentation of image to be appropriate for crack detection in concrete. Initially the image scanned with a crack prospective is refined with the help of the developed line emphasis filter. The processed histogram of gray scale image is subsequently smoothed in by means of an average filter then resolved for perceiving considerable peaks depending upon the offset distance and cross over index which are the dynamic parameters. These criterions are evaluated from the primary peaks detected and updated individual origin, exclusive of any analytical criterion. The result for segmentation is in that case selected for binarization of image.

Song et al. [21] presented algorithm for the crack detection on road surface based on the binary image dual-thresholds and ridge edge detection. The reduction of multi-scale data image utilized for original image contraction to get rid of noise, able to even and improve cracks in an image. From the gray scale image, the main cracks are extracted with the fractional calculus ridge edge detection process. Consequently, applying both the long line and short thresholds, the resulted binary image is further processed for the removing the noise and short curves to get segments of crack area. The slits in cracks associated with a function correlation curve. It can situate and distinguish slight and thin cracks accurately which are complex to classify using other conventional algorithms. Meng et al. [22] proposed algorithm of image processing methods through least squares, which widen the image gray level, also construct more uniform gray values. The defects analyzed through maximum entropy and otsu binarization method. The method introduces the method of least squares binarization depending upon the blocking whereas it takes spatial layout of pixels in concern as well as takes a binarization idea on blockwise, segmenting image with the help of multi-threshold; the cracked regions could exactly located. The major benefits and limitations over various thresholding techniques is illustrated in Table 1 as in the following.

Table 1. Various thresholding techniques comparison for defect detection.

Methods	Strengths	Weaknesses
Histogram methods	• Efficiency in computation • Performs better for different characteristics images	• Sensitive to noise • It becomes difficult in segmenting the images with weak contrast or uneven reflection due to the irregular shape or surface curvature
Adaptive thresholding	• Achieves good detection accuracy • Better Computational metrics • Robust algorithm performance	• Precise selection of window for thresholding is essential
Otsu thresholding	• Detects both huge and tiny defects as of large background	• It fails when histogram is unimodal
Shearlet transform	• The method offers higher directional sensitivity	• The method consumes more time
Wavelet transform	• Procures accurate multi-scale analysis of image • Ease of textural feature extraction and prospect of direct thresholding • Enhanced noise reduction	• Some defect information may be lost • The method fails to detect defect in images having color variance presence and smooth edges
Gabor transform	• Absolute detection for defects on cracks and boundaries • Provides best possible defect identification for frequency as well as spatial domain	• Computational complexity • Parameters selection for optimal filter is relatively complicated
Clustering method	• Applicable on texture images with high background and images of complex structures • It works well in case of bimodal image • More efficient in terms of processing time	• The method fails for somewhat discolored imperfection on light colored surface
Iterative method	• Computationally efficient method with varied complexities • The method is robust against varying characteristics of image	• It fails in cases of detecting micro sized defects
Computer vision	• Robustness in excluding random noise	• It's not suited for some complex images of real-world used in construction
Edge based thresholding	• Identifies the defects effectively and accurately • The method provides superior results	• It is not suitable for defects other than holes and cracks • The method consumes more time
Entropy based thresholding	• High detection accuracy and good versatility	• The algorithm has some issues with the signal-to-noise ratio and sensitivity • Computationally intensive process

3 Conclusion

This paper presents a study of various available thresholding techniques with dissimilar image processing approaches practicable for defect detection in surfaces. Various thresholding approaches along with their stability and fragility have been reviewed in the study. The survey shows that each of the thresholding techniques suffers from some drawbacks. From the analysis, it is clear that each of the thresholding technique is suitable for defect detection on a particular application, as the image intensity level parameter between object and background, image object size, noise and image object position would greatly affect the thresholding performance. Also, the survey reveals the existing thresholding techniques contributions that is useful for additional research on thresholding based defect detection method.

References

1. Bradley, D., Roth, G.: Adaptive thresholding using the integral image. J. Graph. Tools **12**(2), 13–21 (2011)
2. Melnikov, A., Sivagurunathan, K., Guo, X., Tolev, J., Mandelis, A., Ly, K., Lawcock, R.: Non-destructive thermal-wave-radar imaging of manufactured green powder metallurgy compact flaws (cracks). NDT&E Int. **86**, 140–152 (2017)
3. Nacereddine, N., Zelmat, M., BelaYfa, S.S., Tridi, M.: Weld defect detection in industrial radiography based digital image processing. In: 3rd International Conference-Sciences of Electronic, Technologies of Information and Telecommunications, Tunisia (2005)
4. Li, X., Tso, S.K., Guan, X.-P., Huang, Q.: Improving automatic detection of defects in castings by applying wavelet technique. IEEE Trans. Ind. Electron. **53**(6), 1927–1934 (2006)
5. Aslam, Y., Santhi, N., Ramasamy, N., Ramar, K.: An effective surface defect detection method using adaptive thresholding fused with PSO algorithm. Int. J. Simul. Syst. Sci. Technol. **19**(6), 1 (2018)
6. Zhang, X., Krewet, C., Kuhlenkötter, B.: Automatic classification of defects on the product surface in grinding and polishing. Int. J. Mach. Tools Manuf. **46**(1), 59–69 (2006)
7. Zhang, Y.: The design of glass crack detection system based on image pre-processing technology. In: 2014 IEEE 7th Joint International Information Technology and Artificial Intelligence Conference, pp. 39–42, 20–21 (2014)
8. Aminzadeh, M., Kurfess, T.: Automatic thresholding for defect detection by background histogram mode extents. J. Manuf. Syst. **37**, 83–92 (2015)
9. Lee, D.: A new edge-based intra-field interpolation method for deinterlacing using locally adaptive-thresholded binary image. IEEE Trans. Consum. Electron. **54**(1), 110–115 (2008)
10. Aslam, Y., Santhi, N., Ramasamy, N., Ramar, K: A modified adaptive thresholding method using cuckoo search algorithm for detecting surface defects. Int. J. Adv. Comput. Sci. Appl. **10**(5) (2019)
11. Aslam, Y., Santhi, N., Ramasamy, N., Ramar, K.: Metallic surface coating defect detection using firefly based adaptive thresholding and level set method. Int. J. Innov. Technol. Explor. Eng. **8**(1) (2019)
12. Truong, M.T.N., Kim, S.: Automatic image thresholding using Otsu's method and entropy weighting scheme for surface defect detection. Soft Comput. **22**(13), 4197–4203 (2018)
13. Liu, B., Yang, T.: Image analysis for detection of bugholes on concrete surface. Constr. Build. Mater. **137**, 432–440 (2017)

14. Xie, L., Lin, L., Yin, M., Meng, L., Yin, G.: A novel surface defect inspection algorithm for magnetic tile. Appl. Surf. Sci. **375**, 118–126 (2016)
15. Karlekar, V.V.: Fabric defect detection using wavelet filter. In: Proceedings of the 1st International Conference on Computing Communication Control and Automation, ICCUBEA 2015, pp. 26–27 (2015)
16. Vijaykumar, V.R., Sangamithirai, S.: Rail defect detection using Gabor filters with texture analysis. In: 3rd International Conference on Signal Processing, Communication and Networking (ICSCN) (2015)
17. Pham, V.H., Lee, B.R.: An image segmentation approach for fruit defect detection using k-means clustering and graph-based algorithm. J. Comput. Sci. **2**, 25–33 (2015)
18. Ahmed, N.B.C.: Automatic crack detection from pavement images using fuzzy thresholding. In: ICCAD 2017, Hammamet, Tunisia (2017)
19. Lu, R., Yin, D.: Component surface defect detection based on image segmentation method. In: 28th Chinese Control and Decision Conference (CCDC) (2016)
20. Dinh, T.H., Ha, Q.P.: Computer vision-based method for concrete crack detection. In: 14th International Conference on Control, Automation, Robotics & Vision, Phuket, Thailand (2016)
21. Song, H., Wang, W., Wang, F., Wu, L.: Pavement crack detection by ridge detection on fractional calculus and dual-thresholds. Int. J. Multimedia Ubiquit. Eng. **10**(4), 19–30 (2015)
22. Meng, Q., Wang, X., Zhang, Y.: Research on a least squares thresholding algorithm for pavement crack detection. In: Sixth International Conference on Information Science and Technology, Dalian, China (2016)

Machine Learning Approach to Recommender System for Web Mining

Jagdeep Kaur$^{(\boxtimes)}$ and Jatinder Singh Bal$^{(\boxtimes)}$

Department of Computer Science and Engineering,
Sant Baba Bhag Singh University, Jalandhar, Punjab, India
jagdeepkaurrai859@gmail.com,
bal_jatinder@rediffmail.com

Abstract. One of the major challenges face by webmasters is the introduction of numerous choices to the customer, which leads to repetitive and difficulty in locating the correct item or data on the webpage. In the traditional approach, if the data was changed, pooling approach was possible, only if data variation was within the cluster information. In case the data exceeds the limit, classification was difficult to perform. Therefore, we need to have a classification approach that can work under these conditions. In the proposed work we have implemented Hybrid of ANN and KNN approach and found improvement in the recommendation system with greater accuracy.

Keywords: Web mining · Artificial Neural Network · Genetic Algorithm · Artificial Intelligence · Information mining

1 Introduction

Data mining is the procedure which goes under the class of software engineering to explore expansive measure of informational index which has a place with particular example. Here vast informational collection remains for Big Data. The example in the enormous information contains a few strategies or methods at a crossing point of AI that is Artificial Intelligence, database frameworks and measurements. The term information digging remains for the extraction of information from the information distribution centre and furthermore it performs design coordinating and finds helpful learning from the colossal measure of information. It is additionally supplanted or substituted with the procedure of basic leadership frameworks like AI and machine learning and so forth. The languages which are utilized for information mining, for example, information investigation on the extensive scale database, information examination, machine learning and AI are the most proper terms [1]. In the proposed work we have arranged clients on the reproduced dynamic sessions as extracted from testing sessions by gathering dynamic clients' snap stream and it is matched with comparative group or class in the information store, in order to create an arrangement of proposals to the customer in a Real-Time premise.

© Springer Nature Switzerland AG 2020
D. J. Hemanth et al. (Eds.): ICICI 2019, LNDECT 38, pp. 638–646, 2020.
https://doi.org/10.1007/978-3-030-34080-3_72

2 Web Data Mining

It isn't overstated to state that the World Wide Web is a standout amongst the most energized effects to the people and the general public in most recent 10 years. It likewise changes the method for doing work, business, giving and getting great quality training and overseeing different associations. The impact is the aggregate and finish change of data, for example, gathering, passing on, and trade. Along these lines, in the present time web ends up being the best and biggest wellspring of data which is accessible on this planet. Likewise, Web comprises of an assorted, dangerous, tremendous, dynamic and unstructured information archive which supplies mind blowing, immense and substantial measure of data to the clients and in some cases raises multifaceted nature of information that how to manage the data from alternate perspectives. Since, the web mining puts down its foundations profoundly into the information mining, consequently it isn't identical to information mining. The Web mining term deals with applying the information mining methods and to find and concentrate the helpful and vital data from the reports and services of World Wide Web [2].

3 Web Usage Mining

Web usage mining [10] generally finds helpful and critical data from the optional information which is received while the clients are surfing on the Web. The main objective in the area of mining is to examine between the genuine and expected webpage utilization, anticipating the client's conduct inside the website and change of the website as per the interest of the clients. There are as such no clear refinements between the Web use mining and the Web content mining and Web structure mining [3] (Figs. 1 and 2).

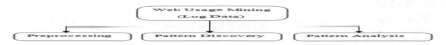

Fig. 1. Contents of Web usage mining [3] (Source: IJANA, pp. 1422–1429)

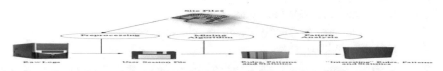

Fig. 2. Web usage mining [4] (Source: IRJET, pp. 1319–1322)

4   Artificial Neural Network

ANN [9] stands for Artificial Neural Network. The neural network serves as a basis for intelligent and complex vision systems for various industrial and research applications.

All neural networks are subjected to a training process, similar to how a child learns from the environment by observing the examples. Neural calculus can never compete with conventional calculation. Instead, it works together to achieve successful solutions.

In general, neural networks are applied when the rules formed for the expert system are not applicable. These are particularly useful for the classification and function approach problem [5] (Fig. 3).

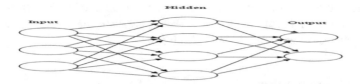

Fig. 3. A simple neural network [6] (Source: IJEERT, pp. 143–148)

5   K Nearest Neighbour (KNN) Classification

KNN is a powerful Machine Learning algorithm. It is easy to interpret and takes less computation time. With increasing values of k in KNN [10], boundaries become more distinct and clear.

5.1 PseudoCode of KNN

1. Load the data set to be used
2. Assign Initial values of k
3. To determine the predicted class, perform iterations starting from 1 to total number of data points in training
    3.1   Using Euclidean distance metric, find out the intermediate distance between data to be tested and every row of training data.
    3.2   Based on the distance values, perform sorting of the calculated distances in ascending order.
    3.3   Retrieve from the sorted array top k rows.
    3.4   Find out the most frequently occurring class of these rows.
    3.5   Return the predicted class.

The closest neighbor's run is the KNN [7] shape when K = 1. In the event that the arrangement of an example is obscure, at that point it might be normal considering the grouping of the closest neighbors. Hence, an obscure example can be ordered by the closest neighbor's grouping. Figure 4 shows KNN choice lead for K = 1 and K = 4 for an arrangement of tests separated into two classes. In Fig. 4(a), an obscure example is ordered utilizing just a single known specimen; in Fig. 4(b) more than one known example is utilized. In the last case, the parameter K is set to 4, with the goal that the four nearest tests are considered to order the obscure. Three of them have a place with a similar class, while just a single has a place with another class. In the two cases, the obscure specimen is named having a place with the class on the left. Figure 4 explains the KNN calculation.

Fig. 4. KNN classification [7] (Source: IJERA 3, pp. 605–610)

The execution of the KNN classifier is essentially controlled by the decision of K, and in addition the separation metric connected.

6 Information Mining

Information Mining analyse different sources of information with the aim of producing self-explanatory results. There are different methods utilized for information mining. Some of them are classified as:

6.1 Statistics.
6.2 Artificial Intelligence technique for data mining.
6.3 Decision Tree approach.
6.4 Genetic Algorithm.
6.5 Visualization [8].

6.1 Statistics

Measurements is an essential procedure to perform for information mining. Insights is performed on the yield of the information mining so as to isolate the important information from insignificant information. Information cleaning is a procedure in which information is cleaned by removing disturbance. Statics can likewise be helpful to recuperate the missing information on the premise of insights. For information investigation, grouping and trial outlining methods are utilized [8].

6.2 Artificial Intelligence (AI) Techniques

Information mining additionally utilizes the idea of AI i.e. Counterfeit consciousness (Artificial Intelligence). The systems like machine learning, neural Networks go under the point of Artificial Intelligence. There is a rundown of some different procedures likewise, for example, design acknowledgment, Knowledge introduction, Knowledge Acquisition additionally a sort of computerized reasoning systems and perform by information mining.

6.3 Decision Tree Approach

Decision tree has tree like structure where test on an attribute is represented with internal node, result of the test is a branch and class label is represented by each leaf node. The most important strength of this approach is its ability to generate rules that can be understood easily. They are able to perform classification with very less computation. Its major weakness is its training which is very expensive.

6.4 Genetic Algorithms

The thought behind the advancement of Genetic Algorithm [11] is Darwin's Theory of development. Inherited counts are arranged as overall chase heuristics. Genetic counts are a particular class of formative estimations (generally called transformative retribution) that use techniques inspired by transformative science, for instance, heritage, change, assurance, and cross breed (furthermore called recombination). Genetic computations are completed as a PC generation in which a masses of dynamic portrayals (usually termed as chromosomes) of contender plans (which are individuals or creatures) to an improvement issue propels toward good arrangements [8]. In every period, the wellbeing of each individual in the population is calculated, different individuals are haphazardly looked over the present population (in perspective of their health). The new population is then used as a piece of the accompanying cycle of the count. Conventionally, the figuring closes when any of the most outrageous number of periods are conveyed, or level of wellbeing has been pursued by the population. If the figuring has finished in view of a most outrageous number of times, an agreeable course of action could have been achieved. Hereditary Algorithms are utilized as a part of information mining to remove the issues that happen by grouping.

6.5 Visualization [12]

Perception alludes to the pictorial portrayal of any issue, framework or method. It gives the better comprehension to the issue or framework. Perception assumes an imperative part in information mining moreover. Information mining by utilizing perception can be accomplished with the assistance of human mind or help [8]. The strategies utilized for representation in multi-dimensional informational indexes are as per the following:

- Project framework,
- Scatter Plot frameworks,
- Co-plot projection.

7 Machine Learning

Machine Learning involves making the computer learn without doing programming. In this we train the system by feeding inputs and getting relevant outputs. At first the system is made to learn or trained, and later on testing is done by feeding the input whose output is not known to the system. The system works based on its past experiences and gets the output, which is then matched with the standard output. It is similar to a human brain, which learns from its past experiences and take decisions based on what it has been taught. İn this work we have used KNN based clustering technique and ANN based Classification technique. In order to find out the distance between the two users, and get the nearest neighbour, KNN based clustering is applied. Hence in traditional system both input and program are feeded to get the desired output while in Machine Learning Input and Output both are feeded and the system develops its own program.

8 Results

The proposed algorithm is implemented in MATLAB to test on the desired dataset. Here are the calculated results from the proposed work (Fig. 5).

Fig. 5. Graph predicting Euclidean Distance of Unknown user to Other user, that is from user 1 to user 6

The above figure shows the Euclidean Distance between Unknown User and the Other user, that is from user 1 to user 6. Here distance from each and every user is taken with other users (Fig. 6).

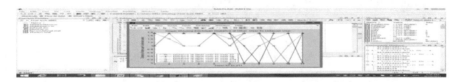

Fig. 6. Graph showing Euclidean Distance of Other user to Unknown user, that is from user 7 to user 11.

The above figure shows the Euclidean Distance of Unknown User and the Other user that is from user 7 to user 11. Here distance from each and every user is taken with other users (Fig. 7).

Fig. 7. Accuracy proposed of hybrid ANN-KNN approach

As shown in the figure, here accuracy percentage of proposed ANN-KNN approach is calculated Here, the calculated accuracy is 90.90% (Figs. 8, 9 and Table 1).

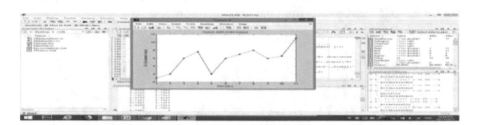

Fig. 8. Hybrid ANN-KNN output

Users	Daily's Name	News Category	Added Required Feed Type	Class
1	CNN News	World	www.*world	World
2	China Daily	Business	www.*business	Business
4	CNN news	Politics	www.*politics	Politics
5	Punch Ng	Entertainment	www.*entertainment	Entertainment
3	Punch Ng	Politics	www.*politics	Politics

Fig. 9. Data set

Table 1. Accuracy of hybrid ANN-KNN approach

S. No	Technique used	Accuracy
1.	Hybrid KNN and ANN model	90.90%

9   Conclusion

The framework performs arrangement of clients on the reproduced dynamic sessions which are extracted from testing sessions by gathering dynamic clients' snap stream and this is matched with comparative group or class in the information store, in order to create an arrangement of proposals to the customer in a Real-Time premise. Accuracy of proposed Hybrid of ANN and K-NN comes out to be 90.90%) which helps to improve the recommendation system. The result of the experiment figures out that an automatic Real-Time recommendation system with K-NN classification model and neural network implemented using the Euclidean distance method is able to produce and deliver accurate results to the client at any time, based on his current requirement instead of the information based on his earlier visit to the site.

10   Future Scope

The proposed technique can be tested on other datasets to check it accuracy and execution time. The proposed work can be compared with the other data mining techniques to check the accuracy of the system in future work. Thus, various other data mining techniques can be compared with one another to check the system performance and accuracy. Besides accuracy other parameters such as entropy, variance, density can also be used.

References

1. Talakokkula, A.: A survey on web usage mining, applications and tools. Comput. Eng. Intell. Syst. **6**, 22–30 (2015)
2. Kosala, R., Blockeel, H.: Web mining research: a survey. ACM SIGKDD Explor. Newsl. **2**, 1–15 (2000)
3. Rana, C.: A study of web usage mining research tools. IJANA **3**, 1422–1429 (2012)
4. Chavda, S.: Recent trends and novel approaches in web usage mining. Int. Res. J. Eng. Technol. (IRJET) **4**, 1319–1322 (2017)
5. Kaur, J., Singh, J.: Particle swarm optimization based neural network. Int. J. Emerg. Technol. Eng. Res. **3**, 5–12 (2015)
6. Kumar, P., Sharma, P.: Artificial neural networks-a study. Int. J. Emerg. Eng. Res. Technol. **2**, 143–148 (2014)
7. Imandoust, S.B., Bolandraftar, M.: Application of K-Nearest Neighbor (KNN) approach for predicting economic events: theoretical background. Int. J. Eng. Res. Appl. **3**, 605–610 (2013)
8. Deepashri, K.S., Kamath, A.: Survey on techniques of data mining and its applications. Int. J. Emerg. Res. Manag. Technol. **6**, 198–201 (2017)
9. Kaur, J.: Recommendation system with Automated Web Usage data mining by using k-Nearest Neighbour (KNN) classification and artificial neural network (ANN) algorithm. Int. J. Res. Appl. Sci. Eng. Technol. **8**, 160–168 (2017)
10. Adeniyi, D.A.: Automated Web usage data mining and recommendation system using K-Nearest Neighbor classification method. Appl. Comput. Inf. **12**, 90–108 (2016)

11. Call, J.M.: Genetic algorithms for modelling and optimization. J. Comput. Appl. Math. **184**, 205–222 (2005)
12. Nazeer, F., Nazeer, N., Akbar, I.: Data visualization techniques – a survey. Int. J. Res. Emerg. Sci. Technol. **4**, 4–8 (2017)

Hybrid Machine Learning Approach for Skin Disease Detection Using Optimal Support Vector Machine

K. Melbin[1]([✉]) and Y. Jacob Vetha Raj[2]

[1] Manonmaniam Sundaranar University, Abishekapatti, Tirunelveli 627012,
Tamil Nadu, India
melbinmean@gmail.com
[2] Nesamony Memorial Christian College, Martandam 629165,
Tamil Nadu, India
jacobvetharaj@gmail.com

Abstract. Dermoscopy is the one of the major discussing topic in medical field to detect the skin diseases. Due to the bad frequencies, low contrast and noises the positive result of the dermoscopy is unpredictable. In this paper, a hybrid machine learning model of Singular Value Decomposition (SVM) based Whale Optimization Algorithm (WOA) is proposed for the identification of skin disease. Initially, the image database is segmented using level set approach. In order to retrieve the feature vectors from the segmented image, extraction of features from those datasets are carried out. Hence, the feature vectors are extracted using histogram and Local binary pattern (LBP) method. Once feature is extracted, the skin disease classification is done using Whale optimization (WOA) based Support Vector Machine (SVM). The simulation results show that the proposed method outperforms many other algorithms in terms of accuracy and other performance measures in the identification of skin disease.

Keywords: Dermoscopy · Skin image · Segmentation · Classifier · WOA-SVM

1 Introduction

Fundamentally, skin is the largest organ of integumentary system, which can cover the entire body. Warm, damage, microscopic tissues, infection, radiation and sensitivity can affect the skin easily [1] and made many skin diseases. Because of the complexity in territories the dermatologist is the standard method. The valuable tool of dermoscopy is used in clinics to identify the skin diseases, which is the integral part of the experiments. Dermoscopy consist of well qualified magnification lens and the arrangements of energetic lights with polarized and non polarized lenses are fixed inside the instrument. In medical field, dermatology method used to identify the skin diseases in the starting stage only but it failed to detect in fatal stage of the skin diseases. Dermatology exposes the architect and features which are imperceptible to the naked eye and this method does not predict or detect the skin diseases correctly. Therefore, it supports further morphologic instructions during the clinical examination

© Springer Nature Switzerland AG 2020
D. J. Hemanth et al. (Eds.): ICICI 2019, LNDECT 38, pp. 647–658, 2020.
https://doi.org/10.1007/978-3-030-34080-3_73

of skin injuries. Nowadays, the dermoscopy method of segmenting and classifying part of the image contains some drawbacks. So in this paper used WOA based SVM classifier algorithm to boost up the quality of the dermoscopy.

The most important objectives of this paper are summarized as follows:

- A hybrid machine learning model combining support vector machine and Whale Optimization algorithm (WOA) is proposed for skin disease classification.
- The color and texture features are examined using Histogram and Local Binary Pattern method. The edge is extracted from the shape.

2   Review of Related Works

The skin is the outer tissue of our body; which is mostly affected by bacteria, fungi and virus. Many researchers can be suggested medical technologies to cure the skin diseases. Barata et al. [2] introduced melanomas determination method, it based on texture classification and color. Hence, the dissimilar identification of melanocytic and melanomas are very difficult. Choudhari and Biday [3] were proposed the skin cell damage identification method, the morbidity and mortality process can be decreased with feature extraction and segmentation method. But the white and black parts are delivered from the results of feature category. Madooei et al. [4] delivered the conversation of grey scale method which is depends on human skin. Hence, it produced usual and lesion of pigmented skin with larger amount of reduction capacity. Song et al. [5] declared boosting of semi supervised method with high accuracy of categorization and multi class holding, because it contains low detection effect in skin cancer with reduced amount of classification. Schwarz et al. [6] presented ultra skin image analysis of optoacoustic in Dermoscopy. It determines the depth of the penetrated skin cells and capacity levels using dermoscopy and it based on multispectral wavelength procedure.

3   Proposed WOA-SVM Based Skin Disease Identification

The proposed section consists of computer based skin disease identification method of WOA-SVM. Initially, the image segmentation process such as color, shape and texture is carried out by means of level set concept. In this, the color and texture features are extracted by using histogram modelling and LBP method also the shape is extracted by mean of edge determination process. The database is used to collect the features vectors and the feature classification is done by using WOA based Support Vector Machine. The step by step process of skin disease identification is shown in Fig. 1.

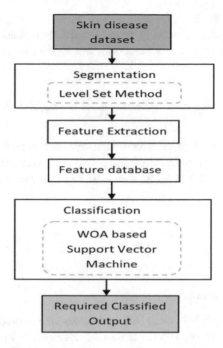

Fig. 1. Flow diagram of proposed method

3.1   Segmentation of Injury Region

Basically, the segmentation process is more helpful to evaluate the image and we use level set method for segmentation process. Thus, the level set method is used to detect the injury portion and it is the instrument to detect the shapes, surface, curves. The level set method easily holds the topological alterations and used to analyze the boundary region. Mathematically, the curve and surface values are initially kept as zero, since the smoothing function is $\lambda(p,q,r)$ and surface as $\lambda(p,q,r)=0$. The planner closed curve C_p is also takes the zero level set value and the function $\lambda(p,q,r)$ is illustrated in Eq. (1) and the evolution ordinary movement of $\lambda(p,q,r)$ is expressed in Eq. (2).

$$C_p(r) = \{(p,q)|\lambda(p,q,r)=0\} \tag{1}$$

$$\frac{\partial \lambda}{\partial r} + R|\nabla \lambda| = 0 \tag{2}$$

Here, gradient operator ∇ and R denote the evolution speed. The general movement of the level set equation is handled by the variational level set, which can minimize the energy level set function λ is expressed in Eq. (3)

$$\frac{\partial \lambda}{\partial r} = -\frac{\partial E}{\partial \lambda} \tag{3}$$

Some information is the representation of various energy expressions and the developing contour is able to modify based on the modification reason. The development speed of R is denoted with geodesic active contour (GAC) model [7] is expressed as follows:

$$\frac{\partial \lambda}{\partial r} = |\nabla \lambda| divg\, x \left(\frac{\nabla \lambda}{|\nabla \lambda|} \right) + k\,x |\nabla \lambda| \tag{4}$$

Hence, k is the coefficient of divergence operator constants and x is the edge criterion function. The Gaussian filter G_σ consists of standard deviation σ and the segmented injury regions are detected using the level set method, which is denoted in Eq. (5).

$$x = \frac{1}{1 + |\nabla G_\sigma * \mathrm{Img}|^2} \tag{5}$$

3.2   Extraction of Features

In this the part, the segmented section of the images are allowed for feature extraction process. Therefore, the color, texture features are extracted by using Histogram and Local Binary Pattern methods. After the segmentation procedure, the disease categorization is correctly identified by using the feature vectors.

- The modelling of color histogram with image vectors are used for the conversion of RGB color element. In this work, the images are initially read. By using various tertiary arrays to accumulate the Red, Green and blue element values and its histogram values are estimated [8]. Hence, the various arrays are more useful to accumulate the tertiary histogram value of the element. The summation process is used to estimate the 256 bins also the mean value is calculated. Each value are computed and stored in different arrays with the Standard deviation of the color histogram is estimated using Eq. (6),

$$\sigma = \sqrt{\frac{1}{M} \left(\sum_{i=1}^{N} hi * (i - \bar{x})^2 \right)} \tag{6}$$

- Hence, the image object boundaries are determined by edge detection techniques also the discontinuous intensity is detected. In the field of image processing, machine and computer eyesight with data extraction and image segmentation are carried out for edge detection. The operator function of canny shape is more useful to detect the edges. Gaussian filter is used to remove the noises and also the smoothening region is obtained. Where x be the kernel, t and u be the pixel coordinates at the row and column respectively.

$$H_{tu} = \frac{1}{2\Pi\sigma^2} \exp\left(\frac{(t-(x+1))^2 + (t-(x+1))^2}{2\sigma^2}\right) \tag{7}$$

Each pixels by its image direction and gradient intensity is estimated by using Eq. (8)

$$G = \sqrt{G_l^2 + G_m^2} \tag{8}$$

Therefore, the horizontal and vertical direction are denoted as 'G$_l$' and 'G$_m$' with the mutual augmentation of rectangular factor 'Rtan2' is expressed in Eq. (9),

$$\Theta = R\tan(G_l, G_m) \tag{9}$$

Based on the maximal gradient with the non maximal suppression is applied. The suppressed direction and magnitude are not exceeding the pixel points. The pixels are filtering out by the using high and low value of threshold. If it is small with the threshold value is low. Finally, the weaker edges only detect the edges.

- The computer eyesight classification done by the kind of visual descriptor in LBP. The difference among centre and nearer pixel is the symbol of LBP texture. For each image, the binary threshold code is subjected to the nearer pixels. The vector LBP function is created as follows [9]: the image window size is classified into 16 * 16 pixels and every pixel is compared with 8 nearer with the circle of clockwise and counter clockwise position. Hence the centre pixel value is more than the neighbour means 0 else 1. These histograms consist of 256-dimensional feature vector and combined the histogram cells, which produce the window of entire feature vector

3.3 WOA Based Support Vector Machine Classification

Once feature is extracted, the classification is done using Whale optimization based support vector machine. To do that, an efficient Machine Learning (ML) method is proposed based on support vector machine (SVM) and Whale Optimization Algorithm (WOA). The prey surrounding, bubble net attacking (exploitation) and prey searching (exploration) are the basic idea of WOA algorithm.

Step 1: Prey surrounding: The humpback whale is used to represent the prey position and the prey surrounding process is initialized. Thereafter, the remaining agents are attempted to update the location towards the present best location. The amount of iteration is denoted as i.

$$\vec{A} = \left|\vec{E}.\vec{Y}^*(i) - \vec{E}(i)\right| \tag{10}$$

$$\vec{Y}(i+1) = \vec{Y}^*(i) - \vec{B}.\vec{C} \tag{11}$$

Where, $\vec{A}$ and $\vec{E}$ are the vector coefficients with Y^* is the location prey and the position of whale is mentioned as $\vec{E}(i)$. The location of Y^* is updated based on the best solution. The vector functions of $\vec{B}$ and $\vec{E}$ is estimated using Eqs. (12) and (13). The function of random vector $\vec{u}$ is uniformly distributed to the interval value of $[0, 1]$.

$$\vec{B} = 2\vec{b}.\vec{u} - \vec{b} \tag{12}$$

$$\vec{E} = 2\vec{u} \tag{13}$$

Step 2: Exploitation Stage (Attacking bubble net process): The method of surrounding shrinking is the bubble net characteristics of whales. Also the humpback movement is determined by using spiral Eq. (14).

$$\vec{Y}(i+1) = \vec{A}'e^{al}\,\cos(2\pi l) + \vec{Y}^*(i) \tag{14}$$

Here, the best obtained solution is denoted by $\vec{A}' = |\vec{Y}^*(i)\vec{Y}(i)|$. Hence, the tth whale and the prey distance is the model of best solution. The interval value of $[-1, 1]$ is for constant a and random numeral l respectively. So the whales are swim in the circular position. Hence, the probability function is distributed to the interval $[0, 1]$.

$$\vec{Y}(i+1) = \begin{cases} \vec{Y}^*(i) - \vec{B}.\vec{C} & \text{if } p < 0.5 \\ \vec{A}'e^{al}\,\cos(2\pi l) + \vec{Y}^*(i) & \text{if } p \geq 0.5 \end{cases} \tag{15}$$

Step 3: Exploration (Prey searching process): Based on the each position by group search randomly proceeded and while contrasted to the next set. Thus, the movement of search agent left from the next one also the vector $\vec{B}$ randomly distributed to $[1, -1]$ interval. Based on the random search agent with its location is updated.

$$\vec{A} = |\vec{E}.\vec{Y}(i)_{rdm} - \vec{Y}(i)| \tag{16}$$

$$\vec{A}(i+1) = \vec{Y}(i)_{rdm} - \vec{B}.\vec{C} \tag{17}$$

Hence, the present iteration of the randomly vector (whale) selected as $\vec{Y}(i)_{rdm}$. The global optimum is determined by the exploration of $|B| > 1$. Significantly, the selected search agent of present best agent position is updated using $|B| < 1$. The unsymmetrical description is done by WOA. The extracted vectors are randomly creating the real number using WOA, the number ranges between 0 to 1. The features are orbited and the extra values are represented using the parameter C and γ. Hence, the orbited value is equal to 0.5 means the new value becomes 1 so the suitable feature value is elected. Therefore, the value is below 0.5 means it rounded to 0 and does not select any features, which is shown in Fig. 2.

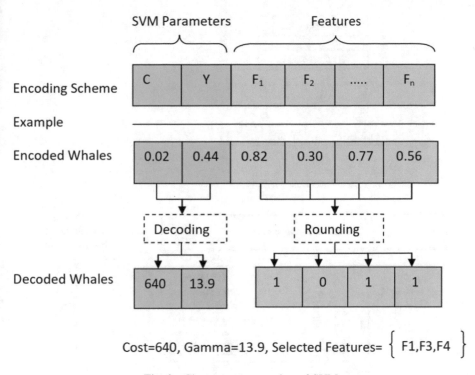

Cost=640, Gamma=13.9, Selected Features= $\left\{$ F1,F3,F4 $\right\}$

Fig. 2. Chrome representation of SVM

The scaled values of C becomes [0.01, 35000.0] and γ[0.0001, 32.0] is explained using Eq. (18). Here U, V is the output, old values with $\max_U$ and $\min_U$ of upper to lower limit interval. The SVM classification accuracy with fitness value is given to the feedback to Whale Optimization Algorithm. The WOA based SVM architecture is illustrated in Fig. 3.

$$U = \frac{V - \min_V}{\max_V - \min_V}(\max_U - \min_U) + \min_U \qquad (18)$$

Significantly, the initial step contains selected features from the SVM classifier, using validation partition in training and testing of dataset. The portioned features only selected for the process and it enters to the SVM classifier. It produce the suitable set of training with cross validation of 3-fold. After that SVM to generate the image accuracy classification of training set with suitable fitness using Whale Optimizer Algorithm. If it does not meet the termination part, again the process is repeated. Hence, it goes through the termination part, the classifier produce the better result compared to all iterations using C and γ. Ultimately, the WOA based SVM classifier delivered the suitable classified output.

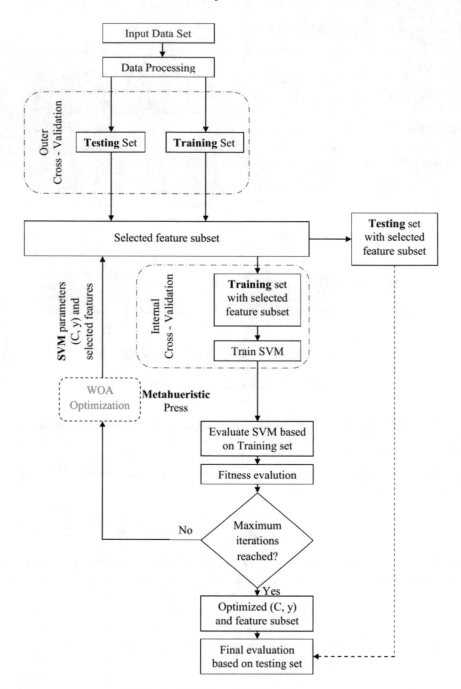

Fig. 3. WOA based SVM architecture

4   Result and Discussion

4.1   Experimental Setup

Numbers of research are performed to determine the skin disease using the Dermoscopy by the computer aided system for identification of skin. This model is implemented by the *Matlab R2017a* in a computer system with *Intel core i5 3.90 GHz* CPU, *8 GB* RAM and with Windows 8 64 bit operating system.

4.2   Segmentation Performance

In the first part of result section, segmentation evaluation metrics such as Sensitivity, Specificity, Positive Predictive Value (PPV), Negative Predictive Value (NPV), False Positive Rate (FPR) and False Negative Rate (FNR) are estimated and segmented skin infections are presented. In Fig. 4, proposed model successfully segmented the desired affected regions of skin of the image, while other existing models fail to segment it.

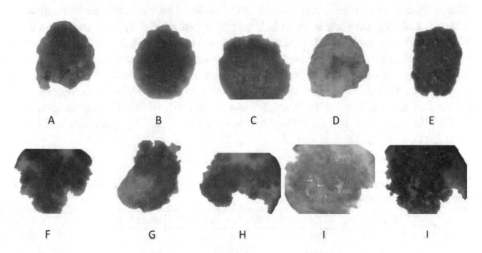

Fig. 4. The segmented dataset of the given input dataset

The main basic evaluation criterion for the comparison of existing approach k- means clustering approach and proposed level set and proposed method provides better segmentation than the existing one. This is used to quantify the quality of the segmentation result. The comparison table is given below in Table 1. It is observed that the segmented image B and F produces 96% of accurate segmentation in various metrics.

Table 1. The comparison of the proposed level set with existing K-mean clustering approach for metrics such as Sensitivity, Specificity, PPV, NPV, FPR and FNR

Segmented images	Level set approach (Proposed)						K-means clustering					
	Sensitivity	Specificity	PPV	NPV	FPR	FNR	Sensitivity	Specificity	PPV	NPV	FPR	FNR
A	0.92	0.95	0.61	0.99	0.07	0.09	0.75	0.85	0.56	0.91	0.11	0.18
B	0.96	0.97	0.77	0.99	0.04	0.05	0.86	0.9	0.73	0.92	0.09	0.1
C	0.93	0.91	0.54	0.99	0.1	0.08	0.89	0.88	0.43	0.9	0.19	0.19
D	0.91	0.98	0.91	0.98	0.02	0.1	0.9	0.91	0.87	0.89	0.08	0.2
E	0.89	0.97	0.76	0.99	0.02	0.12	0.82	0.89	0.68	0.87	0.04	0.26
F	0.96	1.00	0.98	0.99	0	0.05	0.86	0.92	0.84	0.86	0.19	0.17
G	0.91	0.93	0.89	0.91	0.09	0.1	0.85	0.84	0.78	0.88	0.18	0.19
H	0.85	0.90	0.9	0.85	0.09	0.19	0.74	0.83	0.87	0.79	0.2	0.27
I	0.86	0.95	0.69	0.97	0.07	0.16	0.78	0.8	0.58	0.82	0.15	0.28
J	0.80	0.82	0.96	0.43	0.2	0.18	0.76	0.72	0.89	0.38	0.28	0.3

4.3   Classification Performance

The second part is filled with classification result. The performance evaluation of different existing models are given in Table 2, it can be clearly seen that this model WOA based Support Vector Machine has produced better results in terms of evaluation of statistical uncertainty (Accuracy), the proportion of actual negatives that are identified as negatives (Specificity) and the proportion of actual positives that are identified as positives (Sensitivity). The proposed method of SVM-WOA is compared with different classifiers like K-NN, ANN, SVM-GA and SVM-PSO. Finally our proposed SVM-WOA produces better performance in terms of accuracy, specificity and sensitivity, which is illustrated in Fig. 5.

Table 2. Segmentation accuracy, specificity and sensitivity result of WOA based SVM compared with various existing approaches

Classifiers	Specificity (%)	Sensitivity (%)	Accuracy (%)
k-NN	89	76	90
ANN	90	80	85
SVM-GA	96	72	93
SVM-PSO	93	74	94
SVM-WOA	99	84	98

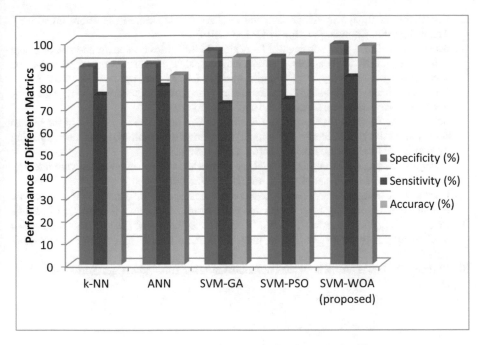

Fig. 5. Performance of proposed SVM-WOA classifier.

5 Conclusion

In this paper, the skin diseases are recognized by means of hybrid machine learning approach of SVM based WOA. The level set concept is widely used to segment skin disease from the image and the feature extraction procedure is carried out by LBP method. Experimentally, the images are taken from the ISIC ARCHIVE dataset and the implementation process is performed in the platform of *Matlab R2017a*. The evaluation metrics of proposed level set methods with sensitivity, specificity, PPV, NPV, FPR and FNR performance ratio is high compared to the K-means clustering. Thus, the proposed SVM-WOA classifier is compared with various kinds of classifiers. But the proposed method delivers the better performance result in accuracy of 99%, sensitivity 84% and specificity 98%. In future, the performance of SVM classifier will be enhanced in terms of Multi kernel approaches.

References

1. Yadav, N., Narang, V.K., Shrivastava, U.: Skin diseases detection models using image processing. Int. J. Comput. Appl. **137**(12), 0034–0039 (2016)
2. Barata, C., Marques, J.S.: A bag of features approach for the classification of melanomas in dermoscopy images. In: Computer Vision Techniques for the Diagnosis of Skin Cancer, pp. 49–69. Springer (2014)

3. Choudhari, S., Biday, S.: Artificial neural network for skin cancer detection. Int. Emerg. Trends Technol. Comput. Sci. **3**(5), 147–153 (2014)
4. Madooei, A., Drew, M.S., Sadeghi, M., Atkins, M.S.: Intrinsic melanin and haemoglobin colour components for skin lesion malignancy detection. In: International Conference on Medical Image Computing and Computer Assisted Intervention, pp. 315–322 (2012)
5. Song, E., Huang, D., Hung, C.C.: Semi-supervised multiclass adaboost by exploiting unlabeled data. Expert Syst. Appl. **38**(6), 6720–6726 (2011)
6. Schwarz, M., Soliman, D., Omar, M., Buehler, A.: Optoacoustic dermoscopy of the human skin: tuning excitation energy for optimal detection bandwidth fast and deep imaging in vivo. IEEE Trans. Med. Imaging **36**(6), 1287–1296 (2017)
7. Zhang, K., Zhang, L., Zhang, D.: A level set approach to image segmentation with intensity in homogeneity. IEEE Trans. Cybern. **46**(2), 546–557 (2015)
8. Prabhu, J., Kumar, J.S.: Wavelet based content image retrieval using color and texture feature extraction by gray level matrix and color matrix. J. Comput. Sci. **10**(1), 15 (2014)
9. Liu, L., Lao, S., Fieguth, P.W., Guo, Y., Wang, X., Pietikäinen, M.: Median robust extended local binary pattern for texture classification. IEEE Trans. Image Process. **25**(3), 1368–1381 (2016)
10. Mirjalili, S., Lewis, A.: The whale optimization algorithm. Adv. Eng. Softw. **95**, 51–67 (2016)

Segmentation Analysis Using Particle Swarm Optimization - Self Organizing Map Algorithm and Classification of Remote Sensing Data for Agriculture

Jagannath K. Jadhav[1]([✉]), Amruta P. Sonavale[2], and R. P. Singh[1]

[1] SSSUTMS, Sehore, Madhya Pradesh, India
jagannathjadhav3030@gmail.com, dr.rp_singh@gmail.com
[2] VTU University, Belagavi, Karnataka, India
amruta.ps713@gmail.com

Abstract. Remote sensing (RS) has become one of the vital approaches to get the information directly from the earth's surface. In recent years, with the event of environmental informatics, RS information has contend a crucial role in several areas of analysis, like atmosphere science, ecology, soil pollution, etc. When monitoring, the multispectral satellite data problem are vital once. Therefore, in our analysis, automatic segmentation has aroused a growing interest of researchers over the past few years within the multispectral RS system. To beat existing shortcomings, we provide automatic semantic segmentation while not losing significant information. So, we use SOM for segmentation functions. Additionally, we've got planned a particle swarm improvement (PSO) algorithmic rule for directly sorting out cluster boundaries from SOM. The most objective of this work is to get a complete accuracy of over eighty fifth (OA> 85%). Deep Learning (DL) could be a powerful image process technique, together with RS image.

Keywords: Deep Learning · Self organizing map · Change detection · Remote sensing · Crop classification

1 Introduction

Remote sensing on image processing has improved a lot in the last decades to detecting and classifying the objects on earth. It is widely used in environmental studies, weather forecasts, agricultural purposes and mapping. But, today RS has used as one of the promising techniques in agriculture for the mapping and identification of crops. Remote sensing is one of the best techniques for acquiring and generating agricultural data [12]. RS uses the visible and invisible electromagnetic spectrum to record data. for example, Crops, which are affected by diseases and insect attacks [1, 2]. Depends on its physical features different object among the earth surface reflect or emits different electromagnetic energy in various wavelength [5]. As far as in DL process, a neural network is used in DL for the best classification process. It uses two classification approaches of supervised NN and unsupervised NN method [6, 7]. Mostly used

© Springer Nature Switzerland AG 2020
D. J. Hemanth et al. (Eds.): ICICI 2019, LNDECT 38, pp. 659–668, 2020.
https://doi.org/10.1007/978-3-030-34080-3_74

satellite to monitor farm data is LANDSAT 8 (L-8), SENTINEL-1A (S-1A). Each satellite image has a high resolution RGB band which has a spatial resolution of 0.41 m and the multispectral scene combines seasonal data [3]. Thermal Infrared Sensor (TIRS) and Operational Land Image (OLI) sensor collects the image from 2 long-wave thermal bands and 9 short-^{wave} bands and measures the 11 bands according to the reflected and emitted energy [4]. JAS is introduced to estimate the cultivated area to collect information on the crop [10, 11] (Fig. 1).

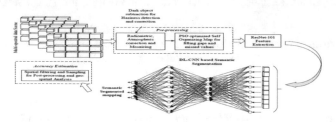

Fig. 1. Architecture of the proposed system

2   Proposed Methodology

The proposed methodology consists of preprocessing, feature extraction and classification which are as follows.

2.1   Pre-processing

The pre-processing functions include the necessary operations before analyzing basic data and consist of geometric correction, radiometric correction and atmospheric correction to improve the ability to qualitatively and quantitatively interpret the components of the image. The raw input images cannot be directly used for segmentation process without the preprocessing [9].

2.1.1   Input Database

For the proposed framework, a multispectral RS image is considered for the input database, i.e. the agricultural images got from the Landsat public dataset and the USGS dataset. These datas are used as input for the current segmentation technique as shown in Fig. 2.

| a (i) | a (ii) | b (i) | b (ii) |

Fig. 2. Input images a(i) image acquired in 2001, a(ii) image acquired in 2009 from Landsat public dataset, b(i) image acquired in 2011, b(ii) image acquired in 2014 images from USGS dataset.

2.2   Radiometric Calibration

The main objective of the radiometric calibration approach is, to recover the real reflectance or radiance of the target image, the visual appearance of the image is increased by Both of corrections, radiometric correction eliminate radiometric errors and geometric correction eliminates geometric distortion in RS images. For each pixel as DN, the sensors collects and records the EM radiation. These numeric numbers of pixels are converted to important real units, such as reflection or brightness. In ENVI, some Landsat data can be directly converted to reflectivity without first having to calculate radiation which is as shown in Fig. 3.

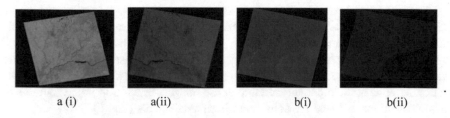

| a (i) | a(ii) | b(i) | b(ii) |

Fig. 3. Radiometric calibration of a(i) image acquired in 2001 and a(ii) image acquired in 2009 from Landsat public dataset, b(i) image acquired in 2011 and b(ii) image acquired in 2014 images from USGS dataset.

2.2.1   Atmospheric Correction

Satellite images require atmospheric correction before moving to any study or undergoing further processing to obtain various surface parameters with spatial images which are used to convert the Digital Numbers (DN) into radiation and reflection since the data provided to the user are in DN. To match multi-time and multi-touch images, DN must be converted to physical units. There are several methods of atmospheric correction.

2.2.1.1. *Dark-Object Subtraction (dos).* The function of DOS is based on the image approaches, assume the satellite image with zero percentage of reflectance such the signals are recorded by the sensor from these features. It is used to removes the component caused by the spread of additives from RS data. The Dark object subtraction is utilized for reducing the haze component in RS data. It tracks all the darkest pixel for each bands. Moreover, DOS approach subtract a value from each pixel for each bands for eliminating the scattering which is as shown in the Fig. 4.

2.2.2   Mosaicing

Mosaicing approach is utilized for combining several separate images in one scene. It can often be impossible to get the image of a huge document as it has a certain size. Then the data are partially scanned, and split the images. This divided images are used for agriculture to monitor the growth of crops, identify the pests and estimate the yield of crops in large agricultural areas.

a(i)	a(ii)	b(i)	b (ii)

Fig. 4. DOS Atmospheric Corrections of image acquired in A(i) 2001 and A(ii) 2009 from Landsat public dataset, B(i) image acquired in 2011 and B(ii) image acquired in 2014 from USGS dataset.

2.2.2.1. *Feature Extraction.* FEs are elements of 2 corresponding input images, while input data are taken from patches to match input images. In the Initial stage, the FE functions were selected for automating registration function based on two approaches to understanding functionality have been created. These approaches are based on the selection of signs of an important structure in the images. Significant points (corners of areas, intersections of lines) are considered here as elements (Fig. 5).

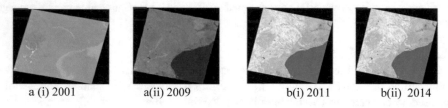

a (i) 2001	a(ii) 2009	b(i) 2011	b(ii) 2014

Fig. 5. Atmospheric Corrections of image acquired in a(i) 2001 and a(ii) 2009 from Landsat public dataset, b(i) image acquired in 2011 and b(ii) image acquired in 2014 from USGS dataset.

2.2.2.2. *Image Fusion (iF).* The remote sensing image fusion is to obtain an image that simultaneously has spatial resolutions and high spectral images, for example pansharpening. This indicate that IF fusion of low and high resolution panchromatic and multi spectral (MS) image, which is used to get the high resolution image.

2.2.2.3. *Image Registration.* Aligns mostly more than 2 images at that time or from a different point of view. To record images, we define a general ordered transformation to organize the images based on the reference of each images.

2.2.2.4. *Homographic Computation.* Homogeneous calculation is mapping the image space that is often among two images at the same part. It used to join several images to create full view of image.

2.2.2.5. *Image Warping (IW).* IW process is used to operate the image digitally, so the shape of the image gets distorted. To decrease the distortion, it is sufficient to deform the images into a plane based on the reference image. It also used to correct the distortion in images.

2.3   Self Organizing Map Segmentation

In the segmentation process, the pre-processing stage is optimized by using the SOM to segment the image and then restore loss of data which are collected from the satellite data. To restore the missing satellite data, the SOM is used. In addition, SOM training data is manually selected because it becomes difficult to use the method in automatic mode. Separate reconstructing of RS images is done for every bands i.e., each spectral bands is trained using separate SOM which is as shown in below Fig. 6.

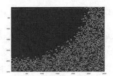

Fig. 6. SOM Segmentation   **Fig. 7.** Clustering of classes

2.3.1   Learning Algorithm

This SOM algorithm don't access the output of other network. In its place, the weight of the node compare the input vector, from this the lattice values are selected for optimization process.

2.3.1.1. *Training Occurs in Several Steps and Over Many Iterations.* In initial s stage, the training process access first. During that processing, the sample data vector is taken from the dataset and find the similarity between sample and each neurons. The BMU is determined: the weight of the conveyor is more related to x. After that, the BMU weight vector values and the best neighbor values are updated. Training usually takes place in two stages.

2.3.1.2. The SOM weight of the connected code project is initialized so the weight value is denoted by $0 < W < 1$. Weights of neurons have geographic properties. In the resulting segmentation, the spectral data are reflect and arranged based on the spatial appearance of the patterns.

2.3.1.3. *BMU.* The Calculation of BMU of all nodes iteratively, then analyze the Euclidean distance in-between input vector and the weight vector of each node. The weight vector with the node is known as BMU.

$$Dist = \sqrt{\sum_{i=0}^{i=n} (U_i - M_i)^2}$$

2.3.1.4. *The Weights Adjusting.* The weight of BMU for each node are adjusted. The closest input vector with node (minimum d) is chosen as the winner or BMU of the map. Then, all weights of neighborhood nodes is defined by the BMU and further it was updated by the formula as $(t + 1) = W(t) + L(t)(V(t) - W(t))$.

2.3.2   Clustering

It is the process of determining the objects class and separate the objects as larger and smaller classes. The group of clusters are formed by SOM tool and provide visual inspection. The performance of each users are in manual inspection or using the hierarchical algorithms, which is used to calculate the cluster limit. Therefore we propose the new heuristic algorithm, to find the automatic limit of the cluster from the output code vectors which is as shown in Fig. 7.

2.3.3   SOM Optimized by PSO

During the training process the fitness function of multi dimension problem has been solved and develops the segmentation process. The steps of this model is as fallows:

Step 1: Initiate the vector velocity and every particles position of input image.
Step 2: Analyze the individual and overall fitness value of every particle. After computing, better value are taken as the best current position.
Step 3: The Best particle position is compared with target to obtain the particles optimal fitness value. If the better value then take it as the current overall optimal position. Then, update the speed and positions of the particles.
Step 4: the termination status is checked if it meets end condition then the process is complete or else get back to step 2 (Fig. 8).

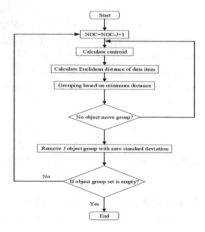

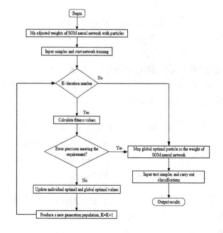

Fig. 8. Flow chart of clustering   **Fig. 9.** Flow chart of PSO optimized SOM

In this document, the precision values of the SOM classification in the NN were chosen for the fitness function of Fig. 9.

2.4   Feature Extraction of ResNet-101

The renued Deep-Lab NN is called Residual Network. ResNet mainly used for image classification, object location and object detection. The main challenge of forming a deeper network is that the accuracy deteriorates with the depth of the network. The ResNet authors have proposed a residual learning approach to facilitate the formation of deeper networks. Based on the design ideas of BN, the small convoluted cores, the fully convoluted network. Then the results of these two branches are added. The Fig. 10 below shows the ResNet-101 Feature Extraction.

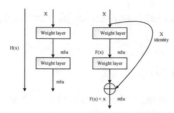

Fig. 10. ResNet-101 Feature Extraction

2.4.1 ResNet Image Training.

This NN may segment objects better by multiple scales. It is necessary to create a residual network by adapting the ResNet DCNN image classification to obtain a better performance of semantic segmentation [8].

3 Experimental Results

The results and the analysis, as well as the corresponding indicators of evaluation of the experimental results are described in this section.

3.1 Change Detection

CD is to use to understand the level of changes occur in the study area. The proper CD technology allows you to obtain and compare pairs or sets of RS images from different period of times. Spatial changes or Statistical can detect in the radiation, geometry or texture of the features (Fig. 12).

A   B

Fig. 11. CD result of agriculture land of (A) Landsat acquired in 2001 and 2009 (B) USGS acquired in 2011 and 2014

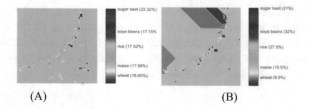

Fig. 12. Semantic classification using (A) ResNet 101 CNN of Landsat RS image; (B) ResNet 101 CNN of USGS RS image.

3.2   Image Classification Using CNN

The classification of the segmented images are used to find the coverage of biophysical territory, the socio-economic use of agriculture and to distinguish between the types of analysis. drawbacks are overcome and provide detailed mapping information of the segmentation using ResNet 101 CNN. Consequently, in this work, we classifies the crops as wheat, sugar beet, soybeans, rice, and corn shown in Fig. 11.

Figure 13 A (Dataset 1) shows the percentage of crops classified using CNN. It maps and classify the crops in percentage, in which levels are the predicted crop classification. The overall area is considered as 100%, and based on the percentage of area the crops are classified. Figure 13 B (Dataset 2) represents the map of Crop classification with percentage, in which levels are the predicted crop classification.

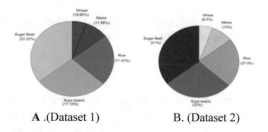

A .(Dataset 1)   **B. (Dataset 2)**

Fig. 13. Classification of crops in PercentageA (Dataset 1) and classification of crops in percentage B (Dataset 2).

Table 1 shows the evaluation of conditions of classified crops in percentage for dataset 1 and dataset 2. The table shows for finding the crops, its growth at different stages for each crop per hectare is identified. Then the total area in hectare in which each crop is grown over the benefited agriculture land is identified. It also explained about the condition of crops, whether it was fertile, moderate fertile or non-fertile.

Table 1. Evaluation of conditions of classified crops in percentage for Dataset 1 and Dataset 2.

	Source	Crop classes	Area	Fields	Percentage	Condition of crops		
						Fertile	Moderate fertile	Non-fertile
1	Dataset 1	Sugar beet	860.7	53	21%		✓	
2		Soya beans	3006.9	49	32%	✓		
3		Rice	296	98	27.5%	✓		
4		Wheat	3695.9	102	9.5%			✓
5		Maize	4329.1	8	10%			✓
1	Dataset 2	Sugar beet	786	89	35.5%	✓		
2		Soya beans	2991	54	28%	✓		
3		Rice	401	77	22%		✓	
4		Wheat	2499	112	5.5%			✓
5		Maize	4461	22	9%			✓

4   Conclusions

In this study, we developed the segmentation of land cover and the classification of crop types. RS images are very high-resolution images that are used to classify agriculture and land cover. our paper introduced the segmentation and classification of significant data without loss. In this document, we use SOM for segmentation purposes. We use a deep residual network to speed up the learning process. Residual deep networks have been demonstrated in the highly efficient RS image classification model. Overall the main definition of research activity is to get an accuracy of over 85%. Therefore CNN can go beyond the best classification of some types of crops and offers assignment accuracy of over 85% for all main crops.

References

1. Lavreniuk, M.S., Skakun, S.V., Shelestov, A.J.: Large-scale classification of land cover using retrospective satellite data. Cybern. Syst. Anal. **52**(1), 127–138 (2016)
2. Chen, Y., Lin, Z., Zhao, X., Wang, G., Gu, Y.: Deep learning-based classification of hyperspectral data. IEEE J. Sel. Top. Appl. Earth Obs. Remote Sens. **7**(6), 2094–2107 (2014)
3. Drusch, M., Bello, D., Colin, O., Fernandez, V.: Sentinel-2: ESA's optical high-resolution mission for GMES operational services. Remote Sens. Environ. **120**, 25–36 (2012)
4. Roy, D.P., Welder, M.A., Loveland, T.R.: Landsat-8: science and product vision for terrestrial global change research. Remote Sens. Environ. **145**, 154–172 (2014)

5. Stokes, G.M., Schwartz, S.E.: The atmospheric radiation measurement (ARM) program: programmatic background and design of the cloud and radiation test bed. Bull. Am. Meteor. Soc. **75**(7), 1201–1222 (1994)
6. Kussul, N., Lavreniuk, M., Skakun, S., Shelestov, A.: Deep learning classification of land cover and crop types using remote sensing data. IEEE Geosci. Remote Sens. Lett. **14**(5), 778–782 (2017)
7. Cheng, G., Zhou, P., Han, J.: Learning rotation-invariant convolutional neural networks for object detection in VHR optical remote sensing images. IEEE Trans. Geosci. Remote Sens. **54**(12), 7405–7415 (2016)
8. Inglada, J.: Automatic recognition of man-made objects in high resolution optical remote sensing images by SVM classification of geometric image features. ISPRS J. photogram. Remote Sens. **62**(3), 236–248 (2007)
9. Mather, P.M., Koch, M.: Computer Processing of Remotely-Sensed Images: an Introduction. John Wiley and Sons, Hoboken (2011)
10. Focareta, M., Marcuccio, S., Votto, C., Ullo, S. L.: Combination of Landsat 8 and Sentinel 1 data for the characterization of a site of interest. a case study: the royal palace of Caserta, October 2015
11. Grant, K., Siegmund, R., Wagner, M., Hartmann, S.: Satellite-based assessment of grassland yields. Int. Arch. Photogram. Remote Sens. Spat. Inf. Sci. **40**(7), 15 (2015)
12. Whitehead, K., Hugenholtz, C.H.: Remote sensing of the environment with small unmanned aircraft systems (UASs), part 1: A review of progress and challenges. J. Unmanned Veh. Syst. **2**(3), 69–85 (2014)

Motion Blur Detection Using Convolutional Neural Network

R. B. Preetham$^{(\boxtimes)}$ and A. Thyagaraja Murthy

Department of E&C, Sri Jayachamarajendra College of Engineering, JSS Science
and Technology University, Mysuru, India
preethamrb1995@gmail.com, trm.sjce@gmail.com

Abstract. In this paper, we identify movement obscure from a solitary, hazy picture. We propose a profound learning way to deal with and anticipate the likelihood dissemination of movement obscure at the fix level by utilizing a Convolutional Neural Network (CNN). The design we follow will moved toward the issue by cutting 100 pictures into $30 \times 3\,0$ fixes and connected our movement obscure calculation to them (with an irregular rate of half). At that point named the hazy and non-foggy patches with 1 s (0 for still, 1 for hazy), and stacked the adjusted pictures as our preparation information. In this Paper, we aim to estimate blurred motion from a single blurry image and propose an in-depth learning approach to predict probabilistic patch level movement blur distribution using a Convolutional Neural Network (CNN).

Keywords: Convolutional Neural Network · Motion deblur · Python · NumPy · OpenCV

1 Introduction

The convolution neural network (CNN) is a class of deep neural networks, most ordinarily connected to examining visual symbolism and imagery. CNNs are regularized forms of multilayer perceptron's. Multilayer perceptron's ordinarily allude to completely associated networks, that is, every neuron in one layer is associated with all neurons in the following layer. The "completely connectedness" of these networks makes them inclined to overfitting information. The representative method of regularization incorporates accumulation of some type of size estimation of loads to the loss work. Be that as it may, CNNs adopt an alternate strategy towards regularization: they exploit the progressive example in information and collect increasingly complex examples utilizing littler and less difficult examples. Consequently, on the size of connectedness and unpredictability, CNNs are on the lower extreme. The Fig. 1 depicts the convolution neural network. The picture deblurring goes for recuperating a sharp picture from an obscured picture in view of camera vibration, entity movement or out-of-center. This paper centres around assessing as well as evacuating spatially shifting movement obscure. Non-uniform deblurring has pulled in much consideration as of late. Techniques in work on erratic haze caused by camera turns, in-plane interpretations or forward out-of-plane interpretations. its compelled for evacuating non uniform haze predictable through these movement suppositions. One more class of methodologies

D. J. Hemanth et al. (Eds.): ICICI 2019, LNDECT 38, pp. 669–676, 2020.
https://doi.org/10.1007/978-3-030-34080-3_75

deals with non-uniform movement obscure brought about by item movement. They gauge obscure parts by examining picture measurements, obscure range, or with a learning approach exploitation carefully assembled alternative.

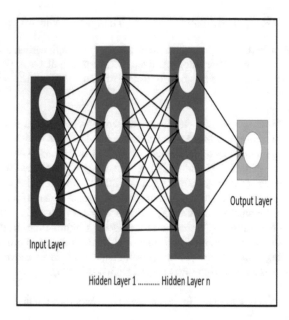

Fig. 1. Convolutional Neural Network.

Different methodologies together gauge the sharp picture and obscure bits exploitation a sparsity earlier. Today is as yet testing to evacuate unequivocally uneven movement obscure caught in complex scenes. This work proposes a innovative profound learning-based way which deals with assessing non-uniform movement obscure, trailed through fix insight based deblurring model adjusted for non-uniform movement obscure. We gauge the conceivable outcomes of movement bits at the fix level utilizing a convolutional neural system (CNN), at that point intertwine the estimations based on fix into a thick area of movement bits utilizing a Markov arbitrary field (MRF) model. To completely use the CNN, this work tends to propose to build applicant movement portion set expected by CNN misuse an image turn method, which fundamentally helps its exhibition to the movement part estimation. Exploiting the solid element learning intensity of CNNs, we can well anticipate the difficult movement obscure that can barely be all around evaluated by the cutting-edge approaches.

2 Literature Survey

This section provides a review of literature to set a foundation of discussing various Motion Blur Detection aspects.

Estimating precise movement blur kernels for non-uniform picture deblurring is essential. In [4, 5, 9, 10, 13], heterogeneous motion haze remains shapely as a worldwide motion of camera, which essentially evaluates a uniform kernel in the camera motion space. Approaches in [6, 7, 12] mutually evaluates sharp image and motion kernels. [6, 7, 12] depend on a sparsity preceding deduce the idle sharp image for higher motion kernel estimation. In this paper we use different approach, we assess obscure moving kernels legitimately utilizing the local patches which do not require a camera motion estimation or an inert sharp picture.

Other kinds of methods [1, 3] approximates dimensionally varying motion blur which supports innate image options. In the Fourier remodel region, the Approach [1] evaluates movement blur aided blur spectrum screening of picture patch. [8] by the use of natural image statistics it examines kernel motion blur and [3] by analyzing the alpha maps of image edges. [2] gives a regression perform which estimates motion blur kernel supported around overseen options. Different to them, in this paper we tend to estimate motion blur kernels using convolutional neural network, accompanied by a fastidiously constructed movement kernel expansion technique, and use MRF model to forecast thick movement kernel fields.

Our approach will well measure confounded and obscure robust motion, which can hardly be evaluated by previous methodologies. Some related research on learning-based deblurring methods has been done since late. [8] introduces a discriminative deblurring strategy to exploitation of standardized CRF models for uniform haze evacuation. [9] proposes a neural system approach for learning a denoiser to stifle noises during deconvolution. [11] styles an image deconvolution neural system for non-daze deconvolution. These above-mentioned methodologies focuses on structuring a superior model of standardized haze evacuation based on learning. Our methodology is used to estimate and expel obscurely uniform movements more effectively and our CNN-based methodology offers a powerful technique for dealing with this issue.

3   Generating the Data for Training

We generated the training data using images from the Pascal Visual Object Classes Challenge 2010 (VOC2010) data set. Our work was done in Python using the PIL, numpy, opencv, and os libraries. Once we had the original images from Pascal, we had to modify them to fit our needs. We needed to have 100 images, each partially blurred and with a corresponding matrix indicating which part of the image is blurred (Fig. 2).

We achieved this by:

- Making a blurred copy of the original image.
- Cutting both images (original and blurry) into 30 × 30 patches.
- Creating a 2D List in Python of size 30 × 30, to represent each image patch We initialize each element to 0 (to represent non-blurry).
- Picking half the patches from the list and marking them as 1 (to represent blurry).

- Putting the final image together to get a partially-blurred, qualifying image (and its corresponding matrix) (Fig. 3).
- Saving the image as "n.jpg" (n is image's serial number), in addition adding matrix to a list (to form a 3D 'list of lists') containing the matrices of all the image (Fig. 4).

Fig. 2. Images of the original to blur process

Fig. 3. Image splitting process

Fig. 4. Final image with its corresponding matrix

4 Learning the Convolutional Neural Network

Once we had the prepared images, we loaded them into our training set. We ran into a problem loading the images into a NumPy array, where our images were of the form (30, 30, 3), while the Keras.Conv2D layer required input to be of the form (3, 30, 30). We solved this by using the numpy.swapexes() function to alter the images' shape in order to fit the convolutional layer.

We Then Apply the CNN Learning Model. First, we apply a Convolution 2Dlayerwith 7×7 filters, followed by a ReLU function. The Conv layer's parameters comprised of a lot of learn capable filters. Each filter is little spatially, however reaches out through the input volume depth.

During the Forward Pass, we slide each filter over the width and height of the volume's input and evaluate dot products between the filter's entries and the input at any position. ReLU is the rectifier function-an activation function that can be used by neurons, just like any other activation function. A node victimization the rectifier activation operate is named a ReLU node. ReLU sets every single negative an incentive in the grid x to 0, and every other value are kept consistent. ReLU is evaluated after the convolution, and in this way a nonlinear actuation work (like tanh or sigmoid).

After that, we add a MaxPooling2D layer by means of pool size of 2×2. MaxPooling is a sample-based discretization method. The goal is to down-example an input depiction, diminishing its dimensionality and considering suspicions to be made about highlights contained in the binned sub-districts.

We then add Dropout layer with dropout rate of 0.2, which makes our learning process faster. Dropout randomly ignoring nodes is useful in CNN models because it prevents interdependencies from emerging between nodes. This permits the network to learn more and form a more robust relationship. We then do the'Conv2D, ReLU, MaxPooling2D, Dropout' circle again. Finally, we add a fully-connected layer with ReLU, then SoftMax the result. SoftMax can be defined as a classifier towards the end of neural network—a logistic regression to regularize outputs to a value between 0 and 1.

We Set Our Model's Learning Rate to be 0.01. This might generally be too big, but we made this decision for the sake of brevity - it was the fastest way to show a result. We chose a batch size of 126 (because we had large training data). We also chose Adam as our optimizer as it's the most efficient optimizer for our model (Fig. 5).

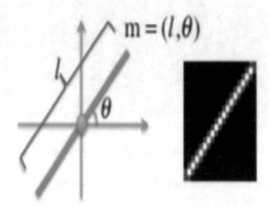

Fig. 5. Motion kernel represented by a motion vector

5   Results

After training with 100 epochs, we had testing accuracy of 92%, which is a very optimal rate for our model. Our training model is saved in an HDF5 file, "motionblur. h5" (Fig. 6).

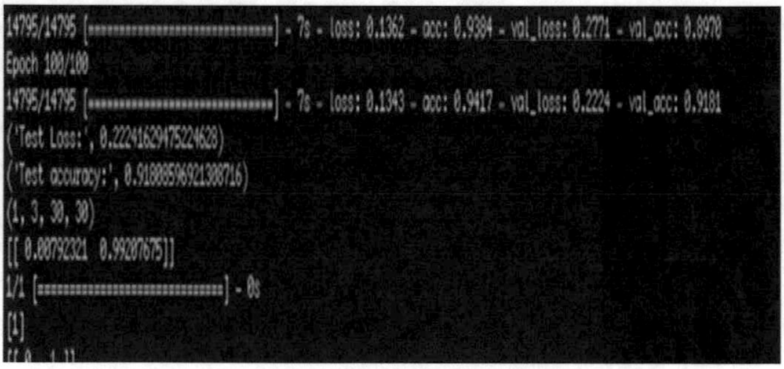

Fig. 6. Accuracy output

In this Paper, the designed model can qualitatively compare the approaches. These findings obviously indicate that the movement blur extraction findings of our strategy can be considerably better than the other referred approaches.

Input Image   Output Image

6   Conclusion

In this paper, we have come up with a novel CNN-based motion deblurring approach. This prompt cutting edge motion deblurring results. We sliced 100 pictures into 30×30 patches and applied our own movement blur algorithm to them at a random pace of 50%. We then marked blurry and non-blurry pictures with 0 s and 1 s (0 for still and 1 for blurry) and then loaded the altered pictures as our training data. In the future, we are keen on structuring a CNN for assessing the general non-uniform haze kernels. We are additionally interested in designing a CNN framework that can gauge and evacuate general hazy spots in a solitary system.

References

1. Chakrabarti, A., Zickler, T., Freeman, W.: Analyzing spatially-varying blur. In: CVPR, pp. 2512–2519 (2010)
2. Couzinie-Devy, F., Sun, J., Alahari, K., Ponce, J.: Learning to estimate and remove the non-uniform image blur. In: Foster, I., Kesselman, C. (eds.) The Grid: Blueprint for a New Computing Infrastructure. CVPR, 2013. Morgan Kaufmann, San Francisco (1999)
3. Dai, S., Wu, Y.: Motion from blur. In: CVPR (2008)
4. Gupta, A., Joshi, N., Lawrence Zitnick, C., Cohen, M., Curless, B.: Single image deblurring using motion density functions. In: ECCV (2010)
5. Hirsch, M., Schuler, C., Harmeling, S., Scholkopf, B. Fast removal of non-uniform camera shake. In: ICCV (2011)
6. Ji, H., Wang, K.: A two-stage approach to blind spatially-varying motion deblurring. In: CVPR (2012)
7. Kim, T.H., Lee, K.M.: Segmentation-free dynamic scene deblurring. In: CVPR (2014)
8. Schmidt, U., Rother, C., Nowozin, S., Jancsary, J., Roth, S.: Discriminative non-blind deblurring. In: CVPR (2013)
9. Tai, Y., Tan, P., Brown, M.: Richardson-lucy deblurring for scenes under a projective motion path. IEEE T. PAMI 33(8), 1603–1618 (2011)
10. Whyte, O., Sivic, J., Zisserman, A., Ponce, J.: Non-uniform deblurring for shaken images. IJCV 98(2), 168–186 (2012)

11. Xu, L., Ren, J.S., Liu, C., Jia, J.: Deep convolutional neural network for image deconvolution. In: NIPS (2014)
12. Xu, L., Zheng, S., Jia, J.: Unnatural l0 sparse representation for natural image deblurring. In: CVPR (2013)
13. Zheng, S., Xu, L., Jia, J.: Forward motion deblurring. In: ICCV (2013)

Machine Learning Based Predictive Models and Its Performance Evaluation on Healthcare Systems

K. Veerasekaran[1]([⊠]) and P. Sudhakar[2]

[1] Department of Computer Science,
Thiru. A. Govindaswamy Government Arts College, Villupuram, India
veeradevihar@hotmail.com
[2] Department of Computer Science and Engineering,
Annamalai University, Chidambaram, India
kar.sudha@gmail.com

Abstract. In previous days, the growth in various technologies and the huge data generation has produced a drastic development in database and sources. Medicinal area characterizes a rich data field. A wide-ranging quantity of medicinal data is presently offered, reaching from details of medical indications to several kinds of medicinal data and creation of image capturing components. The manual extraction of medicinal designs is a hard job due to the nature of medicinal field includes enormous, lively, and difficult information. DM is accomplished to improve the value to extract medicinal designs. This paper outlines the applications of DM on the groups of diseases is projected. The key effort is to examine machine learning (ML) models commonly utilized for the prediction, prognosis and treatment of significant standard diseases like heart diseases, cancer, hepatitis, and diabetes. A set of different classifier models is applied to examine their efficiency. This examination distributes a complete investigation of the recent position of diagnosing diseases by the use of ML models. The attained detection rate of the numerous applications ranges between 70–100% based on diseases, utilized data and methods.

Keywords: Classification · Disease diagnosis · Diabetes · Machine learning

1 Introduction

In the earlier days, growing of data is observed in all fields and nearly all categories. Medicinal data is dramatically improved in the previous days due to a drastic development of data in medicinal field. The volumes of data created by medical field contacts are very difficult and large to be process and analyze by classical approaches. This data frequently hides valued information. Medicinal research people face an issue to find significant information from this enormous volume of resource. So, medical information is a quickly rising method which is afraid of employing computer science to health information.

Healthcare data is the domain of information science concerning with the examination, utilization and diffusion of medicinal information uses computers to several

D. J. Hemanth et al. (Eds.): ICICI 2019, LNDECT 38, pp. 677–684, 2020.
https://doi.org/10.1007/978-3-030-34080-3_76

views of medical field [1]. Health information is described as "every aspect of realizing and boosting the efficient association, investigation, administration, and utilizing data in medical field". It includes the utilization of information for discovering and managing the new data linking to health and disease. A suitable computer-based system and effective systematic procedures can support to learn significant unseen data from massive medical dataset [2]. At present, data mining (DM) is become prevalent in medical field.

DM offers technologies and methodologies to convert huge information into beneficial data to make decisions. DM is described as "a method of nontrivial mining of understood, earlier unidentified and potentially valuable data from the information stored in a database". A wider process is the core step that is known as data discovery in dataset. This procedure contains the purpose of some previous-processing approaches predictable at simplifying the usage of the DM procedure and post processing techniques pointed to filtering and refining the realized information. Developing the predictive approaches from several medical data resources is probable utilizing knowledge detection or DM method built on dissimilar ML models and the estimated precision of the outcome smart system could even reach higher correctness. This method is identifying correlations or designs between diverse areas in huge medical databases. DM is the medical techniques can be employed to analyze and discovery unseen designs private patients' database or medical dataset. Inspired by widely growing of cancer, hepatitis, heart disease and diabetes patients in annually and the accessibility of enormous number of patients' information is utilized by scientists for assisting professional medicinal care in the disease management.

This examination makes investigations being carried out utilizing various ML models. The remaining sections in the study are presented here. Initially, a brief introduction of the methodologies and the diseases concentrated in this paper is provided in Sect. 2. The employed methodologies and the filtered data are provided in Sect. 3. Then, the investigation of experimental outcome is eloborated in Sect. 4. The conclusion is given by Sect. 5.

2   Background Information

2.1   Adopted Diseases

The aim of DM to diverse disease is fastly spreading, in this analysis, heart diseases, Cancer and Hepatitis are attempted as generally spread diseases globally and growing mortality diseases based on World Health Organization.

2.2   Adopted Diseases

CVD is the important reason of deadly people in over world for previous 10 years. The value of 17.5 m public died from heart disease in 2012, represented around a total of 31% of all global deaths. It is determined that 6.7 million were died due to stroke and 7.4 million deaths due CVD. The three portions of CVD death revenue place in lower and middle income countries. Maximum heart diseases banned by resolving personal risk criteria like smoking, physical inactivity, over weightiness, unhealthy food and injurious usage of alcohol. Heart disease is related to an illness of the blood vessels, heart and comprises heart attacks, heart failure and hypertension [3].

2.3 Hepatitis

Hepatitis is the main hazardous disease and death reason globally is particularly Hepatitis C. It is infectious liver diseases which could varies in sternness from a slight infection exists for some days to sever, longlasting illness. The hepatitis C virus is generally banquet in case blood from a diseased people enters into the body to a vulnerable person. Annually, hepatitis C virus is diseased around 3–4 million people. Almost 150 million people are chronically sick and suffering from liver cirrhosis or liver cancer. Annually, at least 3,50,000 people affected from liver diseases correlated hepatitis. A total of six genotypes of hepatitis C are present and it could be answer contrarily to medication [4].

2.4 Cancer

İt is an important reason of illness and death of universe with nearly 14 million new patients and 8.2 million cancer concerned death in 2012. The latest case count is predicted to increase by 70% in the upcoming twenty years. Amongst men, maximum 5 common organs of cancer identified in 2012 are colorectum, stomach, lung, prostate, and liver cancer. It is predictable that case count will be increased to a drastic rate annually. Cancer is a common word for a huge set of diseases which disturbs the organ of the body. An important characteristic of cancer is fast formation of cancer cells which raise outside their normal boundary and then attack detached part of the body and spreads to additional organs, the final method is denoted as metastasizing that is the main reason of deaths from cancer.

2.5 Diabetes

Diabetes is a metabolic illness where a body is incapable to control sugar categorized by hyperglycaemia i.e. higher blood glucose levels resultant to imperfections while secreting insulin. The chronic hyperglycaemia of diabetes is related with disorders of carbohydrates, fats and protein metabolisms will leads to lasting injury, dysfunction and various organ failures, especially the nerves, kidneys, blood vessels, eyes, and heart. The body wants insulin to utilize sugar, fats and proteins from the diet for day to day activities [1]. In case of medical development, still many of diabetes mellitus are unaware as various signs are identical to other diseases. So, it takes extended interval for diagnosis of such a diabetes disease [2]. Hence, there is a necessity to propose a professional system that indicates the risk of disease and also supports to discover the solutions using knowledge. The evaluation paper we focus on prevailing approaches for diabetes detection so as to know the recent progresses in the field of diabetes under healthcare. After analyzing all the methods utilized to diagnose diabetes disease, we create still some effort in diagnosing the disease with high correctness. This process is performed generally on SGPGI dataset or PIMA Indian dataset.

3 Machine Learning Models

Machine learning is used for training the structure over the huge database, where used ML models can be utilized to extract designs or construct a model and utilize to created designs or models for making prediction in the upcoming in definite cases. The dataset utilized to study the models is called as the training dataset. The record producing the training set is denoted as training examples and casually nominated the sample instances. The techniques are made cooperatively using training dataset. As the values or class labels of every individual training instance offered, this stage is called as supervised learning.

The dataset utilized to evaluate the model quality is called as test dataset. It is utilized to evaluate the analytical correctness of the techniques. Test sample is randomly nominated and self-determining of training sample. The outcome of predictive techniques on a specified test set is % of test instances which are properly forecasted by the techniques. When the precision of the sample is measured adequate, the techniques can be utilized to categorize or estimate upcoming information record or objects for that the class labels or values are indefinite. Figure 1 illustrates the various steps of learning. The diverse ML models can be associated and estimated by the following conditions:

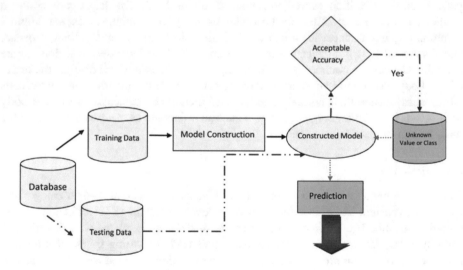

Fig. 1. Various steps of learning process

- Speed: Calculation cost includes creating as well as utilizing the techniques.
- Scalability: Capability to develop the models proficiently provided huge quantity of information.
- Interpretability: The phase of understanding which are delivered by the techniques.
- Predictive accuracy: The model ability for properly predicting the class labels or the values of latest or unknown data.

3.1   K-Nearest Neighbour (K-NN)

K-NN is an easy process which saves ever accessible case and categorizes fresh cases using a comparison metric. It is known as memory based classification since training samples required to be in memory at execution periods for categorizing unlabelled objects, the distance of the object and labelled objects is determined, its K-NN is recognized and the class which looks maximum among the neighbor's area allocated the component to be classified. Diverse formula can be utilized for computation of the distance from test as well as training instances; Euclidean distance is general formula. The construction techniques are inexpensive, but categorizing unidentified objects are moderately expensive. Some main problems disturb the process of K-NN, like the k values. When k is very large, the boundaries among the classes are lesser distinct. When k is very low, the outcome is influenced by noise points (i.e. the neighborhood might contain numerous points from other classes). Under dual categorization, it is good to consider odd values for k, it will avoid similar votes. The precision of K-NN is also influenced by unrelated features or when feature scale is unreliable through their significance, the distance undergoes domination using huge scaled features [5].

3.2   Artificial Neural Network (ANN)

ANN is the important method to technical information, which has ability to hold and denote difficult input or output relationship. The intention of NN lies in the generation of a system which performs proper mapping of inputs to outputs by the use of data so the model could be utilized for creating the output when the anticipated output is not known [6]. ANN is constructed to multilayer of nodes which are connected to one another. It involved an "input" layer, many "hidden" layers, and the "output" layer to characterize outcome. The neuron count in a layer and the layer count is based on difficulty of considered model. The neuron cells in input layer accept the information and handover it to the neuron cells in primary unseen layers over weight link. The information is mathematically treated and outcome is shifted to neuron cells in the subsequent layer. The neuron cells in final layer provides outcome of the network. It is highly difficult to calculate ANN. The absence of description is restriction for ANN, it is similar to "black box". Therefore, so it is complex to understand the methods and conclusions of every node in the network and also the authentication of these network is not a simple process.

3.3   Associative Classification (AC)

This classifier method intends to determine smaller collection of rules in dataset known as class association rules (CAR), to build a classification model. Constructing a predictive technique using AC includes two stages: (1) generation of CARs and (2) construction of a classification model from the created CARs.

3.4   CARs Creation

Creation of CARs is completed by concentrating on distinct subsets of suggestion rules where the right handed side is controlled to categorization of classes. Then, the AC algorithm determines frequent rules substances (attribute values which comes through class labels over the customer defined thresholds) [7]. Using the adequate frequent rule objects, CARs is produced.

3.5   Decision Tree (DT)

DT is a flowchart like structure, where every inside node present in the tree signifies a examination on any attribute, the branches from node label, probable results from test node, and the leaves of the tree represent classes [8]. The root node is uppermost node in a tree. For classifying the unidentified samples, the decision tree will test the sample's attribute value. A route tracking takes place from root to leaf nodes which carry the class prediction of the samples. Overfitting and class overlapping are the two features which restricts the use of DTs. In addition, it is not easier to optimize the DT.

4   Performance Validation

4.1   Dataset

Pima Indian dataset is utilized for building the prediction model of diabetes. The purpose of this database is to analyze the patient for diabetes based on specific diagnostic measures includes in this dataset. While choosing a sample from large dataset, many limitations were accepted. In this case, every patient is female particularly lesser than the age of 21 years of Pima Indian. The dataset includes the many medical predictor variables and one target parameter as output. These variables comprised of age, pregnancy count, BMI, level of insulin, and so on. Table 1 reports the information about the applied dataset. A total of 768 instances are present in the dataset along with the set of 8 attributes. Among them, around 34.90% and 65.10% of instances falls into the categories of positive and negative samples respectively. In addition, a set of two classes are present in the dataset.

Table 1. Dataset details

Description	Dataset-1
No. of instances	768
Attributes count	8
Class count	2
Percentage of positive samples	34.90%
Percentage of negative samples	65.10%
Data sources	[9]

4.2 Results Analysis

This section provides a detailed performance analysis of various classifiers on the applied diabetes dataset. Table 2 gives the results analysis of various classification models in terms of accuracy.

Table 2. Performance comparison of various models

Classifiers	Accuracy
K-means+LR	95.42
Artificial intelligence	94.03
ANFIS+KNN	80.00
ANN+PCA	89.20
CART	78.39

Figure 2 show comparison of the performance of various techniques on the employed dataset in terms of accuracy. The table values indicated that the presented k-means+LR model displays higher classifier outcome by achieving an accuracy of 95.42%. Nonetheless, the Artificial intelligence model showed somewhat better performance by achieving a slightly lower accuracy value of 94.03%. Concurrently, the ANN+PCA model tries to manage the classification procedure by attaining a moderate accuracy value of 89.2%. On the other hand, the ANFIS+KNN model results showed better performance over the CART by getting the accuracy value of 80%. Then, CART model performance is very low by attaining a minimum accuracy of 78.39% compared to other models.

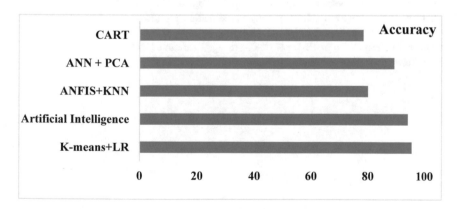

Fig. 2. Comparative results of various classifiers interms of accuracy

5 Conclusion

In this paper, a performance comparison of various classification models namely ANN, K-NN, DT, AC to diagnose, prognose, and treat cancer, heart, hepatitis and diabetic diseases are presented. The simulation outcome depicted that the K-means+LR model shows superior classification with the maximum accuracy of 95.42 whereas the other classifiers namely AI, ANFIS+KNN, ANN+PCA and CART obtains the accuracy values of 94.03, 80, 89.2 and 78.39 respectively. These values indicated that the K-means+LR model is found to be an appropriate tool for disease diagnosis. In future, this model can be implemented in medical organizations for diagnosing the patient's disease in real time.

References

1. National Library of Medicine (2017). http://www.nlm.nih.gov/tsd/acquisitions/cdm/subjects 58.html. Accessed 14 Nov 2017
2. American Medical Informatics Association (2017). http://www.amia.org/informatics/. Accessed 12 Nov 2017
3. World Health Organization: Cardiovascular diseases (CDV's) (2017). http://www.who.int/mediacentre/factsheets/fs317/en/. Accessed 15 May 2017
4. World Health Organization: Cancer (2017). http://www.who.int/mediacentre/factsheets/fs297/en/. Accessed 13 Feb 2017
5. Medjahed, S.A., Saadi, T.A., Benyettou, A.: Breast cancer diagnosis by using k-nearest neighbour with different distances and classification rules. Int. J. Comput. Appl. **62**(1), 1–5 (2013)
6. Amato, F., López, A., Peña-Méndez, E.M., Vanhara, P., Hampl, A., Havel, J.: Artificial neural networks in medical diagnosis. J. Appl. Biomed. **11**(2), 47–58 (2013)
7. Thabtah, F.A., Cowling, P.I.: A greedy classification algorithm based on association rule. Appl. Soft Comput. **7**(3), 1102–1111 (2007)
8. Bramer, M.: Principles of DM, vol. 180. Springer, London (2007)
9. https://www.kaggle.com/uciml/pima-indians-diabetes-databas

A Comprehensive Study on Distributed Denial of Service Attacks in Internet of Things Based Smart Grid

A. Valliammai[1], U. Bavatharinee[1], K. Shivadharshini[1],
N. Hemavathi[1], M. Meenalochani[2]([✉]), and R. Sriranjani[1]

[1] School of EEE, SASTRA Deemed to be University, Thanjavur, India
[2] Department of EEE, Kings College of Engineering, Thanjavur, India
meenalochani79@gmail.com

Abstract. Internet of Things is one of the emerging fields and it is expected that more than 50 million devices are connected through internet by 2020. In addition, extending internet connectivity into various applications will add comfort to our day to-day life. One such application is smart grid which necessitates advanced metering infrastructure for monitoring and control operation. However, Internet of Things poses a serious threat since the data generated from these devices grow longer thereby resulting in big data which is to be stored in database and further analyzed for control. If any malicious activity occurs in either database or during communication, then it would result in major disaster. Therefore, mechanisms for secure data communication are mandate. Furthermore, security mechanism should be cognitive as Internet of Things involves diverse devices. Based on the literature, it is revealed that Distributed Denial of Service is one of the most prevalent attacks in Internet of Things applications leading to unavailability of service to legitimate users. Hence, the key objective of the paper is to explore the distributed denial of service attack, its types and various mitigation methods. Furthermore, the countermeasures to prevent these types of attacks are addressed.

Keywords: Smart grid · Advanced Metering Infrastructure · Internet of Things · Distributed Denial of Service · Cyber attacks

1 Introduction

Smart grid, a type of Cyber Physical System (CPS) is a modern electric network with Advanced Metering Infrastructure (AMI). AMI facilitates computing and communication capability to conventional grids and is one of the emerging fields in power industry. Further, smart grid enables two way communication, i.e. from producers to consumers and vice-versa. To facilitate two way communication, all the devices in the network are equipped with sensors, processor and wireless connectivity. However in case of smart grid, a wide area electric network cannot be supported through wireless network alone due to its own limitations. Hence, internet connectivity is mandate and therefore, Internet of Things (IoT) forms the backbone of smart grid. Further, the objective of the smart grid is to realize monitoring and controlling of power consumption of devices to

© Springer Nature Switzerland AG 2020
D. J. Hemanth et al. (Eds.): ICICI 2019, LNDECT 38, pp. 685–691, 2020.
https://doi.org/10.1007/978-3-030-34080-3_77

achieve Demand Response (DR) or peak shaving. To attain this, the components of AMI such as smart meters, communication networks, data acquisition and data management systems are employed. AMI involves diversified hardware and software components for gathering, maintaining and controlling the energy management system. Further, smart grid can work automatically during power failures through smart detection schemes and also by proper load scheduling, power delivery even during peak hours can be attained. In addition, isolation of faulty parts of the electrical network can be done. In addition, the critical infrastructure with heterogeneous networks and devices facilitate self healing and resilient features to smart grid. As the monitoring and controlling of the entire grid relies on the trustworthiness of data and its delivery, the data obtained through AMI ought to be error free which would otherwise result in erroneous decision making. Further, the data from sensors grows at faster rate resulting in big data which is stored in cloud environment. Thus, the smart grid is prone to severe threats due to the framework of heterogeneous platforms. During transmission of power, mis-configuration of network connectivity may interrupt the service imposing vulnerability. In a similar way, privilege elevation or compromising the administrator due to poor network administration and communication shortcomings may result in unauthorized access to substations that would result in major disaster. In addition, as AMI involves diverse devices, upgrading and integrating new technology and facilities by all the devices may be impossible that would result in vulnerabilities. Remote monitoring and controlling of smart grid infrastructure involves variety of communication networks such as wireless, mobile, radio frequency transceivers and communication over power lines that rely on different protocols would also adds to vulnerabilities.

Physical threats, spreading malware, exploiting control system network are the major vulnerabilities incurred in smart grids of which IP spoofing, distributed denial of service, sybil attacks, botnets and attacks in data centres are gaining significance. In the available pool of vulnerabilities, Distributed Denial of Service (DDoS) attacks are ranked as the highest risk since it results in unavailability of service to authenticated users. Hence, the rest of the paper deals with research work that address cyber security threats.

2   Cyber Security Threats in Advanced Metering Infrastructure

Security solutions of traditional networks cannot be applied to grid networks because security in Information Technology (IT) networks concentrates more protection at the center of the network, while the protection in smart grid is accomplished at the network center and edge. Also, in IT networks rebooting devices in case of failure is allowed, while it is not acceptable in smart grids since services must be available at all times. Different types of cyber physical attacks in smart grid were discussed in the literature. Petrinets used for modeling attacks on the smart grid is investigated [1]. This method is novel where large petrinets can be created from tiny petrinets. The proposal is prescribed for smart meter attacks and its effectiveness is validated using Python. Attacks in electric grid with inaccurate information are identified using graph theory based approach [2]. Susceptibility of data communication to attacks in a network is modeled based on the techniques of discovery, access, possibility, data speed and attacks. Thus, the study of the effect due to integrated cyber-physical attack on smart grid is carried out. A Time Delay Switch (TDS) recovery protocol enhancement without the use of cryptography to sense and progress from the TDS attacks is proposed [3]. Simulation results prove the usefulness of the proposal against attacks which occur in load frequency control components of electrical power grid.

Cyber-attack models consisting of two stages for smart grid based on complete as well as partial information on the network are proposed [4]. The proposed cyber-attack models are validated through complete tests on different IEEE benchmarks. An intrusion detection and prevention system to propose counter measures against cyber attacks is dealt [5]. In this work, if a node bypasses necessary encryption and authorization mechanisms, then this scheme acts as secondary protection system to enable detection and counter measures. Unsupervised anomaly detection for large scale smart grids is attempted to distinguish a normal error and a cyber attack [6]. Attacks which occur due to insertion of wrong data are detected using a combined method of state estimation and cognitive dynamic system in a smart grid environment [7].

Of the existing cyber attacks, Distributed Denial of Service, which is one of the malignant weapons on the cyber attacks is elaborated in detail.

3 Distributed Denial-of-Service (DDoS) Attacks

DDoS is a kind of cyber attack in which the attacker tries to use all the resources of the server so that server is not able to respond to legitimate request as discussed earlier. Now-a-days, DDoS attacks are well targeted, causing only the targeted user to be affected not the entire network. Denial of Service (DoS) is a subversion of DDoS attack where a single attacker jams the network which denies the service to authenticated users. Further, DDoS has its domain for all network connected devices, one such domain which has been infected is smart grid, an AMI. As mentioned, cyber attacks are classified based on component, protocol and topology wise. In case of component based attack, the field components like Remote Terminal Units are attacked through remote access whereas injecting false data/mechanism in the communication protocols with malicious software falls under protocol based attacks. Topology based attacks involve exploitation of vulnerability through network topology. DDoS attack can reduce the speed of data communication in real time. This leads to wrong decision making by the central controller as they are unaware of the complete view of the electrical system.

They are classified based on the quantity of traffic and the defenseless being targeted.

SYN Flood: This is a form of DDoS attack in which the perpetrator sends a SYN requests to the system which is targeted in order to exploit all the server resources to make the system not responsive to the legitimate traffic. Then the victim sends an acknowledgement request as SYN-ACK back to the perpetrator. If the perpetrator responds with an ACK then the connection is established. The perpetrator send multiple SYN packets to the target frequently with spoofed IP address. The targeted victim is uninterrupted leaving the open connections until the ports become available again. This type of attack is known as "HALF OPEN ATTACK".

UDP Flood: UDP is known as the User datagram protocol. It is a continuous networking protocol in which the flood targets the random ports of the network with UDP packets. The victim server searches for the application in the ports but no application is found.

HTTP Flood: Hyper text transfer protocol appears to be a legal POST requests are destroyed by the hacker. It makes use of less bandwidth when compared to other types of DDoS attacks but this flood forces the server to utilize maximum resources.

Ping of Death: In this type of attacks the attackers manipulates the IP protocols by sending a vicious call to the system. Before two decades this ping of death is known as the popular attack but nowadays it becomes less effective.

SMURF Attack: A smurf attack decades the IP and ICMP where IP is referred to as the Internet Protocol and ICMP is the Internet Control Message Protocol by using a malware code called smurf. By using the ICMP it starts spoofing the IP address thereby causing the flood.

Application Layer Attacks: This type of DDoS attacks are designed to attack the application layer of the Open System Interconnection system by mainly focusing on the vulnerabilities or issues. Finally, the contents of the application are not delivered to the user. They attack the specific applications of DDoS most commonly the web services. Perpetrators try to develop new types of attacks and vectors to launch a new wave of attacks.

4   Detection Mechanisms

Detection mechanisms to DDoS attacks in smart grid were proposed in the literature. A handshaking technique using Merkle-tree which is found to improve the network resiliency when a DoS attack occurs is introduced [8]. A DoS attack model is developed and the effectiveness of the proposal against DoS attack is tested and also the performance of the network in terms of overhead and delay is improved. A mathematical model for security of the smart grid is presented [9]. Kalman filter is utilized to obtain the variables of state processes which are further given as input to the $\chi2$-detector for detecting attacks and faults including short-term attacks, DoS attack and long-term attacks. Jamming, a type of DoS attack interrupts the data transfers that happen in real time in smart grid. Using the electricity marketing model, for jamming attack scenario, an active game between the adversary and the target where the optimal policies are chosen by both sides to improve their profits is proposed [10]. Simulation results show that the jamming attack has affected the electricity prices as well as the profits of both sides. The effect of three probable cyber attacks on smart grid using a combination of cyber-power modeling and test bed through simulation is analyzed [11]. Man-in-the-middle and denial-of-service attacks are modeled using the proposed technique. Honey pots for detecting and gathering DDoS attack as a decoy system in the AMI network is presented [12]. The communication between the attackers and the defenders are studied and efficient techniques for both sides are obtained. The results prove that the proposal effectively improves the network efficiency after the deployment of honey pots. The effect of cyber attacks in a smart grid metering is presented [13]. DDoS attacks are modeled in a test bed scenario consisting of smart meters. The results show that during shifting of loads the risk of botnet attacks are reduced. Detection of false data injection attacks and DoS attacks online in a smart grid environment is analyzed [14]. The system is modeled as a discrete-time linear dynamic system and Kalman filter is used to perform state estimation. Results proved that proposed detectors are fast and accurate in detecting random attacks.

Detection of false-data injection and denial-of-service attacks through Signal Temporal Logic (STL) is attempted [15]. STL monitors the output voltages and currents of micro grids for deviation from the predefined values. Based on the deviation, the attack can be detected and intensity of attack can be measured. Denial of sleep, forge, and replay attacks are detected and isolated in wireless sensor network based smart grid applications [16]. Some of the detection techniques are pattern matching, clustering, statistical methods and deviation analysis. Detection in deviation is enhanced by using the past data as the threshold value which will be assigned to an argument by using the above methods mentioned to count the GET request received. Affinity propagation (AP) is a recently proposed clustering algorithm capable of handling massive data sets within a short duration to obtain results that are acceptable. AP is a clustering algorithm which deals with the idea of message passing between data points. With the growing complexity of DDoS attacks it is very important to find versatile methods of mitigation. The nature and working of DDoS attacks differ in each case. There are lot of ways to prevent the systems from these types of dangerous assaults.

5 Defense Mechanism

The main motive of DDoS defense system is to sense the attack as soon as possible. The attack is detected easily but no immediate measures can be taken to stop the attack. It is categorized into source-end, victim-end, core-end and hybrid-end defense mechanisms. In source-end defense mechanism, the defense architecture is applied at the sources to protect the systems from the attack whereas for victim-end defense mechanism, the defense architecture is applied at the routers to detect the vicious traffic of the victim network. The unique use of this mechanism is that the attack can be detected both in the online and offline modes. Further in core-end defense mechanism, individual routers try to distill the malicious packets from the traffic. The accuracy of this defense mechanism is better when compared to the other mechanisms. In addition, hybrid-end defense mechanism indicates many resources are available to detect the attacks.

A communication system with improved ability of recovering when jamming occurs is designed [17]. The proposal is capable of collecting enough readings continuously from smart meters for estimating the states of a smart grid for different attacks. It is proved that the proposal is successful in communicating between smart meters and controllers in the presence of multiple jammers in the network. An authentication protocol based on cloud-trusted authorities-based for providing authentications mutually in a distributed manner is discussed [18]. Results show that the protocol is capable of protecting the network against impersonation attacks, redirection attacks, DoS attacks man-in-the-middle attacks. In addition, it offers a complete protection against flood-based DoS attacks.

A complete and organized survey on the different attacks and protection mechanisms for a smart grid is provided [19]. Important cyber physical attacks which occur in the smart grid and the appropriate protection mechanisms are discussed. A technique for alleviating the DoS, in a microgrid is proposed [20]. The proposal is applied on a 25 kV microgrid and its efficacy is validated through hardware in real-time. An integrated approach to address cyber threats in power systems by interlinking the attack in Information and Communication layer and physical impact in physical layer is addressed [21]. Self-organizing communication architecture to mitigate impact of denial of service attack in smart grid is demonstrated using simulation. A mechanism to ensure multiuser authentication and DoS countermeasure algorithm for smart grid based on wireless sensor network is proposed [22].

6   Conclusion

This paper deals with the glimpse of security threats in internet of things based smart grids. The threats are categorized as physical, network and protocol based attacks. Among the pool of attacks distributed denial of service, one of the prevalent attacks is elaborated. A thorough survey on distributed denial of service attacks, its types, detection and defense mechanisms are dealt.

References

1. Chen, T.M., Sanchez-Aarnoutse, J.C., Buford, J.: Petri net modeling of cyber-physical attacks on smart grid. IEEE Trans. Smart Grid **2**(4), 741–749 (2011)
2. Srivastava, A., Morris, T., Ernster, T., Vellaithurai, C., Pan, S., Adhikari, U.: Modeling cyber-physical vulnerability of the smart grid with incomplete information. IEEE Trans. Smart Grid **4**(1), 235–244 (2013)
3. Sargolzaei, A., Yen, K.K., Abdelghani, M.N., Sargolzaei, S., Carbunar, B.: Resilient design of networked control systems under time delay switch attacks. Appl. Smart Grid **5**, 15901–15912 (2017)
4. Wang, H., Ruan, J., Wang, G., Zhou, B., Liu, Y., Fu, X., Peng, J.: Deep learning-based interval state estimation of AC smart grids against sparse cyber attacks. IEEE Trans. Ind. Inform. **14**(11), 4766–4778 (2018)
5. Radoglou-Grammatikis, P.I., Sarigiannidis, P.G.: Securing the smart grid: a comprehensive compilation of intrusion detection and prevention systems. IEEE Access **7**, 46595–46620 (2019)
6. Karimipour, H., Dehghantanha, A., Parizi, R.M., Choo, K.-K.R., Leung, H.: A deep and scalable unsupervised machine learning system for cyber-attack detection in large-scale smart grids **7**, 80778–80788 (2019)
7. Oozeer, M.I., Haykin, S.: Cognitive dynamic system for control and cyber-attack detection in smart grid. IEEE Access **7**, 78320–78335 (2019)

8. Hu, B., Gharavi, H.: Smart grid mesh network security using dynamic key distribution with merkle tree 4-way handshaking. IEEE Trans. Smart Grid 5(2), 550–558 (2014)
9. Manandhar, K., Cao, X., Hu, F., Liu, Y.: Detection of faults and attacks including false data injection attack in smart grid using Kalman filter. IEEE Trans. Control Netw. Syst. 1(4), 370–379 (2014)
10. Ma, J., Liu, Y., Song, L., Han, Z.: Multiact dynamic game strategy for jamming attack in electricity market. IEEE Trans. Smart Grid 6(5), 2273–2282 (2015)
11. Liu, R., Vellaithurai, C., Biswas, S.S., Gamage, T.T., Srivastava, A.K.: Analyzing the cyber-physical impact of cyber events on the power grid. IEEE Trans. Smart Grid 6(5), 2444–2453 (2015)
12. Wang, K., Du, M., Maharjan, S., Sun, Y.: Strategic honeypot game model for distributed denial of service attacks in the smart grid. IEEE Trans. Smart Grid 8(5), 2474–2482 (2017)
13. Sgouras, K.I., Kyriakidis, A.N., Labridis, D.P.: Short-term risk assessment of botnet attacks on advanced metering infrastructure. IET Cyber-Phys. Syst. Theory Appl. 1–9 (2017)
14. Kurt, M.N., Yılmaz, Y., Wang, X.: Distributed quickest detection of cyber-attacks in smart grid. IEEE Trans. Inf. Forensics Secur. 13(8), 2015–2030 (2018)
15. Beg, O.A., Nguyen, L.V., Johnson, T.T., Davoudi, A.: Signal temporal logic-based attack detection in DC microgrids. IEEE Trans. Smart Grid 10(4), 3585–3595 (2019)
16. Dhunna, G.S., Al-Anbagi, I.: A low power WSNs attack detection and isolation mechanism for critical smart grid applications. IEEE Sens. J. 19(13), 5315–5324 (2019)
17. Liu, H., Chen, Y., Chuah, M.C., Yang, J., Poor, H.V.: Enabling self-healing smart grid through jamming resilient local controller switching. IEEE Trans. Dependable Secure Comput. 14(4), 377–391 (2017)
18. Saxena, N., Choi, B.J.: Integrated distributed authentication protocol for smart grid communications. IEEE Syst. J. 12(3), 2545–2556 (2018)
19. He, H., Yan, J.: Cyber-physical attacks and defences in the smart grid: a survey. IET Cyber-Phys. Syst. Theory Appl. 1(1), 13–27
20. Chlela, M., Mascarella, D., Joós, G., Kassouf, M.: Fallback control for isochronous energy storage systems in autonomous microgrids under denial-of-service cyber-attacks. IEEE Trans. Smart Grid 9(5), 4702–4711 (2018)
21. Cameron, C., Patsios, C., Taylor, P.C., Pourmirza, Z.: Using self-organizing architectures to mitigate the impacts of denial-of-service attacks on voltage control schemes. IEEE Trans. Smart Grid 10(3), 3010–3019 (2019)
22. Afianti, F., Wirawan, W., Suryani, T.: Lightweight and DoS resistant multiuser authentication in wireless sensor networks for smart grid environments. IEEE Access 7, 67107–67122 (2019)

Dual Band Microstrip Patch Ultra-wide Band Antenna for LTE Wireless Applications

V. Deepthi Chamkur[1], C. R. Byrareddy[2], and Saleem Ulla Shariff[3(✉)]

[1] Department of E and C Engineering, VTU, Bangalore Institute
of Technology Research Centre, Bengaluru 560004, Karnataka, India
vdeepthichamkur@gmail.com
[2] Department of E and C Engineering, Bangalore Institute of Technology,
Bengaluru 560004, Karnataka, India
byrareddycr@yahoo.co.in
[3] Research Associate R&D, Bangalore Institute of Technology,
Bengaluru 560004, Karnataka, India
saleem_shariff@yahoo.co.in

Abstract. A Multi slot dual band with an ultra-wide band 'T' and 'G' slots shaped Micro-strip UWB Patch Antenna proposed has been discussed in this paper. The sizes of length & widths of the uwb-patch has been chosen and varied in such a manner that it occupies compact volume of $32 \times 28 \times 1.7$ (1523 mm^3) and it is designed on a substrate FR4-epoxy having a dielectric constant of $\varepsilon r = 4.4$. The antenna is capable to operate in dual band with second band being a wide band ranging from 3.59 GHz to 10 GHz. The first band operates in 2.32 GHz–2.48 GHz range and is obtained by slot size variations in the 'TG' slot geometry. The advantages of the proposed 'T' and 'G' shaped slots design is that of the two band obtained, one ultra-wide band frequency range of operation can be achieved without the slots size variations using the same specifications or dimensions thus overcoming the need for extra enhancement of the surface area while designing the antenna. TG shaped Antenna is covering applications from ISM 2.4 WLAN Band, LTE band No 40 and wide band wireless applications in 5.2/5.8 GHz ISM WLAN, Radio altimeters (4.2 GHz) and Wimax (3.5/5.55 GHz). Direct probe feeding method using a 50 Ω micro-strip line has been used with the width 3 mm for the micro-strip line. The Coupling between the two slots plays a better role for obtaining the wider bandwidth. The analysis of the parameters such as directivity, bandwidth, return loss (dB), gain and VSWR (Voltage Standing Wave Ratio) of the Microstrip patch antenna with 'T' & 'G' slot has been performed using HFSS v15 [24] tool. The obtained return losses (RL) and the radiations patterns are found to be suitable for the LTE operations and moderately omnidirectional in nature. With the variation in the length (L) and width (W) in the 'T' & 'G' slots shaped geometry, the performance of uwb patch antenna has been studied with the comparisons of the simulated results in this paper.

Keywords: RFID · Wireless · Line fed · Dual band · 'TG' shaped ·
Multi slot · Microstrip · FR4 epoxy · Patch · Ground plane · Wi-Fi · Wimax ·
LTE · ISM · Radio altimeters

© Springer Nature Switzerland AG 2020
D. J. Hemanth et al. (Eds.): ICICI 2019, LNDECT 38, pp. 692–703, 2020.
https://doi.org/10.1007/978-3-030-34080-3_78

1 Introduction

The current trend of modernization cum advancements in the technology in mobile devices related to communication, the wireless data connectivity capability for various ranges of applications related with personal and business [16] has improved a lot. It's evident from studies that the Micro-strip Patch antennas are known to have narrow bandwidths [3] which can be an hindrance for them to be opted for uses in some of the advanced modern wireless operations & applications [5–9]. Therefore, there has been an overwhelming increase of demand tremendously for the low-profile, easy to design & fabricate multi bands with ultra wide band broad bandwidth antennas, those can be useful & easily integrated within the devices of current generation having advanced modern communication.

The popular standard such as WiMAX/IEEE 802.16 standard [3] covers the 5.2 GHz and WLAN/IEEE 802.11b standard are known to cover the band for 2.4 GHz [1]. A variety of different techniques have been proposed and studies have shown to achieve the wider bandwidth of operation. Some of the various frequently employed techniques include changing the physical size of the antennas [2], modification of the radiating patch shapes to allow the paths of the current to travel at longer distances [3], and adding gaps or multi layers. The techniques employed also include creating the slots with different shapes such as using U-slot array [3], slots form [7], Y-V Slot [4], shorting wall [6], stacked patch [4], E-shaped or 4G Shaped [4] slots itched on patch mounted on thick substrates with 32 × 30 mm ground plane size [3]. Researchers opt for the Microstrip antennas [4] due to the features such as low cost, compactness, low profile, easy fabrication lightweight & modeling for their research [3]. The focus of current researchers is pretty much dedicated and focussed towards the design and development of multi-bands with ultrawide bandwidth antennas for Long Term Evolution (LTE) wireless related applications [3–17]. In this paper, a direct inline fed dual band wideband antenna [3] using TG shaped itched patch element on a compact rectangular shaped structure has been proposed. The two slots with shapes of 'T' and

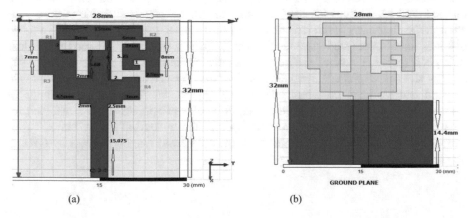

(a)   (b)

Fig. 1. The proposed design with dimensions for the 'T' and 'G' shaped MS patch on FR-4 substrate. (a) Radiating patch (b) Ground plane.

'G' shapes are employed on the micro strip radiating patch side for excitation [4] of the both the bands such a narrow band and wideband region. The proposed dual-band with one of the band containing wideband frequency range are obtained by variation of the slots of the TG shape and itching out the slots with rectangular shapes of specific length as per the Fig. 1(a–b). Thus the designed micro strip antenna can have applications in both narrowband applications [4] (i.e. the 2.4 GHz WLAN and 3.5 GHz WiMAX etc.) and ultrawideband operations in the frequency range from starting with 3.59 GHz to 10 GHz for other various ultra-band wireless standards. The proposed design is covering the 2.4 GHz WLAN/Bluetooth/Wi-Fi band in its first band which is ranging from 2.32 GHz to 2.48 GHz [3] and IEEE operating bands of 5.2 GHz/5.8 GHz WLAN/WiMAX in the wide band range including RFID, Radio altimeter bands [5] 3.5 GHz WiMax band in its second band which is an ultra-wide band. The proposed antenna has a simple design with use of direct line feeding method [19] in the antenna. It has been employed to provide the direct feeding with a 50 Ω (ohm) microstrip line [16]. The 'T' and 'G' slots shaped ultra-wide band antenna has been designed and analyzed using the HFSS – v.15 (High Frequency Structure Simulator – v15) Software design tool [24] which functions based on the principle of Finite Integration Technique [FIT] for analysis of the various parameters such as directivity [4], VSWR (Voltage standing wave ratio), bandwidth, return losses and gain of micro-strip patch antenna [19] along with current and electric field density distribution.

The compact 'T' and 'G' slots shaped dual band ultrawide band microstrip patch antenna has the volume including the ground-plane & the substrate 32 mm × 28 mm × 1.7 mm. The uwb antenna is composed of a substrate [14] having di-electric constant of 4.4 (FR4 epoxy) with the partial ground itched out and 'T' and 'G'-shape slotted radiating patch, which has the resistance and impedance bandwidths improved helping in covering the required frequency bands [6]. It is known fact that fine tuning in the antenna design gives good directivity and polarization [5]. The miniaturization of the sizes of L & W of the 'T' and 'G' slots of radiating patch is done to get flexibility [3] in the desired frequency for ease of operation in the wider band of frequencies of operating ranges and the same has been tuned for the mentioned listed defined applications [17]. Here in this paper, for the compact micro-strip 'T' and 'G' slots shaped line-fed dual band ultra-wide band antenna, the effects pertained to the relative variation in the slots sizes of the radiating patch of the antenna elements has been studied [4]. It is found in the study that, a resonant frequency with amazing good return loss with good radiation patterns has been obtained [25] for the design. The discussed 'T' and 'G' slot shaped slotted ultra-wide band Microstrip patch antenna has improved gain & band width with good return loss having VSWR < 2 for the required range of frequencies of operation. The Voltage Standing Wave Ratio VSWR < 2 obtained for the proposed design of the antenna results in the ultra-wide band to give return loss of almost below −10.0 dB throughout the ultra-wide band ranging from 3.59 GHz to 10 GHz [12].

2   Proposed Antenna Design

The proposed 'T' and 'G' shaped slots structured micro-strip uwb-patch antenna is as displayed in the Fig. 1(a). The proposed antenna consists of a 'T' & 'G' shape slotted radiating patch with a partially itched ground plane which is sandwiched between the FR4 epoxy substrate [3]. The 'T' and 'G' shaped slotted patch is made up of copper which is a conducting material [17] in the radiating patch and the ground plane which is also made of a conducting material (copper) is placed on the other side of the FR4 substrate as shown in the Fig. 1(b).

The substrate used in antenna can be composed of a dielectric material which are made using various type of materials [4], these materials can give varying dielectric constants and it can range between 2.2 to 12. The well-known dielectric materials used as a substrate by the researchers are Rogers RT/duroid 6006 (6.15), FR4 (Flame resistant-4) (4.4), etc. From the Figs. 1(a)–(b), for the proposed 'T' and 'G' slots shaped ultra-wide band patch antenna, as shown in the geometry with the dimensions, we can observe that it has volume of $32 \times 28 \times 1.7$ mm^3. The patch and the ground planes of the wide band antenna are sandwiched between a low profile, low cost substrate such as FR4 epoxy having the di-electric constant $\varepsilon r = 4.4$ with the thickness of substrate $h = 1.6$ mm [19]. The size of the partially itched ground plane with the width $W = 28$ mm and length $L = 14.4$ mm has the antenna being excited with a 50 Ω direct micro-strip line feeding technique [19]. From the study, it is observed that fine tuning in the antenna is required to obtain the desired resonances of frequency for the wider bands; hence the slots have been created of various shapes and dimensions. The dimensions of the Rectangular shapes itched out such as R1 is 3 mm $\times$ 3 mm, R2 is 2.5 mm $\times$ 3mm, R3 is 5 mm $\times$ 2.5 mm and R4 is 4.5 mm $\times$ 3.5 mm. With the variation in the L-> length of the partially itched ground planes [18] and the 'T' and 'G' slots size variation in the radiating patch, the return loss, the band width, gain and the VSWR of the antenna can been varied [3] (Table 1).

Table 1. The proposed antenna dimensions specifications.

Parameter	Specifications	Dimensions (mm)
Patch	T - shape slot	2×8 horizontal, 5.68×2 vertical
	G - shape slot	8×2 vertical
Ground plane	Partial itching	$14.4 \times 28 \times 0.05$
Feeding	Line feeding	15.075×3
Substrate	FR4 ($\varepsilon r = 4.4$)	$32 \times 28 \times 1.6$

3 Results Discussion

The Ansoft HFSS [24] tool has been used to design the 'T' and 'G' slots shaped dual band patch antenna and the simulated results have been discussed in this section. The Plots for Return loss patterns, gain, VSWR, radiation patterns, current & electric field density plots are simulated and extracted from HFSS tool [24] and are shown in respective figures in coming sections.

Figure 3 is the plot of the simulated result for the return losses v/s frequency plot extracted from the HFSS design tool [9]. After extracting the return loss and frequency data, the same plot has been plotted using excel for the analysis in Fig. 2. The ultra-wide band can be seen from 3.59 GHz and extends till 10 GHz. The Fig. 2 is the plot for the same extracted results plotted using Sigma plot or excel [9] software. The peak resonances are observed at 2.4 GHz (m6) and 4 GHz (m7) and moderate resonances at 2.32 GHz (m1), 2.4863 GHz (m2), 3.59 GHz (m3), 7.10 GHz (m4) and 9.9 GHz (m5) and the Return loss of −15.01 dB (m6) and −16.14 dB (m7) are obtained at those peak resonance frequencies respectively. As per the Figs. 2 and 3, we can observe two bands such as first one being a narrow band with a bandwidth of 164.8 MHz starting from 2.32 GHz till 2.4863 GHz and the second band being an ultra-wide band starting from the 3.59 GHz to 10 GHz coverage with a bandwidth BW of 6.40 GHz. The available bandwidth in the proposed antenna covers the frequencies range that can cover various wireless applications bands such as WLAN band/WiMAX band etc. The variations in the dimensions of the length L & the width W of the 'T & G' shaped slots with the variations in the size of length and Width for the shape of ground plane are found to have a good mutual effect with respect to impedance matching, gain, VSWR return loss, etc.

Figures 4 and 5 depicts the 'T' and 'G' slots shaped antenna & there results compared with respect to variations in the length, width and the size in slots shape. We can observe that the minute or small variations in the length L and width W (in mm) added to a lot of variations in the peak resonance & bandwidths of the frequency band of ultra-wide band as shown in Fig. 5. In Fig. 4 the plots for the different shapes as compared in the Fig. 5 has been drawn. It can be observed that the minute variations in the sizes of the structure of the 'TG' shape shown as in Fig. 6 have led to lot of variations in the peak resonance value of the ultra-wide band. As seen from Fig. 4, the comparisons between the return losses of the final structure with respect to the return losses of other four shapes have been studied. We can observe return loss variations happening in the wide band.

Let's study the effects of the variations of the sizes of the shapes from the shapes of Fig. 4. From the shapes & dimensions mentioned in the Fig. 4, shape 2 has given less bandwidth and the peak resonance return loss is quite good. With etching of rectangular shapes of sizes 3 × 3, 2.5 × 3 in the shape 2, the shape 3 has been obtained which further increases the bandwidth of the wide band and but the peak resonance drops a bit. Similar variations has been done in the other shapes as well and finally the final shape gave us two bands, Thus the uwb antenna can operate in dual band with second band being a wide band which ranges from 3.59 GHz to 10 GHz with considerable good return loss.

The first band operates in 2.32 GHz–2.48 GHz range, with peak resonance at 2.4 GHz with −15 db return loss and has been obtained by slot size variations in the 'TG' slot geometry. To get rid of the discontinuity in the wide band in the different shapes, i.e. at some points for the different sizes of shapes shown in Fig. 6, the return loss was approaching less than −10 dB return loss. Hence by trial method and with etching the small variations in the sizes of 'TG' shape has resulted us with the final structure with continuous band of ultra-wide band as shown in Fig. 1(a). Thus with these fine tuning in the shapes of the antenna, we have obtained the final continuous return loss v/s frequency band which has −10 dB of return loss with continuity throughout the ultra-wide frequency band. For further discussions, hence this structure has been chosen as the final structure [3]. Figures 4 and 5 discussed about the 'TG' shaped dual band antenna for the return losses v/s frequency results comparisons with respect to the results from final antenna shape. The minute changes in the sizes of length & width have resulted in the changes on the peak resonances in the dual band containing the ultra-wide band as shown in Fig. 5. Through experimentation, the Shape of Fig. 2 had resulted in more bandwidth but there is a small discontinuity in the return losses v/s frequency band that led to the appearance of two bands with one being the ultra-wide band which is starting from 3.59 GHz up-to 10 GHz. The rectangular slots introduced in the TG shape patch structures along with minute variations to the sizes of the length of the ground plane has resulted in good return loss v/s frequency in the ultra-wide band with one more narrow band resonating at 2.4 GHz.

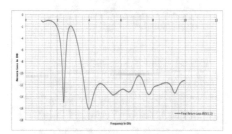 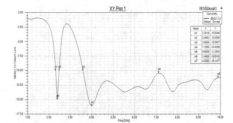

Fig. 2. Graph drawn for the plotting of return loss pattern v/s frequency using excel.

Fig. 3. 2D simulated graph of the return loss patterns v/s frequency extracted from the design tool.

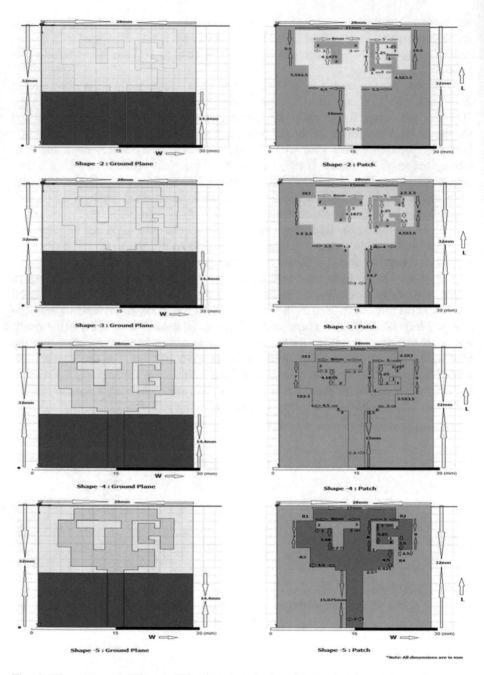

Fig. 4. The proposed 'T' and 'G' slots shaped ultrawide band micro-strip patch antenna structure comparisons with respect to the variations of lengths L and widths W.

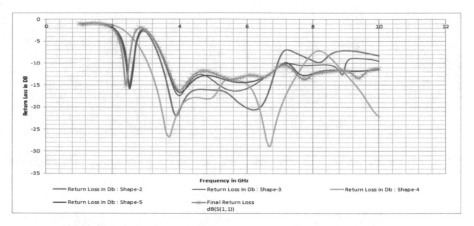

Fig. 5. The graph of the return loss v/s frequency comparisons with respect to the variations in the shapes.

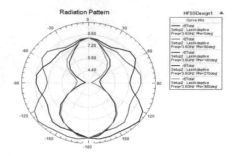

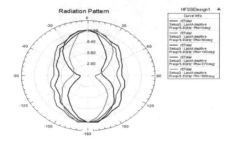

Fig. 6. Graph displaying the plot of radiation patterns plotted at phi = 0, 90, 180 and 270° for freq = 3.5 GHz

Fig. 7. Graph displaying the radiation pattern plotted at phi = 0, 90, 180 and 270° for freq = 5.8 GHz.

Figures 6 and 7 displays the plot of radiation patterns which has been simulated in the E-plane and the H-plane in HFSS tool [24] at 3.5 GHz and 5.8 GHz frequencies at phi = 0, 90, 180, 270° respectively [12]. From the plots, it can be noted and observed that the nature of Radiation patterns are almost omnidirectional and the plots pertaining to the 3D rectangular patterns have been shown in Fig. 8, and it has depicted a very good polarization for proposed 'T' and 'G' slots shaped dual band ultra wide band antenna.

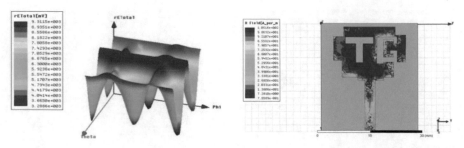

Fig. 8. Plot for the simulated 3D rectangular pattern.

Fig. 9. The graph of magnetic field distribution

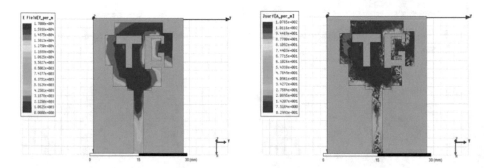

Fig. 10. The graph of electric field distribution.

Fig. 11. Graph of surface current density distribution.

In HFSS tool [24], surface current density, magnetic field and electric field [3] distribution plots has been simulated for the proposed TG shaped ultra-wide band antenna for various frequencies of operation such as 3.5 GHz etc., are shown in Figs. 9, 10 and 11. It can be easily noted from the Fig. 11, that the surface current density for the antenna is maximum [5] at the edges of the radiating patch where as low or minimal at the centre. Figure 12 plots the simulated results for the VSWR. As shown in the fig, we can note that the VSWR (Voltage Standing Wave Ratio) for the proposed 'T' and 'G' slots shaped micro-strip patch antenna is less than 2, which is one of the important conditions to be satisfied to obtain the required bandwidth for the specified frequency ranges [3]. The Fig. 13 is the plot for the stacked gain simulated from the HFSS for the proposed design for freq = 3.5 GHz at phi = 0, 90, 180, 270 and 360°.

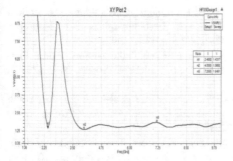
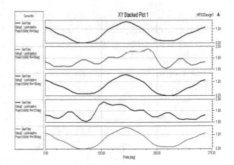
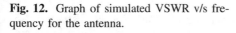

Fig. 12. Graph of simulated VSWR v/s frequency for the antenna.

Fig. 13. Graph of simulated stacked gain v/s frequency at 3.5 GHz.

4 Conclusion

The 'Dual band Microstrip Patch Ultra-wide band Antenna for LTE Wireless Applications' paper discusses about how the proposed antenna designed can be harnessed for various operations related to wireless applications such as Zigbee/Bluetooth, IEEE WLAN (2.4 GHz) Wi-Fi, RFID, Radio altimeter and WLAN/Wimax LTE applications [3]. In this paper, the proposed dual band ultra-wide band patch antenna uses the typical 'TG' shaped patch elements with the etched out rectangular slots etched on the copper plated patch with a compact volume and the variations of the length and width of the slots, these effects have been studied using simulation with the use of Ansoft HFSS [24] tool. The various parameters with the help of the HFSS tool [24] such as VSWR, Gain, Current & Electric field distribution, radiation patterns [9], return loss [18], of the designed 'T' and 'G' slots shaped ultra-wide band patch antenna were validated with the help of simulation. Through the variation of the length L & width W of the slots in the radiating patch and the variation of the shape of ground plane structure, the frequencies with a good bandwidth(BW) range in the ultrawide band antenna can be controlled easily [3]. From the analysis of the results, we can note that the main advantages of the proposed 'TG' shaped wide band antenna is its light compact weight, low profile with low cost, optimal radiation patterns. Good return loss (dB) with VSWR < 2 has helped the proposed design suitable comfortably for its use in various wireless and other applications like WLAN/WiMAX, RFID, Bluetooth, LTE, Radio Altimeter, 5G & others etc.

Acknowledgment. All the authors are thankful and gratefully acknowledge the support encouragement & feedback received from the colleagues from the Electronics & Communication department, Bangalore Institute of Technology Research Centre, Bengaluru, India.

References

1. Kumar, G., Ray, K.P.: Broadband Microstrip Antennas, pp. 18–23. Artech House, Boston (2003)
2. Pozar, D.M., Schaubert, D.H.: Microstrip Antennas. IEEE Press, New York (1995). https://doi.org/10.1109/9780470545270

3. Deepthi Chamkur, V., Byrareddy, C.R.: 4G shaped wide band patch antenna for wireless applications. In: 2018 3rd International Conference on Communication and Electronics Systems (ICCES) (2018). https://doi.org/10.1109/CESYS.2018.8723907

4. Abutarboush, H.F., etal.: A reconfigurable wideband and multiband antenna using dual-patch elements for compact wireless devices. IEEE Trans. Antennas Propag. **60**(1) (2012). https://doi.org/10.1109/TAP.2011.2167925

5. Zhang, X.Y., Zhang, Y., Pan, Y.-M., Duan, W.: Low-profile dual-band filtering patch antenna and its application to LTE MIMO system. IEEE Trans. Antennas Propag. **65**(1) (2017). https://doi.org/10.1109/TAP.2016.2631218

6. Wang, H., Huang, X.B., Fang, D.G.: A single layer wideband U-slot microstrip patch antenna array. IEEE Antennas Wirel. Propag. Lett. **7**, 9–12 (2008). https://doi.org/10.1109/LAWP.2007.914122

7. Laila, D., Sujith, R., Nijas, C.M., Anandan, C.K., Vasudevan, K., Mohanan, P.: Modified CPW fed monopole antenna with suitable radiation pattern for mobile handset. Microwave Rev. (2011). https://doi.org/10.1109/ICCSP.2011.5739340

8. Abutarboush, H.F., Nilavalan, R., Cheung, S.W., Nasr, K.M., Peter, T., Budimir, D., Al-Raweshidy, H.: A reconfigurable wideband and multiband antenna using dual-patch elements for compact wireless devices. IEEE Trans. Antennas Propag. (2012). https://doi.org/10.1109/TAP.2011.2167925

9. Nithya, D., Chitra, M.P.: Performance analysis of filtering antennas. In: 2017 International Conference on Wireless Communications, Signal Processing and Networking (WiSPNET) (2017). https://doi.org/10.1109/WiSPNET.2017.8299711

10. Guo, Y.-X., Khoo, K.-W., Ong, L.C.: Wideband dual-polarized patch antenna with broadband baluns. IEEE Trans. Antennas Propag. **55**(1), 78–83 (2007). https://doi.org/10.1109/TAP.2006.888398

11. Gosh, S.: Ultra wideband planar crossed monopole Antenna as EMI Sensor. Int. J. Microwave Opt. Technol. **7**(4), 247–254 (2012)

12. Ban, Y.-L., Chen, J.-H., Ying, L.-J., Li, J.L.-W.: Ultrawideband antenna for LTE/GSM/UMTS wireless USB dongle applications. IEEE Antennas Wirel. Propag. Lett. **11**, 403–406 (2012). https://doi.org/10.1109/LAWP.2012.2192470

13. Garg, R., Bhartia, P., Bahl, I., Ittipiboon, A.: Microstrip Antenna Design Handbook. Artech House, Boston (2001)

14. Oppermann, I., Hamalainen, M., Inatti, J.: UWB Theory and Applications. Wiley, New york (2004). https://doi.org/10.1002/0470869194

15. Byrareddy, C.R., Reddy, N.C.E., Sridhar, C.S.: A compact dual band planar RMSA for WLAN/WIMAX applications. Int. J. Adv. Eng. Technol. **2**(1), 98–104 (2012). ISSN 2231-1963

16. Balanis, C.A.: Advanced Engineering Electromagnetics. Wiley, Newyork (1989)

17. Krishna, D.D., Gopikrishna, M., Anandan, C.K.: A CPW fed triple band monopole antenna for WiMAX/WLAN applications slot antenna. In: IEEE Proceedings of 38th European Microwave Conference, pp. 897–900 (2008). https://doi.org/10.1109/EUMC.2008.4751598

18. Foudazi, A., Hassani, H.R., Nezhad, S.M.A.: Small UWB planar monopole antenna with added GPS/GSM/WLAN bands. IEEE Trans. Antennas Propag. **60**(6), 2987–2992 (2012). https://doi.org/10.1109/TAP.2012.2194632

19. Lee, S., Park, H., Hong, S., Choi, J.: Design of a multiband antenna using a planner inverted-F structure. In: Proceedings of the 9th International Conference on Advanced Communication Technology, vol. 3, pp. 1665–1668 (2007). https://doi.org/10.1109/ICACT.2007.358690

20. Song, K., Yin, Y.Z., Chen, B.: Triple-band open slot antenna with a slit and a strip for WLAN/wiMAX applications. Prog. Electromamagn. Res. Latt. **22**, 139–146 (2011). https://doi.org/10.2528/PIERL11011406

21. Bhave, M.M., Yelalwar, R.G.: Multiband reconfigurable antenna for cognitive-radio. In: 2014 Annual IEEE India Conference (INDICON) (2014). https://doi.org/10.1109/INDICON.2014.7030609

22. Sigmaplot and Microsoft excel software

23. Ansoft HFSS (High Frequency Structure Simulation) V-15 tool by Ansoft: HFSS user manual. Ansoft Corporation, USA

24. Help Share Ideas. http://www.helpshareideas.com

25. Gurubasavanna, M.G., Shariff, S.U., Mamatha, R., Sathisha, N.: Multimode authentication based electronic voting kiosk using Raspberry Pi. In: 2018 2nd International Conference on I-SMAC (IoT in Social, Mobile, Analytics and Cloud) (I-SMAC) (2018)

Fingerprint Based Cryptographic Key Generation

K. Suresh[1,2(✉)], Rajarshi Pal[2], and S. R. Balasundaram[1]

[1] Department of Computer Applications, National Institute of Technology,
Thiruchirapalli, India
suresh.1088@gmail.com
[2] Centre of Excellence in Cyber Security,
Institute for Development and Research in Banking Technology, Hyderabad, India

Abstract. Secure storage of a cryptographic key is a challenging task. In this paper, a fingerprint based cryptographic key generation approach has been proposed. It can be used in a symmetric key setup. The use of gray code makes the approach interesting, as gray code reduces the mismatch of binary representations of two successive integers. It has been experimentally shown that it is impossible to generate the same key from the fingerprint of an intruder. Moreover, it does not require to store neither the fingerprint of the user nor the cryptographic key. It makes the scheme more secure.

Keywords: Cryptographic key generation · Bio-cryptosystem · Securing cryptographic key

1 Introduction

Cryptographic techniques play an important role in securing information either in transit or storage. Cryptographic keys are at the core of these techniques. Secrecy of these keys need to be maintained. The public key in an asymmetric key system is an exception which is shared publicly. But secure storage of these keys (the symmetric key for a symmetric key system and the private key for an asymmetric key system) is a challenge. Bio-cryptosystem is a field of study where biometrics of a user is used to secure these cryptographic keys. There can be three types of bio-cryptosystem techniques: (i) key release, (ii) key binding and (iii) key generation.

In a cryptographic key release based bio-cryptosystem, access to the cryptographic key is restricted using biometric-based authentication. During the authentication process, the acquired biometric template is compared against a stored reference template which was acquired during enrollment. If the acquired template matches with the stored template, the cryptographic key is released to the user. The major drawback in this scheme is that both the biometric template and the cryptographic key are stored in the database. A database compromise may leak the cryptographic key as well as biometrics.

© Springer Nature Switzerland AG 2020
D. J. Hemanth et al. (Eds.): ICICI 2019, LNDECT 38, pp. 704–713, 2020.
https://doi.org/10.1007/978-3-030-34080-3_79

In cryptographic key binding scheme, the cryptographic key is hidden using biometrics of an user. This binding process results in an unification of the secret key and the biometric data. The result of this fusion is termed as helper data. Hence, it does not divulge either the key or the biometric data. A key retrieval algorithm can extract the same key with the help of helper data. There are multiple ways of binding cryptographic key with biometrics. They are (i) biometric encryption [1] (ii) fuzzy vault [2,3] and (iii) fuzzy commitment [4,5]. The main idea behind the key binding technique is to lock the secret key using biometric information. The privacy of randomly generated key and the complexity of key binding algorithm are the major aspects which govern the security of the key and the biometric template.

In cryptographic key generation scheme, a secret key is generated from biometrics. Several approaches of cryptographic key generation from fingerprint biometrics can be found in [6–13]. There are two ways of generating the secret key. Firstly, the user specific keys are extracted directly from the biometric features [6]. In this approach, authors have proposed a scheme in which the unique code is generated from the extracted fingerprint features using the convolution coding technique. Further, the cryptographic key is generated from the unique code using standard hash function. The use of strong hash algorithms to generate secret key from unique code provides non-repudiation and resistance to brute force attack.

Secondly, the secret key is generated with the use of helper data (obtained from the given biometric data) and the quantized biometric features [7–9]. This helper data is stored in the database during enrollment. A weakness of helper data based scheme is the probability of recuperating original biometric images from stored templates.

In [10], authors proposed a key exchange protocol to integrate the biometric data of two communicating parties (sender and receiver). A secret key is generated from the combined fingerprints of two parties which is further used for secure data communication. A cancellable fingerprint template has been used in this context. Additionally, a cancellable fingerprint based session key generation between sender and receiver is proposed in [11–13]. From the literature, it is clearly shows that if the biometric information is compromised by the attacker then the secret key will also be compromised.

Apart from fingerprint, the cryptographic key can be generated from other biometric traits like face [14], ECG [15,16], handwritten signature [17], etc.

The proposed approach, in this paper, presents a novel technique of generating cryptographic key from fingerprint biometrics. The use of gray code in this approach is an interesting aspect of the proposed method. Gray code reduces the mismatch of bits in binary representations of two consecutive integers. Hence, it reduces the mismatch of bits in the cryptographic keys, which are inherently introduced due to uncertainty in biometrics. An analysis is carried out to reveal that an impostor cannot generate the same cryptographic key as an authentic user.

The rest of the paper is organized as follows. Section 2 describes the proposed method of cryptographic key generation. Section 3 reports experimental results and analysis. Section 4 concludes about the contribution of this paper.

2   Proposed Method

In this paper, an attempt has been made to generate a cryptographic key directly from the unique extracted features of user's fingerprint. The key contribution of this method is that neither cryptographic key nor original biometric data is saved anywhere. So, this method ensures the privacy of user's biometric data as well as solves the cryptographic key storage issue. An overview of the proposed method is schematically shown in Fig. 1.

2.1   Pre-processing

Few pre-processing steps are carried out on the captured fingerprint image such as contrast enhancement, binarization and thinning. Contrast enhancement improves the quality of the fingerprint by reducing the undesired effects of scars, creases, pores, blurs and incipient ridges. It helps in improving the detection of minutiae points. Binarization converts the gray scale fingerprint image into a binary image containing the ridges and valleys. Thinning reduces the width of the binarized ridge lines into 1-pixel. Binarization and thinning processes are required to locate minutiae points.

2.2   Minutiae Extraction

In a fingerprint, there exist variety of minutiae points like ridge ending, ridge bifurcation, island, pore, lake, core point, delta point etc. These minutiae points can be defined as:

- Ridge ending - where the ridge line terminates
- Ridge bifurcation - where the ridge line splits into two ridges
- Island - a ridge line between two deviated ridges
- Pore - a small hole in ridge lines
- Core point - the top most point on the innermost ridge
- Delta point - tri-radial point with three ridges radiating from it

These minutiae points are shown in Fig. 2. The most widely accepted minutiae points are ridge ending and ridge bifurcation. Hence, in this proposed approach, only ridge ending and ridge bifurcation points are considered. The thinned binary image is analyzed to identify the localized pixel pattern where ridge ends or splits. Next, all the false minutiae points like (side minutiae, points detected in poor quality regions, islands, lakes etc.) are removed.

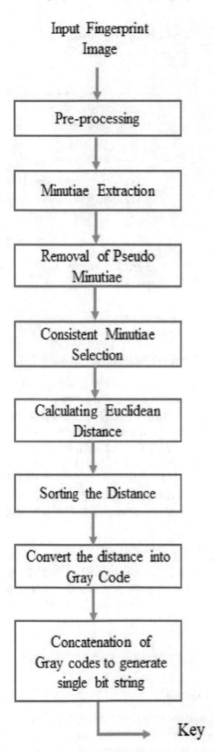

Fig. 1. Block diagram of the proposed cryptographic key generation process.

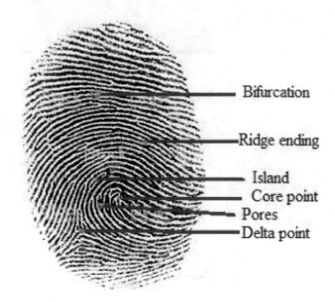

Fig. 2. Minutiae points.

2.3   Consistent Selection of Minutiae Points

After removing false minutiae points, only good quality and the true minutiae points are obtained. Due to inherent uncertainty of biometrics, the acquired fingerprint images of the same subject may not always contain exactly the same set of minutiae points. Since the objective of the proposed approach is to generate a stable cryptographic key, the set of minutiae points being used for key generation must be always the same. So, in this method, a set of minutiae points are manually selected within the common region of all the fingerprint images of the same subject. This ensures that the same set of minutiae points is always selected for the fingerprint images of a particular subject.

2.4   Euclidean Distance Calculation

After selecting the consistent set of minutiae points, the Euclidean distance between the coordinates of each pair of minutiae points are calculated. Let two minutiae points P_i and P_j be located in the coordinates (x_i, y_i) and (x_j, y_j), respectively. The Euclidean distance is computed as:

$$d_{i,j} = \sqrt{(x_i - x_j)^2 + (y_i - y_j)^2} \tag{1}$$

2.5   Sorting of the Distances

If there are n selected minutiae points from a fingerprint, $^nC_2$ such distances are calculated by considering the distance between each pair of minutiae points. Then, these $^nC_2$ distance values are sorted in ascending order.

2.6   Converting Distance into Gray Code

Each of the distance is converted into an m-bit binary number. Using m-bits, a maximum distance value $2^m - 1$ can be represented. Then the binary number is converted into gray code. Gray code is called as unit distance code in which the successive code words differ only by one bit. This provides a major boost to the proposed method as slight change in the distance value will have minimum effect on the generated bit string.

For example, a distance value 11 is represented as '00001011' and a distance value 12 is represented as '00001100' in 8-bit binary. Though the distance changes by 1 from 11 to 12, the corresponding binary strings differ in three positions. The 8-bit gray code representation of 11 and 12 are '00001110' and '00001010', respectively. Hence, these two gray code representations differ in only one position. Hence. the usage of gray code representation reduces the error arising in slight variation in distances of a pair of minutiae points in two different instances of the fingerprint.

Let the m-bit binary string be '$b_0 b_1 b_2 b_3 .... b_{m-1}$'. The corresponding gray code is '$g_0 g_1 g_2 g_3 .... g_{m-1}$'. The conversion of binary to gray code is stated below:

$$g_0 = b_0$$
$$g_i = b_{i-1} \oplus b_i$$

where $\oplus$ represents bit-wise exclusive-OR operation.

2.7   Concatenation of Gray Codes to Generate the Key

Gray codes (in m-bit representation) of the sorted distances are concatenated to obtain a bit string which acts as the cryptographic key. It is to be noted that there are $^nC_2$ distance values for n minutiae points. Hence, the length of the key is: $m \times {^nC_2}$ bits.

3   Experimental Results and Analysis

This section presents the experimental observations. The openly available public database Fingerprint Verification Competition (FVC) 2002 Database DB1 [18] is used to generate cryptographic key. This DB1 consists of 80 fingerprints of 10 persons with 8 instances of each person. Hence, there are total 80(10 × 8) fingerprint images in DB1 database. Each fingerprint image is of size 388 × 374, which were captured at 500 dots per inch (dpi).

In the proposed approach, ridge ending and ridge bifurcation are considered as the minutiae points. For extracting the minutiae points from fingerprint image, NIST's NBIS software tool FpMV (Fingerprint Minutiae Viewer) [19] has been used. This tool takes a fingerprint image as input and gives minutiae points as output with x and y coordinate values, angle, quality and type of minutiae. The sample result of using this tool is shown in Fig. 3. In the proposed method, x and y coordinate values are used to generate the keys. Few more implementational

Fig. 3. Minutiae detection using FpMV tool.

details are as follows: 7 minutiae points are selected from each fingerprint image by observing the point-to-point correspondence among extracted minutiae points from various fingerprint images of same person. Hence $^7C_2 = 21$ distances are obtained. 8-bit representation is used for the gray code. It is experimentally observed that all the distance measures are in the range $[0, 255]$. Hence, 8-bit representation is sufficient to encode the distances. Hence, the length of the obtained key is: $21 \times 8 = 168$ bits.

Due to inherent variation in capturing any biometric signal, there will be difference among the generated key strings from multiple fingerprint instances of the same signal. This may be corrected using error correction codes (such as Reed-Solomon, Hadamard code). But feasibility of the proposed scheme is studied by estimating the mentioned difference among the key strings. Hamming distance is used to estimate the said difference between a pair of key strings obtained from two fingerprint images of the same person. Hamming distance computes the count of dissimilar bits in two binary strings. In the DB1 database, there are 8 fingerprint images for each of 10 persons. Hence, $^8C_2 = 28$ Hamming distances can be obtained between each pair of fingerprints of a person. Hence, total 280 such comparisons (using Hamming distance) are possible. But it has been observed that 12 of those total 80 fingerprint images are either of low quality or contain partial fingerprint. Hence, desired minutiae points could not be extracted from those 12 images. Therefore, a total of 205 pairs (out of 280 possible pairs) of keys are compared using Hamming distance. The observed maximum Hamming distance (as indicated in Table 1) is 20, the average distance being 12 (approximated to the nearest integer).

Table 1. Hamming Distance (HD) comparison between genuine and impostor cryptographic key.

	Minimum HD	Average HD	Maximum HD
Genuine fingerprint	4	12	20
Impostor fingerprint	40	65	81

Moreover, it is also assessed whether there will be resemblance between the generated keys from fingerprints of two different persons (i.e., one is an authentic user and the other is an impostor). Hence, Hamming distance is compared between a pair of keys generated from two fingerprints of two different persons. A total of 5760 such comparisons are possible, as each fingerprint of a person is compared with every fingerprint of every other person in the database (i.e., 8 fingerprints × 9 impostors = 72 experimental comparison for each fingerprint). Hence, maximum number of possible comparisons is 5760 (72 comparisons/fingerprint × 80 fingerprints). But as desired minutiae points could not be extracted from 12 out of 80 available fingerprints, the actual 4146 comparisons are carried out for this impostor category. The minimum Hamming distance (as indicated in Table 1) for this impostor category is observed as 40, average distance being 65 (approximated to the nearest integer).

Distributions of the Hamming distances of two categories (genuine and impostor) are presented in Fig. 4. Here, the genuine category refers to the case

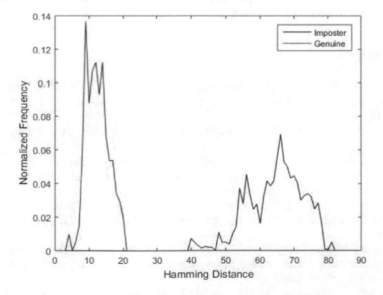

Fig. 4. Distribution of Hamming distances between cryptographic keys for genuine and impostor cases.

that keys from two fingerprints of the same person are compared. The impostor category refers to the case that keys from two fingerprints of two different persons are compared. It can be clearly seen that the distributions of the Hamming distances of these categories are well separated. It indicates the feasibility of the proposed scheme. Because the key generated by the fingerprint of an impostor is very different than a key generated from another fingerprint instance of the same person.

4   Conclusion

Secure storage of the cryptographic key is a challenging task. The proposed method provides a solution to this problem by generating the cryptographic key from fingerprint. Hence, neither the cryptographic key nor the fingerprint template is stored in the database. Hence, it removes the possibility of stealing the cryptographic key by an attacker. As a single key string is obtained in the proposed approach, it can be used in a symmetric key setup. Usage of gray code is an important aspect of the proposed method as it reduces the mismatch between the keys being generated from multiple instances of same fingerprint.

An analysis has been reported to highlight that the mismatch between a pair of keys due to an impostor is higher than the mismatch due to the inherent variation in the acquired fingerprint images. Therefore, an impostor cannot generate the same key as generated by an authentic person. Moreover, the minor variation in the keys from fingerprints of the same person can be corrected using error correction codes. This makes the proposed scheme very interesting.

In future, the research will focus on following two aspects: (1) consistent selection of minutiae points and (2) incorporation of error correction code to obtain a stable cryptographic key.

References

1. Soutar, C., Roberge, D., Stoianov, A., Gilroy, R., Kumar, B.V.: Biomteric encryption - enrollment and verification procedures. In: Proceedings of SPIE, pp. 24–35 (1998)
2. Juels, A., Sudan, M.: A fuzzy vault scheme. Des. Codes Crypt. **38**(2), 237–257 (2006)
3. Jin, Z., Teoh, A.B.J., Goi, B.M., Tay, Y.H.: Biometric cryptosystems: a new biometric key binding and its implementation for fingerprint minutiae-based representation. Pattern Recogn. **56**, 50–62 (2016)
4. Juels, A., Wattenberg, M.: A fuzzy commitment scheme. In: Proceedings of the 6th ACM Conference on Computer and Communications Security, pp. 28–36. ACM (1999)
5. Sasa, A., Milosavljevic, M., Veinovic, M., Sarac, M., Jevremovic, A.: Fuzzy commitment scheme for generation of cryptographic keys based on iris biometrics. IET Biom. **6**(2), 89–96 (2016)
6. Panchal, G., Samanta, D., Barman, S.: Biometric-based cryptography for digital content protection without any key storage. Multimed. Tools Appl. 1–22 (2017)

7. Panchal, G., Samanta, D.: A novel approach to fingerprint biometric-based cryptographic key generation and its applications to storage security. Comput. Electr. Eng. **69**, 461–478 (2018)
8. Panchal, G., Samanta, D.: Comparative features and same cryptographic key generation using biometric fingerprints. In: International Conference on Advances in Electrical, Electronics, Information, Communication and Bio-Informatics, Chennai, India, pp. 691–695 (2016)
9. Sarkar, A., Singh, B.K., Bhaumik, U.: Cryptographic key generation scheme from cancellable biometrics. In: Pattnaik, P., Rautaray, S., Das, H., Nayak, J. (eds.) Progress in Computing, Analytics and Networking. Advances in Intelligent Systems and Computing, vol. 710. Springer, Singapore (2018)
10. Barman, S., Chattopadhyay, S., Samanta, D., Panchal, G.: A novel secure key-exchange protocol using biometrics of the sender and receiver. Comput. Electr. Eng. **64**, 65–82 (2017)
11. Sarkar, A., Singh, B.K.: Cryptographic key generation from cancelable fingerprint templates. In: International Conference on Recent Advances in Information Technology (RAIT), pp. 1–6. IEEE (2018)
12. Sarkar, A., Singh, B.K.: A cancelable biometric based secure session key agreement protocol employing elliptic curve cryptography. Int. J. Syst. Assur. Eng. Manag. 1–20 (2019)
13. Sarkar, A., Singh, B.K.: A cancelable fingerprint biometric based session key establishment protocol. Multimed. Tools Appl. 1–27 (2019)
14. Wu, L., Liu, X., Yuan, S., Xiao, P.: A novel key generation cryptosystem based on face features. In: IEEE 10th International Conference on Signal Processing, pp. 1675–1678. IEEE (2010)
15. Karimian, N., Guo, Z., Tehranipoor, M., Forte, D.: Highly reliable key generation from electrocardiogram (ECG). IEEE Trans. Biomed. Eng. **64**(6), 1400–1411 (2016)
16. Moosavi, S.R., Nigussie, E., Levorato, M., Virtanen, S., Isoaho, J.: Low-latency approach for secure ECG feature based cryptographic key generation. IEEE Access **6**, 428–442 (2017)
17. Yip, W.K., Goh, A., Ngo, D.C.L., Teoh, A.: Generation of replaceable cryptographic keys from dynamic handwritten signatures. In: International Conference on Biometrics, pp. 509–515. Springer, Heidelberg (2006)
18. http://bias.csr.unibo.it/fvc2002/databases.asp
19. https://www.nist.gov/services-resources/software/fingerprint-minutiae-viewer-fpmv

Efficient Deduplication on Cloud Environment Using Bloom Filter and IND-CCA2 Secured Cramer Shoup Cryptosystem

Y. Mohamed Sirajudeen[✉], C. Muralidharan, and R. Anitha

DST Cloud Research Lab, Department of Computer Science and Engineering,
Sri Venkateswara College of Engineering, Sriperumbudur 602 117,
Tamilnadu, India
ducksirajsmilz@gmail.com, murali20infotech@gmail.com,
ranitha@svce.ac.in

Abstract. Cloud Service Providers (CSP) offers several services over the internet to the cloud users. One of the predominant services offered to the cloud users is data storage. With the rapid growth of digital data, redundancy has become the bottleneck issue in cloud computing environment. Storing identical copies of user data increases the storage overhead. The traditional way of performing deduplication is by comparing the cipher texts, which will result in high time complexity and heavy load on users. To utilize the cloud storage effectively and to reduce the work load of cloud users, a novel inline data deduplication method is created based on Bloom filter. The proposed scheme combines the application of bloom filter and Cramer Shoup cryptosystem to perform secure inline deduplication. Two matrices are created such as keyword matrix and file matrix based on the application of the bloom filter to perform deduplication on the cloud storage. The empirical study of the proposed deduplication technique is carried out on the Eucalyptus v4.2.0 open source private cloud on top of the Xeon processor with 64 GB of RAM Speed. It has been observed that the proposed deduplication improves the efficiency and effectiveness of the deduplication compared to the traditional deduplication algorithms.

Keywords: Data deduplication · Redundancy · Cloud computing · Cramer shoup encryption technique

1 Introduction

Cloud Computing offers several services over the internet based on the demand raised by the cloud users ranging from smaller backup services to larger services. The desirable properties of cloud computing like rapid elasticity, measured service, resource pooling and fault tolerance has made it as a promising technology [1]. The important service offered by the cloud provider is storage, in which user can store their data in the cloud and allows to access it from anywhere. The data are resided on different geological locations which makes the system vulnerable against the cyber-attacks. It also opens a gate for internal attacks. Thus, trusting a Cloud Service Provider is not advisable in all scenarios [2]. Encrypting the user data before uploading to the cloud

© Springer Nature Switzerland AG 2020
D. J. Hemanth et al. (Eds.): ICICI 2019, LNDECT 38, pp. 714–723, 2020.
https://doi.org/10.1007/978-3-030-34080-3_80

will make the user data secure. At the same time, performing deduplication becomes very crucial as the data are stored in encrypted form. In case of shared documents, the same data can be encrypted to cloud storage several time by several users. Even though, the cloud has large number of storage space, redundant copies will increase complications in the data management [3].

Deduplication is one of the well-known technique which is used to perform efficient data management operations on cloud storages [4]. Deduplication is an idea used to remove the duplicate copies from the storage. Rather storing multiple copies, it removes all the redundant files and allows only one original copy to get stored in the cloud storage. It has several advantages like, better storage utilization, reduced data transfer and reduced network wastage. Deduplication can be performed in file or block levels. (i.e.) File level deduplication refers to single instance storage (SIS). It compares the entire file with the files that are stored in the cloud storage. If the file doesn't match with any of the file in the cloud storage, then it will be marked as unique file and are uploaded to the cloud storage. Block level deduplication compares each and every block of the file. Block level deduplication is more efficient and time consuming process. It is used in several public cloud services, such as Mozy, Google Drive. In this research work, an efficient in-line deduplication technique is proposed based on Cramer Shoup cryptosystem and bloom filter to manage redundant data in the cloud storage. The proposed work uses keyword extraction and keyword hashing for identifying the redundant data.

Organization of the paper is as follows, Sect. 2, focuses on the related works that has been carried out on deduplication and bloom filter. Section 3, explains the deduplication system architecture and discusses about the design of proposed system. Section 4, illustrates the performance evaluation of the proposed work and Sect. 5 concludes the paper.

2 Related Works

This section explains about the need of deduplication in a pervasive distributed storages and also it reviews the bloom filter and its application on cloud storage systems.

2.1 Data Deduplication

The term deduplication is an evolving idea in the field of distributed storage systems. Cloud Service Providers (CSP) like Google drive, Dropbox, Amazon Web Service, IBM Cloud are using deduplication technique to save storage spaces in the cloud [10, 11]. Convergent encryption [5] is a technique which gives independent master keys to encrypt user data and outsource to cloud. But convergent key encryption mechanism generates large keys as the count of user increases and also it requires the cloud users to dedicatedly protect the master key. The overhead of the key management is high. To reduce the overhead of the convergent key encryption mechanism, Wen et al., proposed a DARE algorithm [6]. The intention of DARE algorithm is to reduce the overhead time in large scale distributed storage. The idea behind the DARE is to introduce, Deduplication Adjacency (DupAdj) to identify the two data chunks with the similarity by inspecting the adjacent data chunks. But it lacks in the data fragmentation while performing deduplication. Jiang et al. [7] proposed a new hybrid authorized dedupli-cation in which they considers the different privileges of user to perform deduplication check inside the cloud storage [12]. Mihir et al. [13] proposed a message-locked encryption to perform secure deduplication. Here, the keys for the encryption and decryption is derived directly from the message itself. Message locked key generation mechanism maps the message 'M' to key 'K' to get encrypted. The resultant values are the cipher text 'C' that is mapped to a randomized tag. These tags are used to perform deduplication in a secure manner. Though the generated keys in ML encryption are short, the time complexity to perform deduplication is also reduced in the shared environment. But while considering the security aspect of all these existing dedupli-cation technique, no algorithm withstands IND-CCA2 attacks.

2.2   Bloom Filter

In the year of 1979, Burton. H. Bloom came up with a design of bloom filters, which is used to identify the members of a particular set. It has been used in various applications like IP table mapping, access control, metadata services, etc. Consider a set array $'m'$, which consists of the meta-information of the sets and $S = \{x_1, x_2..., x_n\}$ where $'n'$ represents the number present in the set $'m'$. Initially, the values of the array $'m'$ are stated zero. Later, $'K'$ number of hash functions are used to set the items $'x'$ to the set array. $Hashfun (x) = 1 \leq i \leq k$. Elements of set array are inserted by, $Hashfun (x)$. To achieve lower rate of false positive error, filter size can be increased accordingly. The optimum number of hash function is calculated by $k = (m/n) \ln 2$. Some of the real-time applications that uses bloom filters are Google BigTable, PostgreSQL.

Later several variants of bloom filter algorithms were developed to improve the latency and reduce the false positive errors. Stable bloom filter (SBF) [8] is an idea to perform deduplication on the streaming data. It is based on the bitmap sketch. Some of the important bloom filters are Scalable Bloom filter [9] and Counting Bloom Filter (CBF) [14].

In Scalable Bloom filter (SBF), the number of the filter can be scaled from two to $'n'$ filters depending upon the growth of the set array. If one bloom filter is filled, then the bit values can be stored in another filters. Counting Bloom Filter (CBF) has a special counter value. These counter values are used to replace the bits on the array (i.e.) if an element is inserted then the counter will be increased and if the element get deleted then the value of the counter will be decreased.

Some of the other variant of bloom filters are Cassandra bloom filter (CSB), Layered bloom filters, Deletable bloom filter, Hierarchical bloom filter and Adaptive bloom filter.

3   System Model

The proposed model intends to perform deduplication on the cloud storage using bloom filter. The major entities in the cloud storages are, (i) Data Owner (ii) Data Holder and (iii) Cloud Service Providers. The proposed deduplication architecture compromises of three layers, First layer namely deduplication layer which performs the inline data-deduplication operation. It examines the incoming data and prevents duplicated writes to the cloud system. The second level of the proposed architecture performs, data encryption and decryption operation. Here, INDCCA2 secured, Cramer Shoup algorithm is used to encrypt the user data. The third layer is the physical storage of the cloud service provider, where the data is stored in the encrypted format.

3.1   Data De-duplication Layer

The proposed deduplication system ensures two layer inline deduplication approach to stop the redundant copies from entering the cloud storage. The first step is to create a keyword for each document in the cloud.

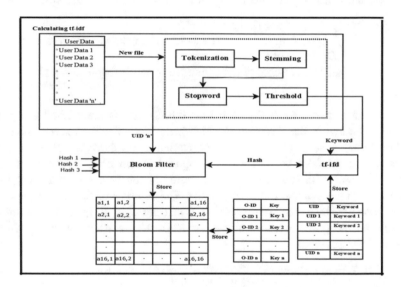

Fig. 1. Proposed in-line deduplication

To identify the suitable keywords for the documents, a separate corpus is maintained. High level pre-processing technique is used to find out the keywords for each files. The architecture of deduplication layer is represented in Fig. 1.

Pre-processing of User Data to Identify the Keyword: Preprocessing unit consists of tokenization, stemming, stop word removal and sets a minimum threshold value to identify the tf-idf (term frequency – inverse document frequency) based keyword. Whenever a user file is uploaded to the cloud, a separate UID and a keyword is extracted from the file and are represented as *'keyword (UID)'*. The below section represents how the keyword is created.

Tokenization and Stop Word Removal of User Data: Initially, the contents of the user file are chopped into pieces and are called as tokens. Tokenization and stop word process removes punctuation, comma, dot and special character from the user uploaded document and creates a temporary file. Later the temporary file is fed into the stemming module.

Stemming module removes the affixes and suffixes of each token. For example if the word 'processing' is said to be a token, then the stemmed output will be 'process'. The proposed work uses porter stemmer algorithm for stemming process.

tf-idf: It is a numerical statistical method to represent the importance of the 'word' in a document stored in a corpus or a collection. Here, tf (term frequency) represent how frequently a word occurs in a particular document and idf (inverse term frequency) says whether the term is common in all documents in the corpus. Value of the tf-idf increases with respect to the importance of the word (i.e.) the number of time a word occurs in a particular document and the entire corpus.

Consider a corpus or collection of document 'D', document 'd' which belongs to corpus 'D', and word 'w' then the tf-idf can be calculated from the following formula,

$$Tf\,idf\,(W_d) = f_{w,d} * log\,(|D|/fw, D);\ Where,$$
$$f_{w,d} = \text{Number of times a word 'w' appears in document 'd'}$$
$$|D| = \text{Size of the corpus}$$
$$f_w, D = \text{Number of documents in which 'w' appears in D}$$

After extracting keywords for each user file, keywords are hashed using bloom filter and are stored in the KW matrix which has a size of $n * n$ matrix. Bloom filter uses three hash functions ('k') to hash the keywords (i.e.) the value of 'k' is set to be three to reduce the false positive value. Proposed bloom filter uses one way MD5, SHA1 and SHA2 hash functions. The keywords of the user's files are fed to hash functions which are independent to each other. The outcome of the hash value changes the corresponding bit values of the KW-matrix from 0 to 1. Instead of hashing the entire file, the proposed work identifies the keyword for each file using tf-idf and hashes only the keyword.

Algorithm: 1: Deduplication verification- KW-Matrix generation
Input   : *New file requested to get stored in the cloud*
Output   : *Allow hashing or call file bloom matrix function*
    1.   If (New File == yes)
    2.   {   {   Calculate the tf-idf value;
    3.   Find the keyword for the newFile;
    4.   Store Keyword as 'new-k'
    5.   }
    6.   Compare function ()
    7.   {
    8.   Compare 'new-k' with K-table
    9.   if (new-k == present in K-table)
    10.   {
    11.   Call 'file-Matrix checking'
    12.   }
    13.   else
    14.   { Allow Hashing;
    15.   Store new-k Hashed value in KW-Matrix
    16.   Store CT in Cloud Storage; }
    17.   }   }

In some scenarios, performing the deduplication using keywords cannot be compromisingly work. Because sometimes two or more different documents can have the same tf-idf keyword. In such cases, comparing the keywords alone will not be sufficient to identify the duplication and it may lead to a false positive error (i.e.) Type I error. To overcome this issue, whenever a keyword match occurs, the proposed deduplication architecture introduces a second level of hash verification (i.e.) the entire user file is hashed and stored in a File Bloom Matrix. As like KW-matrix, MD5, SHA1 and SHA2 are used for hashing. The deduplication architecture is denoted in Fig. 2.

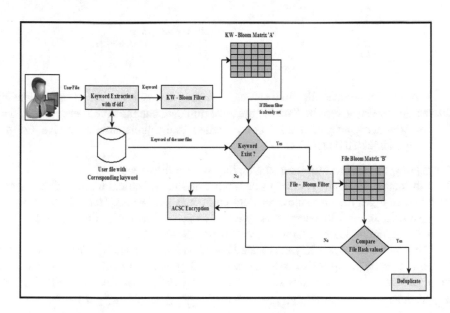

Fig. 2. Proposed in-line deduplication

After creating File Bloom matrix, the second level of deduplication check can be performed. With this secondary check the duplicate files can be removed effectively. The algorithm for second level of deduplication verification is shown below.

Algorithm: 2: Second level deduplication verification – File Matrix
Input   : *User file*
Output   : *Allow ACSC encryption and stores Cipher text in Cloud*

1.	*File-Matrix checking function ( )*	
2.	*{*	*Hash 'User File' using H1, H2, and H3*
3.		*Compare Hash Values with File Bloom Matrix*
4.	*If (Hash Values == Already Set to 1)*	
5.		*{ Reject as de-duplicate value }*
6.	*Else*	
7.		*{*
8.		*Allow Encryption*
9.		*Store CT in Cloud Storage;*
10.	*}*	*}*

After the inline-deduplication process is over, the user file are encrypted using Cramer Shoup cryptosystem. The file are split into data chunks before encrypting. The Cramer Shoup cryptosystem hold good against Indistinguishable Adaptive Chosen Cipher text attack (IND-CCA2).

Encrypting the Data Chunks and Storing in Cloud Storage
Once the data chunks are verified to be unique and non-identical to any of the data in the cloud storage, the encryption of data chunks is performed. Here, Cramer Shoup Cryptosystem is used to encrypt the data. It is proved to be secure under Indistinguishable Adaptive Chosen Cipher text Attack (IND-CCA2).

The public and private key are created based on Pseudo Random Generator (PRG). Sender is a cyclic group *'G'* of order *'q'* with PRG g_1 and g_2. Later, the sender creates a set of random values as privates keys $\{x_1, x_2, y_1, y_2,$ and $z.\}$ of order *'q'*. With the help of private key, creates public key by, $c = g_1^{x1} \cdot g_2^{x2}$; $d = g_1^{y1} \cdot g_2^{y2}$; $h = g_1^z$.

After creating the public and private keys, data chunks are encrypted. Consider a data chunk *'m'* to be encrypted. The sender chooses a random value *'k'*, and generates $u_1 = g_1^k; u_1 = g_2^k$; encrypt $e = h^k.m$ and create $v = c^k.d^{k\alpha}$. Later, the cipher text $\{u_1,$ $u_2, e, v\}$ are uploaded to the cloud storage. The third layer of the proposed scheme is Cloud storage layer. This is where the encrypted user data are stored.

The next section will describe about the performance evaluation of the proposed inline deduplication technique.

4   Implementation and Result Discussion

The proposed deduplication technique is compared with several other existing experiments. A private cloud using Eucalyptus v4.2.0 was installed on Intel Xeon E5 2620 v4, which has a processing speed of 2.1 GHz, 64 GB of RAM memory and 4 TB of storage space.

For testing the proposed deduplication technique, ten random users were chosen. The cloud users make the request to upload *'n'* number of files to the cloud storage. The information of the cloud users are kept confidential.

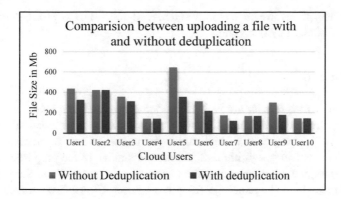

Fig. 3. Comparison between uploading a file with and without deduplication

No user is aware of another user. The experiment is done with and without deduplication. While performing the experiment without deduplication mechanism, several redundant files were found and storage utilization is poor. But when the experiment is carried out with the proposed deduplication technique, the redundant data are stopped from entering into the cloud storage efficiently. The comparison between the data storage with and without deduplication technique is made and represented in Fig. 3. From the above graph and result, it is clear that the proposed technique saves large amount of storage space in the cloud environment.

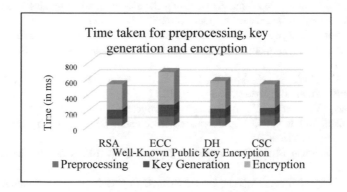

Fig. 4. Time taken to generate public, privates key and encryption

While considering the computation cost of the proposed system, time taken to system initialization, key generation and encryption are considered as major factors. The time taken for the proposed algorithm to generate key and perform encryption is compared with well-known public key encryption is Fig. 4.

The size of the encrypted file is 432 Mb with the fixed data chunk size of 1024 Kb. To make a fair comparison, instead of comparing the algorithm with normal integer public key encryptions, we have modified RSA, Elgamal, DH key exchange and pallier into a matrix public key encryption.

5 Conclusion

This research work presents a novel inline deduplication technique based on bloom filter and asymmetric key Cramer Shoup Cryptosystem on the cloud storage servers. The result of the proposed deduplication technique shows a promising outcome compared to the existing algorithms. The main advantage of the proposed scheme is, it reduces the computation overhead of the client as well as the Cloud Storage Provider. Moreover, it stands well against the Adaptive Chosen Cipher Text Attack (CCA2) in the cloud environment. Though it stand well against CCA2 attacks, internal attacks cannot affect the integrity of the user data. In future, the proposed deduplication scheme will be converted into a form in which it supports homomorphic deduplication operations.

Acknowledgement. The work of this paper is financially sponsored by Science and Engineering Research Board (SERB), Department of Science and Technology (DST), Government of India. The Grant Number is ECR/2016/000546.

References

1. Kenji, E.K., Jonathan, M., John, Z.: Diffusing the cloud: cloud computing and implications for public policy. J. Ind. Compet. Trade **11**, 209–237 (2011)
2. Liu, Y., Sun, Y., Ryoo, J., Rizvi, S., Vasilakos, A.V.: A survey of security and privacy challenges in cloud computing: solutions and future directions. J. Comput. Sci. Eng. **9**, 119–133 (2015)
3. Yan, Z., Ding, W., Yu, X., Zhu, H., Deng, R.H.: Deduplication on encrypted big data in cloud. IEEE Trans. Big Data **2**, 138–150 (2017)
4. Waghmare, V., Kapse, S.: Authorized deduplication: an approach for secure cloud environment. In: International Conference on Information Security and Privacy (2016). Proc. Comput. Sci. **78**, 815–823
5. Diffie, W., Hellman, M.E.: New direction in cryptography. IEEE Trans. Inf. Theory **12**, 644–654 (1976)
6. Xia, W., Jiang, H., Feng, D., Tian, L.: DARE: a deduplication-aware resemblance detection and elimination scheme for data reduction with low overheads. IEEE Trans. Comput. **65**, 1692–1705 (2016)
7. Jiang, T., Chen, X., Wu, Q., Ma, J., Susilo, W., Lou, W.: Secure and efficient cloud data deduplication with randomized tag. IEEE Trans. Inf. Forensic Secur. **12**, 532–545 (2017)
8. Deng, F., Rafiei, D.: Approximately detecting duplicates for streaming data using stable bloom filters. In: Proceedings of the ACM International Conference on Management of Data, pp. 25–36 (2006)
9. Almeida, P., Baquero, C., Preguica, N., Hutchison, D.: Scalable bloom filters. Inf. Process. Lett. **101**, 255–261 (2007)

10. Venish, A., Siva Sankar, K.: Study of chunking algorithm in data deduplication. In: Proceedings of the International Conference on Soft Computing System, Advances in Intelligent System and Computing, pp. 13–20 (2016)
11. Dal Bianco, G., Galante, R., Goncalves, M.A., Canuto, S., Heuser, C.: A practical and effective sampling selection strategy for large scale deduplication. IEEE Trans. Knowl. Data Eng. **27**, 1–15 (2015)
12. Li, J., Li, Y.K., Chen, X., Lee, P.P.C., Lou, W.: A hybrid cloud approach for secure authorized deduplication. IEEE Trans. Parallel Distrib. Syst. **20**, 1–11 (2014)
13. Mihir, B., Sriram, K., Thomas, R.: Message-locked encryption and secure deduplication. In: Proceedings of the Advances in Cryptography, pp. 296–312 (2013)
14. Li, W., Huang, K., Zhang, D., Qin, Z.: Accurate counting bloom filter for large scale data processing. Math. Probl. Eng. **4**, 1–11 (2013)

Biometric Image Encryption Based on Chaotic Sine Map and Information Entropy

Mahendra Patil[1(✉)], Avinash Gawande[2], and Dilendra[1]

[1] Faculty of Engineering, Pacific Academy of Higher Education and Research University, Udaipur, India
apmahendra@yahoo.com
[2] Sipna College of Engineering and Technology, Amravati, Maharashtra, India

Abstract. The Multi-biometric frameworks are utilized in numerous significant biometric applications and more are connected to the UIDAI framework in India. Encryption of biometric pictures is pivotal because of the dynamic increment in the utilization of powerless frameworks and systems for chronicling, transmission and confirmation. This examination proposes a tumultuous sine guide based picture encryption plan utilizing data entropy calculation with different degrees of perplexity and dissemination. The confused sine map with data entropy encryption calculation has a high key space, entropy around equivalent to eight (for dim pictures). The significant preferred position of this strategy is the best qualities for the quantity of pixels and the bound together normal pace of progress it offers. These better qualities obviously demonstrate that the scrambled pictures delivered by the riotous sine map with data entropy calculation plan are arbitrary. Exploratory recreations were performed to acquire the cryptographic parameters and approve the calculation through normal assaults, for example, compression and noise. The proposed calculation is strong because of the pressure assault which is fundamentally required during chronicling and recuperation. Hence, the proposed cryptographic calculation can be connected to the pictures of fingerprints utilized in participation observing frameworks, in biometric gets to, and so forth utilized by most surely understood open and private associations.

Keywords: Attacks · Biometric images · Encryption algorithm · Chaotic sine map · Information entropy · Robustness

1 Introduction

The ongoing flood in web innovation has prompted an expansion in informal communities and online media sharing. Sight and sound information, for example, pictures, sound and video are moved to immense volumes on the Web. Clearly, information and security must be ensured. Data Encryption Standard are not appropriate to sight and sound encryption. Pictures have certain properties, for example, high information excess and huge record sizes, which make these figures unreasonable. The best approach to get around this impediment was to utilize riotous frameworks for picture encryption. Confused frameworks have pseudo-irregularity, solid affectability to beginning conditions and ergodicity these properties make them reasonable for the execution of

© Springer Nature Switzerland AG 2020
D. J. Hemanth et al. (Eds.): ICICI 2019, LNDECT 38, pp. 724–732, 2020.
https://doi.org/10.1007/978-3-030-34080-3_81

clamorous figures [1–3]. Riotous maps are commonly used to actualize confused figures. These maps can be extensively arranged into two sorts: discrete and constant time frameworks. Cryptographic calculations have been widely read by scientists for information already. This standard has created calculations, for example, Data Encryption Standard (DES), Advanced Encryption Standard (AES), RC4 and RC6. Nonetheless, the utilization of these calculations to pictures may not bring about for all intents and purposes invaluable picture encryption plans. Picture encryption plans are not the same as information encryption plans since picture information is commonly enormous, takes more time to scramble and furthermore requires equivalent unscrambling time. Computational overhead makes the customary encryption calculation for scrambling and decoding ongoing pictures unseemly [4–8]. Multi-biometric frameworks are utilized in numerous significant biometric applications or more all connected to the UIDAI framework in India. Be that as it may, multi-biometric frameworks like UIDAI require the capacity of various biometric pictures of finger prints for every client, which involves an extra chance for client protection and PC wrong doings. Among all the bio-measurements, fingerprints are generally utilized in close to home check frameworks, specifically from little scale to enormous scale association. Encryption of biometric pictures is critical because of the dynamic increment in the utilization of defenseless frameworks and systems for filing, transmission and check. Numerous arrangements are proposed dependent on the spatial area and on the recurrence or the methodology of the change space [9–12]. The spatial space system utilizes conventional cryptographic calculations while the recurrence area utilizes changes, for example, Wavelet and Fourier. In the two cases the fundamental thought is to blend the places of the pixels in the biometric pictures and to clamorously produce change orders [13–15]. The primary target of tumultuous cryptanalysis is to reveal data about the mystery key to confused encryption (or secure correspondence) diagram under a wide range of security models: figure content - just assault [5], known-plaintext assault [16, 17], picked assault in clear [18, 19], assault in picked content [20], and outlandish differential association [21]. In the mean time, disordered cryptanalysis additionally gives another viewpoint to ponder the dynamic properties of the hidden tumultuous framework. Since the debasement of any riotous framework psychotic happens in an advanced area a confusion based encryption plan may have some unique security defects that don't exist in non-tumultuous encryption plans. In this way, you need to create a pattern that has progressively spread and disarray. It is noticed that the inherent qualities of biometric pictures speak to the primary bottleneck in the execution of conventional cryptography plans. So these customary plans are not truly appropriate for picture encryption. So the inspiration is to produce plans to scramble biometric pictures productively without giving up inborn highlights. To improve picture security, a biometric cryptosystem approach is utilized that consolidates encryption and biometrics. With this methodology, the biometric picture is scrambled with the assistance of the key. This examination proposes a riotous sine guide based picture encryption plan utilizing data entropy calculation with various degrees of perplexity and dispersion. The tumultuous sine map with data entropy encryption calculation has a high key space, entropy around equivalent to eight (for dark pictures) and significantly less relationship between nearby pixels on a level plane, vertically and askew. The significant preferred position of this strategy is the best qualities for the quantity of pixels and the brought together normal pace of progress it offers. These better qualities unmistakably show that the encoded pictures delivered by the clamorous sine map with data entropy calculation plan are arbitrary. Trial reproductions were performed to acquire the cryptographic parameters and approve the calculation through regular assaults, for example, clamor and pressure. The proposed calculation is powerful because of the pressure assault which is for the most part required during filing and recuperation. Subsequently, the proposed cryptographic calculation can be connected to the pictures of fingerprints utilized in participation observing frameworks, in biometric gets to, and so on utilized by most understood open and private associations. The paper is composed as pursues segment 1 acquaints with encryption, segment 2 depicts encryption and decoding calculation in subtleties. Results are examined in area 3 lastly closed in segment 4.

2 Modified Chaotic Sine Map Based Using Information Entropy Algorithm

In this section chaotic sine map based using information entropy algorithm for encryption of biometric images is discussed along with decryption algorithm. Decryption is exactly the reverse process of encryption.

2.1 Encryption Algorithm

Let I represent the 8-bit gray scale input biometric image of the size m × n. The steps involved in the encryption of the biometric image are as follows

1. Initialize the parameters mu, x0 and y0.
2. Initialize N is the number of iterations, i = 0.
3. Obtain random number using chaotic sine map, input image, iterations and above parameters as follows

$$x(q, p+1) = \sin(pi * mu * (y(q, p) + 3) * x(q, p) * (1 - x(q, p))) \qquad (1)$$

$$y(q, p+1) = \sin(pi * mu * (x(q, p+1) + 3) * y(q, p) * (1 - y(q, p))) \qquad (2)$$

$$x'(q, p) = (x(q, p) + ((s+1)/(s + x(q, p+1) + y(q, p+1) + 1))) \qquad (3)$$

$$x''(q, p) = \mod(x'(q, p), 1) \qquad (4)$$

$$y'(q, p) = (y(q, p) + ((s+2)/(s + x(q, p+1) + y(q, p+1) + 2))) \qquad (5)$$

$$y''(q, p) = \mod(y'(q, p), 1) \qquad (6)$$

where s is the entropy of the input image.
4. Horizontal permutation move, j^{th} column of the input image to n^{th} column of the output image where n is obtained from the jth value of the random number x''.
5. Vertical permutation move, j^{th} row of the input image to n^{th} row of the output image where n is obtained from the jth value of the random number y''.

6. Gray level diffusion using two XOR operations

$$x1(x, y) = bitxor(ic(x, y), 255) \tag{7}$$

$$x2(x, y) = bitxor(x1(x, y), u(1, y)) \tag{8}$$

7. Repeat the process for N number of iterations, i = i + 1.

The encrypted image Ie is obtained and thus the process of encryption is completed. Vectors kr and kc are designated as secret key and obtained randomly from the any source. Also vector kr and kc are randomly hidden into the final encrypted image thus reduces key management. Vectors kr and kc can be again obtained from the encrypted image during decryption for further processing of biometric image.

2.2 Decryption Algorithm

Let Ie represent the 8-bit gray scale input encrypted biometric image of the size m + 1 × n + 1 due to hidden vectors kr and kc. The steps involved in the decryption of the encrypted biometric image are as follows

1. Initialize the parameters mu, x0 and y0.
2. Initialize N is the number of iterations, i = 0.
3. Obtain random number using chaotic sine map, input image, iterations and above parameters as follows

$$x(q, p+1) = \sin(pi * mu * (y(q, p) + 3) * x(q, p) * (1 - x(q, p))) \tag{9}$$

$$y(q, p+1) = \sin(pi * mu * (x(q, p+1) + 3) * y(q, p) * (1 - y(q, p))) \tag{10}$$

$$x'(q, p) = (x(q, p) + ((s+1)/(s+x(q, p+1) + y(q, p+1) + 1))) \tag{11}$$

$$x''(q, p) = mod(x'(q, p), 1) \tag{12}$$

$$y'(q, p) = (y(q, p) + ((s+2)/(s+x(q, p+1) + y(q, p+1) + 2))) \tag{13}$$

$$y''(q, p) = mod(y'(q, p), 1) \tag{14}$$

where s is the entropy of the input image.

4. Gray level diffusion using two XOR operations

$$x2(x, y) = bitxor(x1(x, y), u(1, y)); \tag{15}$$

$$x1(x, y) = bitxor(ic(x, y), 255); \tag{16}$$

5. Vertical permutation move, n^{th} row of the input image to j^{th} row of the output image where n is obtained from the nth value of the random number y''.
6. Horizontal permutation move, n^{th} column of the input image to j^{th} column of the output image where n is obtained from the nth value of the random number x''.
7. Repeat the process for N number of iterations, i = i + 1.

The complete process of encryption and decryption with secret key management policy facilitates security of the biometric images during the process of transmission, storage and verification. Also the modified algorithm overcomes the disadvantage of not having constant values in vectors kr and kc. The complete process is based on defusing pixel position in more confused manner with no secret key generation and management policy.

3   Experimental Results

Exploratory outcomes for visual testing, entropy investigation, pressure assault and clamor assault was acquired on the encoded picture through disorderly sine guide based picture encryption plan utilizing data entropy calculation with numerous degrees of disarray and dissemination.

3.1   Visual Testing

In this analysis, the biometric picture of the 125×125 pixel unique mark was utilized for the visual tests. Figure 1 represents the picture of the biometric info unique finger impression and its encoded yield picture through the proposed calculation. It is obviously comprehended that there is positively no pictographic closeness between these pictures. You need the scrambled picture to be principally not quite the same as its unique information picture. We for the most part get two parameters to figure the contrasts among information and encoded picture.

Table 1. NPCR and UACI values

Biometric images	NPCR	UACI
Finger print 1	99.60	43.29
Finger print 2	99.63	26.86
Finger print 3	99.62	27.88

Input images

Encrypted image

Fig. 1. Experimental results for visual testing

To start with, the measure is the quantity of pixel change rates (NPCR) that determines the level of various pixels between two pictures. The second is the brought together normal variable power (UACI), which estimates the normal force of the distinctions in pixels between two pictures [16]. To accomplish the presentation of a perfect picture encryption calculation, the NPCR esteems must be as huge as could be allowed and the UACI esteems must be around half. Table 1 shows the acquired estimations of NPCR and UACI for the first info picture and their scrambled partner. Subsequently the qualities are extremely near 100% for the NPCR measure and furthermore the UACI esteems are worthy. The high rate estimations of the NPCR measure plainly demonstrate that the pixel positions have been subjectively adjusted. Besides, the UACI esteems demonstrate that for all intents and purposes all the grayscale estimations of the pixels of the encoded picture have been transformed from their qualities in the first info picture.

3.2   Noise Attack

To quantify the strength of the proposed picture encryption plan dependent on the proposed sinusoidal guide utilizing the data entropy calculation, arbitrary clamor was added to the encoded picture. At that point, the boisterous scrambled picture was decoded utilizing the calculation. In this examination, a salt and pepper clamor of thickness 0.1 was added to the scrambled picture. Figure 2 demonstrates the loud info picture and its scrambled and unscrambled partner pictures. The outcomes unmistakably demonstrate that the commotion assault, for example, salt and pepper clamor significantly affects the decoded picture appeared in Table 2.

Table 2. Correlation coefficient after noise and compression attack

Biometric images	Noise attack	Compression attack
	Correlation coefficients	
Finger print 1	0.97	0.88
Finger print 2	0.95	0.88
Finger print 3	0.72	0.42

3.3   Compression Attack

Principally in biometric picture frameworks, the first picture of the unique mark info will be checked, encoded and put away. It is along these lines basic to assess the impact of pressure on the translating procedure. In this investigation, the first encoded picture was compacted to JPEG organization and after that unscrambled.

Along these lines, it is plainly delineated that pressure does not essentially influence the unscrambling procedure and is viable against this procedure. Table 2 shows the estimations of the connection coefficients got between the first picture and the one deciphered after the commotion and pressure assaults. The estimation of the connection coefficient without assault was practically equivalent to one.

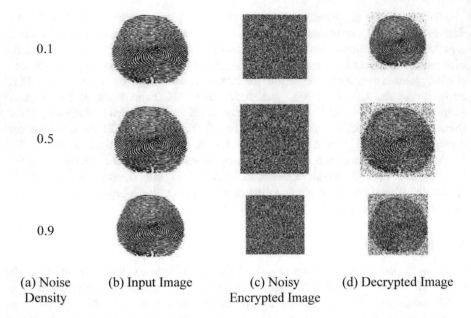

<table>
(a) Noise Density | (b) Input Image | (c) Noisy Encrypted Image | (d) Decrypted Image
</table>

(a) Noise   (b) Input Image   (c) Noisy   (d) Decrypted Image
Density   Encrypted Image

Fig. 2. Experimental results for salt and pepper noise attack at various densities

3.4   Entropy Analysis

For 8-piece dim scale pictures having 256 levels, in the event that each degree of dark is thought to be equiprobable, at that point the entropy of this picture will be hypothetically equivalent to 8 (no. of bits). Therefore it is comprehended that in a perfect world an encryption calculation for bio-metric pictures should give an encoded picture having entropy about equivalent to 8. Table 3 demonstrates the entropy estimations of the three info unique mark pictures and their separate encoded pictures.

Table 3. Entropy values

Biometric images	Input image	Encrypted image
Finger print 1	4.79	7.99
Finger print 2	6.47	7.99
Finger print 3	6.28	7.99

4   Conclusion

We proposed clamorous sine guide based picture encryption plan utilizing data entropy for unique mark biometric pictures in this paper. Trial reenactments were performed to got the encryption parameters and approve the calculation through regular assaults, for example, commotion and pressure. The recreation was finished utilizing MATLAB programming R2014a on Intel i5 processor with 3 GHz speed and 4 GB RAM. Visual testing, entropy examination, clamor assault and pressure assault was performed on the acquired trial results. The proposed calculation is best appropriate for biometric picture encryption particularly unique finger impression. It is powerful to pressure assault and fulfills all encryption parameters, for example, NPCR, UACI and entropy. Further the strength of the proposed calculation should be improved for commotion assault.

References

1. Wen, J., Chang, X.-W.: A modified KZ reduction algorithm. In: Proceedings of IEEE International Symposium on Information Theory, no. 7, pp. 451–455, June 2015
2. Bhatnagar, G., Wu, Q.J.: Chaos-based security solution for fingerprint data during communication and transmission. IEEE Trans. Instrum. Meas. **61**(4), 876–887 (2012)
3. Hassibi, A., Boyd, S.: Integer parameter estimation in linear models with applications to GPS. IEEE Trans. Sig. Process. **46**(11), 2918–2925 (1998)
4. Neelamani, R., Baraniuk, R., de Queiroz, R.: Compression color space estimation of JPEG images using lattice basis reduction. In: Proceedings of IEEE International Conference on Image Processing, vol. 1, pp. 890–893 (2001)
5. Norouzi, B., Seyedzadeh, S.M., Mirzakuchaki, S., Mosavi, M.R.: A novel image encryption based on row-column, masking and main diffusion processes with hyper chaos. Multimed. Tools Appl. 1–31 (2013)
6. Nagar, A., Nandakumar, K., Jain, A.K.: Multibiometric cryptosystems based on feature-level fusion. IEEE Trans. Inf. Forensics Secur. **7**(1), 255–268 (2012)
7. Liu, H., Wang, X.: Color image encryption based on one-time keys and robust chaotic maps. J. Comput. Math. Appl. **59**, 3320–3327 (2010)
8. Murugan, B., Gounder, A.G.N.: Image encryption scheme based on block based confusion and multiple levels of diffusion. IET Comput. Vis. **10**(6), 593–602 (2016)
9. Bringer, J., Chabanne, H., Patey, A.: Privacy-preserving biometric identification using secure multiparty computation: An overview and recent trends. Sig. Process. Mag. **30**(2), 42–52 (2013)
10. Bhatnagar, G., Wu, Q.J., Raman, B.: A new fractional random wavelet transform for fingerprint security. IEEE Trans. Syst. Man Cybern. Part A Syst. Hum. **42**(1), 262–275 (2012)
11. Chen, C.L.P., Zhang, T., Zhou, Y.: Image encryption algorithm based on a new combined chaotic system. In: IEEE International Conference on Systems, Man, and Cybernetics (2012)
12. Wang, Y., et al.: A new chaos-based fast image encryption algorithm. J. Appl. Soft Comput. **11**, 514–522 (2011)
13. Wu, X., Wang, Z.: A new DWT-based lossless chaotic encryption scheme for color images. In: IEEE International Conference on Computer and Computational Sciences (2015)
14. Chen, G., Mao, Y., Chui, C.K.: A symmetric image encryption scheme based on 3D chaotic cat maps. Chaos, Solitons Fractals **21**(3), 749–761 (2004)
15. Sun, J., Liao, X., Chen, X., Guo, S.: Privacy-aware image encryption based on logistic map and data hiding. Int. J. Bifurcation Chaos **27**(5) (2017). Article no. 1750073
16. Li, C., Xie, T., Liu, Q., Chen, G.: Cryptanalyzing image encryption using chaotic logistic map. Nonlinear Dyn. **78**(2), 1545–1551 (2014)
17. Li, C.: Cracking a hierarchical chaotic image encryption algorithm based on permutation. Sig. Process. **118**, 203–210 (2016)

18. Li, S., Li, C., Chen, G., Lo, K.-T.: Cryptanalysis of the RCES/RSES image encryption scheme. J. Syst. Softw. **81**(7), 1130–1143 (2008)
19. Ge, X., Lu, B., Liu, F., Luo, X.: Cryptanalyzing an image encryption algorithm with compound chaotic stream cipher based on perturbation. Nonlinear Dyn. **90**(2), 1141–1150 (2017)
20. Yap, W.-S., Phan, R.C.-W., Yau, W.-C., Heng, S.-H.: Cryptanalysis of a new image alternate encryption algorithm based on chaotic map. Nonlinear Dyn. **80**(3), 1483–1491 (2015)
21. Wang, Q., et al.: Theoretical design and FPGA-based implementation of higher-dimensional digital chaotic systems. IEEE Trans. Circ. Syst. I: Reg. Papers **63**(3), 401–412 (2016)

A Robust Digital Image Watermarking Technique Against Geometrical Attacks Using Support Vector Machine and Glowworm Optimization

Parmalik Kumar[✉] and A. K. Sharma

Shri Venkateshwara University, Gajraula, UP, India
parmalikkumar83@gmail.com,
drarvindkumarsharma@gmail.com

Abstract. The tampering and forgery of digital multimedia data remains as one of the most important challenges all over the internet. Digital image water-marking techniques are used to overcome these challenges. The digital image watermarking techniques provides the security for multimedia data. In this paper, discuss the image features optimization-based digital image watermarking algorithm. The proposed algorithm prevents geometrical attacks during the transmission of data. The proposed algorithm uses support vector machine and glowworm optimization algorithms. The support vector machine classifies the features data of raw image and glowworm optimization used for the reduction cum selection of features for the proceeds of embedded for the extraction of features used SIFT (scale-invariant feature transform) function. The scale-invariant feature transform is major dominated features of critical points and features points. The proposed algorithm implemented in MATLAB software. For the validation and testing purpose used 300 image dataset. For the evaluation of performance measure, these parameters such as PSNR, NC, and SIM. The measured results show better performance instead of SIFT and SVM-GA.

Keywords: Attacks · GA · GSO · SIFT · SVM · Transform function · Watermarking

1 Introduction

The advancement of digital devices and software increases the uses of digital multimedia data. The digital multimedia data can be easily modified and loss their content authentication. The authentication of content and copyright is a new area of challenge over the internet. The digital image watermarking operates in two modes, such as robust and fragile watermarking [1, 2, 4]. The process of watermark proceeds in two sections, such as pixel-based operation and frequency-based operation; the pixel-based methods is called the spatial domain of watermark. Now in the current domain of watermark draw attention in the field of feature-based watermarking. The feature-based watermarking uses the transform function for the extraction of features. By the nature of the Fourier function, the transform function is a major dominated texture feature.

© Springer Nature Switzerland AG 2020
D. J. Hemanth et al. (Eds.): ICICI 2019, LNDECT 38, pp. 733–747, 2020.
https://doi.org/10.1007/978-3-030-34080-3_82

The texture features are transparent features of an image. The extraction of features of the image is the essential phase of digital image watermarking [3, 5, 6]. For the extraction of features used discrete wavelet transform, integer wavelet transform and continuous wavelet transform function. In this paper used SIFT (scale-invariant features transform), the scale-invariant features transform function to extract the features in low key points [11]. The role of SIFT is primarily dominated in image processing for the extraction of features in pattern recognition and watermarking. The classification and machine learning algorithms play an essential role in digital image watermarking. The machine learning algorithms boot the performance of embedding in digital image watermarking. İn this paper support vector machine is used for the categorization of similar features data. The similar categorized features data process for the generation of patterns. Nowadays, optimization algorithms play an important role in image processing and watermarking. Authors [2] used genetic algorithms for the optimization of features data, but the limitation of genetic algorithms decline the value of the similarity index of the watermark image. İnspite of Genetic algorithm used glowworm optimization algorithm (GSO). The generation of patterns gives the randomness of coefficient and increase the value of the number of correlations. The high value of correlation shows a more robust digital image [12, 17–19]. The selection of features, increase the performance of support vector machine for making a pattern. In this paper used a glowworm optimization algorithm for feature optimization. Glowworm optimization reduces the dissimilarity of features point. The reduces features point provides the similarity index of image and increase the value of imperceptibility of watermark image. The strength of the proposed algorithm is analyzed with 300 image data sets. The 300-image data set contains different categories of images. These image data sets also contain a symbol image for the process of the watermark [21, 23]. Moreover, the performance also varies with change in the type of cover image and different group of attacks. The quality of watermark is estimated the parameter PSNR (peak to signal noise ratio). The transform-based function also incorporated with other methods and algorithms such as heuristic function and neural network and enhanced the security strength of watermarking techniques. The current scenario of algorithms cannot generate the process of dynamic pattern binding for the watermarking process. The third-party attacker predicts the weak embedding of the watermark [20, 22]. The organization of the paper, as in Sect. 2, discusses the support vector machine (SVM) and SIFT. In Sect. 3, discuss the glowworm algorithm. In Sect. 4, discuss the proposed algorithm. In Sect. 5 discuss the simulation process of proposed algorithms and finally, conclusion & future scope in Sect. 6.

2 Support Vector Machine and SIFT

The Support vector Machine (SVM)

Support vector machine classified data into two planes +1 and −1. The process of classification used in different area of image processing and pattern recognition. The great advantage of support vector machine is training of sample data for the generation of pattern. The principle of support vector machine is translation of space such as linear to non-linear space. The process of working of support vector machine is describe here

[14]. The shown Fig. 1 the symbol of blacks and white represent the class label of support vector machine. The H indicates the separation plane of two categorizes of data. The value of plane is also called hyperplane, the hyperplane increase the value of margin $2/||w||$. The two planes H1 and H2 is boundary of categorized data and parallel to H plane. The incline data of H1 and H2 is called support vector.

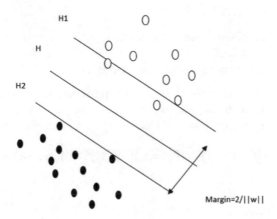

Fig. 1. The process of planes in categorization of class in support vector machine.

The set of sample data (x1, y1),.......(x1, y1), yi $\in \{-1, 1\}$, to lies in hyperplane H, the function of space transform is, $Z = \phi(x)$, is applied for x is linear. The value of weight and constant b describe as

$$\begin{cases} w^T z_i + b \geq 1 & y_i = 1 \\ w^T z_i + b \leq -1, & y_i = -1 \end{cases} \tag{1}$$

i.e. $y_i(w^T z_i + b) \geq 1, \qquad i = 1, 2, \ldots, l$ (2)

The process of SIFT function generates features key point of digital image. The process of features extraction describes here [3, 13, 30–32]:

Step 1 – Scaling of Space:
The process of mapping pf 2D image data into 2D SIFT transform the process of transfer used linear kernel scaling factor the derivation as:

$$f_{out} = K_n * f_{in}$$

Where,

K_n – kernel
f_{in} - input signal
* - convolution operation

Scale-space $S(x, y, \sigma)$ of image $I(x, y)$ is expressed mathematically as:

$$S(x, y, \sigma) = G(x, y, \sigma) * I(x, y)$$

$$G(x, y, \sigma) = \frac{1}{2\pi\sigma^2} e^{-\frac{x^2+y^2}{2\sigma^2}}$$

Where,

$G(x, y, \sigma)$ - scale variable Gaussian function
(x, y) - spatial coordinates
σ - scale-space factor (using image's smoothness)

The feature point value denoted by σ. The maximum value of σ denoted a good feature value and the minimum value of σ extract minimum feature.

The process of *DOG*(*difference of Gaussian*) space describe mathematically as:

$$D(x, y, \sigma) = [G(x, y, k\sigma) - G(x, y, \sigma)] * I(x, y) = S(x, y, k\sigma) - S(x, y, \sigma)$$

Where,

$I(x, y)$ - input image
k - multiple of 2 neighboring scale-spaces
$*$ - convolution operation.

To measure high and low value of $D(x, y, \sigma)$, each feature points matched with its neighbor.
Step 2 – calculate the feature points:
The orientation and location of the feature points, the inner points of orientation is called feature point and left outer point is calculated in step 3 formula.
Step 3 - formula:

$$\begin{cases} \theta(x,y) = \tan^{-1} 2\{[I_L(x, y+1) - I_L(x, y-1)]/[I_L(x+1, y) - I_L(x-1, y)] \\ g(x,y) = \sqrt{[I_L(x+1, y) - I_L(x-1, y)]^2 + [I_L(x, y+1) - I_L(x, y-1)]^2} \end{cases}$$

Where,

$\theta(x, y)$ - orientation of the gradient
$g(x, y)$ - magnitude

The direction of key points is visualized by histogram the maximum gradient direction of feature key points collects as feature descriptor. The collected feature point creates feature matrix. The collected feature matrix shown in Fig. 2.

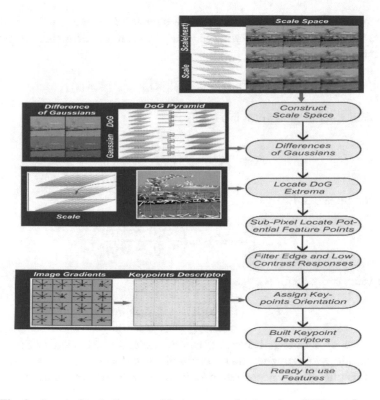

Fig. 2. Process block diagram of feature extraction based on SIFT transform.

3   Glowworm Optimization (GSO)

The feature matrix of source image and symbol image passes through glowworm optimization algorithms.

The process of glowworm optimization maps all features data in terms of glow-worm and define the set of population for further searching of features point for matching the objective function. The process of mapping describes as [7, 8, 10]:

The glowworm map K with objective function J(xk(t))

The position of glowworm defines as xk(t)

The acceleration factor α

The similar features value update with luciferin li

The group of similar glowworm creates set of Nk(t)

The local update function is given in Eq. (3)

$$r_d^k(t+1) = min\{rs, max\{0, r_d^k(t) + \beta(nt - |Nk(t)|)\}\} \qquad (3)$$

The local decision value is $r_d^i(t+1)$ and the range of value of t + 1, rs is the range of similar features nt is similarity range. The total number of optimized glow local update function is in Eq. (4)

$$N_{k(t)} = \left\{ j : \|xk(t) - xk(t)\| < r_d^k;\; lk(t) < lk(t) \right\} \tag{4}$$

The derivation of glowworm in Eq. (5)

$$p_{ij(t)} = \frac{l_{k(t)} - l_{k(t)}}{\sum_{f \in Nk(t)} lf(t) - lk(t)} \tag{5}$$

Position update in Eq. (6)

$$x_{k(t+1)} = xk(t) + s\left(\frac{sj(t) - xk(t)}{\|xj(t) - xk(t)\|} \right) \tag{6}$$

The value of Luciferin is updated in Eq. (7)

$$l_{k(t)} = (1 - \rho)lk(t - 1) + \gamma j(xk(t)) \tag{7}$$

The value of ρ belongs between 0 and 1. the end of iteration of SIFT features matrix with glowworm optimization process gives the optimal features matrix for the process of features point categorization as input of support vector machine [7, 8].

4   Proposed Methods

The proposed algorithm is combination of GSO and support vector machine. The GSO function used for the process of feature optimization and feature selection. The process of feature selection finds the group of similar texture feature for the process of vector in support vector [9]. The increase value of support vector increases the value of margin and generates the better patterns for watermarking. Radial basis kernel function is used for support vector machine. The process of radial kernel function is very fast and training error of pattern is also decrease. For the estimation of pattern difference used person coefficient to measure the pattern difference value. The pattern difference value proceeds the location and position of watermark embedding. The process of algorithm describes here (Fig. 3).

1. Input: a SIFT feature point in GSO and SVM

2. Output: $dynamic\ pattern\ DP_{Pf)}$

3. Compute $D_{(P_t,k)}$ and $k - disimarilty(p_t)$

4. for all $DP \in SVM_{(f_t,k)}$ do

5. estimate local pattern $-Lp(f_t, DP)$

6. end for

7. $W_{update} \leftarrow GSO\ \{the\ set\ of\ glows\}$

8. for all $DP \in W_{update}$ and $FP \in M_{(DG,K)}$ do

9. Update $k - disimarilty(DP)$ and $cluester - ds(GSO, DP)$

10. if $DP_{(FP,k)}$ then

11. $W_{update} \leftarrow W_{update} \cup \{DP\}$

12. end if

13. end for

14. for all $DP \in W_{update}$ do

15. Update $FD(DP)$ and $FD(\{GSO_{o,k}\})$

16. end for

17. return $DP(dynamic\ pattern)$

18. measure   person   coefficient   for   both   pattern   p1.............pn, s1.................sn

19. if value of difference is near about zero.

20. The process of embedding is done

21. Measure value of parameters

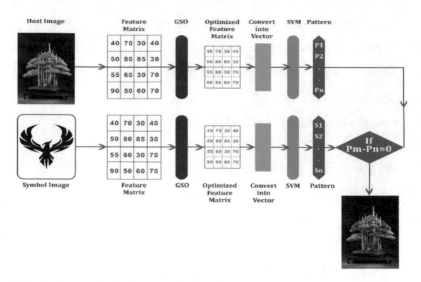

Fig. 3. Process block diagram of proposed algorithm for watermark embedding.

5 Experimental Result Analysis

Validation and analysis of proposed algorithms are performed in 300 color image datasets. The datasets of a color image consist of different categories such as forest, bonsai, pot, etc. these image datasets collected from CVG_UGR datasets [24]. For the analysis of algorithms used following hardware and operating system, (1) the machine equid with intel core seven processor with operating system windows 10. (2) in this operating system used MATLAB software for the implementation of algorithms. Measuring the performance of algorithms used following formula according to us define objective [25–29]. RMSE factor estimate the error deference value of source image and final watermark image.

$$RMSE = \sqrt{\frac{1}{m \times n} \sum_{i=0}^{m-1} \sum_{j=0}^{n-1} \left[I_{image}(i,j) - I''_{image}(i,j) \right]^2} \tag{8}$$

where $m \times n$ is the resolution of I (i, j) is the pixel location, I_{image} is the source image, and I''_{image} is the watermarked image (Table 1). Then PSNR value is calculated as

$$PSNR = 20log_{10} \frac{Max(I_{image})}{RMSE} \tag{9}$$

Table 1. Discuss the process of watermark embedding with the host image and symbol image and finally generated with the watermark image. Also try to attack for watermark image such as cropping, scaling, row-column blanking, rotation, image flapling, noise attack, etc.

Sr. No.	Host	Symbol	Watermarked Image
1.	Forest	Pyramid	
2.	Bonsai	Eagle	

Table 2. The experimental analysis of the watermark image with different types of attacks when we proceed with our simulation. There is five types of attack performed here these name salt-noise attack, gaussian noise attack, cropping attack, scaling attack, and rotation attack. All the above-described attack represented with a host image in this table, when we estimate the SIM, PSNR and NC.

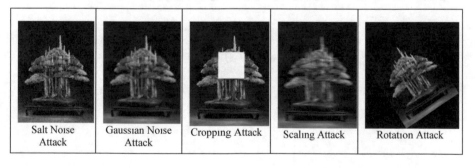

| Salt Noise Attack | Gaussian Noise Attack | Cropping Attack | Scaling Attack | Rotation Attack |

Table 3. The experimental analysis of watermark image with different types of attacks which is discussed in Table 2. We analyzed the simulation with each attack (salt-noise attack, guissian noise attack, cropping attack, scaling attack and rotation attack) and estimate their variation in concern of PSNR, NC and percentage of similarity of the target image. The proposed methods achieve high-performance value and great strength against attacks in the comparison of SIFT technique and SVM-GA technique. İn the SIM and PSNR parameters have a higher numeric result of the proposed technique, but the NC parameter has the lower result value of proposed technique that represented the better performance of the proposed technique.

Attack	Method	SIM (%)	PSNR (dB)	NC
Salt noise	SIFT [12, 15, 16]	95.68	90.1437	0.8626
	SVM-GA [2]	96.54	91.3851	0.9527
	Proposed	97.77	93.4456	0.8023
Cropping	SIFT [12, 15, 16]	68.85	84.2437	0.7524
	SVM-GA [2]	68.75	80.1122	0.8326
	Proposed	70.47	88.1174	0.6427
Rotation	SIFT [12, 15, 16]	82.47	79.2413	0.7965
	SVM-GA [2]	85.24	77.2431	0.8334
	Proposed	89.68	80.2494	0.7662
Gaussian noise	SIFT [12, 15, 16]	65.26	67.2236	0.8903
	SVM-GA [2]	77.34	80.4743	0.8544
	Proposed	79.67	81.4724	0.8147
Scaling	SIFT [12, 15, 16]	75.30	85.1737	0.8124
	SVM-GA [2]	79.43	89.3946	0.7626
	Proposed	80.64	92.8922	0.7427

Calculated value of robustness used this formula as

$$NC = (\frac{\sum_{i=0}^{m-1} \sum_{j=0}^{n-1} W_f(i,j) - W_d(i,j)}{\sqrt{\sum_{i=0}^{m-1} \sum_{j=0}^{n-1} W_o^2(i,j)} \times \sqrt{\sum_{i=0}^{m-1} \sum_{j=0}^{n-1} W_E^2(i,j)}} \tag{10}$$

Where Wf is the source watermark and Wd is decoded watermark Figs. 4, 5 and 6.

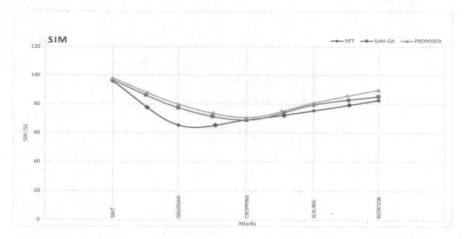

Fig. 4. Window show that the comparative analysis of SIM using SIFT, SVM-GA and proposed techniques with different host image (Almira) and symbol image (authenticate). With the help of performance graph indicate the high performance of proposed technique because it shows high numeric percentage result value of SIM compared to all techniques these are SIFT, SVM-GA. In the case of salt noise attack SIFT have 95.68%, and SVM-GA have 96.54% but the proposed method have 97.77% that means it improved SIM 1.23%. In the case of gaussian attack SIFT have 65.26%, and SVM-GA have 77.34% but the proposed method have 79.67% that means it improved SIM 2.34%. In the case of cropping attack SIFT have 68.85%, and SVM-GA have 68.75% but the proposed method have 70.47% that means it improved SIM 1.27%. In the case of Scaling attack SIFT have 75.3%, and SVM-GA have 79.43% but the proposed method have 80.64% that means it improved SIM 1.07%. In the case of Rotation attack SIFT have 82.47%, and SVM-GA have 85.24% but the proposed method have 89.68% that means it improved SIM 4.44%.

The similarity percentage is estimated as

$$sim = \left(1 - \frac{RMSE}{MAX(Iimage)} \times 100\right)\%$$

(11)

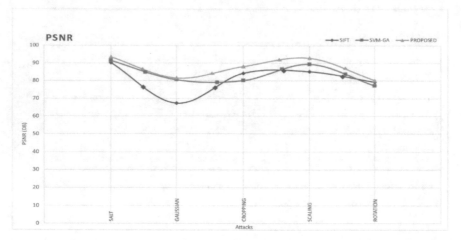

Fig. 5. Window show that the comparative analysis of PSNR using SIFT, SVM-GA and proposed techniques with different host image (Almira) and symbol image (authenticate). With the help of performance graph indicate the high performance of proposed tehnique because it shows high numeric result value of PSNR compared to all techniques these are SIFT, SVM-GA. In the case of salt noise attack SIFT have 90.1437, and SVM-GA have 91.3851 but the proposed method have 93.4456% that means it improved PSNR 2.13 dB. In the case of gaussion attack SIFT have 67.2236, and SVM-GA have 80.4743 but the proposed method have 81.4724 that means it improved PSNR 1.01 dB. In the case of cropping attack SIFT have 84.2437, and SVM-GA have 80.1122 but the proposed method have 88.1174 that means it improved PSNR 8.0052 dB. In the case of Scaling attack SIFT have 85.1737, and SVM-GA have 89.3946 but the proposed method have 92.8922 that means it improved PSNR 3.561 dB. In the case of Rotation attack SIFT have 79.2413, and SVM-GA have 77.2431 but the proposed method have 80.2494 that means it improved PSNR 3.006 dB.

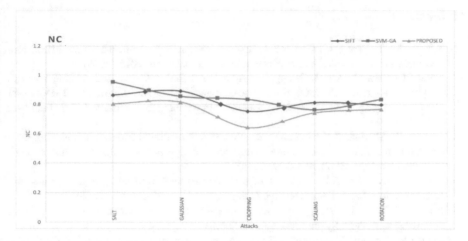

Fig. 6. Window show that the comparative analysis of NC using SIFT, SVM-GA and proposed techniques with different host image (Almira) and symbol image (authenticate). With the help of performance graph indicate the high performance of proposed tehnique because it shows low numeric result value of NC compared to all techniques these are SIFT, SVM-GA. In the case of salt noise attack SIFT have 0.8626, and SVM-GA have 0.9527 but the proposed method have 0.8023 that means it less NC 0.152. In the case of gaussion attack SIFT have 0.8903, and SVM-GA have 0.8544 but the proposed method have 0.8147 that means it less NC 0.403. In the case of cropping attack SIFT have 0.7524, and SVM-GA have 0.8326 but the proposed method have 0.6427 that means it less NC 0.2101. In the case of Scaling attack SIFT have 0.8124, and SVM-GA have 0.7626 but the proposed method have 0.7427 that means it less NC 0.2701. In the case of Rotation attack SIFT have 0.7965, and SVM-GA have 0.8334 but the proposed method have 0.7662 that means it less NC 0.672.

6 Conclusion and Future Work

The enhanced value of NC and SIM shows the better performance of the proposed method for digital image watermarking. The analysis of result shows in Table 3 with compression of SIFT transform function and SVM-GA. The SIFT based watermarking shows the lower value of NC and SIM. However, in the case of SVM-GA value of NC and SIM is increase. The process of glowworm optimization enhances the better patterns selection for embedding in watermarking. The better patterns of embedding enhance the security of digital watermarking. Also, increase the value of PSNR. The point of trust of proposed algorithms is to justify the concept of dynamic pattern generation in the case of cascaded neural network models. The concept of dynamic pattern generation shows more randomness of patterns, and the perception of pattern prediction is decreased in the future, using the proposed algorithms in other fields of multimedia data such as video and audio.

References

1. Sun, L., Xu, J., Zhang, X., Tian, Y.: An image watermarking scheme using Arnold transform and fuzzy smooth support vector machine. Math. Probl. Eng. **2015** (2015)
2. Zhou, X., Cao, C., Ma, J., Wang, L.: Adaptive digital watermarking scheme based on support vector machines and optimized genetic algorithm. Math. Probl. Eng. **2018** (2018)
3. Yen, S.H., Wang, C.J.: SVM based watermarking technique. Tamkang J. Sci. Eng. **9**(2), 141–150 (2006)
4. Kumar, P., Sharma, A.K.: A comprehensive review of digital image watermarking using transform function for the feature extraction and optimization algorithm for dynamic embedding. JETIR. **3**(12), 364–379 (2016)
5. Kumar, P., Sharma, A.K.: Analysis of digital watermarking techniques using transform-based function. Int. J. Futur. Revolut. Comput. Sci. Commun. **3**, 41–48 (2017)
6. Zhou, X., Zhang, H., Wang, C.: A robust image watermarking technique based on DWT, APDCBT, and SVD. Symmetry **10**(3), 77 (2018)
7. Singh, N., Kumari, J., Aggarwal, C.: Feature selection for steganalysis using glow worm algorithm
8. Yepes, V., Martí, J.V., García-Segura, T.: Cost and CO2 emission optimization of precast–prestressed concrete U-beam road bridges by a hybrid glowworm swarm algorithm. Autom. Constr. **1**(49), 123–134 (2015)
9. Marinaki, M., Marinakis, Y.: A glowworm swarm optimization algorithm for the vehicle routing problem with stochastic demands. Expert Syst. Appl. **15**(46), 145–163 (2016)
10. Shrichandran, G.V., Sathiyamoorthy, S., Malarchelvi, P.D.: A hybrid glow-worm swarm optimization with bat algorithm based retinal blood vessel segmentation. J. Comput. Theor. Nanosci. **14**(6), 2601–2611 (2017)
11. Islam, M., Roy, A., Laskar, R.H., et al.: Neural network based robust image watermarking technique in LWT domain. J. Intell. Fuzzy Syst. **34**(3), 1691–1700 (2018)
12. Wang, C., Zhang, Y., Zhou, X.: Robust image watermarking algorithm based on ASIFT against geometric attacks. Appl. Sci. **8**(3), 410 (2018)
13. Nam, S.H., Kim, W.H., Mun, S.M., Hou, J.U., Choi, S., Lee, H.K.: A SIFT features based blind watermarking for DIBR 3D images. Multimed. Tools Appl. **77**(7), 7811–7850 (2018)
14. Singh, K.M.: A robust rotation resilient video watermarking scheme based on the SIFT. Multimed. Tools Appl. **77**(13), 16419–16444 (2018)
15. Li, J., Lin, Q., Yu, C., Ren, X., Li, P.: A QDCT-and SVD-based color image watermarking scheme using an optimized encrypted binary computer-generated hologram. Soft. Comput. **22**(1), 47–65 (2018)
16. Zhou, N.R., Luo, A.W., Zou, W.P., et al.: Secure and robust watermark scheme based on multiple transforms and particle swarm optimization algorithm. Multimed. Tools Appl. **78**(2), 2507–7523 (2019)
17. Zheng, Z., Saxena, N., Mishra, K.K., Sangaiah, A.K., et al.: Guided dynamic particle swarm optimization for optimizing digital image watermarking in industry applications. Futur. Gener. Comput. Syst. **88**, 92–106 (2018)
18. Ali, M., Ahn, C.W., et al.: An optimal image watermarking approach through cuckoo search algorithm in wavelet domain. Int. J. Syst. Assur. Eng. Manag. **9**(3), 602–611 (2018)
19. Ansari, I.A., Pant, M., Ahn, C.W., et al.: Artificial bee colony optimized robust-reversible image watermarking. Multimed. Tools Appl. **76**(17), 18001–18025 (2017)
20. Bahrami, Z., Tab, F.A., et al.: A new robust video watermarking algorithm based on SURF features and block classification. Multimed. Tools Appl. **77**(1), 327–345 (2018)

21. Balasamy, K., Ramakrishnan, S., et al.: An intelligent reversible watermarking system for authenticating medical images using wavelet and PSO. Cluster Comput. 1–2 (2018)
22. Chatterjee, S., Sarkar, S., Hore, S., Dey, N., Ashour, A.S., Balas, V.E., et al.: Particle swarm optimization trained neural network for structural failure prediction of multistoried RC buildings. Neural Comput. Appl. **28**(8), 2005–2016 (2017)
23. Li, D., Deng, L., Gupta, B.B., Wang, H., Choi, C., et al.: A novel CNN based security guaranteed image watermarking generation scenario for smart city applications. Inf. Sci. **479**, 432–447 (2019)
24. Makbol, N.M., Khoo, B.E., Rassem, T.H., Loukhaoukha, K., et al.: A new reliable optimized image watermarking scheme based on the integer wavelet transform and singular value decomposition for copyright protection. Inf. Sci. **417**, 381–400 (2017)
25. Mehta, R., Rajpal, N., Vishwakarma, V.P., et al.: A robust and efficient image watermarking scheme based on Lagrangian SVR and lifting wavelet transform. Int. J. Mach. Learn. Cybern. **8**(2), 379–395 (2017)
26. Mitashe, M.R., Habib, A.R., Razzaque, A., Tanima, I.A., Uddin, J., et al.: An adaptive digital image watermarking scheme with PSO, DWT and XFCM. In: 2017 IEEE International Conference on Imaging, Vision & Pattern Recognition (icIVPR), 13 February 2017, pp. 1–5. IEEE (2017)
27. Saxena, N., Mishra, K.K., et al.: Improved multi-objective particle swarm optimization algorithm for optimizing watermark strength in color image watermarking. Appl. Intell. **47** (2), 362–381 (2017)
28. Shih, F.Y., Zhong, X., Chang, I.C., Satoh, S.I., et al.: An adjustable-purpose image watermarking technique by particle swarm optimization. Multimed. Tools Appl. **77**(2), 1623–1642 (2018)
29. Sun, L., Xu, J., Liu, S., Zhang, S., Li, Y., Shen, C.A., et al.: A robust image watermarking scheme using Arnold transform and BP neural network. Neural Comput. Appl. **30**(8), 2425–2440 (2018)
30. Thakkar, F.N., Srivastava, V.K., et al.: A fast watermarking algorithm with enhanced security using compressive sensing and principle components and its performance analysis against a set of standard attacks. Multimed. Tools Appl. **76**(14), 15191–15219 (2017)
31. Kukenys, I., McCane, B., et al.: Classifier cascades for support vector machines, pp. 1–6. IEEE (2008)
32. Camlica, Z., Tizhoosh, H.R., Khalvati, F., et al.: Medical image classification via SVM using LBP features from saliency-based folded data, pp. 1–5. IEEE (2015)

Robot Operating System Based Path Planning

Ashwin Vasudevan[1], Ajith Kumar[1], Nrithya Theetharapan[1(✉)],
and N. S. Bhuvaneswari[2]

[1] Anna University, Chennai, India
contact@ashwinvasudevan.com, ajithofficials@gmail.com, nrithya17@gmail.com
[2] GKM College of Engineering and Technology, Chennai, India
bhuvaneswarins@rediffmail.com

Abstract. As autonomous cars are catapulted into prominence, path planning has taken the center stage. In this study, We implement the Tangent-bug approach to path planning. We have discretised the Tangent-bug algorithm into a set of independent operations. Transition from one iteration to the next is governed by the value of the range sensor and the convergent criterion. The tangent detection and tangent selection algorithms are tested by implementing on a 2D differential drive robot in the player stage simulator. Contrary to existing solutions, this paper proposes implementation on standard robotics middle-ware ROS (Robot Operating System) allowing effortless deployment on various robotics platforms. Features developed in this module are decentralised, amenable to scaling across multiple robotics platforms with minimal configuration.

Keywords: Tangent-bug · Path planning · Robot Operating System · Obstacle detection · Player stage simulator

1 Introduction

Path planning is the process of identifying which route to take, based on the knowledge of map and sensor data. Path planning can be classified into several types. Some of them are, knowledge based path planning, sensor based path planning, combined path planning and sensor based hierarchical path planning.

In sensor based path planning, the information is not available at the beginning of path finding. The information about the environment is gathered while the robot is in motion. The sensor produces a feedback signal which is a current position of the robot that is compared with the destination set point [9]. Based on the comparison results, path planning is accomplished. Therefore, the sensor based path planning can be depicted as a simple closed loop feedback control mechanism [10].

Some of the popular path planning algorithms are breadth first search [14], depth first search [11], uniform cost search [12], bidirectional search, pure heuristic search [6], A* search [13], D* search and Hill climbing search [8]. Bug algorithms [5] are a simpler approach. Bug algorithms use current state and the goal

© Springer Nature Switzerland AG 2020
D. J. Hemanth et al. (Eds.): ICICI 2019, LNDECT 38, pp. 748–755, 2020.
https://doi.org/10.1007/978-3-030-34080-3_83

state knowledge for path planning. In Bug1 algorithm, if an obstacle is encountered, the robot circumnavigates around the obstacle and moves towards the
goal again. In Bug2 algorithm, the robot is made to move in an m-line which
connects initial point to goal point. If an obstacle is encountered, then the robot
is made to follow the obstacle until it reaches the m-line again.

2   Tangent-Bug Algorithm

Tangent-bug algorithm is an improved version of BUG1 and BUG2 algorithm
and depends only on the range sensor data. Path planning to reach the goal is
achieved by using the output of range sensor and the knowledge of current pose
and goal pose. The algorithm is explained in the section below.

2.1   Algorithm

The range sensor gives the raw distance information from the environment to the
robot covering the 360-degree circumference of the robot. This can be achieved
by emanating laser rays in all directions The algorithm switches between two
planner states: motion to goal and boundary follow. Motion to goal operation
is executed when no obstacle is encountered on the path of the robot or the
heuristic distance is decreasing from a followed obstacle to goal.

$$h = d(x_i, Q_i) + d(Q_i, q_{\text{goal}}) \tag{1}$$

is the heuristic distance to goal calculated by the robot. The robot finds points
that will decrease this heuristic distance and move towards that point. Two
parameters d_{reach} and d_{followed} are calculated

d_{reach} = shortest distance between the goal and the blocking obstacle.
d_{followed} = shortest distance between the boundary and the goal.

The robot starts to move on the tangent line, with the direction selected based
on goal position. Boundary follow operation is executed until $d_{\text{followed}} >= d_{\text{reach}}$,
otherwise, the motion to goal operation starts to execute.

3   State Estimations

To evaluate the current state of the robot in the n-dimensional space, simple
Euclidean transformation can be used. The pose of a robot can be described
by a configuration. The initial pose of the robot can be represented by a vector
$\begin{bmatrix} x_i & y_i & \alpha_i \end{bmatrix}^T$ and goal can be represented as $\begin{bmatrix} x_f & y_f & \alpha_f \end{bmatrix}^T$. At any instant, the robot
in 2D space can be represented by a point $p(x_1, y_1)$. The angular distance can
be calculated as

$$\alpha = \arctan(\frac{y}{x}) \tag{2}$$

By Pythagorean theorem,

$$\Delta l = \sqrt{(\Delta x)^2 + (\Delta y)^2} \tag{3}$$

$$\Delta \alpha = arctan\frac{\Delta y}{\Delta x} - \alpha \tag{4}$$

3.1 Operation Selection

The presence of obstacle in the path of the robot to the goal can be confirmed with the knowledge of Δl and $\Delta \alpha$. Consider *obs* be a parameter whose value becomes equal to 1 when an obstacle is present in the path of the robot. Otherwise, its value becomes zero. Considering the robot's width as w, *obs* will set to one if Δl is greater than current range value and the gap through which the robot can pass is less than or equal to robot's width.

$$obs = \begin{cases} 0 & \text{if } \Delta \alpha > 0 \, and \, (s_{\min} + \Delta \alpha) > \alpha_i \\ 0 & \text{if } \Delta \alpha < 0 \, and \, (s_{\max} + \Delta \alpha) < \alpha_i \\ 1 & \text{if } \Delta l > r_i \, and \, (r_i \times \sin(\alpha_i - \Delta \alpha)) <= w \end{cases} \tag{5}$$

$$\forall i \in \{0, 1, 2 ..... (N_{\text{samples}} - 1)\}$$

where, α_i = goal direction inferred from the i_{th} laser beam and r_i = goal distance inferred from the i_{th} laser beam The operations included are force stop, stop at goal, turn at goal, approach goal, follow tangent and right turn.

3.2 Tangent Detection

The difference between ranges of two successive measurements is Δ which is given by the formula below.

$$\Delta = r_i - r_{i+1} \tag{6}$$

By considering the robot width w, number of tangents can be detected. Consider a vector $T \in R^3$ which contains angle, range, correction factor as its element. The correction factor is selected based on the robot's width.

The number of vectors T that can be obtained is proportional to the number of tangents detected. The following equation represents the two cases of tangent detection based on two successive range measurements.

$$T = \begin{cases} \begin{bmatrix} \alpha_i \\ r_{i+1} \\ -\sin(\frac{W}{r_i}) \end{bmatrix} & if(\Delta > 2W) \vee ((r_i > r_{\max}) \wedge (r_{i+1} < r_{\max})) \\ \begin{bmatrix} \alpha_{i+1} \\ r_i \\ \sin^{-1}(\frac{W}{r_i}) \end{bmatrix} & if(\Delta < -2W) \vee ((r_i < r_{\max}) \wedge (r_{i+1} > r_{\max})) \end{cases} \tag{7}$$

$$N_{\text{tan}} = N_{\text{tan}} + 1$$
$$\forall n \in \{0, 1, 2.....(N_{\text{samples}} - 1)\} \tag{8}$$

3.3   Tangent Selection

A suitable tangent is selected based on $\Delta\alpha$ and angle of all possible tangents. Consider a vector T_s whose parameters are initially zero.

For $N_{\text{tan}} > 0$, Consider the column matrix,

$$T_s = \begin{bmatrix} \alpha_s & r_s & c_s \end{bmatrix} \in R^3$$

which is used to specify the selected tangent. A tangent can be selected according to the following equation.

$$T_s = \left\{ T_i \quad \text{if } (T_{20} - \Delta\alpha) < (\alpha_n - \Delta\alpha) \right. \tag{9}$$

3.4   Distance Angle Update

At each iteration, the value of Δl and Δa is computed. Consider the distance-angle matrix $L \begin{bmatrix} \Delta l \\ \Delta \alpha \end{bmatrix} \in R^2$, whose parameters are updated in every time step.

4   Implementation in ROS

ROS provides hardware abstraction, code reusability, platform independent approach, message passing functionalities between different nodes [1]. Robot operating system also provides interaction with other simulation tools such as stage simulator [2], Gazebo [4] and Rviz [3].

   Our approach to Tangent-bug algorithm can be implemented in any simulator with the knowledge of laser scan sensor data [7] and pose of the robot in the n dimensional space. To get the pose of the robot at any instant, the commonly preferred method is odometry. Gazebo and stage simulators are able to provide odometry data.

   The start and end angle of the laser scan can be adjusted just by the changing values in the message. In the same way, the maximum and minimum range of the laser scan can also be specified. The position field specifies the pose of the robot with x, y, z parameters. The stage simulator is a 2D robot simulator, therefore, the z-parameter will always be equal to zero. The value of x and y are used to calculate the distance and angle to goal from the current position by transforming this Cartesian coordinate information to polar coordinates. The co variance field can be adjusted depending upon the deviation of the model. The results shown below are obtained on the assumption that co-variance field is zero.

   Complete flow of the proposed approach is represented in Algorithm 1.

Algorithm 1. New approach to tangentbug algorithm

1: **procedure** START
2:     Initialize variables   ▷ maximum, minimum velocities maximum and minimum angles etc.
3:     Get laser scan,Odometry ,goal point.
4: **end procedure**
5: **procedure** TANGENT DETECTION
6:     compute the tangent vectors ,T= $\begin{bmatrix} a_n \\ r_n \\ c_n \end{bmatrix}$
7:     $N_{\text{tan}} = N_{\text{tan}} + 1$ ▷ tangent count
8:     $\forall i \in \{0, 1, 2.....N_{\text{samples}} - 1\}$
9: **end procedure**
10: **procedure** TANGENT SELECTION
11:     select a tangent ,T= $\begin{bmatrix} as \\ rs \\ cs \end{bmatrix}$ ▷ best tangent
12:     set $IAT = 1$.
13:     $\forall i \in \{0, 1, 2.....N_{\text{tan}}\}$
14: **end procedure**
15: **procedure** DISTANCE-ANGLE UPDATE
16:     compute L= $\begin{bmatrix} \Delta l \\ \Delta a \end{bmatrix}$
17: **end procedure**
18: **procedure** CONTROLLER   ▷ optional
19:     compute G
20:     compute V_{max}
21:     calculate V= $\begin{bmatrix} vl \\ va \end{bmatrix}$ ▷ velocity vector
22: **end procedure**
23: **procedure** PUBLISH
24:     publish vl, va - linear and angular velocities.
25:     reset Parameters
26:     Goto START
27: **end procedure**
28: **procedure** MAIN
29:     Goto START
30:     calculate $\Delta x, \Delta y, \Delta l, \Delta \alpha$
31:     calculate s_{inc}, α_i
32:     set or reset obs ▷ $obs = 1$,if obstacle is present
33:     compute $OP_{12}, OP_{13}, OP_{14}$ ▷ stop at goal,turn at goal,approach goal
34:     **if** $OP_{12}=OP_{13},=OP_{14} = 0$ **then**
35:         Goto TANGENT DETECTION
36:         **if** more than one tangent is detected **then**
37:             Goto TANGENT SELECTION
38:         **end if**
39:     **end if**
40:     compute OP_{14}, OP_{15}
41:     set or reset W_{TR} ,W_{FT} ,W_{AG}
42:     Goto DISTANCE-ANGLE UPDATE
43:     Goto PUBLISH
44: **end procedure**

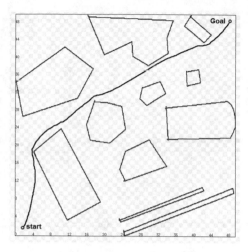

Fig. 1. Trajectory followed by the robot.

5   Simulation Results

Player stage simulator is chosen as the preferred 2D simulation environment.Simulation is carried out in various start and end goal conditions. In all cases, robot was able to plan a trajectory to goal. Figure 1 depicts on the trajectory followed by the robot. Table 1 represents an example of tangent selection strategy. Plot of linear and angular displacement changes in the simulator for some fixed time window are represented in Figs. 2 and 3 respectively. Similarly, linear and angular velocities are represented in Figs. 4 and 5.

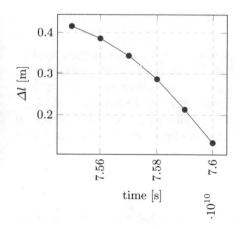

Fig. 2. Plot of Δl over time.

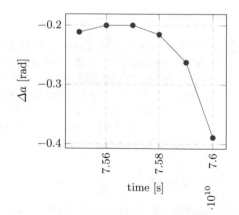

Fig. 3. Plot of Δa over time.

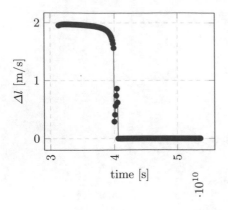

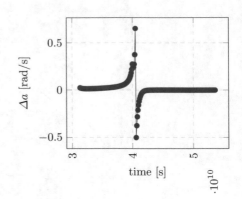

Fig. 4. Plot of linear velocity over time.   **Fig. 5.** Plot of angular velocity over time.

Table 1. Follow tangent: dx: −16.582, dy: −37.839, da: −1.193, dl: 41.313, dl: 1.110, da: −0.156

s.no	T_{11} (rad)	T_{21} (m)	T_{31} (rad)
Tangent: 1	−0.623	1.110	0.467
Tangent: 2	−0.412	9.335	0.054
Tangent: 3	0.921	9.381	−0.053
Tangent: 4	1.220	9.479	0.053
Tangent: 5	1.308	7.807	−0.064
Tangent: 6	1.325	6.406	−0.078
Selected tangent:	−0.623	1.110	0.467

6 Conclusion

An implementation of the Tangent-Bug algorithm, which can be directly applied to robots running on Robot Operating System is presented. Detailed description of each step including estimation of current state, operation selection tangent detection and selection are presented. The implementation details provided, facilitate easy deployment on to a real time system. This algorithm can be used in resource constrained environments where complex optimization based strategies are computationally expensive.

References

1. Cousins, S.: Welcome to ROS topics [ROS topics]. IEEE Robot. Autom. Mag. **17**(1), 13–14 (2010)
2. Gerkey, B., Vaughan, R.T., Howard, A.: The player/stage project: tools for multi-robot and distributed sensor systems. In: Proceedings of the 11th International Conference on Advanced Robotics, vol. 1, pp. 317–323 (2003)

3. Gossow, D., Leeper, A., Hershberger, D., Ciocarlie, M.: Interactive markers: 3-D user interfaces for ROS applications [ROS topics]. IEEE Robot. Autom. Mag. **18**(4), 14–15 (2011)
4. Koenig, N., Howard, A.: Design and use paradigms for Gazebo, an open-source multi-robot simulator. In: Proceedings of the 2004 IEEE/RSJ International Conference on Intelligent Robots and Systems, IROS 2004, vol. 3, pp. 2149–2154. IEEE (2004)
5. Ng, J., Bräunl, T.: Performance comparison of bug navigation algorithms. J. Intell. Robot. Syst. **50**(1), 73–84 (2007)
6. Pohl, I.: Bi-directional and heuristic search in path problems. Ph.D. thesis, Department of Computer Science, Stanford University (1969)
7. Rui, W., Xiaoge, L., Shuo, W.: A laser scanning data acquisition and display system based on ROS. In: 2014 33rd Chinese Control Conference (CCC), pp. 8433–8437. IEEE (2014)
8. Selman, B., Gomes, C.P.: Hill-climbing search. In: Encyclopedia of Cognitive Science (2006)
9. Skogestad, S., Postlethwaite, I.: Multivariable Feedback Control: Analysis and Design, vol. 2. Wiley, New York (2007)
10. Smith, O.J.M.: Feedback Control Systems. McGraw-Hill, New York (1958)
11. Tarjan, R.: Depth-first search and linear graph algorithms. SIAM J. Comput. **1**(2), 146–160 (1972)
12. Verwer, B.J.H., Verbeek, P.W., Dekker, S.T.: An efficient uniform cost algorithm applied to distance transforms. IEEE Trans. Pattern Anal. Mach. Intell. **4**, 425–429 (1989)
13. Yao, J., Lin, C., Xie, X., Wang, A.J., Hung, C.C.: Path planning for virtual human motion using improved a* star algorithm. In: 2010 Seventh International Conference on Information Technology: New Generations (ITNG), pp. 1154–1158. IEEE (2010)
14. Zhou, R., Hansen, E.A.: Breadth-first heuristic search. Artif. Intell. **170**(4), 385–408 (2006)

Efficiency of Naïve Bayes Technique Based on Clustering to Detect Congestion in Wireless Sensor Network

Jayashri B. Madalgi[1]([⊠]) and S. Anupama Kumar[2]

[1] Gogte Institute of Technology, Belgavi, Karnataka, India
jayashri@git.edu
[2] Rashtreeya Vidyalaya College of Engg, Bengaluru, Karnataka, India
anupamakumar@rvce.edu.in

Abstract. Wireless sensor network (WSN) is the network of sensor nodes set up to supervise physical observable fact. Congestion is state in the network when too many packets are present in the network than capacity of network. Congestion can be at node level or link level. Our work is to related to node level congestion. Because of funnel like topology of wireless sensor network, the congestion occurs at the nodes near sink as all the nodes start sending data to sink node whenever an event occurs. Congestion detection is vital as it leads in poor performance of network. In this paper we have implemented the machine learning techniques to detect congestion in wireless sensor network using Ensemble approach of clustering and classification. The Naïve Bayes classification based on the K- means and Expectation- Maximization clustering algorithms are applied to generate the classifier model. The classification model is also generated using only Naïve Bayes algorithm and the performance is compared with classifier of ensemble approach. The analysis of performance parameters of the generated models indicates that EM based Naïve Bayes classifier model is more accurate in detection of the congestion for the generated our network data set.

Keywords: Naïve Bayes · K means clustering · EM clustering · Wireless sensor networks · Congestion detection

1 Introduction

Sensor networks have many applications in various domains [1] like monitoring of humidity, temperature, precipitation etc. for monitoring the environment, forest fire detection or an industry fire detection, object tracking or identification, habitat monitoring, disaster management, military surveillance etc. Some of the researchers have used fuzzy logic, neural network, classification, clustering to detect congestion in wired and wireless networks. In our earlier paper [2] we also have build the congestion detection classifier model using support vector machine (SVM) and the model with best

© Springer Nature Switzerland AG 2020
D. J. Hemanth et al. (Eds.): ICICI 2019, LNDECT 38, pp. 756–764, 2020.
https://doi.org/10.1007/978-3-030-34080-3_84

pair of (c, γ) for the with radial basis function is found for our network traffic data set. In this paper, a novel approach of machine learning model is suggested to detect congestion using ensemble method – clustering based classification.

In Sect. 2 the survey of the work using ensemble approach is detailed, in Sect. 3 Work carried in wireless sensor network and machine learning algorithms is explained. In Sect. 4 the performance measures of the classifier model are compared and the best model with ensemble method is suggested.

2 Related Work

The authors in [3–5] have used the combination of clustering and classification to analyze the accuracy of ensemble classifier with bench mark data sets. The authors in [6] have classified the images using combined approach of classification and clustering. The authors in [7, 8] also have used ensemble approach of clustering with classification in their research work, The authors in [9] combines both approaches to reduce the efforts of tweets classification and the authors in [10] use the combined approach to software quality valuation.

The Naïve Bayes classifier model uses the probabilistic approach. The probabilistic relationship is between the attribute set and the class variable. In our data set is of independent attributes and Naïve Bayes works well with independent attributes.

The clustering technique is used as preprocessing step as it groups similar instances into cluster and cluster index can be used as classification value. The clustering algorithm chosen are K means and Expectation-Maximization (EM) algorithm. The K means is simple, partitional and prototype based that tries to get user specified k number of clusters. EM algorithm is probabilistic approach which uses distributions of various types, can find cluster of different sizes. This algorithm works well with less dimensionality.

3 Work Carried

3.1 WSN Simulation Using NS2 Tool

The WSN simulation with arbitrary placement of 25 sensor nodes is designed using network simulator tool - NS 2.35. The 4 source nodes are chosen which run two FTP (File Transfer Protocol) applications and two CBR (Constant Bit Rate) applications to propel the data packets towards the sink node in the network. The trace file of simulation for 50 s is analyzed. The network configuration parameters are illustrated in the following table.

Table 1. Wireless sensor network simulation parameters

Parameter	Value
Buffer length	50
Number of nodes	25
Network interface	WirelessPhy
Simulation time	50
Sink	1
Protocol	DSDV
Antenna	Omni antenna
Queue	DropTail/PriQueue
Initial energy	30.012010 (energy in joules)
Reception and transmission power	0.3 (power in watts)
Data rate	64 Kb, 128 Kb

The snapshot of network with deployments of sensors randomly is shown below:

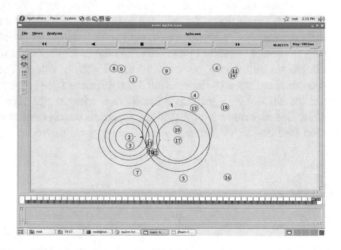

Fig. 1. Sensor network with nodes random deployment

The simulation is experimental for 200 times and in each run nodes are randomly deployed. And for the each execution network trace file is analyzed and the data set is built using the AWK scripts coded to get the required parameters. To get larger data set with more number of instances, resampling without displacement of data method is used. The data set comprises of the following 4 parameters namely packet service time, packet loss, packets delivery ratio and packet inter-arrival time. These values of these parameters signify the congestion state in network and therefore the occurrence of congestion can be identified. The data is CSV (comma separated values) file consist of independent attributes packet service time, packet loss, packet_inter-arrival time and packet delivery ratio. The target variable is the classification attribute as class attribute with values {low, med, high} which stand for the different levels of congestion. All the attributes are of quantitative type. The class value is manually interpreted using the average value of independent attributes.

3.2   Ensemble Classifier Model Generation Using Clustering

3.2.1   Naïve Bayes Classification

The Bayes classifier uses probabilistic approach to solve classification problem, It is based on Bayes theorem which uses conditional probability [11].

$$P(Y|X) = \frac{P(X|Y)P(Y)}{P(X)}$$

The principle Bayes theorem for classification task is explained below:

Assume every attribute and the class attribute as random variables, regard as a record with the attributes $(X_1, X_2, \ldots, X_d)$, the objective is to predict class attribute Y. particularly classification goal is to get the value of Y that maximize $P(Y \mid X_1, X_2, \ldots, X_d)$.

It requires computing posterior probability

$$P(Y|X_1X_2\ldots X_d) = \frac{P(X_1X_2\ldots X_d|Y)P(Y)}{P(X_1X_2\ldots X_d)}$$

Initially, the congestion detection classifier is generated using Naïve Bayes algorithm. The confusion matrix is contingency table indicating the accuracy of classifier. The confusion matrix of the model built is as shown below:

Table 2. Naïve Bayes classifier confusion matrix

High	Medium	Low	Classified as
1397	1385	75	High
0	1008	228	Medium
0	0	2307	Low

The accuracy of model is 73.625%. Clustering as pre processing step helps to group the similar instances. These clusters are assumed as classes and then classification is applied.

3.2.2   K - Means Clustering

This clustering technique has a partition clustering approach. The parameter required is number of clusters – K must be specified. Each data point is assigned with the nearby centroid data point cluster, the process is repeated until centroids do not change.

The common measure to evaluate clusters is SSE (Sum of Squared Error). The error for each point is the distance to its nearest cluster. To get SSE value, take the square of these errors and then sum them [11].

$$SSE = \sum_{i=1}^{K} \sum_{x \in C_i} dist^2(c_i, x)$$

Here x represents a data point in $C_{i \text{ cluster}}$ and the c_i is the centroid data point in the cluster C_i.

In the first approach, the k means clustering technique is applied to our data set and cluster index is added to the data set as an class label. Consequently Naïve Bayes is applied to develop model for congestion detection in WSN.

Table 3. Confusion matrix of Naïve Bayes classifier with k means clustering

Cluster0	Cluster1	Cluster2	Classified as
805	0	0	cluster0
0	1934	0	cluster1
0	1519	2142	cluster2

3.2.3 EM (Expectation - Maximization) Clustering

In the second approach, the EM clustering is applied to our data set and cluster index is added to the data set as an class label. Consequently Naïve Bayes is applied to develop model for congestion detection in WSN.

Table 4. Confusion matrix of Naïve Bayes classifier with EM clustering

Cluster0	Cluster1	Cluster2	Classified as
3609	0	224	cluster0
0	181	0	cluster1
0	0	2386	cluster2

4   Analysis of Results

The machine learning algorithms are applied using open source tool WEKA [12] by Waikato University, New Zealand, which provides support for most of the data mining algorithms. The confusion matrix can be used to get the measures used to evaluate performance such as true positive rate is also known as sensitivity, false positive rate.

The other measures precision (p), F-measure, kappa statistic, area under ROC curve (receiver operating characteristic curve) consider the probability estimations of classes. ROC plots the arc between TPR values and FPR values. The other performance parameters are also observed to analyze the performance of classifier model generated to detect congestion. The 10-fold cross validation system is used to build the model. The following table shows performance parameters of the model generated using the combination of clustering and classification approach.

Table 5. Comparison of other performance parameters

Performance parameter	Naïve Bayes classification	Naïve Bayes classification with K means clustering	Naïve Bayes classification with EM clustering
Accuracy	73.625%	76.2656%	96.5
Precision	0.846	0.867	0.968
Recall	0.736	0.763	0.965

The precision is fraction indicating part of those predicted positive are actually positive. Recall is true positive rate.

The comparison graph of other measures of performance area are shown below.

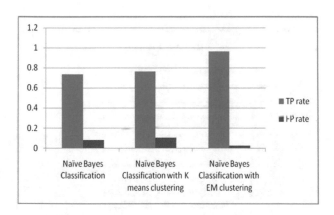

Fig. 2. TP rate and FP rate in different approches

The highest TP rate of model generated using Naïve bayes classification with EM clustering is 96.5%.

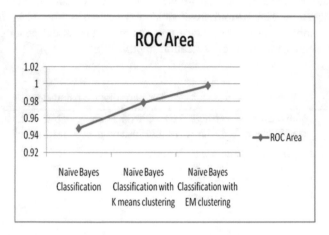

Fig. 3. Area under ROC curve in different approches

The area below the ROC curve known as AUC furthermore the value 1of AUC indicates the classification method is perfect. The our model generated using Naïve bayes clasification with EM clustering has AUC value 0.936.

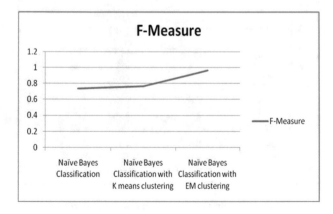

Fig. 4. F-measure in all 3 different approches

The harmonic mean of precision and recall is known as F-measure. This value of it is also more for the model in the 3rd approach.

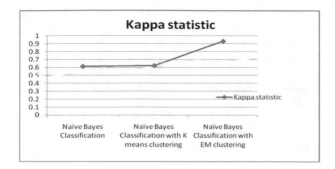

Fig. 5. Kappa statistic in all 3 different approches

Kappa measure compares the observed accuracy with expected accuracy. And the value between 0.81–1.00 indicates that the performance of classifier model is almost perfect.

5   Conclusion

The classification has 2 categories namely the unsupervised and supervised. The supervised use available class information where as unsupervised have set of data objects without class label. We built model using Naïve Bayes classifier and then we have tried to combine both and this approach improves the performance of model generated to detect the congestion. It is observed that clustering as pre processing step helps to reduce the effort of classification and similar objects get appropriate classification labels. Ensemble approach of classification with clustering improves the performance of classifier model than the model with Naïve Bayes classifier only to detect congestion. Experiment is also performed using WEKA tool with the same approach and the results of it also shows that model generated using EM clustering algorithm with Naïve Bayes gives more accurate results for classification compared to Naïve Bayes with the K means algorithm for our network data set.

References

1. Dawhkova, E., Gurtov, A.: Survey on congestion control mechanisms for wireless sensor networks. Center for Wireless Communication, Finland (2013)
2. Madalgi, J.B., Kumar, S.A.: Development of wireless sensor network congestion detection classifier using support vector machine. In: IEEE International Conference on Computational Systems and Information Technology for Sustainable Solutions (CSITSS-2018). RVCE, Bengaluru, IEEE Digital Library (2018). ISBN 978-1-5386-6078-2
3. Karami, H., Taheri, M.: A novel framework to generate clustering algorithms based on a particular classification structure. In: Artificial Intelligence and Signal Processing Conference (AISP), Shiraz, pp. 201–204 (2017). https://doi.org/10.1109/AISP.2017.8324081

4. Chakraborty, T.: EC3: combining clustering and classification for ensemble learning. In: IEEE International Conference on Data Mining (ICDM), New Orleans, LA, pp. 781–786 (2017). https://doi.org/10.1109/ICDM.2017.92
5. Erol, H., Tyoden, B.M., Erol, R.: Classification performances of data mining clustering algorithms for remotely sensed multispectral image data. In: 2018 Innovations in Intelligent Systems and Applications (INISTA), Thessaloniki, pp. 1–4 (2018). https://doi.org/10.1109/INISTA.2018.8466320
6. Zhou, L., Wang, L., Ge, X., Shi, Q.: A clustering-based KNN improved algorithm CLKNN for text classification. In: 2010 2nd International Asia Conference on Informatics in Control, Automation and Robotics, Wuhan, pp. 212–215 (2010). https://doi.org/10.1109/CAR.2010.5456668
7. Ru, X., Liu, Z., Huang, Z., Jiang, W.: Class discovery based on K-means clustering and perturbation analysis. In: 2015 8th International Congress on Image and Signal Processing (CISP), Shenyang, pp. 1236–1240 (2015). https://doi.org/10.1109/CISP.2015.7408070
8. Alapati, Y.K.: Combining clustering with classification: a technique to improve classification accuracy. Int. J. Comput. Sci. Eng. **5**(6), 2319–7323 (2016)
9. Papas, D., Tjortjis, C.: Combining clustering and classification for software quality evaluation. In: Likas, A., Blekas, K., Kalles, D. (eds.) Artificial Intelligence: Methods and Applications, SETN 2014. Lecture Notes in Computer Science, vol. 8445. Springer, Cham (2014)
10. de Oliveira, E., Basoni, H.G., Saúde, M.R., Ciarelli, P.M.: Combining clustering and classification approaches for reducing the effort of automatic tweets classification. In: Proceedings of the International Conference on Knowledge Discovery and Information Retrieval (KDIR-2014), pp. 465–472 (2014). https://doi.org/10.5220/0005159304650472
11. Tan, P.-N., Steinbach, M., Kumar, V.: Introduction to Data Mining (2009)
12. Hall, M., Frank, E., Holmes, G., Pfahringer, B., Reutemann, P., Witten, I.H.: The WEKA data mining software: an update. SIGKDD Explor. **11**(1), 10–18 (2009)

Microcontroller Based Security Protection and Location Identification of Bike Riders

S. Sarath Chandra[1]([✉]), J. Sriram Pavan[1], and Bandla Prasanthi[2]

[1] Department of ECE, QIS College of Engineering and Technology,
Ongole 523272, AP, India
`ssschandra@gmail.com`, `jampani@gmail.com`.
`srirampavan@gmail.com`
[2] Department of ECE, Acharya Nagarjuna University, Guntur 522510, AP, India
`ssanthichandra@gmail.com`

Abstract. As the bikers in our nation are expanding, the street setbacks are additionally expanding step by step. Due to which numerous losses occur, and the vast majority of them are caused by the continuous carelessness and furthermore numerous passings happen because of absence of brief therapeutic consideration required by the harmed individual. This spurs us to develop a framework that guarantees the security of biker, by making it important to wear a head protector/helmet, according to government rules, additionally to get legitimate and brief restorative consideration, in the wake of meeting with a mishap. The proposed framework is a smart protective cap. Every Bike rider should wear helmet to start the bike, if not message will appear on the screen. After a moment it will ask for Password if he/she enters the correct password then ignition will on, next it will check the alcoholic percentage levels of the rider and the acceleration of the bike & alcohol levels will displays on screen. If rider exceeds the alcoholic levels immediately bike will go in a OFF mode and the rider location will sent to a nearest police station.

Keywords: ARM7 microcontroller · Accelerometer · Alcohol sensors · Battery and RF transceiver

1 Introduction

Usually, the road accidents will occur due to the non-wearing of helmets. Particularly in india there is a lot of human injuries & property damages due to the serious accidents. Even if they wore the helmets some accidents are occurring due to drowsy & alcohol levels in human body for this case we have introduced Smart Helmet, which is inter linked with motor vehicle. By using this we are able to identify the alcohol levels of the person and if he/she exceeds the cutoff levels of the alcohol bike will not start. and also by using this technology we can send a SMS to the concerned authorities along with their GPS locations.

© Springer Nature Switzerland AG 2020
D. J. Hemanth et al. (Eds.): ICICI 2019, LNDECT 38, pp. 765–770, 2020.
https://doi.org/10.1007/978-3-030-34080-3_85

2 Literature Survey

2.1 Smart Helmet Using ARM7

There is a disturbing increment in the grimness and mortality because of bike street auto collisions. This has involved incredible concern universally. In India, it is this is a report about a brilliant cap [9] which make bike driving to be even more secured.

2.2 Smart Helmet Using GSM and GPS Technology for Accident Detection and Reporting System [3]

In this Method when a bike rider hits any vehicle or any object and if the helmet hits the ground then the vibration sensor will send a message to the GSM modem in this process GPS [7] also come into activation. With immediate effect relatives or known persons will get a message with location address of injured person.

3 Proposed Method

The proposed framework is a clever protective cap. The framework guarantees the wellbeing [9] of the biker, by making it important to wear the Helmet, according to the administration rules, additionally to get appropriate and brief therapeutic consideration, in the wake of meeting with a mishap. A module is joined in the head protector, to such an extent that, the module will match up and attached to the bicycle. The framework will consist these functionalities like:

1. It will guarantee that the rider has worn the protective cap. In the event that he neglects to do as such, the bicycle won't begin.
2. Likewise guarantee that biker has not devoured liquor. On the off chance that the rider is tanked, the bicycle won't begin.
3. A mishap recognition module will be introduced on the bicycle, which will probably recognize mishap and will almost certainly advise rapidly the mishap to police control room and on the off chance that if the mishap is minor, rider can prematurely end message sending by squeezing the prematurely end switch.

   Mainly it consist of two sections:
   (i) Module on protective cap
   (ii) Module on the bicycle.

3.1 In-Built Vehicle Section

In this project, we design and develop a vehicle headgear for safe ride which will optimize the usage of the Arm7 microcontroller Fig. 1.

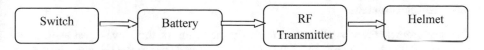

Fig. 1. Block diagram for helmet section

4 System Design

System Design consists of the following.

4.1 Helmet Part

It basically consists [2] of a switch, Microcontroller and Transmitter Fig. 2.

4.2 Switch

Uses for ON & OFF purpose.

4.3 Transmitter

The Transmitter used to transmits the Serial data with the frequency of 434 MHz as a Radio Frequency.

4.4 Bike Part

It comprises of a RF Receiver unit, Microcontroller unit, Abort switch and GSM [8] Module.

4.5 Receiver

The RF receiver [11], receives the transmitter data in a wireless mode Fig. 4.

4.6 Microcontroller

This is the genuine fundamental authority unit [12] of the entire circuit and the ventures will be continued into it [6]. As demonstrated by the data it will get from the module on bike it will control the yield of remarkable parts [1]. In perspective on the yield of both the accelerometers on bike and head defender, it will send information to nearest police home office if there ought to be an event of a setback using GSM module, and reliant on the yields of alcohol sensor and switch, it will send a hand-off respect the engine.

4.7 Alcohol Sensor

Utilized for recognizing liquor fixation in breath. It gives a simple yield dependent on liquor focus [14]. On the off chance that the measure of liquor surpasses the edge esteem it won't enable the bicycle to begin [4].

4.8   Accelerometer

An accelerometer can be utilized to quantify the tilting of the bicycle just as the head protector. The tilt of the protective cap is estimated and sent to the microcontroller. On the off chance that the point of the bicycle is zero (0) as for ground, it will distinguish that mishap has happened. Microcontroller: All the simple yields from every one of the sensors on the head protector are sent to this microcontroller as info. As per the limit set for liquor sensor, accelerometer and the low or high yield of the switch, a choice is made and sent to the module on bicycle remotely.

4.9   GSM Module

It is used as a Mobile phone it can transmits the messages to the destinations and also it can transmits [13] data depends upon the usage [3]. We will utilize SMS use of it to send a SMS to the police headquarters if there should be an occurrence of mishap Fig. 3.

5   System Work Flow

In this if the rider wears a helmet Fig. 5 the system will ask for password if they are entered correct password Fig. 6 it will goes to next step otherwise bike will not move. Next alcohol sensor [10] will sense the particular person condition if he/she is with in a range of alcohol content the bike will start or if he/she crosses the range i.e. cutoff value Figs. 7, 8 and 9 using the RF module the message will be sent to nearby police station and also bike will not start [5].

6   Results

(See Figs. 2, 3, 4, 5, 6, 7, 8, 9)

Fig. 2. Helmet section

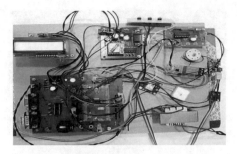

Fig. 3. Vehicle section

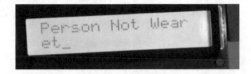

Fig. 4. LCD display shows rider not wearing helmet

Fig. 5. LCD display shows rider wearing helmet

Fig. 6. Security password entered

Fig. 7. Checking the parameters

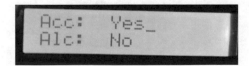

Fig. 8. If accident detects

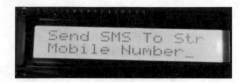

Fig. 9. Sending SMS alert

7   Conclusion

Using Smart Helmet concept we can reduce not only property damage but also the human lives. The main intend of this proposed framework is that it guarantees the wellbeing of the biker, by making them to be aware of wearing the helmet.

References

1. Krishna Chaitanya, V., Praveen Kumar, K.: Shrewd cap utilizing arduino, Hyderabad (2013)
2. Prudhvi Raj, R., Sri Krishna Kanth, Ch., Bhargav Aditya, A., Bharath, K.: Shrewd detective Helmet. AEEE, India
3. Manjesh, N., Raj, S.: Shrewd helmet using GSM and GPS technology for accident detection and reporting system. Int. J. Electr. Electron. Res. **2**(4), 122–127 (2014)
4. Vijayan, S., Govind, V.T., Mathews, M., Surendran, S., Sabah, M.: Liquor identification utilizing brilliant protective cap framework. IJETCSE. 8(1) (2014)
5. Xu, R., Zhou, S., Li, W.J.: MEMS accelerometer based nonspecific-user hand gesture recognition. IEEE Sens. J. **12**(5), 1166–1173 (2011)
6. Mazidi, M.A., Mazidi, J.G.: The 8051 Microcontroller and Embedded Systems
7. Remote mishap data utilizing GPS and GSM. Res. J. Appl. Sci. Eng. Technol. (2012). Maxwell Sci. Organ. **4**(17) (2012)
8. Sarath Chandra, S., Sai Sravanthi, G., Prasanthi, B., Vijaya Rangan, R.: IoT based garbage monitoring system. In: 2019 1st International Conference on Innovations in Information and Communication Technology (ICIICT) (2019)
9. Karthikeyan, S., Singh, S., Jain, H.M., Sailesh Kumar, M., Priya, V.: Smart and assistive driving headgear. In: 2018 3rd International Conference on Communication and Electronics Systems (ICCES) (2018)
10. Patel, D.H., Sadatiya, P., Patel, D.K., Barot, P.: IoT based obligatory usage of safety equipment for alcohol and accident detection. In: 2019 3rd International conference on Electronics, Communication and Aerospace Technology (ICECA) (2019)
11. Sridevi, G., Sarath Chandra, S.: Automated unauthorized parking detector with smart vehicle parking. Indian J. Sci. Technol. **12**(26), 1–5 (2019)
12. https://www.elprocus.com/arm7-based-lpc2148-microcontroller-pin-configuration/
13. https://www.elprocus.com/gsm-architecture-features-working/
14. https://www.sunrom.com/p/alcohol-sensor-module-mq3

Internet of Things Towards Smart Solar Grid Systems for Home Appliances

Shrihariprasath Basuvaiyan[1]([⊠]) and J. Preethi[2]

[1] Anna University, Chennai, India
shrihari_1991@outlook.com
[2] Anna University, Coimbatore, India
preethi17j@yahoo.com

Abstract. Internet of Things (IoT) and Machine to Machine (M2M) Communication are two technologies that become enormously developed in last decades. In the linear perspective of energy saving, the Solar Smart Grid (SSG) is an excellent solution to optimize the energy consumption. This paper demonstrates a design of Smart Solar Power Management (SSPM) System using IoT. The proposed system integrates real time monitoring of data and control the system with precision. This paper consists of three major sections: Real time solar power system monitoring and controlling; Solar home power system engineering database collection and management; system fault analysis. SSPM System is based on a mixture of factors according to solar grid management necessities and operation rules. Sophisticated functions of Smart Solar Power Management System can also be varied via wireless M2 M Communication. The main objective of the system to overcome the field maintenance, system troubleshooting and to provide mean time to repair in existing Solar Home Power Systems. Any device from anywhere using smart phones via Internet can control the proposed SSPM system. The prototype of the system is tested and implemented in several customer's places to analyze implementation issues and challenges.

Keywords: Internet of Things · Machine to machine communication · Solar power system · Smart grid · Real monitoring and control · Maintenance and automation

1 Introduction

For the past few years, Internet of Thing (IoT) has conventional loyalty [1]. In general, the things speaking to each other which are linked to wired and wireless networks is described as IoT [2]. Lately, it is employed not only for the area of custom electronics and gadgets but also in other versatile fields such as a smart home, smart car, smart city, healthcare, smart grid system, and industrial application [3]. At present, a unity of the fundamental renewable energy source is the Solar Photo Voltaic (PV). In future green energy towards solar energy is becoming a potential result [4].

As the contribution of its research plan development, the European Commission (EC) has well-determined IoT as a cohesive part of the Future Internet (FI) where "things having individualities and practical personalities operating in smart spaces employing intelligent ports to link and communicate among social environment" [5].

© Springer Nature Switzerland AG 2020
D. J. Hemanth et al. (Eds.): ICICI 2019, LNDECT 38, pp. 771–778, 2020.
https://doi.org/10.1007/978-3-030-34080-3_86

The principle of photovoltaic effect that changes the solar energy into electricity by solar cells is the technology associated with Solar Power System. The maintenance and management of Solar Power Systems are very complex [6]. Due to fluctuations solar irradiance, temperature and other factors the generation of power from the solar photovoltaic system is erratic in nature. Hence, SSPM systems are required to overcome the field maintenance, system troubleshooting and to provide mean time to repair in existing Solar Home Power Systems [7]. The majority of research studies in monitoring and controlling of solar PV system to improve the performance of solar PV system [8–11]. This paper is designed as follows: Sect. 2.2 presents a problem defined and proposed system feature. Section 3 presents the overall design of the system. The prototype system and implementation setup are presented in Sect. 4. Then Sect. 5 describes the conclusion and future work.

2   System Analysis

2.1   Problem Definition

Solar power system faces four main challenges; mean time to repair is tedious, inflexibility, poor manageability, and difficulty in maintaining the system. The main objectives of proposed system are to design and implement an intelligent system that is capable of managing the solar power system and stores all parameters in the cloud server in order to analyses the performance of the system through an easy manageable web application using Cloud Server. In proposed system IoT protocols is used to interconnect solar power system with smart phone using Android application. The proposed system has a great flexibility by using Wi-Fi technology to interconnect the solar power systems to the internet [12].

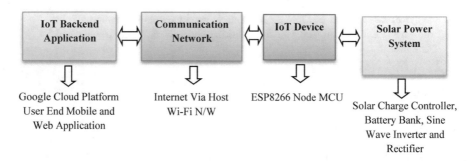

Fig. 1. Proposed system block diagram

2.2   Proposed System Feature

The proposed system is to manage the state of a Solar Power System through an IoT-based network in order to control the system remotely. The data from the Solar Power System is transmitted through the existing internet network. An ESP8266 Node MCU module is employed to send data to the cloud server using MQTT (Message Queuing Telemetry Transport) Protocol. The proposed system consists of four parts. They are Physical Device, IoT device, Communication Network and IoT Backed End Application. In proposed system, Physical device is solar power system, which comprises of Solar panel, Solar Charge Controller, Sinewave Inverter, AC to DC rectifier and Batteries. IoT Device is link layer, which comprises ESP8266 Node MCU (Microcontroller Unit) which communicate with solar power system and links solar power system with Host Network. Internet is the network layer where data logging from the plant for real-time computation is done which includes a database for storage. In the IoT Backend application, refined web-application and Android mobile application are designed using google firebase and Ionic Framework respectively.

3 Overall System Design

The overall design of a proposed system mainly consists of the Solar power system, IoT Device, and cloud server. In the proposed system, the IoT device is responsible for connecting the host network and Solar power system to the internet. In the proposed system, solar power system is interconnected to IoT device using UART serial communication protocol. The combination of the solar power system and IoT device act as MQTT Clients. In the proposed system MQTT remote broker is used as a bridge between cloud server, User End device and solar power system. The block diagram of the proposed system is shown in Fig. 1. In the proposed system we have implemented 10 solar power systems for analyzing the performance of the machine and implementation issues.

3.1 Solar Power System

The solar power systems involve the best transfer of energy generated in the Solar PV Panel array to load offered by the combination of battery bank and load. This best load requirement varies with solar cell's operating temperature and insolation. The load offered by the battery to PV array also varies with its temperature and the state of the charge of the battery. The maximum energy can be generated from the PV array to the battery bank can be achieved by inserting a smart MPPT (Maximum Power Point Tracking) Charger controller. The MPPT controller is a chopper circuit which continuously adjusts the voltage and current level to match both the variable PV output and the load. The solar power system is designed using STM32F030 ARM Cortex M0 microcontroller Unit (MCU) which is 32-bit RISC Architecture core. The overall system operations and functions are controlled by this MCU. The proposed prototype system is designed for a maximum load capacity of 750watts Capacity which used for the domestic home power system. This system includes both solar energy generation and solar energy storage which ensuring the backup power supply during the day and night. The battery is charged through solar alone or solar and AC grid shared mode. Once the battery is fully charged, the Hybrid function will start automatically. The load will run only through solar power depending on its generation. In case of any power failure in night time, battery backup will support. This smart functionality is controlled by ARM STM32F030 MCU.

This system uses IRF3205 MOSFET as power switching element in both H-Bridge circuit and Solar MPPT charger circuit. The Pulse Width Modulation techniques are used in H-Bridge circuit to generate pure sinewave output. In Solar MPPT charger session, we incorporated Perturb and observe MPPT algorithm to achieve maximum solar power from the panel. The MPPT charger circuit consists of DC-DC buck converter topology using IRFB4227 MOSFET as a power switching element. In this system, TLP250 Opto driver is used as driver circuits for switching the Power MOSFETS in H-Bridge circuit and Dc-Dc buck converter circuit. The Input grid voltage, Output grid voltage, Output load current, Battery voltage, Battery current, Solar voltage and solar current information are continuously sampled and measure to calculate power, status and conditions based on functionality using MCU session and displayed in machine locally using 16X2 LCD (Liquid Crystal Display) display. This system software is designed and developed using Embedded C. The program complied and fused to the ARM STM32F030 using IAR workbench and STM Cube. The voltage and current sense signaling circuits are used to measure the voltage and current information's. The IoT device is interconnected with MCU session using the UART interface. Designed solar power system which used in proposed system is shown in the Fig. 2.

Fig. 2. Proposed system prototype

3.2   Working of Proposed System

The proposed system has three power sources, they are a battery, Solar Panel and AC grid. First 12 V 150 AH battery is connected to the battery terminal of the solar power system. In the proposed power supply circuit is used to produce regulated +5 V to power up the MCU session all other signal conditioning circuit. In MCU session +5 Vdc is regulated to 3.3 V supply because the STM32F030 operating voltage is 3–3.3 V. Initially, MCU is configured to all the peripherals and measure all the voltage and current in continuous sampling time. If the battery voltage is above the targeted level only MCU will start operation flow. Now the UART connection and LCD display are initialized and start data communication with ESP8266. The Heart of the IoT Device is an ESP8266 Node Microcontroller Unit is a low-cost MCU constructed with Wi-Fi protocol. The ESP8266 Wi-Fi Module is a self-contained SOC with integrated TCP/IP protocol stack that can be connected with any microcontroller access to Wi-Fi network. The ESP8266 is capable of either hosting an application or offloading all Wi-Fi networking functions from another application processor. In the proposed system firmware for ESP8266 Node, MCU is designed and developed using IoT lightweight operating system called Mongoose OS. Mongoose OS is as IoT firmware development framework. In the proposed system firmware is developed using JavaScript Application program Interface. In the proposed system MQTT protocol is used for communicating with Cloud Server and User End device. MQTT is a publish/subscribe, tremendously simple and lightweight messaging protocol, specially designed for constrained IoT devices and low-bandwidth, high-latency or unreliable networks. Thus the IoT Device is the main bridge between the Solar power system and cloud server. In the proposed system, Google Cloud Platform is used as a cloud server. The user can manage the solar power system with help of web application and mobile application. The interactive web application is developed using Google Firebase and node.js. The mobile application is designed and developed using the Ionic Framework. Web application is mainly designed for administration and management service which mainly used by manufacture and product dealers. The authorized person can also manage the solar power system using web application to provide mean time to service and product maintenance. The proposed system can be monitored and controlled by the customer using smart phone via Android application. When the product is registered with web application and mobile application only it will communicate with cloud server. Each and every system has unique physical address, if the physical address matched with registered application only the user authenticates with mobile application. The proposed solar system automatically select energy source based on the solar power and battery voltage as per the software designed in MCU session.

4 Implementation and Results

The prototype of the proposed system is designed and implemented. The Status and parameters from the Solar power systems are measured and transferred to the cloud server via the internet using ESP8266. It also gives the instant status of the solar power system and fault indication to the customer and authorized dealer or administrator. In the proposed system, with the help of webpage application and smartphone android application, solar power system is managed. Initially proposed system is installed in 10 clients of SS System Coimbatore and analyzed performance and implementation issues. The Experimental results are analyzed to develop a product to the next level [13]. The data from the solar power system are sent to the cloud server via ESP8266 using the Internet. Figure 3 shows the front-end design of web application which will allow the user to monitor and control the solar power system. This interactive web application act as a CRM tool for manufacturer and product distributor. In the proposed system, the mobile application is designed for the customer to manage solar power system. In the

Fig. 3. Front end design of web application

proposed system implementation, the Quality-of Service (QoS) is a primary issue in order to assure its effectiveness and robustness. In IoT Application, we have limited resources that will degrade the QoS. Congestion in IoT will encourage to dilute the expected QoS of the related applications. In this case, to ensure seamless data transmission scarce resources are used efficiently. The power consumption of the IoT Device can be cut down by reducing the rate of packet retransmission, which is caused by congestion [14]. The data is communicated using the Wi-Fi protocol via ESP8266 NodeMCU, the low-cost ESP8266 NodeMCU has been used for the purpose. Wi-Fi protocol has been used as it matched the data rate, distance and power consumption and cost that are aimed for.

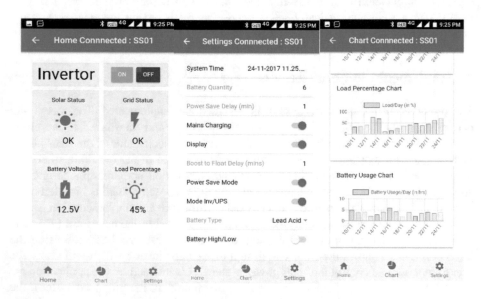

Fig. 4. Front end design of android application

5 Conclusion and Future Work

Use of IoT based SSPM system is an important phase for renewable energy sources. Thus, the automation and intellectualization of solar power system will enhance the future decision-making process for the large-scale solar power plant and grid integration. In this paper, we proposed a design of Smart Solar Power Management System using IoT. The approach is designed, analyzed, implemented and experimentally proven system and works successfully. The designed system not only monitors and control the electrical parameter of the solar system, but it also stores the instant status of solar power system and manages the mean time to repair. The main objective of the system is to provide field maintenance, system troubleshooting and repair. Thus the prototype of the system is tested and implemented in several customer's places to analyze implementation issues and challenges. This approach can be modified in future by using sophisticated technologies and equipped with much more intelligence for faster computation. This system can be further deployed in large scale to enhance the operation and maintenance of the solar power system using Constrained application protocol to increase Quality of Service and reduce congestion.

References

1. Chen, X.S., et al.: Application of internet of things in power-line monitoring. In: 2012 International Conference on Cyber-Enabled Distributed Computing and Knowledge Discovery (CyberC). IEEE (2012)
2. Byeongkwan, K., et al.: IoT-based monitoring system using trilevel context making model for smart home services. In: 2015 IEEE International Conference on Consumer Electronics (ICCE). IEEE (2015)
3. Surie, D., Laguionie, O., Pederson, T.: Wireless sensor networking of everyday objects in a smart home environment. In: International Conference on Intelligent Sensors, Sensor Networks and Information Processing 2008, ISSNIP 2008. IEEE (2008)
4. Ranhotigamage, C., Mukhopadhyay, S.C.: Field trials and performance monitoring of distributed solar panels using a low-cost wireless sensor network for domestic applications. IEEE Sens. J. 11(10), 2583–2590 (2011)
5. Constantin, S., et al.: GPRS based system for atmospheric pollution monitoring and warning. In: 2006 IEEE International Conference on Automation, Quality and Testing, Robotics, vol. 2. IEEE (2006)
6. Reddy, S.R.N.: Design of remote monitoring and control system with automatic irrigation system using GSM-bluetooth. Int. J. Comput. Appl. 47(12) (2012)
7. Belghith, O.B., Sbita, L.: Remote GSM module monitoring and photovoltaic system control. In: International Conference on Green Energy, 2014. IEEE (2014)
8. Ou, Q., et al.: Application of internet of things in smart grid power transmission. In: 2012 Third FTRA International Conference on Mobile, Ubiquitous, and Intelligent Computing (MUSIC). IEEE (2012)
9. Moon, S., Yoon, S.-G., Park, J.-H.: A new low-cost centralized MPPT controller system for multiply distributed photovoltaic power conditioning modules. IEEE Trans. Smart Grid 6(6), 2649–2658 (2015)

10. Jihua, Y., Wang, W.: Research and design of solar photovoltaic power generation monitoring system based on TinyOS. In: 2014 9th International Conference on Computer Science & Education (ICCSE). IEEE (2014)
11. Vineeth, V.V., Radhika, N., Vanitha, V.: Intruder Detection and Prevention in a Smart Grid Communication System. Proc. Technol. **21**, 393–399 (2015)
12. Shrihariprasath, B., Rathinasabapathy, V.: A smart IoT system for monitoring solar PV power conditioning unit. In: World Conference on Futuristic Trends in Research and Innovation for Social Welfare (Startup Conclave). IEEE (2016)
13. Weranga, K.S.K., Chandima, D.P., Kumarawadu, S.P.: Smart metering for next generation energy efficiency & conservation. In: 2012 IEEE Innovative Smart Grid Technologies-Asia (ISGT Asia). IEEE (2012)
14. Halim, N.H.B., Yaakob, N.B., Isa, A.B.A.M.: Congestion control mechanism for Internet-ofThings (IOT) paradigm. In: 2016 3rd International Conference on Electronic Design (ICED). IEEE (2016)

Author Index

© Springer Nature Switzerland AG 2020
D. J. Hemanth et al. (Eds.): ICICI 2019, LNDECT 38, pp. 779–781, 2020.
https://doi.org/10.1007/978-3-030-34080-3

Printed in the United States
By Bookmasters